KU-491-824

Mathematics: Theory & Applications

Series Editor
Nolan Wallach

Armand Borel
Lizhen Ji

Compactifications of Symmetric and Locally Symmetric Spaces

Birkhäuser
Boston • Basel • Berlin

Armand Borel (*deceased*)
Institute for Advanced Study
Princeton, NJ 08540
USA

Lizhen Ji
University of Michigan
Department of Mathematics
Ann Arbor, MI 48109
USA

Mathematics Subject Classification (2000): 22F30, 22E40, 32J05 (primary); 57S30, 11F75, 11F03, 11F23, 11F72, 11H55, 30F25, 31C35 (secondary)

Library of Congress Control Number: 2005934870

ISBN-10 0-8176-3247-6 e-IBSN 0-8176-4466-0
ISBN-13 978-0-8176-3247-2

Printed on acid-free paper.

 Birkhäuser 

Printed in the United States of America. (TXQ/MP)

9 8 7 6 5 4 3 2 1

www.birkhauser.com

To

Gaby Borel, Lan Wang

&

Lena, Emily, and Karen

Contents

Preface

Symmetric spaces and locally symmetric spaces occur naturally in many branches of mathematics as moduli spaces and special manifolds, and are often noncompact. Motivated by different applications, compactifications of symmetric spaces and locally symmetric spaces have been studied intensively by various methods. For example, a typical method to compactify a symmetric space is to embed it into a compact space and take the closure, while a typical method to compactify a locally symmetric space is to attach ideal boundary points or boundary components. In this book, we give uniform constructions of most known compactifications of both symmetric and locally symmetric spaces together with some new compactifications. We also explain how different types of compactifications arise and are used; in particular, why there are so many different kinds of compactifications of one space, and how they are related to each other. We hope to present a comprehensive survey of compactifications of both symmetric and locally symmetric spaces. It should be pointed out that this book emphasizes the geometric and topological aspects of the compactifications; on the other hand, we have tried to provide adequate references to omitted topics.

The book is divided into three parts corresponding to different classes of compactifications. In Part I, we study compactifications of Riemannian symmetric spaces in the usual sense that the symmetric spaces are open and dense subsets. In Part II, we study compact smooth manifolds in which the disjoint union of more than one but finitely many symmetric spaces is contained as an open, dense subset, and the closure of each copy of the symmetric spaces is a manifold with corners. In Part III, we study compactifications of locally symmetric spaces, and their relations to the metric and spectral properties of the locally symmetric spaces.

Though these parts treat different types of compactifications, they are closely related in various ways. In fact, there are several basic, unifying themes in this book:

1. Siegel sets of rational parabolic subgroups and the reduction theory of arithmetic groups have played an important role in the study of compactifications of locally symmetric spaces. We show that compactifications of a symmetric space X can also be studied through a generalization for real parabolic subgroups of

Siegel sets of rational parabolic subgroups and the reduction theory of arithmetic groups. Therefore, compactifications of symmetric and locally symmetric spaces can be studied in parallel using similar methods.

2. Compactifications of a locally symmetric space $\Gamma\backslash X$ can be studied and understood better through compactifications of the homogeneous space $\Gamma\backslash G$, which is a principal fiber bundle over $\Gamma\backslash X$ whose fibers are maximal compact subgroups of G.
3. Certain compactifications of symmetric spaces in Part II can be realized by gluing other compactifications from Part I in a manner analogous to the method of obtaining a closed manifold by doubling a manifold with boundary, which is called self-gluing in this book. In Part II, we describe a new and simpler technique for constructing analytic structures on compactifications by complexifying the symmetric spaces into complex symmetric varieties and using compactifications of the complex symmetric varieties as projective varieties.
4. Although compactifications of locally symmetric spaces are traditionally constructed using compactifications of symmetric spaces, we construct in Part III these compactifications independently of compactifications of symmetric spaces, and hence use only the $\mathbb{Q}$-structure of the spaces.
5. In Part III, we consider metric properties of compactifications of locally symmetric spaces. This gives a new perspective on sizes of compactifications and deepens our understanding of their properties. It also simplifies applications to extension properties of holomorphic maps from the punctured disk to Hermitian locally symmetric spaces. Furthermore, it clarifies relations to the continuous spectrum of the locally symmetric spaces.

The basic plan of the book was worked out and agreed upon by the two authors around December 2002. Part II was mostly written by the first author before his unexpected death on August 11, 2003.[1] The task of finishing this book fell to the second author, who regrets that his style may fail to meet the high standards of the first author, but who must nevertheless take responsibility for any errors or inaccuracies in the text.

This book is partially based on the joint papers [BJ1] [BJ2] [BJ3] [BJ4]. The book project was proposed by N. Wallach, an editor of this series, to the authors near the end of the European summer school on Lie theory in Marseille-Luminy, France, 2001, where the authors gave a joint series of lectures on compactifications of symmetric spaces and locally symmetric spaces.

The second author would like to thank N. Wallach for the book proposal and comments on an earlier version of this book, and J.P. Anker and P. Torasso for inviting the authors to give the lectures at the European summer school on Lie theory in 2001. He would also like to thank N. Mok for arranging and for inviting him to participate in the multiyear program on Lie theory organized by the first author at the University of Hong Kong, where some of the joint work was carried out, and

[1] Except for the introductions to Part II and the chapters there, several comments, and minor changes of notation for consistency, Part II was the version the first author finished and in the form he liked in June 2003.

his teacher M. Goresky for providing the carefully taken notes [Mac] that motivated Chapter 12 in Part III and for many helpful and encouraging conversations on various topics, comments, and suggestions. He thanks G. Prasad for helpful conversations and comments, and S. Zucker and A. Korányi for very helpful correspondence, conversations, specific and general suggestions, and detailed comments on an earlier version of the book. The second author would also like to thank his teacher S.T. Yau for inviting both authors to run a multiyear summer school on *Lie groups and automorphic forms* at the Center of Mathematical Sciences at Zhejiang University, which further encouraged us to work on this book project. It is sad that the first author could not attend any of the activities he planned at the center in Hangzhou.[2]

The second author would also like to thank R. Lazarsfeld for suggestions about the index and the layout of the book, E. Gustafsson, the first author's secretary, for typing up Part II, and A. Kostant for help and many suggestions during the preparation of this book. A seminar talk by W. Fulton on how to write mathematics led to some improvements of the writing and style of this book. The work of this book has been partially supported by NSF grants and an A.P. Sloan research fellowship.

Finally, the second author would like to thank his wife, Lan Wang, and his three daughters, Lena, Emily, and Karen, for their support, understanding, and patience during the intense writing and revising periods of this book.

[2] The second author co-organized an international conference titled *Algebraic groups, arithmetic groups, representation theory and automorphic forms* in memory of the first author at the Center of Mathematical Sciences at Zhejiang University, Hangzhou, July 26–30, 2004. The second author would like to thank the director of the center, S.T. Yau, and its staff members for their efforts and hard work to make the conference a success.

Compactifications of Symmetric and Locally Symmetric Spaces

0

Introduction

Symmetric spaces form an important class of Riemannian manifolds and are divided into three types: flat type, compact type and, noncompact type. Flat symmetric spaces are given by the Euclidean spaces $\mathbb{R}^n$ and theirthe quotients. A typical symmetric space of compact type is the unit sphere S^{n-1} in $\mathbb{R}^n$, and a typical symmetric space of noncompact type is given by a real hyperbolic space, i.e., a simply connected Riemannian manifold of constant negative curvature -1. The hyperbolic plane $\mathbf{H}$ can be identified with the symmetric space $SL(2, \mathbb{R})/SO(2)$. An important and natural generalization of $\mathbf{H}$ is $SL(n, \mathbb{R})/SO(n)$, the space of positive definite $n \times n$ symmetric matrices of determinant 1, which is a symmetric space of noncompact type.

The Euclidean space $\mathbb{R}^n$ is diffeomorphic, by retracting along rays from the origin, to the open unit ball with center at the origin, and hence it can be compactified by adding the unit sphere S^{n-1}. Alternatively, it can be obtained by adding the sphere S^{n-1} at infinity, which is the set of equivalence classes of asymptotic geodesics. This compactification of $\mathbb{R}^{n-1}$ is called the *geodesic compactification* and is useful in Fourier analysis on $\mathbb{R}^n$ (see [Me1]). Even for $\mathbb{R}^n$, there are other compactifications. Briefly, given additional structures on $\mathbb{R}^n$ such as polyhedral cone decompositions, $\mathbb{R}^n$ admits polyhedral compactifications (see §I.18 below), which are very different from the above-mentioned compactification and are responsible for several compactifications of higher-rank symmetric spaces and locally symmetric spaces.

In this book, we are mainly concerned with symmetric spaces of noncompact type and their quotients by arithmetic subgroups, which are referred to here as *locally symmetric spaces*. Quotients of symmetric spaces of noncompact type are abundant. For example, every Riemann surface of genus greater than or equal to 2 is a quotient of the upper half-plane, which becomes the hyperbolic plane when endowed with the Poincaré metric.

Locally symmetric spaces of noncompact type often occur naturally as moduli spaces in algebraic geometry and number theory and are usually noncompact, for example, the moduli space of elliptic curves and more general polarized abelian varieties [Mum2], the moduli space of abelian varieties with certain endomorphism groups [Hu], the moduli space of polarized K-3 surfaces and the related Enriques surfaces [Lo1] [Lo2], and some configuration spaces of points [Yo1] [Yo2].

Symmetric spaces of noncompact type are diffeomorphic to Euclidean spaces and hence noncompact. Clearly this diffeomorphism gives a compactification of the symmetric space via the geodesic compactification of the Euclidean spaces. But other structures of symmetric spaces can produce different compactifications that are important for some applications.

It will be seen below that compactifications of symmetric spaces and locally symmetric spaces are closely related. Even understanding some properties of compact locally symmetric spaces is related to compactifications of symmetric spaces. For example, in the proof of the Mostow strong rigidity [Mos], the *maximal Satake compactification* of the symmetric spaces and their *maximal Furstenberg boundaries* are used crucially.

As explained below, the problem of compactifying symmetric and locally symmetric spaces arises from many branches of mathematics motivated by various applications. In fact, different applications impose different conditions on the sizes of the ideal boundary.

The purposes of this book are the following:

1. To motivate how compactifications were constructed, why they have the given properties and structures of the boundary, and to explain connections between different types of compactifications.
2. To give uniform constructions of almost all compactifications and new compactifications.
3. To examine geometrical, topological properties of compactifications in terms of group structures and from the perspective of applications they lead to.

In the first part of this introduction, we recall briefly the history of compactifications of both symmetric and locally symmetric spaces. In the second part, we summarize new points of view on compactifications presented in this book. Then we give an outline of the contents of the book. Finally, we mention some important topics not covered in this book and the relation between this book and the earlier one [GJT].

Although they are important and have rich structures, symmetric and locally symmetric spaces of flat and compact types with the exception of the Euclidean spaces $\mathbb{R}^n$ are not studied in this book. For example, the study of compact quotients of $\mathbb{R}^n$ is closely related to the geometry of lattices and the geometry of numbers, and understanding them is also very important for many applications (see [Cass1] [CoS] [Gru] [GruL] [Si3]). The classification of such compact quotients of $\mathbb{R}^n$ or equivalently crystallographic groups is nontrivial (see [Wo2] for these and other related results on spherical space forms).

0.1 History of compactifications

We will mention only some of the (major) compactifications in the introduction here and leave others to the various chapters below.

0.1.1 The simplest symmetric space of noncompact type is the hyperbolic plane. One model is given by the Poincaré upper half-plane

$$\mathbf{H} = \mathbf{H}_1 = \{x + iy \mid x \in \mathbb{R}, y > 0\} = \mathrm{SL}(2, \mathbb{R})/\mathrm{SO}(2),$$

with the *Poincaré metric* $ds^2 = \frac{dx^2+dy^2}{y^2}$. The group $\mathrm{SL}(2, \mathbb{R})$ acts isometrically on $\mathbf{H}$ by fractional linear transformations. Another model is given by the open unit disk D with the Poincaré metric, often called the *Poincaré disk*,

$$D = \{z \in \mathbb{C} \mid |z| < 1\} = SU(1, 1)/U(1), \quad ds^2 = \frac{|dz|^2}{(1 - |z|^2)^2},$$

where $SU(1, 1)$ acts isometrically on D also by fractional linear transformations. In fact, the identification between $\mathbf{H}$ and D is given by the *Cayley transformation*

$$\mathbf{H} \to D, \quad z \mapsto \frac{z - i}{z + i}. \tag{0.1.1}$$

The disk D can be compactified by adding the unit circle S^1, and the $SU(1, 1)$-action on D extends continuously to the boundary. In terms of $\mathbf{H}$, this compactification $\overline{\mathbf{H}}$ is homeomorphic to $\mathbf{H} \cup \mathbb{R} \cup \{\infty\}$, which is considered as a subset of $\mathbb{C} \cup \{\infty\}$ The action of $\mathrm{SL}(2, \mathbb{R})$ on $\mathbf{H}$ extends to a continuous action on $\overline{\mathbf{H}}$.

This model of the hyperbolic plane in terms of the disk D was given by Beltrami in 1868, but the compactification $D \cup S^1$ was used by and must have been known to Poisson in the early 1800s in the famous *Poisson integral formula* for harmonic functions u on D with continuous boundary values:

$$u(z) = \int_{S^1} \frac{1 - |z|^2}{|z - \xi|^2} u(\xi) d\xi,$$

where $d\xi$ is the Haar measure on S^1 of total measure equal to 1. This could be the first example of a compactification of a symmetric spaces of noncompact type. (See [Ji7] for discussions of many different constructions of this compactification of D.)

0.1.2 A very important but simple example of locally symmetric spaces is the *modular curve*

$$\mathrm{SL}(2, \mathbb{Z})\backslash\mathbf{H} = \mathrm{SL}(2, \mathbb{Z})\backslash\mathrm{SL}(2, \mathbb{R})/\mathrm{SO}(2),$$

where $\mathbf{H}$ is the Poincaré upper half-plane. More general locally symmetric spaces related to $\mathbf{H}$ are given by quotient spaces $\Gamma\backslash\mathbf{H}$, for subgroups $\Gamma \subset \mathrm{SL}(2, \mathbb{Z})$ of finite index, for example the *principal congruence subgroups*

$$\Gamma(N) = \left\{\gamma \in \mathrm{SL}(2, \mathbb{Z}) \mid \gamma \equiv \begin{pmatrix} 1 & 0 \\ 0 & 1 \end{pmatrix} \mod N\right\}, \quad N \geq 1.$$

It is well known that these quotient spaces are noncompact. In fact, the space $\mathrm{SL}(2, \mathbb{Z})\backslash\mathbf{H}$ is the *moduli space of elliptic curves*. It can be mapped bijectively to

$\mathbb{C}$ under the *j-invariant* of elliptic curves (see [Sil, p. 46]), and hence can be compactified to the complex projective space $\mathbb{C}P^1 = \mathbb{C} \cup \{\infty\}$ by adding a single *cusp point* $\{\infty\}$. In fact, $\mathbb{C}$ has only one end, which is isomorphic to a neighborhood of the origin of $\mathbb{C}$ with the origin removed, i.e., a *punctured neighborhood*; and the compactification is obtained by filling in the *puncture*, or the cusp point. Similarly, other quotients $\Gamma\backslash\mathbf{H}$ can also be compactified to projective curves over $\mathbb{C}$ by adding finitely many cusp points, one cusp point for each end. These results were known at least by the end of the nineteenth century (see [Leh1, Chap. 1]) and have played an important role in the theory of modular forms of one variable (see [Shi]).

Briefly, starting in 1880, Poincaré realized the connection between the theory of modular (or automorphic) forms and the hyperbolic plane $\mathbf{H}$. He introduced the notions of *Fuchsian group* of the first kind and *cusp point*, and he understood the importance of cusp points in controlling the behavior at infinity of modular forms. Though the compactification of $\Gamma\backslash\mathbf{H}$ by adding cusp points was not explicitly mentioned in Poincaré's early papers (see [Leh1, Chap. I] [Po] [Had] [Ma]), it was certainly well understood by him. Related ideas were also developed by Klein in their strong competition around that period (see [Leh1, Chap. I] [Ya, pp. 130–133] [Ja]).

There is a close relation between the compactifications of $\mathbf{H}$ and $\Gamma\backslash\mathbf{H}$. Specifically, instead of the full boundary $\mathbb{R} \cup \{\infty\}$ of the compactification $\overline{\mathbf{H}}$ of $\mathbf{H}$, we pick out only the rational boundary points, $\mathbb{Q} \cup \{\infty\}$. Modify the subset topology of $\mathbf{H} \cup \mathbb{Q} \cup \{\infty\}$ so that horodisks (i.e., the interior of horocycles) near the boundary rational points form a basis of neighborhoods of these boundary points. This topology of the *partial compactification* $\mathbf{H} \cup \mathbb{Q} \cup \{\infty\}$ is often called the *Satake topology*. Then the Γ-action on $\mathbf{H}$ extends to a continuous action on $\mathbf{H} \cup \mathbb{Q} \cup \{\infty\}$ and the quotient $\Gamma\backslash\mathbf{H} \cup \mathbb{Q} \cup \{\infty\}$ is the compactification of $\Gamma\backslash\mathbf{H}$ discussed earlier.

The compactifications of $\mathbf{H}$ and $\Gamma\backslash\mathbf{H}$ and their applications provide the basic framework or paradigm for the general case.

0.1.3 The theory of automorphic forms in several variables was initiated by Siegel in [Si5]. A direct and important generalization of the upper half-plane $\mathbf{H}$ is the *Siegel upper half-space* $\mathbf{H}_n$ of degree n, $n \geq 1$,

$$\mathbf{H}_n = \{X + iY \mid X, Y \in M_n(\mathbb{R}), {}^tX = X, {}^tY = Y, Y > 0\} = Sp(n, \mathbb{R})/U(n),$$

where $Sp(n, \mathbb{R})$ is the group

$$Sp(n, \mathbb{R}) = \left\{ \gamma \in M_{2n}(\mathbb{R}) \mid {}^t\gamma \begin{pmatrix} 0 & I_n \\ -I_n & 0 \end{pmatrix} \gamma = \begin{pmatrix} 0 & I_n \\ -I_n & 0 \end{pmatrix} \right\}.$$

The group $Sp(n, \mathbb{R})$ acts on $\mathbf{H}_n$ by

$$\begin{pmatrix} A & B \\ C & D \end{pmatrix} \cdot (X + iY) = (A(X + iY) + B)(C(X + iY) + D)^{-1},$$

where A, B, C, D are $n \times n$ matrices.

Since the geometry, topology, and compactifications of the modular curve $\Gamma\backslash\mathbf{H}$ have played a fundamental and foundational role in studying the theory of modular forms in one variable, it is a natural problem to understand geometry and compactifications of noncompact quotients of $\mathbf{H}_n$, which will similarly give geometric foundations to the theory of automorphic forms in several variables.

In [Si2] [Si4], Siegel constructed a compactification of a *fundamental domain* in the *Siegel upper half-space* $\mathbf{H}_n$ of the *Siegel modular group* $Sp(n, \mathbb{Z})$, where

$$Sp(n, \mathbb{Z}) = \left\{ \gamma \in M_{2n}(\mathbb{Z}) \mid {}^t\gamma \left| \begin{pmatrix} 0 & I_n \\ -I_n & 0 \end{pmatrix} \gamma = \begin{pmatrix} 0 & I_n \\ -I_n & 0 \end{pmatrix} \right. \right\},$$

where I_n is the identity matrix of size n. This compactification of the fundamental domain defines a compactification of the quotient

$$V_n = Sp(n, \mathbb{Z})\backslash\mathbf{H}_n.$$

The space V_n is the moduli space of *principally polarized abelian varieties* [Mum2, Theorem 4.7] and is hence called the *Siegel modular variety*. This Siegel compactification was used by Christian in [Chr2] [Chr3] to study modular forms but does not seem to be used much by other people. Its relation with other later compactifications is not clear, though it seems to be related to both the Borel–Serre compactification in [BS2] and the toroidal compactifications in [AMRT] in certain ways (see the review of [Si4] in [Leh2]). (See also Remark III.3.17 below for comments on differences between compactifications of a fundamental domain of Γ in a symmetric space X and compactifications of the locally symmetric space $\Gamma\backslash X$.)

In order to relate Siegel modular functions, i.e., modular functions with respect to $Sp(n, \mathbb{Z})$, to meromorphic functions on $Sp(n, \mathbb{Z})\backslash\mathbf{H}_n$, Satake proposed in [Sat6] a procedure to compactify $Sp(n, \mathbb{Z})\backslash\mathbf{H}_n$ into a complex analytic *V-manifold* (or *orbifold*) and constructed such a compactification when $n = 2$.

In [Sat3], Satake defined a compactification V_n^* of V_n by adding the lower-dimensional spaces $V_k = Sp(k, \mathbb{Z})\backslash\mathbf{H}_k$, $0 \leq k < n$,

$$V_n^* = V_n \cup V_{n-1} \cup \cdots \cup V_1 \cup \{\infty\},$$

where the point $\{\infty\}$ corresponds to the space for $k = 0$, i.e., the unique cusp point of $Sp(1, \mathbb{Z})\backslash\mathbf{H}_1 = \mathrm{SL}(2, \mathbb{Z})\backslash\mathbf{H}$. He also showed that the compactification V_n^* is a complex analytic space and conjectured that with this analytic structure, it is a projective variety.

In [Ba1], Baily proved that this Satake compactification V_n^* is a *normal projective variety* and obtained as a corollary that for $n \geq 2$, every meromorphic function on $Sp(n, \mathbb{Z})\backslash\mathbf{H}_n$ is a quotient of two modular forms of the same weight. Later he showed in [Ba2] that any *Hilbert–Siegel modular space* also admits a compactification that is a normal projective variety.

We observe that there are three steps in the construction of the compactification $(V_n)^*$ mentioned in the previous paragraphs:

1. A topological compactification is obtained by adding spaces V_k, $k < n$, of the same type but of lower dimension.

2. A sheaf of analytic functions is constructed on the topological compactification to make it into a compact *normal complex analytic space*.
3. Embedding the compact normal complex space into a complex projective space as a *normal projective subvariety*.

To carry out the second and third steps, the fact that $V_n = Sp(n, \mathbb{Z})\backslash\mathbf{H}_n$ is a Hermitian locally symmetric space, i.e., $\mathbf{H}_n$ has an $Sp(n, \mathbb{R})$-invariant complex structure, is necessary.

On the other hand, there are important non-Hermitian locally symmetric spaces. For example, one such space is

$$\mathrm{SL}(n, \mathbb{Z})\backslash\mathrm{SL}(n, \mathbb{R})/\mathrm{SO}(n), n \geq 3,$$

which is the *moduli space* of equivalence classes of positive definite quadratic forms of determinant 1. Compactifications of such spaces are useful for some applications.

0.1.4 In [Sat1] and [Sat2], Satake started the modern theory of compactifications of general symmetric spaces and locally symmetric spaces, which corresponds to the first step in the construction of V_n^* above.

More specifically, let $X = G/K$ be a Riemannian symmetric space of noncompact type, for example, the Siegel upper half-space

$$\mathbf{H}_n = Sp(n, \mathbb{R})/U(n),$$

or the space of positive definite Hermitian quadratic forms of determinant 1,

$$X = \mathrm{SL}(n, \mathbb{C})/SU(n).$$

Let Γ be an arithmetic subgroup of G and $\Gamma\backslash X$ the associated locally symmetric space, which is assumed to be noncompact in this book. Unlike the case of the Siegel modular variety V_n discussed earlier, there are no obvious choices of lower-dimensional locally symmetric spaces that can be attached at infinity to obtain a topological compactification of $\Gamma\backslash X$ as in Step 1 in §.1.3.

Satake solved this problem in two steps, which correspond to [Sat1] and [Sat2]. To obtain the desired boundary spaces and the topology of how they are attached at infinity of X and $\Gamma\backslash X$, Satake started in [Sat1] by constructing a finite family of Satake compactifications $\overline{X}^S$ of X and decomposing the boundary of the Satake compactifications $\overline{X}^S$ into lower-dimensional symmetric spaces, called *boundary components*. In fact, there are finitely many types of boundary components, and the boundary components of each type form a *G-orbit*. When $X = \mathrm{SL}(n, \mathbb{C})/SU(n)$, the *standard Satake compactification* is obtained by adding *semipositive Hermitian matrices*, and the boundary components are of the form $\mathrm{SL}(k, \mathbb{C})/SU(k)$ for $k \leq n-1$, as expected.

As explained in the example of the compactification of $\Gamma\backslash\mathbf{H}$ in §.1.2, we need only a certain rational part of the boundary of the compactification of $\mathbf{H}$. Instead of rational points, we need some notion of *rational boundary components* of compactifications of X. To pick out certain boundary components of the Satake compactifications of X and use their quotients to compactify locally symmetric spaces

$\Gamma\backslash X$, Satake used in [Sat2] the closure of a suitable fundamental set of Γ in the Satake compactifications of X. It is reasonable to expect that only those boundary components that meet the closure of the fundamental set of Γ in X, which are called Γ*-rational boundary components* in [Sat2] and *Siegel rational boundary components* in this book, are needed to compactify the fundamental set and hence $\Gamma\backslash X$. It should be pointed out that compactifications of a fundamental domain (or set) of Γ in X are different from compactifications of $\Gamma\backslash X$, and there are difficulties in passing from one to the other (see Remark III.3.17 below). Under some assumptions on the fundamental set and the Γ-rational boundary components, he compactified $\Gamma\backslash X$ in two steps:

1. Construct a *partial compactification* of X by adding these Γ-rational boundary components, and endow the partial compactification with a so-called *Satake topology*.
2. Show that Γ acts continuously on the partial compactification with a compact Hausdorff quotient, which defines a Satake compactification of $\Gamma\backslash X$, denoted by $\overline{\Gamma\backslash X}^S$.

Clearly, this compactification of $\Gamma\backslash X$ depends on the Satake compactification $\overline{X}^S$. Since there are finitely many nonisomorphic Satake compactifications of X, there are at most finitely many different Satake compactifications of $\Gamma\backslash X$.

Basically, the assumptions above guarantee that Γ induces a discrete group action on each Γ-rational boundary component of $\overline{X}^S$, and gives a Hausdorff quotient of the Γ-rational boundary component, called a *boundary locally symmetric space*. These boundary symmetric spaces can be attached to the infinity of $\Gamma\backslash X$ and define a compactification of $\Gamma\backslash X$. These assumptions in [Sat2] depend on the action of Γ on the Satake compactifications $\overline{X}^S$ of X and reflect some rational property of the Satake compactifications $\overline{X}^S$; hence, the Satake compactifications of X that satisfy these assumptions are called *geometrically rational* in [Cas2], also called *rational* in this book.

0.1.5 In a different direction, in order to generalize the classical integral Poisson formula in equation (0.1.1) for harmonic functions on the unit disk with continuous boundary values to symmetric spaces of noncompact type, Furstenberg introduced the notion of *Furstenberg boundaries* in [Fu1]. Then he embedded the symmetric space into the *space of probability measures* on the *maximal Furstenberg boundary* and hence obtained the *maximal Furstenberg compactification* as the closure of this embedding [Fu1, Chap. II]. In [Mo1], Moore clarified the structures of all the Furstenberg boundaries, generalized the construction in [Fu1] by defining a Furstenberg compactification for each *faithful Furstenberg boundary*, and showed that these finitely many Furstenberg compactifications $\overline{X}^F$ are isomorphic to the finitely many Satake compactifications $\overline{X}^S$ in [Sat1]. Because of this, the Satake compactifications are often called the *Satake–Furstenberg compactifications* and denoted by $\overline{X}^{SF}$, for example in [GJT]. On the other hand, we will not follow this convention in view of applications of Satake compactifications of X to compactifications of locally symmetric spaces $\Gamma\backslash X$.

Using this identification between the Satake and Furstenberg compactifications of symmetric spaces, Moore showed in [Mo2] that the closure of the *Harish-Chandra canonical realization* of a Hermitian symmetric space of noncompact type as a *bounded symmetric domain* is isomorphic to a minimal Satake compactification, which was verified earlier for *classical domains* by Satake in [Sat1]. Around the same time, Hermann carried out related investigations on compactifications of symmetric spaces and potential theory in these spaces [Her1]–[Her4]. In [Ko1], the problem of intrinsically constructing $\overline{X}^S$ was raised, and an approach using admissible domains and filters was sketched. See also [Ko2–6] for many related results on admissible domains and boundary values of harmonic functions on symmetric spaces.

The closure of a bounded symmetric domain (or Hermitian symmetric space) X is often refereed to as the *Baily–Borel compactification* $\overline{X}^{BB}$ in view of its application to the *Baily–Borel compactification* of Hermitian locally symmetric spaces $\Gamma\backslash X$. In fact, in [BB1], Baily and Borel constructed the compactification $\overline{X}^{BB}$ in three steps:

1. Apply the general procedure in [Sat2] to construct a topological compactification of $\Gamma\backslash X$.
2. Define a sheaf of analytic functions on the topological compactification to make it into a *compact normal complex analytic space*.
3. Embed the compactification in step (2) into a complex projective space as a *normal projective variety* by using automorphic forms.

In other words, they generalized the construction and properties of the compactification V_n^* of the Siegel modular variety V_n in [Sat3] [Ba1] outlined in §0.1.3 above to all Hermitian locally symmetric spaces $\Gamma\backslash X$.

Except for some low-dimensional cases, the boundary $\overline{\Gamma\backslash X}^{BB} - \Gamma\backslash X$ of the Baily–Borel compactification $\overline{\Gamma\backslash X}^{BB}$ is a subvariety of complex codimension at least 2. By the Riemann extension theorem, this immediately implies that the transcendental degree of the field of meromorphic functions on $\Gamma\backslash X$ is equal to $\dim_{\mathbb{C}} \Gamma\backslash X$, which was a problem posed by Siegel, and every such meromorphic function is a quotient of two modular forms of the same weight. In this application, the smallness of the boundary or rather its high codimension is important. It will be seen below that for other applications, we need larger compactifications or rather boundaries.

When X is a *classical domain*, a similar compactification of $\Gamma\backslash X$ was also obtained by Piatetski-Shapiro in [PS] using realizations of X as *Siegel domains of the third kind* (or more special Siegel domains of the first and second kinds) with respect to boundary components in the closure. In fact, such realizations were used to define a topology near the boundary points, called the *cylindrical topology*. Since this approach is different from the general procedure in [Sat2], the cylindrical topology might be different from the Satake topology of the partial compactification of X. On the other hand, it was shown by Kiernan in [Ki] that they define the same topology on the compactication of $\Gamma\backslash X$.

In [Ig3], Igusa showed that the Baily–Borel compactification $\overline{\Gamma\backslash X}^{BB}$ is singular for most $\Gamma\backslash X$ (see also [Chr1]). This is reasonable since in most cases, as pointed out earlier, the boundary of the Baily–Borel compactification $\overline{\Gamma\backslash X}^{BB}$ is a subvariety of complex codimension greater than or equal to 2, rather than a divisor, and the transversal links of the boundary components are not spheres. For example, when $\Gamma\backslash X$ is a *Hilbert modular surface*, the compactification $\overline{\Gamma\backslash X}^{BB}$ is obtained by adding one point to each end. It can be shown easily that the link of each such added point in $\overline{\Gamma\backslash X}^{BB}$ is equal to an $(S^1)^2$-bundle over the circle S^1, and hence not homeomorphic to the sphere S^3, which implies that $\overline{\Gamma\backslash X}^{BB}$ is singular.

By the Hironaka resolution theorem [Hir1] [Hir2], the singularities of $\overline{\Gamma\backslash X}^{BB}$ can be resolved, i.e., there are smooth varieties Y and proper morphisms $Y \to \overline{\Gamma\backslash X}^{BB}$ that are isomorphisms over $\Gamma\backslash X$ (when Γ is torsion free so that $\Gamma\backslash X$ is smooth). But such resolutions are not canonical, and the method is general and works for all varieties over $\mathbb{C}$. The group structures of X and $\Gamma\backslash X$ are not used either in the construction or to pick out some special resolutions by resorting to this general theorem.

Certainly, it is desirable to obtain natural, explicit resolutions of the singularities of $\overline{\Gamma\backslash X}^{BB}$. In [Ig1] [Ig2], Igusa studied a *partial desingularization* of the Baily–Borel compactification of the Siegel modular variety V_n^* by blowing up the singular locus. In [Hir1], Hirzebruch resolved the singularities of the Baily–Borel compactification of the Hilbert modular surfaces by blowing up each ideal point into a finite cycle of rational curves. When X is a *tube domain*, Satake considered some resolutions of the Baily–Borel compactification $\overline{\Gamma\backslash X}^{BB}$ in [Sat5]. Motivated by these results, Mumford and his collaborators constructed in [AMRT] a family of toroidal compactifications $\overline{\Gamma\backslash X}_\Sigma^{tor}$ of $\Gamma\backslash X$ that dominate the Baily–Borel compactification. In general, there are infinitely many such compactifications $\overline{\Gamma\backslash X}_\Sigma^{tor}$ that depend on some auxiliary combinatorial data Σ, and many of them are smooth projective varieties and hence resolve the singularities of $\overline{\Gamma\backslash X}^{BB}$.

The theory of torus embeddings (or toric varieties) developed in [KKMS] plays an essential role in these compactifications, and Σ is a Γ-*admissible family of polyhedral cones,* which are needed to construct torus embeddings used in the toroidal compactifications.

Besides resolving the singularities, the toroidal compactifications are also important for other applications, for example [Mum3] [FC] [Ale] [Nam1–4] [Lo6].

0.1.6 Around the same time, another compactification $\overline{\Gamma\backslash X}^{BS}$ of $\Gamma\backslash X$ was introduced by Borel and Serre in [BS2] in order to obtain a finite $K(\Gamma, 1)$-space (see below for the definition) and use it to study the cohomology groups of Γ. It turns out to give a resolution in the sense of differential topology of the singularities of the Baily–Borel compactification $\overline{\Gamma\backslash X}^{BB}$, or more generally the Satake compactifications $\overline{\Gamma\backslash X}^{S}$.

When Γ is torsion free, $\Gamma\backslash X$ is a $K(\Gamma, 1)$*-space*, i.e., $\pi_1(\Gamma\backslash X) = \Gamma$ and $\pi_i(\Gamma\backslash X) = 0$ for $i \geq 2$.

If $\Gamma\backslash X$ is compact, then $\Gamma\backslash X$ has a finite triangulation and hence is a finite CW-complex, which shows that $\Gamma\backslash X$ is a so-called *finite* $K(\Gamma, 1)$*-space.*

This allows one to study the cohomology groups and other finiteness properties of Γ. For example, it implies that Γ is *finitely presented.*

When $\Gamma\backslash X$ is noncompact, one way to obtain a finite $K(\Gamma, 1)$-space is to construct a compactification $\overline{\Gamma\backslash X}$ of $\Gamma\backslash X$ that has a finite triangulation such that the inclusion $\Gamma\backslash X \to \overline{\Gamma\backslash X}$ is a homotopy equivalence. The *Borel–Serre compactification* $\overline{\Gamma\backslash X}^{BS}$ constructed in [BS2] is a *differential manifold with corners* whose interior is equal to $\Gamma\backslash X$ and hence is a compactification of the desired type.

On the other hand, the Satake compactifications $\overline{\Gamma\backslash X}^{S}$ do not satisfy this homotopy equivalence condition. For example, when $\Gamma\backslash X$ is a Riemann surface $\Gamma\backslash\mathbf{H}$, there is a unique Satake compactification, which is obtained by adding a cusp point to each end. Clearly the nontrivial loops in $\Gamma\backslash X$ in the cusp neighborhoods are homotopic to the cusp points in $\overline{\Gamma\backslash X}^{S}$. Hence the inclusion $\Gamma\backslash X \to \overline{\Gamma\backslash X}^{S}$ is not a homotopy equivalence. In this sense, the Satake compactifications are too small for this purpose. In this example, there is a unique toroidal compactification, which is isomorphic to the Satake compactification and hence does not satisfy this homotopy equivalence condition either. It is worthwhile to point out that the toroidal compactifications of $\Gamma\backslash X$ are often smooth and closed, as in this example, and hence their interior strictly contains $\Gamma\backslash X$, which prevents the inclusion of $\Gamma\backslash X \to \overline{\Gamma\backslash X}$ from being a homotopy equivalence.

Together with the application of the Baily–Borel compactification to the problem of Siegel, the above discussions explain the importance of compactifications with different sizes and properties.

In studying the L^2-cohomology of $\Gamma\backslash X$, a quotient of the Borel–Serre compactification $\overline{\Gamma\backslash X}^{BS}$, called the *reductive Borel–Serre compactification* and denoted by $\overline{\Gamma\backslash X}^{RBS}$, was introduced by Zucker in [Zu1, p. 190]. The basic reason is that $\overline{\Gamma\backslash X}^{BS}$ is too large to support *partition of unity*, which is a basic tool in *De Rham cohomology theory*. It was later used crucially in [GHM] to define and study the *weighted cohomology groups* and to compute the *Lefschetz number* of *Hecke correspondences* (see [GM1] [GM2] and [Go] for a survey of such applications). Roughly, the reductive Borel–Serre compactification is suitable for these applications because it is not too small, so that its singularities are not too complicated. As pointed out earlier, it is not too big either and supports partition of unity.

In order to understand better the local neighborhoods of the Baily–Borel compactification $\overline{\Gamma\backslash X}^{BB}$ at infinity and hence to relate the L^2-cohomology groups of $\Gamma\backslash X$ to other topological invariants such as the *intersection cohomology groups* of $\overline{\Gamma\backslash X}^{BB}$, Zucker showed in [Zu2] that all the Satake compactifications $\overline{\Gamma\backslash X}^{S}$, in particular, the Baily–Borel compactification $\overline{\Gamma\backslash X}^{BB}$, can be realized as quotients of the reductive Borel–Serre compactification $\overline{\Gamma\backslash X}^{RBS}$ and hence also the Borel–Serre compactification $\overline{\Gamma\backslash X}^{BS}$. Since the Borel–Serre compactification $\overline{\Gamma\backslash X}^{BS}$ is a manifold with corners, the dominating map $\overline{\Gamma\backslash X}^{BS} \to \overline{\Gamma\backslash X}^{BB}$ can be considered as a resolution of the singularities in a topological or differential geometric sense as mentioned earlier.

As discussed earlier, the toroidal compactifications $\overline{\Gamma\backslash X}_{\Sigma}^{tor}$ were constructed in [AMRT] to explicitly resolve the singularities of $\overline{\Gamma\backslash X}^{BB}$. A natural problem is to compare them with the Borel–Serre compactification $\overline{\Gamma\backslash X}^{BS}$ and to understand the differences between these two kinds of resolutions.

The methods to construct the Borel–Serre compactification and the toroidal compactifications are very different, and they seem to be unrelated except that they both dominate the Baily–Borel compactification. It was conjectured in [HZ2] that the compactifications $\overline{\Gamma\backslash X}^{BS}$ and $\overline{\Gamma\backslash X}_{\Sigma}^{tor}$ are incompatible except for having the Baily–Borel compactification $\overline{\Gamma\backslash X}^{BB}$ as a *common quotient*. The incompatibility between these compactifications $\overline{\Gamma\backslash X}^{BS}$ and $\overline{\Gamma\backslash X}_{\Sigma}^{tor}$ could be used to study the cohomology groups of Γ [Zu5]. It turns out [Ji3] that the *greatest common quotient* (GCQ) of $\overline{\Gamma\backslash X}^{RBS}$ and $\overline{\Gamma\backslash X}_{\Sigma}^{tor}$ is equal to $\overline{\Gamma\backslash X}^{BB}$, and that the GCQ of $\overline{\Gamma\backslash X}_{\Sigma}^{tor}$ and $\overline{\Gamma\backslash X}^{BS}$ is a new compactification, which is sometimes strictly greater than, i.e., strictly dominates, $\overline{\Gamma\backslash X}^{BB}$. For example, when $\Gamma\backslash X$ is the *Picard modular surface*, there is a unique toroidal compactification $\overline{\Gamma\backslash X}_{\Sigma}^{tor}$, and $\overline{\Gamma\backslash X}^{BS}$ dominates $\overline{\Gamma\backslash X}_{\Sigma}^{tor}$, i.e., the GCQ is equal to $\overline{\Gamma\backslash X}_{\Sigma}^{tor}$. See [Ji3] for precise conditions when the GCQ is equal to $\overline{\Gamma\backslash X}^{BB}$.

Understanding the GCQ is one way to compare two compactifications. Another is to understand the *least common refinement* or *modification* (LCR or LCM) of them, which is equal to the closure of $\Gamma\backslash X$ under the diagonal embedding into the product of the two compactifications. In [GT1], the least common modification (LCM) of $\overline{\Gamma\backslash X}_{\Sigma}^{tor}$ and $\overline{\Gamma\backslash X}^{RBS}$ is shown to be homotopy equivalent to $\overline{\Gamma\backslash X}_{\Sigma}^{tor}$ when the polyhedral cone decomposition Σ is sufficiently fine with respect to Γ. This implies that for any compact subset C of $\Gamma\backslash X$, there is a map $\overline{\Gamma\backslash X}_{\Sigma}^{tor} \to \overline{\Gamma\backslash X}^{RBS}$ that restricts to the identity map on C. Such maps are important in studying characteristic classes of homogeneous vector bundles over $\Gamma\backslash X$ (see also [Zu3], which was motivated by [GT1]).

0.1.7 In the meantime, compactifications of symmetric spaces X have been studied from different points of view. As a noncompact complete Riemannian manifold, X admits a family of *Martin compactifications* $X \cup \partial_\lambda X$, where $\lambda \in (-\infty, \lambda_0)$ and λ_0 is the bottom of the spectrum of X. The Martin compactification $X \cup \partial_\lambda X$ is determined by the asymptotic behavior of the *Green function* of the operator $\Delta - \lambda$ at infinity, where $\Delta \geq 0$ is the Laplace operator of X. Briefly, each point in the *Martin boundary* $\partial_\lambda X$ corresponds to a positive eigenfunction on X with eigenvalue λ, called a *Martin kernel function*, and they span the cone of positive eigenfunctions on X of eigenvalue λ. To remove redundancy (or linear dependence) between these

Martin kernel functions, we need to restrict to the *minimal Martin boundary* $\partial_{\lambda,\min}X$, which gives rise to the *minimal Martin kernel functions*.

A natural problem is to understand these functional-theoretical compactifications and their boundaries geometrically and identify them with other compactifications. When $X = \mathrm{SL}(n, \mathbb{C})/SU(n)$, the Green function can be computed explicitly in terms of elementary functions, and the explicit formula was used by Dynkin in [Dy2] to determine all the Martin kernel functions, the minimal Martin kernels, and their parameter spaces, i.e., the Martin boundaries as sets. On the other hand, the relation of the Martin compactifications to other compactifications was not clarified or studied. For example, the Martin boundaries were not realized as subsets of boundaries of some more geometrically defined compactifications. If G is not a complex group, there is no such explicit formula for the Green function, and the problem is more difficult.

Motivated by this problem, Karpelevič studied the geometry of geodesics in X in great detail in [Ka] and defined the geodesic compactification $X \cup X(\infty)$, where $X(\infty)$ is the set of *equivalence classes of geodesics* in X and homeomorphic to a sphere, often called the *sphere at infinity* of X. He also defined a more refined *Karpelevič compactification* $\overline{X}^K$ whose boundary points correspond to more refined equivalence classes of geodesics. Then he used a suitable subset of the boundary of $\overline{X}^K$ to parametrize the minimal Martin functions, i.e., embedded the minimal Martin boundary $\partial_{\lambda,\min}X$ into the boundary of $\overline{X}^K$.

The geodesic compactification $X \cup X(\infty)$ can also be defined for any simply connected nonpositively curved complete Riemannian manifold in the same way [EO], and has played an important role in the study of geometry of manifolds of nonpositive curvature (see [BGS] [Eb]). It was proved in [And] [Sul] that the Dirichlet problem at infinity on any simply connected Riemannian manifold with negatively pinched sectional curvature is solvable, i.e., given any continuous function f on $X(\infty)$, there is a harmonic function u on X whose boundary value on $X(\infty)$ is given by f. This implies that the vector space of nonconstant bounded harmonic functions on such a manifold is of infinite dimension. (Among symmetric spaces of noncompact type, only those of rank 1 have negatively pinched sectional curvature.) Later in [AS] [Anc], under the same condition on the manifold, the geodesic compactification $X \cup X(\infty)$ was identified with the Martin compactification. Furthermore, it was shown that every boundary point in the Martin compactification is minimal. Since bounded harmonic functions become positive after a sufficiently large positive constant is added, the identification of the Martin boundary for $\lambda = 0$ is a stronger result than the solvability of the Dirichlet problem at infinity.

On the other hand, when the curvature is only nonpositively curved but not negatively pinched, the Martin compactification has not been identified with other more geometric compactifications in general. In fact, there is no conjecture in general on the precise relation between the Martin compactification and the geodesic compactification for this class of manifolds. Symmetric spaces of noncompact type of rank greater than or equal to 2 are important examples of such Riemannian manifolds. Motivated by this problem, the Martin compactifications of symmetric spaces of noncompact type were completely determined in [GJT]. It turns out that both the geodesic compactification and the maximal Satake compactification are needed to determine the Martin compactifications. In fact, for $\lambda < \lambda_0$, the Martin compactification $X \cup \partial_\lambda X$ is equal to the *least common refinement* (or modification) of the geodesic compactification $X \cup X(\infty)$ and the maximal Satake compactification $\overline{X}^S_{\max}$; and for $\lambda = \lambda_0$, $X \cup \partial_\lambda X$ is isomorphic to the maximal Satake compactification $\overline{X}^S_{\max}$ of X. In [GJT], a geometric construction of the maximal Satake compactification, the *dual cell compactification* $X \cup \Delta^*(X)$, was given and used in determining this result. This compactification is related to polyhedral compactifications of Euclidean spaces mentioned at the beginning of the introduction.

0.1.8 In the study of harmonic analysis on symmetric spaces $X = G/K$, in particular, the *Helgason conjecture* on joint eigenfunctions of invariant differential operators on X (see [Hel4] [Hel5]), Oshima constructed in [Os1] a closed real analytic G-manifold $\overline{X}^O$, the *Oshima compactification.*

It contains the disjoint union of finitely many copies of X as an open dense subset. It consists of finitely many G-orbits, and there is a unique compact (or closed) G-orbit, which is equal to the *maximal Furstenberg boundary* of X (or G). Then the invariant differential operators have *regular singularities* along this closed G-orbit, so the theory of differential equations with regular singularities can be applied to study the joint eigenfunctions on X and to prove the Helgason conjecture in [KaK]. (See [Sch] for detailed discussions of the results here.)

To generalize the Helgason conjecture to a class of non-Riemannian symmetric spaces, Oshima and Sekiguchi introduced in [OsS1] a closed real analytic manifold $\overline{X}^{OS}$, the *Oshima–Sekiguchi compactification*, which contains a finite disjoint union of both Riemannian and non-Riemannian symmetric spaces as an open dense subset.

0.1.9 Around the same time, motivated by problems in enumerative algebraic geometry, a different kind of compactification was constructed in [DP1], the *wonderful compactification* of a symmetric variety.

In fact, let $\mathbf{G}$ be a semisimple linear algebraic group, and $\mathbf{H}$ the fixed point set of an involution on $\mathbf{G}$. Then $\mathbf{X} = \mathbf{G}/\mathbf{H}$ is a complex symmetric space (or a symmetric variety). One example of symmetric varieties is obtained from the complexification $\mathbf{X} = X_{\mathbb{C}}$ of a symmetric space $X = G/K$, where G, K are linear, by taking $\mathbf{G}$ to be the complexification $G_{\mathbb{C}}$ of G, $\mathbf{H}$ the complexification $K_{\mathbb{C}}$ of K and $X_{\mathbb{C}} = G_{\mathbb{C}}/K_{\mathbb{C}}$. A particular example is the variety of *nondegenerate quadrics* in $\mathbb{C}P^n$, which is equal to $\mathrm{SL}(n, \mathbb{C})/\mathrm{SO}(n, \mathbb{C})$.

The *variety of complete quadrics* is a smooth compactification of the symmetric variety $\mathrm{SL}(n, \mathbb{C})/\mathrm{SO}(n, \mathbb{C})$; it plays an important role in enumerative algebraic geometry (see [DGMP]). For a general symmetric variety $\mathbf{X}$, the *wonderful compactification* of $\overline{\mathbf{X}}^W$ in [DP1] is a generalization of the variety of *complete quadrics* and has various important applications as well.

0.2 New points of view in this book

0.2.1 There are basically three types of compactifications, and each of the compactifications mentioned above belongs to one of them:

1. Compact spaces that contain a symmetric space X as an open dense subset.
2. Compact smooth analytic manifolds that contain a disjoint union of more than one but finitely many symmetric spaces as an open dense subset.
3. Compact spaces that contain a locally symmetric space $\Gamma\backslash X$ as an open dense subset.

Spaces of type (1) are compactifications of X in the usual sense and include the following: the Satake compactifications $\overline{X}^S$, the Furstenberg compactifications $\overline{X}^F$, the geodesic compactification $X \cup X(\infty)$, the Karpelevič compactification $\overline{X}^K$, the Martin compactification $X \cup \partial_\lambda X$.

Spaces of type (3) are compactifications of $\Gamma\backslash X$ also in the usual sense and include the following: the Satake compactifications $\overline{\Gamma\backslash X}^S$, the Baily–Borel compactification $\overline{\Gamma\backslash X}^{BB}$, the Borel–Serre compactification $\overline{\Gamma\backslash X}^{BS}$, the reductive Borel–Serre compactification $\overline{\Gamma\backslash X}^{RBS}$, the toroidal compactifications $\overline{\Gamma\backslash X}_\Sigma^{tor}$.

Spaces of type (2) include the Oshima compactification $\overline{X}^O$, the Oshima–Sekiguchi compactification $\overline{X}^{OS}$, and the wonderful compactification of complex symmetric spaces $\overline{\mathbf{X}}^W$. Though the complex symmetric space $\mathbf{X}$ is open and dense in the wonderful compactification $\overline{\mathbf{X}}^W$, it will turn out to be better to classify it as a space in type (2), since its real locus is such a space.

An analogue of the Oshima compactification for locally symmetric spaces is the *Borel–Serre–Oshima compactification* of $\Gamma\backslash X$. It does not belong to any of the above three types. Since there is only one such compactification of $\Gamma\backslash X$, we consider it together with compactifications of type (3) for simplicity.

0.2.2 From the brief descriptions of the compactifications in §0.1, compactifications of types (1), (2), and (3) are constructed by very different methods. Specifically, the Satake and the Furstenberg compactifications of the symmetric spaces are obtained by embedding $X = G/K$ into compact G-spaces, though an alternative intrinsic construction of the Satake compactifications was proposed and sketched by Koranyi in [Ko1] using admissible domains and filters converging to ideal boundary points; while the Satake, the Baily–Borel, the Borel–Serre, and the reductive Borel–Serre compactifications of the locally symmetric spaces $\Gamma\backslash X$ are constructed by attaching ideal boundary components that are parametrized by rational parabolic subgroups.

The two basic points of this book are the following:

1. Compactifications of types (1) and (3) can be constructed by a uniform method depending on the reduction theories over $\mathbb{R}$ and $\mathbb{Q}$ respectively, i.e., the finiteness and separation properties of Siegel sets of real parabolic subgroups and

rational parabolic subgroups. The constructions of compactifications of symmetric spaces X and locally symmetric spaces $\Gamma\backslash X$ are parallel and use similar methods.

2. The Oshima compactification $\overline{X}^{O}$ of type (2) can be obtained by gluing up several copies of the maximal Satake compactification $\overline{X}^{S}_{\max}$, a compactification of type (1). The real analytic structure of $\overline{X}^{O}$ can be easily obtained from the real locus $\mathbf{X}^{W}_{\mathbb{C}}(\mathbb{R})$ of the wonderful compactification $\mathbf{X}^{W}_{\mathbb{C}}$ of the complexification $X_{\mathbb{C}}$ of the symmetric space X. The real locus of $\mathbf{X}^{W}_{\mathbb{C}}$ is also related to the Oshima–Sekiguchi compactification $\overline{X}^{OS}$ by a finite-to-one map.

0.2.3 There are five unifying themes in this book.

The *first theme* of this book is that all the compactifications of X in type (1) can be constructed by attaching boundary components of real parabolic subgroups, which is a typical method for constructing compactifications of locally symmetric spaces. Hence compactifications of symmetric and locally symmetric spaces can be studied together in a uniform way.

The *second theme* is that in order to study compactifications of $\Gamma\backslash X$, it is more natural in some sense to study compactifications of the homogeneous space $\Gamma\backslash G$ and then take K-quotients (see §III.13.1), and compactifications of $\Gamma\backslash G$ can also be obtained by embedding it into the compact G-space consisting of closed subgroups of G using lattices in G. Consequently, the reductive Borel–Serre compactification of $\Gamma\backslash X$ can also be obtained by embedding $\Gamma\backslash X$ into a compact space and taking the closure. As mentioned earlier, embedding into compact spaces has been an important way to compactify symmetric spaces. Closely related to this approach is the problem of how to use lattices in Euclidean spaces and other related closed subgroups to study compactifications of locally symmetric spaces when $X = \mathrm{SL}(n, \mathbb{R})/\mathrm{SO}(n)$.

The Oshima compactification of X has two important properties: it is a closed manifold and it has a real analytic structure. The first part of *the third theme* of this book is to treat these two properties separately so that it is easier to understand its structure and relation to other compactifications; the second part of *the third theme* of this book is that for certain questions one can simplify problems by complexifying the symmetric spaces to obtain symmetric varieties and then taking the real locus of compactifications of the symmetric varieties. In fact, the first property corresponds to the self-gluing of the maximal Satake compactification of X, and the existence of the real analytic structure follows easily from the embedding of X into the real locus of the wonderful compactification $\overline{\mathbf{X}}^{W} = \mathbf{X}^{W}_{\mathbb{C}}$ of the complexification $\mathbf{X} = X_{\mathbb{C}}$ and the structure of this real locus. The Oshima–Sekiguchi compactification $\overline{X}^{OS}$ contains symmetric spaces of different types (i.e., some are Riemannian symmetric and others are non-Riemannian symmetric) as open subsets, and can be regarded as the glue-up of compactifications of these symmetric spaces. Unlike the case of self-gluing in the Oshima compactification $\overline{X}^{O}$, it is not obvious why these compactifications of spaces of different types can be glued together along their boundaries. The description of the real locus of the wonderful compactification $\overline{\mathbf{X}}^{W}$ explains naturally this gluing and the compatibility of the boundary components of these spaces of different types. The real locus $\overline{\mathbf{X}}^W(\mathbb{R})$ of the wonderful compactification also occurs naturally in Poisson geometry (see [EL]).

In many papers, for example, [Sat1] [Sat2] [BB1], the construction of compactifications of locally symmetric spaces $\Gamma\backslash X$ depends crucially on compactifications of symmetric spaces X. The uniform approach in this book allows one to construct compactifications of $\Gamma\backslash X$ completely independently of compactifications of symmetric spaces. This is *the fourth theme* of this book.

Compactifications of locally symmetric spaces have not been systematically studied from the point of view of metric spaces in the literature recalled earlier. It is intuitively clear that compactifications are some ways to study the structure of spaces at infinity. It is natural to relate compactifications to other properties which are connected to the geometry at infinity. The *fifth theme* of this book is to understand relations between metrics and compactifications such as the *hyperbolic compactifications* and the geodesic compactification $\Gamma\backslash X \cup \Gamma\backslash X(\infty)$ of $\Gamma\backslash X$, and consider sizes of compactifications according to the metric. A closely related topic concerns relations between the *geometry at infinity*, for example, geodesics in $\Gamma\backslash X$ that go to infinity or boundaries of compactifications of $\Gamma\backslash X$, and the *continuous spectrum* of $\Gamma\backslash X$.

0.3 Organization and outline of the book

The rest of this book is organized according to the three types of compactifications mentioned above in §0.2.

In Part I, we study compactifications that contain a symmetric space X as an open and dense subset. In Part II, we study smooth compactifications that contain a symmetric space X as an open but not dense subset. In Part III, we study compactifications that contain a locally symmetric space $\Gamma\backslash X$ or the associated homogeneous space $\Gamma\backslash G$ as an open, dense subset, together with the Borel–Serre–Oshima compactification and global geometry and spectral theory of $\Gamma\backslash X$. The introductions to the parts and chapters below give more detailed summaries of the contents. We will only give a brief outline here.

In Part I, we first recall the original construction and motivation of each of the classical compactifications of symmetric spaces: the geodesic, the Karpelevič, the Satake, the Furstenberg, and the Martin compactifications. Then we formulate a uniform approach, develop the reduction theory for Siegel sets of real parabolic subgroups and apply the method to give uniform constructions of all these compactifications and compare and relate them. We also give four more constructions of the maximal Satake compactification $\overline{X}^S_{\max}$: the *subgroup compactification* $\overline{X}^{sb}$, the *subalgebra compactification* $\overline{X}^{sba}$, the dual-cell compactification $X \cup \Delta^*(X)$, and a modification $X^S_{\max}$ of the dual-cell compactification suggested by the construction of the Oshima compactification [Os1]. Among these four compactifications, the construction of the last two depend on the structure of closure of flats in the Satake compactifications.

The construction of $\widetilde{X}^S_{\max}$, a variant of the dual-cell compactification, shows that the maximal Satake compactification is a *real analytic manifold with corners,* and the subalgebra construction $\overline{X}^{sba}$ is similar to one construction of the wonderful compactification $\overline{\mathbf{X}}^W$ by embedding $\mathbf{X}$ into a Grassmannian variety. The subgroup compactification of symmetric spaces also motivates the subgroup compactifications of locally symmetric spaces in Part III, Chapter 12.

In Part II, we first recall a general method of self-gluing a manifold with corners into a closed manifold and apply it to reconstruct the Oshima compactification from the maximal Satake compactification. After studying basic facts on semisimple symmetric spaces, we identify the real locus of symmetric varieties (or complex symmetric spaces) through the Galois cohomology, and use these results to determine the real locus of the wonderful compactification of a complex symmetric space. From this we deduce immediately that the maximal Satake compactification is a real analytic manifold with corners. Finally, we relate the Oshima–Sekiguchi compactification to the real locus of the wonderful compactification by a finite-to-one map.

In Part III, after recalling several versions of the reduction theory of arithmetic groups, we recall the original constructions and motivations of the classical compactifications of locally symmetric spaces: the Satake, the Baily–Borel, the Borel–Serre, the reductive Borel–Serre, and the toroidal compactifications. Then we modify the method in [BS2] to formulate a uniform method and apply it to reconstruct the Borel–Serre, the reductive Borel–Serre, and the maximal Satake compactifications. New compactifications of homogeneous spaces $\Gamma\backslash G$ are also constructed. In [Sat2], a general method of passing from a compactification of the symmetric space X to a compactification of $\Gamma\backslash X$ is formulated, and it depends on the *rationality* of the compactification X. On the other hand, in this uniform approach, this subtle question about the rationality of the compactification of X is avoided. We also show that the Borel–Serre compactification $\overline{\Gamma\backslash X}^{BS}$ is a real analytic manifold with corners and can be self-glued into a closed real analytic manifold, the Borel–Serre–Oshima compactification $\overline{\Gamma\backslash X}^{BSO}$.

Then we study metric properties of compactifications, in particular, a simple proof of an *extension theorem of Borel* [Bo6], and a proof of a *conjecture of Siegel* on comparison of two metrics restricted to Siegel sets. Finally, we discuss parametrization of the generalized eigenspaces in terms of the boundaries of $\overline{\Gamma\backslash X}^{RBS}$ and the geodesic compactification $\Gamma\backslash X \cup \Gamma\backslash X(\infty)$, and relations between the geodesics that go to infinity and the continuous spectrum.

0.4 Topics related to the book but not covered and classification of references

It should be pointed out that in this book we emphasize the geometric aspects of compactifications, and do not study their more refined properties, for example, the Baily–Borel compactification $\overline{\Gamma\backslash X}^{BB}$ as a normal projective variety defined over a number field (see [Mi1] [Mi2]) or even over the integers (see [Ch1–4] [FC] [Fa]

[Lar]). We do not discuss in detail the toroidal compactifications either (see [AMRT] [Nam1] [HKW] for comprehensive treatments).

The following is the list of topics that are closely related to this book but not discussed in detail or omitted completely. We hope that the references provided here are adequate.

1. Geometry of numbers, lattices in $\mathbb{R}^n$ and applications. [Cass1–2] [Con] [CoS] [Gru] [GruL] [MM1–2] [MR] [Ma] [PR] [So1] [Si3] [Wo2].
2. Cohomology of arithmetic groups and S-arithmetic groups. [As1–4] [AsB] [AM] [AR] [Bo7] [BS1-3] [BW] [BFG] [Br2] [BuMo] [BuW] [CKS1–3] [GHM] [GHMN] [Go] [GM1–2] [GT1–3] [Har1–6] [HZ1–3] [HW1–3] [KuM] [Kug] [LeS1–2] [LeW1–4] [Leu7] [Li] [LiM] [LiS1–3] [MaM1–2] [Mil1–2] [MiR] [Ra2–3] [Rol1–3] [RSc] [RSg] [RSp1–4] [Sap3] [SaS] [SaZ] [Shw1–5] [Ser2–4] [So2] [Ve1–4] [Wes1–2] [Zu1–12].
3. Topology of compactifications: singular cohomology of Satake compactifications, and the intersection cohomology of the Baily–Borel compactification. [Bes] [BG] [CL] [DP3] [Grs1–3] [Ha] [HaT] [Hat1–5] [Lo3] [LoR] [MM1–2] [Mc1–2] [San3] [SaS] [SaZ] [Tc1–2] [Vd] [Zu1–12].
4. Euclidean Buildings, p-adic symmetric spaces and their compactifications. [BS3] [Br1] [Dr] [Ger1–2] [Kz] [Laf1–3] [Lan1–2] [Mor] [Ni] [Pi1] [RSg] [Sch] [ScT] [Te] [Ti2] [VT1–2] [Wer1–3].
5. The wonderful compactification: topology and other compactifications. [BDP] [DGMP] [DP1-3] [DS] [Kaus] [Kn1–2] [KL] [LP] [Lu1–3] [LV] [MP] [Pro] [Ri1–2] [Sen] [Sp2–3] [Str1–3] [Tc1–2] [Th] [Tim] [Vu1–2].
6. Rigidity of locally symmetric spaces and discrete subgroups. [Ba1–3] [BBE] [BBS] [BGS] [Ben1–2] [BCG] [BuMo] [BuS1–2] [CoG1–2] [Eb] [EO] [Mag] [Mok1–2] [MZ] [Mos] [NR]. See also [Ji9] for more references about rigidity of locally symmetric spaces and complex manifolds.
7. Harmonic analysis and potential theory of symmetric spaces. [Anc] [And] [AS] [AJ] [BFS1-2] [Bet] [BO] [Dy1-2] [Fl] [Fu1–7] [GW] [HeS] [Hel1-5] [Her1–6] [HOW] [Is] [KaS] [Ko1–10] [Me2] [OO] [Os3–4] [Roo] [Ros] [Sch] [Schm1] [ScT] [Sul] [Uz1–2] [Va] [Wal1–2] [Wal9] [War] [Wo1] [Wo4].
8. Large-scale geometry of symmetric spaces and arithmetic groups. [BMP] [BMW] [BuM1–5] [Gol] [Gro1] [Lot] [LMR]. See also [Ji9] for more references.
9. Modular forms and spectral theory of automorphic forms. [Ar1–3] [By] [BMP] [BMW] [BuM1-5] [Chr1–3] [Deg] [DW] [Dei1–2] [DH1–2] [Dim] [Fr] [FrH] [Ga] [HC] [Has1] [HaT] [Ji2] [Ji5] [KuM] [Kug] [La] [Leh1] [Li] [LiM] [LiS1–3] [LoM] [Lot] [MW1–3] [MoW] [Mor] [MS] [Mu1–4] [Ni] [Ol1–2] [OW1-2] [Pe1–4] [Pi2–3] [Sal2] [Sar] [Shi] [St] [Sun] [Vd] [VT1–2] [Wal3–9] [Wi] [Zu4] [Zu6–11].
10. Interpretation of compactifications as moduli spaces. [Ale] [AlB1–2] [Hu] [HKW] [HS1-3] [Ig1-3] [Kaus] [Mum3] [Nam1-4] [NR] [San1–2] [Sha] [VT1–2] [Wan1–3].

11. Variation of Hodge structures and period domains. [BN] [BuW] [Cat1–3] [CK] [CKS1–3] [Schm2] [Gri] [GS] [KaU] [Kug] [Le] [Schm2] [Sha] [Ste1–2] [Uz1–3] [Zu8].
12. Toric varieties. [At] [Ful] [Jur] [KKMS] [Od] [Ta2].
13. Other compactifications of homogeneous spaces, groups, and configuration spaces. [Bet] [Bo] [Bre] [Bri1–3] [BLV] [FM] [HT] [He1–2] [IP1–2] [Jun] [Kan1–4] [KaS] [Kaus] [Kn1–2] [KL] [Kom1–2] [Kus1–4] [Laf1–3] [LMP] [LL] [LM] [Lo1–2] [Lo4–7] [Lu1–3] [LV] [Mas] [MZ] [Nad] [NT] [Ne1–3] [Roo] [Sin] [SiY] [Str1–3] [Tc1–2] [Th] [Tim] [Ts] [Ul] [Xu] [Ye1–2].
14. Monodromy groups of some differential equations [Ho1-2] [Yo1-2].
15. Teichmüller spaces, mapping class groups and compactifications of Teichmüller spaces and moduli spaces[1] [Hae1-2] [Hav1-3] [Iv1-4] [Thu] [Wolf] [Wolp].

To complement the above list of references for omitted topics and to make it more convenient to use the long bibliography at the end of this book, we classify some of the references for the main topics discussed in this book.

a. General results about symmetric spaces and Lie groups. [Bo11] [Eb] [EO] [FH] [Hel1-3] [Ji7] [KW] [Ri1] [Ros] [Sat8] [Sp1] [Ter1-2] [Thor1-2] [Va] [Wal1-2] [War] [Wo2-3] [WK] [Zi]
b. Compactifications of symmetric spaces. [BJ1-2] [Cas2] [DGMP] [DP1-2] [DS] [De] [Dim] [Dy1-2] [EL] [Fu1-7] [GW] [GJT] [GT] [Has2] [He1–2] [Her1–6] [Ho1–2] [Ji1] [JL] [Ka] [KaK] [Kom1–2] [Ko1–10] [KW] [Kus1–4] [LL] [Mar] [Mo1–2] [Ne1–3] [Os1–4] [OsS1] [Sap2] [Sat1] [Sat9–16] [Ta3] [Uz1–3] [WK] [Zu2].
c. Algebraic groups, arithmetic groups, reduction theories, and locally symmetric spaces. [Al1-2] [As1] [As3-4] [Bon1–4] [Bo1–7] [Bo9] [Bo12–14] [BHC] [Con] [Do1–3] [EGM] [GaR] [Gra1–2] [HC] [Ho1–2] [Hu] [Ji6] [Kat] [KM] [LR] [Leh1] [MM1–2] [MR] [Ma] [MT] [OW2] [PR] [Po] [Ra1] [Sap1] [Sap3] [Sel1–3] [Ser2–3] [Si4] [So1] [Sp4] [Vd] [Zi] [Zu3] [Zu5] [Zu8].
d. Compactifications of locally symmetric spaces. [AMRT] [Ba1–7] [BB1–2] [BJ1–4] [BS1] [Cas2] [Ch1–4] [Fa] [FC] [HZ1–3] [Hi1–3] [HKW] [Ig1–3] [IP1–2] [Ji3–5] [KaU] [Ki] [Lar] [LMP] [Lo1–2] [Lo7] [Mac] [Mi1–2] [Mum1] [Mum3] [Nad] [NT] [Nam1–4] [Ni] [NR] [Sat2–7] [Sat9] [Sat11–12] [SiY] [Ts] [Wan1–3] [Wi]
e. Metric properties of locally symmetric spaces and their compactifications. [Abe] [AbM] [Din] [Gu] [Htt1–2] [Ji4] [JM] [JZ1–2] [KK1–2] [KO] [Leu1–5] [Sal2] [Sar] [Si2] [Sun]

Finally, we should mention that this book is complementary to the book [GJT]. For example, [GJT] concentrates on the identification of the Martin compactification, and the structure and applications of the maximal Satake compactification used in

[1] The Teichmüller space of a closed oriented surface of genus can be identified with the upper half-plane **H** and its mapping class group is equal to the modular group $SL(2, \mathbb{Z})$. In general, there are many similarities between Teichmüller spaces and symmetric spaces of noncompact type, and mapping class groups and arithmetic groups.

this identification, but does not discuss compactifications of locally symmetric spaces or the Oshima compactification. Furthermore, Part I of this book was motivated by a question in [GJT] on how to prove directly that the G-action on X extends to a continuous action on the dual-cell compactification $X \cup \Delta^*(X)$ without using its identification with a compactification that admits a continuous G-action, such as the Martin compactification $X \cup \partial_{\lambda_0} X$ at the bottom of the spectrum λ_0 of X or the maximal Satake compactification $\overline{X}^S_{\max}$.

Conventions and notation

Basic notation

Riemannian symmetric spaces are always denoted by X and assumed to be nonpositively curved, i.e., not containing a symmetric space of compact type as a factor, or equivalently equal to products of symmetric spaces of noncompact type and Euclidean spaces. A symmetric pair that gives rise to such a symmetric space X is denoted by (G, K), where G is a noncompact reductive Lie group, and K is a maximal compact subgroup of G, and hence $X = G/K$. For contrast, a *non-Riemannian symmetric space* is denoted by G/H.

A complex symmetric space (or a symmetric variety) is denoted by $\mathbf{X} = \mathbf{G}/\mathbf{H}$. Hence the complexification $X_{\mathbb{C}}$ of $X = G/K$ is $\mathbf{X} = \mathbf{G}/\mathbf{H}$, where $\mathbf{G}$ is the complexification $G_{\mathbb{C}}$ of G, and $\mathbf{H}$ is the complexification $K_{\mathbb{C}}$ of K when G, K are linear.

Arithmetic subgroups or more generally discrete subgroups of G are denoted by Γ, and locally symmetric spaces are denoted by $\Gamma\backslash X = \Gamma\backslash G/K$, which are always assumed to be noncompact unless indicated otherwise.

A compactification of a symmetric space X is denoted by $\overline{X}$ with a superscript consisting of the first letters of the names of the people associated with the compactification together with additional information in the subscript. For example, a Satake compactification of X is denoted by $\overline{X}^S_\tau$, where τ is a faithful projective representation of G used in defining the compactification. Exceptions to this pattern are the geodesic compactification, which is denoted by $X \cup X(\infty)$, and the Martin compactifications which are denoted by $X \cup \partial_\lambda X$ as well.

A compactification of a locally symmetric space $\Gamma\backslash X$ is denoted either by $\overline{\Gamma\backslash X}$ with a superscript consisting of the first letters of the names of the people associated with the compactification, for example, the Borel–Serre compactification $\overline{\Gamma\backslash X}^{BS}$, or with a superscript consisting of an abbreviation of certain properties of the compactification together with a suitable subscript, for example, the toroidal compactification $\overline{\Gamma\backslash X}^{tor}_\Sigma$, where Σ is a certain polyhedral cone decomposition. A compactification of $\Gamma\backslash X$ is also denoted by $\Gamma\backslash\overline{X}$ with a suitable superscript or subscript, which suggests that it is the quotient by Γ of a partial compactification $\overline{X}$ of X. Compactifications of the homogeneous space $\Gamma\backslash G$ are denoted similarly.

By a *G-compactification* of X, we mean a compact Hausdorff space that contains X as an open subset such that the G-action on X extends to a continuous action on the compactification. The compactification may not necessarily contain X as a dense subset.

The Lie algebras of G, K are denoted by $\mathfrak{g}, \mathfrak{k}$, and the Cartan decomposition of $\mathfrak{g}$ by $\mathfrak{g} = \mathfrak{k} \oplus \mathfrak{p}$. A maximal abelian subalgebra of $\mathfrak{p}$ is denoted by $\mathfrak{a}$, and its corresponding subgroup by A. The set of roots of $\mathfrak{a}$ acting on $\mathfrak{g}$ is denoted by $\Phi(\mathfrak{g}, \mathfrak{a})$ or $\Phi(G, A)$, the set of positive roots by $\Phi^+(\mathfrak{g}, \mathfrak{a})$ or $\Phi^+(G, A)$, and the set of simple roots by $\Delta(\mathfrak{g}, \mathfrak{a})$ or $\Delta(G, A)$.

For any two elements $g, h \in G$, define

$$^{g}h = ghg^{-1}, \quad h^{g} = g^{-1}hg.$$

The same notation applies when h is replaced by a subset of G.

Real vs. rational

Since both symmetric and locally symmetric spaces are studied in this book, we need different notations for linear algebraic groups and real Lie groups. In the following, linear algebraic groups are denoted by boldface capital roman letters, and the corresponding Lie groups by capital roman letters. For example, $\mathbf{G}$ denotes a reductive linear algebraic group defined over $\mathbb{Q}$, and G denotes its real locus $\mathbf{G}(\mathbb{R})$; a rational parabolic subgroup is often denoted by $\mathbf{P}$ or $\mathbf{Q}$, while a real parabolic subgroup is denoted by P or Q. In this book, parabolic subgroups are always assumed to be proper unless indicated otherwise.

For a rational parabolic subgroup $\mathbf{P}$, its real locus $P = \mathbf{P}(\mathbb{R})$ has two *Langlands decompositions*: the *real* and the *rational* decompositions,

$$P = N_P A_P M_P, \quad P = N_P A_{\mathbf{P}} M_{\mathbf{P}},$$

where A_P is the *real split component*, and $A_{\mathbf{P}}$ is the *rational split component*, $A_{\mathbf{P}} \subseteq A_P$ with equality if and only if the $\mathbb{Q}$-rank of $\mathbf{P}$ is equal to the $\mathbb{R}$-rank of $\mathbf{P}$. They induce two *boundary symmetric spaces*

$$X_P = M_P/K \cap M_P, \quad X_{\mathbf{P}} = M_{\mathbf{P}}/K \cap M_{\mathbf{P}}.$$

The notation $e(\mathbf{P})$ stands for the boundary component associated with the rational parabolic subgroup $\mathbf{P}$ and depends on the section where it appears and the partial compactification of X under discussion. Similarly, $e(P)$ is the boundary component associated with the real parabolic subgroup P and changes from section to section and depends on the compactification of X under discussion.

The Tits building of a reductive Lie group G is denoted by $\Delta(G)$, and the Tits building of a reductive linear algebraic group $\mathbf{G}$ defined over $\mathbb{Q}$ by $\Delta_{\mathbb{Q}}(\mathbf{G})$.

Topological spaces are always assumed to be Hausdorff, and topologies are often described in terms of convergent sequences.

Numbering

The sections in each part are numbered consecutively and independently of the chapters they belong to and start from 1 in each part. The section number is proceeded by the part number. For example, the first section in Part III is denoted by §III.1. The chapters are also numbered consecutively and independent of the parts. So the chapter numbering does not affect the section numbering.

The numbering of equations in each section starts from 1, and each equation number includes the section number where it appears. The equation numbers appear in parentheses, while the numbers for theorems, propositions, etc. are not so enclosed in order to distinguish these two sets of numbers.

Part I

Compactifications of Riemannian Symmetric Spaces

In this part, we study compactifications of Riemannian symmetric spaces where the symmetric spaces are open and dense. There are two purposes in this part: to review most of the known compactifications and to give a uniform construction of these compactifications. This uniform method in §I.8 is similar to compactifications of locally symmetric spaces in §III.8 in Part III.

Symmetric spaces of noncompact type occur naturally in different contexts, for example as special Riemannian manifolds in differential geometry and special homogeneous spaces in Lie group theory. There are many compactifications of symmetric spaces of higher rank that are of different sizes and have different properties. They are useful for different applications: the Satake compactifications for applications in harmonic analysis, rigidity of discrete subgroups, and global geometry; the Furstenberg compactifications for applications in potential theory and ergodic theory; the geodesic compactification for applications in analysis and differential geometry; the Martin compactification and the Karpelevič compactification for applications in potential theory and Brownian motion; the Oshima compactification for applications in representation theory.

In Chapter 1, we explain the motivations of these compactifications and the original constructions. In Chapter 2, we first propose a uniform, intrinsic construction, where a key concept is the notion of generalized Siegel sets; then we apply this method to reconstruct all the compactifications recalled earlier together with the real Borel–Serre partial compactification, which is important to some global properties of arithmetic subgroups. In Chapter 3, we study relations between these different compactifications and properties of the compactifications as topological spaces and other more refined structures on them. These studies lead to several new constructions of the maximal Satake compactification. They can also be used to prove that $\overline{X}^S_{\max}$ is a real analytic manifold with corners. This property will be used to reconstruct the Oshima compactification from the self-gluing of the maximal Satake compactification $\overline{X}^S_{\max}$.

1

Review of Classical Compactifications of Symmetric Spaces

Compactifications of symmetric spaces are closely related to the geometry at infinity of symmetric spaces, and are useful in studying behavior at infinity of functions on symmetric spaces.

Since the geometry at infinity of symmetric spaces can be described in terms of real parabolic subgroups, we recall the notion of (standard) parabolic subgroups, the real Langlands decomposition of parabolic subgroups, and the induced horospherical decomposition of the symmetric spaces in §I.1. In §I.2, we define an equivalence relation on geodesics and the induced geodesic compactification $X \cup X(\infty)$. Then we relate the structure of geodesics to parabolic subgroups and hence obtain a geometric realization of the spherical Tits building $\Delta(G)$, and introduce the topological Tits building. In §I.3, we explain how a refined classification of geodesics leads to boundary symmetric spaces, and the dimension count of geodesics motivates the Karpelevič compactification $\overline{X}^K$, which is a blow-up of the geodesic compactification. These two compactifications are directly motivated by the geometry, or the structure of geodesics. In §I.4, we recall the Satake compactifications $\overline{X}^S$. Though the original motivation was to produce suitable boundary points and to use them to compactify locally symmetric spaces, the Satake compactifications have also played an important role in understanding the global geometry of and harmonic analysis on symmetric spaces. We emphasize the standard Satake compactification $\overline{\mathcal{P}_n}^S$ of the symmetric space $\mathcal{P}_n$ of positive definite Hermitian matrices of determinant 1 obtained by attaching semipositive definite matrices. In §I.5, we recall basics of Hermitian symmetric spaces and the Baily–Borel compactification $\overline{X}^{BB}$ of Hermitian symmetric spaces and show that it is isomorphic to a minimal Satake compactification. In §I.6, we explain how the Poisson integral formula for harmonic functions on the unit disk leads to the Furstenberg compactifications $\overline{X}^F$. The structure of the cone of positive eigenfunctions, in particular positive harmonic functions, leads to the Martin compactifications $X \cup \partial_\lambda X$ in potential theory in §I.7.

I.1 Real parabolic subgroups

Our main object of interest in Part I is a symmetric space of noncompact type, hence real linear connected semisimple groups play an important role here. However, the construction of parabolic subgroups leads one to a slightly broader class of groups. To avoid repetitions, and prepare for Parts II and III, we shall already add some remarks on them, referring to a later part of the book for more details. For thorough discussions of parabolic subgroups, see [Wal1–2] [War] [Va].

This section is organized as follows. After recalling basic definitions about root systems, parabolic subalgebras, and parabolic subgroups, we introduce the Langlands decomposition of parabolic subgroups (equation I.1.10) and the induced horospherical decomposition of symmetric spaces (equation I.1.14), which plays a basic role in relating the geometry of symmetric spaces to parabolic subgroups. Then we discuss the relative Langlands and horospherical decompositions for a pair of parabolic subgroups when one is contained in the other (equations I.1.21 and I.1.23). Such relative decompositions are important for the purpose of understanding the topology of the boundary components of compactifications of symmetric spaces, for example, relations between the boundary components of a pair of parabolic subgroups P, Q satisfying $P \subset Q$.

I.1.1 To motivate parabolic subgroups and the induced horospherical decompositions, we consider the example of $G = \mathrm{SL}(2,\mathbb{R})$. Fix the maximal compact subgroup $K = \mathrm{SO}(2)$. Let $X = \mathrm{SL}(2,\mathbb{R})/\mathrm{SO}(2)$ be the associated symmetric space. Then X has two important models: the *upper half-plane* $\mathbf{H}$ and the *Poincaré disk D*.

As explained in the introduction (§0.1.1), the disk model is convenient for understanding the compactification of X and its boundary. On the other hand, the model $\mathbf{H}$ is convenient for understanding the structure of neighborhoods of cusps and the compactification of quotients $\Gamma\backslash\mathbf{H}$.

An effective and group-theoretical way to explain this upper-half-plane model $\mathbf{H}$ is to use the Langlands decomposition of parabolic subgroups.

In fact, as explained in §0.1.1, the $\mathrm{SL}(2,\mathbb{R})$-action on $\mathbf{H}$ extends continuously to $\mathbf{H} \cup \mathbb{R} \cup \{\infty\}$. The stabilizer of the point $\{\infty\}$ is the subgroup

$$P = \left\{ \begin{pmatrix} a & b \\ 0 & a^{-1} \end{pmatrix} \mid a \neq 0, b \in \mathbb{R} \right\}.$$

It is a *parabolic subgroup* of $\mathrm{SL}(2,\mathbb{R})$.

The group P contains three subgroups: the *unipotent radical*

$$N_P = \left\{ \begin{pmatrix} 1 & b \\ 0 & 1 \end{pmatrix} \mid b \in \mathbb{R} \right\},$$

the *split component*

$$A_P = \left\{ \begin{pmatrix} a & 0 \\ 0 & a^{-1} \end{pmatrix} \mid a > 0 \right\},$$

and

$$M_P = \left\{ \pm \begin{pmatrix} 1 & 0 \\ 0 & 1 \end{pmatrix} \right\}.$$

They give rise to the Langlands decomposition of P:

$$P = N_P A_P M_P,$$

where the map $N_P \times A_P \times M_P \to N_P A_P M_P = P$ is a diffeomorphism.

It can be shown easily that P acts transitively on $\mathbf{H}$, and the Langlands decomposition of P gives the horospherical decomposition of $\mathbf{H}$:

$$\mathbf{H} = P/P \cap K = P/M_P \cong N_P \times A_P,$$

where the N_P-factor corresponds to the x-coordinate and the A_P-factor to the y-coordinate. In fact, for any $a > 0$, $b \in \mathbb{R}$,

$$\begin{pmatrix} 1 & b \\ 0 & 1 \end{pmatrix} \cdot \begin{pmatrix} a & 0 \\ 0 & a^{-1} \end{pmatrix} \cdot i = b + a^2 i.$$

This example shows the importance of parabolic subgroups in understanding the structure at infinity of X. After defining parabolic subgroups and discussing their structures for general reductive groups G, we illustrate them through the example of $G = \mathrm{SL}(n, \mathbb{R})$.

I.1.2 We first recall that if G is a real Lie group with finitely many connected components, its maximal compact subgroups (which exist) are conjugate under the identity component G^o and meet every connected component of G. If K is one of them, then $X = G/K$ is a manifold diffeomorphic to Euclidean space. Assume now that

(∗) *G is reductive, linear, with finitely many connected components and the center $\mathcal{C}G^o$ of G^o is compact.*

Then $\mathfrak{g} = \mathcal{D}\mathfrak{g} \oplus \mathfrak{c}$, where $\mathfrak{c}$ is the Lie algebra of $\mathcal{C}G^o$, and $\mathcal{D}\mathfrak{g} = [\mathfrak{g}, \mathfrak{g}]$. The group G admits a unique *Cartan involution* θ with fixed point set K. Since the center $\mathcal{C}G^o$ is compact, $\mathfrak{c} \subseteq \mathfrak{k}$.

Let $\mathfrak{p}$ be the (-1)-eigenspace of θ in $\mathfrak{g}$. It is the orthogonal complement to $\mathfrak{k} \cap \mathcal{D}\mathfrak{g}$ in $\mathcal{D}\mathfrak{g}$ with respect to the *Killing form* $B(\ ,\ )$. We have the familiar *Cartan decomposition*

$$\mathfrak{g} = \mathfrak{k} \oplus \mathfrak{p} \text{with } [\mathfrak{k}, \mathfrak{k}] \subset \mathfrak{k}, [\mathfrak{k}, \mathfrak{p}] \subset \mathfrak{p}, [\mathfrak{p}, \mathfrak{p}] \subset \mathfrak{k}. \tag{I.1.1}$$

The restriction of B to $\mathfrak{k} \cap \mathcal{D}\mathfrak{g}$ (resp. $\mathfrak{p}$) is negative (resp. positive) definite. The subspace $\mathfrak{p}$ may be identified with the tangent space $T_{x_0}(X)$ to X at the basepoint $x_0 = K \in G/K = X$. The restriction of the Killing form to $\mathfrak{p}$ defines a *G-invariant metric* on G/K with respect to which it is a simply connected complete Riemannian symmetric space of noncompact type, i.e., of nonpositive sectional curvature. (In the *de Rham decomposition*, it is a direct product of irreducible such spaces, without flat component.)

We shall use repeatedly, without further reference, the fact that if L is a closed subgroup of G, with finitely many connected components, stable under θ, then it is reductive and the restriction of θ to L is a Cartan involution.

I.1.3 The structure theory of G can be viewed from two points of view: a differential-geometric one, which has its origin in E. Cartan's theory of symmetric spaces and restricted roots, and a more algebraic one in the framework of the theory of linear algebraic groups. Both will be used in this book. However, in Part I, it is the former that is predominant and we shall adopt it. We assume familiarity with it (see e.g. [Bo11], [Hel3]) and review it mainly to fix notation. The relations with the algebraic one will be discussed later.

The maximal subalgebras of $\mathfrak{p}$ are abelian and conjugate under K. Let $\mathfrak{a}$ be one. A linear form $\lambda \in \mathfrak{a}^*$ on $\mathfrak{a}$ is a *root* (or a *restricted root*) if it is nonzero and the *root space*

$$\mathfrak{g}_\lambda = \{V \in \mathfrak{g} \mid [H, V] = \lambda(H)V, (H \in \mathfrak{a})\} \neq 0. \tag{I.1.2}$$

The set of roots is a *root system* in $\mathfrak{a}^*$, to be denoted by $\Phi(\mathfrak{g}, \mathfrak{a})$ or simply Φ. Its *Weyl group* $W = W(\mathfrak{g}, \mathfrak{a})$ may be identified with $\mathcal{N}_K(\mathfrak{a})/\mathcal{Z}_K(\mathfrak{a})$, where $\mathcal{N}_k(\mathfrak{a})$ is the normalizer of $\mathfrak{a}$ in K and acts via the adjoint representation, and $\mathcal{Z}_K(\mathfrak{a})$ is the centralizer of $\mathfrak{a}$.

Each root $\alpha \in \Phi$ determines the *root hyperplane* $\mathbf{H}_\alpha$ on which it is zero. The connected components of the complement in $\mathfrak{a}$ of the union of the $\mathbf{H}_\alpha$ are called the *Weyl chambers* and are permuted in a simply transitive manner by W. Fix one, to be called the *positive Weyl chamber* and denoted by $\mathfrak{a}^+$. Its closure is a fundamental domain for the action of W. The choice of $\mathfrak{a}^+$ defines an ordering on $\Phi(\mathfrak{g}, \mathfrak{a})$ and we define

$$\Phi^+(\mathfrak{g}, \mathfrak{a}) = \{\alpha \in \Phi \mid \alpha > 0 \text{ on } \mathfrak{a}^+\},$$

the set of *positive roots*. The set of *simple roots* in Φ^+ is denoted by $\Delta(\mathfrak{g}, \mathfrak{a})$ or Δ. Let

$$\mathfrak{n} = \sum_{\alpha>0} \mathfrak{g}_\alpha \qquad \mathfrak{n}^- = \sum_{\alpha<0} \mathfrak{g}_\alpha.$$

These are nilpotent subalgebras exchanged by θ, normalized by $\mathfrak{a}$, and

$$\mathfrak{g} = \mathfrak{n}^- \oplus \mathfrak{z}(\mathfrak{a}) \oplus \mathfrak{n}, \tag{I.1.3}$$

where $\mathfrak{z}(\mathfrak{a})$ is the centralizer of $\mathfrak{a}$ in $\mathfrak{g}$. Moreover,

$$\mathfrak{z}(\mathfrak{a}) = \mathfrak{m} \oplus \mathfrak{a}, \quad \text{where} \quad \mathfrak{m} = \mathfrak{k} \cap \mathfrak{z}(\mathfrak{a}). \tag{I.1.4}$$

Let $A = \exp \mathfrak{a}$. The exponential is an isomorphism of $\mathfrak{a}$ onto A. Identify A with its orbit Ax_0 in $X = G/K$ through the basepoint $x_0 = K$. Then the conjugates of A under K are the *maximal flat totally geodesic subspaces* of X passing through the basepoint corresponding to K. Any maximal flat totally geodesic subspace of X is a translate of Ax_0 under some element of G.

We shall also view the roots as homomorphisms of A into the multiplicative group of strictly positive numbers $\mathbb{R}^*_{>0}$ by the rule $a^\alpha = \exp \alpha(\log a)$. Strictly speaking this means that our original α is the differential at the origin of the map just defined, but we shall use α for both, unless it leads to confusion. From that global point of view $\Phi = \Phi(\mathfrak{g}, \mathfrak{a})$ is also the *root system* $\Phi(G, A)$ of G with respect to A, and $\mathfrak{g}_\alpha$ may be defined as

$$\mathfrak{g}_\alpha = \{V \in \mathfrak{g} \mid \mathrm{Ad}\, a(V) = a^\alpha V, (a \in A)\}.$$

Parabolic subalgebras. We first define the standard parabolic subalgebras with respect to the ordering of Φ determined by the positive chamber $\mathfrak{a}^+$. For $I \subset \Delta$ let

$$\mathfrak{a}_I = \bigcap_{\alpha \in I} \ker \alpha. \tag{I.1.5}$$

We let Φ^I be the set of roots that are linear combinations of elements in I and $\mathfrak{a}^I$ the orthogonal complement of $\mathfrak{a}_I$ in $\mathfrak{a}$. Then

$$\mathfrak{a} = \mathfrak{a}_I \oplus \mathfrak{a}^I, \quad \mathfrak{a}^I \perp \mathfrak{a}_I. \tag{I.1.6}$$

The *standard parabolic subalgebra* $\mathfrak{p}_I$ is generated by the centralizer $\mathfrak{z}(\mathfrak{a}_I)$ and $\mathfrak{n}$. It can be written as

$$\mathfrak{p}_I = \mathfrak{n}_I \oplus \mathfrak{a}_I \oplus \mathfrak{m}_I,$$

where

$$\mathfrak{n}_I = \sum_{\alpha \in \Phi^+ - \Phi^I} \mathfrak{g}_\alpha, \quad \mathfrak{m}_I = \mathfrak{m} \oplus \mathfrak{a}^I \oplus \sum_{\alpha \in \Phi^I} \mathfrak{g}_\alpha.$$

Note that in the two extreme cases, $I = \Delta, \emptyset$,

$$\mathfrak{p}_\Delta = \mathfrak{g}, \quad \mathfrak{p}_\emptyset = \mathfrak{n} \oplus \mathfrak{a} \oplus \mathfrak{m},$$

so that if we conform to the above notation, we have

$$\mathfrak{m} = \mathfrak{m}_\emptyset, \quad \mathfrak{a} = \mathfrak{a}_\emptyset, \quad \mathfrak{n} = \mathfrak{n}_\emptyset.$$

We note that $\mathfrak{m}_I$ is not necessarily semisimple. It is reductive, and the center is contained in $\mathfrak{m} \cap \mathfrak{m}_I$. It is stable under θ, and $\mathfrak{a}^I$ plays for $\mathcal{D}\mathfrak{m}_I = [\mathfrak{m}_I, \mathfrak{m}_I]$ the same role as $\mathfrak{a}$ for $\mathfrak{g}$. The root system $\Phi(\mathcal{D}\mathfrak{m}_I, \mathfrak{a}^I)$ is Φ^I.

It follows from standard commutation relations that $\mathfrak{z}(\mathfrak{a}_I)$ normalizes $\mathfrak{n}_I$ and $\mathfrak{n}_I^-$, and that the centralizer of $\mathfrak{a}_I$ in either is reduced to zero. Since $\mathfrak{g} = \mathfrak{n}_I^- \oplus \mathfrak{p}_I$, it follows that $\mathfrak{p}_I$ is *self-normalizing*.

A subalgebra $\mathfrak{p}$ of $\mathfrak{g}$ is *parabolic* if it is conjugate to a standard one. If so, it is conjugate to only one, say $\mathfrak{p}_I$, and I is called the *type* of $\mathfrak{p}$. Since the various $\mathfrak{a}$ are conjugate, this class of subalgebras is independent of the choice of $\mathfrak{a}$.

Definition I.1.5 *A subgroup P of G is parabolic if it is the normalizer of a parabolic subalgebra $\mathfrak{p}$ in $\mathfrak{g}$.*

The normalizer of $\mathfrak{p}_I$ is the *standard parabolic subgroup* P_I. In this book, parabolic subgroups are assumed to be proper unless indicated otherwise. Equivalently, for the standard parabolic subgroups P_I, I is assumed to be a proper subset of $\Delta = \Delta(\mathfrak{g}, \mathfrak{a})$.

When $I = \emptyset$, $P_\emptyset$ is a *minimal parabolic subgroup*, and any minimal parabolic subgroup of G is conjugate to $P_\emptyset$. When $\Delta - I$ consists of one element, P_I is a *standard maximal parabolic subgroup*, and every maximal parabolic subgroup of G is conjugate to one of these standard maximal parabolic subgroups. Let r be the number of elements in Δ. Then there are r-conjugacy classes of maximal parabolic subgroups. It is known that any parabolic subgroup containing $P_\emptyset$ is standard, i.e., of the form P_I, and for any two distinct subsets I, I', P_I, $P_{I'}$ are not conjugate under G. Clearly, for any given minimal parabolic subgroup P, the structure of parabolic subgroups containing P is similar.

The subgroup P_I is equal to its normalizer, as follows from its definition and the fact that $\mathfrak{p}_I$ is self-normalizing; hence so are all parabolic subgroups. Let N_I, A_I be the exponentials of $\mathfrak{n}_I$ and $\mathfrak{a}_I$, i.e., the Lie subgroups in G corresponding to the Lie subalgebras $\mathfrak{n}_I$ and $\mathfrak{a}_I$. The group P_I is the semidirect product of its *unipotent radical* N_I and of $\mathcal{Z}(A_I)$, the centralizer of A_I in G. The latter is a "*Levi subgroup*" of P_I, i.e., a maximal reductive subgroup (they are all conjugate under N_I). Moreover,

$$\mathcal{Z}(A_I) = M_I \times A_I. \tag{I.1.7}$$

Here M_I has Lie algebra $\mathfrak{m}_I$, but it is not connected in general. It is stable under θ, hence $K \cap M_I = \mathcal{Z}_K(A_I)$ meets all connected components of M_I and M_I is generated by M_I^o, the identity component, and $\mathcal{Z}_K(A_I)$. It satisfies the assumption $(*)$ of §I.1.1. We have

$$P_I = N_I A_I M_I \cong N_I \times A_I \times M_I. \tag{I.1.8}$$

More precisely, the map

$$(n, a, m) \mapsto nam \qquad (n \in N_I\ ,\ a \in A_I\ ,\ m \in M_I)$$

is an analytic isomorphism of analytic manifolds. It yields a decomposition of P_I in which M_I and A_I are θ-stable, called a *Langlands decomposition* of P_I. To emphasize the dependence on the basepoint $x_0 = K$ or equivalently the associated Cartan involution θ, it is also called the Langlands decomposition with respect to the basepoint x_0.

Since M_I is θ-stable, $K_I = M_I \cap K$ is maximal compact in M_I. It is also maximal compact in P_I. The quotient

$$X_I = M_I/K_I = P_I/K_I A_I N_I \tag{I.1.9}$$

is a symmetric space of noncompact type for M_I, called the *boundary symmetric space* associated to P_I. We remark that X_I can also be identified with a subspace of X as the M_I-orbit of $x_0 = K \in X$, but we emphasize that it is often attached at the infinity of X and hence we call it a boundary symmetric space.

Since $G = P_I K$ (as a consequence of the *Iwasawa decomposition* $G = NAK$, where $N = \exp \mathfrak{n}$ and $NA \subseteq P_I$), P_I acts transitively on X, and the Langlands decomposition in equation (I.1.8) induces a decomposition of X associated to P_I, called the *horospherical decomposition*

$$X \cong N_I \times A_I \times X_I,$$

i.e., the map

$$\mu_0 : N_I \times A_I \times X_I \to X, \quad (n, a, mK_I) \mapsto namK$$

is an analytic diffeomorphism of manifolds.

I.1.6 The parabolic subalgebras of $\mathfrak{m}_I$ are the direct sums of its center with the parabolic subalgebras of $\mathcal{D}\mathfrak{m}_I$. Their normalizers are, by definition, the parabolic subgroups of M_I. They are also the intersections of M_I with the parabolic subgroups of G contained in P_I. The standard ones can be described as above, starting from $A^I = \exp \mathfrak{a}^I$, Φ^I, identified with $\Phi(M_I, A^I)$, and the subsets of I. Any other is conjugate to one and only one standard one. One can also define similarly parabolic subalgebras and subgroups of a group satisfying the assumption §I.1.1(∗).

I.1.7 Example. We illustrate the above concepts by the example of $G = \mathrm{SL}(n, \mathbb{R})$, $n \geq 2$. Fix the maximal compact subgroup $K = \mathrm{SO}(n)$. Then a maximal abelian subalgebra $\mathfrak{a}$ in $\mathfrak{p}$ is given by

$$\mathfrak{a} = \{\mathrm{diag}(t_1, \ldots, t_n) \mid t_1 + \cdots + t_n = 0\}.$$

Its Lie group

$$A = \{\mathrm{diag}(a_1, \ldots, a_n) \mid a_1, \cdots, a_n > 0, a_1 \cdots a_n = 1\}.$$

Choose the positive chamber

$$\mathfrak{a}^+ = \{\mathrm{diag}(t_1, \ldots, t_n) \in \mathfrak{a} \mid t_1 > t_2 > \cdots > t_n\}.$$

Then the nilpotent subalgebra $\mathfrak{n}$ consists of strictly upper triangular matrices (i.e., with 0's on the diagonal), and its Lie group N consists of upper triangular matrices with 1's on the diagonal). Then the standard minimal parabolic subgroup $P_\emptyset$ is the subgroup of upper triangular matrices. For each $k < n$, there is a standard maximal parabolic subgroup P_k given by block upper triangular matrices

$$P_k = \left\{ \begin{pmatrix} A & B \\ 0 & D \end{pmatrix} \in \mathrm{SL}(n, \mathbb{R}) \mid A \in M_{k\times k}, B \in M_{k\times n-k}, D \in M_{n-k\times n-k} \right\}.$$

Hence there are $n-1$ conjugacy classes of maximal parabolic subgroups of $\mathrm{SL}(n, \mathbb{R})$.

Note that each k determines an ordered partition $\{1, \dots, n\} = \{1, \dots, k\} \cup \{k+1, \dots, n\}$. For more general ordered partitions

$$\Sigma : \{1, \dots, n\} = \{1, \dots, i_1\} \cup \{i_1 + 1, \dots, i_2\} \cup \cdots \cup \{i_s + 1, \dots, n\},$$

there are standard parabolic subgroups P_Σ given by upper triangluar matrices whose blocks are determined by the partitions Σ.

The Langlands decomposition of the standard minimal parabolic subgroup $P_\emptyset$ is given by

$$P_\emptyset = NAM,$$

where $M = \{\mathrm{diag}(\pm 1, \dots, \pm 1)\}$. The Langlands decomposition of the standard parabolic subgroups P_Σ is similar if we use block matrices.

The boundary symmetric space of a maximal standard parabolic subgroup P_k is equal to $\mathrm{SL}(k, \mathbb{R})/\mathrm{SO}(k, \mathbb{R}) \times \mathrm{SL}(n-k, \mathbb{R})/\mathrm{SO}(n-k, \mathbb{R})$. The boundary symmetric space of a general standard parabolic subgroup P_Σ is equal to $\mathrm{SL}(i_1, \mathbb{R})/\mathrm{SO}(i_1, \mathbb{R}) \times \cdots \times \mathrm{SL}(n - i_s, \mathbb{R})/\mathrm{SO}(n - i_s, \mathbb{R})$.

Remark I.1.8 Parabolic subgroups have been defined here within the context of real Lie groups, to limit the prerequisites for Part I. It is, however, more natural to look at them from the point of view of algebraic groups, which allows one to give a more intrinsic definition.

An algebraic subgroup $\mathbf{P}$ of a connected linear algebraic group $\mathbf{G}$ defined over an algebraically closed ground field is called a *parabolic subgroup* if the homogeneous space $\mathbf{G}/\mathbf{P}$ is a projective variety.

Our group G is assumed to be linear, hence embedded in some $\mathrm{SL}_n(\mathbb{R})$. Let $G_\mathbb{C}$ be its complexification, i.e., the closure of G in $\mathrm{SL}_n(\mathbb{C})$ with respect to the Zariski topology, equivalently, the smallest algebraic subgroup of $\mathrm{SL}_n(\mathbb{C})$ containing it. Then our parabolic subgroups are the intersections with G of the parabolic subgroups of $G_\mathbb{C}$ that are defined over $\mathbb{R}$. A similar remark is valid for the parabolic-subgroups of M_I.

I.1.9 The group G acts on the set of parabolic subgroups by conjugation, and every parabolic subgroup P is conjugate to a unique standard parabolic subgroup P_I under G and also under K. Choose $k \in K$ such that $P = {}^k P_I$. Define

$$N_P = {}^k N_I, \quad A_P = {}^k A_I, \quad M_P = {}^k M_I.$$

Though the choice of k is not unique, the subgroups N_P, A_P, M_P are well-defined. In fact, N_P is the nilpotent radical of P, and $A_P M_P$ is the unique *Levi subgroup* in P stable under the Cartan involution θ associated with K. The subgroup A_P is called the *split component* of P with respect to the basepoint $x_0 = K$.

The decomposition of P_I in equation (I.1.8) is transported to give the *Langlands decomposition* of P:

$$P = N_P A_P M_P \cong N_P \times A_P \times M_P, \tag{I.1.10}$$

and the map

$$N_P \times A_P \times M_P \to P, \quad (n, a, m) \mapsto nam,$$

is an analytic isomorphism of manifolds. This decomposition is *equivariant* with respect to the following P-action on $N_P \times A_P \times X_P$:

$$n_0 a_0 m_0 (n, a, z) = (n_0\, {}^{a_0 m_0} n, a_0 a, m_0 z), \tag{I.1.11}$$

where $n_0 \in N_P, a_0 \in A_P, m_0 \in M_P$.

Similarly, the boundary symmetric space of a parabolic subgroup can be defined as in the case of standard parabolic subgroups. Specifically, let $K_P = M_P \cap K$. Then K_P is a maximal compact subgroup of M_P, and

$$X_P = M_P / K_P \tag{I.1.12}$$

is a symmetric space of noncompact type of lower dimension, called the *boundary symmetric space* associated with P. Since M_P commutes with A_P and normalizes N_P, X_P can be written as a *homogeneous space*

$$X_P = P/N_P A_P K_P, \quad mK_P \mapsto N_P A_P m K_P = m N_P A_P K_P. \tag{I.1.13}$$

Under this action of P on X_P, A_P and N_P act trivially.

The Langlands decomposition of P induces the *horospherical decomposition* of X associated with P,

$$X \cong N_P \times A_P \times X_P.$$

The map

$$\nu_0 : N_P \times A_P \times X_P \to X, \quad (n, a, mK_P) \mapsto namK, \tag{I.1.14}$$

is an analytic diffeomorphism. It is clear from the above discussions that this horospherical decomposition depends on the basepoint $x_0 = K$. In fact, the subgroups A_P, M_P and the Langlands decomposition of P all depend on the basepoint $x_0 = K$. In the following, we will identify $N_P \times A_P \times X_P$ with X, and for a point $(n, a, z) \in N_P \times A_P \times X_P$, the image $\nu_0(n, a, z) \in X$ will be denoted by either (n, a, z) or naz for simplicity, unless a different basepoint is used.

The K-conjugation on parabolic subgroups transports the Langlands decomposition, and the K-action on X preserves the horospherical decomposition. Specifically, for any $z = mK_P \in X_P, k \in K$, define

$$k \cdot z = {}^k m\, {}^k K_P \in X_{{}^k P}. \tag{I.1.15}$$

Note that ${}^k K_P = K_{^kP}$. Then for $(n, a, z) \in N_P \times A_P \times X_P = X$, $k \in K$, the point $k \cdot \nu_0(n, a, z) = k(n, a, z)$ has the following horospherical coordinates with respect to ${}^k P$:

$$k \cdot (n, a, z) = ({}^k n, {}^k a, kz) \in N_{^kP} \times A_{^kP} \times X_{^kP}. \tag{I.1.16}$$

On the other hand, for $g \in G - K$, the conjugation by g transports the components of the Langlands decomposition with respect to the basepoint x_0 to those in the Langlands decomposition of ${}^g P$ with respect to the new basepoint gx_0 (see the comments before Proposition I.19.25).

I.1.10 For any proper, not necessarily minimal, parabolic subgroup P of G, all the parabolic subgroups containing it can also be described explicitly as above in the case of standard parabolic subgroups containing the minimal parabolic subgroup $P_\emptyset$.

Let $\Phi(P, A_P)$ be the set of roots of the adjoint action of $\mathfrak{a}_P$ on the Lie algebra $\mathfrak{n}_P$. We remark that $\Phi(P, A_P)$ is not a root system. For example, when P is minimal, $\Phi(P, A_P)$ is the set of positive roots with respect to a suitable ordering on $\Phi(\mathfrak{g}, \mathfrak{a}_P)$ rather than a root system.

As in §I.1.2, we also view them as characters of A_P defined by $a^\alpha = \exp \alpha(\log a)$. Then there are linearly independent roots $\alpha_1, \dots, \alpha_r$, $r = \dim A_P$, such that any root is a linear combination of them. These roots are called *simple roots* in $\Phi(P, A_P)$, and the set $\{\alpha_1, \dots, \alpha_r\}$ is denoted by $\Delta(P, A_P)$. In fact, let $P_0 \subseteq P$ be a minimal parabolic subgroup. With respect to a suitable ordering, $\Phi(P_0, A_{P_0})$ is the set of positive roots in $\Phi(\mathfrak{g}, \mathfrak{a}_{P_0}) = \Phi(P_0, A_{P_0})$, and $\Delta(P_0, A_{P_0})$ is the set of simple roots. When $P \neq P_0$, $\Phi(P, A_P)$ consists of nontrivial restrictions to A_P of elements in $\Phi(P_0, A_{P_0})$, and $\Delta(P, A_P)$ those in $\Delta(P_0, A_{P_0})$.

For any subset $I \subset \Delta(P, A_P)$, there is a unique parabolic subgroup P_I containing P such that

$$A_{P_I} = \{a \in A_P \mid a^\alpha = 1, \alpha \in I\}$$

is the split component of P_I with respect to the basepoint x_0, and $\Delta(P_I, A_{P_I})$ is the set of restrictions to A_{P_I} of elements in $\Delta(P, A_P) - I$. Conversely, every parabolic subgroup Q containing P is of this form P_I for a unique subset I of $\Delta(P, A_P)$.

For each such P_I, define $\mathfrak{a}_P^I$ to be the orthogonal complement of $\mathfrak{a}_{P_I}$ in $\mathfrak{a}_P$, and $A_P^I = \exp(\mathfrak{a}_P^I)$ the corresponding subgroup. Then

$$\mathfrak{a}_P = \mathfrak{a}_{P_I} \oplus \mathfrak{a}_P^I, \quad A_P = A_{P_I} A_P^I \cong A_{P_I} \times A_P^I. \tag{I.1.17}$$

There is also a related, but different decomposition of A_P that will be important in studying the corner structures of the maximal Satake compactification later. Let

$$\begin{aligned} \mathfrak{a}_{P,P_I} &= \{e^H \mid H \in \mathfrak{a}_P, \alpha(H) = 0, \alpha \in \Delta(P, A_P)\}, \\ A_{P,P_I} &= \exp \mathfrak{a}_{P,P_I}. \end{aligned}$$

Then

$$\mathfrak{a}_P = \mathfrak{a}_{P,I} \oplus \mathfrak{a}_{P,P_I}, \quad A_P = A_{P_I} A_{P,P_I} = A_{P_I} \times A_{P,P_I}. \tag{I.1.18}$$

The difference between these two decompositions is that $\mathfrak{a}_{P,P_I}$ is not perpendicular to $\mathfrak{a}_{P_I}$ in general.

Let $\alpha_1, \dots, \alpha_r$ be the simple roots in $\Delta(P, A_P)$. Then the map

$$A_P \to \mathbb{R}^r, \quad a \mapsto (a^{-\alpha_1}, \dots, a^{-\alpha_r}), \tag{I.1.19}$$

is a diffeomorphism onto the open quadrant $\mathbb{R}^r_{>0}$, and the closure of A_P in $\mathbb{R}^r$ is the corner $\mathbb{R}^r_{\geq 0}$, and is denoted by $\overline{A_P}$.

Under the identification in equation (I.1.19), the decomposition of A_P in equation (I.1.18) is the decomposition according to the standard coordinates of $\mathbb{R}^r$.

I.1.11 Each parabolic subgroup P has the Langlands decomposition given in equation (I.1.10). As mentioned earlier, it leads to the horospherical decomposition of X in equation (I.1.14). It also leads to a decomposition of G. In fact, $G = PK$, which is equivalent to that P acts transitively on $X = G/K$. It can be shown that $K \cap P = K \cap M_P$, which implies that

$$G = N_P A_P M_P K \cong N_P \times A_P \times M_P K. \tag{I.1.20}$$

For applications to understanding relations between boundary components of different parabolic subgroups, we need the *relative Langlands decomposition* for pairs of parabolic subgroups.

Specifically, for every pair of parabolic subgroups P, Q, $P \subset Q$, there is a unique parabolic subgroup P' of M_Q such that

$$N_P = N_Q N_{P'}, \quad M_{P'} = M_P, \quad A_P = A_Q A_{P'}, \tag{I.1.21}$$

which implies that

$$X_Q = N_{P'} \times A_{P'} \times X_P, \tag{I.1.22}$$

called the *relative horospherical decomposition* for the pair $P \subset Q$.

Since $K_Q = K \cap M_Q$ is a maximal compact subgroup and hence $M_Q = P'(K \cap M_Q)$, the Langlands decomposition of P' implies the relative Langlands decomposition of Q with respect to P:

$$Q = N_P A_P M_P K_Q \cong N_P \times A_P \times M_P K_Q. \tag{I.1.23}$$

In fact, it follows from equation (I.1.21), equation (I.1.20), and

$$\begin{aligned} Q &= N_Q A_Q M_Q = N_Q A_Q (P' K_Q) = N_Q A_Q (N_{P'} A_{P'} (M_{P'} K_Q) \\ &= N_Q N_{P'} A_Q A_{P'} M_P K_Q. \end{aligned}$$

The existence of the parabolic subgroup P' in equation (I.1.21) can be explained as follows. Assume $Q = P_I$ in the notation of the previous subsection. Let $\Phi(\mathfrak{g}, \mathfrak{a}_P)$ be the roots of $\mathfrak{a}_P$ acting on $\mathfrak{g}$,

$$\mathfrak{g} = \mathfrak{g}_0 + \sum_{\alpha \in \Phi(\mathfrak{g}, \mathfrak{a}_P)} \mathfrak{g}_\alpha,$$

where $\mathfrak{g}_0 = \mathfrak{a} \oplus \mathfrak{m}_P$. Let $\Phi^I(\mathfrak{g}, \mathfrak{a}_P)$ be the roots in $\Phi(\mathfrak{g}, \mathfrak{a}_P)$ that are linear combinations of roots in I. Then

$$\mathfrak{m}_Q = \mathfrak{a}_P^I + \sum_{\alpha \in \Phi^I(\mathfrak{g}, \mathfrak{a}_P)} \mathfrak{g}_\alpha + \mathfrak{m}_P.$$

Let $\Phi_I^+(P, A_P)$ be the set of roots in $\Phi(P, A_P)$ that do not vanish on $\mathfrak{a}_{P_I}$, and $\Phi^{I,+}(\mathfrak{g}, \mathfrak{a}_P) = \Phi^I(\mathfrak{g}, \mathfrak{a}_P) \cap \Phi(P, A_P)$. Clearly, $\Phi(P, A_P) = \Phi_I^+(P, A_P) \cup \Phi^{I,+}(\mathfrak{g}, \mathfrak{a}_P)$. Then

$$\mathfrak{n}_Q = \sum_{\alpha \in \Phi_I^+(P, A_P)} \mathfrak{g}_\alpha.$$

Define

$$\mathfrak{p}' = \mathfrak{a}_P^I + \mathfrak{m}_P + \sum_{\alpha \in \Phi^{I,+}(\mathfrak{g}, \mathfrak{a}_P)} \mathfrak{g}_\alpha.$$

Then $\mathfrak{p}'$ is a parabolic subalgebra of $\mathfrak{m}_Q$. Let P' be the normalizer of $\mathfrak{p}'$ in M_Q. Then P' is a parabolic subgroup of M_Q and satisfies the conditions

$$A_{P'} = A_P^I, \quad M_{P'} = M_P, \quad N_{P'} = \exp \sum_{\alpha \in \Phi^{I,+}(P, A_P)} \mathfrak{g}_\alpha. \tag{I.1.24}$$

Therefore P' satisfies the conditions in equation (I.1.21). It can be shown that $P' = M_Q \cap P$. Briefly, it follows from (1) $M_Q = N_{P'} A_{P'} M_{P'} (K \cap M_Q)$, $P = N_P A_P M_P$; and (2) $N_{P'} \subset N_P$, $A_{P'} \subset A_P$, and $M_P = M_{P'}$.

I.1.12 Summary and comments. In this section, we have defined parabolic subgroups by first introducing the standard ones associated with a choice of a positive chamber $\mathfrak{a}^+$ of a maximal abelian subalgebra $\mathfrak{a}$. Two important concepts are the Langlands decomposition of parabolic subgroups and the induced horospherical decomposition of X. To understand these results, it is helpful to keep the example of $G = \mathrm{SL}(n, \mathbb{R})$ in mind.

As mentioned earlier, the horospherical decomposition is basically an analogue of the upper-half-plane model of the symmetric space $\mathrm{SL}(2, \mathbb{R})/\mathrm{SO}(2)$. Results in this section will be used repeatedly in later sections.

In this book, the space $X_P = M_P / K \cap M_P$ is called the boundary symmetric space. The reason is that it often appears in the boundary of compactifications of X. This terminology fits well with the philosophy of this book that parabolic subgroups are related to the geometry at infinity of symmetric spaces. Another point of view is to consider X_P as a submanifold of X. In fact, let $x_0 = K \in G/K$ be a basepoint. Then X_P can be identified with the M_P-orbit through x_0. When we push the basepoint x_0 to infinity in the positive direction determined by P (i.e., through the positive chamber $e^{\mathfrak{a}_P^+} \cdot x_0$), X_P is pushed to infinity (or the boundary) of X. This point of view is convenient for the purpose of studying behaviors at infinity of harmonic functions on symmetric spaces (see [Ko1–7]).

I.2 Geodesic compactification and Tits building

In this section, we discuss the geodesic compactification $X \cup X(\infty)$, which is defined in terms of equivalence classes of geodesics. Classification of geodesics leads to the (spherical) Tits building $\Delta(G)$ and its geometric realization, and establishes a close connection between the Tits building and the boundary of compactifications. The discussions in this section further explain the role of parabolic subgroups in understanding the geometry of symmetric spaces.

In this book, all geodesics are directed and have unit speed. For more details about geometry of simply connected nonpositively curved spaces and geodesics, see [BGS] [Eb] [Ka].

This section is organized as follows. We discuss the geodesic compactification through the examples of $\mathbb{R}^n$ and the Poincaré disk. Then we define an equivalence relation on the set of geodesics in I.2.2, and the related sphere at infinity $X(\infty)$ (I.2.3) and the geodesic compactification $X \cup X(\infty)$ (I.2.4). The continuous extension of G to $X \cup X(\infty)$ is given in I.2.5. Then we show that parabolic subgroups arise naturally as stabilizers of points in the sphere at infinity (I.2.6), and that the Langlands decomposition can be expressed in terms of more refined equivalence relations on geodesics (I.2.15, I.2.29, I.2.31). After recalling briefly the (spherical) Tits building $\Delta(G)$ (I.2.18), we give a geometric realization of the Tits building (I.2.19) and use it to describe the topological Tits building (I.2.21, I.2.22). Then we refine the equivalence relation on geodesics to the N-relation (I.2.26) and use it to explain how the boundary symmetric spaces X_P arise in the parameter spaces of N-equivalence classes (I.2.29), which motivates the definition of the Karpelevič compactification $\overline{X}^K$ in the next section.

I.2.1 The geodesic compactification $X \cup X(\infty)$ can be defined in the same way for any simply connected nonpositively curved Riemannian manifold. To motivate this construction, we consider two special examples.

First, consider the Euclidean space $\mathbb{R}^n$. Two parallel geodesics γ_1, γ_2 in $\mathbb{R}^n$ satisfy the condition that $d(\gamma_1(t), \gamma_2(t))$ is a constant function, in particular,

$$\limsup_{t \to +\infty} d(\gamma_1(t), \gamma_2(t) < +\infty.$$

Let S^{n-1} be the unit sphere in $\mathbb{R}^n$. Then every geodesic in $\mathbb{R}^n$ is parallel to a unique geodesic of the form tV, $V \in S^{n-1}$, and hence the set of parallel classes of geodesics of $\mathbb{R}^n$ can be identified with S^{n-1}.

Let $\pi : [0, +\infty) \to [0, 1)$ be a diffeomorphism whose derivative at $r = 0$ is equal to 1 and higher derivatives vanish at $r = 0$. Then π defines a diffeomorphism

$$\pi : \mathbb{R}^n \to B^n, \quad rV \mapsto \pi(r)V,$$

where B^n is the open unit ball in $\mathbb{R}^n$, $r \geq 0$, and $V \in S^{n-1}$. Hence $\mathbb{R}^n$ can be compactified by adding S^{n-1} to $\pi(\mathbb{R}^n) = B^n$. (See [Me1, p. 13] for a more intrinsic description of this compactification and its applications.)

In this compactification, for any sequence $t_j \to +\infty$ and a geodesic γ parallel to tV, $\gamma(t_j)$ converges to V. This motivates the direct construction of this compactification of $\mathbb{R}^n$ in terms of geodesics to be discussed in the next subsection.

In the case of the Poincaré disk, for any two distinct geodesics γ_1, γ_2, $d(\gamma_1(t), \gamma_2(t))$ is never constant. But either $\limsup_{t\to+\infty} d(\gamma_1(t), \gamma_2(t)) < +\infty$ or $\lim_{t\to+\infty} d(\gamma_1(t), \gamma_2(t)) = +\infty$. In the former case, with a suitable reparametrization on the geodesics, $\lim_{t\to+\infty} d(\gamma_1(t), \gamma_2(t)) = 0$. So in general, we could not consider parallel equivalence classes of geodesics.

I.2.2 Recall from §I.1.2 that $X = G/K$ is a symmetric space of noncompact type. Let $\mathfrak{g} = \mathfrak{k} + \mathfrak{p}$ be the *Cartan decomposition* associated with K. Then the tangent space $T_{x_0}X$ at the basepoint $x_0 = K$ can be canonically identified with $\mathfrak{p}$. The *Killing form* $B(\cdot, \cdot)$ of $\mathfrak{g}$ restricts to a positive definite quadratic form $\langle \cdot, \cdot \rangle$ and hence defines a G-invariant Riemannian metric on $X = G/K$. In this metric, X is a *simply connected, nonpositively curved* Riemannian manifold. In the following, we will fix this Riemannian metric on X. The induced norm on $\mathfrak{p}$ is denoted by $\| \ \|$.

Two geodesics γ_1, γ_2 in X are defined to be *equivalent*, denoted by $\gamma_1 \sim \gamma_2$, if

$$\limsup_{t\to+\infty} d(\gamma_1(t), \gamma_2(t)) < +\infty.$$

It is clearly an equivalence relation. Denote the *set of equivalence classes of geodesics* by $X(\infty)$,

$$X(\infty) = \{\text{all geodesics in } X\}/\sim .$$

For any geodesic γ, its *equivalence class* is denoted by $[\gamma]$.

Proposition I.2.3 *The set $X(\infty)$ can be canonically identified with the unit sphere in the tangent space T_xX at any basepoint x, in particular, the unit sphere in $\mathfrak{p} = T_{x_0}X$.*

Proof. Clearly, the set of geodesics passing through the basepoint x is parametrized by the unit sphere in T_xX. Since X is simply connected and nonpositively curved, comparison with the Euclidean space implies that any two such different geodesics are not equivalent; and it remains to show that every equivalence class contains a geodesic passing through x.

Let γ be any geodesic. Consider the sequence of points $\gamma(n)$, $n \geq 1$. Since X is simply connected and nonpositively curved, there is a unique geodesic γ_n passing through x and $\gamma(n)$. Assume that $\gamma_n(0) = x$. The compactness argument shows that there is a subsequence $\gamma_{n'}$ such that $\gamma_{n'}(t)$ converges uniformly for t in compact subsets to a geodesic $\gamma_\infty(t)$. Clearly, γ_∞ passes through x. Comparison with the Euclidean space again implies that γ_∞ is equivalent to γ.

Because of the identification in the above proposition and the fact that $X(\infty)$ forms the boundary of a compactification, the set $X(\infty)$ is called the *sphere at infinity*, or the *visibility sphere*. The geodesic compactification can be defined for every simply connected, nonpositively curved Riemannian manifold, and Proposition I.2.3 also holds.

I.2.4 The sphere $X(\infty)$ can be attached at the infinity of X to define the *geodesic compactification* $X \cup X(\infty)$. For convenience, we describe its topology by convergent sequences (see [JM, §6] and §I.8 below for details on how to define a topology in terms of convergent sequences).

For any $[\gamma] \in X(\infty)$, an unbounded sequence y_j in X converges to $[\gamma]$ if the geodesic from a basepoint x to y_j converges to a geodesic in $[\gamma]$. Comparison with the Euclidean space shows that this topology is independent of the choice of the basepoint x. For convenience, we will use the basepoint $x_0 = K$. For a geodesic γ passing through x_0, the intersection with X of a fundamental system of neighborhoods of $[\gamma]$ in $X \cup X(\infty)$ is given by a family of truncated cones $C(\gamma, \varepsilon_j, t_j)$ based on the geodesic $\gamma(t)$ and truncated at t_j, where $t_j \to +\infty$, $\varepsilon_j \to 0$. Specifically, for any $\varepsilon > 0$, the cone $C(\gamma, \varepsilon)$ consists of points x such that the angle between γ and the geodesic from x_0 to x is less than ε, and the truncated cone is defined by

$$C(\gamma, \varepsilon, t_j) = C(\gamma, \varepsilon) - B(\gamma(0), t_j),$$

where $B(\gamma(0), t_j)$ is the ball of radius t_j with center $\gamma(0)$. Because of this, the topology on $X \cup X(\infty)$ was called the *conic topology* in [AS], and the compactification $X \cup X(\infty)$ was hence called the *conic compactification* in [GJT]. It can be checked easily that the induced topology on $X(\infty)$ coincides with the subset topology on the unit sphere in $T_x X$ in the above proposition.

Since the name of conic compactification is technical and does not emphasize the role of geodesics, the compactification $X \cup X(\infty)$ will be called the *geodesic compactification* in this book. In some papers, it is also called the *visibility (sphere) compactification*.

Proposition I.2.5 *The isometric action of G on X extends to a continuous action on the compactification $X \cup X(\infty)$.*

Proof. Since both the equivalence relation between geodesics and convergent sequences are preserved under isometries, the action of G on X extends to the compactification $X \cup X(\infty)$, and the extended action is continuous.

One application of this extended action is the following characterization of parabolic subgroups as stabilizers of points in $X(\infty)$.

Proposition I.2.6 *For any point in $X(\infty)$, its stabilizer in G is a parabolic subgroup. Conversely, every proper parabolic subgroup of G is the stabilizer of some point in $X(\infty)$.*

The proof will be given below after several lemmas. It should be pointed out that for a given parabolic subgroup, its fixed points are in general not unique. A more precise version of this proposition will be given in Corollary I.2.17 below after explicit relations between geodesics and parabolic subgroups have been established.

Lemma I.2.7 *If two geodesics γ, δ are equivalent, then there is a continuous family of geodesics γ_s, $s \in [0,1]$, connecting them, $\gamma_0 = \gamma$, $\gamma_1 = \delta$, such that for every s, γ_s is equivalent to γ and δ.*

Proof. Let $p = \gamma(0), q = \delta(0)$. Let $c : [0,1] \to X$ be a continuous curve connecting p, q, $c(0) = p$, $c(1) = q$. The proof of Proposition I.2.3 shows that there exists a unique geodesic γ_s in the equivalence class $[\gamma] = [\delta]$ such that $\gamma_s(0) = c(s)$. Comparison with the Euclidean space shows that γ_s depends continuously on s and hence forms a continuous family of geodesics connecting γ and δ.

Corollary I.2.8 *All the geodesics in each equivalence class of geodesics in X form a continuous family.*

One problem is to identify a parameter space for these continuous families. For this purpose, we need to relate parabolic subgroups to geodesics explicitly.

Let P be a proper parabolic subgroup of G, $A_P = \exp \mathfrak{a}_P$ its split component. The root hyperplanes of the roots in $\Phi(P, A_P)$ divide $\mathfrak{a}_P$ into chambers. Let $\mathfrak{a}_P^+$ be the unique chamber such that the roots in $\Phi(P, A_P)$ are positive on $\mathfrak{a}_P^+$. Define

$$\mathfrak{a}_P^+(\infty) = \{H \in \mathfrak{a}_P^+ \mid \|H\| = 1\}, \tag{I.2.1}$$

which can be identified with the set of equivalence classes of geodesics in $\mathfrak{a}_P$ that have representatives γ such that $\gamma(t) \in e^{\mathfrak{a}_P^+} x_0$ for $t \gg 0$. Note that $\mathfrak{a}_P^+(\infty)$ is an open simplex, and its closure $\overline{\mathfrak{a}_P^+(\infty)}$ in $\mathfrak{a}_P$ is a closed simplex.

Lemma I.2.9 *For each $H \in \mathfrak{a}_P^+(\infty)$, $a \in A_P$, $n \in N_P$, and $z \in X_P$,*

$$\gamma_{n,a,z}(t) = (n, a \exp tH, z) \in N_P \times A_P \times X_P = X \tag{I.2.2}$$

is a geodesic in X.

Proof. Since $A_P x_0$ is a totally geodesic flat submanifold of X, $e^{tH} x_0$ is a geodesic in X. Write $z = mK_P$, where $m \in M_P$. Then $\gamma_{n,a,z}(t) = nam \cdot e^{tH} x_0$ is the image of a geodesic under an isometry of X and is hence a geodesic in X.

Lemma I.2.10 *When H is fixed, for different choices of n, a, z, the geodesics $\gamma_{n,a,z}$ are equivalent.*

Proof. It suffices to prove that $\gamma_{n,a,z}$ is equivalent to $\gamma_{e,e,x_0} = \exp tHx_0$. By the G-invariance of the metric,

$$\begin{aligned} d(\gamma_{n,a,z}(t), \gamma_{e,e,x_0}(t)) &= d(nae^{tH} z, e^{tH} x_0) \\ &= d(e^{-tH} ne^{tH} az, x_0) \to d(az, x_0), \end{aligned}$$

as $t \to +\infty$, since $e^{-tH} ne^{tH} \to e$.

Lemma I.2.11 *If $H_1, H_2 \in \mathfrak{a}_P^+(\infty)$, $H_1 \neq H_2$, then for any $n_1, n_2 \in N_P$, $a_1, a_2 \in A_P$, $z_1, z_2 \in X_P$, the geodesics $\gamma_1(t) = (n_1, a_1 e^{tH_1}, z_1)$, $\gamma_2(t) = (n_2, a_2 e^{tH_2}, z_2)$ are not equivalent.*

Proof. Since $H_1 \neq H_2$, the geodesics $e^{tH_1}x_0$, $e^{tH_2}x_0$ are not equivalent, by Proposition I.2.3. Then the lemma follows from the previous lemma.

Lemma I.2.12 *If P_1, P_2 are different parabolic subgroups, $H_1 \in \mathfrak{a}_{P_1}^+(\infty)$, $H_2 \in \mathfrak{a}_{P_2}^+(\infty)$, then the geodesics $\gamma_1(t) = e^{tH_1}x_0$, $\gamma_2(t) = e^{tH_2}x_0$ are not equivalent.*

Proof. By Proposition I.2.3, it suffices to prove $H_1 \neq H_2$. If $H_1 = H_2$, then $\mathfrak{a}_{P_1} = \mathfrak{a}_{P_2}$ and hence $\mathfrak{a}_{P_1}^+ = \mathfrak{a}_{P_2}^+$, since H_1 (H_2) belongs to the interior of the chamber $\mathfrak{a}_{P_1}^+$ ($\mathfrak{a}_{P_2}^+$ respectively), and $\mathfrak{a}_{P_1}, \mathfrak{a}_{P_2}$ can intersect only along chamber faces if they are not identical. Now there are only finitely many parabolic subgroups that have $\mathfrak{a}_{P_1}$ as the split component, and each corresponds to a unique chamber. This implies that $P_1 = P_2$. This contradicts the assumption and hence implies that $H_1 \neq H_2$.

I.2.13 Proof of Proposition I.2.6. For any point $[\gamma] \in X(\infty)$, let $H \in \mathfrak{p} = T_{x_0}X$ be the unique vector such that $\gamma(t) = \exp tH \in [\gamma]$. Let P_0 be a minimal parabolic subgroup. Then the Cartan decomposition

$$\mathfrak{p} = \cup_{k \in K} Ad(k)\overline{\mathfrak{a}_{P_0}^+} \tag{I.2.3}$$

and Lemma I.2.12 show that there exists a unique parabolic subgroup $P = {}^k P_{0,I}$ such that $H \in \mathfrak{a}_P^+(\infty)$. For any $p \in P$, by the Langlands decomposition of P, write $p = nam$, where $n \in N_P, a \in A_P, m \in M_P$. Then

$$p \cdot \gamma(t) = (n, a \exp tH, mx_0). \tag{I.2.4}$$

By Lemma I.2.10, $[p \cdot \gamma] = [\gamma]$ and hence P stabilizes $[\gamma]$.

On the other hand, for $g \in G \setminus P$, ${}^g P \neq P$. Write $g = kp$, where $k \in K$, $p \in P$. Then by equation (I.2.4) and equation (I.1.16),

$$\begin{aligned} g \cdot \gamma(t) &= k \cdot (n, a \exp tH, mx_0) \\ &= ({}^k n, {}^k a \exp tAd(k)(H), {}^k mx_0) \in N_{{}^k P} \times A_{{}^k P} \times X_{{}^k P}. \end{aligned}$$

Since ${}^k P \neq P$, Lemmas I.2.10 and I.2.12 imply that $g \cdot \gamma$ and γ are not equivalent. This shows that P is the stabilizer of $[\gamma]$.

Conversely, for any proper parabolic subgroup P and any vector $H \in \mathfrak{a}_P^+(\infty)$, define a geodesic $\gamma(t) = \exp tH \cdot x_0$ in X. The above arguments show that P is the stabilizer of $[\gamma]$. This completes the proof of Proposition I.2.6.

Remark I.2.14 For a different proof of Proposition I.2.6, see [GJT, Proposition 3.8].

The parametrization of each geodesic $\gamma(t)$ is unique up to a shift in the parameter t. For each $H \in \mathfrak{a}_P^+(\infty)$, let $\langle H\rangle^\perp$ be the orthogonal complement to the linear subspace $\langle H\rangle$ spanned by H. Then after reparametrization, the component a in the geodesics $\gamma_{n,a,z}$ in equation (I.2.2) can be chosen to lie in $\exp\langle H\rangle^\perp$.

Proposition I.2.15 *For every parabolic subgroup P and $H \in \mathfrak{a}_P^+(\infty)$, let $\gamma_H(t) = e^{tH}x_0$, called the canonical geodesic associated with H. Then the family of geodesics in the equivalence class $[\gamma_H]$ is parametrized by $N_P \times e^{\langle H\rangle^\perp} \times X_P$.*

Proof. The previous three lemmas show that two geodesics $\gamma_i(t) = (n_i, a_i e^{tH_i}, z_i)$ associated with two parabolic subgroups P_i, $i = 1, 2$, as above are equivalent if and only if $P_1 = P_2$ and $H_1 = H_2$. Then the identification of the parameter space follows from the comments above.

Proposition I.2.16 *For every parabolic subgroup P, $\mathfrak{a}_P^+(\infty)$ can be canonically identified with a subset of $X(\infty)$ by the map $H \mapsto [\gamma_H]$, where $\gamma_H(t) = \exp tHx_0$. The sphere at infinity $X(\infty)$ admits a disjoint decomposition*

$$X(\infty) = \coprod_P \mathfrak{a}_P^+(\infty), \tag{I.2.5}$$

where P runs over all (proper) parabolic subgroups.

Proof. By the Cartan decomposition of X, for the minimal parabolic subgroup P_0,

$$X = K \exp \overline{\mathfrak{a}_{P_0}^+} x_0.$$

Let $\overline{\mathfrak{a}_{P_0}^+(\infty)}$ be the closure of $\mathfrak{a}_{P_0}^+(\infty)$ in $\mathfrak{a}_{P_0}$. Then

$$X(\infty) = K \cdot \overline{\mathfrak{a}_{P_0}^+(\infty)}.$$

Since the faces of the simplicial cone $\overline{\mathfrak{a}_{P_0}^+}$ are $\mathfrak{a}_{P_{0,I}}^+$, the faces of the simplex $\overline{\mathfrak{a}_{P_0}^+}(\infty)$ are $\mathfrak{a}_{P_{0,I}}^+(\infty)$. Combined with the fact that every parabolic subgroup P is conjugate under K to a standard parabolic subgroup $P_{0,I}$, it follows that

$$X(\infty) = \cup_P \mathfrak{a}_P^+(\infty).$$

By Lemma I.2.12, for two different parabolic subgroups P_1, P_2, $\mathfrak{a}_{P_1}^+(\infty)$, $\mathfrak{a}_{P_2}^+(\infty)$ are disjoint. Hence, the decomposition in the above equation is disjoint.

The disjoint decomposition of $X(\infty)$ into $\mathfrak{a}_P^+(\infty)$ in the above proposition can be recovered through the G-action on $X(\infty)$.

Corollary I.2.17 *For every parabolic subgroup P, the set of points in $X(\infty)$ whose stabilizer is equal to P is exactly equal to $\mathfrak{a}_P^+(\infty)$, and the set of points in $X(\infty)$ that are fixed by P is equal to the closure of $\mathfrak{a}_P^+(\infty)$ in $X(\infty)$, which can be identified with the closure $\overline{\mathfrak{a}_P^+(\infty)}$ of $\mathfrak{a}_P^+(\infty)$ in $\mathfrak{a}_P$.*

Proof. The proof of Proposition I.2.6 gives the first statement. For the second statement, we note that a parabolic subgroup Q contains P if and only if $\mathfrak{a}_Q^+$ is a face of the polyhedral cone $\mathfrak{a}_P^+$, which is in turn equivalent to that $\mathfrak{a}_Q^+(\infty)$ is a face of $\mathfrak{a}_P^+(\infty)$. Since

$$\overline{\mathfrak{a}_P^+(\infty)} = \coprod_{Q \supseteq P} \mathfrak{a}_Q^+(\infty),$$

the first statement implies the second.

I.2.18 Now we recall several basic facts on Tits buildings and explain how the decomposition in equation (I.2.5) gives a geometric realization of the Tits building of G. For more details about the Tits buildings and proofs of the statements below, see [Ti1] [Ti2] [Br1] (see also [Ji9] for many applications of Tits buildings to geometry and topology).

The Tits building $\Delta(G)$ of the group G is an infinite simplicial complex such that there is a one-to-one correspondence between the set of simplexes and the set of proper parabolic subgroups in G satisfying the following compatibility conditions:

1. For each parabolic subgroup P, denote the corresponding simplex by Δ_P. When P is maximal, Δ_P is a vertex, i.e., a simplex of dimension 0.
2. For every pair of parabolic subgroups P, Q, Δ_P contains Δ_Q as a face if and only if $P \subset Q$.

The group G acts on the set of parabolic subgroups by conjugation, and hence also acts on $\Delta(G)$ simplicially. Since each parabolic subgroup is equal to its normalizer, the stabilizer of each simplex Δ_P is equal to P.

An important feature of the Tits building $\Delta(G)$ is the rich structure of apartments. For any Cartan subalgebra $\mathfrak{a}$ of the symmetric pair $(\mathfrak{g}, \mathfrak{k})$, i.e., a maximal abelian subalgebra in $\mathfrak{p}$, the orthogonal complement of $\mathfrak{k}$ in $\mathfrak{g}$, there are only finitely many parabolic subgroups P whose split component A_P with respect to the basepoint x_0 is contained in $A = \exp \mathfrak{a}$, and they correspond to the chambers and chamber faces in $\mathfrak{a}$; their corresponding simplexes Δ_P form a finite subcomplex $\Sigma(\mathfrak{a})$, called an apartment of $\Delta(G)$. This subcomplex $\Sigma(\mathfrak{a})$ is a *triangulation of the unit sphere* $\mathfrak{a}(\infty)$ in $\mathfrak{a}$ induced by the chamber decomposition. More generally, for any $g \in G$, the subalgebra $Ad(g)\mathfrak{a}$ for the symmetric pair $(\mathfrak{g}, Ad(g)\mathfrak{a})$ also determines an apartment $\Sigma(Ad(g)\mathfrak{a})$ by using the split component with respect to the basepoint gx_0. They exhaust all the apartments in $\Delta(G)$.

The apartments in $\Delta(G)$ satisfy the following conditions:

1. For any two simplexes Δ_1, Δ_2, there exists an apartment Σ containing Δ_1, Δ_2;
2. if Σ' is another apartment containing Δ_1, Δ_2, then there is an isomorphism between Σ and Σ' that fixes Δ_1 and Δ_2 pointwise.

Proposition I.2.19 *For each parabolic subgroup P, the simplex Δ_P in $\Delta(G)$ can be identified with the subset $\overline{\mathfrak{a}_P^+(\infty)} \subset X(\infty)$, the set of points fixed by P (Corollary I.2.17); the underlying space of the building $\Delta(G)$ can be identified with $X(\infty)$.*

Proof. When P is a maximal parabolic subgroup, $\dim \mathfrak{a}_P = 1$, and hence $\mathfrak{a}^+(\infty)$ consists of one point. For a pair of parabolic subgroups P, Q, the inclusion $P \subset Q$ holds if and only if $\overline{\mathfrak{a}_Q^+}$ is a face of $\overline{\mathfrak{a}_P^+}$, which is equivalent to that $\overline{\mathfrak{a}_Q^+(\infty)}$ is a face of $\overline{\mathfrak{a}_P^+(\infty)}$. This proves the first statement. The second statement follows from the decomposition in equation (I.2.5).

The G-action on $\Delta(G)$ canonically extends to an action on the underlying space of $\Delta(G)$, which will also be denoted by $\Delta(G)$.

Lemma I.2.20 *The space $\Delta(G)$ admits a G-invariant metric whose restriction to each apartment gives a sphere. This metric is called the Tits metric.*

Proof. For each parabolic subgroup P, the simplex Δ_P inherits a metric from the norm $\| \ \|$ on $\mathfrak{a}_P$ as the subset $\overline{\mathfrak{a}_P^+(\infty)}$. For every pair of parabolic subgroups P, Q, the inclusion $\Delta_Q \subset \Delta_P$ is isometric. Therefore, these metrics are compatible and define a metric on $\Delta(G)$. Clearly, for any $g \in G$, the map $g : \Delta_P \to \Delta_{gP}$ is isometric, and hence the metric on $\Delta(G)$ is G-invariant.

In the Tits metric, for any $H \in \Delta_P$ and $g \in G - P$, $gH \in \Delta_{gP}$, and the distance $d(H, gH)$ is independent of g. On the other hand, it is reasonable to expect that if g is closer to the identity element, then Δ_{gP} should be closer to Δ_P. This is important for applications to compactifications of the symmetric space X. Briefly, it will be seen that the boundary of compactifications of X is often a cell complex related to or parametrized by the Tits building $\Delta(G)$, and a topology is needed to measure the closeness of these boundary cells (or boundary components). For this purpose, we use the *topological Tits building* from [BuS1, Definition 1.1].

Recall that maximal totally geodesic flat submanifolds of X have the same dimension, called the rank of X and denoted by $rk(X)$, which is also equal to the dimension of the split component A_P of minimal parabolic subgroups P and hence is also equal to the rank of G, denoted by $rk(G)$. In fact, any maximal flat through x_0 is of the form $A_P x_0$ for some minimal parabolic subgroup P. Let $r = rk(X)$. The top dimensional simplexes in $\Delta(G)$ have r vertices. For each $k \leq r$, let Δ_k be the set of simplexes with k vertices, i.e., of dimension $k - 1$. The group G acts on Δ_1, the set of vertices, with r-orbits. Fix an order on the set of these orbits. Then for every $k \leq r$, there is an injective map $\Delta_k \to (\Delta_1)^k$, by mapping each simplex to its ordered vertices.

Definition I.2.21 *A topological Tits building of G is the Tits building $\Delta(G)$ with a Hausdorff topology on the set Δ_1 such that for all $k \leq r$, Δ_k is a closed subset of $(\Delta_1)^k$.*

For each $k \leq r$, Δ_r is given the subset topology of $(\Delta_1)^k$. As mentioned earlier, the topologies of Δ_k for all $k \leq r$ allow one to measure the closeness of the simplexes in $\Delta(G)$, and the condition of the image of Δ_k being closed in the definition imposes a compatibility condition.

Using the above geometric realization of $\Delta(G)$ by subsets of $X(\infty)$, we can realize the topologies as follows.

Let $\mathcal{S}(X(\infty))$ be the space of closed subsets of $X(\infty)$. The identification of $X(\infty)$ with the unit sphere in $T_{x_0}X$ in Proposition I.2.3 defines a metric on $X(\infty)$. The Hausdorff distance defines a metric on $\mathcal{S}(X(\infty))$: For $A, B \in \mathcal{S}(X(\infty))$, the Hausdorff distance $d^H(A, B)$ is defined by

$$d^H(A, B) = \inf\{\varepsilon \mid d(x, B) < \varepsilon, d(A, y) < \varepsilon \text{ for all } x \in A, y \in B\}.$$

Proposition I.2.22 *The geometric realization of $\Delta(G)$ in Proposition I.2.19 induces an injective map*

$$\Delta_k \to \mathcal{S}(X(\infty)), \quad \Delta_P \mapsto \overline{\mathfrak{a}_P^+(\infty)}$$

for every $k = 1, \dots, r$, and the induced subset topology gives the structure of a topological Tits building on $\Delta(G)$.

Proof. The image of each Δ_k in $\mathcal{S}(X(\infty))$ is a compact subset. Since the vertices of $\overline{\mathfrak{a}_P^+(\infty)}$ depend continuously on $\overline{\mathfrak{a}_P^+(\infty)}$, the injective map $\Delta_k \to (\Delta_1)^k$ is continuous and its image is also compact, and hence closed.

Remark I.2.23 As explained in [Ji9], in many applications to geometry, for example, the Mostow strong rigidity [Mos], the rank rigidity of nonpositively curved manifolds [Bal1] [Bal2] [BuS2], and the classification of isoparametric submanifolds of codimension at least 3 [Te1] [Tho1] [Tho2], it is the topological Tits building rather than the usual Tits building that is used. One basic reason is that the topology on the topological Tits building gives a nondiscrete topology on the automorphism group of the building [BuS1].

I.2.24 In the geodesic compactification $X \cup X(\infty)$, we attach one ideal point at infinity to each equivalence class. To motivate more refined compactifications in later sections, it is natural to use the internal structure in each equivalence class and attach boundary points accordingly. If the rank of G is greater than or equal to 2, then there are nonconjugate parabolic subgroups, and hence there are different types of geodesics. This is made precise in the following result.

Proposition I.2.25 *Let P_0 be a minimal parabolic subgroup of G. Then the simplex $\overline{\mathfrak{a}_{P_0}^+(\infty)}$ is a set of representatives of the G-orbits in $X(\infty)$, and hence G acts transitively on $X(\infty)$ if and only if $rk(X)$ is equal to 1.*

Proof. Since $X(\infty) = \coprod_P \mathfrak{a}_P^+(\infty)$, where P runs over all proper parabolic subgroups, and every parabolic subgroup is conjugate to a standard one $P_{0,I}$, it follows that every G-orbit contains a point in $\overline{\mathfrak{a}_{P_0}^+(\infty)}$. On the other hand, for every $H \in \overline{\mathfrak{a}_{P_0}^+(\infty)}$, let $P_{0,I}$ be the unique standard parabolic subgroup such that

$H \in \mathfrak{a}^+_{P_{0,I}}(\infty)$. Then for every $g \in G$, either $g \in P_{0,I}$ and hence $gH = H$; or $g \notin P_{0,I}$, ${}^g P_{0,I} \neq P_{0,I}$, $gH \in \mathfrak{a}^+_{{}^g P_{0,I}}(\infty)$, and hence $gH \notin \mathfrak{a}^+_{P_{0,I}}(\infty)$. This implies that no two points in $\overline{\mathfrak{a}^+_{P_0}(\infty)}$ lie in the same orbit of G, which proves the first statement. The fact that $\overline{\mathfrak{a}^+_{P_0}(\infty)}$ consists of one point if and only if the rank of X is equal to 1 implies the second statement.

I.2.26 To study the internal structure of each equivalence class of geodesics $[\gamma] \in X(\infty)$, we define a more refined relation on geodesics.

Two geodesics γ_1, γ_2 in X are called *N-related* if

$$\lim_{t \to +\infty} d(\gamma_1(t), \gamma_2) = 0,$$

where $d(\gamma_1(t), \gamma_2) = \inf_{s \in \mathbb{R}} d(\gamma_1(t), \gamma_2(s))$.

It should be pointed out that there are pairs of N-related geodesics γ_1, γ_2 such that $d(\gamma_1(t), \gamma_2(t)) \not\to 0$. For example, take $\gamma_2(t) = \gamma_1(t+\delta)$, for a positive constant δ. Then $d(\gamma_1(t), \gamma_2) = 0$, and γ_1, γ_2 are clearly N-related but $d(\gamma_1(t), \gamma_2(t)) = \delta \neq 0$. Therefore, to avoid the problem of different parametrizations of geodesics and to get a simple geometric condition to define the N-equivalence relation, we need to use $d(\gamma_1(t), \gamma_2)$ rather than $d(\gamma_1(t), \gamma_2(t))$ as in the earlier equivalence relation.

Remark I.2.27 In [Ka], an equivalence class $[\gamma]$ of geodesics is called a *finite bundle* (or F-bundle), and an *N-equivalence class* is called a *null bundle* (or N-bundle). Detailed studies of these bundles play a crucial role in [Ka]. In fact, parabolic subgroups and Langlands decomposition were defined there in terms of them.

Lemma I.2.28 *If two geodesics γ_1, γ_2 are N-related, then $\gamma_1 \sim \gamma_2$, i.e., $[\gamma_1] = [\gamma_2] \in X(\infty)$.*

Proof. If γ_1 is N-related to γ_2, then there exists a sequence t_n such that $\varepsilon_n = d(\gamma_1(n), \gamma_2(t_n)) \to 0$ as $n \to +\infty$. We claim that $|t_n - n|$ is bounded, and hence $d(\gamma_1(n), \gamma_2(n))$ is bounded, which implies that $\gamma_1 \sim \gamma_2$. In fact, by the triangle inequality,

$$\begin{aligned} t_n &= d(\gamma_2(t_n), \gamma_2(0)) \leq d(\gamma_2(t_n), \gamma_1(n)) + d(\gamma_1(n), \gamma_1(0)) + d(\gamma_1(0), \gamma_2(0)) \\ &= \varepsilon_n + n + d(\gamma_1(0), \gamma_2(0)) = n + \varepsilon_n + d(\gamma_1(0), \gamma_2(0)). \end{aligned}$$

Similarly,

$$\begin{aligned} t_n &\geq d(\gamma_1(n), \gamma_1(0)) - d(\gamma_2(t_n), \gamma_1(n)) - d(\gamma_1(0), \gamma_2(0)) \\ &= n - \varepsilon_n - d(\gamma_1(0), \gamma_2(0)). \end{aligned}$$

These two inequalities imply the claim.

This lemma shows that the N-relation defines an equivalence relation on each equivalence class $[\gamma] \in X(\infty)$.

Proposition I.2.29 *For every parabolic subgroup P, $H \in \mathfrak{a}_P^+(\infty)$ and its associated geodesic $\gamma_H(t) = e^{tH}x_0$, the set of N-equivalence classes in $[\gamma_H]$ can be parametrized by $\langle H\rangle^\perp \times X_P$. In particular, for any $n \in N_P$, $n \cdot \gamma_H$ is N-related to γ_H.*

Proof. By Proposition I.2.15, every geodesic in $[\gamma_H]$ can be written uniquely, up to a shift in the parameter, in the form

$$\gamma_{n,a,m}(t) = (n, ae^{tH}, mx_0) \in N_P \times A_P \times X_P,$$

where $n \in N_P$, $a \in \langle H\rangle^\perp$, $m \in M_P$. By the G-invariance of the metric, we have

$$\begin{aligned} d(\gamma_{n_1,a_1,m_1}(t), \gamma_{n_2,a_2,m_2}(t)) &= d(n_1a_1e^{tH}m_1x_0, n_2a_2e^{tH}m_2x_0) \\ &= d((a_2e^{tH}m_2)^{-1}(n_2^{-1}n_1)(a_2e^{tH}m_2) \\ &\quad \times a_2^{-1}a_1m_2^{-1}m_1x_0, x_0) \\ &\to d(a_2^{-1}a_1m_2^{-1}m_1x_0, x_0). \end{aligned}$$

This implies that when a, m are fixed, but n changes, the geodesics $\gamma_{n,a,m}$ are N-related. The converse is also true, i.e., if $(a_1, m_1x_0) \neq (a_2, m_2x_0)$, the geodesics $\gamma_{n_1,a_1,m_1}, \gamma_{n_2,a_2,m_2}$ are not N-related. In fact, since M_P commutes with A_P, it can be shown that the map

$$\mathfrak{a}_P \times X_P \cong A_P \times X_P \to X, \quad (V, z) \mapsto e^V z,$$

is an isometric embedding. This implies that $d(a_2^{-1}a_1m_2^{-1}m_1x_0, x_0) = 0$ if and only if $m_2x_0 = m_1x_0$ and $a_2 = a_1$. When $n = e$, $d(\gamma_{e,a_1,m_1}(t), \gamma_{e,a_2,m_2}) = d(\gamma_{e,a_1,m_1}(t), \gamma_{e,a_2,m_2}(t))$. Consequently, γ_{e,a_1,m_1} is N-related to γ_{e,a_2,m_2} if and only if $a_1 = a_2$, $m_1x_0 = m_2x_0$. Together with the earlier result, this implies that $\gamma_{n_1,a_1,m_1}, \gamma_{n_2,a_2,m_2}$ are N-related if and only if $a_1 = a_2, m_1x_0 = m_2x_0$, and hence proves the proposition.

Corollary I.2.30 *If the rank of X is equal to 1, then every pair of equivalent geodesics are N-related.*

Proof. If the rank is equal to 1, then for every parabolic subgroup P, $\dim \mathfrak{a}_P = 1$, X_P consists of one point, and the parameter space $\langle H\rangle^\perp \times X_P$ in Proposition I.2.29 consists of one point.

I.2.31 Proposition I.2.29 and Corollary I.2.30 give a geometric interpretation of the Langlands decomposition of parabolic subgroups. Basically, for a geodesic $\gamma_H(t)$ as in Proposition I.2.29, its image under the action of N_P gives the N-related geodesics, and the further action of A_PM_P sweeps out all the geodesics in the equivalence class of $\gamma_H(t)$.

I.2.32 Summary and comments. In this section, we realized parabolic subgroups as the stabilizers of boundary points of the geodesic compactification $X \cup X(\infty)$ and gave a geometric realization of the Tits building $\Delta(G)$ in terms of simplexes $\mathfrak{a}_P^+(\infty)$, which can be used to give an explicit description of the topological building. We also gave a geometric interpretation of the Langlands decomposition of parabolic subgroups and the induced horospherical decomposition by using the refined N-equivalence relation on geodesics. All these results show that parabolic subgroups are related to the geometry at infinity of X, which is a basic point of this book.

The study of geodesics in negatively curved and simply connected surfaces (or manifolds) was started by Hadamard and Cartan. The geodesic compactification became well known and popular after [EO], though it was defined for symmetric spaces in [Ka].

The Tits buildings were motivated by the problem of giving a uniform geometric interpretation of the exceptional Lie groups. They have turned out to have many applications in geometry and other subjects. In these applications, the enhanced topological spherical Tits building is important. See the survey [Ji9] for details and references.

I.3 Karpelevič compactification

In this section, we describe the Karpelevič compactification $\overline{X}^K$ as a blow-up of the geodesic compactification $X \cup X(\infty)$ by making use of the internal structures of each equivalence class of geodesics, or rather points in $X(\infty)$, obtained in the previous section.

This section is organized as follows. We first use the dimension count to motivate how to blow up the points in $X(\infty)$, or equivalently what kinds of boundary components can be attached at infinity (I.3.3–I.3.5). Then we recall the original inductive definition of the Karpelevič compactification as a set in I.3.7, following [Ka]. The compactification is explained through the example $X = \mathbf{H} \times \mathbf{H}$ in I.3.8. The convergence of interior points to the boundary is described in I.3.9, and the convergence of boundary points is given in I.3.12. Several properties of $\overline{X}^K$ are stated in I.3.15.

I.3.1 When the rank of X is equal to 1, most nontrivial compactifications of X are isomorphic to the geodesic compactification $X \cup X(\infty)$. On the other hand, in the higher-rank case, there are different compactifications. One basic reason is that there are differences between the equivalence relation and the N-equivalence relation between geodesics, and there are different kinds of geodesics (Proposition I.2.25).

The Karpelevič compactification discussed in this section is a good example to illustrate this. For this purpose, a simple but useful example is $X = \mathbf{H} \times \mathbf{H}$, i.e., $G = \mathrm{SL}(2, \mathbb{R}) \times \mathrm{SL}(2, \mathbb{R})$. The geometry is simple but the difference between the (usual) equivalence relation and the N-relation is clear. See §I.3.7 below for details.

I.3.2 The *Karpelevič compactification* $\overline{X}^K$ in [Ka] was motivated by the problem of studying positive eigenfunctions and asymptotic behaviors of bounded harmonic functions on X. To study the behaviors at infinity of bounded harmonic functions, we need boundary spaces at infinity. The problem of positive eigenfunctions will be discussed in more detail in §1.7 below on the Martin compactification. Briefly, two problems that Karpelevič solved are (1) to find a family of linearly independent positive eigenfunctions such that other eigenfunctions can be expressed as unique linear combinations of them, or more precisely, superpositions of them, and (2) to parametrize this family by a subset in the boundary of the compactification $\overline{X}^K$.

To solve these problems, detailed structures of geodesics in X were used to refine the geodesic compactification $X \cup X(\infty)$ and define the compactification $\overline{X}^K$ and to understand the geometry at infinity. Some of the results in [Ka] have been recalled in §I.2.

I.3.3 The Karpelevič compactification $\overline{X}^K$ is a blow-up of $X \cup X(\infty)$ obtained by using the N-equivalence classes in each equivalence class $[\gamma]$ of geodesics.

The choice of the blow-up in [Ka] was motivated by the dimension count. A different explanation was given in [GJT] and will be recalled later in §I.14.

A natural way of blowing up the geodesic boundary $X(\infty)$ is to attach one ideal point to each geodesic instead of collapsing the whole equivalence class to one point. Since it is reasonable to expect that the dimension of the boundary should be of one dimension less than X, the dimension count below will show that the boundary that contains one point for each geodesic will be too large, and we need to collapse the N-equivalence classes of geodesics at infinity.

Lemma I.3.4 *Let P_0 be a minimal parabolic subgroup. For each $H \in \mathfrak{a}_{P_0}^+(\infty)$, let $\gamma_H(t) = e^{tH}x_0$. The set $\{[k\gamma_H] \mid k \in K, H \in \mathfrak{a}_{P_0}^+(\infty)\}$ of equivalence classes of geodesics is of dimension* $\dim X - 1$.

Proof. Let $K_{P_0} = K \cap P_0$. Then $K_{P_0} = K \cap M_{P_0}$, and it is the centralizer in K of each element $H \in \mathfrak{a}_{P_0}^+(\infty)$. The Cartan decomposition shows that $K/K_{P_0} \times \exp \mathfrak{a}_{P_0}^+$ is diffeomorphic to an open subset in X and hence of dimension $\dim X$. This implies that the dimension of $K/K_{P_0} \times \mathfrak{a}_{P_0}^+(\infty)$ is equal to $\dim X - 1$. The Cartan decomposition also implies that for two different cosets $k_1, k_2 \in K/K_{P_0}$, and two different $H_1, H_2 \in \mathfrak{a}_{P_0}^+(\infty)$,

$$Ad(k_1)H_1 \neq Ad(k_2)H_2,$$

and hence by Proposition I.2.3, $[k_1\gamma_{H_1}] \neq [k_2\gamma_{H_2}]$. These two results prove the proposition.

Since each equivalence class $[\gamma_H]$ contains a positive-dimensional family of geodesics $ne^{tH}x_0$, $n \in N_P$, this lemma shows that if we assign one ideal point to each geodesic, then the ideal boundary will have dimension greater than $\dim X$.

For any nonminimal parabolic subgroup P and $H \in \mathfrak{a}_P^+(\infty)$, N-equivalence classes of geodesics in $[\gamma_H]$ need to be identified further. In fact, by Proposition I.2.15, the set of N-equivalence classes of geodesics in $[\gamma_H]$ is parametrized by $\langle H\rangle^\perp \times X_P$. The dimension count in the next lemma shows that the factor $\langle H\rangle^\perp$ needs to be collapsed also.

For each geodesic γ, let $[\gamma]_N$ be the N-equivalence class containing γ. Then we have the following result.

Lemma I.3.5 *For any nonminimal parabolic subgroup P, $K/(K\cap P)\times\mathfrak{a}_P^+(\infty)\times X_P$ is of dimension* $\dim X - 1$, *and the set*

$$\{[k\gamma_{m,H}]_N \mid k \in K, H \in \mathfrak{a}_P^+(\infty), mx_0 \in X_P\}$$

is of dimension $\dim X - 1$, *where* $\gamma_{m,H}(t) = e^{tH}mx_0$.

Proof. Let P_0 be a minimal parabolic subgroup contained in P, and write $P = P_{0,I}$. Then $A_{P_0} = A_P A_{P_0}^I$ (see equation (I.1.17) in §I.1). Since $A_{P_0}x_0 \subset A_P X_P$ and $kA_P X_P = A_{^kP} X_{^kP}$, the Cartan decomposition $X = KA_{P_0}x_0$ implies that

$$X = KA_{P_0}x_0 = K(A_P \cdot X_P) \cong K/K_P \times A_P \times X_P,$$

and hence that $K/(K\cap P)\times \mathfrak{a}_P^+(\infty)\times X_P$ is of dimension $\dim X - 1$. Lemma I.2.12 shows that for different cosets $k_1, k_2 \in K/(K\cap P)$ and any points $H_1, H_2 \in \mathfrak{a}_P^+(\infty)$, $m_1x_0, m_2x_0 \in X_P$, the geodesic classes $k_1\gamma_{m_1,H_1}$, $k_2\gamma_{m_2,H_2}$ are not equivalent. Together with Proposition I.2.29, this implies that the N-equivalence classes $[k\gamma_{m,H}]_N$ are parametrized by $K/(K\cap P)\times\mathfrak{a}_P^+(\infty)\times X_P$.

I.3.6 The above lemma and the comments before it suggest that a natural blow-up is to replace each point $[\gamma_H] \in X(\infty)$ by X_P, where $H \in \mathfrak{a}_P^+(\infty)$, $\gamma_H(t) = e^{tH}x_0$. Since the inverse image of a point should be closed, we need a compactification of X_P. This will be done inductively.

I.3.7 The Karpelevič compactification $\overline{X}^K$ is defined inductively on $rk(X)$ in [Ka, §13].

First we define the Karpelevič compactification $\overline{X}^K$ as a set. For each point $\xi \in X(\infty)$, denote its stabilizer in G by P_ξ, which is a parabolic subgroup by Proposition I.2.17. Denote the Langlands decomposition of P_ξ by $P_\xi = N_\xi A_\xi M_\xi$, and the boundary symmetric space by X_ξ.

Let $r = rk(X)$. When $r = 1$, $\overline{X}^K$ is defined to be the geodesic compactification $X \cup X(\infty)$.

Assume that for every symmetric space Y of noncompact type of rank less than r, the compactification $\overline{Y}^K$ has been defined. For every point $\xi \in X(\infty)$, the boundary symmetric space X_ξ has rank less than r, and hence $\overline{X_\xi}^K$ is defined. Then the Karpelevič compactification $\overline{X}^K$ is defined by

$$\overline{X}^K = X \cup \coprod_{\xi \in X(\infty)} \overline{X_\xi}^K.$$

Clearly, there is a surjective map $\overline{X}^K \to X \cup X(\infty)$ that restricts to the identity map on X. This is exactly the blow-up described in the previous paragraph.

I.3.8 Example. We illustrate the construction $\overline{X}^K$ using the example of $X = \mathbf{H} \times \mathbf{H}$, i.e., $G = \mathrm{SL}(2, \mathbb{R}) \times \mathrm{SL}(2, \mathbb{R})$. There are two types of geodesics in X: (1) generic ones, $\gamma(t) = (\gamma_1(at), \gamma_2(bt))$, $a^2 + b^2 = 1$, $a, b > 0$, γ_1, γ_2 are geodesics in $\mathbf{H}$, (2) nongeneric geodesics, $\gamma(t) = (z_1, \gamma_2(t))$ or $\gamma(t) = (\gamma_1(t), z_2)$, where $z_1, z_2 \in \mathbf{H}$.

If γ is generic, its boundary symmetric space $X_{[\gamma]}$ consists of one point. On the other hand, if γ is nongeneric, then $X_{[\gamma]} = \mathbf{H}$. By definition,

$$\overline{(X_{[\gamma]})}^K = \mathbf{H} \cup \mathbf{H}(\infty) = \mathbf{H} \cup \mathbb{R} \cup \{\infty\}.$$

Hence the fibers of the map $\overline{X}^K \to X \cup X(\infty)$ on the boundary are either points or $\mathbf{H} \cup \mathbf{H}(\infty)$.

One can also work out the case of $X = \mathbf{H} \times \mathbf{H} \times H$. In this space, there are different types of nongeneric geodesics γ, and the spaces $X_{[\gamma]}$ are either $\mathbf{H}$ or $\mathbf{H} \times \mathbf{H}$, and hence the fibers over them are $\mathbf{H} \cup \mathbf{H}(\infty)$ or the compactification $\overline{\mathbf{H} \times \mathbf{H}}^K$ described above.

I.3.9 The topology of $\overline{X}^K$ is also defined inductively on the rank $r = rk(X)$.

When $r = 1$, the topology of $\overline{X}^K$ is equal to the topology of the geodesic compactification $X \cup X(\infty)$ defined in §I.2.4. Assume that for symmetric spaces Y of rank less than r, the topology of $\overline{Y}^K$ has been defined already.

We first describe convergence of interior sequences to boundary points. For any $\xi \in X(\infty)$, the rank of X_ξ is less than r, and hence the topology of $\overline{X_\xi}^K$ has been defined.

Identify $A_\xi \times X_\xi$ with the submanifold $\{e\} \times A_\xi \times X_\xi \subset N_\xi \times A_\xi \times X_\xi \cong X$. Then a sequence $y_j \in X$ converges to a point $z_\infty \in \overline{X_\xi}^K$ if y_j can be written as $y_j = g_j(a_j, z_j)$, where $g_j \in G$, $(a_j, z_j) \in A_\xi \times X_\xi$ satisfy the following conditions: as $j \to +\infty$,

1. $g_j \to e$,
2. $(a_j, z_j) \to \xi$ in $X \cup X(\infty)$,
3. $z_j \to z_\infty$ in $\overline{X_\xi}^K$.

Remark I.3.10 Since $\xi \in \mathfrak{a}_P^+(\infty)$, the second condition implies that

$$d(a_j x_0, x_0) \to +\infty, \quad d(z_j, x_0)/d(a_j x_0, x_0) \to 0,$$

but z_j may be unbounded, which is the case when z_∞ belongs to the boundary of $\overline{X_\xi}^K$.

Since $A_\xi \times X_\xi$ is a lower-dimensional submanifold, the factor g_j is needed to define general interior sequences converging to a boundary point. Furthermore, the factor g_j allows one to show that the G-action on X extends to a continuous action on $\overline{X}^K$.

To discuss convergence of sequences on the boundary, we need the following results.

Lemma I.3.11 *For every pair of points $\xi, \eta \in X(\infty)$, if the corresponding parabolic subgroups $P = P_\xi$, $Q = P_\eta$ satisfy $P \subset Q$, then P determines a parabolic subgroup P' of M_Q as in equation (I.1.21), and ξ determines a unique point $\xi' \in \mathfrak{a}^+_{P'}(\infty)$, and hence X_P and its compactification $\overline{X_P}^K$ can be canonically embedded into $\overline{X_Q}^K$ as the boundary space $\overline{X_{\xi'}}^K$. Denote this map $\overline{X_\xi}^K \to \overline{X_{\xi'}}^K \subset \overline{X_Q}^K$ by $\pi_{\xi,\eta}$.*

Proof. By equation (I.1.21), $X_P = X_{P'}$. It remains to find the point ξ'. Since $Q \supset P$, we can write $Q = P_I$. Then $\mathfrak{a}_P = \mathfrak{a}_Q \oplus \mathfrak{a}^I_P$. By assumption, $P = P_\xi$, and hence $\xi \in \mathfrak{a}^+_P(\infty)$. Write $\xi = \xi_1 + \xi_2$, where $\xi_1 \in \mathfrak{a}_Q$ and $\xi_2 \in \mathfrak{a}^I_P$. Note that $\mathfrak{a}_{P'} = \mathfrak{a}^I_P$, and hence $\xi_2 \in \mathfrak{a}^+_{P'}$. Choose a positive constant c such that $\xi' = c\xi_2 \in \mathfrak{a}^+_{P'}(\infty)$. Then $X_P = X_{\xi'}$. By definition, $\overline{X_Q}^K$ contains the compactification $\overline{X_{P'}}^K$.

I.3.12 Now we are ready to define convergent sequences on the boundary of $\overline{X}^K$. For any $\xi \in X(\infty)$ and a point $z \in \overline{X_\xi}^K$, a sequence $y_j \in \partial\overline{X}^K$ converges to z if and only if the following conditions hold:

1. Let $\xi_j \in X(\infty)$ be the point such that $y_j \in \overline{X_{\xi_j}}^K$. Then ξ_j converges to ξ_∞ in $X(\infty)$, and hence there exists a sequence $g_j \in G$ converging to the identity e such that $P_{g_j\xi_j}$ is either equal to or contained in P_ξ when $j \gg 1$.
2. The identification $X_{\xi_j} \cong X_{g_j\xi_j}$ extends to an isomorphism $\overline{X_{\xi_j}}^K \cong \overline{X_{g_j\xi_j}}^K$. Composed with the map $\pi_{g_j\xi_j,\xi} : \overline{X_{g_j\xi_j}}^K \to \overline{X_\xi}^K$ in Lemma I.3.11, it gives a map $\pi_j : \overline{X_{\xi_j}}^K \to \overline{X_\xi}^K$. Then the image $\pi_j(y_j)$ in $\overline{X_\xi}^K$ converges to z.

Remark I.3.13 In condition (1), the factor g_j is crucial in order to define the map π_j and measures the closeness of the boundary symmetric spaces X_{ξ_j} and $X_{g_j\xi_j}$; Otherwise there is not necessarily any inclusion relation between P_{ξ_j} and P_ξ. As explained in §I.2, this is the basic point of the topological Tits building.

Remarks I.3.14 In [Ka], the topology of $\overline{X}^K$ was defined completely in terms of geodesics. A slightly different formulation of the topology of $\overline{X}^K$ was given in [GJT], where a maximal abelian subalgebra $\mathfrak{a}$ (or equivalently a minimal parabolic subgroup) was fixed throughout. In the above formulation, parabolic subgroups and the relations in equation (I.1.21) between parabolic subgroups play a crucial role.

Since $\overline{X}^K$ is defined inductively, every boundary point of $\overline{X}^K$ can be represented in the form $(X_1, X_2, \ldots, X_k; z)$, where X_{i+1} is a boundary symmetric space of X_i, i.e., there exists a point $\xi \in X_i(\infty)$ such that $X_{i+1} = (X_i)_\xi$, and $z \in X_k$. It would be desirable to give a noninductive description of $\overline{X}^K$ and hence to explain this tower of boundary symmetric spaces X_i. In [GJT, Chapter V], such a noninductive description was given. Basically, for a sequence $H_j \in \overline{\mathfrak{a}_{P_0}^+}$ going to infinity, the limit of $e^{H_j} x_0$ in $X \cup X(\infty)$ does not depend on the root values $\alpha(H_j)$ such that $\alpha(H_j)/\|H_j\| \to 0$, but only on those values $\alpha(H_j)$ that are comparable to $\|H_j\|$. On the other hand, the limit of $e^{H_j} x_0$ in $\overline{X}^K$ depends on all root values $\alpha(H_j)$, and their different rates of going to infinity correspond to the successive boundary symmetric spaces in $(X_1, X_2, \ldots, X_k; z)$.

The basic properties of $\overline{X}^K$ in [Ka] are summarized in the following proposition.

Proposition I.3.15

1. *The space $\overline{X}^K$ is a compact, metrizable Hausdorff space containing X as a dense open subset.*
2. *The isometric G-action on X extends to a continuous action on $\overline{X}^K$.*
3. *For any $\xi \in X(\infty)$, the closure of the boundary symmetric space X_ξ in $\overline{X}^K$ is the Karpelevič compactification $\overline{X_\xi}^K$.*

The proofs of these results were given in [Ka, §13] and are quite long. We will give an alternative construction of $\overline{X}^K$ and prove these properties later in §I.14.

I.3.16 Summary and comments. In this section, we used the dimension count to determine the extent of the blowup of boundary points in $X(\infty)$. The example of $X = \mathbf{H} \times \mathbf{H}$ illustrates how the blowup of boundary points $[\gamma]$ in $X(\infty)$ depends on whether the geodesic γ is generic.

The construction in [Ka] is very complicated due to the inductive nature of the definition. Both the compactification and its construction are interesting and beautiful. Unfortunately it has not been well understood and much appreciated. The compactification $\overline{X}^K$ is briefly discussed in [Eb]. For the case of $G = \mathrm{SL}(n, \mathbb{R})$, it was also described in a different and more direct way in [Ne1]. The first noninductive construction of $\overline{X}^K$ was given in [GJT]. Another one will be given in §I.14 below.

I.4 Satake compactifications

In this section, we recall the Satake compactifications $\overline{X}^S$ of X in [Sat1], which are obtained by embedding X into some compact ambient spaces. As mentioned earlier, [Sat1] started the modern study of compactifications of symmetric and locally symmetric spaces and was motivated to define compactifications of locally symmetric spaces in [Sat2]. Some results in [Sat1] are slightly reformulated and given different proofs in this section.

For simplicity, we assume in this section that G is an adjoint semisimple Lie group, and $X = G/K$ as above. There are two steps in constructing the Satake compactifications $\overline{X}^S$:

1. Compactify the special symmetric space $\mathcal{P}_n = \mathrm{PSL}(n, \mathbb{C})/\mathrm{PSU}(n)$ of positive definite Hermitian matrices of determinant 1 by using semipositive Hermitian matrices to get $\overline{\mathcal{P}_n}^S$, called the *standard Satake compactification*.
2. Embed X into $\mathcal{P}_n$ for some n as a *totally geodesic submanifold* and take the closure of X in $\overline{\mathcal{P}_n}^S$ to get the Satake compactification $\overline{X}^S$ associated with the embedding $X \hookrightarrow \mathcal{P}_n$.

As it will be shown below, Step (2) is equivalent to faithful projective representations of G. The standard compactification $\overline{\mathcal{P}_n}^S$ is only one of the many Satake compactifications of $\mathcal{P}_n$ and corresponds to the *identity (standard) representation* $\mathrm{PSL}(n, \mathbb{C}) \to \mathrm{PSL}(n, \mathbb{C})$. Certainly, $\mathrm{PSL}(n, \mathbb{C})$ admits other *faithful projective representations* $\mathrm{PSL}(n, \mathbb{C}) \to \mathrm{PSL}(m, \mathbb{C})$, which lead to different *totally geodesic embeddings* $\mathcal{P}_n \hookrightarrow \mathcal{P}_m$. The closures of $\mathcal{P}_n$ in $\overline{\mathcal{P}_m}^S$ are often different from $\overline{\mathcal{P}_n}^S$. The major problems in the Satake compactifications of $X = G/K$ are (1) to understand how the compactifications depend on the representations of G or the embeddings of X into $\mathcal{P}_n$, and (2) to determine the G-orbits and boundary components in the compactifications.

This section is organized as follows. First, we discuss the example of $\overline{\mathcal{P}_2}^S$ by identifying $\mathcal{P}_2$ with the real hyperbolic space $\mathbf{H}^3$ (I.4.2). Then we study the compactification $\overline{\mathcal{P}_n}^S$ directly using the spectral theorem of Hermitian matrices and determine its boundary points and boundary orbits explicitly (Propositions I.4.4 and I.4.9), where a crucial part is to determine the limit points of unbounded sequences of diagonal matrices; then we determine the normalizer and centralizer of the boundary components (Propositions I.4.5, I.4.6, and I.4.7), hence completing step (1). Then we show the equivalence between totally geodesic embeddings of X into $\mathcal{P}_n$ and faithful projective representations $\tau : G \to \mathrm{PSL}(n, \mathbb{C})$ (Proposition I.4.12). To determine the structure of the boundary of the Satake compactifications $\overline{X}^S$ by using arguments similar to those in the above special case, we introduce the crucial notion of μ_τ-connected sets of simple roots (Definition I.4.16), where μ_τ is the highest weight of τ. After identifying the closure of the positive Weyl chamber (Proposition I.4.23), we determine the boundary components of the compactification and their stabilizers in terms of these μ_τ-connected sets (Proposition I.4.29), which show that there are only finitely many nonisomorphic Satake compactifications (Proposition I.4.35, see also Corollary I.4.32 and Proposition I.4.38). The structure of boundary components and the closure of the positive chamber naturally lead to the axiomatic characterization of the Satake compactifications in Proposition I.4.33. The relation between the finitely many Satake compactifications is given in Proposition I.4.35. The G-orbits in the boundary of the Satake compactifications are given in Proposition I.4.40 and Corollary I.4.41. The general results on Satake compactifications are illustrated through two examples: the maximal Satake compactification in I.4.42, and the standard Satake compactification of $\mathcal{P}_n$ in I.4.43.

I.4.1 The symmetric space $\mathcal{P}_n = \mathrm{PSL}(n, \mathbb{C})/\mathrm{PSU}(n)$ can be identified with the space of *positive definite Hermitian matrices* of determinant 1 through the map

$$g\mathrm{PSU}(n) \mapsto gg^*.$$

When positive definite matrices degenerate, they become positive semidefinite matrices. Therefore, it is natural to use these matrices to compactify $\mathcal{P}_n$.

Let $\mathcal{H}_n$ be the real vector space of Hermitian $n \times n$ matrices, and $P(\mathcal{H}_n)$ the associated real projective space. For each nonzero matrix A in $\mathcal{H}_n$, let $[A]$ denote the image of A in $P(\mathcal{H}_n)$, i.e., the line $\mathbb{R}A$ spanned by A. The group $\mathrm{PSL}(n, \mathbb{C})$ acts on $\mathcal{H}_n$ by

$$g \cdot A = gAg^*, \quad g \in \mathrm{PSL}(n, \mathbb{C}),\ A \in \mathcal{H}_n.$$

Clearly the action descends to an action on $P(\mathcal{H}_n)$,

$$g \cdot [A] = [gAg^*].$$

Since the matrices in $\mathcal{P}_n$ have determinant 1, the map

$$i : \mathcal{P}_n \to P(\mathcal{H}_n), \quad A \mapsto [A], \tag{I.4.1}$$

is a $\mathrm{PSL}(n, \mathbb{C})$-equivariant embedding.

The closure of $i(\mathcal{P}_n)$ in the compact space $P(\mathcal{H}_n)$ is a $\mathrm{PSL}(n, \mathbb{C})$-equivariant compactification of $\mathcal{P}_n$, called the *standard Satake compactification* of $\mathcal{P}_n$ and denoted by $\overline{\mathcal{P}_n}^S$.

I.4.2 Before studying the general case $\overline{\mathcal{P}_n}^S$, we identify the Satake compactification $\overline{\mathcal{P}_2}^S$. In this example, we can see explicitly how positive definite matrices degenerate to semipositive ones.

Let $\mathbf{H}^3$ be the hyperbolic space of dimension 3, i.e., the three-dimensional simply connected Riemannian manifold of constant curvature -1. An important model of $\mathbf{H}^3$ is given by the upper half-space

$$\mathbf{H}^3 = \{(z, r) \mid z \in \mathbb{C}, r \in (0, +\infty)\}$$

with the metric $ds^2 = \frac{dx^2+dy^2+dr^2}{r^2}$, where $z = x + iy$. Let $\mathbb{H}$ be the algebra of quaternions with the standard basis $1, i, j, k$:

$$i^2 = -1, \quad j^2 = -1, \quad k^2 = -1, \quad k = ij.$$

Then $\mathbf{H}^3$ can be identified with a subset of $\mathbb{H}$ by the map

$$\mathbf{H}^3 \to \mathbb{H}, \quad (x + iy, r) \mapsto x + iy + jr + 0k.$$

Under this identification, $\mathrm{SL}(2,\mathbb{C})$ acts on $\mathbf{H}^3$ by

$$\begin{pmatrix} a & b \\ c & d \end{pmatrix} \cdot w = (aw+b)(cw+d)^{-1},$$

where $w \in \mathbf{H}^3$ and $\begin{pmatrix} a & b \\ c & d \end{pmatrix} \in \mathrm{SL}(2,\mathbb{C})$. It can be shown that $\mathrm{SL}(2,\mathbb{C})$ acts transitively on $\mathbf{H}^3$ and the stabilizer of the point $(0,1) = j$ is equal to $\mathrm{SU}(2)$ (see [EGM, §I.1] for more details). Clearly, the center of $\mathrm{SL}(2,\mathbb{C})$ acts trivially on $\mathbf{H}^3$, and hence $\mathrm{PSL}(2,\mathbb{C})$ also acts on $\mathbf{H}^3$. Therefore,

$$\mathbf{H}^3 = \mathrm{SL}(2,\mathbb{C})/\mathrm{SU}(2) = \mathrm{PSL}(2,\mathbb{C})/\mathrm{PSU}(2). \tag{I.4.2}$$

Composed with the identification $\mathrm{PSL}(2,\mathbb{C})/\mathrm{PSU}(2) = P_2$, this implies that

$$\mathbf{H}^3 \cong P_2.$$

Let $\mathbf{H}^3 \cup \mathbf{H}^3(\infty)$ be the geodesic compactification. We claim that $\overline{\mathcal{P}_2}^S$ is isomorphic to $\mathbf{H}^3 \cup \mathbf{H}^3(\infty)$ under this identification.

To prove this, we need an explicit formula for this identification in equation (I.4.2). Let

$$P = \left\{ \begin{pmatrix} a & b \\ 0 & a^{-1} \end{pmatrix} \mid b \in \mathbb{C}, a > 0 \right\},$$

a parabolic subgroup of $\mathrm{SL}(2,\mathbb{C})$. Then P acts transitively on $\mathbf{H}^3$. In fact, for any $b \in \mathbb{C}$, $a > 0$,

$$\begin{pmatrix} 1 & b \\ 0 & 1 \end{pmatrix} \begin{pmatrix} a & 0 \\ 0 & a^{-1} \end{pmatrix} \cdot j = b + a^2 j.$$

The corresponding Hermitian matrix in $\mathcal{P}_2$ is equal to

$$\begin{pmatrix} 1 & b \\ 0 & 1 \end{pmatrix} \begin{pmatrix} a & 0 \\ 0 & a^{-1} \end{pmatrix} \begin{pmatrix} a & 0 \\ 0 & a^{-1} \end{pmatrix} \begin{pmatrix} 1 & 0 \\ \bar{b} & 1 \end{pmatrix} = \begin{pmatrix} a^2 + |b|^2 a^{-2} & ba^{-2} \\ \bar{b}a^{-2} & a^{-2} \end{pmatrix}. \tag{I.4.3}$$

Note that $\mathbf{H}^3(\infty)$ can be identified with $\mathbb{C} \cup \{\infty\}$. This implies that when a sequence $w_n \in \mathbf{H}^3$ converges to a boundary point $b \in \mathbb{C}$ in $\mathbf{H}^3 \cup \mathbf{H}^3(\infty)$, the corresponding matrix in $\mathcal{P}_2$ converges in the Satake compactification $\overline{\mathcal{P}_2}^S$ to the line in $P(\mathcal{H}_2)$ passing through the semipositive matrix

$$\begin{pmatrix} |b|^2 & b \\ \bar{b} & 1 \end{pmatrix}.$$

Similarly, if a sequence $w_n \in \mathbf{H}^3$ converges to the boundary point $\{\infty\}$ in $\mathbf{H}^3 \cup \mathbf{H}^3(\infty)$, then it converges to the line in $P(\mathcal{H}_2)$ passing through the semipositive matrix

$$\begin{pmatrix} 1 & 0 \\ 0 & 0 \end{pmatrix}.$$

This proves that $\overline{\mathcal{P}_2}^S$ is isomorphic to the geodesic compactification $\mathbf{H}^3 \cup \mathbf{H}^3(\infty)$.

I.4.3 In the general case, to understand the standard Satake compactification $\overline{\mathcal{P}_n}^S$, we need to identify the boundary points and the $\mathrm{PSL}(n, \mathbb{C})$-orbits on the boundary.

Since the subset of semipositive Hermitian matrices is closed, it is clear that the boundary points of $\overline{\mathcal{P}_n}^S$ are of the form $[A]$, where A is a semipositive Hermitian matrix; by adding a small multiple of a positive definite matrix, it is also clear that every such point $[A]$ appears in the boundary of $\overline{\mathcal{P}_n}^S$.

Even though we have determined the whole boundary, its structures such as the G-orbits and lower-dimensional symmetric spaces inside the boundary are not clear. Furthermore, an intrinsic description of the convergence of interior points to the boundary is not clear either.

In the following, we often denote elements in $PGL(n, \mathbb{C})$ by their lifts in $GL(n, \mathbb{C})$, i.e., by matrices. We note that by the spectral theorem, any Hermitian matrix A can be written as

$$A = B \operatorname{diag}(d_1, \dots, d_n) B^*, \tag{I.4.4}$$

where $B \in \mathrm{PSU}(n)$, and $d_1, \dots, d_n \in \mathbb{R}$, $d_1 \geq d_2 \geq \cdots \geq d_n$. Since $\mathrm{PSU}(n)$ is compact, we first concentrate on the limit points of the diagonal matrices.

Let $y_j = \operatorname{diag}(d_{1,j}, \dots, d_{n,j})$ be an unbounded sequence in $\mathcal{P}_n$. Then $d_{1,j} \cdots d_{n,j} = 1$. Assume that $d_{1,j} \geq \cdots \geq d_{n,j}$. Since y_j is not bounded, it follows that

$$d_{1,j} \to +\infty, \quad d_{n,j} \to 0.$$

By passing to a subsequence if necessary, we can assume that for all $i \in \{1, \dots, n\}$, $\lim_{j \to +\infty} d_{i,j}/d_{1,j}$ exists, and denote the limit by a_i. Let i_0 be the largest integer such that

$$a_1, \dots, a_{i_0} > 0, \quad a_{i_0+1} = \cdots = a_n = 0.$$

Since $a_1 = 1$ and $a_n = 0$, such i_0 exists. Then the limit of y_j in $\overline{\mathcal{P}_n}^S$ is given by

$$\begin{aligned} i(y_j) = [Y_j] &= [\operatorname{diag}(a_{1,j}/a_{1,j}, a_{2,j}/a_{1,j}, \dots, a_{n,j}/a_{1,j})] \\ &\to [\operatorname{diag}(a_1, a_2, \dots, a_{i_0}, 0, \dots, 0)]. \end{aligned}$$

Using the decomposition in equation (I.4.4), we obtain that if an unbounded sequence y_j in $\mathcal{P}_n$ converges in $\overline{\mathcal{P}_n}$, its limit point is of the form

$$[B \operatorname{diag}(a_1, a_2, \dots, a_{i_0}, 0, \dots, 0) B^*],$$

where $a_1 \geq a_2 \geq \cdots \geq a_{i_0} > 0$, $B \in \mathrm{PSU}(n)$. Let

$$\pi : \mathrm{PSU}(i_0) \to \mathrm{PSU}(n), \quad C \mapsto \begin{pmatrix} C & 0 \\ 0 & I_{n-i_0} \end{pmatrix}$$

be the natural embedding. Then for any $g \in \mathrm{PSU}(i_0)$,

$$\pi(g) \cdot [\mathrm{diag}(a_1, a_2, \ldots, a_{i_0}, 0, \ldots, 0)] = \left[\begin{pmatrix} g\ \mathrm{diag}(a_1, a_2, \ldots, a_{i_0}) g^* & 0 \\ 0 & 0 \end{pmatrix} \right].$$

Let $b = a_1 \ldots a_{i_0}$. Then $\mathrm{diag}(a_1/b, \ldots, a_{i_0}/b) \in \mathcal{P}_{i_0}$. This implies that when $a_1, \ldots, a_{i_0}$ take all the values satisfying the condition $a_1 \geq a_2 \geq \cdots \geq a_{i_0}$, and g ranges over $\mathrm{PSU}(i_0)$, the set $\{\pi(g) \cdot [\mathrm{diag}(a_1, \ldots, a_{i_0}, 0, \ldots, 0)]\}$ can be identified with $\mathcal{P}_{i_0} = \mathrm{PSL}(i_0, \mathbb{C})/\mathrm{PSU}(i_0)$, a lower-dimensional symmetric space of the same type as $\mathcal{P}_n$. Embed $\mathcal{P}_{i_0}$ into $\overline{\mathcal{P}_n}^S$ through this identification,

$$\mathcal{P}_{i_0} \to \overline{\mathcal{P}_n}^S, \quad A \mapsto \left[\begin{pmatrix} A & 0 \\ 0 & 0 \end{pmatrix} \right]. \tag{I.4.5}$$

Then the above discussions show that

$$\overline{\mathcal{P}_n}^S = \mathcal{P}_n \cup K\mathcal{P}_{n-1} \cup \cdots \cup K\mathcal{P}_1, \tag{I.4.6}$$

where $K = \mathrm{PSU}(n, \mathbb{C})$.

Since matrices in $\mathcal{P}_k$ have rank k, the above decomposition is disjoint. This decomposition describes the K-action on the compactification $\overline{\mathcal{P}_n}^S$. In fact, it also describes the G-orbits structure, where $G = \mathrm{PSL}(n, \mathbb{C})$.

Proposition I.4.4 *Each summand in equation (I.4.6) is a G-orbit, and $\mathcal{P}_n$ admits a disjoint decomposition*

$$\overline{\mathcal{P}_n}^S = \mathcal{P}_n \coprod G\mathcal{P}_{n-1} \coprod \cdots \coprod G\mathcal{P}_1. \tag{I.4.7}$$

Proof. To prove this proposition, we need to relate the boundary symmetric space $\mathcal{P}_i$ to parabolic subgroups. Let

$$P = P_i = \left\{ \begin{pmatrix} A & B \\ 0 & cI_{n-i} \end{pmatrix} \in \mathrm{PSL}(n, \mathbb{C}) \mid A \in GL(i, \mathbb{C}), B \in M_{i \times n-i}(\mathbb{C}), c \in \mathbb{C}^\times \right\}, \tag{I.4.8}$$

a parabolic subgroup in $\mathrm{PSL}(n, \mathbb{C})$. Then the associated boundary symmetric space is given by

$$X_P = \mathrm{PSL}(i, \mathbb{C})/\mathrm{PSU}(i) = \mathcal{P}_i.$$

It can be checked easily that for any $p \in P$, $p\mathcal{P}_i = \mathcal{P}_i$, where $\mathcal{P}_i$ is embedded in $\overline{\mathcal{P}_n}^S$ as in equation (I.4.5). Since $G = KP$, for any $k \in K$ and $g \in G$, we can write $gk = k'p'$, where $k' \in K$, $p' \in P$. Then

$$g(k\mathcal{P}_i) = k'(p'\mathcal{P}_i) = k'\mathcal{P}_i \subset K\mathcal{P}_i.$$

Hence $K\mathcal{P}_i$ is G-invariant. Since P acts transitively on $\mathcal{P}_i$, G acts transitively on $K\mathcal{P}_i$. This implies that the decomposition in equation (I.4.7) is the disjoint decomposition into G-orbits.

To determine structures of these G-orbits, we need to identify the *stabilizer (or normalizer)* $\mathcal{N}(\mathcal{P}_i)$ and the *centralizer* $\mathcal{Z}(\mathcal{P}_i)$ of the lower dimensional symmetric spaces $\mathcal{P}_i$ in the boundary, where

$$\mathcal{N}(\mathcal{P}_i) = \{g \in \mathrm{PSL}(n, \mathbb{C}) \mid g\mathcal{P}_i = \mathcal{P}_i\}, \tag{I.4.9}$$

$$\mathcal{Z}(\mathcal{P}_i) = \{g \in \mathcal{N}(\mathcal{P}_i) \mid g|_{\mathcal{P}_i} = Id\}. \tag{I.4.10}$$

Define a parabolic subgroup Q_i of $\mathrm{PSL}(n, \mathbb{C})$ by

$$Q_i = \left\{\begin{pmatrix} \alpha & \beta \\ 0 & \delta \end{pmatrix} \in \mathrm{PSL}(n, \mathbb{C}) \mid \alpha \in GL(i, \mathbb{C}), \beta \in M_{i\times n-i}(\mathbb{C}), \delta \in GL(n-i, \mathbb{C})\right\}. \tag{I.4.11}$$

Proposition I.4.5 *For the boundary symmetric space $\mathcal{P}_i$ in $\overline{\mathcal{P}_n}^S$, the normalizer is given by $\mathcal{N}(\mathcal{P}_i) = Q_i$.*

Proof. Since

$$\begin{pmatrix} \alpha & \beta \\ \gamma & \delta \end{pmatrix}\begin{pmatrix} A & 0 \\ 0 & 0 \end{pmatrix}\begin{pmatrix} \alpha^t & \gamma^t \\ \beta^t & \delta^t \end{pmatrix} = \begin{pmatrix} \alpha A\alpha^t & \alpha A\gamma^t \\ \gamma A\alpha^t & \gamma A\gamma^t \end{pmatrix}, \tag{I.4.12}$$

it is clear that Q_i stabilizes $\mathcal{P}_i$, and $Q_i \subseteq \mathcal{N}(\mathcal{P}_i)$. On the other hand, when $\gamma \neq 0$,

$$\gamma A\gamma^t \neq 0,$$

since A is positive definite, and hence $\begin{pmatrix} \alpha & \beta \\ \gamma & \delta \end{pmatrix} \cdot \left[\begin{pmatrix} A & 0 \\ 0 & 0 \end{pmatrix}\right] \notin \mathcal{P}_i$. This implies that $\mathcal{N}(\mathcal{P}_i) = Q_i$.

The computation in equation (I.4.12) also gives the following two results.

Proposition I.4.6 *For the boundary symmetric space $\mathcal{P}_i$ in $\overline{\mathcal{P}_n}^S$, the centralizer $\mathcal{Z}(\mathcal{P}_i)$ is equal to the subgroup*

$$\left\{\begin{pmatrix} \alpha & \beta \\ 0 & \delta \end{pmatrix} \in Q_i \mid \alpha = cI_i, c \in \mathbb{C}\right\}.$$

Proposition I.4.7 *For the basepoint $z_0 = [\mathrm{diag}(1, \dots, 1, 0, \dots, 0)] \in \mathcal{P}_i$ in the boundary space, its stabilizer in* $\mathrm{PSL}(n, \mathbb{C})$ *is given by*

$$\left\{\begin{pmatrix} \alpha & \beta \\ 0 & \gamma \end{pmatrix} \in Q_i \mid \alpha \in U(n)\right\}.$$

These results completely describe the structure of the Satake compactification $\overline{\mathcal{P}_n}^S$. In the above notation, for $g \in \mathrm{PSL}(n, \mathbb{C})$ and $i \in \{1, \dots, n-1\}$, the spaces $g \cdot \mathcal{P}_i$ are called the *boundary components*. It can be seen easily that the closure of $\mathcal{P}_i$ in $\overline{\mathcal{P}_n}^S$ is isomorphic to the standard compactification of $\mathcal{P}_i$. Hence this compactification is inductive in a certain sense, which also holds for general Satake compactifications.

I.4.8 Though the boundary component $\mathcal{P}_i$ is equal to the boundary symmetric space X_{P_i} of the parabolic subgroup P_i in equation (I.4.8), the space $\mathcal{P}_i$ does not uniquely determine the group P_i, rather it determines Q_i in equation (I.4.11) as its stabilizer. Clearly, Q_i is a maximal parabolic subgroup, and every maximal parabolic subgroup arises this way. We call $\mathcal{P}_i$ the boundary component associated with Q_i and denoted by $e(Q_i)$:

$$e(Q_i) = \mathcal{P}_i.$$

Similarly, there is a boundary component $e(Q)$ for every maximal parabolic subgroup Q, which is defined as follows. Since every maximal parabolic subgroup Q is conjugate to Q_i in equation (I.4.11) for some i, write $Q = {}^g Q_i$. Then the decomposition

$$X_{Q_i} = \mathcal{P}_i \times \mathcal{P}_{n-i} = e(Q_i) \times \mathcal{P}_{n-i} \tag{I.4.13}$$

is transported to a decomposition of X_Q,

$$X_Q = g \cdot \mathcal{P}_i \times g \cdot \mathcal{P}_{n-i} = e(Q) \times g \cdot \mathcal{P}_{n-i}. \tag{I.4.14}$$

Though the element g in $Q = {}^g Q_i$ is not unique, the decomposition in equation (I.4.14) is unique, and hence

$$e(Q) = g \cdot \mathcal{P}_i \tag{I.4.15}$$

is well-defined.

Proposition I.4.9 *The standard Satake compactification $\overline{\mathcal{P}_n}^S$ is decomposed into the disjoint union of $\mathcal{P}_n$ and the boundary components*

$$\overline{\mathcal{P}_n}^S = \mathcal{P}_n \cup \coprod_{Q} e(Q), \tag{I.4.16}$$

where Q ranges over proper maximal parabolic subgroups of $\mathrm{PSL}(n, \mathbb{C})$, *and $e(Q)$ is given in equation (I.4.15).*

In the compactification $\overline{\mathcal{P}_n}^S$, all boundary components isomorphic to $\mathcal{P}_i$ belong to one G-orbit. In this sense, it is a *minimal Satake compactification*. In fact, as shown below, there are finitely many partially ordered nonisomorphic Satake compactifications of any symmetric space X, and $\overline{\mathcal{P}_n}^S$ is a minimal element in this partial ordering of the Satake compactifications of $\mathcal{P}_n$.

I.4.10 The second step of the Satake compactifications of X is to embed X into $\mathcal{P}_n$ as *totally geodesic submanifolds*. Such embeddings correspond to faithful projective representations of G.

For any irreducible faithful projective representation τ of G,

$$\tau : G \to \mathrm{PSL}(n, \mathbb{C})$$

satisfying the condition

$$\tau(\theta(g)) = (\tau(g)^*)^{-1}, \tag{I.4.17}$$

where θ is the Cartan involution on G associated with K, and $g \mapsto (g^*)^{-1}$ is the Cartan involution on $\mathrm{PSL}(n, \mathbb{C})$ associated with $\mathrm{PSU}(n)$, the map

$$i_\tau : X = G/K \to \mathcal{P}_n = \mathrm{PSL}(n, \mathbb{C})/\mathrm{PSU}(n), \quad gK \mapsto \tau(g)\tau(g)^*,$$

is well-defined. In fact, the condition in equation (I.4.17) implies that $\tau(K) \subset \mathrm{PSU}(n)$, and hence for any $k \in K$, $\tau(k)\tau(k)^* = \mathrm{Id}$.

Lemma I.4.11 *The map $i_\tau : X \to \mathcal{P}_n = \mathrm{PSL}(n, \mathbb{C})/\mathrm{PSU}(n)$ is an embedding.*

Proof. Let $\mathfrak{g} = \mathfrak{k} \oplus \mathfrak{p}$ be the Cartan decomposition of the Lie algebra $\mathfrak{g}$. Then

$$\mathfrak{p} \cong \exp \mathfrak{p} \cong X, \quad H \mapsto \exp H \mapsto e^H x_0.$$

Denote the derivative of τ by $d\tau$. Clearly, $d\tau : \mathfrak{g} \to \mathfrak{sl}(n, \mathbb{C})$ is injective. For any $x = \exp H \in \exp \mathfrak{p}$,

$$i_\tau(x) = \exp 2d\tau(H) \in \mathcal{P}_n. \tag{I.4.18}$$

This implies that i_τ is an embedding.

Recall that a submanifold M of $\mathcal{P}_n$ is called *totally geodesic* if for any two points $p, q \in M$, the unique geodesic in $\mathcal{P}_n$ passing through p, q belongs to M. (Recall that $\mathcal{P}_n$ is simply connected and nonpositively curved, and hence any two points can be connected by a unique geodesic.)

Proposition I.4.12 *The image $i_\tau(X)$ under the embedding i_τ is a totally geodesic submanifold of $\mathcal{P}_n$, and any irreducible, totally geodesic embedding $i(X)$ of X in $\mathcal{P}_n$ is of this form, where the irreducibility of $i(X)$ means that the set of $n \times n$ matrices (or linear transformations on $\mathbb{C}^n$) $i(x)$, $x \in X$, is irreducible.*

Proof. Let $x_0 = K \in X = G/K$ be the distinguished basepoint. It suffices to consider the case $p = i_\tau(x_0) = Id \in \mathcal{P}_n$. Write $q = i(\exp H x_0)$, where $H \in \mathfrak{p}$. By equation (I.4.17), $d\tau(H)$ is a Hermitian matrix, and hence $e^{td_\tau(H)} Id$, $t \in \mathbb{R}$, is a geodesic in $\mathcal{P}_n$ and passes through p and q. Since

$$e^{td_\tau(H)} Id = i_\tau(e^{\frac{1}{2}tH} x_0),$$

this geodesic belongs to $i_\tau(X)$.

Conversely, if $i(X)$ is totally geodesic, then for any point $p \in i(X)$, the geodesic symmetry of $\mathcal{P}_n$ based at p preserves $i(X)$ and restricts to the geodesic symmetry of $i(X)$ at p, and hence any isometry in $G = Is^0(X)$ of $i(X)$ extends to an isometry of $\mathcal{P}_n$. Therefore, for any $g \in G$, there exists a matrix $T \in \mathrm{SL}(n, \mathbb{C})$ such that the extended isometry of g on $\mathcal{P}_n$ is given by the action

$$T : \mathcal{P}_n \to \mathcal{P}_n, \quad A \mapsto TAT^* \quad (A \in \mathcal{P}_n).$$

Since $i(X)$ is a set of irreducible matrices, the Schur lemma implies that T is uniquely determined by g up to a scalar multiple. Denote the image of T in $\mathrm{PSL}(n, \mathbb{C})$ by $\tau(g)$. It can be checked that τ is a faithful, irreducible projective representation of G (see [Sat1, p. 85] for details).

I.4.13 Note that for any irreducible faithful projective representation τ of G into $\mathrm{PSL}(n, \mathbb{C})$, there exists a suitable norm on $\mathbb{C}^n$ such that the condition in equation (I.4.17) is satisfied. Since irreducible faithful projective representations of G into $\mathrm{PSL}(n, \mathbb{C})$ correspond bijectively to irreducible faithful representations of $\mathfrak{g}$ into $\mathfrak{sl}(n, \mathbb{C})$ and are classified by the theory of highest weights, there exist infinitely many irreducible faithful projective representations $\tau : G \to \mathrm{PSL}(n, \mathbb{C})$ (see [GJT, Chapter IV]).

Definition I.4.14 For any such totally geodesic embedding $i_\tau : X \to \mathcal{P}_n$, the closure of $i_\tau(X)$ in $\overline{\mathcal{P}_n}^S$ is called the Satake compactification associated with the representation $\tau : G \to \mathrm{PSL}(n, \mathbb{C})$, and denoted by $\overline{X}^S_\tau$.

It should be pointed out that for $G = \mathrm{PSL}(n, \mathbb{C})$, we need to consider representations besides the identity representation in order to obtain other Satake compactifications of P_n.

Since G acts on $\mathcal{P}_n$ via the representation τ,

$$g \cdot A = \tau(g) A \tau(g)^*, \quad g \in G, A \in \mathcal{P}_n,$$

the embedding $i_\tau : X \to \mathcal{P}_n$ is G-equivariant, and hence the Satake compactification $\overline{X}^S_\tau$ is a *G-compactification*, i.e., the G-action on X extends to a continuous action on $\overline{X}^S_\tau$. Since there are infinitely many irreducible faithful projective representations τ, there are infinitely many corresponding Satake compactifications. It will be shown below that there are only finitely many nonisomorphic Satake compactifications as G-topological spaces.

I.4.15 To understand the structures of $\overline{X}^S_\tau$ such as G-orbits and boundary components, and the dependence on the representation τ, we follow the methods used above in describing the standard Satake compactification $\overline{\mathcal{P}_n}^S$.

Let $\mathfrak{g} = \mathfrak{k} \oplus \mathfrak{p}$ be the Cartan decomposition. Let $\mathfrak{a} \subset \mathbf{P}$ be a maximal abelian subalgebra in $\mathfrak{p}$. Choose a suitable basis of $\mathbb{C}^n$ such that for $H \in \mathfrak{a}$, $\tau(e^H)$ is a diagonal matrix

$$\tau(e^H) = \mathrm{diag}(e^{\mu_1(H)}, \dots, e^{\mu_n(H)}), \tag{I.4.19}$$

where $\mu_1, \dots, \mu_n$ are the *weights* of τ listed with multiplicity. Choose a positive chamber $\mathfrak{a}^+$ in $\mathfrak{a}$. Then the Cartan decomposition gives

$$X = K \exp \overline{\mathfrak{a}^+} x_0.$$

Since K is compact, we first consider limits of sequences of the form $i_\tau(e^{H_j} x_0)$ in $\overline{\mathcal{P}_n}^S$, where $H_j \in \overline{\mathfrak{a}^+}$ is unbounded. By equation (I.4.19), we get

$$i_\tau(e^{H_j} x_0) = [\mathrm{diag}(e^{2\mu_1(H_j)}, \dots, e^{2\mu_n(H_j)})]. \tag{I.4.20}$$

There are different ways for the sequence H_j to go to infinity. For example, it can go to infinity within a bounded distance of a Weyl chamber face $\mathfrak{a}_I$ of $\mathfrak{a}^+$, or strictly inside the Weyl chamber $\mathfrak{a}^+$ and the distance to all the faces of $\mathfrak{a}^+$ goes to infinity. To understand the effect of these differences on the limits of $i_\tau(e^{H_j} x_0)$, we need to introduce some concepts for the weights of representations.

Let $\Delta = \Delta(\mathfrak{g}, \mathfrak{a})$ be the set of simple roots in $\Phi(\mathfrak{g}, \mathfrak{a})$ determined by the positive chamber $\mathfrak{a}^+$. Let μ_τ be the *highest weight* of τ with respect to the partial ordering determined by Δ. Then other weights μ_i of τ are of the form

$$\mu_i = \mu_\tau - \sum_{\alpha \in \Delta} c_{\alpha,i}\, \alpha, \tag{I.4.21}$$

where $c_{\alpha,i}$ are nonnegative integers.

Definition I.4.16 For each weight μ_i, define its support $\mathrm{Supp}(\mu_i) = \{\alpha \in \Delta \mid c_{\alpha,i} > 0\}$.

Definition I.4.17 A subset I of Δ is called μ_τ-connected if $I \cup \{\mu_\tau\}$ is connected, i.e., not the union of two subsets orthogonal with respect to the Killing form.

When the Dynkin diagram is joined with $\{\mu_\tau\}$ by edges to all the vertices of the simple roots not perpendicular to μ_τ, then $I \cup \{\mu_\tau\}$ is a connected subset if and only if I is μ_τ-connected. This is the reason for the above definition. These two concepts are related by the following result [Sat1, Lemma 5 in §2.3].

Proposition I.4.18 *For any weight* μ_i*, its support* $\mathrm{Supp}(\mu_i)$ *is a* μ_τ*-connected subset of* Δ*. Conversely, any* μ_τ*-connected subset* I *of* Δ *is equal to* $\mathrm{Supp}(\mu_i)$ *for some weight* μ_i*.*

Let I be a μ_τ-connected (proper) subset of Δ. Suppose the sequence $H_j \in \overline{\mathfrak{a}^+}$ satisfies the following conditions:

1. for $\alpha \in \Delta - I$, $\alpha(\mu_j) \to +\infty$,
2. for $\alpha \in I$, $\lim_{j \to +\infty} \alpha(H_j)$ exists and is finite.

Recall from §I.1.3 that $\mathfrak{a}^I$ is the orthogonal complement of $\mathfrak{a}_I$ in $\mathfrak{a}$. Define

$$\mathfrak{a}^{I,+} = \{H \in \mathfrak{a}^I \mid \alpha(H) > 0, \alpha \in I\},$$

a positive chamber in $\mathfrak{a}_I$. Let $H_\infty \in \overline{\mathfrak{a}^{I,+}}$ be the unique vector such that $\alpha(H_\infty) = \lim_{j\to+\infty} \alpha(H_j)$ for $\alpha \in I$.

By reordering the weights if necessary, we can assume that $\mu_1 = \mu_\tau, \mu_2, \dots, \mu_k$ are all the weights with support contained in I, and $\mu_{k+1}, \dots, \mu_n$ the other weights. Since H_j is not bounded, $i_\tau(e^{H_j} x_0)$ is not bounded, and hence $k \leq n-1$. By equation (I.4.20) and equation (I.4.21),

$$\begin{aligned} i_\tau(e^{H_j} x_0) &= [e^{2\mu_\tau(H_j)} \operatorname{diag}(1, e^{-2\sum_{\alpha\in\Delta} c_{\alpha,2}\alpha(H_j)}, \dots, e^{-2\sum_{\alpha\in\Delta} c_{\alpha,n}\alpha(H_j)})] \\ &= [\operatorname{diag}(1, e^{-2\sum_{\alpha\in\Delta} c_{\alpha,2}\alpha(H_j)}, \dots, e^{-2\sum_{\alpha\in\Delta} c_{\alpha,n}\alpha(H_j)})]. \end{aligned}$$

For each $\ell \geq k+1$, there exists $\alpha \notin I$ such that $c_{\alpha,\ell} > 0$, and hence

$$e^{-2\sum_{\alpha\in\Delta} c_{\alpha,\ell}\alpha(H_j)} \to 0.$$

This implies that

$$i_\tau(e^{H_j} x_0) \to [\operatorname{diag}(1, e^{-2\sum_{\alpha\in I} c_{2,\alpha}\alpha(H_\infty)}, \dots, e^{-2\sum_{\alpha\in I} c_{k,\alpha}\alpha(H_\infty)}, 0, \dots, 0)].$$

I.4.19 In general, for any sequence $H_j \in \overline{\mathfrak{a}^+}$ going to infinity, by passing to a subsequence if necessary, we can assume that there exists a proper subset $J \subset \Delta$ such that

1. for all $\alpha \in \Delta \setminus J$, $\alpha(H_j) \to +\infty$,
2. for all $\alpha \in J$, $\alpha(H_j)$ is bounded and $\lim_{j\to+\infty} \alpha(H_j)$ exists.

Let $I \subset J$ be the largest μ_τ-connected subset contained in J. Since the union of two connected μ_τ-subsets is μ_τ-connected, and the empty set is μ_τ-connected, such I exists and is unique.

Let $H_\infty \in \overline{\mathfrak{a}^{I,+}}$ be the unique vector such that $\alpha(H_\infty) = \lim_{j\to\infty} \alpha(H_j)$ for $\alpha \in I$. Note that the vector H_∞ does not depend on the limits $\lim_{j\to+\infty} \alpha(H_j)$ for $\alpha \in J \setminus I$. Using Proposition I.4.18 and the fact that I is the largest μ_τ-connected subset in J, we can show as above that for any $\ell \geq k+1$,

$$e^{-2\sum_{\alpha\in\Delta} c_{\alpha,\ell}\alpha(H_j)} \to 0,$$

and hence

$$i_\tau(e^{H_j} x_0) \to [\operatorname{diag}(1, e^{-2\sum_{\alpha\in I} c_{2,\alpha}\alpha(H_\infty)}, \dots, e^{-2\sum_{\alpha\in I} c_{k,\alpha}\alpha(H_\infty)}, 0, \dots, 0)]. \tag{I.4.22}$$

I.4.20 On the other hand, for any $H_\infty \in \overline{\mathfrak{a}^{I,+}}$, there exists an unbounded sequence $H_j \in \overline{\mathfrak{a}^+}$ such that (1) for $\alpha \in I$, $\alpha(H_j) \to \alpha(H_\infty)$, (2) for $\alpha \in \Delta \setminus I$, $\alpha(H_j) \to +\infty$. This implies that there exists a map

$$i_I : e^{\overline{\mathfrak{a}^{+,I}}} \to \overline{X}^S_\tau, \tag{I.4.23}$$

$$H_\infty \mapsto [\mathrm{diag}(1, e^{-2\sum_{\alpha\in I} c_{2,\alpha}\alpha(H_\infty)}, \dots, e^{-2\sum_{\alpha\in I} c_{k,\alpha}\alpha(H_\infty)}, 0, \dots, 0)]. \tag{I.4.24}$$

We can show that the map i_I is an embedding. For each weight μ_i of the representation $\tau : G \to \mathrm{PSL}(n, \mathbb{C})$, let V_{μ_i} be the corresponding weight space. Let I be a μ_τ-connected subset and $\mu_1, \dots, \mu_k$ the weights with support contained in I. Define

$$V_I = V_{\mu_1} \oplus \cdots \oplus V_{\mu_k}. \tag{I.4.25}$$

Let P_0 be the minimal parabolic subgroup whose split component is equal to $\mathfrak{a}$ and whose associated chamber $\mathfrak{a}^+_{P_0}$ is equal to $\mathfrak{a}^+$. Let $P_{0,I}$ be the standard parabolic subgroup associated with I.

Lemma I.4.21 *The subspace V_I is invariant under $\tau(g)$, $g \in M_{P_{0,I}}$, and the induced representation of $M_{P_{0,I}}$ on V_I, denoted by τ_I, is a multiple of an irreducible faithful one.*

See [Sat1, §2.4, Lemma 8] for proof. Let $\mathcal{H}_I$ be the space of Hermitian matrices on V_I, and $P(\mathcal{H}_I)$ the associated real projective space. As in equation (I.4.18), we can show that the map

$$i_{\tau_I} : X_{P_{0,I}} = M_{P_{0,I}}/K \cap M_{P_{0,I}} \to P(\mathcal{H}_I), \quad gK \cap M_{P_{0,I}} \mapsto \tau_I(g)\tau_I(g)^*, \tag{I.4.26}$$

is an embedding. Extending by zero on the orthogonal complement of V_I in $\mathbb{C}^n$, we can embed $\mathcal{H}_I$ into $\mathcal{H}_n$, and hence $P(\mathcal{H}_I)$ into $P(\mathcal{H}_n)$. Therefore, we have an embedding

$$i_\tau : X_{P_{0,I}} \to P(\mathcal{H}_n). \tag{I.4.27}$$

Note that $e^{\mathfrak{a}^I} \cdot K \cap M_{P_{0,I}} \subset X_{P_{0,I}}$. Then

$$i_\tau(e^{H_\infty} \cdot K \cap M_{P_{0,I}}) = i_I(e^{H_\infty}).$$

This implies that the map i_I in equation (I.4.23) is an embedding.

Lemma I.4.22 *Let I, I' be two different μ_τ-connected subsets of Δ. Then $i_I(\overline{\mathfrak{a}^{I,+}})$, $i_{I'}(\overline{\mathfrak{a}^{I',+}})$ are disjoint subsets of $\overline{X}^S_\tau$.*

Proof. Since $I \neq I'$, $V_I \neq V_{I'}$. Since for any $H \in \overline{\mathfrak{a}^{I,+}}$, $i_{\tau_I}(H)$ is a line spanned by a positive definite matrix on V_I, this implies that $P(\mathcal{H}_I)$ and $P(\mathcal{H}_{I'})$ as subsets in $P(\mathcal{H}_n)$ are disjoint, which proves the lemma.

Proposition I.4.23 *Let $H_j \in \overline{\mathfrak{a}^+}$ be an unbounded sequence. Then $e^{H_j}x_0$ converges in $\overline{X}_\tau^S$ if and only if there exists a μ_τ-connected subset I of Δ such that*

1. *for all $\alpha \in I$, $\lim_{j\to+\infty} \alpha(H_j)$ exists and is finite,*
2. *for any μ_τ-connected subset I' properly containing I, there exists $\alpha \in I' \setminus I$ such that $\alpha(H_j) \to +\infty$.*

Let H_∞ be the unique vector in $\overline{\mathfrak{a}^{I,+}}$ such that $\alpha(H_\infty) = \lim_{j\to+\infty}\alpha(H_j)$ for all $\alpha \in I$. Then

$$\begin{aligned} i_\tau(e^{H_j}x_0) &\to i_I(e^{H_\infty}) \\ &= [\mathrm{diag}(1, e^{-2\sum_{\alpha\in I} c_{2,\alpha}\alpha(H_\infty)}, \dots, e^{-2\sum_{\alpha\in I} c_{k,\alpha}\alpha(H_\infty)}, 0, \dots, 0)]. \end{aligned}$$

Therefore, the closure of the positive chamber $e^{\overline{\mathfrak{a}^+}}x_0$ in $\overline{X}_\tau^S$ is given by

$$\begin{aligned} \overline{i_\tau(e^{\overline{\mathfrak{a}^+}}x_0)} &= i_\tau(e^{\overline{\mathfrak{a}^+}}x_0) \cup \coprod_{\mu-connected\ I\subset\Delta} i_I(e^{\overline{\mathfrak{a}^{I,+}}}) \\ &\cong \overline{\mathfrak{a}^+} \cup \coprod_{\mu-connected\ I\subset\Delta} \overline{\mathfrak{a}^{I,+}}. \end{aligned}$$

Proof. Suppose the sequence H_j satisfies the condition in the proposition. For any subsequence $H_{j'}$, by passing to a further subsequence if necessary, we can assume that there exists $J \subset \Delta$ such that (1) for any $\alpha \in J$, $\lim_{j\to+\infty}\alpha(H_{j'})$ exists and is finite, and (2) for $\alpha \in \Delta \setminus J$, $\alpha(H_{j'}) \to +\infty$. Then I is the largest μ_τ-connected subset of J, and $\alpha(H_\infty) = \lim_{j\to+\infty}\alpha(H_{j'})$ for $\alpha \in I$. By equation (I.4.22), this implies that $i_\tau(e^{H_{j'}}x_0)$ converges to the limit $i_I(e^{H_\infty})$. Since $H_{j'}$ is an arbitrary subsequence, this implies that $i_\tau(e^{H_j}x_0)$ converges to $i_I(e^{H_\infty})$.

By Lemma I.4.22, different choices of I and H_∞ give different limit points, and hence any sequence $i_\tau(e^{H_j}x_0)$ converging to $i_I(e^{H_\infty})$ must satisfy the properties in the proposition.

The last statement on the closure of the positive chamber follows from the discussions above Lemma I.4.22.

The above proposition completely describes the closure of the positive chamber $e^{\overline{\mathfrak{a}^+}}x_0$ in $\overline{X}_\tau^S$. In $\overline{\mathcal{P}_n}^S$, the boundary is decomposed into boundary components of the form $g \cdot \mathcal{P}_k$, $k \leq n-1$. Similar results hold for $\overline{X}_\tau^S$ and are given in the next proposition.

Proposition I.4.24 *For any μ_τ-connected subset $I \subset \Delta$, the image $i_\tau(X_{P_{0,I}})$ of $X_{P_{0,I}}$ under the embedding in equation (I.4.27) is contained in $\overline{X}_\tau^S$, and*

$$\overline{X}_\tau^S = i_\tau(X) \cup \bigcup_{\mu_\tau\text{-connected } I\subset\Delta} K \cdot i_\tau(X_{P_{0,I}}).$$

Proof. We note that the embedding of $X_{P_{0,I}}$ in equation (I.4.27) is $K \cap M_{P_{0,I}}$-equivariant and $X_{P_{0,I}} = (K \cap M_{P_{0,I}}) \exp \overline{\mathfrak{a}^{I,+}} x_0$, where $x_0 = (K \cap M_{P_{0,I}})$ is the basepoint in $X_{P_{0,I}}$. Then the proposition follows from Proposition I.4.23.

We will show that the decomposition in the proposition is the disjoint decomposition into G-orbits. The group G acts on $\overline{X}_{\tau}^{S}$ through the representation τ: for $g \in G, x = [A] \in \overline{X}_{\tau}^{S}$,

$$g \cdot x = \tau(g) \cdot x, \text{ where } \tau(g) \cdot x = [\tau(g) A \tau(g)^*].$$

For convenience, τ is sometimes dropped from the action.

Lemma I.4.25 *For any $g \in P_{0,I}$, $\tau(g) \cdot i_\tau(X_{P_{0,I}}) = i_\tau(X_{P_{0,I}})$.*

Proof. Since $P_{0,I} = N_{P_{0,I}} A_{P_{0,I}} M_{P_{0,I}}$, it suffices to check that each factor stabilizes $i_\tau(X_{P_{0,I}})$. By equations (I.4.26) and (I.4.27), $M_{P_{0,I}}$ clearly stabilizes it. To show that $N_{P_{0,I}}$ stabilizes $i_\tau(X_{P_{0,I}})$, recall that $i_\tau(X_{P_{0,I}}) \subset P(\mathcal{H}_n)$ is a set of lines spanned by positive definite matrices on V_I in equation (I.4.25). Since V_I is the sum of all weight spaces μ_i with $\mathrm{supp}(\mu_i) \subset I$, if $g \in N_{P_{0,I}}$, $\tau(g)$ stabilizes V_I and acts trivially on it, and hence it also acts trivially on $i_\tau(X_{P_{0,I}}) \subset P(\mathcal{H}_I)$ and $\tau(g) \cdot i_\tau(X_{P_{0,I}}) = i_\tau(X_{P_{0,I}})$. Since all the roots in I vanish on $\mathfrak{a}_{P_{0,I}}$, $A_{P_{0,I}}$ acts trivially on the projective space of V_I, and hence trivially on $P(\mathcal{H}_I)$. This implies that $A_{P_{0,I}}$ stabilizes $i_\tau(X_{P_{0,I}})$.

Remark I.4.26 Another proof might explain more directly why $P_{0,I}$ stabilizes $i_\tau(X_{P_{0,I}})$, and $N_{P_{0,I}}, A_{P_{0,I}}$ act trivially on it. Given any point $z = ke^{H_\infty} x_0 \in X_{P_{0,I}}$, where $k \in K \cap M_{P_{0,I}}$, $H_\infty \in \overline{\mathfrak{a}^{I,+}}$, pick a sequence $H_j \in \mathfrak{a}_I$ such that $\alpha(H_j) \to +\infty$ for $\alpha \in \Delta \setminus I$. Then $i_\tau(ke^{H_j + H_\infty} x_0)$ converges to $i_\tau(z)$. For any $g \in P_{0,I}$, $gk \in P_{0,I}$. Write $gk = nam$, where $n \in N_{P_{0,I}}, a \in A_{P_{0,I}}, m \in M_{P_{0,I}}$. Then

$$gke^{H_j + H_\infty} x_0 = name^{H_j + H_\infty} x_0 = e^{H_j}(e^{-H_j} n e^{H_j}) a (me^{H_\infty}) x_0.$$

Since $e^{-H_j} n e^{H_j} \to e$, it can be checked easily that

$$i_\tau(gke^{H_j + H_\infty} x_0) \to [\tau(me^{H_\infty}) \tau(me^{H_\infty})^*] \in i_\tau(X_{P_{0,I}}).$$

Therefore, $\tau(g) \cdot z \in i_\tau(X_{P_{0,I}})$.

Proposition I.4.27 *For each μ_τ-connected subset I of Δ, $K \cdot i_\tau(X_{P_{0,I}})$ is stable under G, and hence the G-orbits in $\overline{X}_{\tau}^{S}$ are given by*

$$\overline{X}_{\tau}^{S} = i_\tau(X) \cup \bigcup_{\mu\text{-connected } I \subset \Delta} G\, i_\tau(X_{P_{0,I}}).$$

Proof. Since $G = KP_{0,I}$, for any $g \in G$, $k \in K$, write $gk = k'p'$, where $k' \in K$, $p' \in P_{0,I}$. By Lemma I.4.25, $p' i_\tau(X_{P_{0,I}}) = i_\tau(X_{P_{0,I}})$, and hence

$$gki_\tau(X_{P_{0,I}}) = k'\tau(X_{P_{0,I}}) \subset Ki_\tau(X_{P_{0,I}}).$$

To show that the G-orbits in $\overline{X}_\tau^S$ in the above proposition are disjoint, we need to determine the *normalizer*

$$\mathcal{N}(X_{P_{0,I}}) = \{g \in G \mid gX_{P_{0,I}} = X_{P_{0,I}}\},$$

and the *centralizer*

$$\mathcal{Z}(X_{P_{0,I}}) = \{g \in \mathcal{N}(X_{P_{0,I}}) \mid g|_{X_{P_{0,I}}} = Id\}.$$

Definition I.4.28 For any subset $I \subset \Delta$, let I' be the union of roots in Δ that are orthogonal to $I \cup \{\mu_\tau\}$, and $J = I \cup I'$. Then J is called the μ_τ*-saturation* of I, and the parabolic subgroup $P_{0,J}$ is called the μ_τ-saturation of $P_{0,I}$.

Proposition I.4.29 *For a μ_τ-connected subset I, let J be μ_τ-saturation of I as in the above definition, $I' = J - I$. Then the normalizer $\mathcal{N}(X_{P_{0,I}})$ of $X_{P_{0,I}}$ is equal to $P_{0,J}$, and the centralizer $\mathcal{Z}(X_{P_{0,I}})$ is equal to $N_{P_{0,J}} A_J M_{P_{0,I'}}$. For any $g \in G \setminus P_{0,J}$,*

$$g \cdot i_\tau(X_{P_{0,I}}) \cap i_\tau(X_{P_{0,I}}) = \emptyset.$$

Therefore, for any $g \in G$, either

$$g \cdot i_\tau(X_{P_{0,I}}) = i_\tau(X_{P_{0,I}})$$

or

$$g \cdot i_\tau(X_{P_{0,I}}) \cap i_\tau(X_{P_{0,I}}) = \emptyset.$$

Proof. The stabilizer of V_I in G is a parabolic subgroup Q. As explained in Lemma I.4.25, $P_{0,I}$ stabilizes V_I, and hence $Q \supseteq P_{0,I}$. This implies that $Q = P_{0,J'}$, where $J' \supseteq I$. If I is not the largest μ_τ-connected subset in J', then $\tau(P_{0,J'})V_I$ contains weight spaces V_{μ_i} such that $\mathrm{Supp}(\mu_i) \not\subseteq I$, and hence Q does not stabilize V_I. This contradiction implies that $Q \subseteq P_{0,J}$.

Since I is perpendicular to I', $M_{P_{0,J}} = M_{P_{0,I}} M_{P_{0,I'}}$. Then the Langlands decomposition of $P_{0,J}$ can be written as

$$P_{0,J} = N_{P_{0,J}} A_{P_{0,J}} M_{P_{0,I'}} M_{P_{0,I}}.$$

As in the proof of Lemma I.4.25, it can be shown that the first three factors act trivially on V_I, and the last factor stabilizes V_I. Hence, $P_{0,J} \subseteq Q$. Combined with the previous paragraph, it implies that $Q = P_{0,J}$, i.e., the stabilizer of V_I is equal to $P_{0,J}$.

It follows that for any $g \in G \setminus P_{0,J}$,

$$\tau g V_I \neq V_I. \tag{I.4.28}$$

Since $N_{P_{0,J}}$, $A_{P_{0,J}}$ and $M_{P_{0,I'}}$ act trivially on V_I and hence on $i_\tau(X_{P_{0,I}})$, and $M_{P_{0,I}}$ clearly stabilizes and $M_{P_{0,I}}/(M_{P_{0,I}} \cap M_{P_{0,I'}})$ acts effectively on $i_\tau(X_{P_{0,I}})$, we obtain that $\mathcal{N}(X_{P_{0,I}}) = P_{0,J}$, and $\mathcal{Z}(X_{P_{0,I}}) = N_{P_{0,J}} A_{P_{0,I}} M_{P_{0,I'}}$. Since any point z in $i_\tau(X_{P_{0,I}})$ is represented by a line spanned by a positive definite matrix on V_I, it follows that $g \cdot z$ is represented by a line spanned by a positive definite matrix on $\tau(g)V_I$. Together with equation (I.4.28), this implies that

$$g \cdot z \notin i_\tau(X_{P_{0,I}}),$$

and hence

$$g \cdot i_\tau(X_{P_{0,I}}) \cap i_\tau(X_{P_{0,I}}) = \emptyset.$$

Proposition I.4.30 *For any two μ_τ-connected subsets I_1, I_2 and $g \in G$, either $g \cdot i_\tau(X_{P_{0,I_1}}) \cap i_\tau(X_{P_{0,I_2}}) = \emptyset$ or $g \cdot i_\tau(X_{P_{0,I_1}}) = i_\tau(X_{P_{0,I_2}})$.*

Proof. By Proposition I.4.29, it suffices to prove that if $I_1 \neq I_2$, then for any $g \in G$,

$$g \cdot i_\tau(X_{P_{0,I_1}}) \cap i_\tau(X_{P_{0,I_2}}) = \emptyset.$$

This is equivalent to showing that

$$\tau(g) V_{I_1} \neq V_{I_2}. \tag{I.4.29}$$

Let J_1 be the μ_τ-saturation of I_1, and J_2 the μ_τ-saturation of I_2. Then $J_1 \neq J_2$. Note that the stabilizer of V_{I_1} is P_{0,J_1} and the stabilizer of V_{I_2} is equal to P_{0,J_2}. Since $J_1 \neq J_2$, P_{0,J_1} is not conjugate to P_{0,J_2}, and hence equation (I.4.29) holds.

Definition I.4.31 For any $g \in G$ and any μ_τ-connected subset I, the subset $g \cdot i_\tau(X_{P_{0,I}})$ is called a boundary component of the Satake compactification $\overline{X}_\tau^S$.

The above proposition and its proof give the following result.

Corollary I.4.32 *Any two different boundary components of $\overline{X}_\tau^S$ are disjoint, and the G-orbit decomposition of $\overline{X}_\tau^S$ is given by*

$$\overline{X}_\tau^S = i_\tau(X) \cup \coprod_{\mu\text{-connected } I \subset \Delta} G\, i_\tau(X_{P_{0,I}}).$$

The Satake compactification $\overline{X}_\tau^S$ can be characterized in the next proposition.

Proposition I.4.33 *As a topological space, $\overline{X}_\tau^S$ is the unique compactification of X satisfying the following conditions:*

1. *The symmetric space X is a dense open subset of $\overline{X}_\tau^S$, and the G-action on X extends to a continuous action on $\overline{X}_\mu^S$.*
2. *For any μ_τ-connected subset $I \subset \Delta$, the boundary symmetric space $X_{P_{0,I}}$ is embedded in $\overline{X}_\tau^S$. The boundary components $g \cdot X_{P_{0,I}}$, $g \in G$, are disjoint, and the disjoint decomposition of $\overline{X}_\tau^S$ into G-orbits is given by*
$$\overline{X}_\tau^S = X \cup \coprod_{\mu\text{-connected } I \subset \Delta} G X_{P_{0,I}}.$$
3. *For each μ_τ-connected I, let J be the μ_τ-saturation of I. Then the normalizer (i.e., stabilizer or isotropy subgroup) of $X_{P_{0,I}}$ is equal to the μ_τ-saturation $P_{0,J}$ of $P_{0,I}$.*
4. *The closure of the positive chamber $e^{\overline{\mathfrak{a}^+}} x_0$ in $\overline{X}_\tau^S$ is given by*
$$\overline{e^{\mathfrak{a}^+} x_0} = e^{\overline{\mathfrak{a}^+}} x_0 \cup \coprod_{\mu\text{-connected } I \subset \Delta} \overline{\mathfrak{a}^{I,+}}$$
with the following topology: for any unbounded sequence $H_j \in \overline{\mathfrak{a}^+}$, $e^{H_j} x_0$ converges to a point $H_\infty \in \overline{\mathfrak{a}^{I,+}}$ if and only if (a) for all $\alpha \in I$, $\alpha(H_j) \to \alpha(H_\infty)$, (b) for any μ_τ-connected subset I_1 properly containing I, there exists a root $\alpha \in I_1 \setminus I$ such that $\alpha(H_j) \to +\infty$.

Proof. From the above discussions (I.4.23–I.4.32), it is clear that $\overline{X}_\tau^S$ satisfies the conditions in the proposition. We need to show that $\overline{X}_\tau^S$ is uniquely determined by these conditions. (2) determines $\overline{X}_\tau^S$ completely as a G-set. Let $\overline{\overline{\mathfrak{a}^+} x_0}$ be the closure of $\overline{\mathfrak{a}^+} x_0$ in $\overline{X}_\tau^S$. By Proposition I.4.23, it is clearly compact. Since $X = K \exp \overline{\mathfrak{a}^+} x_0$ and the K-action is continuous, the map from the compact space $K \times \exp \overline{\overline{\mathfrak{a}^+} x_0} \to \overline{X}_\tau^S$ is continuous and surjective. Hence there exists an equivalence relation $\sim$ on $K \times \exp \overline{\overline{\mathfrak{a}^+} x_0}$ such that the quotient is isomorphic to $\overline{X}_\tau^S$. But the equivalence relation is determined by (2). Therefore, $\overline{X}_\tau^S$ is completely determined by the conditions (1), (2), (3).

I.4.34 The above proposition shows that as a G-topological space, the Satake compactification $\overline{X}_\tau^S$ depends only on the set of μ_τ-connected subsets of $\Delta = \Delta(\mathfrak{g}, \mathfrak{a})$, the set of simple roots. Let $\mathfrak{a}^*$ be the dual of $\mathfrak{a}$, and $\mathfrak{a}^{*,+}$ the positive chamber dual to $\mathfrak{a}^+$ under the Killing form. Then $\mu_\tau \in \overline{\mathfrak{a}^{*,+}}$, and there is a unique chamber face C of $\overline{\mathfrak{a}^{*,+}}$ such that C contains μ_τ as an interior point. Clearly, C determines the set of μ_τ-connected subsets I of Δ and vice versa; and hence $\overline{X}_\tau^S$ depends only on the chamber face C.

It is known that for a faithful irreducible representation τ of G, the highest weight μ_τ is connected to some root in every connected component of $\Delta(\mathfrak{g}, \mathfrak{a})$. Conversely, for any nonzero integral weight μ in $\overline{\mathfrak{a}^{*,+}}$ that is connected to every connected component of $\Delta(\mathfrak{g}, \mathfrak{a})$, a suitable multiple of μ is the highest weight of a finite-dimensional irreducible faithful representation τ. If X is irreducible, or equivalently G is simple, then a suitable multiple of any nonzero integral weight in $\overline{\mathfrak{a}^{*,+}}$ is the highest weight of some τ.

Proposition I.4.35 (Relations between Satake compactifications) *The isomorphism class of the Satake compactifications $\overline{X}^S_\tau$ only depends on the Weyl chamber face of $\overline{\mathfrak{a}^{*,+}}$ containing μ_τ, and hence there are only finitely many nonisomorphic Satake compactifications $\overline{X}^S_\tau$ of X.*

When X is irreducible, they correspond bijectively to the set of all faces of $\overline{\mathfrak{a}^{*,+}}$ of positive dimension. These compactifications are partially ordered. Specifically, let $\overline{X}^S_{\tau_1}$, $\overline{X}^S_{\tau_2}$ be two Satake compactifications and C_1, C_2 the corresponding chamber faces. If μ_{τ_1} is more regular than μ_{τ_2}, i.e., C_2 is a face of C_1, then $\overline{X}^S_{\tau_1}$ dominates $\overline{X}^S_{\tau_2}$, i.e., the identity map on X extends to a continuous, surjective map $\overline{X}^S_{\tau_1} \to \overline{X}^S_{\tau_2}$. In particular, when μ_τ is generic, i.e., belongs to $\mathfrak{a}^{*,+}$, $\overline{X}^S_\tau$ is maximal, i.e., it dominates all other Satake compactifications, and hence is denoted by $\overline{X}^S_{\max}$. On the other hand, if τ_μ is contained in an edge of $\mathfrak{a}^{*,+}$, $\overline{X}^S_\tau$ is a minimal compactification.

When X is reducible, $X = X_1 \times \cdots \times X_n$, where $X_i, \ldots, X_n$ are irreducible. Then

$$\overline{X}^S_\tau = \overline{X_1}^S_{\tau_1} \times \cdots \times \overline{X_n}^S_{\tau_n}.$$

In this case, the isomorphism classes of the Satake compactifications $\overline{X}^S$ can also be parametrized by some chamber faces. Specifically, $\mathfrak{a}^{*,+} = \mathfrak{a}_1^{*,+} \times \cdots \times \mathfrak{a}_n^{*,+}$, corresponding to the above decomposition of X; and any chamber face of $\mathfrak{a}^{*,+}$ has a similar decomposition. The set of chamber faces that have no trivial factor from any $\mathfrak{a}_i^{*,+}$ parametrizes the isomorphism classes of the Satake compactifications of X.

This proposition was first proved by Zucker in [Zu2]. We will describe the map $\overline{X}^S_{\tau_1} \to \overline{X}^S_{\tau_2}$ here and give a different proof later in Remark I.11.19 using the uniform constructions of the Satake compactifications in §§I.10, I.11. For this purpose, we decompose the Satake compactifications $\overline{X}^S_\tau$ into disjoint unions of boundary components.

I.4.36 To illustrate the Satake compactifications of reducible symmetric spaces, consider the case $X = \mathbf{H} \times \mathbf{H}$. Then $\mathfrak{a}^* = \mathbb{R} \times \mathbb{R}$, and $\mathfrak{a}^{*,+} = \mathbb{R}_{>0} \times \mathbb{R}_{>0}$. The chamber $\mathfrak{a}^{*,+}$ is the only chamber face that has no trivial factor from either of the two copies of $\mathbb{R}_{>0}$. Hence there is a unique Satake compactification of X given by

$$\overline{X}^S = \overline{\mathbf{H}} \times \overline{\mathbf{H}} = (\mathbf{H} \cup \mathbf{H}(\infty)) \times (\mathbf{H} \cup \mathbf{H}(\infty)).$$

I.4.37 Let $I \subset \Delta$ be a μ_τ-connected subset, and $J = I \cup I'$ the μ_τ-saturation of I. Then any parabolic subgroup P that is conjugate to $P_{0,I}$, $P = {}^g P_{0,I}$ for some $g \in G$, is called a *μ_τ-connected parabolic subgroup*, and $Q = {}^g P_{0,J}$ the *μ_τ-saturation of P*. Any parabolic subgroup Q of the form ${}^g P_{0,J}$ is called a *μ_τ-saturated parabolic subgroup*. For each μ_τ-saturated parabolic subgroup Q, the boundary symmetric space X_Q admits an isometric splitting

$$X_Q = X_P \times X_{P'}, \tag{I.4.30}$$

where $Q = {}^g P_{0,J}$, $P = {}^g P_{0,I}$, $P' = {}^g P_{0,I'}$. It should be pointed out that though P, P' depend on the choice of g, the splitting of X_Q in equation (I.4.30) is canonical. (Compare with equation (I.4.15) and the comments before it.) The subspace X_P is a boundary component of $\overline{X}^S_\tau$, and its normalizer is equal to Q. Denote X_P by $e(Q)$ and call it the *boundary component of Q*.

Proposition I.4.38 *The Satake compactification $\overline{X}^S_\tau$ admits a disjoint decomposition into boundary components:*

$$\overline{X}^S_\tau = X \cup \coprod_{\mu\text{-saturated } Q} e(Q).$$

Proof. It follows from Proposition I.4.33 that the boundary components of $\overline{X}^S_\tau$ are disjoint and that the normalizer of each boundary component is a μ_τ-saturated parabolic subgroup.

I.4.39 The map from $\overline{X}^S_{\tau_1}$ to $\overline{X}^S_{\tau_2}$ in Proposition I.4.35 is obtained as follows. Since μ_{τ_1} is more regular than μ_{τ_2}, every μ_{τ_2}-connected parabolic subgroup is also μ_{τ_1}-connected, but the converse is not true.

For any μ_{τ_1}-connected subset I, let I_1 be the largest μ_{τ_2}-connected subset in I. Let $I' = I \setminus I_1$. Then

$$X_{P_{0,I}} = X_{P_{0,I_1}} \times X_{P_{0,I'}}.$$

Denote the projection $X_{P_{0,I}} \to X_{P_{0,I_1}}$ by π. Let J be the μ_{τ_1}-saturation of I, and J_1 the μ_{τ_2}-saturation of I_1. Clearly, $J \subseteq J_1$ and hence $P_{0,J} \subseteq P_{0,J_1}$. The boundary component $e(P_{0,J})$ in $\overline{X}^S_{\tau_1}$ is $X_{P_{0,I}}$, and the boundary component $e(P_{0,J_1})$ in $\overline{X}^S_{\tau_2}$ is $X_{P_{0,I_1}}$, and hence there is a canonical map

$$\pi : e(P_{0,J}) \to e(P_{0,J_1}).$$

In general, for any μ_{τ_2}-saturated parabolic subgroup Q, there is a unique μ_{τ_1}-saturated parabolic subgroup Q_1 containing Q such that there is a canonical projection π from the boundary component $e(Q)$ in $\overline{X}^S_{\tau_1}$ to the boundary component $e(Q_1)$ in $\overline{X}^S_{\tau_2}$. Combined with the identity map on X, this defines the map

$$\pi : \overline{X}^S_{\tau_1} \to \overline{X}^S_{\tau_2}$$

in Proposition I.4.35. To prove it, it suffices to check the continuity of π, which will be given in Remark I.11.19.

To determine the structure of the G-orbits in $\overline{X}^S_\tau$ as G-homogeneous spaces, we need to determine the stabilizers of points in the boundary components.

Proposition I.4.40 *Let I be a μ_τ-connected subset of Δ, and J is μ_τ-saturation. Let x_0 be the basepoint $x_0 = K \cap M_{P_{0,I}}$ in $X_{P_{0,I}}$. Then the stabilizer of x_0 is $N_{P_{0,J}} A_{P_{0,J}} M_{P_{0,J\setminus I}} (K \cap M_{P_{0,I}})$, and hence the G-orbit through x_0 is given by*

$$Gx_0 = GX_{P_{0,I}} \cong G/N_{P_{0,J}} A_{P_{0,J}} M_{P_{0,J\setminus I}} (K \cap M_{P_{0,I}}).$$

Proof. Since the normalizer of $X_{P_{0,I}}$ is equal to $P_{0,J}$ and $N_{P_{0,J}} A_{P_{0,J}} M_{P_{0,J\setminus I}}$ fixes x_0, the proposition follows from the Langlands decomposition of $P_{0,J}$ and the fact that $K \cap M_{P_{0,I}}$ is the stabilizer of x_0 in $M_{P_{0,I}}$.

Corollary I.4.41 *The G-orbit $GX_{P_{0,I}}$ in $\overline{X}^S_\tau$ is a fiber bundle over $G/P_{0,J}$ with fibers equal to the boundary symmetric space $X_{P_{0,I}}$.*

Proof. Since $P_{0,J}/N_{P_{0,J}} A_{P_{0,J}} M_{P_{0,J\setminus I}} (K \cap M_{P_{0,I}}) \cong X_{P_{0,I}}$, the result follows from the above proposition.

The quotient space $G/P_{0,J}$ is compact and called a (real) *flag space (variety)*. It is also a *Furstenberg boundary* to be defined in §I.6.

I.4.42 We illustrate the above general structures of the Satake compactifications $\overline{X}^S_\tau$ using two examples. The first example is the maximal Satake compactification $\overline{X}^S_{\max}$.

When τ is a *generic representation*, i.e., the highest weight μ_τ is *generic* (or belonging to the interior of the positive chamber $\mathfrak{a}^{*,+}$), every parabolic subgroup P is μ_τ-connected, and is hence μ_τ-saturated. The boundary component $e(P)$ is equal to X_P, and the Satake compactification $\overline{X}^S_\tau$ is the maximal Satake compactification $\overline{X}^S_{\max}$, which can be decomposed into a disjoint union

$$\overline{X}^S_{\max} = X \cup \coprod_P X_P, \tag{I.4.31}$$

where P ranges over all parabolic subgroups of G. The G-orbits in $\overline{X}^S_{\max}$ are given by

$$\overline{X}^S_{\max} = X \cup \coprod_{I \subset \Delta} GX_{P_{0,I}},$$

where P_0 is the minimal parabolic subgroup corresponding to the simple roots $\Delta = \Delta(\mathfrak{g}, \mathfrak{a})$.

I.4.43 The second example concerns the standard Satake compactification $\overline{\mathcal{P}_n}^S$. It is the Satake compactification associated with the identity representation

$$\tau = Id : G = \mathrm{PSL}(n, \mathbb{C}) \to \mathrm{PSL}(n, \mathbb{C}).$$

Let the maximal abelian subspace $\mathfrak{a}$ be given by

$$\mathfrak{a} = \{\mathrm{diag}(t_1, \dots, t_n) \mid t_i \in \mathbb{R}, t_1 + \cdots + t_n = 0\},$$

and the positive chamber by

$$\mathfrak{a}^+ = \{\mathrm{diag}(t_1, \dots, t_n) \in \mathfrak{a} \mid t_1 > t_2 > \cdots > t_n\}.$$

Then the set of simple roots is given by

$$\Delta = \Delta(\mathfrak{g}, \mathfrak{a}) = \{\alpha_1 = t_1 - t_2, \dots, \alpha_{n-1} = t_{n-1} - t_n\}.$$

The highest weight μ_τ is equal to t_1, and the only simple root connected to μ_τ is α_1. Then *μ_τ-connected* (proper) subsets of Δ are of the form $I = \{\alpha_1, \dots, \alpha_{k-1}\}$, where $k \leq n - 1$. The *μ_τ-saturation* I is $J = \{\alpha_1, \dots, \alpha_{k-1}\} \cup \{\alpha_{k+1}, \dots, \alpha_{n-1}\}$. The minimal parabolic subgroup P_0 corresponding to the positive chamber $\mathfrak{a}^+$ is the subgroup of upper triangular matrices, and the subgroup $Q = P_{0,J}$ is the μ_τ-saturation of $P_{0,I}$ consisting of upper triangular matrices corresponding to the partition $\{1, \dots, k-1\} \cup \{k, \dots, n\}$. The boundary symmetric spaces are given by

$$X_P = \mathcal{P}_k, \quad X_Q = \mathcal{P}_k \times \mathcal{P}_{n-k}.$$

The boundary component $e(Q)$ of Q is equal to $\mathcal{P}_k$. Since $\mu_\tau = t_1$ belongs to the edge of $\mathfrak{a}^{*,+}$ defined $\{\lambda \in \mathfrak{a}^{*,+} \mid \langle \lambda, \alpha_2 \rangle = \cdots = \langle \lambda, \alpha_{n-1} \rangle = 0\}$, the Satake compactification $\overline{\mathcal{P}_n}^S$ is a minimal one in the partially ordered set of Satake compactifications of $\mathcal{P}_n$, as mentioned earlier.

The Satake compactification $\overline{\mathcal{P}_n}^S$ is special in the sense that its boundary components are of the same form $\mathcal{P}_k$, $k \leq n - 1$, as $\mathcal{P}_n$. The Baily–Borel compactification discussed in the next section has the same property.

Remarks I.4.44 If X is irreducible, i.e., the Dynkin diagram is connected, there is a one-to-one correspondence between the minimal Satake compactifications $\overline{X}_\tau^S$ and the set of simple roots (or the vertices of the Dynkin diagram).

In general, the normalizer of a boundary component of such a Satake compactification is not necessarily a maximal parabolic subgroup. For example, when $n \geq 5$, μ_τ a suitable weight only connected to α_3, then $I = \{\alpha_3\}$ is μ_τ connected and its μ_τ-saturation is $J = \{\alpha_1, \alpha_3, \alpha_5, \dots, \alpha_{n-1}\}$, and hence the saturated parabolic subgroup $P_{0,J}$ is not a maximal parabolic subgroup.

I.4.45 We conclude this section by mentioning several applications of the Satake compactifications of X. In [Sat2], the Satake compactifications of symmetric spaces were used to construct compactifications of locally symmetric spaces (see §III.3 below). For applications to geometry and harmonic analysis, the maximal Satake compactification is the most useful one among the family of all Satake compactifications. For example, it is used in the proof of the Mostow strong rigidity in [Mos], and the determination of the Martin compactification of X in [GJT] (see §I.7 below), and its boundaries are used in the harmonic analysis on X (see [KaK] [Ko1–2]).

I.4.46 Summary and Comments. In this section, we motivated the Satake compactifications by the example of the space $\mathcal{P}_n$ of positive definite Hermitian matrices of determinant one. The importance of this example is that it explains naturally why $\mathcal{P}_n$ is noncompact and how it can be compactified by adding semipositive matrices. This is the original approach in [Sat1]. A slightly different procedure using spherical representations was given by Casselman in [Cas2]. The axioms in Proposition I.4.33 explain clearly the nature of the Satake compactifications as G-spaces and hence can be used to determine whether other compactifications are isomorphic to the Satake compactifications. See [Mo1], [Ko1] and Corollary I.5.29, for example.

I.5 Baily–Borel compactification

In this section, we recall basic properties of Hermitian symmetric spaces and the Baily–Borel compactification. Most results are recalled without proofs, which can be found in the references [Wo3], [Mok1], [Hel3], [BB1], and [Sat8]. In view of applications to the Baily–Borel compactification of locally symmetric spaces, we will follow the notation in [BB1].

In this section, we will assume that X is a Hermitian symmetric space unless indicated otherwise. The Baily–Borel compactification is one of the minimal Satake compactifications as a topological compactification, and it has the additional property that every boundary component is also a Hermitian symmetric space. In fact, among all the Satake compactifications of Hermitian symmetric spaces, it is the only one having this property.

This section is organized as follows. First we recall bounded symmetric domains (I.5.1) and the Bergman metric of bounded domains (I.5.4). Then we discuss Hermitian symmetric spaces X of noncompact type and their characterization in terms of the center of a maximal compact subgroup (Proposition I.5.9), and their realization as bounded symmetric domains via the Harish-Chandra embedding, whose closure gives the Baily–Borel compactification $\overline{X}^{BB}$ (Definition I.5.13). In I.5.20, we introduce the notion of analytic boundary components of domains in $\mathbb{C}^n$. After describing the analytic boundary components of X or rather $\overline{X}^{BB}$ (Proposition I.5.27) and their normalizer (Proposition I.5.28), we give a proof that the Baily–Borel compactification $\overline{X}^{BB}$ is isomorphic to a minimal Satake compactification $\overline{X}^{S}_{\mu}$ (Corollary I.5.29), which avoids using identification through the Furstenberg compactifications and is hence different from the original proof in [Mo2].

Definition I.5.1 A bounded domain $\Omega \subset \mathbb{C}^n$ is called symmetric if for any $z \in \Omega$, there exists an involutive holomorphic automorphism with z as an isolated fixed point.

At first sight, it is not obvious that such an involution, if it exists, is unique. As shown below, every bounded domain Ω has a canonically associated Hermitian metric, the Bergman metric, with respect to which any holomorphic automorphism is an isometry. This implies that the involutive automorphism, if it exists, is unique and is the geodesic symmetry in the Bergman metric.

Definition I.5.2 A connected complex manifold M with a Hermitian structure is called a Hermitian symmetric space if every point $z \in M$ is an isolated fixed point of an involutive holomorphic isometry s_z.

It will be seen below that the relation between these symmetric domains and Hermitian symmetric spaces is that any bounded symmetric domain with the Bergman metric is a Hermitian symmetric space of noncompact type, and any Hermitian symmetric space of noncompact type can be canonically realized as a bounded symmetric domain.

I.5.3 We briefly recall the definition of the Bergman metric of a bounded domain Ω in $\mathbb{C}^n$. Let

$$H(\Omega) = \left\{ f \mid f \text{ is a holomorphic function on } \Omega, \int_\Omega |f|^2 < +\infty \right\}.$$

It can be shown that $H(\Omega)$ is a closed subspace of $L^2(\Omega)$ and is hence a Hilbert space with the inner product given by

$$(f, g) = \int_\Omega f(z)\overline{g(z)}dz, \quad f, g \in H^2(\Omega).$$

Let $\varphi_1, \dots, \varphi_j, \dots$ be any orthonormal basis of $H(\Omega)$. Then

$$K(z, \bar{\zeta}) = \sum_{j=1}^{\infty} \varphi_j(z)\overline{\varphi_j(\zeta)}$$

converges uniformly for z and w in compact subsets of Ω and satisfies the *reproducing property*: for any $f \in H(\Omega)$,

$$f(z) = \int_\Omega K(z, \bar{\zeta}) f(\zeta) d\zeta.$$

This reproducing property shows that $K(z, \bar{\zeta})$ is independent of the choice of the orthonormal basis; $K(z, \bar{\zeta})$ is called the *Bergman kernel* of the domain Ω.

Proposition I.5.4 *The Bergman kernel $K(z, \bar{\zeta})$ defines a Kähler metric on Ω by the following formula:*

$$g_{ij}(z) = \frac{\partial^2}{\partial z_i \overline{\partial z_j}} \log K(z, \bar{z}),$$

where the norm of a holomorphic vector $X = \sum_j \xi_j \frac{\partial}{\partial z_j}$ *is given by*

$$g(X, X) = \sum_{i,j} g_{ij} \xi_i \bar{\xi_j}.$$

The Hermitian metric in the above proposition is called the *Bergman metric* and has the following important invariance property.

Proposition I.5.5 *Let Ω, Ω' be two bounded domains in $\mathbb{C}^n$, and $\phi : \Omega \to \Omega'$ a holomorphic diffeomorphism. Then ϕ is an isometry with respect to the Bergman metric.*

Corollary I.5.6 *Any bounded symmetric domain Ω in $\mathbb{C}^n$ endowed with the Bergman metric is a Hermitian symmetric space.*

Since bounded holomorphic functions separate points on Ω, it follows that a bounded symmetric domain Ω is a Hermitian symmetric space of noncompact type. In fact, there is no nontrivial bounded holomorphic function on the flat Hermitian symmetric spaces ($\mathbb{C}^n$) or compact Hermitian symmetric spaces.

I.5.7 Consider the example of the unit ball in $\mathbb{C}^n$:

$$B^n = \{z \in \mathbb{C}^n \mid \|z\| < 1\}.$$

Its Bergman kernel $K_{B^n}(z, \bar{\zeta})$ is

$$K_{B^n}(z, \bar{\zeta}) = \frac{n!}{\pi^n} \frac{1}{(1 - z \cdot \bar{\zeta})},$$

and the Bergman metric is

$$g_{ij}(z) = \frac{n+1}{(1 - |z|^2)^2} ((1 - |z|^2)\delta_{ij} + \bar{z}_i z_j).$$

When $n = 1$, it is equal to the Poincaré metric

$$g_{11}(z) = \frac{2}{(1 - |z|^2)^2}.$$

See [Kr, Proposition 1.4.23] for details.

I.5.8 Besides the unit ball B_n, there are other *classical domains*, which are obtained basically by replacing vectors by matrices. There are four types of domains. Let $M(p, q, \mathbb{C})$ be the space of complex $p \times q$ matrices.

The type I domain $D^I_{p,q}$ is defined by

$$D^I_{p,q} = \{z \in M(p,q,\mathbb{C}) \mid {}^t\bar{z} \cdot z - I_q < 0\} \cong SU(p,q)/S(U(p) \times U(q)). \quad \text{(I.5.1)}$$

When $q = 1$, $D^I_{p,1}$ is the unit ball B^p in $\mathbb{C}^p$. For any p, q, $D^I_{p,q}$ is biholomorphic to $D^I_{q,p}$. It may not be trivial to check this directly. For example, $D^+_{1,p}$ is a subset of $M(1,p,\mathbb{C}) = \mathbb{C}^p$ defined by more than one inequality, while $D^I_{p,1}$ is the unit ball defined by only one inequality. To see this biholomorphism, we note that $SU(p,q)$ is isomorphic to $SU(q,p)$, and $S(U(p) \times U(q))$ corresponds to $S(U(q) \times U(p))$ under the isomorphism. Hence $D^I_{p,q}$ is biholomorphic to $D^I_{q,p}$.

The type II and III domains are cut out of the type I domain as follows. The type II domain D^{II}_n is defined by

$$D^{II}_n = \{z \in D^I_{n,n} \mid {}^t z = -z\} \cong \mathrm{SO}^*(2n)/U(n), \quad \text{(I.5.2)}$$

where $\mathrm{SO}^*(2n)$ is the subgroup of $\mathrm{SO}(2n,\mathbb{C})$ consisting of elements that preserve the skew Hermitian form

$$-z_1\bar{z}_{n+1} + z_{n+1}\bar{z}_1 - \cdots - z_n\bar{z}_{2n} + z_{2n}\bar{z}_n,$$

and is also the noncompact Lie group dual to $\mathrm{SO}(2n) = \mathrm{SO}(2n,\mathbb{R})$ with respect to the maximal compact subgroup $K = \mathrm{SO}(2n) \cap Sp(n) \cong U(n)$ (this explains the superscript * for the duality), and the compact subgroup $U(n)$ is embedded into $\mathrm{SO}^*(2n)$ by

$$A + iB \mapsto \begin{pmatrix} A & B \\ -B & A \end{pmatrix}, \quad \text{(I.5.3)}$$

where $A + iB \in U(n)$, with A, B real.

The type III domain D^{III}_n is defined by

$$D^{III}_n = \{z \in D^I_{n,n} \mid {}^t z = z\} \cong Sp(n,\mathbb{R})/U(n), \quad \text{(I.5.4)}$$

where $U(n)$ is embedded into $Sp(n,\mathbb{R})$ by equation (I.5.3).

The type IV domain D^{IV}_n is defined by

$$D^{IV}_n = \left\{z \in \mathbb{C}^n \mid \|z\|^2 < 2, \|z\|^2 < 1 + \frac{1}{2}|z_1^2 + \cdots + z_n^2|\right\}$$
$$\cong \mathrm{SO}(n,2)/\mathrm{SO}(n) \times \mathrm{SO}(2).$$

The domain D^{IV}_n is the only one of the double-indexed family of Riemannian symmetric spaces $\mathrm{SO}(n,m)/\mathrm{SO}(n) \times \mathrm{SO}(m)$ that is Hermitian symmetric. The reason is that only when either n or m is equal to 2 does the center of $\mathrm{SO}(n) \times \mathrm{SO}(m)$ have positive dimension, and hence other symmetric spaces are not Hermitian (see Proposition I.5.9 below).

Besides these classical domains, there are two *exceptional bounded symmetric domains*, one in (complex) dimension 16, and another in dimension 27. These exhaust all the irreducible bounded symmetric domains. A general bounded symmetric domain is the product of such irreducible ones.

Proposition I.5.9 (Characterization of Hermitian symmetric spaces) *An irreducible symmetric space $X = G/K$ is Hermitian if and only if the center of K has positive dimension, which is equal to 1; and any Hermitian symmetric space of noncompact type can be canonically realized as a bounded symmetric domain in the holomorphic tangent space at $x_0 = K$ in X.*

This embedding of X is called the *Harish-Chandra embedding* and can be described as follows. See [Hel3] for a complete proof of this proposition.

Let $\mathfrak{k}$ be the Lie algebra of K, $\mathfrak{g} = \mathfrak{k} \oplus \mathfrak{p}$ the Cartan decomposition, and $\mathfrak{k}^{\mathbb{C}} = \mathfrak{k} \otimes \mathbb{C}$, $\mathfrak{p}^{\mathbb{C}} = \mathfrak{p} \otimes \mathbb{C}$. Let $\mathfrak{h}$ be a maximal abelian subalgebra of $\mathfrak{k}$. Then $\mathfrak{h}$ is also a maximal abelian subalgebra of $\mathfrak{g}$, and $\mathfrak{h}^{\mathbb{C}} = \mathfrak{h} \otimes \mathbb{C}$ is a Cartan subalgebra of $\mathfrak{g}^{\mathbb{C}} = \mathfrak{g} \otimes \mathbb{C}$. Let $\Phi(\mathfrak{g}^{\mathbb{C}}, \mathfrak{h}^{\mathbb{C}})$ be the associated root system, and denote the root spaces by $\mathfrak{g}_{\alpha}^{\mathbb{C}}$. Since

$$[\mathfrak{h}^{\mathbb{C}}, \mathfrak{k}^{\mathbb{C}}] \subset \mathfrak{k}^{\mathbb{C}}, \quad [\mathfrak{h}^{\mathbb{C}}, \mathfrak{p}^{\mathbb{C}}] \subset \mathfrak{p}^{\mathbb{C}},$$

for each $\alpha \in \Phi(\mathfrak{g}^{\mathbb{C}}, \mathfrak{h}^{\mathbb{C}})$, either $\mathfrak{g}_{\alpha}^{\mathbb{C}} \subset \mathfrak{k}^{\mathbb{C}}$ or $\mathfrak{g}_{\alpha}^{\mathbb{C}} \subset \mathfrak{p}^{\mathbb{C}}$. In the former case, the root α is called a compact root; and in the latter case, the root α is called a noncompact root. Hence,

$$\mathfrak{k}^{\mathbb{C}} = \sum_{\text{compact } \alpha} \mathfrak{g}_{\alpha}^{\mathbb{C}}, \quad \mathfrak{p}^{\mathbb{C}} = \sum_{\text{noncompact } \alpha} \mathfrak{g}_{\alpha}^{\mathbb{C}}.$$

Note that $\mathfrak{p}$ can be identified with the tangent space of X at the basepoint $x_0 = K$, and $\mathfrak{p}^{\mathbb{C}}$ can be identified with the complexification of the tangent space.

All the roots in $\Phi(\mathfrak{g}^{\mathbb{C}}, \mathfrak{h}^{\mathbb{C}})$ take real values on the vector space $i\mathfrak{h}$. Choose an ordering on $\Phi(\mathfrak{g}^{\mathbb{C}}, \mathfrak{h}^{\mathbb{C}})$ such that every positive noncompact root is larger than every compact root. For example, this can be arranged as follows. Choose a basis $v_1, \dots, v_n$ of $i\mathfrak{h}$ such that iv_1 belongs to the center of $\mathfrak{k}$. Then the associated lexicographic order has the desired property. The reason is that compact roots vanish on v_1, but noncompact roots do not. The set of positive roots is denoted by $\Phi^+ = \Phi^+(\mathfrak{g}^{\mathbb{C}}, \mathfrak{h}^{\mathbb{C}})$.

As mentioned in §I.1, G is a linear Lie group. Let $G_{\mathbb{C}}$ be the linear group with Lie algebra $\mathfrak{g}^{\mathbb{C}} = \mathfrak{g} \otimes \mathbb{C}$, i.e., the *complexification* of G. Define

$$\mathfrak{p}^+ = \sum_{\text{noncompact } \alpha \in \Phi^+} \mathfrak{g}_{\alpha}^{\mathbb{C}}, \quad \mathfrak{p}^- = \sum_{\text{noncompact } \alpha \in -\Phi^+} \mathfrak{g}_{\alpha}^{\mathbb{C}}. \tag{I.5.5}$$

Then $\mathfrak{p}^+, \mathfrak{p}^-$ are abelian subalgebras of $\mathfrak{g}^{\mathbb{C}}$. Let P^+, P^- be the corresponding subgroups of $G_{\mathbb{C}}$. The exponential maps

$$\exp : \mathfrak{p}^+ \to P^+, \quad \exp : \mathfrak{p}^- \to P^-$$

are diffeomorphisms. Let $K_{\mathbb{C}}$ be the subgroup of $G_{\mathbb{C}}$ corresponding to $\mathfrak{k}^{\mathbb{C}}$.

There are two possible choices of orderings on $\Phi(\mathfrak{g}^{\mathbb{C}}, \mathfrak{h}^{\mathbb{C}})$ satisfying the above property. Choose the one such that $\mathfrak{p}^+$ becomes the *holomorphic tangent space* at the basepoint x_0.

Proposition I.5.10 *In the above notation, the multiplication*

$$P^+ \times K_{\mathbb{C}} \times P^- \to G_{\mathbb{C}}$$

is injective, holomorphic, and regular with an open image in $G_{\mathbb{C}}$. *The subset* $GK_{\mathbb{C}}P^- \subset G_{\mathbb{C}}$ *is open, and there exists a bounded domain* $\Omega \subset \mathfrak{p}^+$ *containing the zero element such that*

$$GK_{\mathbb{C}}P^- = \exp(\Omega)K_{\mathbb{C}}P^-.$$

Let

$$\pi^+ : P^+K_{\mathbb{C}}P^- = P^+ \times K_{\mathbb{C}} \times P^- \to \mathfrak{p}^+, \quad (p^+, k, p^-) \mapsto \log p^+, \tag{I.5.6}$$

be the projection map to the first factor composed with $\log : P^+ \to \mathfrak{p}^+$. Define an action of G on Ω by

$$g \cdot H = \pi^+(g \exp H), \quad H \in \mathfrak{p}^+.$$

Since $\exp(\Omega)K_{\mathbb{C}}P^- = GK_{\mathbb{C}}P^-$ is invariant under G on the left, $g \cdot H \in \Omega$ for any $g \in G$, $H \in \Omega$.

Proposition I.5.11 *The action of* G *on* Ω *is holomorphic and the stabilizer in* G *of the zero element in* Ω *is equal to* K, *and hence the map* $g \mapsto \pi^+(g)$ *in equation (I.5.6) defines a biholomorphism*

$$\pi^+ : X = G/K \to \Omega.$$

See [Wo3, p. 281] for proofs of the above two propositions.

Lemma I.5.12 *The bounded domain* Ω *in the above proposition is a bounded symmetric domain in* $\mathfrak{p}^+$, *which can be identified with the holomorphic tangent bundle at* x_0 *of* X *when considered as a complex manifold.*

Proof. Since X is Hermitian symmetric, every point of X is an isolated fixed point of an involutive holomorphic map. Hence every point of Ω is also an isolated fixed point of an involutive holomorphic map.

Definition I.5.13 *The embedding* $\pi^+ : X = G/K \to \Omega$ *in the above proposition is called the Harish-Chandra embedding. The closure of* $\pi^+(X)$ *in* $\mathfrak{p}^+$ *is called the Baily–Borel compactification of* X, *denoted by* $\overline{X}^{BB}$.

To show that $\overline{X}^{BB}$ is topologically one of the minimal Satake compactifications and to understand the structure of the boundary components, we need to study the root system $\Phi(\mathfrak{g}^{\mathbb{C}}, \mathfrak{h}^{\mathbb{C}})$ in more detail.

Two roots μ, ν in $\Phi(\mathfrak{g}^{\mathbb{C}}, \mathfrak{h}^{\mathbb{C}})$ are called *strongly orthogonal* if $\mu \pm \nu$ are not roots, which implies that μ, ν are orthogonal in the usual sense. Then there exists a maximal collection $\Psi = \{\mu_1, \dots, \mu_r\}$ of strongly orthogonal roots in Φ^+. For example, let μ_1 be a lowest positive root, and μ_2 a lowest root among all the roots strongly orthogonal to μ_1, and so on ([BB1, §1] [Wo3, p. 279]).

Proposition I.5.14 *Given a maximal collection of strongly orthogonal roots* $\Psi = \{\mu_1, \dots, \mu_r\}$, *we can choose root vectors* $E_\mu \in \mathfrak{g}^{\mathbb{C}}_\mu$ *such that*

$$(E_\mu + E_{-\mu}) \in \mathfrak{p}, \quad E_\mu - E_{-\mu} \in i\mathfrak{p}, \quad B(E_\mu, E_{-\mu}) = \frac{2}{\langle \mu, \mu \rangle}, \tag{I.5.7}$$

where B *is the Killing form of* $\mathfrak{g}^{\mathbb{C}}$, *and the subalgebra*

$$\mathfrak{a} = \sum_{j=1}^{r} \mathbb{R}(E_{\mu_j} + E_{-\mu_j})$$

is a maximal abelian subalgebra in $\mathfrak{p}$.

For each $\mu \in \Psi$, let

$$H_\mu = [E_\mu, E_{-\mu}] \in \mathfrak{h}^{\mathbb{C}}.$$

Define

$$\mathfrak{g}^{\mathbb{C}}[\mu] = \mathbb{C}H_\mu + \mathfrak{g}^{\mathbb{C}}_\mu + \mathfrak{g}^{\mathbb{C}}_{-\mu},$$

and a real form

$$\mathfrak{g}[\mu] = \mathfrak{g}^{\mathbb{C}}[\mu] \cap \mathfrak{g}.$$

Let $G[\mu]$ be the subgroup of G corresponding to the subalgebra $\mathfrak{g}[\mu]$. More generally, for any subset $I \subseteq \Psi$, we get a subalgebra

$$\mathfrak{g}^{\mathbb{C}}[I] = \sum_{\mu \in I} \mathfrak{g}^{\mathbb{C}}[\mu], \quad \mathfrak{g}[I] = \sum_{\mu \in I} \mathfrak{g}[\mu].$$

Let $G[I]$ be the subgroup of G with Lie algebra $\mathfrak{g}[I]$. See [Wo3, p. 280] for the proof of the next result.

Proposition I.5.15 *Let* $x_0 = K$ *be the basepoint in* X. *For each* $\mu \in \Psi$, *the orbit* $G[\mu]x_0$ *in* X *is a totally geodesic submanifold biholomorphic to the unit disk* $D = \{z \in \mathbb{C} \mid |z| < 1\}$ *in* $\mathbb{C}$. *In fact, the image* $\pi^+(G[\mu]x_0)$ *under the Harish-Chandra embedding is a disk in* $\mathbb{C}E_\mu \subset \mathfrak{p}^+$. *For any subset* $I \subset \Psi$, *the orbit* $G[I]x_0$ *in* X *of the subgroup* $G[I]$ *is a totally geodesic submanifold, and the image* $\pi^+(G[I]x_0)$ *in* $\mathfrak{p}^+$ *is the polydisk* $\prod_{\mu \in I} \pi^+(G[\mu]x_0)$ *in* $\prod_{\mu \in I} \mathbb{C}E_\mu$. *When* $I = \Psi$, $G[\Psi]x_0$ *is product of* r *disks, where* r *is equal to the rank of* X, *and*

$$X = KG[\Psi]x_0, \tag{I.5.8}$$

i.e., the compact subgroup K *sweeps out the polydisk* $G[\Psi] \cong D^r$ *to yield the whole symmetric space* X.

Note that the polydisk $G[\Psi]x_0$ contains the maximal flat $e^{\mathfrak{a}}x_0$ in X, and hence the Cartan decomposition gives that

$$Ke^{\mathfrak{a}}x_0 = X.$$

This implies that there is some overlap (or redundancy) in the decomposition in equation (I.5.8). One obvious reason is that the polydisk is a Hermitian symmetric space and has both radial and angular parts, while the flat $e^{\mathfrak{a}}x_0$ has no complex structure at all and consists only of the radial part of the polydisk. (Note that the angular part is contained in K.)

Corollary I.5.16 *Identify the maximal flat* $e^{\mathfrak{a}}x_0 \cong \mathfrak{a}$ *of the polydisk with* $(-1, 1)^r$ *under the map*

$$t_1(E_{\mu_1} + E_{-\mu_1}) + \cdots + t_r(E_{\mu_r} + E_{-\mu_r}) \mapsto (\tanh t_1, \ldots, \tanh t_r).$$

Then the closure $\overline{e^{\mathfrak{a}}x_0}$ *of* $e^{\mathfrak{a}}x_0$ *in* $\overline{X}^{BB}$ *is equal to* $[-1, 1]^r$.

Proof. By Proposition I.5.15, the closure of the polydisk $\pi^+(G[\Psi]x_0)$ in $\mathfrak{p}^+$ is $\prod_{\mu \in I} \overline{\pi^+(G[\mu]x_0)}$, where $\overline{\pi^+(G[\mu]x_0)}$ is the closed disk. This implies that the closure $\overline{e^{\mathfrak{a}}x_0}$ is a cube. The explicit form of the map follows from the Harish-Chandra embedding in equation (I.5.6) and the following equality:

$$\exp t(E_\mu + E_\mu) = \exp(\tanh t \ E_\mu) \ \exp(\log(\cosh t)H_\mu) \ \exp(\tanh t \ E_{-\mu}),$$

where $\exp(\tanh t \ E_\mu) \in P^+$, $\exp(\tanh t \ E_\mu) \in K_{\mathbb{C}}$, and $\exp(\log(\cosh t)H_\mu) \in P^-$.

I.5.17 To study the real roots in $\Phi(\mathfrak{g}, \mathfrak{a})$ using the complex roots in $\Phi(\mathfrak{g}^{\mathbb{C}}, \mathfrak{h}^{\mathbb{C}})$, we need a transform from the Cartan subalgebra $\mathfrak{h}^{\mathbb{C}}$ of $\mathfrak{g}^{\mathbb{C}}$ to another Cartan subalgebra of $\mathfrak{g}^{\mathbb{C}}$ containing the Cartan subalgebra $\mathfrak{a}$ of the symmetric pair $(\mathfrak{g}, \mathfrak{k})$. This is given by the Cayley transform ([BB1, 1.6] [Wo3, p. 281]).

Define

$$\mathfrak{h}^- = \sum_{\mu \in \Psi} i\mathbb{R}H_\mu,$$

and $\mathfrak{h}^+$ the orthogonal complement in $\mathfrak{h}$ equal to $\{X \in \mathfrak{h} \mid [X, \mathfrak{a}] = 0\}$. For any subset $I \subset \Psi$, define

$$\mathfrak{h}^-_{\Psi - I} = \sum_{\mu \in \Psi - I} i\mathbb{R}H_\mu,$$

$$\mathfrak{a}_I = \sum_{\mu \in I} \mathbb{R}(E_\mu + E_{-\mu}).$$

Define a *partial Cayley transform*

$$C_I = \prod_{\mu \in I} C_\mu, \quad C_\mu = \exp \frac{\pi}{4}(E_{-\mu} - E_\mu) \in G_{\mathbb{C}}.$$

Let $\mathfrak{h}_I = \mathfrak{g} \cap Ad(C_I)\mathfrak{h}^{\mathbb{C}}$. Then

$$\mathfrak{h}_I = \mathfrak{h}^+ + \mathfrak{h}^-_{\Psi-I} + \mathfrak{a}_I. \tag{I.5.9}$$

Therefore, the Cayley transform C_I maps the subspace $\mathfrak{h}^-_I \subset \mathfrak{h}^-$ into a subspace $\mathfrak{a}_I \subset \mathfrak{a}$. In particular, when $I = \Psi$, C_Ψ is the *full Cayley transform*, and we get

$$Ad(C_\Psi) : \mathfrak{h}^- = \mathfrak{h}^-_\Psi \to \mathfrak{a}.$$

Let $\gamma_1, \dots, \gamma_r$ be the roots in $\Phi(\mathfrak{g}, \mathfrak{a})$ corresponding to the restriction of $\mu_1, \dots, \mu_r$ to $\mathfrak{h}^-$.

Proposition I.5.18 *Suppose that G is simple, i.e., the Hermitian symmetric space X is irreducible. Then $\Phi = \Phi(\mathfrak{g}, \mathfrak{a})$ belongs to one of the following two types:*

(1) Type C_r,

$$\Phi = \left\{ \pm \frac{\gamma_s + \gamma_t}{2}, 1 \le s, t \le r; \pm \frac{\gamma_s - \gamma_t}{2}, 1 \le s \ne t \le r \right\}.$$

If we order the roots such that $\gamma_1 > \dots > \gamma_r$, then

$$\Delta = \left\{ \alpha_1 = \frac{1}{2}(\gamma_1 - \gamma_2), \dots, \alpha_{r-1} = \frac{1}{2}(\gamma_{r-1} - \gamma_r), \alpha_r = \gamma_r \right\}$$

is the set of simple roots.

(2) Type BC_r,

$$\Phi = \left\{ \pm \frac{\gamma_s + \gamma_t}{2}, 1 \le s, t \le r; \pm \frac{\gamma_s - \gamma_t}{2}, 1 \le s \ne t \le r; \pm \frac{\gamma_s}{2}, 1 \le s \le r \right\}.$$

If we order the roots such that $\gamma_1 > \dots > \gamma_r$, then

$$\Delta = \left\{ \alpha_1 = \frac{1}{2}(\gamma_1 - \gamma_2), \dots, \alpha_{r-1} = \frac{1}{2}(\gamma_{r-1} - \gamma_r), \alpha_r = \frac{\gamma_r}{2} \right\}$$

is the set of simple roots. In both cases, the Weyl group consists of all the signed permutations of $\{\gamma_1, \dots, \gamma_r\}$.

In Type C_r, the simple root α_r is longer than other simple roots; while in Type BC_r, α_r is shorter than other simple roots. In either case, we call it the *distinguished root*. See [BB1, 1.2] [Wo3, p. 284–285] for more discussions about the root systems.

I.5.19 Before discussing the boundary components of $\overline{X}^{BB}$, we illustrate the above discussions using the simplest example $X = SU(1,1)/U(1)$, the unit disk D in $\mathbb{C}$; the groups

$$SU(1,1) = \left\{ \begin{pmatrix} a & \bar{b} \\ b & \bar{a} \end{pmatrix} \mid |a|^2 - |b|^2 = 1 \right\}, \quad K = \left\{ \begin{pmatrix} e^{it} & 0 \\ 0 & e^{-it} \end{pmatrix} \mid t \in \mathbb{R} \right\},$$

the Lie algebras

$$\mathfrak{h} = \mathfrak{k} = \left\{ \begin{pmatrix} it & 0 \\ 0 & -it \end{pmatrix} \mid t \in \mathbb{R} \right\},$$

$$\mathfrak{a} = \left\{ \begin{pmatrix} 0 & -it \\ it & 0 \end{pmatrix} \mid t \in \mathbb{R} \right\},$$

and the corresponding subgroup

$$A = \left\{ \begin{pmatrix} \cosh t & -i \sinh t \\ i \sinh t & \cosh t \end{pmatrix} \mid t \in \mathbb{R} \right\}.$$

The rank of $SU(1,1)$ is equal to 1. Let μ be the only positive root in $\Phi(\mathfrak{g}, \mathfrak{a})$. Then we can take

$$E_\mu = \begin{pmatrix} 0 & -i \\ 0 & 0 \end{pmatrix}, \quad E_{-\mu} = \begin{pmatrix} 0 & 0 \\ i & 0 \end{pmatrix}.$$

It can be checked easily that

$$\mathfrak{a} = \mathbb{R}(E_\mu + E_{-\mu}) = i\mathbb{R} \begin{pmatrix} 0 & -1 \\ 1 & 0 \end{pmatrix},$$

and the Cayley transform is given by

$$C = \exp \frac{\pi}{4}(E_{-\mu} - E_\mu) = \begin{pmatrix} \cos \frac{\pi}{4} & i \sin \frac{\pi}{4} \\ i \sin \frac{\pi}{4} & \cos \frac{\pi}{4} \end{pmatrix} = \frac{1}{\sqrt{2}} \begin{pmatrix} 1 & i \\ i & 1 \end{pmatrix}.$$

Since $\mathfrak{h}$ is of dimension 1,

$$\mathfrak{h}^- = \mathfrak{h}.$$

To show that

$$Ad(C)(i\mathfrak{h}) = \mathfrak{a},$$

we note that

$$C^{-1} = \frac{1}{\sqrt{2}} \begin{pmatrix} 1 & -i \\ -i & 1 \end{pmatrix}, \quad C \begin{pmatrix} 1 & 0 \\ 0 & -1 \end{pmatrix} C^{-1} = i \begin{pmatrix} 0 & -1 \\ 1 & 0 \end{pmatrix}.$$

The group $\mathrm{SL}(2, \mathbb{C})$ and the Cayley transform $C \in \mathrm{SL}(2, \mathbb{C})$ act on $\mathbb{C}P^1 = \mathbb{C} \cup \{\infty\}$. The effect of this transform is given in the next result and can be checked easily.

Proposition I.5.20 *The Cayley transform C maps $0 \in D$ to i, a point on the boundary of the unit disk D, and maps the unit disk D onto the upper half-plane $\{x + iy \mid x \in \mathbb{R}, y > 0\}$.*

I.5.21 To show that $\overline{X}^{BB}$ is a Satake compactification, we need to study the boundary components of the boundary $\partial \overline{X}^{BB} \cong \overline{\pi^+(X)} \subset \mathfrak{p}^+$ as a subset of the complex vector space and determine the normalizer of the boundary components.

In general, let V be a complex manifold, in particular a complex vector space, and $S \subset V$ an open subset. A *holomorphic arc* in S is a holomorphic map

$$f : D = \{z \in \mathbb{C} \mid |z| < 1\} \to V$$

with image contained in S. By a *chain of holomorphic arcs* in S, we mean a finite sequence of holomorphic arcs $f_1, \dots, f_k$ such that the image of f_j meets the image of f_{j+1}. This defines an equivalence relation on S as follows. Two points $p, q \in S$ are defined to be *equivalent* if there exists a chain of holomorphic arcs $f_1, \dots, f_k$ in S such that p belongs to the image of f_1, and q belongs to the image of f_k. An equivalence class is called an *analytic arc* in S. If S is an open subset in V with the (topological) boundary ∂S, then each connected component of S is an *analytic arc component*; the analytic arc components in ∂S are called the *analytic boundary components*. From the definition, it is clear that the boundary ∂S is decomposed into a disjoint union of analytic boundary components; and it is also clear that if $\varphi : V \to W$ is a biholomorphism, then the analytic boundary components of S are mapped to those of $\varphi(S)$.

In the rest of this section, we determine the boundary components of a Hermitian symmetric space X under the Harish-Chandra embedding in $\mathfrak{p}^+$ and their normalizers in G, and then use the characterization of the Satake compactifications in Proposition I.4.33 to show that $\overline{X}^{BB}$ is a Satake compactification.

I.5.22 An illuminating example for the structure of analytic components is the case of the bidisk $D \times D$. For any points $p, q \in \partial D$, $\{p\} \times D$, $D \times \{q\}$, $\{p\} \times \{q\}$ are analytic boundary components of $D \times D$, and any analytic boundary component of $D \times D$ is of such a form.

In fact, if $p = i$, the boundary component $\{i\} \times D$ can be obtained as follows. The second factor D determines a complex submanifold $\{0\} \times D$ of $\mathbb{C}P^1 \times D$, which is an analytic boundary component of $\mathbf{H} \times D$ by the maximal principle for holomorphic functions. Under the Cayley transform C on the first factor, $\{0\} \times D$ gives the analytic boundary component $\{i\} \times D$. The action under elements of $K = U(1) \times U(1)$ gives the general boundary components $\{p\} \times D$. The same method works for the boundary component $D \times \{q\}$.

To get the boundary component $\{i\} \times \{i\}$, we start with the submanifold consisting only of the origin $(0, 0)$ in $\mathbb{C}P^1 \times \mathbb{C}P^1$, which is a boundary component of $\mathbf{H} \times \mathbf{H}$ by the maximal principle, and apply the Cayley transform (C, C) to $(0, 0)$. The K-action gives other boundary components (p, q) of dimension 0.

Clearly, the analytic boundary components of polydisks $D^r = D \times \cdots \times D \subset \mathbb{C}^r$ have similar structures.

I.5.23 To obtain the boundary components of general X, we use the result in Proposition I.5.15 that

$$X = KG[\Psi] \cdot x_0 \cong K \cdot D^r$$

and apply the (partial) Cayley transform to suitable Hermitian submanifolds in X containing the origin as in the previous paragraph, or equivalently, apply appropriate subgroups of K to sweep out the boundary components of the polydisk D^r.

Briefly, for any $I \subset \Psi$, $G[I]x_0$ is a subpolydisk D^I in $D^\Psi = D^r$. There is a unique totally geodesic Hermitian symmetric submanifold X_I in X that contains $G[I]x_0$ as a maximal polydisk. Let C_I be the (partial) Cayley transform associated with I. Then $C_I \cdot G[I] \cdot x_0$ is a boundary component of D^Ψ under the Harish-Chandra embedding. One obvious difference with the case of polydisks is that we need to enlarge the polydisk $G[I]x_0$ to a Hermitian symmetric space X_I so that $C_I \cdot X_I$ is defined and gives a boundary component of X.

But there is one problem with the above approach. The Cayley transform C_I is in $G_\mathbb{C}$, and $X \subset \mathfrak{p}^+$, but $G_\mathbb{C}$ does not act on X or $\mathfrak{p}^+$. In the case of $X = D$, we have $\mathfrak{p}^+ = \mathbb{C}$, and we embedded both of them into $\mathbb{C}P^1$, on which $G_\mathbb{C} = \mathrm{SL}(2, \mathbb{C})$ acts. The analogue of this embedding for general X is given by the Borel embedding.

Since $\mathfrak{k}^\mathbb{C}$ is the centralizer of $\mathfrak{h}^\mathbb{C}$ in $\mathfrak{g}^\mathbb{C}$, $\mathfrak{p}^- + \mathfrak{k}^\mathbb{C}$ is a subalgebra. The corresponding subgroup of $G_\mathbb{C}$ is $P^- K_\mathbb{C}$. Let

$$\mathfrak{g}_u = \mathfrak{k} + i\mathfrak{p}$$

be the compact dual of $\mathfrak{g} = \mathfrak{k} + \mathfrak{p}$, and G_u the compact subgroup of $G_\mathbb{C}$ *dual* to G. Let $X_u = G_u/K$, the compact symmetric space dual to $X = G/K$.

Proposition I.5.24

(1) The compact group G_u acts transitively on $G_\mathbb{C}/K_\mathbb{C}P^-$ with the stabilizer of the coset $K_\mathbb{C}P^-$ equal to K, and hence the map

$$X_u \cong G_\mathbb{C}/K_\mathbb{C}P^-, \quad gK \mapsto gK_\mathbb{C}P^-,$$

is a G_u-equivariant holomorphic surjective diffeomorphism. Therefore, $G_\mathbb{C}$ acts holomorphically on X_u; and $G_\mathbb{C}/K_\mathbb{C}P^-$ is compact, and $K_\mathbb{C}P^-$ is a parabolic subgroup of $G_\mathbb{C}$, and $\mathfrak{p}^+$ is embedded in $G_\mathbb{C}/K_\mathbb{C}P^-$ under the map

$$\mathfrak{p}^+ \to G_\mathbb{C}/K_\mathbb{C}P^- = X_u, \quad H \mapsto \exp H\ K_\mathbb{C}P^-.$$

(2) The symmetric space X is embedded into X_u under the map

$$X = G/K \to X_u = G_\mathbb{C}/K_\mathbb{C}P^-, \quad gK \to gK_\mathbb{C}P^-,$$

and every isometry on X extends to a holomorphic map on X_u. The image of X in X_u is included in the subset $\mathfrak{p}^+$ defined in part (1), and the inclusion $X \hookrightarrow \mathfrak{p}^+$ is the Harish-Chandra embedding.

The embedding of X into its compact dual X_u is called the *Borel embedding* (see [Wo3, p. 278]). The basic example is the embedding of the disk $X = D$ into the complex projective space $X_u = \mathbb{C}P^1$, where $\mathfrak{p}^+$ is identified with the subset $\mathbb{C}$.

Now we resume the discussion about the boundary components of $\overline{X}^{BB}$. For any subset $I \subset \Psi = \{\mu_1, \dots, \mu_r\}$, let $\langle \Psi - I \rangle^{\perp}$ be the subset of roots in $\Phi(\mathfrak{g}^{\mathbb{C}}, \mathfrak{h}^{\mathbb{C}})$ that are perpendicular to $\Psi - I$. Let $\mathfrak{g}_I^{\mathbb{C}}$ be the derived Lie subalgebra of $\mathfrak{h}^{\mathbb{C}} + \sum_{\mu \in \langle \Psi - I \rangle^{\perp}} \mathfrak{g}_{\mu}^{\mathbb{C}}$. The complex Lie subalgebra $\mathfrak{g}_I^{\mathbb{C}}$ has two dual real forms:

$$\mathfrak{g}_I = \mathfrak{g}_I^{\mathbb{C}} \cap \mathfrak{g}, \quad \mathfrak{g}_{I,u} = \mathfrak{g}_I^{\mathbb{C}} \cap \mathfrak{g}_u.$$

Let $G_I^{\mathbb{C}}, G_I, G_{I,u}$ be the corresponding subgroups in $G_{\mathbb{C}}$. Let $x_0 = K$ be the basepoint in X_u, and denote the orbits through x_0 of G_I and $G_{I,u}$ by

$$X_{I,u} = G_{I,u} x_0 = G_I^{\mathbb{C}} x_0 \subset X_u,$$
$$X_I = G_I x_0 \subset X.$$

They are Hermitian symmetric spaces of rank $|I|$. See [Wo3, p. 287–289] for proof of the following result.

Proposition I.5.25

(1) The inclusions $X_{I,u} \hookrightarrow X_u$, $X_I \hookrightarrow X$ are totally geodesic.
(2) The submanifold X_I is contained in $X_{I,u}$, and the inclusion $X_I \hookrightarrow X_{I,u}$ is the Borel embedding of the Hermitian symmetric space X_I of noncompact type.
(3) Let $\mathfrak{p}_I^+ = \mathfrak{p}^+ \cap \mathfrak{g}_I^{\mathbb{C}}$. Then under the Harish-Chandra embedding $\pi^+ : X \to \mathfrak{p}^+$, the image $\pi^+(X_I)$ is contained in $\mathfrak{p}_I^+$, and the embedding $X_I \subset \mathfrak{p}_I^+$ is the Harish-Chandra embedding of X_I.
(4) The boundary components of $\overline{X}^{BB}$ in $\mathfrak{p}^+$ are the subsets of the form

$$kC_{\Psi - I} X_I,$$

where $k \in K$, and $C_{\Psi-I}$ is the (partial) Cayley transform associated with the subset $\Psi - I$, and $\mathfrak{p}^+$ is identified with a subset of X_u.

I.5.26 To relate these analytic (arc connected) boundary components X_I to the boundary components in the Satake compactifications of X, we need to express X_I in terms of the boundary symmetric spaces X_P of parabolic subgroups P.

For the rest of this section, we assume that X is irreducible, and hence the root system $\Phi(\mathfrak{g}, \mathfrak{a})$ is irreducible. Since the Weyl group of G consists of the signed permutations of the roots in $\Psi = \{\mu_1, \dots, \mu_r\}$, we can assume that $I = \{\mu_\ell, \mu_{\ell+1}, \dots, \mu_r\}$ for some ℓ. Let P_0 be the minimal parabolic subgroup of G containing $\exp \mathfrak{a}$ such that $\Delta = \{\alpha_1, \dots, \alpha_r\}$, defined in Proposition I.5.18, is the set of simple roots in $\Phi(P_0, A_{P_0})$. Let $I' = \{\alpha_\ell, \dots, \alpha_r\}$ be the subset of Δ corresponding to $I \subset \Psi$. Let $P_{0,I'}$ be the standard parabolic subgroup associated with I.

Proposition I.5.27 *The analytic boundary component X_I can be identified with the boundary symmetric space of the parabolic subgroup $P_{0,I'}$,*

$$X_I = X_{P_{0,I'}}. \tag{I.5.10}$$

Proof. Let $e^{\mathfrak{a}} x_0$ be the flat in X contained in the polydisk $G[\Psi] \cdot x_0 \subset X$. Then the flat $\exp \mathfrak{a}^{I'} \cdot K \cap P_{0,I'}$ of the boundary symmetric space $X_{P_{0,I'}}$ can be identified with a subflat $\exp \mathfrak{a}^{I'} \cdot x_0$ of $e^{\mathfrak{a}} \cdot x_0$ and is included in the subpolydisk $G[I] \cdot x_0$,

$$\exp \mathfrak{a}^{I'} \cdot x_0 \subset G[I] \cdot x_0.$$

Since I is a maximal collection of strongly orthogonal roots for $G[I]$, the polydisk theorem (I.5.15) for X_I and the Cartan decomposition of $X_{P_{0,I'}}$ implies the equation (I.5.10).

Therefore, every boundary component of $\overline{X}^{BB}$ is of the form $kX_{P_{0,I'}}$, where I' is a connected subset of Δ containing the last longest (or shortest) root α_r, i.e., the distinguished root. Note that the boundary components X_I that meet the closure of $\exp \mathfrak{a}^+ x_0$ in $\overline{X}^{BB}$ satisfy the property that $I = \{\mu_\ell, \mu_{\ell+1}, \dots, \mu_r\}$ for some ℓ.

Proposition I.5.28 *The normalizer $\mathcal{N}(X_{P_{0,I'}})$ of the boundary component $X_I = X_{P_{0,I'}}$ in $\overline{X}^{BB}$ is equal to the maximal parabolic subgroup $P_{0,\Delta-\{\alpha_{\ell-1}\}}$.*

See [BB1, 1.3, 1.5] [Wo3, p. 295] for more details and proof of this proposition.

Corollary I.5.29 *As a G-topological space, $\overline{X}^{BB}$ is isomorphic to the minimal Satake compactification $\overline{X}_\tau^S$ when the highest weight μ_τ is connected only to the distinguished root α_r.*

Proof. Let μ_τ be the highest weight of a representation τ such that it is connected only to α_r. Then the Satake compactification $\overline{X}_\tau^S$ is minimal. It suffices to check that the conditions in Proposition I.4.33 are satisfied by $\overline{X}^{BB}$. By Proposition I.5.25 and equation (I.5.10), every boundary component of $\overline{X}^{BB}$ is of the form X_P for a μ_τ-connected parabolic subgroup P; and for every such P, X_P is contained in $\overline{X}^{BB}$ as a boundary component. By Proposition I.5.28, the normalizer of $X_P \subset \overline{X}^{BB}$ for a μ-connected parabolic subgroup P is the μ-saturation of P. By Corollary I.5.16, the conditions on the closure of the positive chamber $\exp \mathfrak{a}^+ x_0$ are also satisfied. This implies that $\overline{X}^{BB}$ is isomorphic to the minimal Satake compactification $\overline{X}_\tau^S$.

Remark I.5.30 The above corollary was first proved by Moore in [Mo2] using the Furstenberg compactifications, which will be described in the next section. In fact, using the boundary values of bounded holomorphic functions on the bounded symmetric domain $\pi^+(X) \subset \mathfrak{p}^+$, Furstenberg proved in [Fu1, p. 370] that the maximal

Furstenberg compactification $\overline{X}^F_{\max}$ is mapped surjectively and continuously onto $\overline{X}^{BB}$. Moore proved in [Mo2, p. 365–369] that this map factors through a minimal Furstenberg compactification of X, and $\overline{X}^{BB}$ is isomorphic to this minimal Furstenberg compactification of X. In [Mo1, Theorem 8], he proved that the Furstenberg compactifications satisfy the properties of the Satake compactifications in Proposition I.4.33 and hence are isomorphic to the Satake compactifications. Combining these two results, he concluded that $\overline{X}^{BB}$ is isomorphic to the minimal Satake compactification of X in the above proposition. In a certain sense, the proof here is more direct since it avoids the Furstenberg compactifications.

I.5.31 Summary and comments. From many different points of view, Hermitian symmetric spaces are the most interesting symmetric spaces, and the Baily–Borel compactification of a bounded symmetric domain is one of the most interesting Satake compactifications. We have summarized the boundary components and their normalizers of bounded symmetric domains, and the identification of the Baily–Borel compactification with a minimal Satake compactification.

For detailed results on the structures of bounded symmetric domains, see [Wo3] [WK] [KW] [Sat8]. For applications to representation theory, see [Wo1] [Wo4] and the references there.

I.6 Furstenberg compactifications

In this section, we recall the Furstenberg compactifications $\overline{X}^F$ and their relations to the Satake compactifications $\overline{X}^S$. The basic references are [Fu1] [Fu2] [Ko1–3] and [Mo1]. A natural generalization of the problem here motivates the Martin compactification in the next section.

This section is organized as follows. First we recall the Poisson integral formula for harmonic functions on the unit disk D and use the representing measure on the unit circle to define the Furstenberg compactification of D (Lemma I.6.3). Then we recall the notion of boundary spaces of Lie groups (Definitions I.6.6 and I.6.8) and define the Furstenberg compactifications of symmetric spaces by replacing the unit circle with the so-called *faithful Furstenberg boundaries* (Definition I.6.17 and Definition I.6.19). Finally we identify the Furstenberg compactifications $\overline{X}^F$ with the Satake compactifications $\overline{X}^S$ (Proposition I.6.21).

I.6.1 We first consider the example of $X = D$, the Poincaré disk in $\mathbb{C}$. Let Δ be the Laplace operator. The *Dirichlet problem*

$$\begin{cases} \Delta u = 0 & \text{in } D, \\ u = f & \text{on } \partial D, \ f \in C^0(S^1), \end{cases} \tag{I.6.1}$$

is solved by

$$u(z) = \int_{S^1} \frac{1 - |z|^2}{|z - \xi|^2} f(\xi) d\xi, \tag{I.6.2}$$

where the Haar measure $d\xi$ on S^1 is normalized so that its total measure is equal to 1. Each point $z \in D$ determines a measure

$$\mu_z(\xi) = \frac{1 - |z|^2}{|z - \xi|^2} d\xi$$

on S^1. By taking $f = 1$ in equation (I.6.2), we conclude that μ_z is a probability measure on S^1. Let $\mathcal{M}_1(S^1)$ be the space of probability measures on S^1. Then we get a map

$$i : D \to \mathcal{M}_1(S^1), \quad z \mapsto \mu_z. \tag{I.6.3}$$

The space $\mathcal{M}_1(S^1)$ is given the weak-$\star$ topology: a sequence of measures μ_j in $\mathcal{M}_1(S^1)$ converges to a limit μ_∞ if and only if for any $f \in C^0(S^1)$,

$$\int_{S^1} f d\mu_j \to \int_{S^1} f d\mu_\infty.$$

Since S^1 is compact, the space $\mathcal{M}_1(S^1)$ is compact.

Lemma I.6.2 *The map $i : D \to \mathcal{M}_1(S^1)$ is an embedding.*

Proof. First we show that i is injective. For any two points $z_1, z_2 \in D$, if $\mu_{z_1} = \mu_{z_2}$, then for any harmonic function u on D with continuous boundary values on S^1,

$$u(z_1) = u(z_2). \tag{I.6.4}$$

Let x, y be the real and imaginary parts of z. Since x, y are clearly harmonic functions on D and have continuous boundary values, equation (I.6.4) implies that $z_1 = z_2$. To show that i is an embedding, we need to show that μ_{z_j} converges to μ_{z_∞} if and only if $z_j \to z_\infty$. It is clear from the explicit form of μ_z and the fact that S^1 is compact that $z_j \to z_\infty$ implies that μ_{z_j} converges to μ_{z_∞}. By taking $f = x, y$ again, we conclude that $z_j \to z_\infty$ if μ_{z_j} converges to μ_{z_∞}.

The closure of $i(D)$ in $\mathcal{M}_1(S^1)$ is called the *Furstenberg compactification* of D, denoted by $\overline{D}^F$.

Lemma I.6.3 *The compactification $\overline{D}^F$ is isomorphic to the closed unit disk $D \cup S^1$.*

Proof. If a sequence z_j in D converges to a boundary point $\xi \in S^1$, then μ_{z_j} converges to the delta measure δ_ξ supported at ξ. This implies that there is a well-defined continuous map from $D \cup S^1$ to $\overline{D}^F$. Clearly this map is bijective, and is hence a homeomorphism.

I.6.4 To generalize the construction to symmetric spaces X, we need to find an analogue of the boundary S^1. In the case of $X = D$, the boundary S^1 is obtained as the natural boundary in the embedding of D in $\mathbb{C}$. But such an embedding is not obvious in general. For this purpose, we need to characterize S^1 intrinsically in terms of G.

In fact, $D = SU(1,1)/U(1)$, where $SU(1,1)$ acts on D by fractional linear transformations. The action extends continuously to the boundary S^1, and $SU(1,1)$ acts transitively on S^1. The stabilizer of any point $\xi \in S^1$ is a parabolic subgroup of $SU(1,1)$.

This can be seen easily in the model of the upper half-plane

$$\mathbf{H} = \{z \in \mathbb{C} \mid \mathrm{Im}(z) > 0\} = \mathrm{SL}(2,\mathbb{R})/\mathrm{SO}(2).$$

The boundary S^1 of D corresponds to $\mathbb{R} \cup \{i\infty\}$. The stabilizer of $i\infty$ in $\mathrm{SL}(2,\mathbb{R})$ is the parabolic subgroup P consisting of upper triangular matrices, and hence the stabilizers of other points are conjugates of P.

I.6.5 Let G be a connected semisimple Lie group, and $X = G/K$ the associated symmetric space. Now we recall the notion of Furstenberg boundaries of G and X, and relations to harmonic functions on X.

Definition I.6.6 A compact homogeneous space Y of G is called a *boundary* of G, or a *G-boundary*, if for every probability measure μ on Y, there exists a sequence $g_j \in G$ such that $g_j \cdot \mu$ converges to the delta measure δ_x at some point $x \in Y$. A G-boundary is also called a boundary of X.

Remark I.6.7 There is a more general notion of boundaries for any locally compact, σ-compact topological group H. Specifically, a compact H-space Y is called an H-boundary if (1) for every probability measure μ on Y, there exists a sequence $g_j \in H$ such that $g_j \cdot \mu$ converges to the delta measure δ_x at some point $x \in Y$, and (2) Y is a minimal H-space, i.e., Y does not contain any proper H-invariant closed subset. (Clearly, a homogeneous space of H is minimal.)

It is known that if H is amenable, for example, solvable, then the only boundary of H consists of one point. The existence of nontrivial H-boundaries and their sizes measure the extent to which the group H fails to be amenable.

Definition I.6.8 A homogeneous space M is called a universal (or maximal) G-boundary if it is a G-boundary and every other boundary of G is a G-equivariant image of M.

It is known that an equivariant image of a G-boundary is also a G-boundary, and the universal G-boundaries are isomorphic and hence unique up to isomorphism, to be denoted by $\mathcal{F} = \mathcal{F}(G)$ and called the *maximal Furstenberg boundary* of G.

Proposition I.6.9 *Let P_0 be a minimal parabolic subgroup of G. Then G/P_0 is the maximal Furstenberg boundary $\mathcal{F}(G)$. Other boundaries of G are of the form $G/P_{0,I}$, where $P_{0,I}$ are standard parabolic subgroups containing P_0.*

See [Mo2], [Ko3, §2] and [GJT, Lemma 4.48] for complete proofs. We illustrate the idea through the example of $G = \mathrm{SL}(2, \mathbb{R})$.

Proposition I.6.10 *When $G = \mathrm{SL}(2, \mathbb{R})$, and $P_0 = \left\{ \begin{pmatrix} a & b \\ 0 & a^{-1} \end{pmatrix} \mid a \in \mathbb{R}^\times, b \in \mathbb{R} \right\}$, G/P_0 is a boundary of G.*

Proof. Identify G/P_0 with $\mathbb{R} \cup \{\infty\} = \mathbf{H}(\infty)$. Let $\mu \in \mathcal{M}_1(\mathbb{R} \cup \{\infty\})$. Then there exists $k \in \mathrm{SO}(2)$ such that the measure $k\mu$ satisfies

$$\lim_{t \to +\infty} k \cdot \mu(\{x \in \mathbb{R} \cup \{\infty\} \mid x > t \text{ or } x < -t) = 0, \tag{I.6.5}$$

in particular, $k \cdot \mu(\{\infty\}) = 0$. Let $g_t = \begin{pmatrix} t^{-1} & 0 \\ 0 & t \end{pmatrix} \in \mathrm{SL}(2, \mathbb{R})$. Then for any subset $E \subset \mathbb{R} \cup \{\infty\}$,

$$g_t k \cdot \mu(E) = k\mu(t^2 E).$$

This, together with equation (I.6.5), implies that as $t \to +\infty$,

$$g_t k \cdot \mu \to \delta_0.$$

Remark I.6.11

(1) If μ is a smooth measure, then the condition in equation (I.6.5) is satisfied without the action of a suitable element $k \in \mathrm{SO}(2)$. By convolving with a smooth probability measure ν on G with compact support, we get a smooth probability measure $\mu * \nu$. For any sequence g_j such that $g_j \mu * \nu \to \delta_0$, we can get a sequence g'_j such that $g'_j \mu \to \delta_0$. This gives another proof of Proposition I.6.10. See [Ko3, p. 385] for details.

(2) The decomposition $\mathrm{SL}(2, \mathbb{R})/P_0 = \mathbb{R} \cup \{\infty\}$ is given by the Bruhat decomposition of $\mathrm{SL}(2, \mathbb{R})$ with respect to P_0, and the scaling $E \to t^2 E$ corresponds to the adjoint action of A_{P_0} on $N_{\overline{P_0}}$, which is the unipotent radical of the parabolic subgroup opposite to P_0. The proof of Proposition I.6.9 follows a similar line as above.

I.6.12 Since $G = KP_0$, K acts transitively on $\mathcal{F}(G)$. Hence $\mathcal{F}(G)$ admits a unique K-invariant probability measure μ_0.

The notion of Furstenberg boundaries is related to harmonic functions by a generalization of the Poisson integral formula in equation (I.6.2). Since $X = G/K$, a function on X can be identified with a right K-invariant function on G.

Proposition I.6.13 *For every bounded harmonic function u on X, there exists an element $f \in L^\infty(\mathcal{F})$ such that*

$$u(g) = \int_{\mathcal{F}} f(g\xi) d\mu_0(\xi) = \int_{\mathcal{F}} f(\xi) P(x, \xi) d\mu_0(\xi),$$

where

$$P(x, \xi) = \frac{d\mu_0(g^{-1}\xi)}{d\mu_0(\xi)} = \frac{dg \cdot \mu_0(\xi)}{d\mu_0(\xi)}$$

is the Poisson kernel associated with the boundary point ξ.

See [Fu1] [Fu2] for proofs and applications of this proposition.

I.6.14 The boundedness assumption in Proposition I.6.13 is crucial, and the conclusion is false otherwise. In fact, the boundedness of u implies that u is *strongly harmonic*, i.e., a joint eigenfunction of all invariant differential operators on X.

A closely related notion is that of μ-*harmonic functions* [Fu4, §§12–13], where μ is a probability measure on G. There are similar integral formulas for bounded μ-harmonic functions using the so-called μ-boundaries [Fu4, Theorem 12.2]. (See [Fu1-7] for many related results.)

The theory of μ-boundaries is very important to superrigidity of lattices. In fact, it can be used to obtain some equivariant maps from the boundaries to certain projective spaces, which are used in the proof of the superrigidity of lattices [Mag, Chap. VI, Theorem 4.3; Chap VII] (see also [Fu2] [Fu3]).

Remark I.6.15 By definition, the Furstenberg boundaries are also called boundaries of X. But none of them is equal to the whole boundary of a compactification of X if the rank of X is strictly greater than 1. For example, when $G = \mathrm{SL}(2,\mathbb{R}) \times \mathrm{SL}(2,\mathbb{R})$ and $X \cong D \times D$, the maximal Furstenberg boundary is $S^1 \times S^1$, the distinguished part, i.e., the corner of the boundary $(\overline{D} \times S^1) \cup (S^1 \times \overline{D})$ of the compactification $\overline{D} \times \overline{D}$, where $\overline{D} = D \cup S^1$.

For any proper subset $I \subset \Delta = \Delta(P_0, A_{P_0})$, denote the boundary $G/P_{0,I}$ by $\mathcal{F}_I$. Then K also acts transitively on $\mathcal{F}_I$.

Let $\mathcal{M}_1(\mathcal{F}_I)$ be the space of probability measures on $\mathcal{F}_I$. Denote the unique K-invariant probability measure on $\mathcal{F}_I$ by μ_I. Then the map

$$i_I : X = G/K \to \mathcal{M}_1(\mathcal{F}_I), \quad gK \mapsto g\mu_I,$$

is well-defined and G-equivariant.

Proposition I.6.16 *The map i_I is an embedding if and only if $\Delta - I$ meets every connected component of Δ, or the associated Dynkin diagram, which is equivalent to $P_{0,I}$ containing no simple factor of G.*

See [Mo2], [Ko3], and [GJT, Proposition 4.49] for proofs of this proposition.

Definition I.6.17 A Furstenberg boundary $\mathcal{F}_I$ such that the map i_I is an embedding is called a *faithful boundary*.

Note that the conditions on $\Delta - I$ in Proposition I.6.16 are equivalent to the fact that there exists a faithful projective representation τ of G such that the set of simple roots connected to the highest weight μ_τ is equal to $\Delta - I$. In particular, the maximal Furstenberg boundary G/P_0 is always faithful.

I.6.18 If G is simple or equivalently Δ is connected, then X is irreducible, and every Furstenberg boundary is faithful.

If X is reducible, the situation is more complicated. For example, when $G = \mathrm{SL}(2,\mathbb{R}) \times \mathrm{SL}(2,\mathbb{R})$, we can take $P_0 = P \times P$, and the maximal Furstenberg boundary is equal to $S^1 \times S^1$, the distinguished (or corner) part of the boundary of $\overline{X} \cong \overline{D} \times \overline{D}$, where P is the parabolic subgroup of $\mathrm{SL}(2,\mathbb{R})$ consisting of upper triangular matrices. The minimal parabolic subgroup P_0 is contained in two proper maximal parabolic subgroups $P \times \mathrm{SL}(2,\mathbb{R})$ and $\mathrm{SL}(2,\mathbb{R}) \times P$, and their corresponding Furstenberg boundaries $S^1 \times q$, $p \times S^1$, where $p, q \in S^1$, are not faithful.

In general, if X is reducible, then $X = X_1 \times \cdots \times X_n$, with each X_i irreducible. Then a boundary $\mathcal{F}$ is also a product $\mathcal{F} = \mathcal{F}_1 \times \cdots \times \mathcal{F}_n$, where $\mathcal{F}_i$ is a boundary of X_i; and $\mathcal{F}$ is faithful if and only if for every $i = 1, \dots, n$, $\mathcal{F}_i \neq X_i$.

As in the case of $\mathcal{M}_1(S^1)$, $\mathcal{M}_1(\mathcal{F}_I)$ is compact with the weak-$\star$ topology.

Definition I.6.19 When $\mathcal{F}_I$ is faithful, the closure of $i_I(X)$ in $\mathcal{M}_1(S^1)$ is a G-compactification of X, called the *Furstenberg compactification* associated with the boundary $\mathcal{F}_I$ and denoted by $\overline{X}_I^F$.

The space $\mathcal{M}_1(\mathcal{F}_I)$ is a compact convex subset of the dual of the Banach space of continuous functions on $\mathcal{F}_I$. An interesting feature of this embedding $i_I : X \to M_1(\mathcal{F}_I)$ is the following property.

Proposition I.6.20 *The restriction to $X = i_I(X)$ of a continuous affine function of $\mathcal{M}_1(\mathcal{F}_I)$ is a bounded harmonic function on X.*

The converse is also true under some assumption (see [Fu2, §4] for details). In this sense, the embeddings $i_I : X \to \mathcal{M}_1(\mathcal{F}_I)$ are harmonic, and hence geometric and special.

The Furstenberg compactifications $\overline{X}_I^F$ have another interesting feature that the Furstenberg boundary $\mathcal{F}_I$ appears explicitly in the boundary of $\overline{X}_I^F$ as the set of delta measures. In fact, $\mathcal{F}_I$ is the only closed G-orbit in $\overline{X}_I^F$. This distinguished boundary is important for problems about boundary behavior of harmonic functions on X and the Poisson transform. These problems have been extensively studied. See [Ko1] [Ko2] [Sch] for details. It is also important in the Mostow strong rigidity of locally symmetric spaces [Mos].

Proposition I.6.21

(1) The Furstenberg compactifications $\overline{X}_I^F$ are partially ordered: if $I \subseteq J$, then $\overline{X}_I^F$ dominates $\overline{X}_J^F$, i.e., the identity map on X extends to a continuous surjective map. When $I = \emptyset$, $\overline{X}_\emptyset^F$ is the maximal Furstenberg compactification, which is also denoted by $\overline{X}_{\max}^F$.

(2) *The Furstenberg compactification* $\overline{X}_I^F$ *is isomorphic to the Satake compactification* $\overline{X}_\tau^S$ *such that the subset of simple roots in* Δ *that are connected to the highest weight* μ_τ *is equal to* $\Delta - I$.

When $I \subset J$, $P_{0,I} \subset P_{0,J}$, and hence there is a surjective G-equivariant map $G/P_{0,I} \to G/P_{0,J}$, under which the K-invariant measure μ_I is pushed down to μ_J. The second part is proved by checking that $\overline{X}_I^F$ satisfies the conditions in Proposition I.4.33 that characterize the Satake compactifications. See [Mo2] for details.

Due to the identification in Proposition I.6.21, the Furstenberg compactifications are often called *Satake–Furstenberg compactifications*.

I.6.22 Summary and comments. In this section, we recall the boundary of the Poincaré disk D and the Poisson integral formula to motivate general boundaries of a semisimple Lie group G, or the associated symmetric space X. When the symmetric space X is not irreducible, there is the important notion of faithful Furstenberg boundaries. This can be illustrated by the example of $D \times D$. Closely related to the original application to bounded harmonic functions are the notions of μ-harmonic functions and μ-boundaries.

For other notions of boundary and applications to geometric topology, see [Bes] [BG].

I.7 Martin compactifications

In this section, we discuss the Martin compactifications of the symmetric space X. The Martin compactifications are defined for any complete Riemannian manifold, and the basic problem is to understand the structure of the cone of positive eigenfunctions on the Riemannian manifold. For a detailed, self-contained exposition about the Martin compactifications of domains in $\mathbb{R}^n$ and general Riemannian manifolds, see [Ta1], and for the proofs of the results for symmetric spaces stated below, see [GJT].

This section is organized as follows. First, we recall conditions on the eigenvalue λ under which the cone $C_\lambda(X)$ is nonempty (I.7.1). Generalizing the Poisson integral formula for the unit disk, we give a complete description of such cones for the Poincaré disk (I.7.2). Then we recall the general construction of the Martin compactification of a complete Riemannian manifold and show that an analogue of the Poisson integral formula over the Martin boundaries holds for eigenfunctions (Proposition I.7.6). To get a unique integral representation, we introduce the *minimal Martin boundary* (Proposition I.7.9). In the case of bounded harmonic functions, we introduce the *Poisson boundary* (Definition I.7.11). Then bounded harmonic functions can be represented in terms of bounded measurable functions on the Poisson boundary (Proposition I.7.12). Finally we recall results in [GJT] on identification of the Martin compactifications of a symmetric space X in terms of the maximal Satake compactification $\overline{X}_{\max}^S$ and the geodesic compactification $X \cup X(\infty)$ (Proposition I.7.15). We also describe explicitly the minimal Martin boundary (I.7.15–I.7.16) and the Poisson boundary (I.7.17).

I.7.1 As mentioned earlier, the theory of Martin compactifications holds for any complete Riemannian manifold. For simplicity, we also use X to denote such a Riemannian manifold in this section.

Let Δ be the Laplace–Beltrami operator of X. Normalize Δ such that $\Delta \geq 0$. For each $\lambda \in \mathbb{R}$, consider the space

$$C_\lambda(X) = \{u \in C^\infty(X) \mid \Delta u = \lambda u, u > 0\}.$$

For simplicity, functions in $C_\lambda(X)$ are called eigenfunctions of eigenvalue λ, and $C_\lambda(X)$ the eigenspace of eigenvalue λ. When $\lambda = 0$, they are harmonic functions.

Let $\lambda_0 = \lambda_0(X)$ be the bottom of the spectrum of Δ:

$$\lambda_0(X) = \inf_{\varphi \in C_0^\infty(X), \varphi \neq 0} \frac{\int_X |\nabla \varphi|^2}{\int_X |\varphi|^2}.$$

Then a result of Cheng and Yau [CY] says that $C_\lambda(X)$ is nonempty if and only if $\lambda \leq \lambda_0(X)$.

If nonempty, $C_\lambda(X)$ is clearly a convex cone. A basic problem in *potential theory* is to identify the set of *extremal elements* (or edges) of $C_\lambda(X)$ and express other functions as linear combinations (or superpositions) of them. This problem is solved by determining the Martin compactifications of X. In fact, for each $\lambda \leq \lambda_0(X)$, there is a Martin compactification $X \cup \partial_\lambda X$. It will turn out that in many cases, all the compactifications for $\lambda < \lambda_0$ are isomorphic to each other but not necessarily to $X \cup \partial_{\lambda_0} X$.

I.7.2 To motivate the Martin compactifications of X, we recall the Poisson integral formula for the unit disk D stated earlier.

Let $D = \{z \in \mathbb{C} \mid |z| < 1\}$. Then its natural compactification is given by $\overline{D} = D \cup S^1$. For each $\xi \in S^1$, define the *Poisson kernel*

$$K(z, \xi) = \frac{1 - |z|^2}{|\xi - z|^2}.$$

Then each $K(z, \xi)$ is harmonic in z, i.e., $\Delta K(\cdot, \xi) = 0$. For any positive harmonic function u on D, there exists a unique nonnegative measure μ on S^1 such that

$$u(z) = \int_{S^1} K(z, \xi) d\mu(\xi).$$

Since positive constants are harmonic functions, any bounded harmonic function can be written as the difference of two positive harmonic functions, and hence admits a similar integral representation by a signed measure.

With respect to the Poincaré metric $ds^2 = \frac{|dz|^2}{(1-|z|^2)^2}$ on D, $\lambda_0(D) = \frac{1}{4}$. For each $s \leq \frac{1}{2}$, $\lambda = s(1-s) \leq \frac{1}{4}$. Define

$$K_\lambda(z, \xi) = K(z, \xi)^s.$$

Let Δ be the Laplace–Beltrami operator of D with respect to the Poincaré metric. Then $K_\lambda(z, \xi)$, $\xi \in S^1$, are all the *extremal elements* of $C_\lambda(D)$, and any function u in $C_\lambda(D)$ can be represented uniquely in the form

$$u(z) = \int_{S^1} K_\lambda(z, \xi) d\mu(\xi), \tag{I.7.1}$$

where μ is a nonnegative measure on S^1. It should be pointed out that if $\lambda \neq 0$, there are in general no bounded nonzero functions in $C_\lambda(D)$.

I.7.3 The example of the Poincaré disk suggests the following problem. For each $\lambda \leq \lambda_0(X)$, construct a compactification $X \cup \partial_\lambda X$ such that

1. for each $\xi \in \partial_\lambda X$, there is a function $K_\lambda(\cdot, \xi) \in C_\lambda(X)$;
2. for each $u \in C_\lambda(X)$, there is a nonnegative measure $d\mu$ on $\partial_\lambda X$ such that an integral formula similar to that in equation (I.7.1) holds.

It will be seen below that in general the *representing measure* μ on the whole boundary $\partial_\lambda X$ of the compactification is not unique, i.e., not all functions $K_\lambda(\cdot, \xi)$, $\xi \in \partial_\lambda X$, are extremal, i.e., some $K_\lambda(\cdot, \xi)$ can be expressed as superpositions of other $K_\lambda(\cdot, \xi)$ and are hence redundant.

If one compactification $\overline{X}^1$ of X satisfies the above conditions, then any other compactification $\overline{X}^2$ that dominates $\overline{X}^1$ also satisfies the conditions. Clearly it is desirable to find as small as possible a compactification $X \cup \partial_\lambda X$ satisfying the above conditions.

If $\dim C_\lambda(X) = 1$, we can use the one-point compactification $X \cup \{\infty\}$, with a generator corresponding to $\{\infty\}$. Otherwise, we need a bigger compactification. One such compactification is given by the Martin compactification.

I.7.4 For each $\lambda \leq \lambda_0(X)$, let $G^\lambda(x, y)$ be a positive *Green function* of $\Delta - \lambda$ that vanishes at infinity, which is unique if it exists. It always exists if $\lambda < \lambda_0(X)$. For $\lambda = \lambda_0(X)$, if $G^\lambda(x, y)$ does not exist, then $\dim C_\lambda(X) = 1$ and define the Martin compactification $X \cup \partial_\lambda X$ to be the one-point compactification as mentioned above.

Assume that $G^\lambda(x, y)$ exists. Let x_0 be a basepoint in X. For any $y \neq x_0$, define the normalized Green function

$$K^\lambda(x, y) = G^\lambda(x, y) / G^\lambda(x_0, y).$$

Then $K^\lambda(x, y)$ is smooth on $X - \{y\}$, and

$$\Delta K^\lambda(x, y) = \lambda K^\lambda(x, y), \quad K^\lambda(x_0, y) = 1.$$

An unbounded sequence y_j in X is called a *fundamental sequence* in the Martin compactification for λ if the sequence of functions $K^\lambda(x, y_j)$ converges uniformly for x in compact subsets to a function $K^\lambda(x)$. Clearly, $K^\lambda(x) \in C_\lambda(X)$. Two fundamental sequences y_j and y'_j are called equivalent if $K^\lambda(x, y_j)$ and $K^\lambda(x, y'_j)$ converge to the same limit function.

Denote the set of equivalence classes of fundamental sequences by $\partial_\lambda X$. For each $\xi \in \partial_\lambda X$, denote the common limit function by $K^\lambda(\cdot, \xi)$.

Then the Martin compactification of X for λ is the set $X \cup \partial_\lambda X$ with the following topology. An unbounded sequence y_j in X converges to a boundary point $\xi \in \partial_\lambda X$ if and only if the function $K^\lambda(x, y_j)$ converges uniformly over compact subsets to $K^\lambda(x, \xi)$; a sequence of boundary points $\xi_j \in \partial_\lambda X$ converges to $\xi_\infty \in \partial_1 X$ if and only if the function $K^\lambda(x, \xi_j)$ converges to $K^\lambda(x, \xi_\infty)$ uniformly over compact subsets.

Proposition I.7.5 *The topological space $X \cup \partial_\lambda X$ is a metrizable compactification of X, and any isometry of X extends to the compactification.*

Proof. The Harnack inequality for positive solutions of $\Delta u = \lambda u$ shows that every sequence in $X \cup \partial_\lambda X$ has a convergent subsequence, and hence $X \cup \partial_\lambda X$ is compact. The Harnack inequality also implies that the function

$$d(y_1, y_2) = \int_{B(x_0;1)} \frac{|K^\lambda(x, y_1) - K^\lambda(x, y_2)|}{1 + |K^\lambda(x, y_1) - K^\lambda(x, y_2)|}$$

defines a distance function on $X \cup \partial_\lambda X$ that induces the topology defined earlier.

Proposition I.7.6 *For any $u \in C_\lambda(X)$, there exists a nonnegative measure μ on the Martin boundary $\partial_\lambda X$ such that*

$$u(x) = \int_{\partial_\lambda X} K^\lambda(x, \xi) d\mu(\xi).$$

Proof. The basic idea of the proof is to use *Riesz's representation theorem* for positive superharmonic functions. Let $L = \Delta - \lambda$. Then functions u satisfying $Lu = 0$ are called L-harmonic, in particular, elements in $C_\lambda(X)$ are all L-harmonic. In terms of L-harmonic functions, we can define L-superharmonic, L-subharmonic functions.

Let $G^\lambda(x, y)$ be the Green function of L as above. Then Riesz's representation theorem states that every positive superharmonic function w is a sum of the form

$$w(x) = \int_X G^\lambda(x, y) d\mu(y) + u(x),$$

where u is a nonnegative harmonic function, and μ is a nonnegative measure. A positive superharmonic function v is called a *potential* if every harmonic function u bounded from above by v, i.e., $u \leq v$, must be nonpositive. Then Riesz's representation theorem implies that the potential v can be written as

$$v(x) = \int_X G^\lambda(x, y) d\mu(y),$$

for some nonnegative measure $d\mu$ on X. Hence, Riesz's representation theorem decomposes a positive superharmonic function into a sum of a potential and a harmonic function.

To prove the proposition, take an exhausting family Ω_n of relatively compact domains with smooth boundary of X, $X = \cup_n \Omega_n$. For any $u \in C_\lambda(X)$, let $H_n u$ be the solution to the exterior Dirichlet problem on $X \setminus \Omega_n$. Define a function u_n by $u_n = u$ on Ω_n, and $u_n = H_n u$ on $X \setminus \Omega_n$. Then u_n is superharmonic and does not bound any nonzero nonnegative harmonic function. Then Riesz's representation theorem gives

$$u_n(x) = \int_X G^\lambda(x, y) d\nu_n(y).$$

Extending the measure $d\nu_n$ trivially outside of X to define a measure on the compactification $X \cup \partial_\lambda X$, we obtain

$$u_n(x) = \int_{X \cup \partial_\lambda X} K^\lambda(x, y) d\mu_n(y), \tag{I.7.2}$$

where $d\mu_n = G^\lambda(x_0, y) d\nu_n$. Since $u(x_0) > 0$, assume $u(x_0) = 1$ for simplicity. Then the normalization $K(x_0, y) = 1$ implies that $d\mu_n$ is a probability measure on $X \cup \partial_\lambda X$. Since $X \cup \partial_\lambda X$ is compact, there is a subsequence $d\mu_{n'}$ that weakly converges to a probability measure $d\mu$. Since the support of $d\mu_n$ is contained in the complement of Ω_n, the measure $d\mu$ is supported on $\partial_\lambda X$. Since the function $K^\lambda(x, y)$ is continuous in the y variable on $X \cup \partial_\lambda X$ and $u_n \to u$, equation (I.7.2) implies that

$$u(x) = \int_{\partial_\lambda X} K^\lambda(x, \xi) d\mu(\xi).$$

Remarks I.7.7

1. If $X \cup \partial X$ is any compactification where the functions $K^\lambda(x, y)$ extend continuously in the y-variable, then the above proof works and gives a representation

   $$u(x) = \int_{\partial X} K^\lambda(x, \xi) d\mu(\xi).$$

   If $X \cup \tilde{\partial} X$ is a compactification that dominates $X \cup \partial X$, then $K^\lambda(x, y)$ also extends continuously onto $X \cup \tilde{\partial} X$. It is clearly desirable to get as small as possible a compactification satisfying this condition.
2. The smallest compactification satisfying the above condition is characterized by the second condition that the extended functions $K^\lambda(x, \xi)$ separate the points in the boundary $\partial_\lambda X$. The Harnack inequality shows that such a compactification exists and is exactly the Martin compactification $X \cup \partial_\lambda X$. This is the reason that in some works, for example in [GJT], the Martin compactification is defined to be a compactification of X satisfying the two conditions (a) the functions $K^\lambda(x, y)$ extend continuously to the boundary in the y-variable, (b) the extended functions separate the boundary points.

I.7.8 Due to the similarity to the Poisson integral formula for the Poincaré disk D recalled earlier, the functions $K^\lambda(x, \xi)$ are called the *Martin kernel functions*, and the integral representation in Proposition I.7.6 is called the *Martin integral representation*.

In general, the measure $d\mu$ in the above proposition is not unique. To get a unique measure, we need to restrict the support of the measure.

A function $u \in C_\lambda(X)$ is called *minimal* or *extremal* if every function $v \in C_\lambda(X)$ satisfying $v \leq u$ is a multiple of u, or equivalently, the ray $\{cu \mid c > 0\}$ is an edge of the cone $C_\lambda(X)$, or an extremal ray of the cone.

Define the *minimal Martin boundary*

$$\partial_{\lambda,\min} X = \{\xi \in \partial_\lambda X \mid K^\lambda(\cdot, \xi) \text{ is minimal}\}.$$

In general, $\partial_{\lambda,\min} X$ is a proper subset of $\partial_\lambda X$. For example, it will be seen later that $\partial_{\lambda,\min} X$ is a proper subset of $\partial_\lambda X$ when X is a symmetric space of rank greater than or equal to 2; on the other hand, if X is of rank 1, $\partial_{\lambda,\min} X = \partial_\lambda X$.

By restricting the support of the measures to the minimal Martin boundary, we have the following uniqueness result.

Proposition I.7.9 *For any $u \in C_\lambda(X)$, there exists a unique measure $d\nu$ on the minimal Martin boundary $\partial_{\lambda,\min} X$ such that*

$$u(x) = \int_{\partial_{\lambda,\min} X} K^\lambda(x, \xi) d\nu(\xi).$$

The unique measure $d\nu$ in this proposition is called the *representing measure* of u. One consequence of this uniqueness of the representing measure is the following domination result.

Proposition I.7.10 *For any two functions $u, v \in C_\lambda(X)$, let $d\nu_u$, $d\nu_v$ be their representing measures. If $v \leq u$, then $d\nu_u$ dominates $d\nu_v$, i.e., $d\nu_v$ is absolutely continuous with respect to $d\nu_u$ and hence $d\nu_v(\xi) = f(\xi) d\nu_u$ for some measurable function f on $\partial_{\lambda,\min} X$ with $|f| \leq 1$ a.e.*

When $\lambda = 0$, $C_\lambda(X)$ consists of harmonic functions, and the constant function $u = 1$ belongs to $C_\lambda(X)$. The support of the representing measure $d\nu_1$ of $u = 1$ is called the *Poisson boundary* of X and denoted by Π.

Definition I.7.11 The measure space $(\Pi, d\nu_1)$ is also called the *Poisson boundary*.

It should be pointed out that the Poisson boundary is not the boundary of a compactification; rather it is a distinguished subset of the Martin boundary $\partial_0 X$. The structure of $(\Pi, d\nu_1)$ as a measure space rather than a topological space is emphasized. Any measure space satisfying the conditions in the proposition below is also called a Poisson boundary.

Proposition I.7.12 *The space of bounded harmonic functions on X is isomorphic to the space $L^\infty(\Pi, d\nu_1)$ under the map*

$$f \in L^\infty(\Pi, d\nu_1) \to u(x) = \int_\Pi K^\lambda(x, \xi) f(\xi) d\nu_1(\xi).$$

Proof. Since any bounded harmonic function can be written as the difference of two positive bounded harmonic functions, the isomorphism follows from the previous proposition.

I.7.13 One application of the Poisson boundary (or rather the inclusion in the Martin compactification) is that almost surely the *Brownian paths* in X converge in the Martin compactification $X \cup \partial_0 X$ to points in the Poisson boundary Π (see [GJT, Remarks 8.31]. In this application, both the topology of the Martin compactification and the embedding of Π in $\partial_{\lambda_0} X$ are important.

I.7.14 By definition, the Martin compactification is determined by the behavior at infinity of the Green function and is a purely function-theoretical compactification. An important problem is to understand and identify it with other geometric and more explicit compactifications.

In [AS], Anderson and Schoen proved that if X is a simply connected Riemannian manifold with the curvature pinched by two negative numbers, then the Martin compactification $X \cup \partial_\lambda X$ when $\lambda = 0$ is isomorphic to the geodesic compactification $X \cup X(\infty)$. Ancona showed in [Anc] that the same result holds for all $\lambda < \lambda_0(X)$. (Note that under the above assumption on X, $\lambda_0(X) > 0$.)

These results motivated a lot of work on the Martin compactifications of simply connected manifolds with nonpositive curvature. Symmetric spaces of noncompact type of rank greater than or equal to 2 form an important class of simple connected Riemannian manifolds whose curvature is nonpositive but not pinched by negative constants.

After partial results by various people, the Martin compactifications of symmetric spaces of noncompact type were completely determined by Guivarch, Ji, and Taylor in [GJT].

Proposition I.7.15 *Let X be a symmetric space of noncompact type. Then $\lambda_0(X) > 0$. When $\lambda = \lambda_0(X)$, the Martin compactification $X \cup \partial_\lambda X$ is isomorphic to the maximal Satake compactification $\overline{X}^S_{\max}$.*

When $\lambda < \lambda_0(X)$, the Martin compactification $X \cup \partial_\lambda X$ is the least compactification dominating both the geodesic compactification $X \cup X(\infty)$ and the maximal Satake compactification $\overline{X}^S_{\max}$, i.e., it is equal to the closure of X under the diagonal embedding $X \to (X \cup X(\infty)) \times \overline{X}^S_{\max}$.

Furthermore, the Karpelevič compactification $\overline{X}^K$ dominates the Martin compactification $X \cup \partial_\lambda X$ and is isomorphic to the latter if and only if the rank of X is less than or equal to 2.

I.7.16 For $\lambda < \lambda_0(X)$, each parabolic subgroup P of G determines a subset $\overline{\mathfrak{a}_P^+(\infty)} \times X_P$ in the Martin boundary, which is the inverse image of $X_P \subset \overline{X}^S_{\max}$

under the dominating map $X \cup \partial_\lambda X \to \overline{X}^S_{\max}$, and

$$\partial_\lambda X = \coprod_P \overline{\mathfrak{a}_P^+(\infty)} \times X_P, \tag{I.7.3}$$

where P runs over all proper parabolic subgroups. For each point $\xi = (H, z) \in \overline{\mathfrak{a}_P^+(\infty)} \times X_P$, the Martin kernel function $K^\lambda(x, \xi)$ is independent of the N_P-factor in the horospherical decomposition $X = N_P \times A_P \times X_P$ and is product of an exponential function on A_P and a spherical function on X_P. When P is minimal, X_P consists of one point. The minimal Martin boundary $\partial_{\lambda,\min} X$ is given by

$$\partial_{\lambda,\min} X = \coprod_{\text{minimal } P} \overline{\mathfrak{a}_P^+(\infty)}, \tag{I.7.4}$$

a disjoint union over all minimal parabolic subgroups P.

It is important to note that the decomposition of the sphere at infinity $X(\infty)$ in terms of the closed simplexes $\overline{\mathfrak{a}_P^+(\infty)}$,

$$X(\infty) = \cup_{\text{minimal } P} \overline{\mathfrak{a}_P^+(\infty)},$$

is not disjoint. In fact, for any two minimal parabolic subgroups P_1, P_2 contained in a proper parabolic subgroup Q, $\overline{\mathfrak{a}_{P_1}^+(\infty)} \cap \overline{\mathfrak{a}_{P_2}^+(\infty)} \supseteq \mathfrak{a}_Q^+(\infty)$. Because of this, the geodesic sphere is too small to parametrize the minimal functions of $C_\lambda(X)$. In [Ka], Karpelevič determined the set of minimal functions in $C_\lambda(X)$ and used a part of the boundary of the Karpelevič compactification $\overline{X}^K$ to parametrize them, which was his motivation to define the compactification $\overline{X}^K$. On the other hand, the above proposition also shows that the Karpelevič compactification is larger than the Martin compactification if the rank is greater than 3.

I.7.17 For $\lambda = \lambda_0$, the Martin boundary is given by $\partial_\lambda X = \coprod_P X_P$, where P runs over all proper parabolic subgroups, and the minimal Martin boundary by

$$\partial_{\lambda_0,\min} X = \coprod_{\text{minimal } P} X_P \cong G/P_0,$$

where P runs over all minimal parabolic subgroups. In fact, for each minimal P, X_P consists of one point, and hence the union can be identified with G/P_0, where P_0 is any fixed minimal parabolic subgroup.

I.7.18 When $\lambda = 0$, the Poisson boundary Π can be described as follows. For each minimal parabolic subgroup P, recall that $\Phi(P, A_P)$ is the set of roots of the adjoint action of $\mathfrak{a}_P$ on $\mathfrak{n}_P$. Let ρ_P be the half sum of roots in $\Phi(P, A_P)$ with multiplicity equal to the dimension of the root spaces, and scale ρ_P to a unique vector $H_P \in \mathfrak{a}_P^+(\infty)$. Then the *Poisson boundary* is given by

$$\Pi = \coprod_{\text{minimal } P} H_P \subset \partial_{0,\min} X = \coprod_{\text{minimal } P} \overline{\mathfrak{a}_P^+(\infty)}. \tag{I.7.5}$$

It can be shown as above that for any fixed minimal parabolic subgroup P_0,

$$\Pi \cong G/P_0.$$

For the constant function 1, its representing measure is the unique K-invariant measure μ_0 on G/P_0. Hence the *Poisson boundary* of X is $(G/P_0, \mu_0)$.

I.7.19 As mentioned earlier, the Brownian paths almost surely converge to points in the Poisson boundary Π. Let $\mathfrak{a}$ be a Cartan subalgebra in $\mathfrak{p}$, $\mathfrak{a}^+$ a positive chamber, and ρ the half sum of positive roots. Then the Cartan decomposition of G gives the polar decomposition of X:

$$X = K \exp \overline{\mathfrak{a}^+} x_0, \quad x = ke^H x_0,$$

where H is uniquely determined by x. In polar coordinates, the H-component of the Brownian paths almost surely goes to infinity in the direction of ρ. This is basically the center of the open simplex $\mathfrak{a}^+(\infty)$. For details of the above statements, see [GJT].

I.7.20 Summary and comments. Here we defined the Martin compactifications in terms of the asymptotic behaviors of the Green function. Then we introduced related notions such as the minimal Martin boundary and the Poisson boundary. We concluded with the identification of the Martin compactification of symmetric spaces.

An important property of the Martin compactifications is the Martin integral formula for positive eigenfunctions. Since bounded harmonic functions become positive when a sufficiently large positive constant is added, the determination of the Martin compactification is a strengthening of identification of the Poisson boundary. The Furstenberg boundary is essentially a measure space, but the Martin compactification has a topology. This topology is important for the application to long-time behaviors of Brownian paths.

The Martin compactification of general Riemannian manifolds has not been identified geometrically, not even for nonpositively curved simply connected Riemannian manifolds. For graphs and more generally simplicial complexes, we can also define analogues of Laplace operators and hence harmonic functions. The Euclidean buildings are analogues of symmetric spaces for p-adic semisimple Lie groups, but their Martin compactifications have not been identified with other more geometric compactifications, for example analogues of the geodesic compactification and the maximal Satake compactification.

2

Uniform Construction of Compactifications of Symmetric Spaces

Among the compactifications discussed in Chapter 1, the Satake compactifications and the Furstenberg compactifications are obtained by embedding the symmetric spaces into some compact ambient spaces and taking the closure. In this chapter, we propose a *uniform, intrinsic approach* to compactifications of X. It is called *intrinsic* since no embedding into compact ambient spaces is used, and both the ideal boundary points and the topology on the compactifications are defined in terms of points of the symmetric spaces. This method was suggested by a modification in §III.8 of the method in [BS2] that was used to compactify locally symmetric spaces. Then we will construct all the compactifications recalled in Chapter 1 using this method.

Since compactifications arise from different situations and are constructed by different methods, this uniform construction allows one to understand relations between them easily. It also relates directly to compactifications of locally symmetric spaces, and hence shows similarities of compactifications of these two classes of spaces.

In §I.8, we formulate the general intrinsic approach and recall from [JM] how to describe a topology in terms of convergent sequences, which is more intuitive and convenient for defining topologies of many compactifications in later sections.

A key notion in the uniform, intrinsic approach is that of generalized Siegel sets, which is used to demonstrate the continuity of the group action on the compactifications. The Hausdorff property of the compactifications follows from the strong separation of the generalized Siegel sets. These results are given in §I.9. In §I.10, we apply this intrinsic uniform approach to construct the maximal Satake compactification $\overline{X}^S_{\max}$. Since this is the first case to which this method is applied and is also the simplest case, the construction is described in detail to illustrate the steps in the general method. In §I.11, we construct the nonmaximal Satake compactifications and describe explicitly the neighborhoods of the boundary points. In §I.12, we apply this method to construct the geodesic compactification $X \cup X(\infty)$. In the process, the geometric realization of the Tits building $\Delta(G)$ comes in naturally. In §I.13, we combine the construction of the maximal Satake compactification and the geodesic compactification to construct the Martin compactification. In §I.14, we apply the method to construct the Karpelevič compactification $\overline{X}^K$. In §I.15, we use the uniform method to construct the real Borel–Serre partial compactification in [BS2], which is useful for some applications in topology of arithmetic groups.

I.8 Formulation of the uniform construction

Recall from §I.4 that the Satake compactifications $\overline{X}^S$ are defined by embedding X equivariantly into the compact space $P(\mathcal{H}_n)$. On the other hand, the geodesic compactification $X \cup X(\infty)$ is obtained by defining ideals points and the convergence of points of X to them completely in terms of the points in X. These two examples represent two different approaches to compactifications of X. The former is called the *embedding method* and the latter is called the *intrinsic method*.

The purpose of Chapter 2 is to give a uniform and intrinsic method to construct most of the compactifications of X discussed earlier in Chapter 1. As explained in §I.2, the stabilizers of the boundary points of the geodesic compactification $X \cup X(\infty)$ are parabolic subgroups, and additional structures on $X(\infty)$ can also be understood in terms of parabolic subgroups. The boundary components of the Satake compactifications and their stabilizers in §I.4 are also given by parabolic subgroups. Therefore, parabolic subgroups are naturally associated with the geometry at infinity of symmetric spaces. Certainly, it is desirable to use parabolic subgroups directly to construct compactifications of symmetric spaces X intrinsically, i.e., defining the ideal boundary points and sequences converging to them without embedding X into some compact spaces.

I.8.1 As suggested by the construction of the Borel–Serre compactification of locally symmetric spaces in [BS2] and a modification in §III.8 below, we propose a uniform and intrinsic construction of compactifications of X. The uniform, general method consists of the following steps:

1. Choose a suitable collection of parabolic subgroups of G that is invariant under conjugation by elements of G.
2. For every parabolic subgroup P in the collection, define a boundary component $e(P)$ by making use of the Langlands decomposition of P and its refinements, or the induced horospherical decomposition of X.
3. For every parabolic subgroup in the collection, attach the boundary component $e(P)$ to X via the horospherical decomposition of X to obtain $X \cup \coprod_P e(P)$, and show that the induced topology on $X \cup \coprod_P e(P)$ is compact and Hausdorff, and the G-action on X extends continuously to the compactification, i.e., the compactification is a G-space.

All the known compactifications can be constructed this way by varying the choices of the collection of parabolic subgroups and their boundary components. In fact, the maximal Satake compactification $\overline{X}^S_{\max}$, the geodesic (or conic) compactification $X \cup X(\infty)$, the Martin compactification $X \cup \partial_\lambda X$, and the Karpelevič compactification $\overline{X}^K$ will be obtained by choosing the full collection of parabolic subgroups. On the other hand, for the nonmaximal Satake compactifications, in particular the Baily–Borel compactification $\overline{X}^{BB}$, we can specify a subcollection of parabolic subgroups according to a dominant weight vector.

I.8.2 The boundary components $e(P)$ often satisfy some compatibility conditions. For simplicity, we choose the collection of all parabolic subgroups. There are two obvious types of compatibility conditions:

1. for any pair of parabolic subgroups P, Q, $P \subseteq Q$ if and only if $e(P) \subseteq \overline{e(Q)}$;
2. or for any pair of parabolic subgroups P, Q, $P \subseteq Q$ if and only if $e(Q) \subseteq \overline{e(P)}$.

Usually, each boundary component $e(P)$ is a cell. If the condition in Type (1) is satisfied, the boundary $\coprod_P e(P)$ is a cell complex parametrized by the set of parabolic subgroups, and the incidence relation between its cells is dual to the incidence relation for the simplexes of the Tits building. The topology of the topological building puts a topology on the set of boundary components. In this case, we also require that for any parabolic subgroup Q, the closure of $e(Q)$ be given by

$$\overline{e(Q)} = e(Q) \cup \coprod_{P \subset Q} e(Q).$$

As will be seen below in §I.10, the boundary components of the maximal Satake compactification satisfy this type of compatibility condition.

If the condition in Type (2) is satisfied, the boundary $\coprod_P e(P)$ is a cell complex over the set of parabolic subgroups whose incidence relation is the same as the incidence relation of the Tits building, and is hence a cell complex homotopic to the Tits building. As in case (1), we need the topology of the topological Tits building to put a topology on the set of boundary components and hence make it into the topological boundary of some compactification. For any parabolic subgroup P, the closure of the boundary component $e(P)$ is given by

$$\overline{e(P)} = e(P) \cup \coprod_{Q \supset P} e(Q).$$

Since there are only finitely many Q containing any given P, the closure $\overline{e(P)}$ is a finite cell-complex. As will be seen below in §I.12, the boundary components of the geodesic compactification $X \cup X(\infty)$ satisfies this compatibility condition.

But there are also other types of mixed compatibility conditions. For example, take a compactification $\overline{X}^1$ whose boundary components satisfy the Type (1) compatibility condition, and a compactification $\overline{X}^2$ whose boundary components satisfy the Type (2) compatibility condition. Let $\overline{X}$ be the closure of X in $\overline{X}^1 \times \overline{X}^2$ under the diagonal embedding, i.e., $\overline{X}$ is the least common refinement of $\overline{X}^1$ and $\overline{X}^2$. For such a compactification, the boundary components will not satisfy either the Type (1) or (2) compatibility condition. For any pair of parabolic subgroups P, Q, $\overline{e(P)} \cap \overline{e(Q)} \neq \emptyset$ if and only if there is an inclusion relation between them, i.e., either $P \subseteq Q$ or $P \supseteq Q$. It should be pointed out that in general, $\overline{e(P)} \cap \overline{e(Q)} \neq \emptyset$ does not imply an inclusion relation between them. In this case, for any parabolic subgroup P,

$$\overline{e(P)} = e(P) \cup \cup_Q (\overline{e(P)} \cap \overline{e(Q)}),$$

where Q ranges over parabolic subgroups Q that either contain P or are contained in P.

By the results in §I.7, the Martin compactification $X \cup \partial_\lambda X$ for $\lambda < \lambda_0$ is the least common refinement of $X \cup X(\infty)$ and $\overline{X}^S_{\max}$. As seen below, its boundary components satisfy this type of mixed compatibility condition. Though the Karpelevič compactification $\overline{X}^K$ is not the least common refinement of some other compactifications whose boundary components satisfy either Type (1) or (2) compatibility conditions, its boundary components also satisfy the mixed compatibility condition. In both the Martin and the Karpelevič compactifications, the boundary $\cup_P e(P)$ is a cell complex whose incidence relation is not directly related to the incidence relation on the Tits building.

I.8.3 There are several general features of this approach that will become clearer later.

1. It gives an explicit description of neighborhoods of boundary points in the compactifications of X and sequences of interior points converging to them, which clarifies the structure of the compactifications and is also useful for applications (see [Zu2] [GHMN]). In [Sat1] and other works [GJT], the G-orbits in the Satake compactifications $\overline{X}^S_\tau$ and convergent sequences in a maximal totally geodesic flat submanifold in X through the basepoint x_0 are fully described, but there do not seem to be explicit descriptions of neighborhoods of the boundary points in $\overline{X}^S_\tau$.
2. It relates compactifications of symmetric spaces X directly to compactifications of locally symmetric spaces $\Gamma\backslash X$; in fact, for locally symmetric spaces, the method of [BS2] modified in §III.8 below consists of similar steps by considering only boundary components associated with rational parabolic subgroups instead of real parabolic subgroups for symmetric spaces, and both constructions depend on the reduction theory, in particular, the separation property of Siegel sets in Propositions I.9.8, I.9.11, and III.2.19.
3. By decomposing the boundary into boundary components associated with parabolic subgroups, its relation to the spherical Tits building $\Delta(G)$ of G becomes transparent, and the combinatorial structure of the boundary components is described in terms of the Tits building $\Delta(G)$. It can be seen below that for many known compactifications, the boundary components are cells, and hence the whole boundary of the compactifications is a cell complex parametrized by the Tits building, a fact emphasized in [GJT].
4. Since all the compactifications of X are constructed uniformly, relations between them can easily be determined by comparing their boundary components.
5. Due to the gluing procedure using the horospherical decomposition and generalized Siegel sets, the continuous extension of the G-action to the compactifications can be obtained easily. In [GJT], the continuity of the extension of the G-action to the dual-cell compactification $X \cup \Delta^*(X)$ is obtained through iden-

tification with the Martin compactification, where the continuity of the extended action is clear, rather than directly. This fact is one of the motivations of this uniform construction.

6. Due to the definition of the topology at infinity, one difficulty is to prove the Hausdorff property of the topology. This will follow from the strong separation property of the generalized Siegel sets in Proposition I.9.8 and Proposition I.9.11.

I.8.4 As mentioned earlier, the construction of the Satake compactifications and the Furstenberg compactifications uses embeddings into compact ambient spaces $P(\mathcal{H}_n)$ and $\mathcal{M}_1(G/P)$ respectively, where $\mathcal{H}_n$ is the real vector space of Hermitian matrices and $P(\mathcal{H}_n)$ the associated real projective space, and G/P is a faithful Furstenberg boundary and $\mathcal{M}_1(G/P)$ the space of probability measures on G/P. In general, given a compact G-space Z and G-equivariant embedding $i : X \to Z$, the closure $\overline{i(X)}$ is a G-compactification of X.

This embedding method is direct and the extension of the G-action to the compactification is immediate. On the other hand, it is not always easy to find the compact G-spaces Z and equivariant embeddings $X \to Z$. For example, there is no obvious choice of Z for the Karpelevič compactification, though we can construct the geodesic compactification $X \cup X(\infty)$ using the embedding method through the Gromov compactification §I.17. By combining with the maximal Satake compactification, we can similarly obtain the Martin compactification $X \cup \partial_\lambda X$ using the embedding method.

The main difficulty with the embedding method is that it is often not easy to understand the structure of the boundary, for example, G-orbits and boundary components, and neighborhoods of boundary points or equivalently characterization of unbounded sequences in X converging to boundary points. As seen in §I.4, these questions form the major part of the Satake compactifications in [Sat1].

I.8.5 The question of intrinsically constructing the Satake compactifications was first raised in [Ko1, pp. 25–26], and a possible construction was outlined there by using admissible regions. An expansion of this construction will be given in [Ko7].

The admissible regions are similar to generalized Siegel sets introduced in the next section (see Remark I.9.13 for more details). They are basically the same when the rank of X is equal to 1. But the Siegel sets or *admissible regions* (or *nontangential regions*) are not large enough to form neighborhoods (or rather intersection with X of neighborhoods) of the boundary points. Therefore, the *strong separation property* of generalized Siegel sets in §I.9 plays an important role in showing the Hausdorff property of the compactifications.

Admissible regions are related to bounded neighborhoods of geodesics and hence are important to understanding the limit behaviors along geodesics of harmonic functions on X. Admissible regions were motivated by the problem of generalizing convergence of harmonic functions on symmetric spaces of Fatou type. They were first introduced in [Ko10]. When X is equal to the Poincaré disk, they are basically the nontangential regions. For other rank-1 symmetric spaces, they are different from nontangential regions but slightly larger. An important property of admissible regions is the invariance under the isometry group of the symmetric space [Ko10, Theorem 6], which is not enjoyed by the collection of nontangential regions. For higher-rank symmetric spaces, there are several types of admissible regions, for example the restricted admissible regions and unrestricted admissible regions. The regions in [Ko1] are unrestricted. For detailed applications of admissible regions and various equivalent definitions, see [Ko1–6] [Ko8–10].

The intrinsic approach outlined in [Ko1, pp. 25–26] is as follows (some details will be provided in the forthcoming paper [Ko7]):

1. Define ideal boundary points to be suitable equivalence classes of certain filters defined by admissible regions.
2. Define the topology of the closure of the Weyl chamber $e^{\mathfrak{a}^+} \cdot x_0$ by the axioms of the Satake compactifications.
3. Use the G-action to define a topology on the union of X and the ideal boundary points and show that it is a G-compactification of X.

Step (1) is natural since by [Ko1, Lemma 2.2], certain filters defined by admissible regions converge to boundary points of Satake compactifications.

As will be seen below, there is some difference between this intrinsic approach and the approach in this book. In fact, in this book, the topology on the Satake compactifications is defined directly on the whole space in terms of the horospherical decomposition, rather than starting from the closure of the Weyl chamber $e^{\mathfrak{a}^+} \cdot x_0$ and then passing to the whole space. The boundary points are explicitly given rather than in terms of filters of subsets of X.

I.8.6 On the other hand, Siegel sets are fundamental to the reduction theory of arithmetic groups (see [Bo3] [Bo4] and §III.1 below). This surprising connection, pointed out by A. Koranyi, makes their applications to compactifications more interesting.

It should be pointed out that the generalized Siegel sets are used to show that the G-action on X extends continuously to compactifications. It sems that admissible regions can be used for the same purpose, and it might be possible to simplify some of the arguments in this chapter (see [Ko7]). Since Siegel sets will be used again for locally symmetric spaces, we will emphasize generalized Siegel sets rather than admissible regions, which have been crucial in [Ko1–6] [Ko8–10].

I.8.7 Another approach to an intrinsic construction of the maximal Satake compactification $\overline{X}^S_{\max}$ was given in [GJT, Chapter III]. In fact, the *dual-cell compactification* $X \cup \Delta^*(X)$ was based on the Cartan decomposition of X: $X = Ke^{\mathfrak{a}}x_0$ and the *polyhedral compactification* of the flat $e^{\mathfrak{a}}x_0$, which is determined by the Weyl chamber decomposition of $\mathfrak{a}$. Unfortunately, this construction does not allow one to show easily that the G-action on X extends continuously to $X \cup \Delta^*(X)$. As mentioned earlier, this problem was the starting point of the uniform approach proposed above.

I.8.8 It seems that the first nontrivial intrinsic compactification of X was the geodesic compactification $X \cup X(\infty)$, and it was probably studied by Hadamard in the context of simply connected, nonpositively curved Riemannian manifolds, though it was first formally constructed for symmetric spaces in [Ka] and for more general simply connected and nonpositively curved spaces in [EO]. Such a compactification has been generalized to certain nonsimply connected manifolds in [JM].

I.8.9 In constructing compactifications of X using the intrinsic method, it it often more convenient and intuitive to describe a topology in terms of convergent sequences.

We recall some conditions that need to be satisfied by convergent sequences from [JM, §6].

A topology on a space can be defined using a *closure operator* [Ku].

Definition I.8.10 *A closure operator for a space X is a function that assigns to every subset A of X a subset $\overline{A}$ satisfying the following properties:*

1. *For the empty set $\emptyset$, $\overline{\emptyset} = \emptyset$.*
2. *For any two subsets $A, B \subset X$, $\overline{A \cup B} = \overline{A} \cup \overline{B}$.*
3. *For any subset $A \subset X$, $A \subset \overline{A}$.*
4. *For any subset $A \subset X$, $\overline{\overline{A}} = \overline{A}$.*

Once a closure operator is given, then a subset A of X is defined to be closed if $\overline{A} = A$, and a subset B of X is open if its complement is closed. Using the four properties listed above, we can check easily that the open subsets define a topology on X.

A closure operator on a space can be defined using the class of convergent sequences. Let X be a space, and $\mathcal{C}$ a class of pairs $(\{y_n\}_1^\infty, y_\infty)$ of a sequence $\{y_n\}$ and a point y_∞ in X. If a pair $(\{y_n\}_1^\infty, y_\infty)$ is in $\mathcal{C}$, we say that y_n $\mathcal{C}$-converges to y_∞ and denote it by $y_n \stackrel{\mathcal{C}}{\to} y_\infty$; otherwise, $y_n \stackrel{\mathcal{C}}{\not\to} y_\infty$.

Motivated by the convergence class of nets in [Ke2, Chapter 4], we introduce the following.

Definition I.8.11 *A class $\mathcal{C}$ of pairs $(\{y_n\}_1^\infty, y_\infty)$ is called a convergence class of sequences if the following conditions are satisfied:*

1. *If $\{y_n\}$ is a constant sequence, i.e., there exists a point $y \in X$ such that $y_n = y$ for $n \geq 1$, then $y_n \stackrel{\mathcal{C}}{\to} y$.*
2. *If $y_n \stackrel{\mathcal{C}}{\to} y_\infty$, then so does every subsequence of y_n.*
3. *If $y_n \stackrel{\mathcal{C}}{\not\to} y_\infty$, then there is a subsequence $\{y_{n_i}\}$ of $\{y_n\}$ such that for any further subsequence $\{y'_{n_i}\}$ of $\{y_{n_i}\}$, $y'_{n_i} \stackrel{\mathcal{C}}{\not\to} y_\infty$.*
4. *Let $\{y_{m,n}\}_{m,n=1}^\infty$ be a double sequence. Suppose that for each fixed m, $y_{m,n} \stackrel{\mathcal{C}}{\to} y_{m,\infty}$; and the sequence $y_{m,\infty} \stackrel{\mathcal{C}}{\to} y_{\infty,\infty}$. Then there exists a function $n : \mathbb{N} \to \mathbb{N}$ such that* $\lim_{m\to\infty} n(m) = \infty$*; and the sequence $y_{m,n(m)} \stackrel{\mathcal{C}}{\to} y_{\infty,\infty}$.*

Lemma I.8.12 *Suppose $\mathcal{C}$ is a convergence class of sequences in a space X. For any subset A of X, define*

$$\overline{A} = \left\{ y \in X \mid \text{there exists a sequence } \{y_n\} \text{ in } A \text{ such that } y_n \stackrel{\mathcal{C}}{\to} y \right\}.$$

Then the operator $A \to \overline{A}$ is a closure operator.

Proof. The properties (1), (2), and (3) in Definition I.8.10 follow directly from the definition. For property (4), we need to show that for any point $y_\infty \in \overline{\overline{A}}$, there exists a sequence y_m in A such that $y_m \stackrel{\mathcal{C}}{\to} y$. By definition, there exists a sequence $y_{m,\infty}$ in $\overline{A}$ such that $y_{m,\infty} \stackrel{\mathcal{C}}{\to} y_\infty$. Since $y_{m,\infty} \in \overline{A}$, there exists a sequence $\{y_{m,n}\}_{n=1}^{\infty}$ in A such that $y_{m,n} \stackrel{\mathcal{C}}{\to} y_{m,\infty}$. Then Definition I.8.11.4 shows that there exists a sequence $y_{m,n(m)}$ in A such that $y_{m,n(m)} \stackrel{\mathcal{C}}{\to} y_\infty$, and hence $y_\infty \in \overline{A}$.

Proposition I.8.13 *A convergence class of sequences $\mathcal{C}$ in X defines a unique topology on X such that a sequence $\{y_n\}_1^\infty$ in X converges to a point $y \in X$ with respect to this topology if and only if $(\{y_n\}_1^\infty, y) \in \mathcal{C}$. The topological space X is Hausdorff if and only if every convergent sequence has a unique limit, and X is compact if and only if every sequence in X has a convergent subsequence.*

Proof. The statement that the convergence class $\mathcal{C}$ defines a unique topology follows from Lemma I.8.12 and the discussion after Lemma I.8.10. For the rest, see [Ke2, Chapter 2].

I.8.14 Summary and comments. In this section, we outlined the general steps in the uniform and intrinsic construction of compactifications of symmetric spaces. To simplify definitions and discussions of topologies of compactifications in this chapter, we introduced the notion of convergence sequences.

I.9 Siegel sets and generalizations

In this section, we discuss Siegel sets and generalized Siegel sets, in particular, strong separation properties of generalized Siegel sets in Propositions I.9.8 and I.9.11, which play a crucial role in applying the uniform method in §I.8 to construct compactifications of symmetric spaces.

In the reduction theory of arithmetic groups, two basic results are the finiteness property of Siegel sets and separation of sufficiently small Siegel sets of rational parabolic subgroups (see §III.2, Proposition III.2.19). Some results for Siegel sets of real parabolic subgroups were developed in [Bo4, §12]. But we need the larger generalized Siegel sets for applications to compactifications of X in this chapter. In this sense, results in this section can be considered as a reduction theory over $\mathbb{R}$.

The Siegel sets for real parabolic subgroups are direct generalizations of the Siegel sets of $\mathbb{Q}$-parabolic subgroups and are defined in equation (I.9.1). They satisfy a separation property similar to Siegel sets associated with $\mathbb{Q}$-parabolic subgroups (Proposition I.9.2). Since we need to put a nondiscrete topology on the set of boundary components, this separation is not strong enough. Instead, we need Proposition I.9.3. To prove the continuous extension of the G-action on X to compactifications, we need generalized Siegel sets (equation I.9.2), which contain the usual Siegel sets (Lemma I.9.4). The strong separation of Siegel sets are given in Propositions I.9.8 and I.9.11.

I.9.1 Now we define Siegel sets associated with proper parabolic subgroups of G. For any $t > 0$, define

$$A_{P,t} = \{a \in A_P \mid a^\alpha > t, \ \alpha \in \Delta(P, A_P)\},$$

where $\Delta(P, A_P)$ is the set of simple roots defined in §I.1.10.

When $t = 1$,

$$A_{P,t} = A_P^+ = \exp \mathfrak{a}_P^+,$$

the positive chamber in A_P. Hence, $A_{P,t}$ is the shift of the positive chamber A_P^+. For bounded sets $U \subset N_P$ and $V \subset X_P$, the set

$$\mathcal{S}_{P,U,t,V} = U \times A_{P,t} \times V \subset N_P \times A_P \times X_P \tag{I.9.1}$$

is identified with the subset $\nu_0(U \times A_{P,t} \times V)$ of X by the horospherical decomposition of X in equation (I.1.14) and called a *Siegel set* in X associated with the parabolic subgroup P.

An important property of Siegel sets is the following separation property.

Proposition I.9.2 (Separation of Siegel sets) *Let P_1, P_2 be two parabolic subgroups of G and let $\mathcal{S}_i = U_i \times A_{P,t_i} \times V_i$ be a Siegel set for P_i $(i = 1, 2)$. If $P_1 \neq P_2$ and $t_i \gg 0$, then*

$$\mathcal{S}_1 \cap \mathcal{S}_2 = \emptyset.$$

Proof. This is a special case of [Bo4, Proposition 12.6]. In fact, let P be a fixed minimal parabolic subgroup. Then P_1, P_2 are conjugate to standard parabolic subgroups P_{I_1}, P_{I_2} containing P,

$$P_1 = {}^{k_1}P_{I_1}, \qquad P_2 = {}^{k_2}P_{I_2},$$

for some $k_1, k_2 \in K$. If for all $t_i > 0$,

$$U_1 \times A_{P_1,t_1} \times V_1 \cap U_2 \times A_{P_2,t_2} \times V_2 \neq \emptyset,$$

then [Bo4, Proposition 12.6] implies that $I_1 = I_2$ and $k_2^{-1}k_1 \in P_{I_1}$. This implies that $P_1 = P_2$.

A special case of this proposition concerns rational parabolic subgroups and their Siegel sets. This separation property for rational parabolic subgroups plays an important role in reduction theory for arithmetic subgroups and compactifications of locally symmetric spaces (see §III.2). For compactifications of symmetric spaces, we need stronger separation properties.

Proposition I.9.3 (Strong separation of Siegel sets) *Let P_1, P_2, S_1, S_2 be as in the previous proposition and let C be a compact neighborhood of the identity element in K. Assume that $P_1^k \neq P_2$ for every $k \in C$. Then there exists $t_0 > 0$ such that $kS_1 \cap S_2 = \emptyset$ for all $k \in C$ if $t_1, t_2 \geq t_0$.*

This proposition follows from the even stronger separation property of Proposition I.9.8 below, which is needed in applying the uniform construction in §I.8 to construct compactifications of X.

Let $B(\cdot, \cdot)$ be the Killing form on $\mathfrak{g}$ as above, θ the Cartan involution on $\mathfrak{g}$ associated with K. Then

$$\langle X, Y \rangle = -B(X, \theta Y), \quad X, Y \in \mathfrak{g},$$

defines an inner product on $\mathfrak{g}$ and hence a left-invariant Riemannian metric on G and the subgroup N_P. Let $B_{N_P}(\varepsilon)$ be the ball in N_P of radius ε with center the identity element.

For a bounded set V in X_P and $\varepsilon > 0$, $t > 0$, define

$$S_{P,\varepsilon,t,V} = \{(n, a, z) \in N_P \times A_P \times X_P = X \mid z \in V, a \in A_{P,t}, n^a \in B_{N_P}(\varepsilon)\}. \tag{I.9.2}$$

We shall call $S_{P,\varepsilon,t,V}$ a *generalized Siegel set* associated with P, and P will be omitted and $S_{P,\varepsilon,t,V}$ denoted by $S_{\varepsilon,t,V}$ when the reference to P is clear.

Lemma I.9.4 *For any bounded set $U \subset N_P$ and $\varepsilon > 0$, when $t \gg 0$,*

$$U \times A_{P,t} \times V \subset S_{\varepsilon,t,V}.$$

Proof. Since the action of $A_{P,t}^{-1}$ by conjugation on N_P shrinks N_P toward the identity element as $t \to +\infty$, it is clear that for any bounded set $U \subset N_P$ and $\varepsilon > 0$, when $t \gg 0$, $a \in A_{P,t}$,

$$U^a \subset B_{N_P}(\varepsilon),$$

and the lemma follows.

I.9.5 On the other hand, when $\dim A_P \geq 2$, $S_{\varepsilon,t,V}$ is not contained in the union of countably infinitely many Siegel sets defined above. In fact, for any strictly increasing sequence $t_j \to +\infty$ and a sequence of bounded sets $U_j \subset N_P$ with $\cup_{j=1}^{\infty} U_j = N_P$, we claim that

$$S_{\varepsilon,t,V} \not\subset \cup_{j=1}^{n} U_j \times A_{P,t_j} \times V.$$

First, we note that

$$S_{\varepsilon,t,V} = \cup_{a \in A_{P,t}} {}^{a}B_{N_P}(\varepsilon) \times \{a\} \times V.$$

For every j such that $t_{j+1} > t_j$, $A_{P,t_j} \setminus A_{P,t_{j+1}}$ is not bounded, and hence there is an unbounded sequence $a_k \in A_{P,t_j} \setminus A_{P,t_{j+1}}$. Fix such a j and a sequence a_k. Then ${}^{a_k}B_{N_P}(\varepsilon)$ is not bounded, and hence when $k \gg 1$,

$${}^{a_k}B_{N_P}(\varepsilon) \times \{a_k\} \times V \not\subset \cup_{l=1}^{j} U_l \times A_{P,t_l} \times V.$$

On the other hand, since $a_k \notin A_{P,t_l}$ for all $l \geq j+1$,

$${}^{a_k}B_{N_P}(\varepsilon) \times \{a_k\} \times V \not\subset U_l \times A_{P,t_l} \times V,$$

and the claim follows.

I.9.6 To cover $S_{\varepsilon,t,V}$, we need to define Siegel sets slightly differently. For any $T \in A_P$, define

$$A_{P,T} = \{a \in A_P \mid a^{\alpha} > T^{\alpha}, \alpha \in \Delta(P, A_P)\}, \tag{I.9.3}$$

and

$$S_{P,\varepsilon,T,V} = \{(n, a, z) \in N_P \times A_P \times X_P = X \mid z \in V, a \in A_{P,T}, n^{a} \in B_{N_P}(\varepsilon)\}, \tag{I.9.4}$$

which is also to be denoted by $S_{\varepsilon,T,V}$.

Siegel sets of the form $U \times A_{P,T} \times V$ are needed for the precise reduction theory of arithmetic subgroups (see [OW] [Sap1] and §III.2 below for more details) and will also be used in §I.11 below to describe the topology of nonmaximal Satake compactifications. An analogue of Lemma I.9.4 holds for the Siegel sets $S_{P,\varepsilon,T,V}$, and the converse inclusion is given in the next lemma.

Proposition I.9.7 *There exist sequences $T_j \in A_{P,t}$ and bounded sets $U_j \subset N_P$ such that*

$$S_{\varepsilon,t,V} \subset \cup_{j=1}^{\infty} U_j \times A_{P,T_j} \times V.$$

Proof. In fact, T_j could be any sequence in $A_{P,t}$ such that every point of $A_{P,t}$ belongs to a δ-neighborhood of some T_j, where δ is independent of j, and $U_j = U$ is a sufficiently large subset.

Proposition I.9.8 (Strong separation of generalized Siegel sets) *For any two different parabolic subgroups P, P' and generalized Siegel sets $S_{\varepsilon,t,V}$, $S_{\varepsilon,t,V'}$ associated with them, and a compact neighborhood C of the identity element in K such that for every $k \in C$, ${}^{k}P \neq P'$, if $t \gg 0$ and ε is sufficiently small, then for all $k \in C$,*

$$kS_{\varepsilon,t,V} \cap S_{\varepsilon,t,V'} = \emptyset.$$

Proof. Let $\tau : G \to \mathrm{PSL}(n, \mathbb{C})$ be a faithful irreducible projective representation whose highest weight is generic. Since G is semisimple of adjoint type, such a representation exists. Choose an inner product on $\mathbb{C}^n$ such that $\tau(\theta(g)) = (\tau(g)^*)^{-1}$, where θ is the Cartan involution on G associated with K, and $A \to (A^*)^{-1}$ is the Cartan involution on $\mathrm{PSL}(n, \mathbb{C})$ associated with $\mathrm{PSU}(n)$. Then $\tau(K) \subset \mathrm{PSU}(n)$.

Let $M_{n\times n}$ be the vector space of complex $n\times n$ matrices, and $P_{\mathbb{C}}(M_{n\times n})$ the associated complex projective space. Composed with the map $\mathrm{PSL}(n, \mathbb{C}) \to P_{\mathbb{C}}(M_{n\times n})$, τ induces an embedding

$$i_\tau : G \to P_{\mathbb{C}}(M_{n\times n}).$$

For each Siegel set $S_{\varepsilon,t,V}$ in X associated with P, its inverse image in G under the map $G \to X = G/K$, $g \mapsto gx_0$, is $\{(n, a, m) \in N_P \times A_P \times M_P K = G \mid m \in VK, a \in A_{P,t}, n^a \in B_{N_P}(\varepsilon)\}$ and denoted by $S_{\varepsilon,t,V}K$.

We claim that the images $i_\tau(kS_{\varepsilon,t,V}K)$ and $i_\tau(S_{\varepsilon,t,V'}K)$ are disjoint for all $k \in C$ under the above assumptions.

Let P_0 be a minimal parabolic subgroup contained in P. Then $P = P_{0,I}$ for a subset $I \subset \Delta(P_0, A_{P_0})$. Let

$$\mathbb{C}^n = V_{\mu_1} \oplus \cdots \oplus V_{\mu_k}$$

be the weight space decomposition under the action of A_{P_0}. Let μ_τ be the highest weight of τ with respect to the positive chamber $\mathfrak{a}^+_{P_0}$. Then each weight μ_i is of the form

$$\mu_i = \mu_\tau - \sum_{\alpha\in\Delta} c_\alpha \alpha,$$

where $c_\alpha \geq 0$. The subset $\{\alpha \in \Delta \mid c_\alpha \neq 0\}$ is called the *support* of μ_i, and denoted by $\mathrm{Supp}(\mu_i)$. For $P = P_{0,I}$, let V_P be the sum of all weight spaces V_{μ_i} whose support $\mathrm{Supp}(\mu_i)$ is contained in I. Since τ is generic, V_P is nontrivial. In fact, $P_{0,I}$ leaves V_P invariant and is equal to the stabilizer of V_P in G, and the representation of M_P on V_P is a multiple of an irreducible, faithful one, and hence τ induces an embedding

$$\tau_P : M_P \to \mathrm{PSL}(V_P). \tag{I.9.5}$$

The group $\mathrm{PSL}(V_P)$ can be canonically embedded into $P_{\mathbb{C}}(M_{n\times n})$ by extending each matrix in $\mathrm{PSL}(V_P)$ to act as the zero linear transformation on the orthogonal complement of V_P. Under this identification, for every $A \in \mathrm{PSL}(V_P)$,

$$A(\mathbb{C}^n) = V_P.$$

Denote the composed embedding $M_P \to \mathrm{PSL}(V_P) \hookrightarrow P_{\mathbb{C}}(M_{n\times n})$ also by τ_P,

$$\tau_P : M_P \to P_{\mathbb{C}}(M_{n\times n}).$$

Similarly, for P', we get a subspace $V_{P'}$ invariant under P' and hence under $M_{P'}$, a subset $\mathrm{PSL}(V_{P'})$ in $P_{\mathbb{C}}(M_{n\times n})$ and an embedding

$$\tau_{P'} : M_{P'} \to \mathrm{PSL}(V_{P'}) \subset P_{\mathbb{C}}(M_{n\times n}).$$

For any $k \in C$, ${}^kP \neq P'$, and hence

$$V_{{}^kP} \neq V_{P'}.$$

Since for any $m \in M_P$, $m' \in M_{P'}$, and any $g \in G$,

$$\tau_P(m)\tau(g)(\mathbb{C}^n) = \tau_P(m)(\mathbb{C}^n) = V_P,$$
$$\tau_{P'}(m')\tau(g)(\mathbb{C}^n) = \tau_{P'}(m')(\mathbb{C}^n) = V_{P'},$$

it follows that for any $g, g' \in G$, $m \in M_P$, $m' \in M_{P'}$, and $k \in C$,

$$\tau(k)\tau_P(m)\tau(g) \neq \tau_{P'}(m')\tau(g'). \tag{I.9.6}$$

If the claim is false, then there exists a sequence g_j in G such that

$$g_j \in k_j S_{\varepsilon_j,t_j,V} K \cap S_{\varepsilon_j,t_j,V'} K,$$

where $k_j \in C$, $\varepsilon_j \to 0$, $t_j \to +\infty$. Since $g_j \in k_j S_{\varepsilon_j,t_j,V} K$, g_j can be written as

$$g_j = k_j n_j a_j m_j c_j,$$

where $n_j \in N_P$, $a_j \in A_{P,t_j}$, $m_j \in M_P$, and $c_j \in K$ satisfy (1) for all $\alpha \in \Delta(P, A_P)$, $a_j^\alpha \to +\infty$, (2) $n_j^{a_j} \to e$, (3) $m_j \in V$. By passing to a subsequence, we can assume that $k_j \to k_\infty \in C$, m_j converges to some $m_\infty \in M_P$, and c_j converges to some c_∞ in K.

By choosing suitable coordinates, we can assume that for $a \in A_{P_0}$, $\tau(a)$ is a diagonal matrix,

$$\tau(a) = \mathrm{diag}(a^{\mu_1}, \dots, a^{\mu_n}),$$

where the weights μ_i with support contained in I are $\mu_1, \dots, \mu_l$ for some $l \geq 1$, and μ_1 is the highest weight μ_τ. Since τ is faithful and I is proper, $l < n$. Recall that $P = P_{0,I}$, and

$$A_P = \{a \in A_{P_0} \mid a^\alpha = 1, \alpha \in I\}.$$

Then

$$\begin{aligned}\tau(a_j) &= \mathrm{diag}(a_j^{\mu_1}, \dots, a_j^{\mu_l}, a_j^{\mu_{l+1}}, \dots, a_j^{\mu_n}) \\ &= \mathrm{diag}(a_j^{\mu_\tau}, \dots, a_j^{\mu_\tau}, a_j^{\mu_\tau - \sum_\alpha c_{l+1,\alpha}\alpha}, \dots, a_j^{\mu_\tau - \sum_\alpha c_{n,\alpha}\alpha}),\end{aligned}$$

where for each $j \in \{l+1, \dots, n\}$, there exists at least one $\alpha \in \Delta - I$ such that $c_{j,\alpha} > 0$. Then as $j \to +\infty$, the image of $\tau(a_j)$ in $P_{\mathbb{C}}(M_{n\times n})$ satisfies

$$\begin{aligned}i_\tau(a_j) &= [\mathrm{diag}(1, \dots, 1, a_j^{-\sum_\alpha c_{l+1,\alpha}\alpha}, \dots, a_j^{-\sum_\alpha c_{n,\alpha}\alpha})], \\ &\to [\mathrm{diag}(1, \dots, 1, 0, \dots, 0)],\end{aligned}$$

where the image of an element $A \in M_{n\times n} \setminus \{0\}$ in $P_{\mathbb{C}}(M_{n\times n})$ is denoted by $[A]$. This implies that

$$i_\tau(g_j) = \tau(k_j) i_\tau(a_j) \tau(n_j^{a_j}) \tau(m_j) \tau(c_j)$$
$$\to \tau(k_\infty)[\mathrm{diag}(1, \ldots, 1, 0, \ldots, 0)] \tau(m_\infty) \tau(c_\infty) = \tau(k_\infty) \tau_P(m_\infty) \tau(c_\infty),$$

since $k_j \to k_\infty$, $n_j^{a_j} \to e$, $m_j \to m_\infty$, $c_j \to c_\infty$, and the image of $\mathbb{C}^n$ under $\mathrm{diag}(1, \ldots, 1, 0, \ldots, 0)$ is equal to V_P. Using $g_j \in S_{\varepsilon_j, t_j, V'} K$, we can similarly prove that

$$i_\tau(g_j) \to \tau_{P'}(m'_\infty) \tau(c'_\infty)$$

for some $m'_\infty \in M_{P'}$ and $c'_\infty \in K$. This contradicts equation (I.9.6), and the claim and hence the proposition are proved.

Remark I.9.9 It seems that the proof of [Bo4, Proposition 12.6] does not apply here. Assume that P, P' are both minimal. Then there exists an element $g \in G$, $g \notin P$ such that $P' = {}^g P$. In the proof of [Bo4, Proposition 12.6], g is written in the Bruhat decomposition $uwzv$, where $w \in W(\mathfrak{g}, \mathfrak{a}_P)$, $w \neq Id$, $u, v \in N_P$, $z \in A_P$. For each fixed g, the components u, v, z are bounded. This is an important step in the proof. If w is equal to the element w_0 of longest length, then for a sufficiently small neighborhood C of g in G (or K), every $g' \in C$ is of the form $u' w_0 z' v'$ with the same Weyl group element W_0 and the components u', v', z' are uniformly bounded, and the same proof works. On the other hand, if w is not equal to w_0, then any neighborhood C of g contains elements g' of the form $u' w_0 z' v'$ whose components u', z', v' are not uniformly bounded, and the method in [Bo4, Proposition 12.6] does not apply directly. The reason for the unboundedness of the components is that $N_P w_0$ is mapped to an open dense subset of G/P.

Remark I.9.10 The above proof of Proposition I.9.8 was suggested by the Hausdorff property of the maximal Satake compactification $\overline{X}^S_{\max}$. In fact, Proposition I.9.8 follows from the Hausdorff property of $\overline{X}^S_{\max}$, by computations similar to those in the proof of Proposition I.10.10. But the point here is to prove this separation property without using any compactification, so that it can be used to construct other compactifications.

Proposition I.9.8 gives the separation property for different parabolic subgroups. For any parabolic subgroup P, separation of Siegel sets for disjoint neighborhoods in X_P is proved in the next proposition.

Proposition I.9.11 (Strong separation of generalized Siegel sets) *For any given parabolic subgroup P and two different boundary points $z, z' \in X_P$, let V, V' be compact neighborhoods of z, z' with $V \cap V' = \emptyset$. If ε is sufficiently small, t is sufficiently large, and C is a sufficiently small compact neighborhood of e in K, then for all $k, k' \in C$, the generalized Siegel sets $kS_{\varepsilon,t,V}$, $k'S_{\varepsilon,t,V'}$ are disjoint.*

Proof. We prove this proposition by contradiction. If it is false, then for all $\varepsilon > 0, t > 0$, and any neighborhood C of e in K,

$$kS_{\varepsilon,t,V} \cap k' S_{\varepsilon,t,V'} \neq \emptyset,$$

for some $k, k' \in C$. Therefore, there exist sequences $k_j, k'_j \in K$, $n_j, n'_j \in N_P$, $a_j, a'_j \in A_P$, $m_j \in VK_P$, $m'_j \in V'K_P$ such that

1. $k_j, k'_j \to e$,
2. $n_j^{a_j} \to e$, $n_j'^{a'_j} \to e$,
3. For all $\alpha \in \Delta(P, A_P)$, $a_j^\alpha, a_j'^\alpha \to +\infty$,
4. $k_j n_j a_j m_j K = k'_j n'_j a'_j m'_j K$.

Since VK_P, $V'K_P$ are compact, after passing to a subsequence, we can assume that both m_j and m'_j converge. Denote their limits by m_∞, m'_∞. By assumption, $VK_P \cap V'K_P = \emptyset$, and hence

$$m_\infty K_P \neq m'_\infty K_P.$$

We claim that the conditions (1), (2), and (3) together with $m_\infty K_P \neq m'_\infty K_P$ contradict the condition (4).

As in the proof of Proposition I.9.8, let $\tau : G \to \mathrm{PSL}(n, \mathbb{C})$ be a faithful representation whose highest weight is generic and $\tau(\theta(g)) = (\tau(g)^*)^{-1}$, where θ is the Cartan involution associated with K. Let $\mathcal{H}_n$ be the real vector space of $n \times n$ Hermitian matrices and $P(\mathcal{H}_n)$ the associated projective space. As in §I.4, τ defines an embedding

$$i_\tau : G/K \to \mathcal{P}(\mathcal{H}_n), \quad gK \mapsto [\tau(g)\tau^*(g)],$$

where $[\tau(g)\tau^*(g)]$ denotes the line determined by $\tau(g)\tau^*(g)$. We will prove the claim by determining the limits of $i_\tau(k_j n_j a_j m_j)$ and $i_\tau(k'_j n'_j a'_j m'_j)$.

Let P_0 be a minimal parabolic subgroup contained in P. Then $P = P_{0,I}$ for a unique subset $I \subset \Delta(P_0, A_{P_0})$. As in the proof of Proposition I.9.8, we can assume that for $a \in A_{P_0}$, $\tau(a)$ is diagonal,

$$\tau(a) = \mathrm{diag}(a^{\mu_1}, \dots, a^{\mu_n}),$$

and the weights $\mu_1, \dots, \mu_l$ are the weights whose supports are contained in I. Then

$$\begin{aligned} i_\tau(k_j n_j a_j m_j) &= [\tau(k_j)\tau(a_j)\tau(n_j^{a_j})\tau(m_j)\tau(m_j)^*\tau(n_j^{a_j})^*\tau(a_j)^*\tau(k_j)^*] \\ &\to [\mathrm{diag}(1, \dots, 1, 0, \dots, 0)\tau(m_\infty)\tau(m_\infty)^* \\ &\quad \times \mathrm{diag}(1, \dots, 1, 0 \dots, 0)^*] \\ &= [\tau_P(m_\infty)\tau_P(m_\infty)^*], \end{aligned} \tag{I.9.7}$$

where

$$\tau_P : M_P \to \mathrm{PSL}(V_P) \hookrightarrow P_{\mathbb{C}}(M_{n\times n}), \quad m \mapsto [\mathrm{diag}(1, \dots, 1, 0, \dots, 0)\tau(m)],$$

is the map in equation (I.9.5) in the proof of Proposition I.9.8. Since τ_P is a faithful representation,

$$\tau_P\tau_P^* : X_P \to P(\mathcal{H}_n), \quad mK_P \mapsto [\tau_P(m)\tau_P(m)^*]$$

is an embedding.

Similarly, we get

$$i_\tau(k'_j n'_j a'_j m'_j) \to [\tau_P(m'_\infty)\tau_P(m'_\infty)^*]. \tag{I.9.8}$$

Since $m_\infty, m'_\infty \in M_P$ and $m_\infty K_P \neq m'_\infty K_P$, we get

$$[\tau_P(m_\infty)\tau_P(m_\infty)^*] \neq [\tau_P(m'_\infty)\tau_P(m'_\infty)^*].$$

Then the condition (4) implies that equation (I.9.8) contradicts equation (I.9.7) and the claim is proved.

Remark I.9.12 As mentioned earlier, the separation property and the finiteness property of Siegel sets for rational parabolic subgroups is a crucial result in the reduction theory of arithmetic subgroups of algebraic groups (see [Bo4] and §III.2 below) and plays an important role in compactifications of locally symmetric spaces $\Gamma\backslash X$. One of the main points of Part I is that the above (stronger) separation properties of the generalized Siegel sets for real parabolic subgroups in Propositions I.9.8 and I.9.11 will play a similar role in compactifications of X.

Remark I.9.13 For a parabolic subgroup P, let P^- be the opposite parabolic subgroup, i.e., P, P^- share a common split component A_P, and the roots in $\Phi(P^-, A_P)$ are equal to the negatives of the roots in $\Phi(P, A_P)$. Denote the Langlands decomposition of P^- by

$$P^- = N_P^- A_P M_P.$$

Since N_P^- is a normal subgroup, this can also be written in the form

$$P^- = A_P \times N_P^- \times M_P \cong A_P \times N_P^- \times M_P,$$

which induces the following decomposition of X:

$$A_P \times N_P^- \times X_P \to X, \quad (a, n, z) \to anz. \tag{I.9.9}$$

This decomposition is different from the horospherical decomposition $X = N_P^- \times A_P \times X_P$ associated with P^-. Let $V \subset X_P$ be a bounded subset as above, and U' a compact neighborhood of the identity element in N_P^-. Then the subet $A_{P,t} \times U' \times V$ in X is an *admissible region* in [Ko1]. It can be checked easily that such a subset is not contained in any Siegel set $U \times A_{P,t} \times V$, and hence the decomposition in equation (I.9.9) is different from the horospherical decomposition.

For any element $g \in P$, in particular for $g \in N_P$, $gS_{P,\varepsilon,t,V}$ is contained in another Siegel set. This invariance under P of the collection of generalized Siegel sets associated with P is important for applications to compactifications of X. The reason is that we expect P to stabilize the boundary component of P and N_P to act trivially on it. If we want to describe the neighborhoods of boundary points in terms of certain sets, it is certainly desirable that the collection of these sets be invariant under P. Furthermore, it is needed to prove the continuity of the extended action. The admissible regions in [Ko1] have similar properties [Ko1, Lemma 2.1].

I.9.14 Summary and comments. Siegel sets associated with $\mathbb{Q}$-parabolic subgroups are basic to the reduction theory of arithmetic subgroups, and this is important for understanding the geometry of locally symmetric spaces. There are several aspects of the reduction theory, but a crucial one concerns separation of Siegel sets.

In this section, we introduced Siegel sets of real parabolic subgroups and larger sets, called the generalized Siegel sets. Then we developed several versions of the separation property of the generalized Siegel sets. The strong separation property will play a fundamental role in the intrinsic construction of compactifications in this chapter.

The reduction theory in [Bo4] [Bo3] deals mainly with arithmetic subgroups, but some parts deal with real parabolic subgroups and their Siegel sets. Parts of [BS2] also deal with parabolic subgroups defined over a field lying between $\mathbb{Q}$ and $\mathbb{R}$. The generalized Siegel sets are strictly larger than Siegel sets and are important for applications below.

I.10 Uniform construction of the maximal Satake compactification

In this section, we apply the uniform method in §I.8 to construct the maximal Satake compactification $\overline{X}^S_{\max}$, which, in a certain sense, is the simplest compactification to which this method applies, and all the general steps in §I.8 can be explained easily.

First we define the boundary component of parabolic subgroups (§I.1.10.1) and describe the convergence of interior points to them in terms of the horospherical decomposition. The generalized Siegel sets are used here in order to show that the G-action extends continuously to the compactification (Proposition I.10.8). To prove the Hausdorff property of the compactification (Proposition I.10.6), we use the strong separation property of the generalized Siegel sets. The compactness of the constructed space is given in Proposition I.10.7. Finally, we show (Proposition I.10.11) that this construction gives the maximal Satake compactification $\overline{X}^S_{\max}$ defined in §I.4.

I.10.1 The first step is to choose a G-invariant collection of parabolic subgroups. In this case, we use the collection of all proper parabolic subgroups.

For every (proper) parabolic subgroup P, define its maximal Satake boundary component by

$$e(P) = X_P,$$

the boundary symmetric space defined in equation (I.1.12) in §I.1. As pointed out in equation (I.1.13), X_P is also a homogeneous space of P, which will be needed below to define the G-action on the compactification. Define

$$\overline{X}_{\max} = X \cup \coprod_P X_P. \tag{I.10.1}$$

To put a topology on $\overline{X}_{\max}$, we need a topology on the collection of boundary components. This can be given in terms of the topology of the topological spherical Tits building recalled in Definition I.2.21 and Proposition I.2.22, or described more directly using the K-action on the parabolic subgroups and the related decompositions.

Specifically, by the K-action on the horospherical decomposition in equation (I.1.16) in §I.1, K acts on $\overline{X}_{\max}$ as follows: for $k \in K$, $z = mK_P \in X_P$,

$$k \cdot z = {}^k m \in X_{{}^k P}. \tag{I.10.2}$$

If we realize X_P as a submanifold $M_P \cdot x_0 = M_P \cdot K \subset X = G/K$, then this action is exactly the restriction of the action of K on X.

The G-action on $\overline{X}_{\max}$ will be defined later.

I.10.2 The topology of $\overline{X}_{\max}$ is defined as follows. First we note that X and X_P have a topology defined by the invariant metric. We need to define convergence of sequences of interior points in X to boundary points and convergence of sequences of boundary points:

1. For a boundary component X_P and a point $z_\infty \in X_P$, an unbounded sequence y_j in X converges to z_∞ if and only if y_j can be written in the form $y_j = k_j n_j a_j z_j$, where $k_j \in K, n_j \in N_P, a_j \in A_P, z_j \in X_P$ such that
   (a) $k_j \to e$, where e is the identity element.
   (b) For all $\alpha \in \Phi(P, A_P)$, $a_j^\alpha \to +\infty$.
   (c) $n_j^{a_j} \to e$.
   (d) $z_j \to z_\infty$.
2. Let Q be a parabolic subgroup containing P. For a sequence $k_j \in K$ with $k_j \to e$, and a sequence $y_j \in X_Q$, the sequence $k_j y_j \in X_{k_j Q}$, which is defined in equation (I.10.2), converges to $z_\infty \in X_P$ if the following conditions are satisfied. Let P' be the unique parabolic subgroup in M_Q corresponding to P as in equation (I.1.21) in §I.1, and write $X_Q = N_{P'} \times A_{P'} \times X_{P'}$. The sequence y_j can be written as $y_j = k'_j n'_j a'_j z'_j$, where $k'_j \in K_Q, n'_j \in N_{P'}, a'_j \in A_{P'}, z'_j \in X_{P'} = X_P$ satisfy the same condition as part (1) above when K, N_P, A_P, X_P are replaced by $K_Q, N_{P'}, A_{P'}, X_{P'}$. Note that if $Q = P$, then $P' = M_Q$, and $N_{P'}, A_{P'}$ are trivial.

These are special convergent sequences, and combinations of them give general convergent sequences. By a combination of these special sequences, we mean a sequence $\{y_j\}$, $j \in \mathbb{N}$, and a splitting $\mathbb{N} = A_1 \coprod \cdots \coprod A_s$ such that for each infinite A_i, the corresponding subsequence y_j, $j \in A_i$, is a sequence of type either (1) or (2). It can be shown easily that these convergent sequences satisfy the conditions in Definition I.8.11. In fact, the main condition to check is the double sequence condition, and this condition is satisfied by double sequences of either type (1) or type (2) above, and hence by general double sequences. Therefore these convergent sequences form a convergence class of sequences, and define a unique topology on $\overline{X}_{\max}$.

I.10.3 It is easy to see that for every P, X_P is in the closure of X. More generally, for a pair of parabolic subgroups P, Q, $P \subseteq Q$ if and only if X_P is contained in the closure of X_Q. Clearly, each boundary component X_P is a cell; and therefore, the boundary components satisfy the type (1) compatibility condition in §I.8.2, and the boundary $\cup_P X_P$ is a cell complex dual to the Tits building $\Delta(G)$.

I.10.4 The description of the topology in terms of convergent sequences is convenient and intuitive for many purposes. On the other hand, neighborhood systems of the boundary points of $\overline{X}^{\max}$ can be given explicitly.

For every parabolic subgroup P, let P_I, $I \subset \Delta(P, A_P)$, be all the parabolic subgroups containing P. For every P_I, X_{P_I} contains X_P as a boundary component. For any point $z \in X_P$, let V be a neighborhood of z in X_P. For $\varepsilon > 0, t > 0$, let $S_{\varepsilon,t,V}$ be the generalized Siegel set in X defined in equation (I.9.2) in §I.9, and let $S^I_{\varepsilon,t,V}$ be the generalized Siegel set of X_{P_I} associated with the parabolic subgroup P' in M_{P_I}, which is associated with P as in equation (I.1.21) in §I.1. Let C be a (compact) neighborhood of e in K. Then the union

$$C\left(S_{\varepsilon,t,V} \cup \coprod_{I \subset \Delta} S^I_{\varepsilon,t,V}\right) \tag{I.10.3}$$

is a neighborhood of z in $\overline{X}_{\max}$. For sequences of $\varepsilon_i \to 0$, $t_i \to +\infty$, a basis V_i of neighborhoods of z in X_P and a basis of compact neighborhoods C_j of e in K, the above union in equation (I.10.3) forms a countable basis of the neighborhoods of z in $\overline{X}_{\max}$.

It can be checked easily that these neighborhoods define a topology on $\overline{X}^{\max}$ whose convergent sequences are exactly those given above.

I.10.5 When a point $y_j \in X$ is written in the form $k_j n_j a_j z_j$ with $k_j \in K, n_j \in N_P, a_j \in A_P, z_j \in X_P$, none of these factors is unique, since $X = N_P \times A_P \times X_P$ and the extra K-factor causes nonuniqueness. Then a natural question is the uniqueness of the limit of a convergent sequence y_j in $\overline{X}_{\max}$, or equivalently, whether the topology on $\overline{X}_{\max}$ is Hausdorff. Since the K-factor is required to converge to e, it is reasonable to expect that the Hausdorff property still holds.

Proposition I.10.6 *The topology on $\overline{X}_{\max}$ is Hausdorff.*

Proof. We need to show that every pair of different points $x_1, x_2 \in \overline{X}_{\max}$ admit disjoint neighborhoods. This is clearly the case when at least one of x_1, x_2 belongs to X.

Assume that both belong to the boundary and let P_1, P_2 be the parabolic subgroups such that $x_1 \in X_{P_1}, x_2 \in X_{P_2}$. There are two cases to consider: $P_1 = P_2$ or not.

For the second case, let C be a sufficiently small compact neighborhood of e in K such that for $k_1, k_2 \in C$,

$$^{k_1}P_1 \neq {}^{k_2}P_2.$$

By definition, $C(S_{\varepsilon,t,V_i} \cup \coprod_I S^I_{\varepsilon,t,V_i})$ is a neighborhood of x_i, $i = 1, 2$. Proposition I.9.8 implies that

$$CS_{\varepsilon,t,V_1} \cap CS_{\varepsilon,t,V_2} = \emptyset.$$

For all pairs of $I_1, I_2, k_1, k_2 \in C$, either

$$^{k_1}P_{1,I_1} \neq {}^{k_2}P_{2,I_2},$$

and hence

$$k_1 S^{I_1}_{\varepsilon,t,V_1} \cap k_2 S^{I_2}_{\varepsilon,t,V_2} = \emptyset,$$

or

$$^{k_1}P_{1,I_1} = {}^{k_2}P_{2,I_2}.$$

In the latter case, P_1, $({}^{k_2}P_2)^{k_1}$ are contained in P_{1,I_1} and correspond to two parabolic subgroups of $M_{P_{1,I_1}}$ by equation (I.1.21). As in the case above for general Siegel sets in X, we get

$$k_1 S^{I_1}_{\varepsilon,t,V_1} \cap k_2 S^{I_2}_{\varepsilon,t,V_2} = \emptyset.$$

This implies that the neighborhoods of x_1 and x_2 are disjoint.

In the first case, $P_1 = P_2$. Since $x_1 \neq x_2$, we can choose compact neighborhoods V_1, V_2 in X_{P_1} such that $V_1 \cap V_2 = \emptyset$. Then Proposition I.9.11 together with similar arguments as above imply that x_1, x_2 admit disjoint neighborhoods. This completes the proof of this proposition.

Proposition I.10.7 *The topological space $\overline{X}_{\max}$ is compact and contains X as a dense open subset.*

Proof. Let P_0 be a minimal parabolic subgroup, and let $P_{0,I}$ $I \subset \Delta = \Delta(P_0, A_{P_0})$ be all the standard parabolic subgroups. Since every parabolic subgroup is conjugate to the standard parabolic subgroup $P_{0,I}$ under K, we have

$$\overline{X}_{\max} = X \cup \coprod_{I \subset \Delta} K X_{P_{0,I}}.$$

Since K is compact, it suffices to show that every sequence in X and $X_{P_{0,I}}$ has a convergent subsequence. First, we consider a sequence in X. If y_j is bounded, it clearly has a convergent subsequence in X. Otherwise, writing $y_j = k_j a_j x_0$, $k_j \in K, a_j \in \overline{A^+_{P_0}}$, we can assume, by passing to a subsequence, that the components of y_j satisfy the following conditions:

1. $k_j \to k_\infty$ for some $k_\infty \in K$,
2. there exists a subset $I \subset \Delta(P_0, A_{P_0})$ such that for $\alpha \in \Delta - I, \alpha(\log a_j) \to +\infty$, while for $\alpha \in I, \alpha(\log a_j)$ converges to a finite number.

Decompose

$$\log a_j = H_{I,j} + H_j^I, \quad H_{I,j} \in \mathfrak{a}_{P_{0,I}}, H_j^I \in \mathfrak{a}_{P_0}^I.$$

Since $\Delta(P_0, A_{P_0}) - I$ restricts to $\Delta(P_{0,I}, A_{P_{0,I}})$, it follows from the definition that $k_j^{-1} y_j = a_j x_0$ converges to $e^{H_\infty^I} x_0 \in X_{P_{0,I}}$ in $\overline{X}_{\max}$, where x_0 also denotes the basepoint $K_{P_{0,I}}$ in $X_{P_{0,I}}$, and H_∞^I is the unique vector in $\mathfrak{a}_{P_0}^I$ such that for all $\alpha \in I, \alpha(H_\infty^I) = \lim_{j\to+\infty} \alpha(\log a_j)$. Together with the action of K on parabolic subgroups and the Langlands decomposition in equation (I.1.16), this implies that $y_j = (k_j k_\infty^{-1})\, {}^{k_\infty} a_j x_0$ converges to a point belonging to $X_{({}^{k_\infty} P_{0,I})}$ in the topology of $\overline{X}_{\max}$.

For a sequence in $X_{P_{0,I}}$, we can similarly use the Cartan decomposition $X_{P_{0,I}} = K_{P_{0,I}} \exp \overline{\mathfrak{a}_{P_0}^{I,+}} x_0$ to extract a convergent subsequence in $\overline{X}_{\max}$.

Proposition I.10.8 *The action of G on X extends to a continuous action on $\overline{X}_{\max}$.*

Proof. First we define a G-action on the boundary $\partial \overline{X}_{\max} = \coprod_P X_P$, then show that this gives a continuous G-action on $\overline{X}_{\max}$.

For $g \in G$ and a boundary point $z \in X_P$, write

$$g = kman,$$

where $k \in K, m \in M_K, a \in A_P, n \in N_P$. Define

$$g \cdot z = k \cdot (mz) \in X_{{}^k P},$$

where $k \cdot (mz)$ is defined in equation (I.1.15) in §I.1. We note that k, m are determined up to a factor in K_P, but km is uniquely determined by g, and hence this action is well-defined. Clearly, this action extends the K-action defined in equation (I.10.2), in particular, $k \cdot X_P = X_{{}^k P}$.

To prove the continuity of this G-action, we first show that if $g_j \to g_\infty$ in G and a sequence $y_j \in X$ converges to $z_\infty \in X_P$, then $g_j y_j \to g_\infty z_\infty$.

By definition, y_j can be written in the form $y_j = k_j n_j a_j z_j$ such that (1) $k_j \in K$, $k_j \to e$, (2) $a_j \in A_P$, and for all $\alpha \in \Phi(P, A_P)$, $\alpha(\log a_j) \to +\infty$, (3) $n_j \in N_P$, $n_j^{a_j} \to e$, and (4) $z_j \in X_P, z_j \to z_\infty$. Write

$$g_j k_j = k'_j m'_j a'_j n'_j,$$

where $k'_j \in K, m'_j \in M_P, a'_j \in A_P, n'_j \in N_P$. Then a'_j, n'_j are uniquely determined by $g_j k_j$ and bounded, and $k'_j m'_j$ converges to the $K M_P$-component of g. By choosing suitable factors in K_P, we can assume that $k'_j \to k$ and $m'_j \to m$, where $g = kman$ as above. Since

$$g_j y_j = k'_j m'_j a'_j n'_j n_j a_j z_j = k'_j {}^{m'_j a'_j}(n'_j n_j) a'_j a_j m'_j z_j$$

and

$$(n'_j n_j)^{a'_j a_j} \to e, \quad m'_j \to m,$$

it follows from the definition of convergence of sequences that $g_j y_j$ converges to $k \cdot (m z_\infty) \in X_{kP}$ in $\overline{X}_{\max}$, which is equal to $g \cdot z_\infty$ as defined above.

The same proof works for a sequence y_j in X_Q for any parabolic subgroup $Q \supset P$. A general sequence in $\overline{X}_{\max}$ follows from combinations of these two cases, and the continuity of this extended G action on $\overline{X}_{\max}$ is proved.

Remark I.10.9 In the above proof, the horospherical decomposition allows us to decompose the g-action into two components. The K-component is easy and the difficulty lies in the P-component. Since it is easy to compute the action of elements of P in terms of the horospherical decomposition associated with P, this difficulty can be overcome. In this step, the N_P-factor in the definition of convergence to boundary points is crucial. We want the N_P-factor to be large enough to absorb bounded elements in N_P but still to be negligible. This motivates the definition of generalized set in equation (I.9.2).

On the other hand, in [GJT], convergence to boundary points in the dual-cell compactification $X \cup \Delta^*(X)$ is defined in terms of the Cartan decomposition $X = K\overline{A^+}x_0$, and it is difficult to compute the Cartan decomposition of $g(ke^H)$ in terms of $g, k \in K$ and $e^H \in \overline{A^+}$. Because of this difference, the continuous extension of the G-action to $X \cup \Delta^*(X)$ is not easy. In fact, the continuity of the G-action is not proved directly there. As mentioned earlier, this is one of the motivations of Part I of this book.

Next we identify this compactification with the maximal Satake compactification $\overline{X}^S_{\max}$ in §I.4.

Proposition I.10.10 *For any Satake compactification $\overline{X}^S_\tau$, the identity map on X extends to a continuous G-equivariant surjective map $\overline{X}_{\max} \to \overline{X}^S_\tau$.*

Proof. Since every boundary point of $\overline{X}_{\max}$ is the limit of a sequence of points in X, by [GJT, Lemma 3.28], it suffices to show that for any unbounded sequence y_j in X that converges in $\overline{X}_{\max}$, then y_j also converges in $\overline{X}^S_\tau$. By definition, there exists a parabolic subgroup P such that y_j can be written as $y_j = k_j n_j a_j m_j K_P$, where $k_j \in K, n_j \in N_P, a_j \in A_P, m_j K_P \in X_P$ satisfy the conditions (1) $k_j \to e$, (2) $n_j^{a_j} \to e$, (3) for all $\alpha \in \Phi(P, A_P)$, $\alpha(\log a_j) \to +\infty$, (4) $m_j K_P$ converges to $m_\infty K_P$ for some m_∞. Then under the map $i_\tau : X \to P(\mathcal{H}_n)$,

$$i_\tau(y_j) = [\tau(k_j n_j a_j m_j)\tau(k_j n_j a_j m_j)^*]$$
$$= [\tau(k_j)\tau(a_j)\tau(n_j^{a_j})\tau(m_j)\tau(m_j)^*\tau(n_j^{a_j})^*\tau(a_j)\tau(k_j)^*].$$

Let P_0 be a minimal parabolic subgroup contained in P. Write $P = P_I$. As in the proof of Proposition I.9.8 (or I.9.11), we can assume, with respect to a suitable basis, that $\tau(a_j) = \mathrm{diag}(a_j^{\mu_1}, \dots, a_j^{\mu_n})$ and that $\mu_1, \dots, \mu_l$ are all the weights whose supports are contained in I. Then as $j \to +\infty$,

$$[\mathrm{diag}(a_j^{\mu_1}, \dots, a_j^{\mu_n})] \to [\mathrm{diag}(1, \dots, 1, 0, \dots, 0)];$$

and hence

$$i_\tau(y_j) \to [\mathrm{diag}(1, \dots, 1, 0, \dots, 0)\tau(m_\infty)\tau(m_\infty)^*\mathrm{diag}(1, \dots, 1, 0, \dots, 0)^*].$$

Proposition I.10.11 *For the maximal Satake compactification $\overline{X}^S_{\max}$, the map $\overline{X}_{\max} \to \overline{X}^S_{\max}$ in Proposition I.10.10 is a homeomorphism.*

Proof. Since both $\overline{X}_{\max}$ and $\overline{X}^S_{\max}$ are compact and Hausdorff, it suffices to show that the continuous map $\overline{X}_{\max} \to \overline{X}^S_{\max}$ is injective. By equation (I.4.31), $\overline{X}^S_{\max} = X \cup \coprod_P X_P$. By the proof of the previous proposition, a sequence $y_j = k_j n_j a_j m_j K_P$ in X satisfying the conditions above with $m_j K_P \to m_\infty K_P$ in $\overline{X}_{\max}$ converges to the same limit as the sequence $a_j m_j K_P$. Under the above identification of $\overline{X}^S_{\max}$, $a_j m_j K_P$ converges to $m_\infty K_P \in X_P$ in $\overline{X}^S_{\max}$. This implies that the map $\overline{X}_{\max} \to \overline{X}^S_{\max}$ is the identity map under the identification $\overline{X}_{\max} = X \cup \coprod_P X_P = \overline{X}^S_{\max}$, and hence is injective.

I.10.12 Summary and comments. In this section, we followed the uniform, intrinsic method to construct the maximal Satake compactification. This construction explains the motivation and definition of generalized Siegel sets, and how they are used in proving the Hausdorff property of the compactification and the continuous extension of the G-action to the compactification.

I.11 Uniform construction of nonmaximal Satake compactifications

In this section we apply the uniform method in §I.8 to construct the nonmaximal Satake compactifications. In contrast to the case of the maximal Satake compactification, we choose a proper G-invariant collection of parabolic subgroups of G and attach their boundary components.

We first motivate and define the collection of these parabolic subgroups and their boundary components in I.11.2–I.11.7. Then we describe the convergence of interior points to these boundary components in I.11.10. Neighborhood bases of boundary points are given in I.11.11–I.11.13. Both steps are more complicated than in the maximal Satake compactification $\overline{X}^S_{\max}$. One reason is that the boundary components are not parametrized by all parabolic subgroups but rather by saturated parabolic subgroups.

After proving the Hausdorff property (Proposition I.11.14) and relations between these compactifications and the maximal Satake compactification, we use them to prove the compactness (Proposition I.11.16) and the continuous extension of the G-action on X (Proposition I.11.17). Finally we identify them with the Satake compactifications $\overline{X}^S_\tau$ in §I.4 (Proposition I.11.18).

This construction of the nonmaximal Satake compactifications is more complicated than the original one in [Sat1]. But the gain is the explicit description of neighborhoods of boundary points and sequences converging to them, which are useful for applications.

I.11.1 Let P_0 be a minimal parabolic subgroup, and $\mu \in \overline{\mathfrak{a}^{*+}_{P_0}}$ a *dominant weight*, both of which will be fixed for the rest of this section. For each such μ, we will construct a compactification $\overline{X}_\mu$ such that the Satake compactification $\overline{X}^S_\tau$ is isomorphic to $\overline{X}^{\mu_\tau}$. Before defining the boundary components, we need to choose the collection of parabolic subgroups and to refine the horospherical decomposition of X, which are needed to attach the boundary components at infinity.

To motivate the choice of the collection of parabolic subgroups, we recall the decomposition of $\overline{X}^S_\tau$ into boundary components in Proposition I.4.38. The boundary components are parametrized by their normalizers, which are given by μ_τ-saturated parabolic subgroups.

First we introduce the analogous definitions. A subset $I \subset \Delta(P_0, A_{P_0})$ is called *μ-connected* if the union $I \cup \{\mu\}$ is connected, i.e., it cannot be written as a disjoint union $I_1 \coprod I_2$ such that elements in I_1 are perpendicular to elements in I_2 with respect to a positive definite inner product on $\mathfrak{a}^*_{P_0}$ invariant under the Weyl group.

A standard parabolic subgroup $P_{0,I}$ is called *μ-connected* if I is μ-connected. Let I' be the union of roots in $\Delta(P_0, A_{P_0})$ that are perpendicular to $I \cup \{\mu\}$. Then $J = I \cup I'$ is called the *μ-saturation* of I, and $P_{0,J}$ is called the *μ-saturation* $P_{0,I}$. Any parabolic subgroup P conjugate to a μ-connected standard parabolic subgroup $P_{0,I}$ is called μ-connected. Any parabolic subgroup Q conjugated to $P_{0,J}$ for a μ-saturated J is called a μ-saturated parabolic subgroup.

For any standard parabolic subgroup $P_{0,J}$, let P_{0,I_J} be the unique maximal one among all the μ-connected standard parabolic subgroups contained in $P_{0,J}$, i.e., I_J is the largest μ-connected subset contained in J. Then I_J is called the *μ-reduction* of J, and the parabolic subgroup P_{0,I_J} is called the *μ-reduction* of $P_{0,J}$. In general, for any parabolic subgroup $Q = {}^g P_{0,J}$, we define a μ-reduction of Q by ${}^g P_{0,I_J}$, a μ-connected parabolic subgroup contained in Q. In fact, it is a maximal element in the collection of μ-connected parabolic subgroups contained in Q. It should be pointed out that this μ-reduction of Q is not unique. In fact, for any $q \in Q - {}^g P_{0,I_J}$, ${}^{qg} P_{0,I_J}$

is also a μ-reduction. On the other hand, the μ-saturation of a parabolic subgroup is unique.

Clearly, the collection of μ-saturated parabolic subgroups is invariant under the conjugation of G, and will be used to construct the compactification of $\overline{X}_{\mu}$.

I.11.2 Next we define the boundary components of μ-saturated parabolic subgroups in two steps: (1) for the standard ones containing the fixed minimal parabolic subgroup P_0, (2) the general case.

Let Q be a μ-saturated standard parabolic subgroup, $Q = P_{0,J}$. Let $Q = N_Q A_Q M_Q$ be the Langlands decomposition associated with the basepoint $x_0 = K \in X$, and $X_Q = M_Q/(K_Q)$ the boundary symmetric space. Let $I = I_J$ be the μ-reduction of J, $J = I \cup I'$. Let $P_{0,I}$ be the standard μ-reduction of $P_{0,J}$, and $X_{P_{0,I}}$ its boundary symmetric space. Then

$$X_Q = X_{P_{0,I}} \times X_{P_{0,I'}}.$$

Define the boundary component $e(Q)$ of Q by

$$e(Q) = X_{P_{0,I}}. \tag{I.11.1}$$

For any μ-saturated parabolic subgroup Q, write $Q = {}^k P_{0,J}$, where J is μ-saturated. Then ${}^k P_{0,I}$ is a μ-reduction of $P_{0,J}$. Define the boundary component $e(Q)$ of Q by

$$e(Q) = X_{{}^k X_{P_{0,I}}}.$$

To show that this definition is independent of the choice of k, we define it directly without using conjugation to the standard parabolic subgroups.

Let Q be any μ_τ-saturated parabolic subgroup. Let $\mathfrak{g} = \mathfrak{k} \oplus \mathfrak{p}$ be the Cartan decomposition of $\mathfrak{g}$, and $\mathfrak{a}$ a maximal subalgebra of $\mathfrak{p}$ contained in the Lie algebra of Q. Then $A = \exp \mathfrak{a}$ is a maximal split torus contained in Q and stable under the Cartan involution θ associated with the basepoint x_0.

Let P be a minimal parabolic subgroup contained in Q. Then A is a split component of P and $Q = P_J$, where $J \subset \Delta(P, A)$. Since P is conjugate to P_0, μ defines a dominant weight on $\mathfrak{a}$, still denoted by μ. Let $I \subset J$ be the maximal μ-connected subset of J, and $I' = J \setminus I$ its orthogonal complement. Let $X_I = X_{P_I}$, $X_{I'} = X_{P_{I'}}$.

Proposition I.11.3 *The boundary symmetric space of Q splits as*

$$X_Q = X_I \times X_{I'}, \tag{I.11.2}$$

and the splitting is independent of the choice of the minimal parabolic subgroup P and the splitting component A. Define the boundary component $e(Q)$ of Q by

$$e(Q) = X_I.$$

Proof. Clearly, this splitting of X_Q does not depend on the choice of the minimal parabolic subgroup P contained in Q. In fact, in the Langlands decomposition of P, $P = N_P A M_P$, M_P is the same for all the different choices of P.

To show that the factors X_I, $X_{I'}$, and the splitting in equation (I.11.2) do not depend on the choice of the subalgebra $\mathfrak{a}$, we determine all such choices of $\mathfrak{a}$. Note that

$$\mathfrak{a} = \mathfrak{a}_J \oplus \mathfrak{a}^I \oplus \mathfrak{a}^{I'},$$

where $\mathfrak{a}^I$ is the orthogonal complement of $\mathfrak{a}_I$ in $\mathfrak{a}$, and $\mathfrak{a}^{I'}$ is the orthogonal complement of $\mathfrak{a}_{I'}$ in $\mathfrak{a}$. Since $\exp \mathfrak{a}^I$ is a maximal split torus in M_{P_I} stable under the Cartan involution associated with K_{P_I} and $\exp \mathfrak{a}^{I'}$ is a maximal split torus in $M_{P_{I'}}$ stable under the Cartan involution associated with $K_{P_{I'}}$, for any $k \in K_{P_I}$ and $k' \in K_{P_{I'}}$,

$$\mathfrak{a}' = \mathfrak{a}_J \oplus Ad(k)\mathfrak{a}^I \oplus Ad(k')\mathfrak{a}^{I'}$$

is a maximal subalgebra of $\mathfrak{p}$ contained in the Lie algebra of Q, and any such abelian subalgebra of the Lie algebra of Q is of this form $\mathfrak{a}'$ for some k, k'. Since $Ad(k')$ acts as the identify on M_{P_I} and $Ad(k)$ also acts as the identify on $M_{P_{I'}}$, the factors X_I, $X_{I'}$ and the splitting of X_Q induced from the subgroup $A' = \exp \mathfrak{a}'$ are the same. Hence, the splitting in equation (I.11.2) is canonical.

I.11.4 The discussion of the standard Satake compactification $\overline{\mathcal{P}_n}^S$ in I.4 is a good example illustrating these definitions. Let P_0 be the minimal parabolic subgroup consisting of upper triangular matrices in $G = \mathrm{PSL}(n, \mathbb{C})$. Then

$$\begin{aligned} \mathfrak{a}_{P_0} &= \{\mathrm{diag}(t_1, \dots, t_n) \mid t_i \in \mathbb{R}, t_1 + \cdots + t_n = 0\}, \\ \Phi(P_0, A_{P_0}) &= \{t_i - t_j \mid 1 \leq i < j \leq n\}, \\ \Delta(P_0, A_{P_0}) &= \{\alpha_1 = t_1 - t_2, \dots, \alpha_{n-1} = t_{n-1} - t_n\}. \end{aligned}$$

Let $\mu = i_1$. Then μ-connected subsets of Δ are of the form $I = \{\alpha_1, \dots, \alpha_{k_1}\}$. The μ-saturation of $P_{0,I}$ is the maximal parabolic subgroup $P_{0,\Delta-\{\alpha_k\}}$. Hence, a parabolic subgroup is μ-saturated if and only if it is a maximal parabolic subgroup. For the parabolic subgroup $Q = P_{0,\Delta-\{\alpha_k\}}$, the canonical splitting of X_Q in equation (I.11.2) is given by

$$X_Q = \mathcal{P}_k \times \mathcal{P}_{n-k},$$

and the boundary component by

$$e(Q) = \mathcal{P}_k.$$

I.11.5 Another example is the *Baily–Borel compactification* of Hermitian symmetric spaces. Assume that X is an irreducible Hermitian symmetric space of noncompact type. By Proposition I.5.18, the root system $\Phi(\mathfrak{g}, \mathfrak{a})$ is either of type C_r or BC_r.

If μ is connected only to the last distinguished root α_r, then μ-connected subsets of $\Delta = \Delta(\mathfrak{g}, \mathfrak{a})$ are of the form $I = \{\alpha_\ell, \dots, \alpha_r\}$. Let P_0 be the minimal parabolic subgroup containing $\exp \mathfrak{a}$ such that Δ is the set of simple roots in $\Phi(P_0, \exp \mathfrak{a})$. Then the μ-saturation of $P_{0,I}$ is equal to $P_{0,\Delta-\{\alpha_{\ell-1}\}}$, which is a maximal parabolic subgroup. Let $Q = P_{0,\Delta-\{\alpha_{\ell-1}\}}$. Then the boundary symmetric space X_Q is not Hermitian, but the boundary component $e(Q) = X_I$ is a Hermitian symmetric space of lower rank. In fact, the root system of X_I is generated by I and has the same type as X. Since the Baily–Borel compactification of X was used to compactify locally symmetric spaces $\Gamma\backslash X$ into projective varieties, in particular complex analytic spaces, it is important that the boundary components be complex spaces as well. This explains the necessity of the splitting of the boundary symmetric space X_Q in equation (I.11.2).

Remark I.11.6 In the above two examples, the μ-saturated parabolic subgroups are maximal parabolic subgroups. It should be pointed out that this is not always the case. In fact, in the above examples, assume the rank $r \geq 5$, and the dominant μ is exactly connected to the two end roots α_1 and α_r. Let $I = \{\alpha_1, \alpha_r\}$. Then the μ-saturation of $P_{0,I}$ is equal to $P_{0,\Delta-\{\alpha_2,\alpha_{r-1}\}}$, which is not a maximal parabolic subgroup of G.

I.11.7 For any parabolic subgroup Q, by equation (I.1.13), the boundary symmetric space X_Q is a homogeneous space of Q. When Q is μ-saturated, the boundary component $e(Q) = X_I$ is also a homogeneous space of Q. In fact, we can define an action of Q on X_I as follows. Let

$$\pi : X_Q = X_I \times X_{I'} \to X_I$$

be the projection onto the first factor. For any $q \in Q$ and $z = mK_{P_{0,I}} \in X_I$, define the action

$$q \cdot z = \pi(q \cdot mK_Q), \tag{I.11.3}$$

where mK_Q is a point in X_Q and $q \cdot mK_Q$ is given by the action of Q on X_Q.

It can be shown that this defines a group action of Q on $e(Q)$. Briefly, for any $q \in Q$, there exist $m_q \in M_{P_{0,I}}, m'_q \in M_{P_{0,I'}}$ such that for any $(z, z') \in X_I \times X_{I'} = X_Q$,

$$q \cdot (z, z') = (m_q z, m'_q z') \in X_I \times X_{I'}.$$

Hence, for $q_1, q_2 \in Q$,

$$q_1 q_2 \cdot (z, z') = (m_{q_1} m_{q_2} z, m'_{q_1} m'_{q_2} z'),$$

which implies that

$$q_1 q_2 \cdot z = q_2 \cdot (q_1 \cdot z).$$

The stabilizer of the basepoint $x_0 = K_{P_{0,I}} \in X_I$ can also be described explicitly (see [BJ2, §5]).

I.11.8 Once the boundary components are defined, we are ready to define the compactification $\overline{X}_\mu$. As a set,

$$\overline{X}_\mu = X \cup \coprod_{\mu\text{-saturated Q}} e(Q) = X \cup \coprod_{\mu\text{-saturated Q}} X_I. \tag{I.11.4}$$

As in the previous section on $\overline{X}_{\max}$, to define the topology on the set of boundary components of $\overline{X}_\mu$, we need to define a K-action on the set of boundary components. Let Q be any μ-saturated parabolic subgroup, and $e(Q) = X_I = X_{P_I}$ its boundary component. For any $k \in K$ and $z = mK_{P_I}$, define

$$k \cdot z = {}^k m K_{{}^k P_I} \in e({}^k Q). \tag{I.11.5}$$

To describe the convergence of interior points of X to the boundary points in $\overline{X}_{\max}$, the horospherical decomposition was used. For the compactification $\overline{X}_\mu$, we need the following more refined decomposition.

Lemma I.11.9 *Let Q be a μ-saturated parabolic subgroup, P a minimal parabolic subgroup contained in Q. Let $Q = P_J$, and P_I the μ-connected reduction of Q. For any parabolic subgroup R satisfying $P_I \subseteq R \subseteq Q$, write $R = P_{J'}$ and $J' = I \cup I'$, where I' is perpendicular to I. Let $X_R = X_I \times X_{P_{I'}}$ be the decomposition similar to that in equation (I.11.2). Then the horospherical decomposition of X with respect to R,*

$$X = N_R \times A_R \times X_R,$$

can be refined to

$$X = N_R \times A_R \times X_I \times X_{P_{I'}}. \tag{I.11.6}$$

Proof. When $R = Q$, the decomposition $X_Q = X_I \times X_{P_{I'}}$ was described in equation (I.11.2). The general case is similar. As in the case $R = Q$, the decomposition $X_R = X_I \times X_{P_{I'}}$ is independent of the choice of the minimal parabolic subgroup P contained in R. For simplicity, $X_{I'}$ is also denoted by $X_{I'}$ and the decomposition in equation (I.11.6) written as

$$X = N_R \times A_R \times X_I \times X_{I'}. \tag{I.11.7}$$

I.11.10 A topology on $\overline{X}_\mu$ is given as follows:

1. For a μ-saturated parabolic subgroup Q, let $P_0 \subseteq Q$ be a minimal parabolic subgroup, and $P_{0,I}$ a μ-connected reduction of Q. Then an unbounded sequence y_j in X converges to a boundary point $z_\infty \in e(Q) = X_I$ if there exists a parabolic subgroup R that is contained in Q and contains $P_{0,I}$ such that in the refined

horospherical decomposition $X = N_R \times A_R \times X_I \times X_{I'}$ in equation (I.11.6), y_j can be written in the form $y_j = k_j n_j a_j z_j z'_j$ such that the factors $k_j \in K$, $n_j \in N_R$, $a_j \in A_R$, $z_j \in X_I$, $z'_j \in X_{I'}$ satisfy the following conditions:

(a) the image of k_j in the quotient $K/K \cap \mathcal{Z}(e(Q))$ converges to the identity coset,

(b) for $\alpha \in \Delta(R, A_R)$, $a_j^{\alpha} \to +\infty$,

(c) $n_j^{a_j} \to e$,

(d) $z_j \to z_\infty$,

(e) z'_j is bounded.

2. For a pair of μ-saturated parabolic subgroups Q_1 and Q_2 such that a μ-connected reduction of Q_1 is contained in a μ-connected reduction of Q_2, let Q'_1 be the unique parabolic subgroup in M_{Q_2} determined by $Q_1 \cap Q_2$. For a sequence $k_j \in K$ whose image in $K/K \cap \mathcal{Z}(e(Q_1)$ converges to the identity coset, and a sequence y_j in $e(Q_2) = X_{I(Q_2)}$, the sequence $k_j y_j$ in $\overline{X}_\mu$ converges to $z_\infty \in e(Q_1) = X_{I(Q_1)}$ if y_j satisfies the same condition as in part (1) above when G is replaced by the subgroup $G_{I(Q_2)}$ of M_{Q_2} whose symmetric spaces of maximal compact subgroups is $X_{I(Q_2)}$, and Q by $Q'_1 \cap G_{I(Q_2)}$.

These are special convergent sequences, and their (finite) combinations give general convergent sequences. We note that all the μ-connected reductions of Q are conjugate under $\mathcal{Z}(e(Q))$. Hence in (1), it does not matter which μ-connected reduction is used, and we can fix a minimal parabolic subgroup $P_0 \subseteq Q$ and the μ-connected reduction $P_{0,I}$ of Q. Then there are only finitely many parabolic subgroups R with $P_{0,I} \subseteq R \subseteq Q$.

The definition of the convergent sequences is motivated by the fact that the maximal Satake compactification $\overline{X}^S_{\max}$ of X dominates all nonmaximal Satake compactifications and the characterizations of convergent sequences of $\overline{X}^S_{\max}$ in §I.10, together with the observation that the action by elements in the centralizer $\mathcal{Z}(e(Q))$ will not change the convergence of sequences to points in $e(Q)$.

I.11.11 Neighborhoods of boundary points can be given explicitly.

For any μ-saturated parabolic subgroup Q and a boundary point $z_\infty \in e(Q)$, let V be a neighborhood of z_∞ in $e(Q)$.

Fix a minimal parabolic subgroup P_0 contained in Q. Let $\mathfrak{a}_{P_0}$ be the Lie algebra of its split component A_{P_0}, W the Weyl group of $\mathfrak{a}_{P_0}$, and $T \in \mathfrak{a}^+_{P_0}$ a regular vector. The convex hull of the Weyl group orbit $W \cdot T$ is a convex polytope Σ_T in $\mathfrak{a}_{P_0}$, whose faces are in one-to-one correspondence with parabolic subgroups R whose split component $\mathfrak{a}_R$ is contained in $\mathfrak{a}_{P_0}$. Denote the closed face of Σ corresponding to R by σ_R. Then

$$\mathfrak{a}_{P_0} = \Sigma_T \cup \coprod_R (\sigma_R + \mathfrak{a}^+_R), \tag{I.11.8}$$

where $\mathfrak{a}_R \subseteq \mathfrak{a}_{P_0}$ as above. Define

$$\mathfrak{a}_{\mu,Q,T} = \cup_{P \subseteq R \subseteq Q} (\sigma_R + \mathfrak{a}^+_R), \tag{I.11.9}$$

where P ranges over all the μ-connected reductions of Q with $\mathfrak{a}_P \subseteq \mathfrak{a}_{P_0}$, and for each such P, R ranges over all the parabolic subgroups lying between P and Q. Then for any such pair P and R, $\mathfrak{a}_{\mu,Q,T} \cap \mathfrak{a}_P$ is a connected open subset of $\mathfrak{a}_P$; when $R \neq P$, it has the property that when a point moves out to infinity along $\mathfrak{a}_R$ in the direction of the positive chamber, its distance to the boundary of $\mathfrak{a}_{\mu,Q,T} \cap \mathfrak{a}_P$ goes to infinity.

More precisely, write $R = P_J$, $J \neq \emptyset$. Recall from equation (I.1.17) that $\mathfrak{a}_P = \mathfrak{a}_{P_J} \oplus \mathfrak{a}_P^J$ and $\mathfrak{a}_P^J$ is the split component of the parabolic subgroup of M_{P_J} corresponding to P. Then for a sequence $H_j \in \mathfrak{a}_R^+$ with $\alpha(H_j) \to +\infty$ for all $\alpha \in \Delta(R, A_R)$ and any bounded set $\Omega \subset \overline{\mathfrak{a}_P^{J,+}}$, when $j \gg 1$,

$$(H_j + \Omega) \subset \mathfrak{a}_{\mu,Q,T} \cap \mathfrak{a}_P. \tag{I.11.10}$$

The positive chamber $\mathfrak{a}_R^+$ intersects the face σ_R at a unique point T_R, and

$$\exp(\mathfrak{a}_R^+ \cap (\sigma_R + \mathfrak{a}_R^+)) = A_{R,\exp T_R} = \exp(\mathfrak{a}_R^+ + T_R),$$

the shifted chamber defined in equation (I.9.3). The face σ_R is contained in the shift by T_R of the orthogonal complement of $\mathfrak{a}_R$ in $\mathfrak{a}_{P_0}$, and $K_R \exp \sigma_R \cdot x_0$ is a codimension-0 set in X_R. Denote the image of $K_R \exp \sigma_R \cdot x_0$ in $X_{I'}$ under the projection $X_R = X_I \times X_{I'} \to X_{I'}$ by $W_{R,T}$.

For each such parabolic subgroup R that is contained in Q and contains a μ-connected reduction P with $\mathfrak{a}_P \subseteq \mathfrak{a}_{P_0}$, define

$$\begin{aligned} S_{R,\varepsilon,e^{T_R},V\times W_{R,T}} = \{&(n,a,z,z') \in N_R \times A_R \times X_I \times X_{I'} \\ &\mid a \in A_{R,e^{T_R}}, n^a \in B_{N_R}(\varepsilon), z \in V, z' \in W_{R,T}\}, \end{aligned} \tag{I.11.11}$$

a generalized Siegel set in X associated with R defined in equation (I.9.4). Then

$$\exp \mathfrak{a}_{\mu,Q,T} \cdot x_0 \subset \cup_{P\subseteq R\subseteq Q} S_{R,\varepsilon,e^{T_R},V\times W_{R,T}}, \tag{I.11.12}$$

where P ranges over all the μ-connected reductions of Q with $\mathfrak{a}_P \subseteq \mathfrak{a}_{P_0}$, and for each such P, R ranges over all the parabolic subgroups lying between P and Q.

Define

$$S^{\mu}_{\varepsilon,T,V} = \cup_{P\subseteq R\subseteq Q} S_{R,\varepsilon,e^{T_R},V\times W_{R,T}}, \tag{I.11.13}$$

where P ranges over all μ-connected reductions of Q with $\mathfrak{a}_P \subseteq \mathfrak{a}_{P_0}$, and for any such P, R ranges over all parabolic subgroups lying between P and Q. It is important to note that there are only finite parabolic subgroups R in the above union.

For each μ-saturated parabolic subgroup Q' such that one of its μ-connected reductions contains a μ-connected reduction P of Q with $\mathfrak{a}_P \subseteq \mathfrak{a}_{P_0}$, we get a similar set $S^{Q',\mu}_{\varepsilon,T,V}$ in $e(Q')$.

For a compact neighborhood C of the identity coset in $K/K \cap \mathcal{Z}(e(Q))$, let $\tilde{C}$ be the inverse image in K of C for the map $K \to K/K \cap \mathcal{Z}(e(Q))$. Then

$$\tilde{C}(S^{\mu}_{\varepsilon,T,V} \cup_{\mu\text{-saturated}\, Q'} S^{Q',\mu}_{\varepsilon,T,V}) \tag{I.11.14}$$

is a neighborhood of z_∞ in $\overline{X}_\mu$.

Proposition I.11.12 *With the above notation, for a basis of neighborhoods C_i of the identity coset in $K/K \cap \mathcal{Z}(e(Q))$, a sequence of points $T_i \in \mathfrak{a}_{P_0}^+$ with $\alpha(T_i) \to +\infty$ for all $\alpha \in \Delta(P_0, A_{P_0})$, a sequence $\varepsilon_i \to 0$, and a basis V_i of neighborhoods of z_∞ in $e(Q)$, the associated sets in equation (I.11.14) form a neighborhood basis of z_∞ in $\overline{X}_\mu$ with respect to the topology defined by the convergent sequences above.*

See [BJ2, Proposition 5.3] for a proof.

Remark I.11.13 It was shown in [Cas2] and [Ji1] that the closure of the flat $\mathfrak{a}_{P_0} = \mathfrak{a}_{P_0}x_0$ in the Satake compactification $\overline{X}_\tau^S$ is canonically homeomorphic to the closure of the convex hull of the Weyl group orbit $W\mu_\tau$ of the highest weight μ_τ. When μ_τ is generic, we can take $T = \mu_\tau$. For a nongeneric μ_τ, there is a collapsing from Σ_T to the convex hull of $W\mu_\tau$, and all the faces σ_R for $P \subseteq R \subseteq Q$ collapse to the face σ_P. The domain $S^\mu_{\varepsilon,T,V}$ was suggested by this consideration.

Proposition I.11.14 *The topology on $\overline{X}_\mu$ defined above is Hausdorff.*

Proof. It suffices to show that if an unbounded sequence y_j in X converges in $\overline{X}_\mu$, then it has a unique limit. Suppose y_j has two different limits $z_{1,\infty} \in e(Q_1)$, $z_{2,\infty} \in e(Q_2)$, where Q_1, Q_2 are two μ-saturated parabolic subgroups. By passing to a subsequence, we can assume that there exist two parabolic subgroups R_1, R_2, $R_1 \subseteq Q_1$, $R_2 \subseteq Q_2$, such that y_j satisfies the condition (1) in the definition of convergent sequences with respect to both R_1 and R_2. By passing to a further subsequence, we can assume that

$$y_j \in C_j S_{R_1,\varepsilon_j,t_j,V_1} \cap C_j S_{R_2,\varepsilon_j,t_j,V_2}, \tag{I.11.15}$$

where $C_j \subset K$ is a sequence of compact neighborhoods of e converging to e, $\varepsilon_j \to 0$, $t_j \to +\infty$, and V_1, V_2 are compact neighborhoods of $z_{1,\infty}$, $z_{2,\infty}$ respectively. If $R_1 \neq R_2$, equation (I.11.15) contradicts the separation property of general Siegel sets for R_1, R_2 in Proposition I.9.8. If $R_1 = R_2$, we can take V_1, V_2 to be disjoint since $z_{1,\infty} \neq z_{2,\infty}$. Then equation (I.11.15) contradicts the separation property in Proposition I.9.11. These contradictions show that y_j must have a unique limit in $\overline{X}_\mu$.

Proposition I.11.15 (Relations between Satake compactifications) *For any two dominant weights $\mu_1, \mu_2 \in \overline{\mathfrak{a}_{P_0}^{*+}}$, if μ_2 is more regular than μ_1, i.e., if $\mu_i \in \mathfrak{a}_{P_0,I_i}^{*+}$ and $I_2 \subseteq I_1$, then the identity map on X extends to a continuous surjective map from $\overline{X}_{\mu_2} \to \overline{X}_{\mu_1}$. If μ_1, μ_2 belong to the same Weyl chamber face, then the extended map from $\overline{X}_{\mu_1}$ to $\overline{X}_{\mu_2}$ is a homeomorphism. When μ is generic, $\overline{X}_\mu$ is isomorphic to the maximal Satake compactification $\overline{X}^S_{\max}$. Hence, for any $\overline{X}_\mu$, the identity map on X extends to a continuous surjective map from $\overline{X}_{\max}$ to $\overline{X}_\mu$.*

Proof. Since μ_2 is more regular than μ_1, every μ_1-connected parabolic subgroup is also μ_2-connected, and every subset of Δ perpendicular to μ_2 is also perpendicular to

μ_1. This implies that for any μ_1-connected parabolic subgroup P, its μ_2-saturation Q_{μ_2} is contained in its μ_1-saturation Q_{μ_1}.

For every μ_2-saturated parabolic subgroup Q, let P_{μ_2} be a μ_2-connected reduction, and P_{μ_1} a μ_1-connected reduction contained in P_{μ_2}. Then P_{μ_1} is also a maximal μ_1-connected parabolic subgroup contained in P_{μ_2}. The decomposition $X_R = X_I \times X_{P_{0,I'}}$ in equation (I.11.7) for $R = P_{\mu_2}$ gives

$$X_{P_{\mu_2}} = X_{P_{\mu_1}} \times X_{P_{0,I'}}.$$

Let Q_{μ_1} be the μ_1-saturation of P_{μ_1}. Then $Q_{\mu_1} \subset Q$. Define a map $\pi : \overline{X}_{\mu_2} \to \overline{X}_{\mu_1}$ such that it is equal to the identity on X, and on the boundary component $e(Q) = X_{P_{\mu_2}}$, a point $(z, z') \in X_{P_{\mu_1}} \times X_{P_{0,I'}} = X_{P_{\mu_2}}$ is mapped to $z \in X_{P_{\mu_1}} = e(Q_{\mu_1})$. Clearly, π is surjective.

Since the convergence in $\overline{X}_{\mu_1}$ is determined in terms of the refined horospherical coordinates decomposition, using $A_{Q_{\mu_1}} \subset A_Q$, and $X_{P_{\mu_2}} = X_{P_{\mu_1}} \times X_{P_{0,I'}}$, it can be checked easily that any unbounded sequence y_j in X that converges to $(z, z') \in X_{P_{\mu_1}} \times X_{P_{0,I'}} = X_{P_{\mu_2}} \subset \overline{X}_{\mu_2}$ also converges to $z \in X_{P_{\mu_1}} \subset \overline{X}_{\mu_1}$. By [GJT, Lemma 3.28], this proves that the map π is continuous.

Proposition I.11.16 *For any dominant weight $\mu \in \overline{\mathfrak{a}_{P_0}^{*+}}$, $\overline{X}_\mu$ is compact.*

Proof. It was shown in Proposition I.10.7 that $\overline{X}_{\max}$ is compact. By Proposition I.11.15, $\overline{X}_\mu$ is the image of a compact set under a continuous map and hence compact.

Proposition I.11.17 *The G-action on X extends to a continuous action on $\overline{X}_\mu$.*

Proof. First, we define the G-action on $\overline{X}_\mu$. For any μ-saturated parabolic subgroup Q and any point $z \in e(Q)$, and any $g \in G$, write $g = kq$, where $k \in K$ and $q \in Q$. Define

$$g \cdot z = k \cdot (qz),$$

where $qz = q \cdot z$ is defined in equation (I.11.3). Then it can be shown as in the case of $\overline{X}_{\max}$ that the G-action on $\overline{X}_\mu$ is continuous, by using the refined horospherical decomposition in equation (I.11.7).

Proposition I.11.18 *For any Satake compactification $\overline{X}^S_{\max}$, the identity map on X extends to a homeomorphism $\overline{X}_{\mu_\tau} \to \overline{X}^S_\tau$.*

The proof is similar to the proof of Proposition I.10.10. See [BJ2, §5] for details.

Remark I.11.19 Propositions I.11.15 and I.11.18 give a more explicit proof of Proposition I.4.35. In [Zu2], Proposition I.4.35 was proved by comparing the closures of a flat in these compactifications of X. On the other hand, the proof here gives an explicit map and its surjectivity is immediate.

I.11.20 Summary and comments. In this section, we followed the uniform method to construct nonmaximal Satake compactifications. It might be more complicated than the original construction in [Sat1]. On the other hand, the explicit description of convergent sequences to boundary points and a neighborhood base makes it easier to understand the compactifications and is also useful for other purposes.

I.12 Uniform construction of the geodesic compactification

In this section, we construct the geodesic compactification $X \cup X(\infty)$ using the uniform method in §I.8. Though this construction is not as direct as the geometric definition in terms of geodesics in §I.2, it is needed for the uniform construction of the Martin compactification $\overline{X}^M$ in the next section. It also illustrates the use of the geometric realization of the Tits building $\Delta(G)$ in §I.2.

This section is organized as follows. The boundary components are defined in I.12.1. The topology is described in terms of convergent sequences in I.12.2, and neighborhood bases are given in I.12.3. The compactness and Hausdorff property are proved in Proposition I.12.4, and the continuous extension of the G-action on X to the compactification is given in Proposition I.12.5. The identification with the geodesic compactification is given in Proposition I.12.6.

I.12.1 As in the compactification $\overline{X}^S_{\max}$, we use the whole collection of parabolic subgroups. For every parabolic subgroup P, define the geodesic boundary component to be

$$e(P) = \mathfrak{a}_P^+(\infty),$$

an open simplex.

Define

$$\overline{X}^c = X \cup \coprod_P \mathfrak{a}_P^+(\infty).$$

The superscript c stands for the conic compactification (in fact, the geodesic compactification $X \cup X(\infty)$ is called the conic compactification in [GJT]), while $\overline{X}^G$ denotes the Gromov compactification in §I.17.

To describe the topology of the boundary components, we need a K-action on the boundary points, which is described as follows: for $k \in K$, $H \in \mathfrak{a}_P^+(\infty)$,

$$k \cdot H = Ad(k)H \in \mathfrak{a}_{{}^kP}^+(\infty). \qquad \text{(I.12.1)}$$

I.12.2 The topology on $\overline{X}^c$ is given in terms of convergent sequences as follows.

1. A unbounded sequence y_j in X converges to $H_\infty \in \mathfrak{a}_P^+(\infty)$ if and only if y_j can be written as $y_j = k_j n_j a_j z_j$ with $k_j \in K, n_j \in N_P, a_j \in A_P, z_j \in X_P$ satisfying the following conditions:

   (a) $k_j \to e$,

(b) for all $\alpha \in \Delta(P, A_P)$, $\alpha(\log a_j) \to +\infty$ and $\log a_j/\|\log a_j\| \to H_\infty$,
(c) $n_j^{a_j} \to e$,
(d) $d(z_j, x_o)/\|\log a_j\| \to 0$, where $x_0 \in X_P$ is the basepoint K_P, and $d(z_j, x_0)$ is the Riemannian distance on X_P for the invariant metric.

2. For a sequence $k_j \in K$, $k_j \to e$, and a parabolic subgroup P' contained in P and a sequence $H_j \in \mathfrak{a}_{P'}^+(\infty)$, $k_j \cdot H_j$ converges to $H_\infty \in \mathfrak{a}_P^+(\infty)$ if and only if $H_j \to H_\infty \in \overline{\mathfrak{a}_{P'}^+(\infty)}$. (Note that $\mathfrak{a}_P^+(\infty) \subset \overline{\mathfrak{a}_{P'}^+(\infty)}$.)

Combinations of these two special types of convergent sequences give general convergent sequences. It can be checked that they define a convergence class of sequences in Definition I.8.11 and hence define a topology on $\overline{X}^c$.

I.12.3 Neighborhoods of boundary points can be given as follows. Note that for any parabolic subgroup P, the closure of $\mathfrak{a}_P^+(\infty)$ in $\overline{X}^c$ is equal to

$$\overline{\mathfrak{a}_P^+(\infty)} = \coprod_{Q \supseteq P} \mathfrak{a}_Q^+(\infty).$$

Let P_0 be a minimal parabolic subgroup contained in P. Then $\mathfrak{a}_P^+(\infty) \subset \overline{\mathfrak{a}_{P_0}^+(\infty)}$. For $H \in \mathfrak{a}_P^+(\infty)$ and $\varepsilon > 0$, let

$$U_{\varepsilon,H} = \{H' \in \overline{\mathfrak{a}_{P_0}^+(\infty)} \mid \|H' - H\| < \varepsilon\},$$

a neighborhood of H in $\overline{\mathfrak{a}_{P_0}^+(\infty)}$. For $t > 0$, let

$$\begin{aligned} V_{\varepsilon,t,H} = \{&(n, a, z) \in N_P \times A_P \times X_P = X \\ &\mid a \in A_{P,t},\ \log a/\|\log a\| \in U_{\varepsilon,H},\ n^a \in B_{N_P}(\varepsilon),\ d(z, x_0)/\|\log a\| < \varepsilon\}. \end{aligned}$$

Let C be a compact neighborhood of e in K. Then the set

$$C V_{\varepsilon,t,H} \cup C U_{\varepsilon,H}$$

is a neighborhood of H in $\overline{X}^c$.

For sequences $\varepsilon_i \to 0$, $t_i \to +\infty$, and a basis C_i of neighborhoods of e in K, $C_i V_{\varepsilon_i,t_i,H} \cup C U_{\varepsilon_i,H}$ forms a basis of neighborhoods of H in $\overline{X}^c$.

Proposition I.12.4 *The space $\overline{X}^c$ is a compact Hausdorff space.*

Proof. Since every parabolic subgroup is conjugate under K to a standard parabolic subgroup $P_{0,I}$, we have

$$\overline{X}^c = K\left(\exp \overline{\mathfrak{a}_{P_0}^+} x_0 \cup \coprod_{\emptyset \subseteq I \subset \Delta} \mathfrak{a}_{P_{0,I}}^+(\infty)\right) = K(\exp \overline{\mathfrak{a}_{P_0}^+} x_0 \cup \overline{\mathfrak{a}_{P_0}^+(\infty)}),$$

where P_0 is a minimal parabolic subgroup. Since K and $\overline{\mathfrak{a}^+_{P_0}(\infty)}$ are compact, to prove that $\overline{X}^c$ is compact, it suffices to show that every unbounded sequence of the form $\exp H_j x_0$, $H_j \in \overline{\mathfrak{a}^+_{P_0}}$, has a convergent subsequence. Replacing by a subsequence, we can assume that $H_j/\|H_j\|$ converges to $H_\infty \in \mathfrak{a}^+_{P_{0,I}}(\infty)$ for some I. By decomposing

$$H_j = H_{j,I} + H_j^I, \quad H_{j,I} \in \mathfrak{a}_{P_0,I}, \ H_j^I \in \mathfrak{a}^I_{P_0},$$

it follows immediately that $\exp H_j x_0 = \exp H_{j,I}(\exp H_j^I x_0)$ converges to H_∞ in $\overline{X}^c$.

To prove the Hausdorff property, let $H_1, H_2 \in \overline{X}^c$ be two distinct points. Clearly they admit disjoint neighborhoods if at least one of them belongs to X. Assume that $H_i \in \mathfrak{a}^+_{P_i}(\infty)$ for some parabolic subgroups P_1, P_2.

First consider the case that $P_1 = P_2$. Let P_0 be a minimal parabolic subgroup contained in P_1, and let $U_{\varepsilon,1}, U_{\varepsilon,2}$ be two neighborhoods of H_1, H_2 in $\overline{\mathfrak{a}^+_{P_0}(\infty)}$ with $\overline{U_{\varepsilon,1}} \cap \overline{U_{\varepsilon,2}} = \emptyset$. Let C be a small compact neighborhood of e in K such that for all $k_1, k_2 \in C$, $k_1 U_{\varepsilon,1} \cap k_2 U_{\varepsilon,2} = \emptyset$. We claim that the neighborhoods $CV_{\varepsilon,t,H_1} \cup CU_{\varepsilon,1}$, $CV_{\varepsilon,t,H_2} \cup CU_{\varepsilon,2}$ are disjoint when t is sufficiently large, ε, and C are sufficiently small.

By the choice of C,

$$CU_{\varepsilon,1} \cap CU_{\varepsilon,2} = \emptyset.$$

We need to show that

$$CV_{\varepsilon,t,H_1} \cap CV_{\varepsilon,t,H_2} = \emptyset.$$

If not, there exist sequences $\varepsilon_j \to 0$, $t_j \to +\infty$, $C_j \to e$ such that

$$C_j V_{\varepsilon_j,t_j,H_1} \cap C_j V_{\varepsilon_j,t_j,H_2} \neq \emptyset.$$

Let $y_j \in C_j V_{\varepsilon_j,t_j,H_1} \cap C_j V_{\varepsilon_j,t_j,H_2}$. Since $y_j \in C_j V_{\varepsilon_j,t_j,H_1}$, then y_j can be written as $y_j = k_j n_j a_j z_j$ with the components $k_j \in K, n_j \in N_{P_1}, a_j \in A_{P_1}, z_j \in X_{P_1}$ satisfying (1) $k_j \to e$, (2) $\|\log a_j\| \to +\infty$, $\log a_j/\|\log a_j\| \to H_1$, (3) $n_j^{a_j} \to e$, (4) $d(z_j, x_0)/\|\log a_j\| \to 0$. Similarly, y_j can be written as $y_j = k'_j n'_j a'_j z'_j$ with k'_j, n'_j, a'_j, z'_j, satisfying similar properties and $\log a'_j/\|\log a'_j\| \to H_2$.

The idea of the proof is that we can ignore the K-, N_P-components and use the separation of geodesics with different directions to get a contradiction. Specifically,

$$d(y_j, x_o) = (1 + o(1))\|\log a_j\| = (1 + o(1))\|\log a'_j\|;$$

hence

$$\|\log a'_j\| = (1 + o(1))\|\log a_j\|$$

and

$$\begin{aligned} d(y_j, k_j a_j x_0) &= d(k_j n_j a_j z_j, k_j a_j x_0) = d(n_j a_j z_j, a_j x_0) \\ &= d(n_j^{a_j} z_j, x_0) = o(1)\| \log a_j \|, \\ d(y_j, k'_j a'_j x_0) &= o(1)\| \log a'_j \|. \end{aligned}$$

Since X is simply connected and nonpositively curved, $H_1 \neq H_2$, and $k_j a_j x_0, k'_j a'_j x_0$ lie on two geodesics from x_0 with a uniform separation of angle between them, comparison with the Euclidean space gives

$$d(k_j a_j x_0, k'_j a'_j x_0) \geq c_0 \| \log a_j \|$$

for some positive constant c_0. This contradicts the inequality

$$d(k_j a_j x_0, k'_j a'_j x_0) \leq d(y_j, k_j a_j x_0) + d(y_j, k'_j a'_j x_0) = o(1)\| \log a_j \|.$$

The claim is proved.

The case $P_1 \neq P_2$ can be proved similarly. In fact, for suitable neighborhoods U_{ε, H_1}, U_{ε, H_2} and a neighborhood C of e in K such that for all $k_1, k_2 \in C$,

$$k_1 U_{\varepsilon, H_1} \cap k_2 U_{\varepsilon, H_2} = \emptyset,$$

the same proof works.

Proposition I.12.5 *The G-action on X extends to a continuous action on $\overline{X}^c$.*

Proof. For $g \in G$ and $H \in \mathfrak{a}_P^+(\infty)$, write $g = kp$ with $k \in K$ and $p \in P$. Define

$$g \cdot H = Ad(k) H \in \mathfrak{a}_{{}^k P}.$$

Since k is uniquely determined up to a factor in K_P and K_P commutes with A_P, this action is well-defined and extends the K-action in equation (I.12.1). Clearly, P fixes $\mathfrak{a}_P^+(\infty)$.

To show that it is continuous, by [GJT, Lemma 3.28], it suffices to show that for any unbounded sequence y_j in X, if y_j converges to H_∞ in $\overline{X}^c$, then $g y_j$ converges to $g H_\infty$. By definition, y_j can be written as $y_j = k_j n_j a_j z_j$ with (1) $k_j \in K, k_j \to e$, (2) $a_j \in A_P$, for all $\alpha \in \Delta(P, A_P)$, $\alpha(\log a_j) \to +\infty$, $\log a_j / \| \log a_j \| \to H_\infty$, (3) $n_j \in N_P$, $n_j^{a_j} \to e$, and (4) $z_j \in X_P$, $d(z_j, x_0)/\| \log a_j \| \to 0$. Write

$$g k_j = k'_j m'_j a'_j n'_j,$$

where $k'_j \in K, m'_j \in M_P, n'_j \in N_P$ and $a'_j \in A_P$. Then m'_j, n'_j, a'_j are bounded, and ${}^{k'_j}P$ converges to ${}^k P$. Since

$$g y_j = k'_j m'_j a'_j n'_j n_j a_j z_j = k'_j \, {}^{m'_j a'_j}(n'_j n_j) a'_j a_j (m'_j z_j),$$

it is clear that $g y_j$ converges to $Ad(k) H_\infty \in \mathfrak{a}_{{}^k P}^+(\infty)$, which is equal to $g H_\infty$ by definition.

Proposition I.12.6 *The identity map on X extends to a continuous map $\overline{X}^c \to X \cup X(\infty)$, and this map is a homeomorphism.*

Proof. To prove that the identity map extends to a continuous map $\overline{X}^c \to X \cup X(\infty)$, by [GJT, Lemma 3.28], it suffices to prove that if an unbounded sequence in X converges in $\overline{X}^c$, then it also converges in $X \cup X(\infty)$. For a unbounded sequence y_j in X that converges to $H_\infty \in \mathfrak{a}_P^+(\infty)$ in $\overline{X}^c$, y_j can be written as $y_j = k_j n_j a_j z_j$, where the components satisfy (1) $k_j \in K, k_j \to e$, (2) $\| \log a_j \| \to +\infty$, $\log a_j / \| \log a_j \| \to H_\infty$, (3) $n_j^{a_j} \to e$, (4) $d(z_j, x_0)/\| \log a_j \| \to 0$.

By the definition of $X \cup X(\infty)$ in §I.2, the geodesic passing through $a_j x_0$ and x_0 clearly converges to the geodesic $\exp t H_\infty x_0$. Since $k_j \to e$, the geodesic passing through $k_j a_j x_0$ and x_0 also converges to $\exp t H_\infty x_0$. We claim that the geodesic passing through y_j and x_0 also converges to $\exp t H_\infty x_0$.

Since

$$d(k_j n_j a_j z_j, k_j n_j a_j x_0) = d(z_j, x_0),$$

and hence

$$d(k_j n_j a_j z_j, k_j n_j a_j x_0)/\| \log a_j \| \to 0,$$

comparison with the Euclidean space shows that both sequences y_j and $k_j n_j a_j x_0$ will converge to the same limit if $k_j n_j a_j x_0$ converges in $X \cup X(\infty)$. Since

$$d(k_j n_j a_j x_0, k_j a_j x_0) = d(n_j^{a_j} x_0, x_0) \to 0$$

and the geodesic passing through $k_j a_j x_0$ and x_0 clearly converges to the geodesic $\exp t H_\infty x_0$, it follows that $k_j n_j a_j x_0$ and hence y_j converges to $H_\infty \in X \cup X(\infty)$. Therefore, the identity map on X extends to a continuous map $\overline{X}^c \to X \cup X(\infty)$.

To show that this extended map is a homeomorphism, it suffices to prove that it is bijective, since $\overline{X}^c$ and $X \cup X(\infty)$ are both compact and Hausdorff. By Proposition I.2.16,

$$X \cup X(\infty) = X \cup \coprod_P \mathfrak{a}_P^+(\infty).$$

Under this identification, the map $\overline{X}^c \to X \cup X(\infty)$ becomes the identity map and is hence bijective.

I.12.7 Summary and comments. By Propositions I.2.16 and I.2.19, the sphere at infinity $X(\infty)$ is the underlying space of the spherical Tits building $\Delta(G)$. The point here is to describe the topology of the geodesic compactification, or the convergence of interior points to the boundary points, in terms of the horospherical decomposition. For the maximal Satake compactification $\overline{X}^S_{\max}$, the boundary complex is dual to the Tits building $\Delta(G)$. For the geodesic compactification $X \cup X(\infty)$, the boundary complex is exactly the Tits building $\Delta(G)$.

I.13 Uniform construction of the Martin compactification

In this section, we apply the uniform method in §I.8 to construct a compactification $\overline{X}^M$ that is isomorphic to the Martin compactification $X \cup \partial_\lambda X$ for any λ below the bottom of the spectrum $\lambda_0(X)$ as recalled in §I.7. Note that for different values of λ, $X \cup \partial_\lambda X$ are isomorphic to each other. In $\overline{X}^S_{\max} = \overline{X}_{\max}$, the boundary is a cell-complex dual to the Tits building $\Delta(G)$, while in $X \cup X(\infty) = \overline{X}^c$, the boundary is a cell-complex equal to $\Delta(G)$. On the other hand, it will be seen below that the boundary of $\overline{X}^M$ is a combination of the boundaries of $\overline{X}_{\max}$ and $\overline{X}^c$ and hence is a cell-complex of more complicated nature.

This section is organized as follows. First we define the boundary component of parabolic subgroups in §I.13.1 and explain the reason for this choice of boundary components in §I.13.2. In fact, there are two choices, and the other one leads to an inductive, complicated construction. The topology in terms of convergent sequences is given in I.13.4. Neighborhood bases are given in Proposition I.13.6. The continuous extension of the G-action on X to the compactification is given in Proposition I.13.7. The compactness is given in Proposition I.13.8. In Proposition 1.13.9, $\overline{X}^M$ is shown to be isomorphic to the least common refinement of $X \cup (\infty)$ and the $\overline{X}^S_{\max}$. The compactification $\overline{X}^M$ is illustrated via the example of $X = D \times D$ in I.13.10. Then we show that the uniform construction gives a Hausdorff compactification $\overline{X}^M$. Finally we show that the compactification $\overline{X}^M$ is isomorphic to the Martin compactification $X \cup \partial_\lambda X$, $\lambda < \lambda_0(X)$.

I.13.1 As in the case of $\overline{X}_{\max}$, we choose the full collection of parabolic subgroups. For any parabolic subgroup P, define its *Martin boundary component* $e(P)$ by

$$e(P) = \overline{\mathfrak{a}_P^+(\infty)} \times X_P,$$

where $\mathfrak{a}^+(\infty)$ is the intersection of the unit sphere in $\mathfrak{a}_P$ with $\mathfrak{a}_P^+$ (see equation (I.2.1)), and $\overline{\mathfrak{a}_P^+(\infty)}$ the closure in $\mathfrak{a}_P$. Define

$$\overline{X}^M = X \cup \coprod_P \overline{\mathfrak{a}_P^+(\infty)} \times X_P,$$

where P ranges over all proper parabolic subgroups.

I.13.2 As recalled in Proposition I.7.15 in §I.7, the Martin compactification $X \cup \partial_\lambda X$ for $\lambda < \lambda_0(X)$ is the least compactification dominating both the geodesic compactification $X \cup X(\infty)$ and the maximal Satake compactification $\overline{X}^S_{\max}$. It is natural to expect that the boundary components of $X \cup \partial_\lambda X$ are obtained by combining those of $X \cup X(\infty)$ and $\overline{X}^S_{\max}$.

In $X \cup X(\infty) = \overline{X}^c$, the boundary component of a parabolic subgroup P is equal to $\mathfrak{a}_P^+(\infty)$; while in $\overline{X}^S_{\max} = \overline{X}_{\max}$, the boundary component of P is X_P. An immediate guess for the boundary component of P in $X \cup \partial_\lambda X$ would be $\mathfrak{a}_P^+(\infty) \times X_P$. This turns out to be incorrect for the following reason. Let

$$\pi : X \cup \partial_\lambda X \to \overline{X}^S_{\max}$$

be the extension of the identity map on X. The fiber over every boundary point in X_P contains $\mathfrak{a}_P^+(\infty)$ and should be compact. Hence it should be some compactification of $\mathfrak{a}_P^+(\infty)$. The choice of $\mathfrak{a}_P^+(\infty) \times X_P$ as the boundary component will lead to the noncompact fiber. It turns out that $\overline{\mathfrak{a}_P^+(\infty)} \times X_P$ is the right choice.

On the other hand, the identity map on X also extends to a dominating map

$$\pi' : X \cup \partial_\lambda X \to X \cup X(\infty),$$

and the fiber over a point in $\mathfrak{a}_P^+(\infty)$ contains X_P, which turns out to be $\overline{(X_P)}^S_{\max}$. Therefore, another possible choice for the boundary component of P would be $\mathfrak{a}_P^+(\infty) \times \overline{(X_P)}^S_{\max}$. It is true that

$$X \cup \partial_\lambda X = X \cup \coprod_P \mathfrak{a}_P^+(\infty) \times \overline{(X_P)}^S_{\max},$$

but the description of the topology using this choice of the boundary component is inductive and complicated, similar to the original construction of the Karpelevič compactification $\overline{X}^K$ in §I.3.

I.13.3 In $\overline{X}^S_{\max}$, each boundary component is a cell, and the boundary complex $\cup_P e(P)$ is a cell complex dual to the Tits building $\Delta(G)$ of G in the sense that for any pair of parabolic subgroups P and Q, $e(P)$ is contained in the closure of $e(Q)$ if and only if $P \subseteq Q$. In $X \cup X(\infty) = \overline{X}^C$, the boundary complex is isomorphic to the Tits building $\Delta(G)$, in particular, for any pair of parabolic subgroups P and Q, $e(P)$ is contained in the closure of $e(Q)$ if and only if $P \supseteq Q$.

On the other hand, the boundary components in $\overline{X}^M$ are of mixed type, and the boundary complex is not isomorphic or dual to the Tits building of G. In fact, for any pair of (proper) parabolic subgroups P and Q, neither $e(P)$ nor $e(Q)$ is contained in the closure of the other.

I.13.4 To define a topology on the set of boundary components of $\overline{X}^M$, we need a K-action on the set of boundary components. For $k \in K$, $(H, z) \in \overline{\mathfrak{a}_P^+(\infty)} \times X_P = e(P)$,

$$k \cdot (H, z) = (Ad(k)H, k \cdot z) \in \overline{\mathfrak{a}_{{}^kP}^+(\infty)} \times X_{{}^kP}\,. \tag{I.13.1}$$

A topology on $\overline{X}^M$ is given in terms of convergent sequences as follows:

1. For a boundary point $(H_\infty, z_\infty) \in \overline{\mathfrak{a}_P^+(\infty)} \times X_P$, an unbounded sequence y_j in X converges to (H_∞, z_∞) if y_j can be written in the form $y_j = k_j n_j a_j z_j$ with the components $k_j \in K, n_j \in N_P, a_j \in A_P, z_j \in X_P$ satisfying the following conditions:

(a) $k_j \to e$,
(b) for all $\alpha \in \Delta(P, A_P)$, $\alpha(\log a_j) \to +\infty$, $\log a_j/\|\log a_j\| \to H_\infty$,
(c) $n_j^{a_j} \to e$,
(d) $z_j \to z_\infty$.

2. For a pair of parabolic subgroups P, Q, $P \subset Q$, let P' be the unique parabolic subgroup of M_Q determined by P in equation (I.1.21). For a sequence $k_j \in K$ with $k_j \to e$ and a sequence $y_j = (H_j, z_j) \in \overline{\mathfrak{a}_Q^+(\infty)} \times X_Q$, the sequence $k_j \cdot y_j$ converges to $(H_\infty, z_\infty) \in \overline{\mathfrak{a}_P^+(\infty)} \times X_P$ if $H_\infty \in \overline{\mathfrak{a}_Q^+(\infty)}$, $H_j \to H_\infty$, and z_j can be written in the form $z_j = k'_j n'_j a'_j z'_j$, where $k'_j \in K_Q$, $n'_j \in N_{P'}$, $a'_j \in A_{P'}$, $z'_j \in X_{P'}$ satisfy the same condition as part (1) above when the pair (G, P) is replaced by (M_Q, P'), except that (b) is replaced by (b'): for all $\alpha \in \Delta(P', A_{P'})$, $\alpha(\log a'_j) \to +\infty$, i.e., z_j converges in $\overline{(X_Q)_{\max}}$ in §I.10.

These are special convergent sequences, and their combinations give the general convergent sequences. It can be checked that they form a convergence class of sequences, and hence define a topology on $\overline{X}^M$.

I.13.5 Neighborhoods of boundary points can be given explicitly as follows. For a parabolic subgroup P and a point $(H, z) \in \overline{\mathfrak{a}_P^+(\infty)} \times X_P$, there are two cases to consider: $H \in \mathfrak{a}_P^+(\infty)$ or not.

In the first case, let U be a neighborhood of H in $\mathfrak{a}_P^+(\infty)$ and V a neighborhood of z in X_P. Let

$$S^M_{\varepsilon,t,U,V} = \{(n, a, z) \in N_P \times A_P \times X_P = X \mid \\ a \in A_{P,t}, \log a/\|\log a\| \in U, n^a \in B_{N_P}(\varepsilon), z \in V\}.$$

For a neighborhood C of e in K, the set

$$C(S^M_{\varepsilon,t,U,V} \cup U \times V)$$

is a neighborhood of (H, z) in $\overline{X}^M$. The reason is that (H, z) is contained in the closure of other boundary components only when $H \in \partial\overline{\mathfrak{a}_P^+(\infty)}$, and hence there is no contribution from parabolic subgroups of different types, i.e., not conjugate under K.

In the second case, $H \in \partial\overline{\mathfrak{a}_P^+(\infty)}$. Let Q be the unique parabolic subgroup containing P such that H is contained in $\mathfrak{a}_Q^+(\infty)$. Let $Q = P_J$. Then P_I with $I \subseteq J$ are all the parabolic subgroups containing P such that $H \in \overline{\mathfrak{a}_{P_I}^+(\infty)}$ and X_P is a boundary symmetric space of X_{P_I}. By equation (I.1.21), P determines a parabolic subgroup of M_{P_I}. Let $S^I_{\varepsilon,t,V}$ be the generalized Siegel set in X_{P_I} associated with P' as defined in equation (I.9.2), where V is a bounded neighborhood of z. Let U be a neighborhood of H in $\overline{\mathfrak{a}_P^+(\infty)}$. Then

$$(\overline{\mathfrak{a}_{P_I}^+(\infty)} \cap U) \times S^I_{\varepsilon,t,V}$$

is the intersection of a neighborhood of (H, z) in $\overline{X}^M$ with the boundary component $\overline{\mathfrak{a}_{P_I}^+(\infty)} \times X_{P_I}$. Let C be a neighborhood of e in K. Then

$$C\left(S_{\varepsilon,t,U,V}^M \cup \coprod_{I \subseteq J} (\overline{\mathfrak{a}_{P_I}^+(\infty)} \cap U) \times S_{\varepsilon,t,V}^I\right)$$

is a neighborhood of (H, z) in $\overline{X}^M$.

Proposition I.13.6 *The topology on $\overline{X}^M$ is Hausdorff.*

Proof. We need to show that every pair of distinct points $x_1, x_2 \in \overline{X}^M$ admit disjoint neighborhoods. If at least one of them belongs to X, it is clear. Assume that they both lie on the boundary, that is $x_i = (H_i, z_i) \in \overline{\mathfrak{a}_{P_i}^+(\infty)} \times X_{P_i}$ for a pair of parabolic subgroups P_1, P_2. There are two cases to consider: $P_1 = P_2$ or not.

In the first case, $(H_1, z_1), (H_2, z_2) \in \overline{\mathfrak{a}_{P_1}^+(\infty)} \times X_{P_1}$. If $z_1 \neq z_2$, existence of the disjoint neighborhoods follows from the corresponding results for $\overline{X}_{\max}$ in Proposition I.10.6. If $z_1 = z_2$, then $H_1 \neq H_2$, and the existence of disjoint neighborhoods follows from the similar result of $\overline{X}^c$ in Proposition I.12.4.

In the second case, $P_1 \neq P_2$, existence of the disjoint neighborhoods follows similarly from the results for $\overline{X}_{\max}$ and $\overline{X}^c$.

Proposition I.13.7 *The G-action on X extends to a continuous action on $\overline{X}^M$.*

Proof. First we define a G-action on the boundary of $\overline{X}^M$. For $(H, z) \in \overline{\mathfrak{a}_P^+(\infty)} \times X_P$, $g \in G$, write $g = kman, k \in K, m \in M_P, a \in A_P, n \in N_P$. Define

$$g \cdot (H, z) = (Ad(k)H, k \cdot mz) \in \overline{\mathfrak{a}_{^kP}^+(\infty)} \times X_{^kP},$$

where $k \cdot mz \in X_{^kP}$ is defined in equation (I.1.15) in §I.1.

To show that this is a continuous extension of the G-action on X, by [GJT, Lemma 3.28], it suffices to show that if an unbounded sequence y_j in X converges to a boundary point (H, z), then gy_j converges to $g \cdot (H, z)$.

By definition, y_j can be written in the form $y_j = k_j n_j a_j z_j$, where $k_j \in K$, $n_j \in N_P$, $a_j \in A_P$, $z_j \in X_P$ satisfy (1) $k_j \to e$, (2) for all $\alpha \in \Delta(P, A_P)$, $\alpha(\log a_j) \to +\infty$, $\log a_j / \| \log a_j \| \to H$, (3) $n_j^{a_j} \to e$, (4) $z_j \to z$. Write

$$gk_j = k_j' m_j' a_j' n_j',$$

where $k_j' \in K$, $m_j' \in M_P$, $n_j' \in N_P$, $a_j' \in A_P$. Then n_j', a_j' are bounded and $k_j' m_j' \to km$. The components k_j', m_j' are not uniquely determined, but determined up to an element in K_P. By choosing this element suitably, we can assume that k_j', m_j' converge to k, m respectively. Then

$$gy_j = k'_j m'_j a'_j n'_j n_j a_j z_j = k'_j \; {}^{m'_j a'_j}(n'_j n_j) \; a'_j a_j m'_j z_j$$
$$= {}^{k'_j m'_j a'_j}(n'_j n_j) \; {}^{k'_j}(a'_j a_j) \; (k_j{}' m'_j z_j) = (k'_j k^{-1}) \; {}^{k m'_j a'_j}(n'_j n_j) \; {}^{k}(a'_j a_j)(k m'_j z_j).$$

From the last expression it can be checked easily that the conditions for convergence in $\overline{X}^M$ are satisfied, and gy_j converges to $(Ad(k)H, kmz) \in \overline{\mathfrak{a}^+_{kP}(\infty)} \times X_{kP}$.

Proposition I.13.8 *The space $\overline{X}^M$ is compact.*

Proof. We need to show that every sequence in $\overline{X}^M$ has a convergent subsequence. First we consider sequences y_j in X. If y_j is bounded, it clearly has a convergent subsequence. Otherwise, we can assume that y_j goes to infinity. Using the Cartan decomposition $X = K \exp \overline{\mathfrak{a}^+_{P_0}} x_0$, $y_j = k_j \exp H_j x_0$ and replacing H_j by a subsequence, we can assume that

1. k_j converges to some $k \in K$,
2. there exists a subset $I \subset \Delta(P_0, A_{P_0})$ such that for $\alpha \in I$, $\alpha(H_j)$ converges to a finite number, while for $\alpha \in \Delta - I$, $\alpha(H_j) \to +\infty$.
3. $H_j / \|H_j\| \to H_\infty \in \overline{\mathfrak{a}^+_{P_{0,I}}(\infty)}$.

Writing $H_j = H_{j,I} + H_j^I$, where $H_{j,I} \in \mathfrak{a}_{P_{0,I}}$, $H_j^I \in \mathfrak{a}^I_{P_0}$, then $\exp H_j^I x_0$ converges to a point $z_\infty \in X_{P_{0,I}}$, and $H_{j,I}/\|H_{j,I}\| \to H_\infty$. From this, it is clear that $y_j = k_j \exp H_{j,I} H_j^I x_0$ converges to $k(H_\infty, z_\infty) \in \overline{\mathfrak{a}^+_{kP_{0,I}}(\infty)} \times X_{kP_{0,I}}$ in $\overline{X}^M$.

Let P_0 be a minimal parabolic subgroup. Since every parabolic subgroup is conjugate to a standard parabolic subgroup $P_{0,I}$ under some elements in K, we have

$$\overline{X}^M = X \cup \coprod_I K(\overline{\mathfrak{a}^+_{P_{0,I}}(\infty)} \times X_{P_{0,I}}).$$

Since K is compact and the closure of $X_{P_{0,I}}$ in $\overline{X}^M$ is $\overline{(X_{P_{0,I}})}_{\max}$ and hence compact; together with the previous paragraph, this implies that $\overline{X}^M$ is compact.

Proposition I.13.9 *The compactification $\overline{X}^M$ is isomorphic to the least common refinement $\overline{X}_{\max} \bigvee \overline{X}^C$ of $\overline{X}_{\max}$ and $\overline{X}^C$.*

Proof. By [GJT, Lemma 3.28], it suffices to show that an unbounded sequence y_j in X converges in $\overline{X}^M$ if and only if y_j converges in both $\overline{X}_{\max}$ and $X \cup X(\infty)$.

If y_j in X converges in $\overline{X}^M$ to $(H, z) \in \overline{\mathfrak{a}^+_P(\infty)} \times X_P$, then it can be written in the form $y_j = k_j n_j a_j z_j$ with $k_j \in K$, $n_j \in N_P$, $a_j \in A_P$ and $z_j \in X_P$ satisfying (1) $k_j \to e$, (2) $\alpha(\log a_j) \to +\infty$, $\alpha \in \Delta(P, A_P)$, $\log a_j / \|\log a_j\| \to H$, (3) $z_j \to z$. Since these conditions are stronger than the convergence conditions of $\overline{X}_{\max}$, it is clear that y_j converges in $\overline{X}_{\max}$ to $z \in X_P$. On the other hand, let P_I be the unique parabolic subgroup containing P such that $\mathfrak{a}_P^+(\infty)$ contains H as an interior point. By decomposing $\log a_j$ according to $\mathfrak{a}_P = \mathfrak{a}_{P,I} \oplus \mathfrak{a}_P^I$, it can be seen easily that y_j converges to $H \in \mathfrak{a}_{P_I}^+(\infty)$ in $\overline{X}^c$.

Conversely, suppose that y_j converges in both $\overline{X}_{\max}$ and $\overline{X}^c$. Let $z \in X_P$ be the limit in $\overline{X}_{\max}$. Then y_j can be written as $y_j = k_j n_j a_j z_j$ with the components satisfying similar conditions as above except (2) is replaced by (2′): $\alpha(\log a_j) \to +\infty$, $\alpha \in \Delta(P, A_P)$. We claim that the second part of (2) is also satisfied, i.e., $\log a_j / \| \log a_j \| \to H$ for some $H \in \overline{\mathfrak{a}_P^+(\infty)}$. In fact, since y_j converges in $\overline{X}^c$, the proof of Proposition I.12.4 shows that $a_j x_0$ also converges in $\overline{X}^c$. This implies that $\log a_j / \| \log a_j \|$ converges in $\mathfrak{a}_P(\infty)$ and the limit clearly belongs to $\overline{\mathfrak{a}_P^+(\infty)}$. Since all the conditions (1)–(4) are satisfied, y_j converges in $\overline{X}^M$ to $(H, z) \in \overline{\mathfrak{a}_P^+(\infty)} \times X_P$.

I.13.10 Example. When the rank of X is equal to 1, $\overline{X}^M$ is isomorphic to both $X \cup X(\infty)$ and $\overline{X}^M$. They are different when the rank is at least 2. We illustrate their difference through the example $X = \mathbf{H} \times \mathbf{H}$. For each geodesic γ in X, the inverse image in $\overline{X}^M$ of the equivalence class $[\gamma]$ under the map $\overline{X}^M \to X \cup X(\infty)$ is either a point or the geodesic compactification of $\mathbf{H}$ depending on whether γ is nonsingular or not. (Recall that γ is nonsingular if $\gamma(t) = (\gamma_1(at), \gamma_2(bt))$, where γ_1, γ_2 are geodesics in $\mathbf{H}$, and $a, b > 0$, $a^2 + b^2 = 1$. On the other hand, the inverse images in $\overline{X}^M$ of the boundary points in $\overline{X}_{\max}^S$ are either points or closed 1-dimensional simplexes. For example, for all $a, b > 0$, as $t \to +\infty$, $\gamma(t)$ converges to the same point in $\overline{X}_{\max}^S$, but for different values of a, b, they converge to different boundary points in $\overline{X}^c$ and hence in $\overline{X}^M$. These different choices of $a, b > 0$ give the 1-dimensional simplex.

Corollary I.13.11 *The compactification $\overline{X}^M$ is isomorphic to the Martin compactification $X \cup \partial_\lambda X$, $\lambda < \lambda_0(X)$.*

Proof. By Proposition I.7.15, $X \cup \partial_\lambda X$, $\lambda < \lambda_0(X)$, is isomorphic to the least common refinement $\overline{X}_{\max}^S \bigvee X \cup X(\infty)$. By Proposition I.10.11, $\overline{X}_{\max}^S \cong \overline{X}_{\max}$, and by Proposition I.12.6, $\overline{X}^c = X \cup X(\infty)$. Then the corollary follows from Proposition I.13.9.

I.14 Uniform construction of the Karpelevič compactification

In this section we apply the uniform method in §I.8 to construct a compactification $\overline{X}_K$ that is isomorphic to the Karpelevič compactification $\overline{X}^K$. We emphasize that in order to distinguish the new construction from the original one, we use the subscript in $\overline{X}_K$.

The construction of $\overline{X}_K$ is a generalization of the construction of $\overline{X}^M$ in the previous section. The basic issue is the choice of the boundary components. It is known that the Karpelevič compactification $\overline{X}^K$ dominates the Martin compactification and hence both the maximal Satake compactification $\overline{X}^S_{\max}$ and the geodesic compactification $X \cup X(\infty)$. As pointed out in §I.13.2, each of the dominating maps $\overline{X}^K \to \overline{X}^S_{\max}$, $\overline{X}^K \to X \cup X(\infty)$ leads to a choice of the boundary component, and we will use the first map. Hence, the boundary component $e(P)$ of P is $\overline{X}_K$ is of the form

$$e(P) = \overline{\mathfrak{a}_P^+(\infty)}^K \times X_P,$$

where $\overline{\mathfrak{a}_P^+(\infty)}^K$ is a compactification of $\mathfrak{a}_P^+(\infty)$. Since the boundary component of P in $\overline{X}^M$ is $\overline{\mathfrak{a}_P^+(\infty)} \times X_P$ and $\overline{X}^K$ dominates $\overline{X}^M$, $\overline{\mathfrak{a}_P^+(\infty)}^K$ is a refinement (or blowup) of $\overline{\mathfrak{a}_P^+(\infty)}$.

I.14.1 This section is organized as follows. We define the boundary of flats in I.14.1, and its topology is given in I.14.2. The relation to the closure in the geodesic compactification is given in Proposition I.14.4. The Karpelevič boundary components are given in I.14.5. The topology of the Karpelevič compactification of flats is given in I.14.6. The topology of $\overline{X}^K$ is described in terms of both convergent sequences and a neighborhood base, which are given in I.14.7. The Hausdorff property of $\overline{X}_K$ is proved in Proposition I.14.9 the compactness in Proposition I.14.10 and the continuous extension of the G-action on X in Proposition I.14.11.

The fibers of the map from $\overline{X}_K$ to $\overline{X}^c = X \cup X(\infty)$ are described in Proposition I.14.12. The relation to the Martin compactification is determined in Proposition I.14.13. In I.14.13, we discuss examples of polydisks $X = D \times \cdots \times D$. Finally in I.14.13, we identify $\overline{X}_K$ with the original Karpelevič compactification $\overline{X}^K$ in Proposition I.14.15.

I.14.2 The simplicial faces in the compactification $\overline{\mathfrak{a}_P^+(\infty)}$ are parametrized by subsets of the set $\Delta = \Delta(P, A_P)$ of simple roots in $\Phi(P, A_P)$. The basic idea in defining the compactification $\overline{\mathfrak{a}_P^+(\infty)}^K$ is to construct faces for all decreasing filtrations of Δ, or equivalently, all ordered partitions of Δ. If we pick out only the first set in the filtration, we get the map to $\overline{\mathfrak{a}_P^+(\infty)}$.

For a pair $J, J' \subset \Delta(P, A_P)$, $J \subset J'$, let

$$\mathfrak{a}_J^{J'} = \mathfrak{a}_{P_J} \cap \mathfrak{a}_P^{J'}. \tag{I.14.1}$$

The restriction of the roots in $J' - J$ yields a homeomorphism $\mathfrak{a}_J^{J'} \cong \mathbb{R}^{J'-J}$. Since

$$\mathfrak{a}_P = \mathfrak{a}_{P_J} \oplus \mathfrak{a}_P^J, \quad \mathfrak{a}_P = \mathfrak{a}_{P_{J'}} \oplus \mathfrak{a}_P^{J'},$$

and

$$\mathfrak{a}_{P_{J'}} \subset \mathfrak{a}_{P_J}, \quad \mathfrak{a}_P^J \subset \mathfrak{a}_P^{J'},$$

we have

$$\mathfrak{a}_P = \mathfrak{a}_{P_{J'}} \oplus \mathfrak{a}_J^{J'} \oplus \mathfrak{a}_P^J.$$

Define

$$\mathfrak{a}_J^{J',+}(\infty) = \{H \in \mathfrak{a}_J^{J'} \mid \|H\| = 1, \alpha(H) > 0, \alpha \in J' - J\}. \tag{I.14.2}$$

For any *ordered partition*

$$\Sigma : I_1 \cup \cdots \cup I_k = \Delta,$$

let $J_i = I_i \cup \cdots \cup I_k$, $1 \leq i \leq k$, $J_{k+1} = \emptyset$, be the induced decreasing filtration. Then we have a decomposition

$$\mathfrak{a}_P = \mathfrak{a}_{J_2} \times \mathfrak{a}_{J_3}^{J_2} \times \cdots \times \mathfrak{a}_{J_{k+1}}^{J_k},$$

where $\dim \mathfrak{a}_{J_{i+1}}^{J_i} = |I_i|$. Define

$$\mathfrak{a}_P^{\Sigma,+}(\infty) = \mathfrak{a}_{J_2}^+(\infty) \times \mathfrak{a}_{J_3}^{J_2,+}(\infty) \times \cdots \times \mathfrak{a}_{J_{k+1}}^{J_k,+}(\infty).$$

Note that $\mathfrak{a}_{J_{k+1}}^{J_k,+}(\infty) = \mathfrak{a}_P^{J_k,+}(\infty)$. If we use the improper parabolic subgroup $P_\Delta = G$, $\mathfrak{a}_{J_2}^+(\infty) = \mathfrak{a}_{P_{J_2}}^+(\infty)$ can be identified with $\mathfrak{a}_{J_2}^{J_1,+}(\infty)$. When Σ is the trivial partition consisting of only Δ, then $\mathfrak{a}_P^{\Sigma,+}(\infty) = \mathfrak{a}_P^+(\infty)$, the interior of $\overline{\mathfrak{a}_P^+(\infty)}$. Other pieces are blow-ups of the boundary faces of $\mathfrak{a}_P^+(\infty)$ as shown in Proposition I.14.4 below.

Define

$$\overline{\mathfrak{a}}_P^{K,+}(\infty) = \coprod_{\Sigma} \mathfrak{a}_P^{\Sigma,+}(\infty), \tag{I.14.3}$$

where Σ runs over all the partitions of Δ.

I.14.3 The space $\overline{\mathfrak{a}}_P^{K,+}(\infty)$ is given the following topology:

1. For every partition Σ, $\mathfrak{a}_P^{\Sigma,+}(\infty)$ is given the product topology.
2. For two partitions Σ, Σ', $\mathfrak{a}_P^{\Sigma,+}(\infty)$ is contained in the closure of $\mathfrak{a}_P^{\Sigma',+}(\infty)$ if and only if Σ is a refinement of Σ', i.e., every part in Σ' is a union of parts of Σ, or equivalently, the filtration of Σ' is a subfiltration of that of Σ. Specifically, the convergence of a sequence of points in $\mathfrak{a}_P^{\Sigma',+}(\infty)$ to limits in $\mathfrak{a}_P^{\Sigma,+}(\infty)$ is given as follows. Assume $\Sigma : I_1 \cup \cdots \cup I_k$, $\Sigma' : I_1' \cup \cdots \cup I_{k'}'$. For any part I_m' in Σ', write $I_m' = I_{n_1} \cup \cdots \cup I_{n_s}$, where the indexes $n_1, \ldots, n_s$ are strictly increasing. Then it suffices to describe how a sequence in $\mathfrak{a}_{J_m'+1}^{J_m',+}(\infty)$ converges to a limit in $\mathfrak{a}_{J_{n_1}+1}^{J_{n_1},+}(\infty) \times \cdots \times \mathfrak{a}_{J_{n_s}+1}^{J_{n_s},+}(\infty)$. Let H_j be a sequence in $\mathfrak{a}_{J_m'+1}^{J_m',+}(\infty)$, and $(H_{n_1,\infty}, \ldots, H_{n_s,\infty}) \in \mathfrak{a}_{J_{n_1}+1}^{J_{n_1},+}(\infty) \times \cdots \times \mathfrak{a}_{J_{n_s}+1}^{J_{n_s},+}(\infty)$. Then H_j converges to $(H_{n_1,\infty}, \ldots, H_{n_s,\infty})$ if and only if the following conditions are satisfied:

(a) For $\alpha \in I_{n_1}$, $\alpha(H_j) \to \alpha(H_{n_1,\infty})$, in particular, $\alpha(H_j) \not\to 0$.
(b) For $\alpha \in I'_m - I_{n_1}$, $\alpha(H_j) \to 0$.
(c) For $\alpha \in I_{n_a}$, $\beta \in I_{n_b}$, $a < b$,

$$\frac{\beta(H_j)}{\alpha(H_j)} \to 0.$$

(d) For $\alpha, \beta \in I_{n_a}$, $1 \leq a \leq s$,

$$\frac{\beta(H_j)}{\alpha(H_j)} \to \frac{\beta(H_{n_a,\infty})}{\alpha(H_{n_a,\infty})} \neq 0.$$

For $\varepsilon > 0$, define a subset $U_\varepsilon^{\Sigma'}(H_\infty)$ in $\mathfrak{a}_P^{\Sigma',+}(\infty)$ as follows:

$$\begin{aligned} U_\varepsilon^{\Sigma'}(H_\infty) = \{(H_1, \dots, H_{k'}) \in \mathfrak{a}_P^{\Sigma',+}(\infty) \mid & \text{ for } 1 \leq m \leq k', I'_m = I_{n_1} \cup \dots \cup I_{n_s}, \\ & (1) \text{ for } \alpha \in I_{n_1}, |\alpha(H_m) - \alpha(H_{n_1,\infty})| < \varepsilon, \\ & (2) \text{ for } \alpha \in I'_m - I_{n_1}, |\alpha(H_m)| < \varepsilon, \\ & (3) \text{ for } \alpha \in I_{n_a}, \beta \in I_{n_b}, a < b, \left|\frac{\beta(H_m)}{\alpha(H_m)}\right| < \varepsilon, \\ & (4) \text{ for } \alpha, \beta \in I_{n_a}, 1 \leq a \leq s, \left|\frac{\beta(H_m)}{\alpha(H_m)} - \frac{\beta(H_{n_a,\infty})}{\alpha(H_{n_a,\infty})}\right| < \varepsilon\}. \end{aligned} \tag{I.14.4}$$

Proposition I.14.4 *There exists a continuous surjective map* $\pi : \overline{\mathfrak{a}}_P^{K,+}(\infty) \to \overline{\mathfrak{a}_P^+(\infty)}$. *This map is a homeomorphism if and only if* $\dim A_P \leq 2$.

Proof. Recall that

$$\overline{\mathfrak{a}_P^+(\infty)} = \coprod_{I \subset \Delta} \mathfrak{a}_{P_I}^+(\infty).$$

For each partition $\Sigma : I_1 \cup \dots \cup I_k = \Delta$, define a map by projecting to the first factor:

$$\begin{aligned} \pi : \mathfrak{a}_P^{\Sigma,+}(\infty) = \mathfrak{a}_{J_2}^+(\infty) \times \mathfrak{a}_{J_3}^{J_2,+}(\infty) \times \dots \times \mathfrak{a}_{J_{k+1}}^{J_k,+}(\infty) \\ \to \mathfrak{a}_{J_2}^+(\infty) = \mathfrak{a}_{P_{J_2}}^+(\infty), \end{aligned}$$

$$\pi(H_{1,\infty}, \dots, H_{k,\infty}) = H_{1,\infty}$$

where $J_i = I_i \cup \dots \cup I_k$ as above. This gives a surjective map

$$\pi : \overline{\mathfrak{a}}_P^{K,+}(\infty) \to \overline{\mathfrak{a}_P^+(\infty)}.$$

If $H_j \in \mathfrak{a}_P^+(\infty)$ converges to $(H_{1,\infty}, \dots, H_{k,\infty}) \in \mathfrak{a}_P^{\Sigma,+}(\infty)$, it is clear from the description of the topology of $\overline{\mathfrak{a}}_P^{K,+}(\infty)$, in particular conditions (a) and (b), that H_j

converges to $H_{1,\infty}$ in $\overline{\mathfrak{a}_P^+(\infty)}$. Since $\mathfrak{a}_P^+(\infty)$ is dense in $\overline{\mathfrak{a}}_P^{K,+}(\infty)$ and hence every point in $\overline{\mathfrak{a}}_P^{K,+}(\infty)$ is the limit of a sequence of points in $\mathfrak{a}_P^+(\infty)$, this proves the continuity of π.

When $\dim A_P = 1$, this map π is clearly bijective. When $\dim A_P = 2$, there are only two nontrivial ordered partitions, $\Sigma_1 : I_1 \cup I_2$, $\Sigma_2 : I_2 \cup I_1$ of Δ, and each of $\mathfrak{a}_P^{\Sigma_1,+}(\infty)$, $\mathfrak{a}_P^{\Sigma_2,+}(\infty)$ consists of one point, corresponding to the two endpoints of the 1-simplex $\mathfrak{a}_P^+(\infty)$, and hence the map π is bijective. On the other hand, if $\dim A_P \geq 3$, there are nontrivial ordered partitions $\Sigma : \Delta = I_1 \cup I_2$ where $|I_1| = 1, |I_2| \geq 2$. For such a Σ, $\mathfrak{a}_P^{\Sigma,+}(\infty)$ has positive dimension and is mapped to the zero-dimensional space $\mathfrak{a}_{J_2}^+(\infty)$, and hence π is not injective.

I.14.5 For each parabolic subgroup P, define its *Karpelevič boundary component* to be

$$e(P) = \overline{\mathfrak{a}}_P^{K,+}(\infty) \times X_P.$$

Define

$$\overline{X}_K = X \cup \coprod_P \overline{\mathfrak{a}}_P^{K,+}(\infty) \times X_P,$$

where P runs over all parabolic subgroups.

To put a topology on $\overline{X}^K$, we need a topology on the set of boundary components, which is given by the K-action on the boundary components. For $k \in K$, $(H, z) \in \overline{\mathfrak{a}}_P^{K,+}(\infty) \times X_P$,

$$k \cdot (H, z) = (Ad(k)(H), k \cdot z) \in \overline{\mathfrak{a}}_{^kP}^{K,+}(\infty) \times X_{^kP}, \tag{I.14.5}$$

where for $H = (H_1, \ldots, H_j) \in \mathfrak{a}_P^{\Sigma,+}(\infty) \subset \overline{\mathfrak{a}}_P^{K,+}(\infty)$ and $\Sigma : I_1 \cup \cdots \cup I_j$,

$$Ad(k)H = (Ad(k)H_1, \ldots, Ad(k)H_j)) \in \mathfrak{a}_{^kP}^{\Sigma,+}(\infty),$$

where Σ induces an ordered partition of $\Delta(^kP, A_{^kP})$ by $Ad(k) : \mathfrak{a}_P \to \mathfrak{a}_{^kP}$, which is denoted by Σ also in the above equation.

I.14.6 Before defining a topology on $\overline{X}_K$, we need to define a topology of $\mathfrak{a}_P \cup \overline{\mathfrak{a}}_P^{K,+}(\infty)$. Given an ordered partition $\Sigma : I_1 \cup \cdots \cup I_k$ and a point $H_\infty = (H_{1,\infty} \ldots, H_{k,\infty}) \in \mathfrak{a}_P^{\Sigma,+}(\infty)$, an unbounded sequence $H_j \in \mathfrak{a}_P$ converges to $(H_{1,\infty} \ldots, H_{k,\infty})$ if and only if

1. For all $\alpha \in \Delta$, $\alpha(H_j) \to +\infty$.
2. For every pair $m < n$, $\alpha \in I_m$, $\beta \in I_n$, $\beta(H_j)/\alpha(H_j) \to 0$.
3. For every m, $\alpha, \beta \in I_m$, $\beta(H_j)/\alpha(H_j) \to \beta(H_{m,\infty})/\alpha(H_{m,\infty})$.

Neighborhoods of boundary points in $\mathfrak{a}_P \cup \overline{\mathfrak{a}}_P^{K,+}(\infty)$ can be given explicitly. For any $H_\infty = (H_{1,\infty} \ldots, H_{k,\infty}) \in \mathfrak{a}_P^{\Sigma,+}(\infty)$ and $\varepsilon > 0$, define

$$
\begin{aligned}
U_\varepsilon^X(H_\infty) =\{H \in \mathfrak{a}_P \mid\ & (1) \text{ for } \alpha \in \Delta, \alpha(H) > \frac{1}{\varepsilon}, \\
& (2) \text{ for every pair } m < n, \alpha \in I_m, \beta \in I_n, |\beta(H)/\alpha(H)| < \varepsilon, \\
& (3) \text{ for every } m, \alpha, \beta \in I_m, |\beta(H)/\alpha(H) - \beta(H_{m,\infty})/\alpha(H_{m,\infty})| < \varepsilon\}.
\end{aligned} \tag{I.14.6}
$$

Combining this with the set $U_\varepsilon^{\Sigma'}(H_\infty)$ in equation (I.14.4), set

$$U_\varepsilon(H_\infty) = U_\varepsilon^X(H_\infty) \cup \cup_{\Sigma'} U_\varepsilon^{\Sigma'}(H_\infty), \tag{I.14.7}$$

where Σ' runs over all the ordered partitions for which Σ is a refinement. This is a neighborhood of H_∞ in $\mathfrak{a}_P \cup \overline{\mathfrak{a}}_P^{K,+}(\infty)$.

I.14.7 The topology of the space $\overline{X}_K$ is given in terms of convergent sequences as follows:

1. An unbounded sequence y_j in X converges to $(H_\infty, z_\infty) \in \overline{\mathfrak{a}}_P^{K,+}(\infty) \times X_P$ if y_j can be written as $y_j = k_j n_j a_j z_j$ with $k_j \in K, n_j \in N_P, a_j \in A_P, z_j \in X_P$ satisfying the following conditions:
   (a) $k_j \to e$.
   (b) $\log a_j \to H_\infty$ in $\mathfrak{a}_P \cup \overline{\mathfrak{a}}_P^{K,+}(\infty)$ in the topology described above.
   (c) $n_j^{a_j} \to e$.
   (d) $z_j \to z_\infty$.
2. Let Q be a parabolic subgroup containing P, and let P' be the parabolic subgroup in M_Q corresponding to P. We note that a partition Σ_Q of $\Delta(Q, A_Q)$ and a partition of $\Sigma^{P'}$ of $\Delta(P', A_{P'})$ combine to form a partition $\Sigma_P = \Sigma_Q \cup \Sigma^{P'}$ of $\Delta(P, A_P)$, where the roots in $\Delta(Q, A_Q)$ and $\Delta(P', A_{P'})$ are identified with the roots in $\Delta(P, A_P)$ whose restrictions are equal to them.
   For a sequence $k_j \in K$ with $k_j \to e$, and a sequence $y_j = (H_j, z_j)$ in $\overline{\mathfrak{a}}_Q^{K,+}(\infty) \times X_Q$, the sequence $k_j y_j$ converges to $(H_\infty, z_\infty) \in \overline{\mathfrak{a}}_P^{K,+}(\infty) \times X_P$ if and only if z_j can be written as $z_j = k'_j n'_j a'_j z'_j$ with $k'_j \in K_Q, n'_j \in N_{P'}$, $a'_j \in A_{P'}, z'_j \in X_{P'}$, and these components and H_j satisfy the following conditions:
   (a) There exists a partition Σ of the form $\Sigma = \Sigma_Q \cup \Sigma^{P'}$, i.e., a combination of two partitions $\Sigma_Q, \Sigma^{P'}$, such that $H_\infty \in \mathfrak{a}_P^{\Sigma,+}$. Write $H_\infty = (H_{\infty,Q}, H'_\infty) \in \mathfrak{a}_Q^{\Sigma_Q,+}(\infty) \times \mathfrak{a}_{P'}^{\Sigma^{P'},+}(\infty)$. The components k'_j, n'_j, a'_j, z'_j satisfy the same condition as in part (1) above when the pair X, P is replaced by X_Q, P' and the limit by $(H'_\infty, z_\infty) \in \overline{\mathfrak{a}}_{P'}^{K,+}(\infty) \times X_{P'}$.
   (b) $(H_j, H'_\infty) \to H_\infty$ in $\mathfrak{a}_P \cup \overline{\mathfrak{a}}_P^{K,+}(\infty)$.

These are special convergent sequences and their combinations give general convergent sequences. It can be checked that they form a convergence class of sequences and hence define a topology on $\overline{X}^K$.

I.14.8 Neighborhood systems of boundary points in $\overline{X}^K$ can be described as follows. For a point $(H_\infty, z_\infty) \in \overline{\mathfrak{a}}_P^{K,+}(\infty) \times X_P$, let V be a bounded neighborhood of z_∞ in X_P, and $U_\varepsilon = U_\varepsilon(H_\infty)$ a neighborhood of H_∞ in $\mathfrak{a}_P \cup \overline{\mathfrak{a}}_P^{K,+}(\infty)$ defined in equation (I.14.7) above. For $\varepsilon > 0, t > 0$, define

$$S^K_{\varepsilon,t,V} = \{(n, a, z) \in N_P \times A_P \times X_P = X \mid \tag{I.14.8}$$

$$\log a \in U_\varepsilon, a \in A_{P,t}, n^a \in B_{N_P}(\varepsilon), z \in V\}. \tag{I.14.9}$$

Let Σ be the partition of $\Delta(P, A_P)$ such that $H_\infty \in \mathfrak{a}_P^{\Sigma,+}(\infty)$. For a parabolic subgroup P_I containing P, let P' be the parabolic subgroup of M_{P_I} corresponding to P in equation (I.1.21). If a partition Σ is of the form $\Sigma_{P_I} \cup \Sigma^{P'}$ in the above notation, we call P_I a Σ-admissible parabolic subgroup. For example, when $\Sigma = \Delta$, the only Σ-admissible parabolic subgroup is P.

For each Σ-admissible parabolic subgroup P_I, write $H_\infty = (H_{\infty,I}, H^I_\infty) \in \overline{\mathfrak{a}}_{P_I}^{K,+}(\infty) \times \overline{\mathfrak{a}}_{P'}^{K,+}(\infty)$. For $\varepsilon > 0$, Let $U_{I,\varepsilon} = U_\varepsilon(H_{\infty,I}) \cap \overline{\mathfrak{a}}_{P_I}^{K,+}(\infty)$, a neighborhood of $H_{\infty,I}$ in $\overline{\mathfrak{a}}_{P_I}^{K,+}(\infty)$. (Recall that $U_\varepsilon(H_{\infty,I})$ is a neighborhood of $H_{\infty,I}$ in $\mathfrak{a}_{P_I} \cup \overline{\mathfrak{a}}_{P_I}^{K,+}(\infty)$ as defined in equation (I.14.7) above.) Let $S^{K,I}_{\varepsilon,t,V}$ be the corresponding neighborhood of H^I_∞ in X_{P_I} defined as in equation (I.14.8). For a neighborhood C of e in K, the set

$$C\left(S^K_{\varepsilon,t,V} \cup \coprod_{\Sigma\text{-admissible } P_I} U_{I,\varepsilon} \times S^{K,I}_{\varepsilon,t,V}\right) \tag{I.14.10}$$

is a neighborhood of (H_∞, z_∞) in $\overline{X}_K$.

Proposition I.14.9 *The topology on $\overline{X}_K$ is Hausdorff.*

Proof. We need to show that any two distinct points x_1, x_2 in $\overline{X}_K$ admit disjoint neighborhoods. This is clearly the case when at least one of them belongs to X. Assume that $x_i = (H_i, z_i) \in \overline{\mathfrak{a}}_{P_i}^{K,+}(\infty) \times X_{P_i}$. There are two cases to consider: $P_1 = P_2$ or not.

In the first case, if $z_1 \neq z_2$, it follows from the proof of Proposition I.10.6 that when $t \gg 0$, ε is sufficiently small, C is a sufficiently small neighborhood of e, and V_i a sufficiently small neighborhood of z_i, $i = 1, 2$, the neighborhoods $C(S^K_{\varepsilon,t,V_i} \cup \coprod U_{I_i,\varepsilon_j} \times S^{K,I}_{\varepsilon,t,V_i})$ of x_i are disjoint. On the other hand, if $z_1 = z_2$, then $H_1 \neq H_2$. We claim that the same conclusion holds.

If not, there exists a sequence y_j in the intersection of $C_j(S^K_{\varepsilon_j,t_j,V_{i,j}} \cup \coprod U_{I_i,\varepsilon_j} \times S^{K,I}_{\varepsilon_j,t_j,V_{i,j}})$ for sequences $\varepsilon_j \to 0$, $t_j \to +\infty$, C_j that shrinks to e, and $V_{i,j}$ shrinks to z_i, $i = 1, 2$.

Assume first that $y_j \in X$. Then $y_j = k_j n_j a_j z_j$ with $k_j \in K, n_j \in N_P, a_j \in A_P$, $z_j \in X_P$ satisfying the conditions (1) $k_j \to e$, (2) $\log a_j \to H_1 \in \overline{\mathfrak{a}}_P^{K,+}(\infty)$, (3) $n_j^{a_j} \to e$, (4) $z_j \to z_1$. Similarly, $y_j = k'_j n'_j a'_j z'_j$ with the components satisfying the same condition except $\log a'_j \to H_2$, $z'_j \to z_2$. Since $H_1 \neq H_2$,

$$\|\log a_j - \log a'_j\| \to +\infty. \tag{I.14.11}$$

Since

$$d(k_j n_j a_j z_j, k_j a_j z_j) = d(n_j^{a_j} z_j, z_j) \to 0$$

and

$$d(k_j a_j z_j, k_j a_j x_0) = d(z_j, x_0)$$

is bounded, it follows that

$$d(k_j a_j x_0, k'_j a'_j x_0) \leq c$$

for some constant c. By [AJ, Lemma 2.1.2],

$$d(k_j a_j x_0, k'_j a'_j x_0) \geq \|\log a_j - \log a'_j\|,$$

and hence

$$\|\log a_j - \log a'_j\| \leq c.$$

This contradicts equation (I.14.11), and the claim is proved. The case that y_j belongs to the boundary of $\overline{X}_K$ can be handled similarly.

In the second case, $P_1 \neq P_2$, and we can use the fact that $S^K_{\varepsilon,t,V}$ is contained in the generalized Siegel set $S_{\varepsilon,t,V}$ defined in equation (I.9.2) in §I.1 and the separation result in Proposition I.9.8 to prove that x_1, x_2 admit disjoint neighborhoods.

Proposition I.14.10 *The space $\overline{X}_K$ is compact.*

Proof. Since $X = K \exp \overline{\mathfrak{a}^+_{P_0}} x_0$, we have $\overline{X}_K = K \overline{\exp \overline{\mathfrak{a}^+_{P_0}} x_0}$, where P_0 is a minimal parabolic subgroup, and $\overline{\overline{\exp \mathfrak{a}^+_{P_0}} x_0}$ is the closure of $\exp \overline{\mathfrak{a}^+_{P_0}} x_0$ in $\overline{X}_K$. Since K is compact, it suffices to prove the compactness of $\overline{\exp \overline{\mathfrak{a}^+_{P_0}} x_0}$, which follows easily from the definition. In fact, for any unbounded sequence $H_j \in \mathfrak{a}^+_{P_0}$, there exists an ordered partition $\Sigma' : I_1 \cup \cdots \cup I_k \cup J = \Delta$ of $\Delta(P_0, A_{P_0})$, where J could be empty, such that after being replaced by a subsequence, H_j satisfies the conditions:

1. For all $\alpha \in J$, $\alpha(H_j)$ converges to a finite limit.
2. For all $\alpha \notin J$, $\alpha(H_j) \to +\infty$.
3. For $\alpha, \beta \in I_m$, $\alpha(H_j)/\beta(H_j)$ converges to a finite positive number.
4. For $\alpha \in I_m$, $\beta \in I_n$, $m < n$, $\alpha(H_j)/\beta(H_j) \to +\infty$.

Then it follows from the definition that $\exp H_j x_0$ converges to a point in $\mathfrak{a}^{\Sigma,+}_{P_J}(\infty) \times X_{P_J} \subset \overline{\mathfrak{a}}^{K,+}_{P_J}(\infty) \times X_{P_J}$, where Σ is the partition $I_1 \cup \cdots \cup I_k$ of $\Delta \setminus J = \Delta(P_{0,J}, A_{P_{0,J}})$.

Proposition I.14.11 *The G-action on X extends to a continuous action on $\overline{X}_K$.*

Proof. For any $g \in G$ and $(H, z) \in \overline{\mathfrak{a}}_P^{K,+}(\infty) \times X_P$, write $g = kman$ with $k \in K, m \in M_P, a \in A_P, n \in N_P$. Define

$$g \cdot (H, z) = (Ad(k)H, k \cdot mz) \in \overline{\mathfrak{a}}_{{}^kP}^{K,+}(\infty) \times X_{{}^kP},$$

where k canonically identifies $\overline{\mathfrak{a}}_P^{K,+}(\infty)$ with $\overline{\mathfrak{a}}_{{}^kP}^{K,+}(\infty)$ in equation (I.14.5), and the K-action on X_P is defined in equation (I.1.15) in §I.1. This defines an extended action of G on $\overline{X}_K$. Arguments similar to those in the proof of Proposition I.13.7 show that this extended action is continuous.

Proposition I.14.12 *The identity map on X extends to a continuous surjective map $\pi : \overline{X}_K \to \overline{X}^c$, and for every point $H \in \mathfrak{a}_P^+(\infty)$, the fiber $\pi^{-1}(H)$ is equal to $\overline{(X_P)_K}$, in particular,*

$$\overline{X}_K = X \cup \coprod_P \mathfrak{a}_P^+(\infty) \times \overline{(X_P)_K},$$

where P runs over all parabolic subgroups. Equivalently, for any $z \in X(\infty)$, let G_z be the stabilizer of z, which is a parabolic subgroup of G, and $X_z = X_{G_z}$, the boundary symmetric space. Then

$$\overline{X}_K = X \cup \coprod_{z \in X(\infty)} \overline{(X_z)_K}.$$

Proof. For any unbounded sequence y_j in X, if it converges to $(H_\infty, z) \in \overline{\mathfrak{a}}_P^{K,+}$ in $\overline{X}_K$, then it follows from the definition of convergence of sequences that y_j converges to $\pi(H_\infty)$ in $\overline{X}^c$, where π is the map in Proposition I.14.4. By [GJT, Lemma 3.28], this defines a continuous map

$$\pi : \overline{X}_K \to \overline{X}^c = X \cup X(\infty).$$

For any point $H \in \partial \overline{X}^c = \coprod_Q \mathfrak{a}_Q^+(\infty)$, let Q be the unique parabolic subgroup such that $H \in \mathfrak{a}_Q^+(\infty)$. For any parabolic subgroup P contained in Q, let P' be the corresponding parabolic subgroup in M_Q. Let $J \subset \Delta(P, A_P)$ such that $Q = P_J$. For any partition $\Sigma : I_1 \cup \cdots \cup I_k$ of $\Delta(P, A_P)$ satisfying $I_1 = \Delta - J$, Σ induces a partition $\Sigma' : I_2 \cup \cdots \cup I_k$ of $\Delta(P', A_{P'})$. For every point $H' \in \mathfrak{a}_{P'}^{\Sigma',+}(\infty)$, $z \in X_P$, then $(H, H') \in \mathfrak{a}_P^{\Sigma,+}(\infty)$, $((H, H'), z) \in \overline{\mathfrak{a}}_P^{K,+}(\infty) \times X_P$, and the fiber $\pi^{-1}(H)$ consists of the union

$$\cup_{P \subseteq Q} \cup_\Sigma \{((H, H'), z) \mid H' \in \mathfrak{a}_{P'}^{\Sigma',+}(\infty), z \in X_P\},$$

where for each $P \subseteq Q$, write $Q = P_J$ as above, the second union being taken over all the partitions Σ with $I_1 = \Delta - J$. This set can be identified with

$$\cup_{P' \subseteq M_Q} \cup_{\Sigma'} \mathfrak{a}_{P'}^{\Sigma',+}(\infty) \times X_{P'} = X_Q \cup \cup_{P' \subset M_Q} \overline{\mathfrak{a}}_{P'}^{K,+}(\infty) \times X_{P'},$$

which is equal to $\overline{(X_Q)_K}$ by definition.

Proposition I.14.13 *The identity map on X extends to a continuous map $\overline{X}_K \to \overline{X}^M$, and this map is a homeomorphism if and only if the rank of X is less than or equal to 2.*

Proof. It is clear from the definitions that if an unbounded sequence y_j in X converges to $(H_\infty, z_\infty) \in \overline{\mathfrak{a}}_P^{K,+}(\infty) \times X_P$ in $\overline{X}_K$, then y_j also converges in $\overline{X}^M$ to $(\pi(H_\infty), z_\infty) \in \overline{\mathfrak{a}_P^+(\infty)} \times X_P$, where π is the map defined in Proposition I.14.4. By [GJT, Lemma 3.28], this defines a continuous map $\overline{X}_K \to \overline{X}^M$. By Proposition I.14.4, this map is bijective if and only if $rk(X) \leq 2$. In this case, it is a homeomorphism.

I.14.14 Example. When $X = D \times D$, $\overline{X}^K$ is the same as the Martin compactification $\overline{X}^M$. On the other hand, when $X = D \times D \times D$, they are different. To see this, let γ be a singular geodesic such that the boundary symmetric space $X_{[\gamma]}$ is equal to $D \times D$. For example, we can take $\gamma(t) = (\gamma_1(t), 0, 0)$, where γ_1 is a geodesic in D. Then $D \times D$ is contained in the boundaries of both $\overline{X}^M$ and $\overline{X}^K$. The closure of $D \times D$ in $\overline{X}^M$ is equal to $\overline{D} \times \overline{D}$, where $\overline{D}$ is the closed unit disk, but its closure in $\overline{X}^K$ is $\overline{D \times D}^M$, the Martin compactification, which strictly dominates $\overline{D} \times \overline{D}$, the maximal Satake compactification.

A maximal flat subspace in $D \times D \times D$ is isometric to $\mathbb{R} \times \mathbb{R} \times \mathbb{R}$. Its closure in the maximal Satake compactification is a cube. Its closure in the Martin compactification is a blow-up of the cube by rounding off the corners. To get the closure in the Karpelevič compactification, we need to blow up further each boundary face in the closure of the Martin compactification into a cell of dimension 2, i.e., of codimension 1.

Proposition I.14.15 *Let $\overline{X}^K$ be the Karpelevič compactification, and $\overline{X}_K$ the compactification defined in this section. Then the identity map on X extends to a homeomorphism $\chi : \overline{X}^K \to \overline{X}_K$*

Proof. The idea is to prove that any unbounded sequence y_j in X that converges in $\overline{X}^K$ also converges in $\overline{X}_K$, and hence we need only to describe the intersection with X of neighborhoods of boundary points in $\overline{X}^K$ which is given in [Ka, §13.8] [GJT, §5.5], and show that it is contained the intersection with X of the neighborhoods of boundary points in $\overline{X}_K$. For details, see [BJ2, §8].

I.14.16 Summary and comments. We give a noninductive construction of $\overline{X}^K$ following the general method in §I.8. A key point is to understand the closure of the maximal flat subspaces. The original definition in [Ka] is inductive on the rank of the symmetric spaces. An important observation in [GJT] is that for an unbounded sequence of points in a positive chamber, the limit point to which the sequence converges depends on the relative rates at which the coordinates of the sequence go to infinity with respect to the simple roots. Then compactifications of flats in X are glued into the compactification of X. As in the case of dual-cell compactification, it is not easy to show directly that the G-action on X extends to a continuous action on the resulting compactification.

The approach here is more global in some sense, and the continuous extension of the G-action is easy to obtain.

I.15 Real Borel–Serre partial compactification

In this section, we use the uniform method in §I.8 to give a slightly different construction of the *partial Borel–Serre compactification* ${}_{\mathbb{R}}\overline{X}^{BS}$ of X over the real numbers in [BS2]. It is well known that [BS2] constructs a partial compactification over the rational numbers, called the partial Borel–Serre compactification of X, whose quotient by an arithmetic subgroup Γ gives the Borel–Serre compactification $\overline{\Gamma\backslash X}^{BS}$ of $\Gamma\backslash X$ in [BS2] (see §III.5 below). Actually, the construction in [BS2] works over any subfield of $\mathbb{R}$. Compactifications of this real Borel–Serre partial compactification ${}_{\mathbb{R}}\overline{X}^{BS}$ of X are useful for some applications concerning global properties of arithmetic subgroups such as the *Novikov conjectures* (see [Gol]).

As suggested by the reductive Borel–Serre compactification of locally symmetric spaces, we also construct in this section a quotient of ${}_{\mathbb{R}}\overline{X}^{BS}$, called the real reductive Borel–Serre partial compactification of X and denoted by ${}_{\mathbb{R}}\overline{X}^{RBS}$.

This section is organized as follows. In I.15.1, we define the Borel–Serre boundary components, and define the topology via convergent sequences. Neighborhood bases are given in I.15.2. The Hausdorff property of ${}_{\mathbb{R}}\overline{X}^{BS}$ is proved in Proposition I.15.3. Though the G-action on X extends to ${}_{\mathbb{R}}\overline{X}^{BS}$, the extension is not continuous (Proposition I.15.6). It is noncompact (Proposition I.15.4) and hence called a *partial compactification*. In I.15.6, the difficulty of compactifying this partial compactification in the case of $X = D$ is discussed. The real reductive Borel–Serre partial compactification ${}_{\mathbb{R}}\overline{X}^{RBS}$ is defined in I.15.9. It is shown in Proposition I.15.11 that it is noncompact and Hausdorff. Its relation to the maximal Satake compactification is discussed in Proposition I.15.13.

I.15.1 We follow the uniform method to construct this partial compactification ${}_{\mathbb{R}}\overline{X}^{BS}$. There is a boundary component $e(P)$ for each real (proper) parabolic subgroup P, defined by

$$e(P) = N_P \times X_P.$$

Let

$$_{\mathbb{R}}\overline{X}^{BS} = X \cup \coprod_{P} e(P)$$

with the following topology. First, we describe some special convergent sequences.

1. An unbounded sequence y_j in X converges to a boundary point $y_\infty = (n_\infty, z_\infty)$ if in the horospherical decomposition $y_j = (n_j, a_j, z_j) \in N_P \times A_P \times X_P$, the components satisfy the following conditions:
   (a) $n_j \to n_\infty$,
   (b) $z_j \to z_\infty$,
   (c) and for any $\alpha \in \Phi(P, A_P)$, $a_j^\alpha \to +\infty$.
2. For two parabolic subgroups P, Q with $P \subseteq Q$, $e(P)$ is contained in the closure of $e(Q)$. Let P' be the parabolic subgroup of M_Q corresponding to P in equation (I.1.21). Then

$$X_Q = N_{P'} \times A_{P'} \times X_{P'} = N_{P'} \times A_{P'} \times X_P,$$

and hence

$$e(Q) = N_Q \times X_Q = N_P \times A_{P'} \times X_P, \tag{I.15.1}$$

where $N_P = N_Q \times N_{P'}$ is used. Under the above notation, a sequence $y_j \in e(Q)$ converges to any given point $(n_\infty, z_\infty) \in e(P)$ if in the decomposition in equation (I.15.1), the coordinates of $y_j = (n_j, a'_j, z_j)$ satisfy the following conditions:
   (a) $n_j \to n_\infty$ in N_P,
   (b) $z_j \in z_\infty$,
   (c) for all $\alpha \in \Phi(P', A_{P'})$, $a_j'^\alpha \to +\infty$.

A general convergent sequence is a combination of these special ones. It can be checked easily that they form a convergence class of sequences and hence define a topology on $_{\mathbb{R}}\overline{X}^{BS}$.

I.15.2 Neighborhoods of boundary points can be given explicitly. For any $(n_\infty, z_\infty) \in e(P)$, let U a neighborhood of n_∞ in N_P, and V be a neighborhood of z_∞ in X_P. For $t > 0$, $U \times A_{P,t} \times V$ is a Siegel set in X associated with P, and

$$U \times A_{P',t} \times V \subset N_P \times A_{P'} \times X_P = X_Q.$$

The union

$$U \times A_{P,t} \times V \cup \coprod_{Q \supseteq P} U \times A_{P',t} \times V$$

is a neighborhood of (n_∞, z_∞) in $_{\mathbb{R}}\overline{X}^{BS}$.

Proposition I.15.3 *The space* $_{\mathbb{R}}\overline{X}^{BS}$ *is Hausdorff.*

Proof. This follows from the separation of Siegel sets in X in Propositions I.9.8 and I.9.11 and its generalization to subsets $U \times A_{P'} \times V$ in $e(Q)$ associated with $P \subset Q$.

Proposition I.15.4 *The space* $_{\mathbb{R}}\overline{X}^{BS}$ *is noncompact.*

Proof. Fix any parabolic subgroup P and consider a sequence $y_j = (n_j, a_j, z_j)$ whose coordinates satisfy the conditions (1) $a_j^{\alpha} \to +\infty$ for all $\alpha \in \Sigma(P, A_P)$, (2) n_j is unbounded, $n_j^{a_j} \to e$, (3) $z_j \to z_\infty$ for some $z_\infty \in X_P$. From the separation of generalized Siegel sets in Proposition I.9.11, it can be shown that for any boundary point $(n, z) \in e(P)$, $z \neq z_\infty$, y_j does not belong to any sufficiently small neighborhood of (n, z) when $j \gg 1$. Since n_j is not bounded, y_j also does not belong to any neighborhood of (n, z_∞) when $j \gg 1$. Similarly, Proposition I.9.8 implies that it does converge to any point in $e(Q)$ for any $Q \neq P$. Therefore, y_j does not converge to any point in $_R\overline{X}^{BS}$.

I.15.5 The G-action on X can be extended to $_{\mathbb{R}}\overline{X}^{BS}$ as follows. For any $g \in G$ and a boundary point $(n, z) \in N_P \times X_P$, write $g = k m_0 a_0 n_0$, where $k \in K$, $m_0 \in M_P$, $a_0 \in A_P$, and $n_0 \in N_P$. Define

$$g \cdot (n, z) = ({}^{k m_0 a_0}(n_0 n), Ad(k) m_0 z) \in N_{gP} \times X_{gP}.$$

Proposition I.15.6 *This extended action G-action on* $_{\mathbb{R}}\overline{X}^{BS}$ *is not continuous.*

Proof. In fact, any sequence $y_j = (n, a_j, z) \in N_P \times A_P \times X_P$ with $a_j^{\alpha} \to +\infty$ for all $\alpha \in \Phi(P, A_P)$ converges to $(n, z) \in e(P)$. On the other hand, y_j can also be written as $g_j \cdot (n_j, a_j', z_j)$, where $g_j \to e$ in G, $n_j \to e$ in N_P, $(a_j')^{\alpha} \to +\infty$ for all $\alpha \in \Phi(P, A_P)$, and z_j converges to z. If the G-action is continuous, then y_j will also converge to $(e, z) \in e(P)$ in $_{\mathbb{R}}\overline{X}^{BS}$. Since it will be shown below that the space $_{\mathbb{R}}\overline{X}^{BS}$ is Hausdorff, this gives a contradiction when $n \neq e$.

On the other hand, for some applications to topology, only the action of discrete subgroups Γ on compactifications of $_{\mathbb{R}}\overline{X}^{BS}$ is required.

I.15.7 The above proof shows that the noncompactness of the factor N_P in the boundary component $e(P)$ causes the noncompactness of $_{\mathbb{R}}\overline{X}^{BS}$. To get a compactification of $_{\mathbb{R}}\overline{X}^{BS}$, we need to compactify the spaces N_P suitably. This turns out to be subtle since there are compatibility conditions of the boundary components of different parabolic subgroups to be satisfied.

I.15.8 Consider the example of $G = SU(1, 1)$ and $X = D$, the unit disk. For each parabolic subgroup P, the boundary component $e(P)$ is equal to $N_P \cong \mathbb{R}$. Compactify N_P by adding two points at infinity $-\infty, +\infty$:

$$\overline{e(P)} = \{-\infty\} \cup e(P) \cup \{+\infty\} \cong [-\infty, +\infty].$$

The topology of

$$\overline{D}^* = D \cup \coprod_P \overline{e(P)}$$

is given as follows:

1. Fix any point $n_\infty \in e(P)$. An unbounded sequence y_j in X converges to n_∞ if in the horospherical coordinate decomposition $X = N_P \times A_P$ (note that X_P consists of one point and $\dim A_P = 1$), $y_j = (n_j, a_j)$, $a_j \to +\infty$ and $n_j \to n_\infty$. A sequence y_j in the boundary $\coprod_Q \overline{e(Q)}$ converges to n_∞ if $y_j \in e(P)$ for $j \gg 1$ and $y_j \to n_\infty$ in $e(P) = N_P$.
2. Identify N_P with the open cell in G/P^- using the *Bruhat decomposition*,

$$G/P^- = N_P \cup \{\infty\},$$

where P^- is the parabolic subgroup opposite to P. Each point $gP^- \in G/P^-$ corresponds to the parabolic subgroup gP^-, and the point ∞ corresponds to P. For the point $\xi = -\infty \in \overline{e(P)}$, an unbounded sequence y_j in X converges to ξ if in the horospherical decomposition $X = N_P \times A_P$, $y_j = (n_j, a_j)$, $n_j \to -\infty$ but there is no condition on a_j; a sequence y_j on the boundary $\coprod_Q \overline{e(Q)}$ converges to $\xi = -\infty$ if $y_j = {}^{g_j}P^-$, where $g_j \in N_P$ for $j \gg 1$, and $g_j \to -\infty$ in $\overline{N_P} = \overline{e(P)} = \{-\infty\} \cup e(P) \cup \{+\infty\}$. Convergence to the point $+\infty \in \overline{e(P)}$ can be described similarly.

It can be shown that $\overline{D}^*$ is a compactification of ${}_{\mathbb{R}}\overline{D}^{BS}$.

Remark I.15.9 This compactification $\overline{X}^*$ was first defined in [Gol], and the topology was described differently. The above definition of the topology emphasizes the role of A_P-, N_P-actions, and the A_P-action is the geodesic action in [BS2]. In a certain sense, the N_P-action on X takes precedence over the A_P-action.

I.15.10 If we drop the factor N_P and take the boundary component to be X_P, then we get the *reductive Borel–Serre partial compactification*

$${}_{\mathbb{R}}\overline{X}^{RBS} = X \cup \coprod_P X_P$$

with the following topology. Fix any point $z_\infty \in X_P$. There are two special types of sequences converging to z_∞, and the general convergent sequences are combinations of these.

1. An bounded sequence $y_j \in X$ converges to z_∞ if in the horospherical decomposition $y_j = (n_j, a_j, z_j) \in N_P \times A_P \times X_P$, the coordinates satisfy (1) $z_j \to z_\infty$, (2) $a_j^\alpha \to +\infty$ for all $\alpha \in \Sigma(P, A_P)$, (3) $n_j^{a_j} \to e$.
2. For any pair of parabolic subgroups P, Q, $e(P)$ is contained in the closure of $e(Q)$ if and only if $P \subseteq Q$, and a sequence y_j in the boundary component $e(Q)$ converges to z_∞ if it satisfies the same condition as in (1) when X is replaced by X_Q and P by the parabolic subgroup P' of M_Q corresponding to P.

Proposition I.15.11 *The partial compactification* ${}_{\mathbb{R}}\overline{X}^{RBS}$ *is a noncompact Hausdorff space.*

Proof. The Hausdorff property follows from the separation properties of generalized Siegel sets in Propositions I.9.8 and I.9.11 . To show that it is noncompact, take the sequence $k_j \in K, k_j \to e$ but $k_j \notin P$. For any point $z_\infty \in X_P$, then $k_j z_\infty$ does not converge to any limit point in ${}_{\mathbb{R}}\overline{X}^{RBS}$.

I.15.12 Set theoretically, ${}_{\mathbb{R}}\overline{X}^{RBS}$ is equal to the maximal Satake compactification $\overline{X}^S_{\max}$. In fact, ${}_{\mathbb{R}}\overline{X}^{RBS}$ is the analogue of the partial reductive Borel–Serre compactification ${}_{\mathbb{Q}}\overline{X}^{RBS}$, whose quotient by arithmetic groups Γ gives the reductive Borel–Serre compactification $\overline{\Gamma\backslash X}^{RBS}$ discussed later in §III.6.

Proposition I.15.13 *The identity map on* $X \cup \coprod X_P$ *gives a continuous map from* ${}_{\mathbb{R}}\overline{X}^{RBS}$ *to* $\overline{X}^S_{\max}$, *but this map is not a homeomorphism.*

Proof. Since both ${}_{\mathbb{R}}\overline{X}^{RBS}$ and $\overline{X}^S_{\max}$ are equal to $X \cup \coprod_P X_P$, they can be identified with each other as sets. From the definition of the topologies of ${}_{\mathbb{R}}\overline{X}^{RBS}$ and $\overline{X}^S_{\max}$, it is clear that if a sequence in ${}_{\mathbb{R}}\overline{X}^{RBS}$, in particular an unbounded sequence in X, converges in ${}_{\mathbb{R}}\overline{X}^{RBS}$, it also converges in $\overline{X}^S_{\max}$. This implies that the identity map on X extends to a continuous map from ${}_{\mathbb{R}}\overline{X}^{RBS}$ to $\overline{X}^S_{\max}$.

The proof of Proposition I.15.11 shows that there is no convergence between points of boundary components in ${}_{\mathbb{R}}\overline{X}^{RBS}$ of different conjugate parabolic subgroups, i.e., the topology on the set of boundary components is discrete. For example, in the case $G = SU(1,1)$, $X = D$, ${}_{\mathbb{R}}\overline{X}^{RBS} = D \cup S^1$ with the discrete topology on the boundary. Since $\overline{X}^S_{\max}$ admits a continuous G-action, $\overline{X}^S_{\max}$ and ${}_{\mathbb{R}}\overline{X}^{RBS}$ are not isomorphic.

Remark I.15.14 This proposition explains that for compactifications of symmetric spaces with a continuous G-action, it is the topological Tits building rather than the (usual) Tits building that is used to parametrize the boundary components.

Proposition I.15.15 *The action of G on X extends to actions on* ${}_{\mathbb{R}}\overline{X}^{BS}$ *and* ${}_{\mathbb{R}}\overline{X}^{RBS}$, *but the extended actions are not continuous.*

Proof. The results on ${}_{\mathbb{R}}\overline{X}^{BS}$ were given in Proposition I.15.6. Now we define the extended action on ${}_{\mathbb{R}}\overline{X}^{RBS}$. For any $g \in G$, $z_\infty \in X_P$, write $g = kman$, where $k \in K, m \in M_P, a \in A_P$, and $n \in N_P$. Define

$$g \cdot z_\infty = k \cdot (m z_\infty) \in X_{kP} = X_{gP}.$$

To show that this extended action is not continuous, we note that for any sequence $k_j \in K$, $k_j \to e$, but $k_j \notin P$. Then for any boundary point $(n_\infty, z_\infty) \in z_\infty$, $k_j \cdot (n_\infty, z_\infty)$ does not converge to (n_∞, z_∞).

I.15.16 Summary and comments. In this section, we applied the general method to construct two partial compactifications of X, the Borel–Serre and the reductive Borel–Serre partial compactifications. Since they compactify X in certain directions but are not compact, they are called *partial compactifications*. Though the terminology of partial compactification is not ideal, it seems to be difficult to choose a good one. Some people call the Borel–Serre partial compactification over $\mathbb{Q}$ defined in §III.5 the *Borel–Serre bordification*, since it is a manifold with corners, and hence some borders have been added. But the term bordification is probably not suitable for the reductive Borel–Serre partial compactification over $\mathbb{Q}$.

3

Properties of Compactifications of Symmetric Spaces

In the previous chapters, we have recalled and constructed many different compactifications of symmetric spaces. In this chapter, we study relations between these compactifications and their more refined properties such as analytic structures and the global shape as topological manifolds.

In §I.16, we introduce the notions of dominating maps between compactifications, common quotients, and common refinements of compactifications of symmetric spaces. Then we study domination relations between all the compactifications of symmetric spaces mentioned in Chapter 1. To motivate the subgroup compactification of locally symmetric spaces in Part III, we discuss in §I.17 two other constructions of the maximal Satake compactification $\overline{X}^S_{\max}$, one using the space of closed subgroups and another using a Grassmannian variety associated with the Lie algebra $\mathfrak{g}$. We also give a different construction of the geodesic compactification $X \cup X(\infty)$ by embedding X into ambient spaces and taking the closure, which gives the Gromov compactification of X. In §I.18, we recall Atiyah's convexity theorem on the moment map and use it to identify the closure of flats in the Satake compactifications and use it to show that the Satake compactifications are topological balls. In §I.19, we recall the dual-cell compactification $X \cup \Delta^*(X)$ in [GJT] and a modification $\widetilde{X}^S_{\max}$ suggested by the Oshima construction in [Os1] to show that $\overline{X}^S_{\max}$ is a real analytic manifold with corners.

I.16 Relations between the compactifications

In the previous sections, we have studied many different compactifications that arise from various problems. In this section, we study relations between them.

First we introduce several notions that describe different relations between compactifications of the same space. Then we prove the existence of the greatest common quotient (GCQ) and the least common refinement (LCR) of two compactifications (Proposition I.16.2). After illustrating these concepts through the example of $X = D \times D$, the bidisk, in I.16.3, we state the general result in I.16.5.

I.16.1 As mentioned earlier, in this book, all compactifications are assumed to be Hausdorff. Recall that given two compactifications $\overline{X}^1, \overline{X}^2$ of X, $\overline{X}^1$ is said to *dominate* $\overline{X}^2$ if the identity map on X extends to a continuous map $\overline{X}^1 \to \overline{X}^2$, which is automatically surjective and called the *dominating map*. The compactification $\overline{X}^1$ is called a *refinement* of $\overline{X}^2$, and $\overline{X}^2$ is called a *quotient* of $\overline{X}^1$. Two compactifications $\overline{X}^1, \overline{X}^2$ are isomorphic if the identity map on X extends to a homeomorphism between them. If both $\overline{X}^1$ and $\overline{X}^2$ are G-compactifications, the extended map is clearly G-equivariant. A *common quotient* of $\overline{X}^1$ and $\overline{X}^2$ is a compactification $\overline{X}$ that is dominated by both $\overline{X}^1$ and $\overline{X}^2$. On the other hand, if a compactification $\overline{X}$ dominates both $\overline{X}^1$ and $\overline{X}^2$, it is called a *common refinement*.

Proposition I.16.2 *Any two compactifications $\overline{X}^1$ and $\overline{X}^2$ admit a unique greatest common quotient (GCQ), denoted by $\overline{X}^1 \wedge \overline{X}^2$ or $GCQ(\overline{X}^1, \overline{X}^2)$, and a least common refinement (LCR) denoted by $\overline{X}^1 \vee \overline{X}^2$ or $LCR(\overline{X}^1, \overline{X}^2)$.*

Proof. The set of compactifications of X is partially ordered with respect to the relation of domination. It can be checked easily that for any ordered chain in this partially ordered set, there is a maximal element and a minimal element. By Zorn's lemma, both GCQ and LCR exist.

The LCR can be realized as the closure of X under the diagonal embedding into $\overline{X}^1 \times \overline{X}^2$. On the other hand, there is no such simple procedure to obtain $\overline{X}^1 \wedge \overline{X}^2$.

I.16.3 **Example.** We consider several examples to illustrate these concepts. Let $X = D \times D$. It has the one-point compactification $X \cup \{\infty\}$, the Satake compactification $\overline{X}^S = \overline{D} \times \overline{D}$ (all the Satake compactifications of D and hence of $D \times D$ are isomorphic to each other), the geodesic compactification, and the Martin compactification $\overline{X}^M$, which is isomorphic to the Karpelevič compactification $\overline{X}^K$.

The one-point compactification $X \cup \{\infty\}$ is a quotient of all the others, and $\overline{X}^K$ dominates all other compactifications. The GCQ of $X \cup X(\infty)$ and $\overline{X}^S$ is $\overline{X}^M$, and their LCR is $X \cup \{\infty\}$.

I.16.4 Recall that $\mathfrak{g} = \mathfrak{k} + \mathfrak{p}$ is the Cartan decomposition of $\mathfrak{g}$ associated with K, and $\mathfrak{a} \subset \mathfrak{p}$ a maximal abelian subspace. Let $\mathfrak{a}^+$ be a positive chamber, and $\Delta = \Delta(\mathfrak{g}, \mathfrak{a})$ the set of simple roots in $\Phi(\mathfrak{g}, \mathfrak{a})$. Let P_0 be the minimal parabolic subgroup containing $A = \exp \mathfrak{a}$ such that $\Delta(P_0, A) = \Delta(G, A) = \Delta(\mathfrak{g}, \mathfrak{a})$.

Proposition I.16.5 *The relations between the compactifications: the geodesic compactification $X \cup X(\infty)$, the Satake compactifications $\overline{X}^S_\tau$, the Baily–Borel compactification $\overline{X}^{BB}$, the Furstenberg compactifications $\overline{X}^F_I$, the Martin compactification $\overline{X}^M = X \cup \partial_\lambda X$, $\lambda < \lambda_0(X)$, the Karpelevič compactification $\overline{X}^K$, and the real Borel–Serre partial compactification ${}_{\mathbb{R}}\overline{X}^{BS}$, ${}_{\mathbb{R}}\overline{X}^{RBS}$ are given as follows:*

1. *When the rank of X is equal to 1, all these compactifications are isomorphic to each other.*

2. *The Karpelevič compactification* $\overline{X}^K$ *dominates the Martin compactification* $\overline{X}^M$, *and they are isomorphic if and only if the rank of* X *is less than or equal to* 2.
3. $\overline{X}^M$ *dominates both the geodesic compactification* $X \cup X(\infty)$ *and the maximal Satake compactification* $\overline{X}^S_{\max}$ *and is equal to* $X \cup X(\infty) \vee \overline{X}^S_{\max}$, *i.e., the LCR of* $X \cup X(\infty)$ *and* $\overline{X}^S_{\max}$. *On the other hand, the GCQ of* $X \cup X(\infty)$ *and* $\overline{X}^S_{\max}$ *is the one-point compactification if the rank of* X *is strictly greater than* 1.
4. *The maximal Satake compactification* $\overline{X}^S_{\max}$ *dominates all other Satake compactifications. More generally, for any two Satake compactifications* $\overline{X}^S_{\tau_1}$ *and* $\overline{X}^S_{\tau_2}$, $\overline{X}^S_{\tau_1}$ *dominates* $\overline{X}^S_{\tau_2}$ *if and only if the highest weight* μ_{τ_1} *of* τ_1 *is more regular than* μ_{τ_2} *in the sense that if* C_i *is the Weyl chamber face of* $\overline{\mathfrak{a}^{*,+}}$ *containing* μ_{τ_i} *as an interior point, then* C_2 *is a face of* C_1. *In particular, if* τ *is a regular representation,* $\overline{X}^S_\tau$ *is the maximal Satake compactification* $\overline{X}^S_{\max}$.
5. *When* X *is a Hermitian symmetric space (or equivalently a bounded symmetric domain), the Baily–Borel compactification* $\overline{X}^{BB}$ *is isomorphic, as a topological* G*-compactification, to the minimal Satake compactification* $\overline{X}^S_\tau$, *whose highest weight* μ_τ *is connected only to the distinguished simple root (in the sense being either the longest or the shortest) in the Dynkin diagram of* $\Phi(\mathfrak{g}, \mathfrak{a})$.
6. *The Furstenberg compactifications are isomorphic to the Satake compactifications, where* $\overline{X}^F_{P_{0,I}}$, $I \subset \Delta$, *is isomorphic to a Satake compactification* $\overline{X}^S_\tau$, *where the highest weight* μ_τ *is connected exactly to the simple roots in* $\Delta - I$. *In particular, the maximal Furstenberg compactification* $\overline{X}^F_{\max}$ *corresponds to* $I = \emptyset$ *and is isomorphic to the maximal Satake compactification* $\overline{X}^S_{\max}$.
7. *The real Borel–Serre partial compactification* ${}_{\mathbb{R}}\overline{X}^{BS}$ *dominates the real reductive Borel–Serre partial compactification* ${}_{\mathbb{R}}\overline{X}^{RBS}$, *which in turn dominates the maximal Satake compactification* $\overline{X}^S_{\max}$. *The dominating map* ${}_{\mathbb{R}}\overline{X}^{RBS} \to \overline{X}^S_{\max}$ *is continuous and bijective but is not a homeomorphism.*

Proof. The result basically follows from the uniform construction of these compactifications by comparing the sizes of the boundary components. Specifically, (1) follows from Propositions I.14.13 and I.14.15, (2) follows from Proposition I.7.15, (3) is contained in Proposition I.4.35, (4) follows from Corollary I.5.29 and the definition of the distinguished root, (5) follows from Proposition I.6.21, the first statement of (6) follows from the definitions of ${}_{\mathbb{R}}\overline{X}^{BS}$ and ${}_{\mathbb{R}}\overline{X}^{RBS}$, and the second statement from Proposition I.15.13.

Remark I.16.6 In this book, we have not discussed functorial properties of compactifications of X, i.e., if $X_1 \to X_2$ is an embedding, and $\overline{X_2}$ is a compactification of X_2, the question is what the induced compactification of X_1 is. Similarly, we can also ask whether a compactification $\overline{X}$ preserves the product, i.e., if $X = X_1 \times X_2$, then $\overline{X} = \overline{X_1} \times \overline{X_2}$.

It can be shown easily that the geodesic compactification $X \cup X(\infty)$ has the functorial property. In fact, for any isometric embedding $X_1 \to X_2$, every equivalence class of geodesics in X_1 is mapped into an equivalence class in X_2. This defines an injective map from $X_1 \cup X_1(\infty)$ to $X_2 \cup X_2(\infty)$. It can be shown easily that it is an embedding.

The example of $X = D \times D$ shows easily that it does not preserve the product structure, i.e., the geodesic compactification of $D \times D$ is not isomorphic to $(D \cup D(\infty)) \times (D \cup D(\infty)) = \overline{D} \times \overline{D}$.

On the other hand, it can be shown that the Satake compactifications preserve the product structure.

I.17 More constructions of the maximal Satake compactification

In this section, we recall two more realizations of the maximal Satake compactification $\overline{X}^S_{\max}$: the *subgroup compactification* $\overline{X}^{sb}$, the *subalgebra compactification* $\overline{X}^{sba}$. Two more realizations will be given in §I.19, the *dual-cell compactification* $X \cup \Delta^*(X)$, and a modification $\widetilde{X}^S_{\max}$ suggested by the Oshima compactification.

The subgroup compactification $\overline{X}^{sb}$ is related to the compactifications of locally symmetric spaces in Part III, Chapter 12, while the subalgebra compactification $\overline{X}^{sba}$ allows us to embed the maximal Satake compactification $\overline{X}^S_{\max}$ into the real locus of the wonderful compactification $\overline{X_{\mathbb{C}}}^W$ of the complex symmetric space $X_{\mathbb{C}} = G_{\mathbb{C}}/K_{\mathbb{C}}$ in Part II, Chapter 7, and the modified dual-cell compactification $\widetilde{X}^S_{\max}$ shows that $\overline{X}^S_{\max}$ is a real analytic manifold with corners. We also give another realization of the geodesic compactification $X \cup X(\infty)$ by the *Gromov compactification* $\overline{X}^G$ and its variant. The Gromov compactification applies to any complete, locally compact metric space, and the Gromov compactification of locally symmetric spaces will be determined and identified with the geodesic compactification in §III.20. As an application, we reconstruct the Martin compactifications $X \cup \partial_\lambda X$ via embeddings into compact ambient spaces in I.17.18.

I.17.1 Let $\mathcal{S}(G)$ be the *space of closed subgroups* in G. It has the following topology. Let $C \subset G$ be a compact subset, and $U \subset G$ a neighborhood of e in G. For any closed subgroup $H \subset G$, define

$$V(H; U, C) = \{H' \in \mathcal{S}(G) \mid H' \cap C \subset U(H \cap C), H \cap C \subset U(H' \cap C)\}.$$

Then $V(H; U, C)$ is a neighborhood of H in $\mathcal{S}(G)$. When C_j is an exhausting family of subsets of G, $\cup_j C_j = G$, and U_j ranges over a neighborhood basis of e, then $V(H; U_j, C_j)$ forms a neighborhood basis of H. It can be checked easily that this defines a topology on $\mathcal{S}(G)$. Intuitively, a sequence H_j converges to H if within any fixed compact subset of G, H_j approximates H uniformly.

Clearly, G acts continuously on $\mathcal{S}(G)$ by conjugation. Hence $\mathcal{S}(G)$ is a G-space.

Proposition I.17.2 *The space $\mathcal{S}(G)$ is a compact Hausdorff space.*

Clearly, the limit of a sequence of subgroups is a closed subgroup. The point is to prove that any sequence of subgroups has a limit. For details, see [Bu3, Chapter 8, §5].

Define a map

$$i : X = G/K \to \mathcal{S}(G), \quad gK \mapsto gKg^{-1},$$

which is G-equivariant. The injectivity of the map i follows from the next result.

Lemma I.17.3 *The maximal compact subgroup K is equal to its own normalizer $\mathcal{N}(K)$.*

Proof. Fix any $g \in \mathcal{N}(K)$. By definition, for any $k \in K$, there exists $k' \in K$ such that $kg = gk'$. Let $x_0 = K \in X$ be the basepoint. Then

$$k(gx_0) = g(k'x_0) = gx_0,$$

and K fixes gx_0. If $gx_0 \neq x_0$, then the geodesic $\gamma(t)$ from x_0 passing through gx_0 is fixed by K. But this is impossible. In fact, let $\mathfrak{g} = \mathfrak{k} + \mathfrak{p}$ be the Cartan decomposition. There exists a maximal abelian subspace $\mathfrak{a}$ in $\mathfrak{p}$ and a nonzero element $H \in \mathfrak{a}$ such that $\gamma(t) = \exp tH$. Since the Weyl group of $\mathfrak{a}$ can be realized as a subgroup of K and the Weyl group of $\mathfrak{a}$ does not fix any nonzero element of $\mathfrak{a}$, the claim follows.

Proposition I.17.4 *The map $i : X \to \mathcal{S}(G)$ is an embedding.*

Proof. Clearly, by the above lemma, the map i is injective. We need to show that for any sequence $g_j K$ in X, $g_j K g_j^{-1}$ converges to $g_\infty K g_\infty^{-1}$ in $\mathcal{S}(G)$ if and only if $g_j K$ converges to $g_\infty K$ in X.

If $g_j K \to g_\infty K$ in X, then

$$g_j K g_j^{-1} = (g_j K)(g_j K)^{-1} \to (g_\infty K)(g_\infty K)^{-1} = g_\infty K g_\infty^{-1}.$$

On the other hand, assume that $g_j K g_j^{-1}$ converges to $g_\infty K g_\infty^{-1}$. We claim that the orbit $g_j K g_j x_0$, where $x_0 = K \in X$, is contained in the sphere in X with center $g_j x_0$ and radius $d(g_j x_0, x_0)$ with respect to the invariant metric d on X. In fact, for any $k \in K$,

$$d(g_j k g_j^{-1} x_0, g_j x_0) = d(k g_j^{-1} x_0, x_0) = d(g_j^{-1} x_0, x_0) = d(g_j x_0, x_0).$$

This implies that as the center of the sphere, $g_j x_0 \to g_\infty x_0$, and hence $g_j K \to g_\infty K$.

Remark I.17.5 See [GJT, p. 132] for other proofs of Lemma I.17.3. The proof of the above proposition is similar to the proof of [GJT, Proposition 9.3].

The closure of $i(X)$ in $\mathcal{S}(G)$ is a G-compactification of X and called the *subgroup compactification* of X and denoted by $\overline{X}^{sb}$.

Proposition I.17.6 *The subgroup compactification $\overline{X}^{sb}$ is isomorphic to the maximal Satake compactification $\overline{X}^{S}_{\max}$.*

This proposition is proved in [GJT, Chapter IX]. The basic idea is as follows. Let $\mathfrak{a}$ be a maximal abelian subspace in $\mathfrak{p}$ as before, and $\mathfrak{a}^+$ a positive chamber. Recall from [GJT, Definition 3.35] that a sequence y_j in X is called *fundamental* if y_j can be written in the form $y_j = k_j \exp H_j x_0$ such that $k_j \in K$, $H_j \in \overline{\mathfrak{a}^+}$ satisfy the following conditions:

1. k_j converges to some k_∞ in K,
2. there exists a subset $I \subset \Delta = \Delta(\mathfrak{g}, \mathfrak{a})$ such that for all $\alpha \in \Delta - I$, $\alpha(H_j) \to +\infty$, and for $\alpha \in I$, $\alpha(H_j)$ converges to a finite limit.

Then a characterization of $\overline{X}^{\max}$ is given in terms of convergence behavior of fundamental sequences in [GJT, Theorem 3.38]. The basic point of the proof of the above proposition is to identify the limit of $i(y_j)$ in $\mathcal{S}(G)$ for fundamental sequences y_j, which is done in [GJT, Proposition 9.14]. In this process, the Furstenberg boundaries play an important role.

I.17.7 There is a variant of the above construction. Let $h = \dim \mathfrak{k}$ and $Gr(\mathfrak{g}, h)$ be the Grassmannian of h-dimensional subspaces in $\mathfrak{g}$. Using the fact that the normalizer of K in G is equal to K in Lemma I.17.3, it can be shown that the normalizer of $\mathfrak{k}$ in G is equal to K under the adjoint action,

$$\mathcal{N}(\mathfrak{k}) = \{g \in G \mid Ad(g)\mathfrak{k} = \mathfrak{k}\} = K.$$

Then the map

$$\iota : X = G/K \to Gr(\mathfrak{g}, h), \quad gK \mapsto Ad(g)\mathfrak{k},$$

is injective.

Proposition I.17.8 *The map $\iota : X \to Gr(\mathfrak{g}, h)$ is an embedding.*

Proof. We need to show that for any sequence $g_j K$ in X, $g_j K$ converges to $g_\infty K$ if and only if $Ad(g_j)\mathfrak{k}$ converges to $Ad(g_\infty)\mathfrak{k}$. By Proposition I.17.4, $g_j K \to g_\infty K$ implies $g_j K g_j^{-1} \to g_\infty K g_\infty^{-1}$, which in turn implies that $Ad(g_j)\mathfrak{k} \to Ad(g_\infty)\mathfrak{k}$. Conversely, since $g_j K g_j^{-1}$ and $g_\infty K g_\infty^{-1}$ are maximal compact subgroups, the convergence of their Lie algebras,

$$Ad(g_j)\mathfrak{k} \to Ad(g_\infty)\mathfrak{k},$$

implies that

$$g_j K g_j^{-1} \to g_\infty K g_\infty^{-1},$$

which, by Proposition I.17.4 again, implies that $g_j K \to g_\infty K$.

The closure of $\iota(X)$ in $Gr(\mathfrak{g}, h)$ is a G-compactification of X, called the *subalgebra compactification* of X, denoted by $\overline{X}^{sba}$.

Proposition I.17.9 *The subalgebra compactification $\overline{X}^{sba}$ is isomorphic to $\overline{X}^{S}_{\max}$.*

This proposition is proved in [JL, Theorem 1.1]. The idea is as follows. Let $\mathfrak{a}$ be a maximal abelian subspace of $\mathfrak{p}$ as above. Using the explicit action of $\mathfrak{a}$ on the root spaces $\mathfrak{g}_\alpha$, it is easy to show that for a fundamental sequence $y_j = g_j K \in X$ in the sense [GJT, Definition 3.35] as recalled earlier, the limit $Ad(g_j)\mathfrak{k}$ exists and can be described explicitly. Once the stabilizers of these limit subalgebras are determined, one can apply the characterization of $\overline{X}^{S}_{\max}$ in Proposition I.4.33 to finish the proof.

Remark I.17.10 The subalgebra compactification $\overline{X}^{sba}$ is suggested by the wonderful compactification $X_{\mathbb{C}} = G_{\mathbb{C}}/K_{\mathbb{C}}$ discussed in §II.9 later. In fact, De Concini and Procesi [DP1] defined the wonderful compactification by a construction similar to the maximal Satake compactification, and Demazure [De] realized it via embedding $X_{\mathbb{C}}$ into the Grassmannian $Gr(\mathfrak{g}\otimes\mathbb{C}, h)$ of complex subspaces in $\mathfrak{g}\otimes\mathbb{C}$ of dimension h,

$$X_{\mathbb{C}} = G_{\mathbb{C}}/K_{\mathbb{C}} \to Gr(\mathfrak{g}\otimes\mathbb{C}, h), \quad gK_{\mathbb{C}} \mapsto Ad(g)\mathfrak{k}\otimes\mathbb{C},$$

where $h = \dim_{\mathbb{C}} \mathfrak{k}\otimes\mathbb{C}$.

I.17.11 As mentioned earlier, the geodesic compactification $X \cup X(\infty)$ of X is intrinsically defined, unlike the Satake and the Furstenberg compactifications, which are obtained by embedding into compact ambient spaces. We review a compactification of Gromov that allows us to realize $X \cup X(\infty)$ as the closure of X embedded into an ambient space (I.17.14). Then we discuss a modification to obtain a compact ambient space in I.17.16.

Let $C^0(X)$ be the space of continuous functions on X with the topology of uniform convergence on compact subsets. Let $\tilde{C}(X)$ be the quotient of $C^0(X)$ by the one dimensional subspace of constant functions. The quotient topology of $\tilde{C}(X)$ is characterized as follows: a sequence $\tilde{f}_j \in \tilde{C}(X)$ converges to $\tilde{f} \in \tilde{C}(X)$ if and only if there exist lifts $f_j, f_\infty \in C^0(X)$ such that $f_j \to f_\infty$ uniformly over compact subsets.

For each point $x \in X$, let $d_x(\cdot) = d(\cdot, x)$ be the distance function measured from the point x. Denote the image of $d(\cdot, x)$ in $\tilde{C}(X)$ by $[d(\cdot, x)]$ or $[d_x(\cdot)]$.

Lemma I.17.12 *The map $i : X \to \tilde{C}(X)$, $x \mapsto [d(\cdot, x)]$, is an embedding.*

Proof. First we show that i is injective. For any two points $y_1, y_2 \in X$, if $i(y_1) = i(y_2)$, then $d_{y_1} - d_{y_2}$ is a constant function, i.e., there exists a constant c such that for any $x \in X$, $d_{y_1}(x) - d_{y_2}(x) = c$. Setting $x = y_1$, we get $-d(y_2, y_1) = c$. Setting $x = y_2$, we get $d(y_1, y_2) = c$. Therefore, $d(y_1, y_2) = 0$ and $y_1 = y_2$. This proves that i is injective.

Since d_y depends continuously on $y \in X$, the map i is continuous. To finish the proof, we need to show that for any sequence $y_j \in X$, if $[d_{y_j}]$ converges to $[d_{y_0}]$ for some $y_0 \in X$, then $y_j \to y_0$.

By definition, there exists a sequence of constants c_j such that

$$d_{y_j}(x) + c_j \to d_{y_0}(x) \tag{I.17.1}$$

uniformly for x in compact subsets of X.

Suppose that $y_j \not\to y_0$, i.e., $d(y_0, y_j) \not\to 0$ as $j \to \infty$. For simplicity, we assume that there exists a positive constant d_0 such that $d(y_0, y_j) \geq d_0$ when $j \geq 1$ and the distance-minimizing geodesic segments $\gamma_j(t)$ from y_0 to y_j converge to a segment $\gamma_0(t)$ with $\gamma_0(0) = y_0$. Fix a positive t_0 such that $t_0 < d_0$.

Setting $x = y_0$ in equation (I.17.1), we get

$$d_{y_j}(y_0) + c_j \to d_{y_0}(y_0) = 0. \tag{I.17.2}$$

Since $\gamma_j(t_0) \to \gamma_0(t_0)$, we obtain that as $j \to +\infty$,

$$\begin{aligned}
|d_{y_j}(\gamma_0(t_0)) - d_{y_j}(\gamma_j(t_0))| &\leq d(\gamma_0(t_0), \gamma_j(t_0)) \to 0, \\
d_{y_j}(\gamma_0(t_0)) + c_j &= d_{y_j}(\gamma_j(t_0)) + c_j + o(1) \\
&= d_{y_j}(\gamma_j(0)) + c_j - t_0 + o(1) \\
&= d_{y_j}(y_0) + c_j - t_0 + o(1) \\
&\to -t_0.
\end{aligned}$$

In the second equality, we have used the assumption that $t_0 < d_0 < d(y_j, y_0)$, and in the last inequality, we have used equation (I.17.2).

On the other hand, by equation (I.17.1),

$$d_{y_j}(\gamma_0(t_0)) + c_j \to d_{y_0}(\gamma_0(t_0)) = t_0.$$

This implies that $-t_0 = t_0$, and hence $t_0 = 0$. This contradicts the positivity of t_0. Therefore $y_j \to y_0$.

Proposition I.17.13 *The closure of $i(X)$ in $\tilde{C}(X)$ is compact.*

Proof. For any $y \in X$, d_y satisfies the following inequality:

$$|d_y(y_1) - d_y(y_2)| \leq d(y_1, y_2).$$

For any representative of the class $[d_y]$, the same inequality holds. Fix a basepoint $x_0 \in X$. In every equivalence class $\bar{d}_y$, choose a representative $\hat{d}_y$ such that $\hat{d}_y(x_0) = 0$. Then the family $\{\hat{d}_y \mid y \in X\}$ is equicontinuous on any compact subsets on X. It follows that for any sequence y_j in X, there exists a subsequence $y_{n'}$ such that $\hat{d}_{y_{j'}}$ converges uniformly over compact subsets to a continuous function on X. That is, any sequence in $i(X)$ has a subsequence converging to a point on the closure of $i(X)$. Therefore, the closure of $i(X)$ in $\tilde{C}(X)$ is compact.

The closure of X in $\tilde{C}(X)$ is called the *Gromov compactification*, denoted by $\overline{X}^G$. This construction works for any complete locally compact manifold.

Proposition I.17.14 *The geodesic compactification $X \cup X(\infty)$ is isomorphic to the Gromov compactification $\overline{X}^G$, i.e., the closure of $i(X)$ in the compact space $\tilde{C}(X)$.*

For the proof, see [BGS]. The basic idea is to characterize the limit functions and to use the characterization to show that for any unbounded sequence y_j in X, y_j converges in $X \cup X(\infty)$ if and only if the image $[d_{y_j}]$ of d_{y_j} converges in $\tilde{S}(X)$.

I.17.15 The spaces $C^0(X)$ and $\tilde{C}(X)$ are both noncompact. This is one difference between the Gromov compactification and the Satake compactifications and the Furstenberg compactifications, since the ambient spaces are compact in the latter cases. But $\tilde{C}(X)$ has removed some noncompactness. In fact, let y_j be an unbounded sequence in X, and $f_j(x) = d(x, y_j)$. Then f_j is not bounded on any compact subset and hence cannot converge uniformly over compact subsets to a function in $C^0(X)$. Therefore, $C^0(X)$ is noncompact. On the other hand, the proof of the above proposition shows that the image $\tilde{f_j} \in \tilde{C}(X)$ will converge after passing to a subsequence. To show that $\tilde{C}(X)$ is noncompact, we take $g_j(x) = f_j^2(x) = d^2(x, y_j)$. Let c_j be the sequence of constant functions such that $f_j(x) - c_j$ converges uniformly over compact subsets to a function $h(x)$. Clearly the limit $h(x)$ is not a constant function. Since

$$g_j(x) = (f_j(x) - c_j)^2 + 2(f_j(x) - c_j)c_j + c_j^2,$$

the fact that $h(x)$ is nonconstant and $c_j \to +\infty$ implies that the image of g_j in $\tilde{C}(X)$ does not converge. This implies that $\tilde{C}(X)$ is noncompact.

I.17.16 By using a suitable compact subspace of $\tilde{C}(X)$, we can embed X into a compact ambient space and hence its closure is automatically compact. This will fit in the same set-up as the Satake and Furstenberg compactifications.

One such subspace is the Sobolev-type space

$$\tilde{S}(X) = \{f \in \tilde{C}(X) \mid |f(x) - f(y)| \leq d(x, y) \text{ for all } x, y \in X\}.$$

It can be shown as in the proof of Proposition I.17.13 that the image of X in $\tilde{C}(X)$ belongs to $\tilde{S}(X)$ and $\tilde{S}(X)$ is compact. Hence, the closure of $i(X)$ in $\tilde{S}(X)$ is compact. Clearly, these embeddings are G-equivariant, and the G-action on X extends to the compactification $\overline{X}^G$. Therefore, the next proposition gives a realization of the geodesic compactification $X \cup X(\infty)$ via the embedding method, i.e., by embedding into a compact ambient G-space.

Proposition I.17.17 *The closure of $i(X)$ in the compact space $\tilde{S}(x)$ is isomorphic to the geodesic compactification $X \cup X(\infty)$.*

I.17.18 The Martin compactifications $X \cup \partial_\lambda X$ were originally defined in an intrinsic way using asymptotic behaviors of Green's function. By combining Proposition I.17.17, and Proposition I.7.15, we can obtain every $X \cup \partial_\lambda X$ as the closure of X under an embedding into some compact ambient G-space.

In fact, let $\tau : G \to \mathrm{PSL}(n, \mathbb{C})$ be a generic representation of G, and $i_\tau : X \to P(\mathcal{H}_n)$ the associated embedding. Denote the embedding of X into $\tilde{S}(X)$ in the previous paragraph by i. Then $P(\mathcal{H}_n) \times \tilde{S}(X)$ is a compact G-space, and the closure of X under the diagonal embedding

$$X \to P(\mathcal{H}_n) \times \tilde{S}(X), \quad x \mapsto (i_\tau(x), i(x))$$

is isomorphic to the Martin compactification $X \cup \partial_\lambda X$ for $\lambda < \lambda_0$. On the other hand, for $\lambda = \lambda_0$, $X \cup \partial_\lambda X$ is isomorphic to the closure of X under the embedding i_τ.

I.17.19 Summary and comments. In this section, we discussed two new constructions of the maximal Satake compactification. Two more will be given in §I.19. These results show that the maximal Satake compactification occurs more naturally than other Satake compactifications in some sense. We also gave another construction of the geodesic compactification by embedding X into some ambient space and taking the closure instead of the original intrinsic approach.

I.18 Compactifications as a topological ball

In this section, we study topological properties of compactifications of X. The basic reference is [Ji1].

Since X is a symmetric space of noncompact type, X is diffeomorphic to, and hence homeomorphic to, the open ball in the tangent space $T_{x_0}X$. A natural question is whether this diffeomorphism or the homeomorphism extends to compactifications of X. A weaker question is whether compactifications are diffeomorphic or homeomorphic to the closed unit ball in $T_{x_0}X$ without preserving the interior diffeomorphism to the open ball.

In this section, we show that many compactifications are topological balls. In the next section, we discuss the differential and analytic structures on compactifications, which show that if the rank of X is greater than or equal to 2, the compactifications are not differential manifolds with boundary and hence not diffeomorphic to the closed ball.

This section is organized as follows. First we discuss the example of $X = D \times D$. Then we show that the Baily–Borel compactification of a bounded symmetric domain is a ball. This motivated the general result that Satake compactifications of symmetric spaces are topological balls. To show that the closure of a flat in the Satake compactification is a ball, we recall the moment map in I.18.6 and Atiyah's convexity theorem in Proposition I.18.7. To determine the shape of the convex polytope in Atiyah's result, we discuss the case of the standard torus action on $\mathbb{C}P^n$ in I.18.8. Then the closure of flats in the Satake compactifications is determined in Proposition I.18.11.

I.18.1 For $X = D \times D$, we have four different compactifications: the one-point compactification $X \cup \{\infty\}$, the geodesic compactification $X \cup X(\infty)$, and the Martin compactification $\overline{X}^M$, which is isomorphic to the Karpelevič compactification.

The one-point compactification is too small and is homeomorphic to the sphere of dimension 4. The geodesic compactification is clearly homeomorphic to the closed unit ball in the tangent space $T_{(0,0)}X$ at $(0,0)$. The Satake compactifications are all isomorphic to $\overline{D} \times \overline{D}$, which can be easily shown to be homeomorphic to a unit ball, since a square is homeomorphic to a disk.

Let $\pi_1 : X \cup X(\infty) \to T_{(0,0)}X$, $\pi_2 : X \cup X(\infty) \to T_{(0,0)}X$ be the homeomorphisms to the unit ball. Then under the diagonal embedding (π_1, π_2), the image is a bounded star-shaped region in the diagonal subspace and is hence homeomorphic to a closed unit ball in the diagonal subspace.

I.18.2 When X is a symmetric space of noncompact type, it is simply connected and nonpositively curved, and hence the exponential map gives a diffeomorphism from the tangent space $T_{x_0}X$ to X. This fact can also be seen explicitly from the Cartan decomposition

$$G = e^{\mathfrak{p}} K, \quad X = G/K \cong \mathfrak{p}.$$

Since $[0, \infty)$ is diffeomorphic to $[0, 1)$ by $t \mapsto \tanh t$, the radial contraction gives a diffeomorphism from X to the unit ball in $T_{x_0}X = \mathfrak{p}$.

I.18.3 As proved in Proposition I.5.29, when X is a Hermitian symmetric space of noncompact type, $\overline{X}^{BB}$ is, as a topological G-compactification, isomorphic to a minimal Satake compactification. We will concentrate on the Satake compactifications.

Proposition I.18.4 *The Baily–Borel compactification* $\overline{X}^{BB}$ *is a topological ball.*

Proof. Embed X into the holomorphic tangent space $\mathfrak{p}^+$ at x_0 as a bounded symmetric domain,

$$\pi^+ : X \to \mathfrak{p}^+.$$

Identify the holomorphic tangent space $\mathfrak{p}^+$ with the tangent space $T_{x_0}X$. Then by Corollary I.5.16, the closure of any flat $e^{\mathfrak{a}}x_0$ in X is a cube $\Omega = [-1, 1]^r$ contained in $\mathfrak{a}$, where $r = \dim \mathfrak{a}$. Let $B_{\mathfrak{a}}$ be the unit ball in $\mathfrak{a}$. Under suitable scaling, the cube Ω is mapped homeomorphically to $B_{\mathfrak{a}}$ such that for any Weyl chamber face $\mathfrak{a}_I$ and $H \in \mathfrak{a}_I \cap \Omega$, the image of H also belongs to $\mathfrak{a}_I$. This implies that these homeomorphisms for the flats $e^{\mathfrak{a}}x_0$ can be glued into a homeomorphism from $\overline{X}^{BB}$ to the closed unit ball in $T_{x_0}X$.

In this proposition, the bounded realization of $\overline{X}^{BB}$ as a bounded set in $\mathfrak{p}^+$ is crucial. In fact, by the Herman convexity theorem (see [Wo3, p. 286]), under the Harish-Chandra embedding π^+, the image $\pi^+(X)$ in $\mathfrak{p}^+$ is the open ball with respect to a Banach norm on $\mathfrak{p}^+$, which of course implies that $\overline{X}^{BB}$ is a topological ball. Note that the norm is not differentiable, and hence the closure is not differentially a unit ball.

I.18.5 The proof of the above proposition follows 2 steps:

1. show that the closure of a flat can be mapped homeomorphically to a unit ball,
2. show that these homeomorphisms on the flats can be glued into a homeomorphism between the compactification of X and the unit ball in the tangent space at x_0.

For general Satake compactifications $\overline{X}_{\tau}^{S}$, the idea is to get a bounded realization of the closure of each flat $e^{\mathfrak{a}}x_0$ in $\overline{X}_{\tau}^{S}$ in $\mathfrak{a} \subset T_{x_0}X$ and to glue up the homeomorphisms of the flats. Such a bounded realization is given by the moment map in the symplectic geometry associated with the torus action and the convexity result.

I.18.6 We first recall several facts about the moment maps in symplectic geometry. An even-dimensional manifold M is a *symplectic manifold* if there exists a closed nondegenerate 2-form ω on M. An important class of symplectic manifolds consists of Kähler manifolds, whose Kähler form is closed and positive definite, hence nondegenerate. Let $T = (S^1)^n$, $n \geq 1$, be a compact torus. Assume that T acts symplectically on M, i.e., preserves the symplectic form ω, and M is simply connected. For any $v \in \mathfrak{t}$, the Lie algebra of T, induces a vector field on M which is also denoted by v. Then $\omega(v, \cdot)$ is a closed 1-form on M, and hence there exists a function φ^v on M such that

$$d\varphi^v = \omega(v, \cdot).$$

The functions φ^v, $v \in \mathfrak{t}$, can be chosen up to a constant such that the map

$$\mathfrak{t} \to C^\infty(M), \quad v \mapsto \varphi^v,$$

becomes a Lie algebra homomorphism when $C^\infty(M)$ is given the Poisson structure. These functions φ^v define the *moment map* of the T-action on M:

$$\Phi : M \to \mathfrak{t}^*, \quad \Phi(x)(v) = \varphi^v(x), x \in M, \quad v \in \mathfrak{t}. \tag{I.18.1}$$

Assume that M is a Kähler manifold and T acts on M symplectically and holomorphically. Then the T-action on M extends to a holomorphic action of the complex torus $T_{\mathbb{C}}$, the complexification of T. Let Y be an orbit of $T_{\mathbb{C}}$,

$$Y = T_{\mathbb{C}} \cdot y_0, \quad y_0 \in M,$$

and $\overline{Y}$ its closure in M. The moment map Φ in equation (I.18.1) restricts to a map

$$\Phi : \overline{Y} \to \mathfrak{t}^*.$$

The following result of Atiyah [At, Theorem 2] describes this map.

Proposition I.18.7 *The image $\Phi(\overline{Y})$ is a bounded convex polytope in $\mathfrak{t}^*$ whose vertices are the image of the fixed points of T in $\overline{Y}$. The map Φ is constant on the T-orbits and induces a homeomorphism from the quotient $\overline{Y}/T$ to the convex polytope $\Phi(\overline{Y})$.*

Since $T_{\mathbb{C}}/T = e^{it} \cong \mathbb{R}^n$ is the noncompact part e^{it} of the complex torus $T_{\mathbb{C}}$, $\overline{Y}/T$ is homeomorphic to the closure $\overline{e^{it}y_0}$, and the proposition identifies this closure with a convex polytope. We will apply this result to identify the closure of flats in the Satake compactifications $\overline{X}^S_\tau$. For this purpose, we need to compute explicitly the moment map.

I.18.8 We start with the example of the torus action on $\mathbb{C}P^{n-1}$. The compact torus $T = (S^1)^n$ acts on $P^{n-1}(\mathbb{C})$ as follows. Let $[z_1, \dots, z_n]$ be the homogeneous coordinates of $\mathbb{C}P^{n-1}$. Then for any $t = (e^{it_1}, \dots, e^{it_n}) \in T$,

$$t \cdot [z_1, \dots, z_n] = [e^{it_1} z_1, \dots, e^{it_n} z_n].$$

Clearly, T acts holomorphically and preserves the Kähler form. Identify $\mathbb{C}P^{n-1}$ with the quotient $\{(z_1, \dots, z_n) \mid \frac{1}{2}\sum_{j=1}^n |z_j|^2 = 1\}/\sim$, where points $(z_1, \dots, z_n)$, $(z_1', \dots, z_n')$ are equivalent if there exists $\theta \in \mathbb{R}$ such that $(z_1, \dots, z_n) = e^{i\theta}(z_1', \dots, z_n')$. Then the map

$$\Phi : \mathbb{C}P^{n-1} \to \mathfrak{t}^* = \mathbb{R}^n, \quad (z_1, \dots, z_n) \mapsto \left(\frac{1}{2}|z_1|^2, \dots, \frac{1}{2}|z_n|^2\right),$$

is the moment map of T acting on $\mathbb{C}P^{n-1}$. The image $\Phi(\mathbb{C}P^{n-1})$ is the standard simplex of dimension $n-1$ with vertices $(1, 0, \dots, 0), \dots, (0, \dots, 0, 1) \in \mathbb{R}^n$. Clearly, the fixed points of T on $\mathbb{C}P^{n-1}$ are the points corresponding to the coordinate axes $[1, 0, \dots, 0], \dots, [0, \dots, 0, 1]$, and their images under Φ give these vertices.

The projective spaces admit other torus actions defined through representations of T. Let $T : T \to \mathrm{SL}(N, \mathbb{C})$ be a representation with image $\tau(T) \subset SU(N)$. Then T acts on $\mathbb{C}P^{N-1}$ by

$$t \cdot [z_1, \dots, z_N] = \tau(t)[z_1, \dots, z_N].$$

Clearly, T acts holomorphically on $\mathbb{C}P^{N-1}$. Since $\tau(T) \subset SU(N)$, T also preserves the Kähler metric.

Lemma I.18.9 *Denote the weights of τ by $\mu_1, \dots, \mu_N$. Then the image of the moment map $\Phi(\mathbb{C}P^{N-1})$ of the T-action associated with τ as above is the convex hull of the weights $\mu_1, \dots, \mu_N$ in $\mathfrak{t}^*$.*

Proof. Decompose $\mathbb{C}^N$ into the weight subspaces

$$\mathbb{C}^N = V_1 \oplus \cdots \oplus V_N,$$

where V_j is the weight space of μ_j. Let the torus $(S^1)^N$ act on $\mathbb{C}P^{N-1}$ by

$$(e^{it_1}, \dots, e^{it_N}) \cdot [v_1, \dots, v_N] = [e^{it_1} v_1, \dots, e^{it_N} v_N],$$

where $v_j \in V_j$. Then the T-action on $\mathbb{C}P^{N-1}$ is the composition of the map

$$T \to (S^1)^N, \quad e^H \mapsto (e^{\mu_1(H)}, \dots, e^{\mu_N(H)})$$

and the action of $(S^1)^N$ on $\mathbb{C}P^{N-1}$. This implies that the moment map of T on $\mathbb{C}P^{N-1}$ is obtained by composing the moment map of $(S^1)^N$ with the projection $\mathbb{R}^N \to \mathfrak{t}^*$. It is clear that the vertices $(1, 0, \dots, 0), \dots, (0, \dots, 0, 1)$ in $\mathbb{R}^N$ are projected to $\mu_1, \dots, \mu_N$ in $\mathfrak{t}^*$. It follows from the above discussion on the image of the moment map $(S^1)^N$ that the image of the moment map $\Phi(\mathbb{C}P^{N_1})$ associated with T is the convex hull of the weights $\mu_1, \dots, \mu_N$.

I.18.10 Now we are ready to apply the Atiyah convexity result to identify the closure of flats in the Satake compactification $\overline{X}^S_\tau$.

Let $\mathfrak{g} = \mathfrak{k} \oplus \mathfrak{p}$ be the Cartan decomposition as above, and $\mathfrak{a} \subset \mathfrak{p}$ a maximal abelian subalgebra. Then $e^{\mathfrak{a}} x_0$ is a maximal totally geodesic flat submanifold in X passing through the basepoint x_0. Conversely, any such flat passing through x_0 is of this form.

Let $\tau : G \to \mathrm{PSL}(n, \mathbb{C})$ be an irreducible faithful projective representation and $\overline{X}^S_\tau$ the associated Satake compactification as in §I.4. Denote the weights of τ in $\mathfrak{a}^*$ by $\mu_1, \dots, \mu_n$. Choose and fix a positive Weyl chamber $\mathfrak{a}^+$ of $\mathfrak{a}$, and let μ_τ be the highest weight of μ. Let W be the Weyl group of $\mathfrak{a}$.

Proposition I.18.11 *The closure of the flat $e^{\mathfrak{a}} x_0$ in the Satake compactification $\overline{X}^S_\tau$ is canonically homeomorphic to the convex hull of the weights $2\mu_1, \dots, 2\mu_n$ in $\mathfrak{a}^*$. This convex hull is stable under the Weyl group, and is equal to the convex hull of the Weyl orbit through τ_μ.*

Proof. Let $\mathcal{H}_n$ be the real vector space of Hermitian matrices, and $\mathcal{H}_n \otimes \mathbb{C}$ its complexification. Let $P(\mathcal{H}_n \otimes \mathbb{C})$ be the associated complex projective space. Then the real projective space $P(\mathcal{H}_n)$ is canonically embedded into $P(\mathcal{H}_n \otimes \mathbb{C})$.

Choose a basis of $\mathbb{C}^n$ such that for any $H \in \mathfrak{a}$, $\tau(e^H)$ is a diagonal matrix,

$$\tau(e^H) = (e^{\mu_1(H)}, \dots, e^{\mu_n(H)}).$$

Then the restriction of the G-action on $\mathcal{H}_n$ to $e^{\mathfrak{a}}$ is given as follows: for any $H \in \mathfrak{a}$, $M = (m_{jk}) \in \mathcal{H}_n$,

$$e^H \cdot M = \tau(e^H) M \tau(e^H) = (e^{\mu_j(H)+\mu_k(H)} m_{jk}).$$

This action clearly extends to a holomorphic action of the complex torus $e^{\mathfrak{a}+i\mathfrak{a}}$ on $\mathcal{H}_n \otimes \mathbb{C}$ and hence on $P(\mathcal{H}_n \otimes \mathbb{C})$. The compact torus $e^{i\mathfrak{a}}$, which is contained in the compact dual of G, acts on $P(\mathcal{H}_n \otimes \mathbb{C})$ as follows: for $e^{iH} \in e^{i\mathfrak{a}}$, $[M] = [(a_{jk})] \in P(\mathcal{H}_n \otimes \mathbb{C})$,

$$e^{iH} \cdot [M] = [(e^{i\mu_j(H)+i\mu_k(H)} m_{jk})].$$

This action of $e^{i\mathfrak{a}}$ clearly preserves the Kähler form of $P(\mathcal{H}_n \otimes \mathbb{C})$. Denote its moment map by

$$\Phi : P(\mathcal{H}_n \otimes \mathbb{C}) \to i\mathfrak{a}^* \cong \mathfrak{a}.$$

Let Id be the image in $P(\mathcal{H}_n \otimes \mathbb{C})$ of the identify matrix. Then the closure of the orbit $e^{\mathfrak{a}} Id$ in $P(\mathcal{H}_n \otimes \mathbb{C})$ is the closure of the flat $e^{\mathfrak{a}} x_0$ in the Satake compactification $\overline{X}_\tau^S$. We will use the moment map Φ to identify it.

Let Y be the orbit $e^{\mathfrak{a}+i\mathfrak{a}} Id$ of the complex torus $e^{\mathfrak{a}+i\mathfrak{a}} Id$ in $P(\mathcal{H}_n \otimes \mathbb{C})$, and $\overline{Y}$ its closure. Then the closure of the orbit $e^{\mathfrak{a}} Id$ can be identified with the quotient of $\overline{Y}$ by the compact torus $e^{i\mathfrak{a}}$. Therefore, it follows from Proposition I.18.7 that the closure of the flat $e^{\mathfrak{a}} x_0$ in $\overline{X}_\tau^S$ is homeomorphic to the convex polytope $\Phi(\overline{Y})$.

We claim that $\Phi(\overline{Y})$ is the convex hull of the weights $2\mu_1, \ldots, 2\mu_n$. To prove the claim, let D_n be the complex subspace of diagonal matrices in $\mathcal{H}_n \otimes \mathbb{C}$ and $P(D_n)$ its associated complex projective space. Clearly the complex torus $e^{\mathfrak{a}+i\mathfrak{a}}$ preserves D_n and the orbit $e^{\mathfrak{a}+i\mathfrak{a}} Id$ is contained in $P(D_n)$. The weights of the representation of $e^{\mathfrak{a}+i\mathfrak{a}}$ on D_n are $2\mu_1, \ldots, 2\mu_n$, and hence by Lemma I.18.9, $\Phi(P(D_n))$ is equal to the convex hull of $2\mu_1, \ldots, 2\mu_n$, which implies that $\Phi(\overline{Y})$ is contained in this convex hull.

To finish the proof, we need to show that this convex hull is contained in $\Phi(\overline{Y})$. Let $2\mu_j$ be any vertex of the convex hull of $2\mu_1, \ldots, 2\mu_n$. Without loss of generality, we can assume that $j = 1$. Then $2\mu_1$ is the image of $[1, 0, \ldots, 0] \in P(D_n)$ under Φ. It suffices to show that $[1, 0, \ldots, 0] \in \overline{Y}$.

Since $2\mu_1$ is an extremal weight, there exists a positive chamber $\mathfrak{a}^+$ such that $2\mu_1$ is the highest weight of the representation on D_n. Then for any $j \geq 2$,

$$2\mu_1 - 2\mu_j = \sum_{\alpha \in \Delta(\mathfrak{g},\mathfrak{a})} c_{\alpha,j} \alpha,$$

where $\Delta(\mathfrak{g}, \mathfrak{a}^+)$ is the set of simple roots determined by the positive chamber $\mathfrak{a}^+$, $c_{\alpha,j} \geq 0$, and $\sum_{\alpha \in \Delta(\mathfrak{g},\mathfrak{a})} c_{\alpha,j} > 0$. For any $H \in \mathfrak{a}^+$, as $t \to +\infty$,

$$\begin{aligned} e^{tH} \cdot Id &= [e^{2\mu_1(tH)}, e^{2\mu_2(tH)}, \ldots, e^{2\mu_n(tH)}] \\ &= [1, e^{(2\mu_2 - 2\mu_1)(tH)}, \ldots, e^{(2\mu_n - 2\mu_1)(tH)}] \\ &\to [1, 0, \ldots, 0] \text{ in } P(D_n). \end{aligned}$$

This proves the claim and hence the first statement of the proposition. The second statement follows from the known fact that the set of weights $\mu_1, \ldots, \mu_n$ is invariant under the Weyl group W, and the their convex hull is equal to the convex hull of the Weyl group orbit $W\mu_\tau$ of the highest weight μ_τ (see [FH, p. 204]).

I.18.12 Let ϕ be the map in Proposition I.18.11 from the closure $\overline{e^{\mathfrak{a}} x_0}$ in $\overline{X}_\tau^S$ to $\mathfrak{a}^*$ induced from the moment map Φ. Then its restriction to the interior $e^{\mathfrak{a}} x_0$ can be given explicitly as follows (see [Od, p. 94] and [Ju]):

$$\phi(e^H x_0) = \sum_{s \in W} \frac{e^{2s\mu_\tau(H)}}{\sum_{s \in W} e^{2s\mu_\tau(H)}} 2s\mu_\tau. \tag{I.18.2}$$

This function is clearly equivariant under the action of the Weyl group. In particular, for any subset $I \subset \Delta(\mathfrak{g}, \mathfrak{a})$ of simple roots, the Weyl wall $\mathfrak{a}_I$ is mapped into its dual $\mathfrak{a}_I^*$,

$$\phi(e^{\mathfrak{a}_I} x_0) \subset \mathfrak{a}_I^*. \tag{I.18.3}$$

Proposition I.18.13 *Every Satake compactification $\overline{X}_\tau^S$ is canonically homeomorphic to the closed unit ball in the tangent space $T_{x_0}X$ of X at the basepoint x_0.*

Proof. For any maximal abelian subspace $\mathfrak{a}$ in $\mathfrak{p}$, denote the map in Proposition I.18.11 by

$$\phi_\mathfrak{a} : \overline{e^\mathfrak{a} x_0} \to \mathfrak{a}^*.$$

We claim that for any two flats $e^\mathfrak{a} x_0$, $e^{\mathfrak{a}'} x_0$, the maps $\phi_\mathfrak{a}$, $\phi_{\mathfrak{a}'}$ agree on their intersection $\overline{e^\mathfrak{a} x_0} \cap \overline{e^{\mathfrak{a}'} x_0}$. By continuity, it suffices to prove that they agree on $e^\mathfrak{a} x_0 \cap e^{\mathfrak{a}'} x_0 = e^{\mathfrak{a} \cap \mathfrak{a}'} x_0$. Note that on each flat $e^\mathfrak{a} x_0$, the moment map is equivariant with respect to the Weyl group, and hence for any Weyl wall $\mathfrak{a}_I$, where $I \subset \Delta(\mathfrak{g}, \mathfrak{a})$,

$$\phi_\mathfrak{a}(e^{\mathfrak{a}_I} x_0) \subset \mathfrak{a}_I^*.$$

This map coincides with the moment map of the compact torus $e^{i\mathfrak{a}_I}$ on $e^{\mathfrak{a}_I + i\mathfrak{a}_I} x_0$. (See also the formula in equation I.18.2.) It is known that the intersection $\mathfrak{a} \cap \mathfrak{a}'$ is of the form $\mathfrak{a}_I$ for some I, and hence the claim follows.

For any maximal abelian subspace $\mathfrak{a}$ in $\mathfrak{p}$, identify its dual $\mathfrak{a}^*$ with $\mathfrak{a}$ under the Killing form. Then the map $\phi_\mathfrak{a}$ induces a map from $\overline{e^\mathfrak{a} x_0}$ to $\mathfrak{a}$, still denoted by $\phi_\mathfrak{a}$. Any ray from the origin of $\mathfrak{a}$ to the boundary of the convex polytope $\phi_\mathfrak{a}(\overline{e^\mathfrak{a} x_0})$ can be scaled down to a vector of unit length. This scaling gives a homeomorphism from $\overline{e^\mathfrak{a} x_0}$ to the closed unit ball $B(\mathfrak{a})$ in $\mathfrak{a}$:

$$\Phi_\mathfrak{a} : \overline{e^\mathfrak{a} x_0} \to B(\mathfrak{a}).$$

By the claim in the previous paragraph, these homeomorphisms for different flats agree on their intersection and hence define a global homeomorphism

$$\Phi : \overline{X}_\tau^S \to B(\mathfrak{p}) = B(T_{x_0} X),$$

where $B(\mathfrak{p})$ is the closed unit ball in $\mathfrak{p}$, which is canonically identified with the closed unit ball in the tangent space $T_{x_0} X$.

Remarks I.18.14

(1) A result similar to the above proposition was also obtained in [Cas2], where it was shown that the decomposition of the closure of the flat $e^a x_0$ in $\overline{X}_\tau^S$ into the $e^\mathfrak{a}$-orbits is combinatorially equivalent to the faces of the convex hull in Proposition I.18.11.

(2) The polytope of the convex hull of the weights $2\mu_1, \dots, 2\mu_n$ appears naturally in *Arthur's trace formula* through the weighted integrals (see [Ar1]). Related decompositions of the flats $e^{\mathfrak{a}}x_0 = \mathfrak{a}$ also occur in the precise reduction theory (see [Sap1] [OW1] [Ar3, §3.2]).

Proposition I.18.15 *The Martin compactifications $X \cup \partial_\lambda X$ are homeomorphic to the unit ball in the tangent space $T_{x_0}X$.*

Proof. By Proposition I.7.15, when $\lambda = \lambda_0$, $X \cup \partial_\lambda X$ is isomorphic to the maximal Satake compactification $\overline{X}^S_{\max}$, and hence homeomorphic to the unit ball by Proposition I.18.13. On the other hand, when $\lambda < \lambda_0(X)$, $X \cup \partial_\lambda X$ is homeomorphic to the closure of X under the diagonal embedding $X \hookrightarrow X \cup X(\infty) \times X \cup \partial_\lambda X$. Since $X \cup X(\infty)$ is clearly homeomorphic to the unit ball, it can be combined with Proposition I.18.13 to show that $X \cup \partial_\lambda X$ is homeomorphic to the unit ball in $T_{x_0}X$.

I.18.16 Proposition I.18.15 and related results were motivated by a question of Dynkin in his ICM talk [Dy1] on relations between the geometry of Riemannian manifolds and the Martin boundaries. His question basically amounts to that if M is simply connected and nonpositively curved, then for any $\lambda < \lambda_0(M)$, the Martin compactification $M \cup \partial_\lambda M$ is homeomorphic to the closed unit ball in T_{x_0}, and hence the Martin boundary $\partial_\lambda M$ has dimension $\dim M - 1$.

When M is negatively pinched, the answer was positive and given in [An] [AS]. Proposition I.18.15 affirms the case when M is a symmetric space of noncompact type. See [Ji1] for more details.

Remark I.18.17 In [Kus], Kusner proved that the Karpelevič compactification $\overline{X}^K$ is also homeomorphic to a closed ball.

By Proposition I.16.5, the Karpelevič compactification $\overline{X}^K$ dominates the Satake compactifications $\overline{X}^S_\tau$. It is conceivable that any G-compactification of X that dominates a Satake compactification is homeomorphic to the closed unit ball in $T_{x_0}X$. It should be emphasized that by a G-compactification of X, we mean that the isometric G-action on X extends to a continuous action to the compactification. The continuity of the G-action is important since it allows one to reduce the problem basically to the closure of flats through the basepoint x_0 using the Cartan decomposition of G (or the polar decomposition of X). In fact, if the continuity assumption is dropped, the conclusion may not hold, and the compactification in [Gol] gives such an example. See [Ji1] for more details about related issues.

I.18.18 Summary and comments. The basic result in this section is that the Satake compactifications are topological balls. It should be emphasized that they are not differential manifolds with boundary if the rank of X is strictly greater than 1. When the rank is equal to 1, they are real analytic manifolds with boundary and diffeomorphic to the closed unit ball. The maximal Satake compactification is a real analytic manifold with corners (see the next section), but other Satake compactifications are more singular. It is not clear whether nonmaximal Satake compactifications have natural canonical analytic structures.

As will be seen in Part II, the analogue of the maximal Satake compactification $\overline{X}^S_{\max}$ for a complex symmetric space $X_{\mathbb{C}}$ is the wonderful compactification $\overline{X_{\mathbb{C}}}^W$ in [DP1] [DP2]. The analogues of nonmaximal Satake compactifications have been defined for complex symmetric spaces (see [Lu1] [Lu2] [Lu3] [LV] [Vu1] [Vu2]), and the Satake compactifications of X can be embedded into these compactifications of the complexification $X_{\mathbb{C}}$ (in fact, into the real locus if they are defined over $\mathbb{R}$). But a problem is that the analytic structure of these nonmaximal compactifications of complex symmetric spaces may depend on the representation used in the definition rather than only on the relative position (or degeneracy) of the highest weight of the representation. Recall that as topological G-spaces, the Satake compactifications $\overline{X}^S_{\tau}$ depends only on the Weyl chamber face that contains the highest weight μ_τ (Proposition I.4.35).

I.19 Dual-cell compactification and maximal Satake compactification as a manifold with corners

In this section, we recall two more constructions of the maximal Satake compactification $\overline{X}^S_{\max}$ using the structure of the closure of maximal flats[1] obtained in the previous section: the dual-cell compactification $X \cup \Delta^*(X)$ in [GJT] and a modification $\widetilde{X}^S_{\max}$ suggested by the Oshima compactification in [Os1]. The dual-cell compactification will be presented in a slightly different way from [GJT, Chapter 3] so that it motivates the modification $\widetilde{X}^S_{\max}$ of the construction in [Os1].

In $X \cup \Delta^*(X)$, the basic idea is to start with a compactification of flats through the basepoint x_0 using the combinatorial data given by the Weyl chamber decomposition and to glue these compactifications naturally along the Weyl chamber walls. In the modification $\widetilde{X}^S_{\max}$ of the Oshima compactification, we start with compactifications of all flats whether they pass through the basepoint x_0 or not and define a suitable equivalence relation on the union of these compactifications of flats in X. Considering all flats gives more flexibility and allows one to show that the maximal Satake compactification $\overline{X}^S_{\max}$ is a real analytic manifold with corners, but also makes the equivalence relation on the compactified flats more complicated.

This section is organized as follows. First we recall polyhedral cone decompositions of vector spaces in I.19.1 and the associated polyhedral compactifications of the vector spaces in I.19.2 and Lemma I.19.5. Then we discuss the Weyl chamber decompositions in I.19.8 and the induced polyhedral compactification of the flats. Due to the symmetry of the Weyl chamber decomposition, the dual-cell compactifications

[1] For convenience, in the following, a flat in X means a maxmal totally geodesic submanifold in X.

of flats are compatible and glue up into the dual-cell compactification $X \cup \Delta^*(X)$ of the symmetric space X (Proposition I.19.13). It is difficult to see why the G-action on X extends continuously to the dual-cell compactification, though it is easy to see that the K-action extends continuously. To overcome this problem, we combine the constructions of the dual-cell compactification and the Oshima compactification to construct the modification $\widetilde{X}^S_{\max}$ and use it to show that $\overline{X}^S_{\max}$ is a real analytic manifold with corners in I.19.16, I.19.26, and I.19.27. An important step in this construction is to extend the corner structure of closure of flats to corners of the compactification of symmetric spaces.

I.19.1 Let V be a vector space over $\mathbb{R}$. A convex cone C in V with the vertex at the origin of V is called a (convex) *polyhedral cone* if there exist linear functions $\lambda_1, \dots, \lambda_n, \gamma_1, \dots, \gamma_m$ on V such that

$$C = \{H \in V \mid \lambda_1(H) = \cdots = \lambda_n(H) = 0, \gamma_1(H) > 0, \dots, \gamma_m(H) > 0\}, \tag{I.19.1}$$

where n could be equal to 0, which implies that $\dim C = \dim V$. We always require that C be proper, i.e., it does not contain any line passing through the origin, which is equivalent to that $m > 0$ unless C consists of the origin. The linear span $\mathrm{Span}(C)$ is the vector subspace generated by $H \in C$, i.e.,

$$\mathrm{Span}(C) = \{H \in V \mid \lambda_1(H) = \cdots = \lambda_n(H) = 0\}.$$

Clearly, C is an open subset of $\mathrm{Span}(C)$.

The origin is clearly a face of every polyhedral cone. For any proper subset $I \subset \{1, \dots, m\}$, there is a nontrivial face

$$C_I = \{H \in V \mid \lambda_1(H) = \cdots = \lambda_n(H) = 0, \gamma_i(H) = 0, i \in I; \gamma_i(H) > 0, i \notin I\}.$$

Any nontrivial proper face of C is of this form C_I for some I.

A *polyhedral cone decomposition* of V consists of a collection Σ of polyhedral cones C such that

1. V admits the disjoint decomposition
$$V = \{0\} \cup \coprod_{C \in \Sigma} C.$$
2. Every face of any cone $C \in \Sigma$ is also a cone in Σ.
3. For any two cones $C_1, C_2 \in \Sigma$, if $C_1 \cap \overline{C_2} \neq \emptyset$, then C_1 is a face of C_2.

Remarks I.19.2

(1) We emphasize that all the cones here are open in their linear spans in order to get the disjoint decomposition and to be consistent with the Weyl chamber decomposition discussed later. In some places, the closed cones are used. In that case, faces of a (closed) cone are contained in the cone.

(2) In torus embeddings or toric varieties, *partial polyhedral cone decompositions* of vector spaces are often used. These are collections of polyhedral cones satisfying only the conditions (2) and (3) above. Our decompositions here are called *complete polyhedral decompositions* (see [Od] [Jur] [Ful]).

I.19.3 Given a polyhedral cone decomposition Σ of a vector space V, there is a *polyhedral compactification* $\overline{V}_\Sigma$ defined as follows. For each nontrivial cone C in Σ, define its boundary component

$$e(C) = V/\mathrm{Span}(C).$$

Then the *polyhedral compactification* $\overline{V}_\Sigma$ is obtained by adding these boundary components:

$$\overline{V}_\Sigma = V \cup \coprod_{C \in \Sigma, C \neq \{0\}} e(C). \tag{I.19.2}$$

To describe convergence of points on the boundary $\coprod e(C)$, we need to describe an induced polyhedral cone decomposition on the boundary component $e(C)$ for every nontrivial cone $C \in \Sigma$. Let Σ_C be the collection of all the cones in Σ for which C is a face. If $\dim C = \dim V$, there is no such cone. But in this case, the boundary component $e(C)$ consists of one point. If $\dim C < \dim V$, there exist such cones. In fact, the union of such cones Σ_C is an open subset containing C. Let

$$\pi : V \to V/\mathrm{Span}(C) = e(C)$$

be the projection. Then for any $C' \in \Sigma_C$, the image $\pi(C')$ is a polyhedral cone in $e(C)$, and the collection of $\pi(C')$, $C' \in \Sigma_C$, gives a polyhedral cone decomposition of $e(C)$, which is also denoted by Σ_C for simplicity. For any cone C' in V satisfying $C' \in \Sigma_C$, the boundary component $e(C')$ can be identified with the boundary component of $\pi(C')$ in the lower-dimensional vector subspace $e(C)$.

I.19.4 Now we are ready to describe the topology of $\overline{V}_\Sigma$.

1. For any C and a boundary point $\xi \in e(C)$, a sequence $x_j \in V$ converges to $\xi \in e(C)$ if the projection $\pi(x_j)$ in $e(C)$ converges to ξ, and in the notation of equation (I.19.1), for any $k = 1, \dots, m$,

$$\gamma_k(x_j) \to +\infty.$$

2. For any two cones C_1, C_2 in Σ, $e(C_1)$ is contained in the closure of $e(C_2)$ if and only if C_2 is a face of C_1, and the convergence of sequences of points in $e(C_2)$ to points in $e(C_1)$ is the same as in (1) using the polyhedral cone decomposition Σ_{C_2} of $e(C_2)$, where $e(C_1)$ is identified with a boundary component of $e(C_2)$ as above.

These are special convergent sequences, and combinations of them give the general convergent sequences. It can be checked that they form a convergence class of sequences and define a topology on $\overline{V}_\Sigma$.

Lemma I.19.5 *The space $\overline{V}_\Sigma$ is Hausdorff and compact.*

Proof. It can be checked easily that every unbounded sequence in V has a convergent subsequence in $\overline{V}_\Sigma$, and every convergent sequence has a unique limit. By induction, it can be shown that the same conclusion holds for each of the finitely many boundary components, and hence for every sequence in $\overline{V}_\Sigma$, which implies that it is Hausdorff and compact.

I.19.6 Let $V(\infty)$ be the unit sphere with respect to any inner product. For any cone C in Σ, the intersection $C \cap V(\infty)$ is a cell in $V(\infty)$ and these cells form a cell complex, denoted by $\Sigma(\infty)$. It can be shown that the boundary components of $\overline{V}_\Sigma$ is a cell complex dual to $\Sigma(\infty)$.

Remarks I.19.7 The construction of the polyhedral compactification $\overline{V}_\Sigma$ has been discussed in [Ta2] [GJT] [Ger1] in different ways. It is also closely related to the *torus embeddings* when the polyhedral cones in Σ are rational with respect to a lattice in V. In fact, in this case, $\overline{V}_\Sigma$ is the noncompact part of the torus embedding, or equivalently, the quotient of the torus embeddings by the compact part of the complex torus $(\mathbb{C}^\times)^d$, where $d = \dim V$ (see [Od] [Ful] [Jur]).

I.19.8 Let $\mathfrak{a}$ be a maximal abelian subspace of $\mathfrak{p}$. Each root $\alpha \in \Phi(\mathfrak{g}, \mathfrak{a})$ defines a *root hyperplane*

$$\mathbf{H}_\alpha = \{H \in \mathfrak{a} \mid \alpha(H) = 0\}.$$

The connected components of

$$\mathfrak{a} - \cup_{\alpha \in \Phi} \mathbf{H}_\alpha$$

are called Weyl chambers. These Weyl chambers together with their faces form a polyhedral cone decomposition of $\mathfrak{a}$, called the *Weyl chamber decomposition* and denoted by Σ_{wc}. In fact, since a face of a *Weyl chamber face* is a Weyl chamber face, it can be checked easily that the conditions in (I.19.1) are satisfied.

Since this polyhedral cone decomposition Σ_{wc} is canonical, and the boundary of the induced polyhedral cone decomposition $\overline{\mathfrak{a}}_{\Sigma_{wc}}$ is a cell complex dual to this canonical decomposition, the boundary of $\overline{\mathfrak{a}}_{\Sigma_{wc}}$ is also denoted by $\Delta^*(\mathfrak{a})$, called the *dual-cell complex*, and the compactification

$$\overline{\mathfrak{a}}_{\Sigma_{wc}} = \mathfrak{a} \cup \Delta^*(\mathfrak{a}) \tag{I.19.3}$$

is called the *dual-cell compactification* of $\mathfrak{a}$ in [GJT, Chapter 3]. For simplicity, the compactification $\mathfrak{a} \cup \Delta^*(\mathfrak{a})$ is also denoted by $\overline{\mathfrak{a}}^*$.

Remark I.19.9 By definition, the boundary components in the dual-cell $\Delta^*(\mathfrak{a})$ are attached at infinity of $\mathfrak{a}$. A bounded realization of these boundary components and the dual-cell compactification $\mathfrak{a} \cup \Delta^*(\mathfrak{a})$ can be realized as follows. Let ρ be a nonsingular element in $\mathfrak{a}$ or $\mathfrak{a}^*$ if $\mathfrak{a}^*$ is identified with $\mathfrak{a}$ under the Killing form, for example, the half sum of positive roots in $\Phi(\mathfrak{g}, \mathfrak{a})$ with multiplicity. Then the convex hull of the Weyl group orbit $W \cdot \rho$ is a bounded polytope homeomorphic to the dual-cell compactification of $\mathfrak{a}$ under a suitable map.

I.19.10 There are many symmetries of this Weyl chamber decomposition Σ_{wc}. In fact, let W be the Weyl group of $\mathfrak{a}$. Then W acts on Σ_{wc}. Furthermore, for any collection of roots $I \subset \Phi(\mathfrak{g}, \mathfrak{a})$, if the intersection

$$\mathbf{H}_I = \cap_{\alpha \in I} \mathbf{H}_\alpha$$

is nonempty, it is called a Weyl wall. The Weyl chamber faces contained in the wall $\mathbf{H}_I$ form a polyhedral cone decomposition of $\mathbf{H}_I$. In fact, they consist of the connected components of

$$\mathbf{H}_I - (\mathbf{H}_\beta \cap \mathbf{H}_I),$$

where β runs over all roots such that $\mathbf{H}_\beta \not\supseteq \mathbf{H}_I$, and their faces.

An important corollary of these symmetries is the following.

Corollary I.19.11 *For any other maximal abelian subspace $\mathfrak{a}'$ in $\mathfrak{p}$, the intersection $\mathfrak{a} \cap \mathfrak{a}'$ of the two maximal abelian subspaces $\mathfrak{a}, \mathfrak{a}'$ of $\mathfrak{p}$ is a Weyl wall of both of them, and the Weyl chamber decompositions of $\mathfrak{a}$ and $\mathfrak{a}'$ induce the same polyhedral cone decomposition on $\mathfrak{a} \cap \mathfrak{a}'$.*

Corollary I.19.12 *For any two maximal abelian subalgebras $\mathfrak{a}, \mathfrak{a}'$, the dual-cell compactifications $\overline{\mathfrak{a}}^*, \overline{\mathfrak{a}'}^*$ induce the same compactification on the intersection $\mathfrak{a} \cap \mathfrak{a}'$, which is the polyhedral compactification $\overline{\mathfrak{a} \cap \mathfrak{a}'}^*$ of $\mathfrak{a} \cap \mathfrak{a}'$ with respect to the polyhedral cones in the previous corollary.*

This compatibility of the polyhedral compactifications $\overline{\mathfrak{a}}^*$ allows us to glue them into a compactification of X.

As recalled earlier, every maximal flat in X passing through x_0 is of the form $e^{\mathfrak{a}} x_0$. Since $e^{\mathfrak{a}} x_0$ can be canonically identified with $\mathfrak{a}$, the dual-cell compactification $\overline{\mathfrak{a}}^*$ of $\mathfrak{a}$ induces a *dual-cell compactification* $\overline{e^{\mathfrak{a}} x_0}^*$ of the flat $e^{\mathfrak{a}} x_0$.

Define

$$\overline{X}^* = \coprod_{\mathfrak{a}} \overline{e^{\mathfrak{a}} x_0}^* / \sim \cong \coprod_{\mathfrak{a}} \overline{\mathfrak{a}}^* / \sim, \tag{I.19.4}$$

where $\mathfrak{a}$ runs over all maximal abelian subspaces of $\mathfrak{p}$, and the equivalence relation $\sim$ is given by the following: for any two flats $\mathfrak{a}$ and $\mathfrak{a}'$, the dual-cell compactifications $\overline{\mathfrak{a}}^*$ and $\overline{\mathfrak{a}'}^*$ are identified along $\overline{\mathfrak{a} \cap \mathfrak{a}'}^*$ in Corollary I.19.12. Equivalently, $\overline{X}^*$ is the union of the dual-cell compactifications $\overline{\mathfrak{a}}^*$ identified along the closure of the intersection

$\mathfrak{a} \cap \mathfrak{a}'$. (See [FK, pp. 31–34] for general results on quotients, gluing, and unions.) As a set,

$$\overline{X}^* = X \cup \bigcup_{\mathfrak{a}} \Delta^*(\mathfrak{a}) / \sim,$$

where $\mathfrak{a}$ runs over all maximal abelian subspaces of $\mathfrak{p}$, and $\sim$ is defined by the identification along $\Delta^*(\mathfrak{a} \cap \mathfrak{a}')$.

Clearly, the space $\overline{X}^*$ has the quotient topology, which induces the original topology on the dual-cell compactifications of the flats. But $\overline{X}^*$ is not compact in this topology.

To make $\overline{X}^*$ into a compactification of X, we fix a maximal abelian subspace $\mathfrak{a}_0$ of $\mathfrak{p}$, or equivalently a maximal flat $e^{\mathfrak{a}_0} x_0$ through x_0. Then any maximal abelian subalgebra $\mathfrak{a}$ of $\mathfrak{p}$ is of the form $Ad(k)\mathfrak{a}_0$ for any $k \in K$. Since $Ad(k)$ maps the Weyl chamber decomposition of $\mathfrak{a}_0$ to the Weyl chamber decomposition of $\mathfrak{a}$, the map $Ad(k)$ extends to a homeomorphism

$$Ad(k) : \overline{\mathfrak{a}_0}^* \to \overline{Ad(k)\mathfrak{a}_0}^*. \tag{I.19.5}$$

Therefore, we have a map

$$K \times \overline{\mathfrak{a}_0}^* \to \coprod_{\mathfrak{a}} \Delta^*(\mathfrak{a}), \quad (k, H) \mapsto Ad(k)H \in \overline{Ad(k)\mathfrak{a}_0}^*,$$

which induces a map

$$K \times \overline{\mathfrak{a}_0}^* \to \overline{X}^*. \tag{I.19.6}$$

The space $\overline{X}^*$ endowed with the quotient topology from $K \times \overline{\mathfrak{a}_0}^*$ is also denoted by $X \cup \Delta^*(X)$. Clearly, the subset topology on X coincides with the original topology of X.

Proposition I.19.13 *The space $X \cup \Delta^*(X)$ is a compact Hausdorff compactification of X, called the dual-cell compactification, and is isomorphic to the maximal Satake compactification $\overline{X}^S_{\max}$.*

Proof. Let $\mathcal{R}$ be the equivalence relation on $K \times \overline{\mathfrak{a}_0}^*$ that defines the quotient $\overline{X}^*$. Since every equivalence class of $\mathcal{R}$ is a homogeneous space of K, it is compact and hence closed. More generally, the $\mathcal{R}$-saturation of any closed set is closed, which implies that the induced partition is closed. Since $K \times \overline{\mathfrak{a}_0}^*$ is metrizable and hence normal, by [FK, p. 33], $X \cup \Delta^*(X)$ is a compact Hausdorff space. To prove the second statement, we note that the compactification $X \cup \Delta^*(X)$ satisfies the conditions in [GJT, Theorem 3.39] and hence is isomorphic to $\overline{X}^S_{\max}$ by Theorem 4.43 in [GJT].

I.19.14 The *dual-cell compactification* $X \cup \Delta^*(X)$ was constructed in [GJT, Chapter 3] and the above construction is a variant. In fact, the construction in [GJT] is given via the Cartan decomposition. Specifically, let $A^+ = A^+_{P_0}$ be a positive Chamber, where P_0 is a minimal parabolic subgroup. Then a sequence y_j in X is called *fundamental* if y_j admits a decomposition $y_j = k_j a_j x_0$, where $k_j \in K, a_j \in \overline{A^+}$ satisfy the following conditions:

1. k_j converges to some k_∞.
2. There exists a subset I of $\Delta(P_0, A_{P_0})$ such that for $\alpha \in I$, a_j^α converges to a finite number, while for $\alpha \in \Delta - I$, $a_j^\alpha \to +\infty$.

Let $H_\infty \in \mathfrak{a}^I$ be the unique element such that for all $\alpha \in I$, $\alpha(H_\infty) = \lim_{j\to+\infty} \alpha(\log a_j)$. Then the *formal limit* of such a fundamental sequence y_j is defined to be $k_\infty \cdot e^{H_\infty} K_{P_{0,I}} \in k_\infty \cdot X_{P_{0,I}}$. The dual-cell compactification $X \cup \Delta^*(X)$ is characterized as the unique compactification of X such that every fundamental sequence converges to its formal limit.

By construction, the K-action on X extends to a continuous action on the dual-cell compactification $X\cup\Delta^*(X)$. In [GJT, p. 45], a G-action on $X\cup\Delta^*(X)$ was given, but it does not seem to be easy to prove directly the continuity of this extension of the G-action. In fact, in [GJT], the continuity was proved by identifying $X \cup \Delta^*(X)$ with the Martin compactification $X \cup \partial_{\lambda_0} X$ [GJT, Theorem 7.33] (or equivalently $\overline{X}^S_{\max}$) and using the continuous G-action on the latter. As mentioned earlier, this issue about the continuity of the G-action is one of the motivations of the uniform method for constructing compactifications in §I.8.

I.19.15 Another method to solve this problem of the continuity of the G-action is to build in the G-action in gluing up the compactifications of flats X. Clearly, we need to use flats besides those passing through the basepoint x_0 as well. Roughly, instead of taking the quotient of $K \times \overline{\mathfrak{a}_0}^*$, we construct a compactification of X as a quotient of $G \times \overline{\mathfrak{a}_0}^*$. The rest of this section is motivated by this, which will be used for the problem of obtaining a different construction of the Oshima compactification $\overline{X}^O$ [Os1] using the self-gluing method in §II.1 below.

I.19.16 For this purpose, we need a more concrete realization of the boundary components of the dual-cell compactification $\overline{\mathfrak{a}}^*$ and a partial compactification contained in $\overline{\mathfrak{a}}^*$.

The Weyl group W of $\mathfrak{a}$ acts on the Weyl chamber decomposition Σ_{wc} and hence on the compactification $\overline{\mathfrak{a}}^*$, carrying one boundary component to another. Let $\mathfrak{a}^+$ be a positive Weyl chamber of $\mathfrak{a}$. Then any cone C in Σ_{wc} is mapped under some element of W to a Weyl chamber face $\mathfrak{a}_I^+$, for some $I \subset \Delta(\mathfrak{g}, \mathfrak{a})$, and the corresponding boundary component $e(C)$ to $e(\mathfrak{a}_I^+)$, where

$$\mathfrak{a}_I = \{H \in \mathfrak{a} \mid \alpha(H) = 0, \alpha \in I\}, \quad \mathfrak{a}_I^+ = \{H \in \mathfrak{a}_I \mid \alpha(H) > 0, \alpha \in \Delta - I\}.$$

It suffices to give a concrete realization of $e(\mathfrak{a}_I^+)$, the boundary component of the Weyl chamber face $\mathfrak{a}_I^+$.

In the notation of §I.1, let $\mathfrak{a}^I$ be the orthogonal complement of $\mathfrak{a}_I$ in $\mathfrak{a}$ with respect to the Killing form. Then

$$e(\mathfrak{a}_I^+) = \mathfrak{a}/\mathfrak{a}_I \cong \mathfrak{a}^I.$$

By adding only the boundary components $e(\mathfrak{a}_I^+) = \mathfrak{a}^I$ for $I \subset \Delta(\mathfrak{g}, \mathfrak{a})$, we get a partial compactification, called the *partial dual-cell compactification*,

$$\overline{\mathfrak{a}}_+ = \mathfrak{a} \cup \coprod_{I \subset \Delta} \mathfrak{a}^I. \tag{I.19.7}$$

Clearly, $\overline{\mathfrak{a}}_+ \subset \overline{\mathfrak{a}}^*$, and any boundary point of $\overline{\mathfrak{a}}^*$ is mapped to a point in $\overline{\mathfrak{a}}_+$ under some element of W. Hence,

$$W\overline{\mathfrak{a}}_+ = \overline{\mathfrak{a}}^*. \tag{I.19.8}$$

This relation is related to the fact that the closure of the positive chamber $\mathfrak{a}^+$ is a fundamental domain of the W-action on $\mathfrak{a}$. But $\overline{\mathfrak{a}}_+$ strictly contains a fundamental domain of the W-action on $\overline{\mathfrak{a}}^*$.

I.19.17 The choice of the partial compactification $\overline{\mathfrak{a}}_+$ is natural also in connection to the Oshima compactification $\overline{X}^O$ [Os1] to be discussed later in §II.2.

Let $\alpha_1, \dots, \alpha_r$ be the set of simple roots in $\Delta(\mathfrak{g}, \mathfrak{a})$. Then $\mathfrak{a}$ can be identified with $\mathbb{R}^\Delta_{>0} = \mathbb{R}^r_{>0}$ by

$$i : \mathfrak{a} \to \mathbb{R}^r_{>0}, \quad H \to (e^{-\alpha_1(H)}, \dots, e^{-\alpha_r(H)}).$$

The closure of $i(\mathfrak{a})$ in $\mathbb{R}^r$ defines a partial compactification of $\mathfrak{a}$,

$$\overline{i(\mathfrak{a})} \cong \mathbb{R}^r_{\geq 0}. \tag{I.19.9}$$

Proposition I.19.18 *The partial compactification $\overline{i(\mathfrak{a})}$ of $\mathfrak{a}$ is the same as the partial dual-cell compactification $\overline{\mathfrak{a}}_+$ in equation (I.19.7), i.e., the closure in the dual-cell compactification $\overline{\mathfrak{a}}^*$.*

Proof. By [GJT, Lemma 3.28], it suffices to show that an unbounded sequence y_j in $\mathfrak{a}$ converges in $\overline{\mathfrak{a}}_+$ if and only if it converges in $\overline{i(\mathfrak{a})}$. Suppose that $H_j \to H_\infty \in \mathfrak{a}^I$ in $\overline{\mathfrak{a}}_+$. Since

$$\mathfrak{a}_I^+ = \{H \in \mathfrak{a} \mid \alpha(H) = 0, \alpha \in I; \alpha(H) > 0, \alpha \in \Delta - I\},$$

by the definition of the topology of $\overline{\mathfrak{a}}^*$ and hence of $\overline{\mathfrak{a}}_+$, for $\alpha \in \Delta - I$,

$$\alpha(H_j) \to +\infty, \tag{I.19.10}$$

and the image of H_j in $\mathfrak{a}/\mathfrak{a}_I \cong \mathfrak{a}^I$ converges to H_∞. Since the roots $\alpha \in I$ vanish on $\mathfrak{a}_I$ and give coordinates of $\mathfrak{a}^I$, the latter condition is equivalent to

$$\alpha(H_j) \to \alpha(H_\infty), \quad \alpha \in I. \tag{I.19.11}$$

The conditions in equations (I.19.10, I.19.11) are exactly the conditions for H_j to converge to a boundary point in $\overline{i(\mathfrak{a})} \subset \mathbb{R}^r_{\geq 0}$.

I.19.19 By the above proposition and equation (I.19.9),

$$\begin{aligned} \overline{\mathfrak{a}}_+ &\cong \mathbb{R}^r_{\geq 0}, \\ e(\mathfrak{a}^+) &\cong \{0\}, \\ e(\mathfrak{a}^+_I) = \mathfrak{a}^I &\cong \{(x_1, \dots, x_l) \mid x_j \in \mathbb{R}, \alpha_j \in I; x_j = 0, \alpha_j \in \Delta - I\}. \end{aligned}$$

In particular, $\overline{\mathfrak{a}}_+$ is a *corner*. This will be used to show that the maximal Satake compactification $\overline{X}^S_{\max}$ is a real analytic manifold with corners of codimension r.

Let P be the minimal parabolic subgroup containing $A = \exp \mathfrak{a}$ and corresponding to the positive chamber $\mathfrak{a}^+$, i.e., in the Langlands decomposition $P = N_P A_P M_P$, $A_P = A$, and the Lie algebra of N_P is given by

$$\mathfrak{n}_P = \sum_{\alpha \in \Phi^+(\mathfrak{g}, \mathfrak{a})} \mathfrak{g}_\alpha.$$

Let

$$\mathfrak{n}_P^- = \sum_{\alpha \in -\Phi^+(\mathfrak{g}, \mathfrak{a})} \mathfrak{g}_\alpha,$$

the sum of the root spaces of the negative roots, and let N_P^- be the corresponding Lie subgroup, which is the unipotent radical of the parabolic subgroup $P^- = N_P^- A_P M_P$ opposite to P.

Proposition I.19.20 *The map*

$$N_P^- \times \mathfrak{a} \to X, \quad (n, H) \mapsto ne^H x_0,$$

is an analytic homeomorphism and extends to a homeomorphism

$$N_P^- \times \overline{\mathfrak{a}}_+ \to \overline{X}^S_{\max},$$

whose image is an open dense subset in $\overline{X}^S_{\max}$.

Proof. By Proposition I.19.13, the closure of the flat $e^{\mathfrak{a}} x_0 \cong \mathfrak{a}$ in $\overline{X}^S_{\max}$ is the dual-cell compactification $\overline{\mathfrak{a}}^* = \mathfrak{a} \cup \Delta^*(\mathfrak{a})$. This implies that the map

$$\mathfrak{a} \to \overline{X}^S_{\max}, \quad H \mapsto e^H x_0,$$

extends to an embedding

$$i : \overline{\mathfrak{a}}_+ \to \overline{X}^S_{\max}.$$

For any subset $I \subset \Delta$, the boundary component $e(\mathfrak{a}_I^+)$ is equal to $\mathfrak{a}^I$, and

$$i(e(\mathfrak{a}_I^+)) \subset X_{P_I} \subset \overline{X}^S_{\max} - X.$$

By Proposition I.4.40, the stabilizer in G of a point $mK_{P_I} \in X_{P_I}$ in $\overline{X}^S_{\max}$ is equal to $N_{P_I} A_{P_I}{}^m K_{P_I}$. Note that the *Bruhat decomposition* (see [GJT, Corollary 2.21]) implies that

$$N_P^- \times e^{\mathfrak{a}^I} \times N_{P_I} A_{P_I}{}^m K_{P_I} \to G, \quad (n^-, e^H, nak) \mapsto n^- e^H nak \qquad \text{(I.19.12)}$$

is an analytic diffeomorphism onto an open subset of G. In fact, by the Bruhat decomposition, $N_{P_I}^- N_{P_I} A_{P_I} M_{P_I}$ is open dense in G. Let P' be the parabolic subgroup in M_{P_I} corresponding to P in equation (I.1.21). Then $N_{P'}^- A_{P'}{}^m K_{P_I} = M_{P_I}$, and equation (I.19.12) follows from $N_P^- = N_{P_I}^- N_{P'}^-$ and $A_{P'} = e^{\mathfrak{a}^I}$.

Then equation (I.19.12) implies that the map

$$N_P^- \times \overline{\mathfrak{a}}_+ = N_P^- \times \left(\mathfrak{a} \cup \coprod_I \mathfrak{a}^I \right) \to \overline{X}^S_{\max}$$

is an embedding. Since

$$\overline{X}^S_{\max} = X \cup \coprod_{I \subset \Delta} G X_{P_I},$$

the image of $N_P^- \times \overline{\mathfrak{a}}_+$ is open and dense in $\overline{X}^S_{\max}$.

Remark I.19.21 By Proposition I.19.18, $\overline{\mathfrak{a}}_+ \cong \mathbb{R}^r_{\geq 0}$ is an analytic corner, and hence $N_P^- \times \overline{\mathfrak{a}}_+$ is an analytic manifold with corners of codimension r. By Proposition I.19.20, $\overline{X}^S_{\max}$ is covered by these corners. It will be shown later in Proposition I.19.27 that the analytic structures of these corners are compatible, and hence $\overline{X}^S_{\max}$ is a compact manifold with corners of codimension r.

I.19.22 Now we are ready to construct the compactification $\widetilde{X}^S_{\max}$ with a built-in continuous G-action, which will turn out to be isomorphic to $\overline{X}^S_{\max}$.

Define an equivalence relation $\sim$ on

$$G \times \overline{\mathfrak{a}}_+ = G \times \left(\mathfrak{a} \cup \coprod_{I \subset \Delta} \mathfrak{a}^I \right)$$

by

$$(g_1, H_1) \sim (g_2, H_2)$$

if and only if

1. either $H_1, H_2 \in \mathfrak{a}$, and

$$g_1 e^{H_1} K = g_2 e^{H_2} K, \tag{I.19.13}$$

2. or there exists $I \subset \Delta$ such that $H_1, H_2 \in \mathfrak{a}^I$, and

$$g_1 e^{H_1} N_{P_I} A_{P_I} K_{P_I} = g_2 e^{H_2} N_{P_I} A_{P_I} K_{P_I}. \tag{I.19.14}$$

Define $\widetilde{X}^S_{\max}$ to be the equivalence classes of $\sim$ on $G \times \overline{\mathfrak{a}}_+$:

$$\widetilde{X}^S_{\max} = G \times \overline{\mathfrak{a}}_+ / \sim$$

with the quotient topology. The group G acts on $G \times \overline{\mathfrak{a}}_+$ by

$$g(g_1, H_1) = (gg_1, H_1).$$

Clearly, the action preserves the equivalence relation $\sim$ and induces a continuous action on the quotient $\widetilde{X}^S_{\max}$.

It will be shown that $\widetilde{X}^S_{\max}$ is a Hausdorff compactification isomorphic to the maximal Satake compactification $\overline{X}^S_{\max}$. This isomorphism is the motivation for the definition of the equivalence relation $\sim$ above. In fact, the group $N_{P_I} A_{P_I}{}^m K_{P_I}$ is the stabilizer of the point $mK_{P_I} \in X_{P_I} \subset \overline{X}^S_{\max}$ (see the proof of Proposition I.19.20).

I.19.23 Let

$$\pi : G \times \overline{\mathfrak{a}}_+ \to \widetilde{X}^S_{\max}$$

be the quotient map. To determine the quotient topology of $\widetilde{X}^S_{\max}$, it suffices to find subsets U of $G \times \overline{\mathfrak{a}}_+$ such that the inclusion $U \hookrightarrow G \times \overline{\mathfrak{a}}$ is continuous, $\pi : U \to \widetilde{X}^S_{\max}$ is injective, and the $\sim$-saturation of open subsets V of U are open subsets of $G \times \overline{\mathfrak{a}}_+$, i.e., $\pi^{-1}(\pi(V))$ is open. Then the map $\pi : U \to \widetilde{X}^S_{\max}$ is a homeomorphism onto $\pi(U)$. We need to find such sets $\pi(U)$ that cover $\widetilde{X}^S_{\max}$.

Let P be a minimal parabolic subgroup containing $A = \exp \mathfrak{a}$ as the split component and corresponding to the positive chamber $\mathfrak{a}^+$ as above. Let

$$U_P = N_P^- \times \overline{\mathfrak{a}}_+ \subset G \times \overline{\mathfrak{a}}_+. \tag{I.19.15}$$

For any $g \in G$, then

$$gU_P = gN_P^- \times \overline{\mathfrak{a}}_+.$$

Lemma I.19.24 *For any $g \in G$, the image of $\pi : gU_P \to \widetilde{X}^S_{\max}$ is an open subset, and π is a homeomorphism onto its image.*

Proof. As in the proof of Proposition I.19.20, the Bruhat decomposition (see equation I.19.12) shows that the map $\pi : gU_P \to \widetilde{X}^S_{\max}$ is injective.

Clearly, the inclusion $gU_P \to G \times \overline{\mathfrak{a}}_+$ is continuous and hence the map $\pi : gU_P \to \widetilde{X}^S_{\max}$ is continuous, by the definition of the quotient topology. We need to show that it is an open map, i.e., the $\sim$-saturation of any open subset of gU_P is an open subset in $G \times \overline{\mathfrak{a}}_+$. Let $W \subset N_P^-$, $V \subset \overline{\mathfrak{a}}_+$ be any two open subsets. It suffices to show that for any point $(q, H) \in W \times V$, the $\sim$-saturation of $gW \times V$ contains a neighborhood of $g(q, H)$ in $G \times \overline{\mathfrak{a}}_+$.

Without loss of generality, we assume that $g = e$ and V is of the form

$$V = V_\Delta \cup \coprod_{I \subset \Delta} V_I,$$

where $V_\Delta \subset \mathfrak{a}$, and $V_I \subset \mathfrak{a}^I$. We will construct the desired neighborhood using this decomposition.

Assume first that $H \in \mathfrak{a}$. Then we can assume that $V_I = \emptyset$ for all $I \subset \Delta$. Let U_Δ be a small neighborhood of e in $e^{\mathfrak{a}}$ and let V'_Δ be a neighborhood of e^H such that

$$WU_\Delta e^H K e^{-H} V'_\Delta \subset WV_\Delta K.$$

Such neighborhoods exist since

$$e^H K e^{-H} e^H = e^H K \subset V_\Delta K$$

and $WV_\Delta K$ is an open subset containing $We^H K$. Hence, $WU_\Delta e^H K e^{-H} \times V'_\Delta$ is contained in the $\sim$-saturation of $W \times V_\Delta$. Since $G = N_P^- A_P K$, $WU_\Delta e^H K e^{-H}$ is an open subset of G. This proves that $WU_\Delta e^H K e^{-H} \times V'_\Delta$ is a neighborhood of (q, H) in $G \times \overline{\mathfrak{a}}_+$.

In general, let $H \in \mathfrak{a}^I$, where $I \subset \Delta$. In this case, we can assume that for $J \subset I$, V_J is empty. On the other hand, for $J \supset I$, V_J is nonbounded. Fix a $J \supset I$. Let U_J be a (small) neighborhood of e in $\mathfrak{a}^J$ to be determined later. Let $V' \subset V$ be a smaller neighborhood of H in $\overline{\mathfrak{a}}_+$, and let $V'_J = V' \cap \mathfrak{a}^I$. Let H_j be a sequence in V'_J such that

$$V'_J \subset \cup_j e^{H_j} U_J.$$

Then

$$WU_J N_{P_J} A_{P_J} e^{H_j} K_{P_J} e^{-H_j} e^{H_j} U_J \subset WN_{P_J} A_{P_J} V_J N_{P_J} A_{P_J} K_{P_J}.$$

This holds when U_J is sufficiently small. Now

$$\cup_{J \supset I} U_j WU_J N_{P_J} A_{P_J} e^{H_j} K_{P_J} e^{-H_j} \times e^{H_j} U_j$$

is a neighborhood of (q, H) in $G \times \overline{\mathfrak{a}}_+$. This shows that the $\sim$-saturation of $W \times V$ contains a neighborhood of (q, H). This completes the proof of the lemma.

Remark I.19.25 The sets $\pi(gU_P)$ clearly cover $\widetilde{X}^S_{\max}$. The open subset U_P is the canonical corner associated with the parabolic subgroup P and depends on the horospherical decomposition of X with respect to P and the basepoint x_0. The set gU_P can also be interpreted this way. In the definition of U_P, we used the Langlands decomposition of P with respect to the basepoint x_0, and this Langlands decomposition induces the associated horospherical decomposition of X associated with P. Under the conjugation by g, the Langlands decomposition becomes the Langlands decomposition of ${}^gP$ with respect to the basepoint gx_0,

$$^gP = N_{^gP} A_{^gP} M_{^gP},$$

and the corresponding horospherical decomposition is obtained by left multiplication by g,

$$X = N_{^gP} \times A_{^gP} \times M_{^gP}/({}^gK \cap M_{^gP}).$$

Therefore, gU_P is the canonical corner associated with ${}^gP$ with respect to the basepoint gx_0.

Proposition I.19.26 *The space $\widetilde{X}^S_{\max}$ is a Hausdorff G-compactification of X.*

Proof. Clearly, $\widetilde{X}^S_{\max}$ contains X as an open dense subset. To show that it is Hausdorff, we note that for any point $x \in \widetilde{X}^S_{\max}$, the set of points g such that $x \in gU_P$ is open and dense in G. This basically follows from the Bruhat decomposition (see [Os1, p. 127]). This implies that for any two points $x, y \in \widetilde{X}^S_{\max}$, there exists an element $g \in G$ such that $x, y \in gU_P$. Since gU_P is an open Hausdorff subset, x, y are separated disjoint open subsets.

To show that it is compact, we note that the image $\pi(G \times \mathfrak{a}) \cong X$ is open and dense in $\widetilde{X}^S_{\max}$. By the Cartan decomposition, $\pi(G \times \mathfrak{a}) = \pi(K \times cl(\mathfrak{a}+))$, where $cl(\mathfrak{a}^+)$ is the closed Weyl chamber, and $\widetilde{X}^S_{\max}$ is equal to the closure of $\pi(K \times cl(\mathfrak{a}^+))$. Since the closure of $cl(\mathfrak{a}^+)$ in $\overline{\mathfrak{a}}_+$ is compact and K is also compact, the closure of $\pi(K \times cl(\mathfrak{a}^+))$ is compact, hence $\widetilde{X}^S_{\max}$ is compact.

Proposition I.19.27 *The space $\widetilde{X}^S_{\max}$ is a compact real analytic manifold with corners of codimension $r = rk(X)$.*

Proof. The proof of the analyticity is the main technical part of this construction. The basic idea is as follows. As pointed out earlier, the corner U_P associated with P has a canonical real analytic structure. For any $g \in G$, the subset gU_P is the corner associated with the parabolic subgroup ${}^gP$ and the basepoint gx_0. These corners are charts of $\widetilde{X}^S_{\max}$. The only problem is to show that their analytic structures are compatible.

When restricted to the interior X, these analytic structures are clearly compatible, since the horospherical decomposition of X is real analytic. The issue is at the boundary of $\widetilde{X}^S_{\max}$. It is not easy to compute explicitly the transition function from U_P to gU_P, which is equivalent to expressing the horospherical decomposition of $x \in X$ with respect to ${}^gP$ and the basepoint gx_0 in terms of the horospherical decomposition with respect to P and the basepoint x_0.

For a point $x \in X$, its horospherical coordinates with respect to P and basepoint x_0 are conjugated by g to the horospherical coordinates of gx with respect to ${}^gP$

and the basepoint gx_0. So if we know how to compute the horospherical coordinates of $g^{-1}x$ with respect to P and the basepoint x_0 in terms of the horospherical coordinates of x with respect to P and the basepoint x_0, we can compute the desired horospherical coordinates of x with respect to ${}^gP$ and the basepoint gx_0 in terms of the horospherical coordinates of x with respect to P and the basepoint x_0.

To show that the horospherical coordinates of $g^{-1}x$, $x \in X$, with respect to P and the basepoint x_0 are given by real analytic functions of the horospherical coordinates of x with respect to P and the basepoint x_0, and can be extended to real analytic functions, it suffices to show that when g belongs to one parameter family $\exp tY$, $Y \in \mathfrak{g}$, the associated vector field on X extends to an analytic field on $\widetilde{X}^S_{\max}$. It turns out that the explicit computation of this vector field over X is easier and given in [Os1, Lemma 3] (since it essentially deals with only one horospherical decomposition), and the formula [Os1, equation (2.3)] shows that it extends to a real analytic vector field on $\widetilde{X}^S_{\max}$.

Proposition I.19.28 *The identity map on X extends to a homeomorphism $\widetilde{X}^S_{\max} \to \overline{X}^S_{\max}$, i.e., the compactification $\widetilde{X}^S_{\max}$ is isomorphic to the maximal Satake compactification $\overline{X}^S_{\max}$. Hence $\overline{X}^S_{\max}$ is a real analytic compact manifold with corners of codimension $r = rk(X)$. In particular, when $rk(X) = 1$, $\overline{X}^S_{\max}$ is a compact manifold with boundary.*

Proof. Under the identification

$$\mathfrak{a} \cong e^{\mathfrak{a}}x_0, \quad H \mapsto e^H x_0,$$

the inclusion $\mathfrak{a} \to X$ extends to a map

$$\overline{\mathfrak{a}}_+ \to \overline{X}^S_{\max},$$

which is a homeomorphism onto its image. Hence it induces a continuous map

$$G \times \overline{\mathfrak{a}}_+ \to \overline{X}^S_{\max}.$$

Since the equivalence relation $\sim$ on $G \times \overline{\mathfrak{a}}_+$ is defined in terms of stabilizers in G of points in $\overline{X}^S_{\max}$, this induces a continuous bijective map

$$\widetilde{X}^S_{\max} = G \times \overline{\mathfrak{a}}_+ / \sim \to \overline{X}^S_{\max},$$

which extends the identity map on X. Since both $\widetilde{X}^S_{\max}$ and $\overline{X}^S_{\max}$ are compact and Hausdorff, they are homeomorphic.

I.19.29 The above construction of $\widetilde{X}^S_{\max}$ was suggested by the construction of the Oshima compactification $\overline{X}^O$ in [Os1], which is a closed (i.e., compact without boundary) analytic manifold. The Oshima compactification $\overline{X}^O$ was motivated by the study of boundary behaviors at the unique closed G-orbit G/P in $\overline{X}^S_{\max}$ of joint eigenfunctions of all the invariant differential operators on X, where P is a minimal parabolic subgroup.

Now we recall briefly the Oshima construction (see §II.2 in Part II for more details). Instead of $\mathbb{R}^r_{\geq 0} \cong \overline{\mathfrak{a}}_+$, we use $\mathbb{R}^r$ and define

$$\overline{X}^O = G \times \mathbb{R}^r / \sim, \tag{I.19.16}$$

where the equivalence relation $\sim$ is a generalization of the equivalence relation for $\widetilde{X}^S_{\max}$ in equations (I.19.13, I.19.14). Briefly, for any real number x, define its sign

$$sgn(x) = 0 \text{ if } x = 0, \quad sgn(x) = \frac{x}{|x|} \text{ if } x \neq 0.$$

For any $t = (t_1, \dots, t_r) \in \mathbb{R}^r$, define its signature

$$\varepsilon_t = (sgn(t_1), \dots, sgn(t_r)).$$

A signature $\varepsilon = (\varepsilon_1, \dots, \varepsilon_r)$ is called *proper* if $\varepsilon_i \neq 0$ for all $i = 1, \dots, r$. For each proper signature ε, let

$$\mathbb{R}^r_\varepsilon = \{(t_1, \dots, t_r) \mid \varepsilon_t = \varepsilon\}.$$

Then

$$\overline{\mathbb{R}^r_\varepsilon} \cong \overline{\mathfrak{a}}_+,$$

and

$$\mathbb{R}^r = \cup_\varepsilon \overline{\mathbb{R}^r_\varepsilon} = \cup_\varepsilon \overline{\mathfrak{a}}_+,$$

where ε runs over all proper signatures.

Using the above identification, the equivalence relation $\sim$ on $G \times \mathbb{R}^r$ is defined as follows. Two points $(g, t), (g', t') \in G \times \mathbb{R}^r$ are equivalent if and only if

1. $\varepsilon_t = \varepsilon_{t'}$, and hence there exists a proper signature ε such that $t, t' \in \overline{\mathbb{R}^r_\varepsilon}$.
2. Under the identification $\mathbb{R}^r_\varepsilon \cong \overline{\mathfrak{a}}_+$, (g, t) and (g', t') are mapped to the same point in $\widetilde{X}^S_{\max}$.

By the same arguments as above (see [Os1] [Sch]), we can prove the following proposition.

Proposition I.19.30 *The space $\overline{X}^O$ is a compact and Hausdorff real analytic G-manifold. It contains 2^r disjoint copies of X, and its analytic structure extends the natural analytic structure of X, and the closure of each X is isomorphic to the maximal Satake compactification $\widetilde{X}^S_{\max} \cong \overline{X}^S_{\max}$.*

The space $\overline{X}^O$ is called the *Oshima compactification.*

I.19.31 In Part II, we will give an alternative construction of $\overline{X}^S$ (see Remark II.10.12) in two steps:

1. Show that $\overline{X}^S_{\max}$ is a real analytic (and semialgebraic) manifold with corners by embedding X into the real locus of the wonderful compactification $\overline{X_{\mathbb{C}}}^W$ of $X_{\mathbb{C}} = G_{\mathbb{C}}/K_{\mathbb{C}}$, the complexification of X.
2. Self-glue $\overline{X}^S_{\max}$ into a closed real analytic manifold, which gives $\overline{X}^O$.

As discussed earlier in Proposition I.19.27, the main, technical step in [Os1] is to show that $\overline{X}^O$ has a real analytic structure. In this alternative approach, the real analytic structure is obtained rather easily from the analytic structure of the smooth projective variety $\overline{X_{\mathbb{C}}}^W$ (note that $X_{\mathbb{C}}$ is a quasi-projective variety).

I.19.32 Summary and comments. In this section, we gave two more constructions of the maximal Satake compactification $\overline{X}^S_{\max}$. The basic idea is to pass from the compactifications of maximal flats in X to the whole symmetric space X. The polyhedral compactification or rather the dual-cell compactification of a vector space explains the underlying duality between the maximal Satake compactification $\overline{X}^S_{\max}$ and the spherical Tits building $\Delta(G)$. The Oshima compactification $\overline{X}^O$ will be discussed more in §II.2. From the point of view of this section, one of the key points in [Os1] is to build in a G-action in the construction of $\overline{X}^O$.

Part II

Smooth Compactifications of Semisimple Symmetric Spaces

In Part I, we have discussed many compactifications of a Riemannian symmetric space in the classical sense that the symmetric space is an open dense subset. Under this restriction, none of these compactifications is a closed smooth manifold.

Part II is chiefly devoted to compactifications of a given semisimple symmetric space in which the symmetric space is an open, not dense subset, but the compactification is a closed real analytic manifold. Therefore, there are two features of the compactifications in Part II that are different from those in Part I:

1. Symmetric spaces are not necessarily Riemannian.
2. Compactifications are closed smooth manifolds, but the symmetric spaces are not dense.

In Chapter 4, we study the first such compactification, the Oshima compactification $\overline{X}^{O}$, of a Riemannian symmetric space X. The basic point is to apply a general self-gluing procedure to obtain it from the maximal Satake compactification $\overline{X}^{S}_{\max}$, using the fact that $\overline{X}^{S}_{\max}$ is a compact manifold with corners.

In Chapter 5, we study basic facts on semisimple symmetric spaces, in particular, those arising from the real form of semisimple linear algebraic groups defined over $\mathbb{R}$. In Chapter 6, we identify the real locus of complex symmetric spaces using the Galois cohomology.

In Chapter 7, we apply the general facts developed in the earlier chapters to determine the real locus of the wonderful compactification of $\overline{\mathbf{X}}^{W}$ of a complex symmetric space $\mathbf{X}$. As a corollary, we obtain easily the existence of the real analytic structure on the maximal Satake compactification $\overline{X}^{S}_{\max}$. Combined with the self-gluing result in Chapter 4, this gives a simple proof of the existence of real analytic structure on the Oshima compactification $\overline{X}^{O}$. The results of this chapter also explain naturally why the study of compactifications of Riemannian symmetric spaces X leads to compactifications of non-Riemannian semisimple symmetric spaces X_{ε}.

In Chapter 8, we relate the Oshima–Sekiguchi compactification $\overline{X}^{OS}$ to the real locus of the wonderful compactification $\overline{\mathbf{X}}^{W}(\mathbb{R})$ of $\mathbf{X} = X_{\mathbb{C}}$, and explain naturally why compactifications of different semisimple symmetric spaces and Riemannian symmetric spaces can be glued together into a closed smooth manifold.

4

Smooth Compactifications of Riemannian Symmetric Spaces G/K

The Oshima compactification $\overline{X}^O$ of a Riemannian symmetric space X is a closed real analytic manifold that contains the union of 2^r copies of X as an open dense subset, where $r = rk(X)$. The closure of each X contains a unique compact G-orbit in its boundary, which is isomorphic to the maximal Furstenberg boundary of G or X and hence the Poisson boundary of X.

The Oshima compactification $\overline{X}^O$ has the crucial property that G-invariant differential operators on X can be extended to differential operators on $\overline{X}^O$ with regular singularities on the Poisson boundary. This property depends on the real analytic structure of $\overline{X}^O$ and can be used to study the asymptotic behaviors at the Poisson boundary of joint eigenfunctions on X of the G-invariant differential operators. Asymptotic behaviors and boundary values of eigenfunctions played an important role in solving the Helgason conjecture in [KaK] (see also [Sch]), which says roughly that joint eigenfunctions on X are the Poisson transform of hyperfunctions on the Poisson boundary. The Oshima compactification $\overline{X}^O$ is also important for other problems in the representation theory of G (see [Os4]).

As briefly recalled in Part I, §I.19, the Oshima compactification $\overline{X}^O$ is closely related to the maximal Satake compactification $\overline{X}^S_{\max}$. In §II.1, we study a procedure of self-gluing multiple copies of a compact manifold with corners into a closed manifold, which is a generalization of doubling a compact manifold with boundary into a closed manifold. It turns out that there is an obstruction to the self-gluing using the minimal number of copies. In §II.2, we apply this procedure to self-glue the maximal Satake compactification $\overline{X}^S_{\max}$ into the Oshima compactification $\overline{X}^O$. The result that the maximal Satake compactification $\overline{X}^S_{\max}$ is a real analytic manifold with corners can be obtained easily by embedding X into the real locus of the wonderful compactification of the corresponding complex symmetric space $X_{\mathbb{C}} = G_{\mathbb{C}}/K_{\mathbb{C}}$. Hence, this approach gives a more direct and conceptional construction of the Oshima compactification $\overline{X}^O$, avoiding the difficult proof of the existence of real analytic structure in [Os1].

Besides the application to the Oshima compactification $\overline{X}^O$, the self-gluing procedure will also be used in §III.16 to self-glue the Borel–Serre compactification $\overline{\Gamma\backslash X}^{BS}$ of the locally symmetric space $\Gamma\backslash X$ into a closed analytic manifold $\overline{\Gamma\backslash X}^{BSO}$.

II.1 Gluing of manifolds with corners

The notion of manifold with corners is assumed to be known. We review only some facts and notation. (See the appendix in [BS] by Douady–Herault).

This section is organized as follows. In II.1.1, we introduce notions of the rank, boundary faces, and boundary hypersurfaces of a manifold with corners. An important concept for the self-gluing of manifolds with corners is that of embedded hypersurfaces. To carry out the self-gluing, we introduce a partition of the set $\mathcal{H}_M$ of boundary hypersurfaces of M. Then the self-gluing of M into $\widetilde{M}$ is given in Proposition II.1.2. Properties of $\widetilde{M}$ are given in Propositions II.1.4 and II.1.5. In the special case that the rank is equal to the number of subsets in the partition of $\mathcal{H}_M$, a more direct construction of $\widetilde{M}$ is given in Proposition II.1.9.

II.1.1 Let M be a connected manifold with corners, m its dimension. Every point $p \in M$ has a local chart of the form $\mathbb{R}^{m-i} \times \mathbb{R}^i_{\geq 0}$, where $R^i_{\geq 0}$ is a positive (closed) quadrant in $\mathbb{R}^i$ and p is sent to the origin. The integer i is called the *rank of* p, and the maximum of i is called *the rank* of M, denoted by $rk(M)$.

The manifold M has a stratification such that each stratum consists of points of the same rank. Every connected component of a stratum is called an *open boundary face* of M, and its closure in M is called a *boundary face*. If a boundary face is of codimension 1, it is called a *boundary hypersurface.*

The boundary ∂M of M is the union of boundary hypersurfaces, which are themselves manifolds with corners of rank strictly less than $rk(M)$. We shall assume that they are all of rank equal to $rk(M) - 1$ and embedded (no self-intersection). More precisely, a boundary hypersurface H is *embedded* if for every point p of rank i and belonging the boundary of H, there exist $i-1$ boundary hypersurfaces $H_1, \dots, H_{i-1}$ different from H such that p belongs to the intersection $H \cap H_1 \cdots \cap H_{i-1}$ and the intersection has codimension i, which is automatically satisfied if all $H_1, \dots, H_{i-1}$ are different.

If all the boundary hypersurfaces are embedded, the intersection of two boundary hypersurfaces is a manifold with corners of rank equal to $rk(M) - 2$ (if not empty), and is the union of boundary hypersurfaces of each of them, considered as manifolds with corners.

For any boundary face of M of codimension i, its boundary hypersurfaces are also embedded if all the boundary hypersurfaces of M are embedded, and they are intersections of $i + 1$ boundary hypersurfaces of M.

An example of a manifold with corners whose boundary hypersurfaces are not embedded is a two-dimensional manifold with one corner point and one boundary hypersurface. It is clear that this two-dimensional manifold cannot be self-glued into a closed smooth manifold. Therefore, the assumption that boundary hypersurfaces are embedded is crucial.

Our aim here is to glue M to a certain number of copies of itself so as to get a smooth manifold, and to give an alternative formulation in the case where it is possible to use the smallest possible number of copies of M, namely $2^{rk(M)}$.

For any connected manifold M with corners, we assume that the set $\mathcal{H}_M$ of boundary hypersurfaces is locally finite in the sense that each point has a neighborhood that meets only finitely many of them.

For the gluing purpose, we need to assume that the set $\mathcal{H}_M$ admits a *finite partition*

$$\mathcal{H}_M = \coprod_{j=1}^{N} \mathcal{H}_{M,j}$$

such that the elements of each $\mathcal{H}_{M,j}$ are disjoint ($1 \leq j \leq N$). If M is compact, then $\mathcal{H}_M$ is finite and such a partition always exists. This is the case considered in [Me2]. The following proposition is an obvious generalization.

Proposition II.1.2 *Suppose that M is a manifold with corners, and the set $\mathcal{H}_M$ of boundary hypersurfaces admits a finite partition as above. Then it is possible to construct a closed manifold $\widetilde{M}$ by gluing 2^N copies of M along boundary hypersurfaces.*

Proof. It is by induction on N. Let M' be a copy of M with the same partition of the set $\mathcal{H} = \mathcal{H}_M = \mathcal{H}_{M'}$ of boundary hypersurfaces. Glue M and M' along the elements of $\mathcal{H}_1$. We claim that $M \cup M'$ is a manifold with corners. In fact, the interior points of the $H \in \mathcal{H}_1$ are manifold points of $M \cup M'$, i.e., have Euclidean neighborhoods. We need to check that boundary points of these hypersurfaces in $\mathcal{H}_1$, i.e., corner points of M, are also corner points of $M \cup M'$. Let $H \in \mathcal{H}_1$ and p in the boundary of H. Suppose that p is of rank i in M. Then $i \geq 2$. Since all the boundary hypersurfaces of M are embedded, there exist i different hypersurfaces $H_1 = H, \dots, H_i$ such that $p \in H_1 \cap \cdots \cap H_i$. By the assumption on the partition, the hypersurfaces in $\mathcal{H}_1$ are disjoint. This implies that $H_2, \dots, H_i$ do not belong to $\mathcal{H}_1$. Then it is clear that after the gluing along H_1, p has a chart in $M \cup M'$ of the form $\mathbb{R}^{m-i+1} \times \mathbb{R}^{i-1}_{\geq 0}$ and becomes a point of rank $i - 1$.

We claim that $\mathcal{H}_{M \cup M'}$ admits a partition in $N - 1$ subsets, each consisting of disjoint boundary hypersurfaces.

For every $j > 1$, divide $\mathcal{H}_{M,j}$ into two subsets:

$$\mathcal{H}_{M,j} = \mathcal{H}_{j,1} \amalg \mathcal{H}_{j,2},$$

where

$$\mathcal{H}_{j,1} = \{H \in \mathcal{H}_j \mid H \cap Z = \emptyset \text{ for all } Z \in \mathcal{H}_1\},$$

and $\mathcal{H}_{j,2}$ is the complement. The elements of $\mathcal{H}_{j,1}$ and their homologues in M' form a set of disjoint boundary hypersurfaces of $M \cup M'$, say $\mathcal{H}''_{j,1}$.

On the other hand, if $H \in \mathcal{H}_{j,2}$, there exists $Z \in \mathcal{H}_1$ such that $H \cap Z \neq \phi$. For any such Z, $H \cap Z$ is a boundary hypersurface of H and Z, and is equal to $H \cap H'$, where H' is the homologue of H on M'. As observed earlier, the assumption on embeddedness of the boundary hypersurfaces of M implies that the boundary hypersurfaces of H are also embedded. Then the gluing of $M \cup M'$ induces one of H and H' along their intersection, which is similarly a manifold with corners, locally Euclidean around an interior point of $H \cap H'$. In particular, $H \cup H'$ is a boundary hypersurface of $M \cup M'$. Let $\mathcal{H}''_{j,2}$ be the set of these glued-up boundaries of $M \cup M'$. They are disjoint since two elements of $\mathcal{H}_{M,j}$ are disjoint. Let $\mathcal{H}''_{M,j} = \mathcal{H}''_{j,1} \cup \mathcal{H}''_{j,2}$. Clearly hypersurfaces in $\mathcal{H}''_j$ are disjoint.

Since every boundary hypersurface of $M \cup M'$ belongs to a unique $\mathcal{H}''_j$ for $j \geq 2$, we have a partition of $\mathcal{H}_{M \cup M'}$ in $N - 1$ subsets:

$$\mathcal{H}_{M \cup M'} = \coprod_{2 \leq j \leq N} \mathcal{H}''_j.$$

If $N = 1$, then M is a manifold with boundary and the previous construction provides the desired manifold $\widetilde{M} = M \cup M'$. We can now use an induction hypothesis, which implies that we can glue 2^{N-1} copies of $M \cup M'$ to obtain a closed manifold $\widetilde{M}$. Altogether, $\widetilde{M}$ is constructed by gluing 2^N copies of M.

Remark II.1.3 In a corner of rank r, there are r boundary hypersurfaces with a nonempty intersection, hence $N \geq rk(M)$. The number N depends on the partition. When M is compact, the maximum value of N is the number N' of boundary hypersurfaces; hence $rk(M) \leq N \leq N'$. We can try to minimize N by a suitable choice of the partition. For instance, start with a maximal set $\mathcal{H}_1$ of disjoint boundary hypersurfaces. Then let $\mathcal{H}_2$ be a maximal set of disjoint boundary hypersurfaces in $\mathcal{H} - \mathcal{H}_1$, and so on. However, the number N may depend on the successive choices, as simple examples show.

Proposition II.1.4 *If M is C^∞ (resp. real analytic), then so is $\widetilde{M}$. Moreover, if a group H acts on M, then this action extends to one on $\widetilde{M}$. The extended action is smooth (resp. real analytic) if H is a Lie group and so is the given action on M.*

Proof. This follows from the construction: around a smooth point x of $H \cap H'$, the local charts in M and M' are obtained from one another by a "reflection principle" with respect to $H \cap H'$. These charts glue into a neighborhood of x in $\widetilde{M}$, which is C^∞ (resp. real analytic) if M is so. To see that a group action on M extends to $\widetilde{M}$, we note that for any two copies of M in $\widetilde{M}$, the group actions on them agree on their intersection, and the combined action on $\widetilde{M}$ gives the extension.

Proposition II.1.5 *The closed manifold $\widetilde{M}$ constructed in Proposition II.1.2 admits a $(\mathbb{Z}/2\mathbb{Z})^N$-action such that the quotient of $\widetilde{M}$ by $(\mathbb{Z}/2\mathbb{Z})^N$ is equal to M. If M admits a group action by H as in Proposition II.1.2, then the extended H-action commutes with this $(\mathbb{Z}/2\mathbb{Z})^N$-action.*

Proof. We prove this by induction. When $N = 1$, M is a manifold with boundary, and $\widetilde{M}$ is obtained from M by doubling across the boundary and clearly admits a $\mathbb{Z}/2\mathbb{Z}$-action.

As in the proof of Proposition II.1.2, $\mathcal{H}_{M \cup M'}$ admits a partition in $N-1$ subsets, and $\widetilde{M}$ is glued from 2^{N-1} copies of $M \cup M'$. By induction, $\widetilde{M}$ admits a $(\mathbb{Z}/2\mathbb{Z})^{N-1}$-action, and the quotient by this group is equal to $M \cup M'$. By Proposition II.1.2, the $\mathbb{Z}/2\mathbb{Z}$-action on $M \cup M'$ extends to an action on $\widetilde{M}$. This $\mathbb{Z}/2\mathbb{Z}$-action commutes with the $(\mathbb{Z}/2\mathbb{Z})^{N-1}$-action on $\widetilde{M}$ by induction. Hence $\widetilde{M}$ admits a $(\mathbb{Z}/2\mathbb{Z})^N$-action, and the quotient of $\widetilde{M}$ by $(\mathbb{Z}/2\mathbb{Z})^N$ is equal to the quotient $M \cup M'$ by $\mathbb{Z}/2\mathbb{Z}$ and hence to M.

To show that the extended H-action on $\widetilde{M}$ commutes with $(\mathbb{Z}/2\mathbb{Z})^N$, we note that $(\mathbb{Z}/2\mathbb{Z})^N$ interchanges different copies of M. Since the H-actions on the all the copies of M are the same, the extended H-action commutes with the $(\mathbb{Z}/2\mathbb{Z})^N$-action.

II.1.6 A special case of gluing. The most important cases for us are some in which $N = rk(M)$ and a further assumption of homogeneity is assumed. In the rest of this section, an alternative construction of $\widetilde{M}$ is given under the assumption that $N = rk(M)$.

The standard example of dimension m is that of a closed quadrant in $\mathbb{R}^m$, say the positive quadrant where all coordinates x_i are nonnegative. A first gluing to the quadrant

$$\{(x_1, \ldots, x_m) \mid x_1 \leq 0,\ x_i \geq 0,\ i \geq 2\}$$

produces the manifold M_1 with corners of rank $m - 1$:

$$\{(x_1, \ldots, x_m) \mid x_i \geq 0,\ i \geq 2\}.$$

A second gluing of M_1 provides the manifold with corners of rank $m - 2$:

$$\{(x_1, \ldots, x_m) \mid x_i \geq 0,\ i \geq 3\},$$

etc. After m steps, we get $\mathbb{R}^m$, a smooth manifold without corners.

We want to reformulate this construction so as to carry it out in one step rather than m and then apply the procedure to general manifolds.

II.1.7 First we need to define a partition of $\mathbb{R}^m$ into 3^m subsets, the open quadrants in $\mathbb{R}^n$ and the lower-dimensional coordinate planes. We follow the conventions of [Os1] and [OsS1]. The natural set of indices for the applications to follow is a set Δ of simple roots, so we shall label the coordinates by elements of Δ and speak of a partition of $\mathbb{R}^\Delta$. The value of $a \in \Delta$ on $t \in \mathbb{R}^\Delta$ is denoted by t^a.

Definition.

1. The signature of $t \in \mathbb{R}$, denoted by sgn t, is 0 if $t = 0$ and $t/|t|$ otherwise.

2. A signature ε on a finite set Δ is a map $\varepsilon : \Delta \to \{1, 0, -1\}$. Its support $s(\varepsilon)$ is the set

$$s(\varepsilon) = \{a \in \Delta \mid \varepsilon(a) \neq 0\}.$$

3. A signature ϵ is called proper if $s(\varepsilon) = \Delta$.

We let $\mathcal{E}(\Delta)$ be the set of all signatures on Δ and $\mathcal{E}^o(\Delta)$ the subset of proper signatures. They have cardinalities $3^{|\Delta|}$ and $2^{|\Delta|}$ respectively, where $|\Delta|$ is the cardinality of Δ. If $J \subset \Delta$, an element of $\mathcal{E}(J)$ is identified with the signature of Δ, which is equal to ε on J and is zero outside J. If $I \subset J$, and $\varepsilon \in \mathcal{E}(I)$, $\varepsilon' \in \mathcal{E}(J)$, we write $\varepsilon \subset \varepsilon'$ if ε and ε' coincide on I. For $\varepsilon \in \mathcal{E}(\Delta)$, let

$$\mathbb{R}^{\Delta,\,\varepsilon} = \{t \in \mathbb{R}^{\Delta} \mid \operatorname{sgn} t^a = \varepsilon(a)\}. \tag{II.1.1}$$

It can be identified with $\mathbb{R}^{J,\,\varepsilon'}$, where $J = s(\varepsilon)$ and ε' is the restriction of ε to J. We have

$$\mathbb{R}^{\Delta} = \coprod_{\varepsilon \in \mathcal{E}(\Delta)} \mathbb{R}^{\Delta,\,\varepsilon}.$$

This can also be written as

$$\mathbb{R}^{\Delta} = \coprod_{J \subset \Delta,\ \varepsilon \in \mathcal{E}^o(J)} \mathbb{R}^{J,\,\varepsilon}.$$

Note that J can be empty, in which case $\mathbb{R}^{J,\varepsilon}$ is the origin.

The closed quadrants in $\mathbb{R}^{\Delta}$ are exactly the subspaces $\mathbb{R}^{\Delta}(\delta)$ defined by

$$\mathbb{R}^{\Delta}(\delta) = \coprod_{\varepsilon \subset \delta} \mathbb{R}^{\Delta,\,\varepsilon},$$

where $\delta \in \mathcal{E}^o(\Delta)$, and $\mathbb{R}^{\Delta}$ may be viewed as the space obtained by gluing them along their intersections. They are all isomorphic and so $\mathbb{R}^{\Delta}$ is the manifold obtained by gluing $2^{|\Delta|}$ copies of $\mathbb{R}^{\Delta}(\delta)$.

II.1.8 Let M be a manifold with corners, r its rank, and Δ a set of cardinality r. We assume that $\mathcal{H}_M$ has a partition

$$\mathcal{H}_M = \coprod_{a \in \Delta} \mathcal{H}_a,$$

where the elements of $\mathcal{H}_a$ are disjoint boundary hypersurfaces of rank $r - 1$. For $J \subset \Delta$, and $|J|$ boundary hypersurfaces H_a, with $H_a \in \mathcal{H}_a$, $a \in J$, the intersection Z of these H_a is either empty or a manifold with corners of codimension and rank $|J|$.

Let $J' \subset \Delta$ and Z' be similarly constructed. If $J \cap J' = \emptyset$, then $Z \cap Z'$ is a manifold with corners of rank and codimension $r - (|J| + |J|')$ or is empty. Assume now that $I = J \cap J'$ is not empty. If for some $a \in I$, the hypersurfaces H_a and H'_a are distinct, then they are disjoint, by the definition of $\mathcal{H}_a$, hence $Z \cap Z'$ is empty. If $H_a = H'_a$ for all $a \in I$, then we are back to the previous case with J' replaced by $J'' = J' - I$.

To be consistent with the notation of Definition II.1.7 on signatures that are used to parametrize different parts of $\mathbb{R}^\Delta$, we will change slightly the one just used. For $J \subset \Delta$, let us denote by $\mathcal{H}_J$ the set of nonempty intersections of elements H_a, where a runs through $\Delta - J$. Thus our previous $\mathcal{H}_a$ becomes $\mathcal{H}_{\Delta-\{a\}}$.

Given a manifold with corners N, we let N^o be its interior. Let $\mathcal{H}^o_J$ be the set of interiors of the elements of $\mathcal{H}_J$. Then

$$M = \coprod_{J \subset \Delta} \mathcal{H}^o_J. \tag{II.1.2}$$

Here it is understood that if $J = \Delta$, then $\mathcal{H}_J = M$ and $\mathcal{H}^o_J = \{M^o\}$. If $J = \emptyset$ and $Z \in \mathcal{H}_J$ then $Z = Z^o$ is a closed manifold.

The elements of the $\mathcal{H}^o_J$ are the strata of a stratification of M in which the closed subspace of codimension i $(0 \leq i \leq r)$ is the union of the $\mathcal{H}^o_J$, where J runs through the subsets of Δ of cardinality $\leq r - i$, or simply $\bigcup_{|J|=|\Delta|-i} \mathcal{H}_J$.

We consider now objects (Z, ε), where $Z \in \mathcal{H}^o_J$ and ε is a signature on Δ with support equal to J, or, equivalently, a proper signature on J. Let

$$\widetilde{M} = \coprod_{J \subset \Delta} \quad \coprod_{Z \in \mathcal{H}^o_J,\ \varepsilon \in \mathcal{E}^o(J)} (Z, \varepsilon).$$

Proposition II.1.9 *There exists a suitable topology on $\widetilde{M}$ with respect to which $\widetilde{M}$ is the manifold $\widetilde{M}$ constructed in Proposition II.1.2.*

Proof. Fix $\delta \in \mathcal{E}^o(\Delta)$. Define a subset $\widetilde{M}(\delta)$ of $\widetilde{M}$ by

$$\widetilde{M}(\delta) = \coprod_{\varepsilon \subset \delta} (Z, \varepsilon).$$

For each Z, there is only one possibility for ε; hence we see from (II.1.2) that $\widetilde{M}(\delta)$ is, set-theoretically, a copy of M. We endow it with the topology of M. We have

$$\widetilde{M} = \bigcup_{\delta \in \mathcal{E}^o(\Delta)} \widetilde{M}(\delta).$$

Let $\delta' \in \mathcal{E}^o(\Delta)$. We want to describe $\widetilde{M}(\delta) \cap \widetilde{M}(\delta')$. Let

$$J(\delta, \delta') = \{a \in \Delta \mid \delta(a) = \delta'(a)\},$$

and let $\varepsilon(\delta, \delta')$ be the common restriction of δ and δ' to $J(\delta, \delta')$. Then

$$\widetilde{M}(\delta) \cap \widetilde{M}(\delta') = \coprod_{J \subset J(\delta, \delta')} \quad \coprod_{Z \in \mathcal{H}^0_J,\ \varepsilon \subset \varepsilon(\delta, \delta')} (Z, \varepsilon).$$

The topologies of $\widetilde{M}(\delta)$ and $\widetilde{M}(\delta')$ induce the same topology on the intersection $\widetilde{M}(\delta) \cap \widetilde{M}(\delta')$. We then endow $\widetilde{M}$ with the sum topology of the topologies on the $\widetilde{M}(\delta)$.

We next show that $\widetilde{M}$ is a smooth manifold without corners. For any point o in a corner of $\widetilde{M}(\delta)$ of codimension $|\Delta|$, a neighborhood of o in $\widetilde{M}(\delta)$ is the same as a neighborhood of the origin in $\mathbb{R}^{n-|\Delta|} \times \mathbb{R}^{|\Delta|}(\delta)$, where $n = \dim M$. From the fact that $\mathbb{R}^{|\Delta|}$ is the manifold obtained by gluing the closed quadrants $\mathbb{R}^{|\Delta|}$, we conclude that these identical neighborhoods of o in $\widetilde{M}(\delta)$ glue into a smooth neighborhood of o in $\widetilde{M}$. Similarly, for any point o in a boundary face (Z, ε) of $\widetilde{M}(\delta)$, the neighborhoods of o in $\widetilde{M}(\delta)$ for all the δ equal to ε on the support $s(\varepsilon)$ glue into a smooth neighborhood.

Since any two $\widetilde{M}(\delta)$ intersect only on their common boundary faces, it can be seen by induction on the rank that $\widetilde{M}$ is the manifold constructed in Proposition II.1.2. Furthermore, the $(\mathbb{Z}/2\mathbb{Z})^{|\Delta|}$ action on $\widetilde{M}$ corresponds to changing the proper signatures δ.

II.1.10 Summary and comments. In this section, we gave two constructions of self-gluing several copies of a manifold M with corners into a closed manifold $\widetilde{M}$. The self-gluing depends on the choice of a partition of the set $\mathcal{H}_M$ of boundary hypersurfaces. When the partition consists of one subset, M is a manifold with boundary, and the self-gluing is the familiar doubling of a manifold with boundary into a closed manifold. The first method is inductive and depends inductively on the number of subsets in $\mathcal{H}_M$. The second method is modeled on obtaining $\mathbb{R}^n$ from the self-gluing of $\mathbb{R}^n_{\geq 0}$ and applies when the number of subsets in the partition of $\mathcal{H}_M$ is equal to the rank of M.

II.2 The Oshima compactification of G/K

In [Os1], Oshima constructs a smooth analytic compactification of $X = G/K$ that contains 2^r open orbits isomorphic to G/K, where $r = rk_{\mathbb{R}}(G)$. We shall first construct it as an application of (II.1.8), using the fact that the maximal Satake compactification of X is a manifold with corners (Proposition I.19.27), and then we recall Oshima's construction.

II.2.1 In order to apply (II.1.8) to G/K, we have to exhibit the structure of a manifold with corners on the maximal Satake compactification $\overline{X}^S_{\max}$.

Let $P = MAN$ be the minimal standard parabolic subgroup, P^- the opposite standard minimal parabolic subgroup, and N^- the unipotent radical of P^-. Then

$$X = N^- \cdot A \text{ and } N^- \cdot \overline{A} \text{ is an open chart in } \overline{X}^S_{\max}. \qquad \text{(II.2.1)}$$

Here $\overline{A} = R^{\Delta}_{\geq 0}$ with the identification

$$a \mapsto (a^{-\alpha_1}, \ldots, \alpha^{-\alpha_r}), \quad (\Delta = \{\alpha_1, \ldots, \alpha_r\}) \qquad \text{(II.2.2)}$$

of A with $\mathbb{R}^{\Delta}_{>0}$. (See Propositions I.19.20 and I.19.27 for proofs of these assertions.)

The orbits of G in $\overline{X}^S_{\max}$ are the subspaces

$$O_J = G/K_J A_J N_J, \tag{II.2.3}$$

where $J \subset \Delta = \Delta(\mathfrak{g}, \mathfrak{a})$, O_J fibers over G/P_J, the fibers being the boundary symmetric spaces conjugate to X_J. The subgroup $K_J A_J N_J$ is denoted by Q_J in the following. In the terminology of II.1.1, the O_J are open boundary faces and their closures are the boundary faces. $\overline{O}_J$ has codimension $rk(X) - |J|$. The boundary hypersurfaces are the sets

$$\overline{O}_{(\alpha)} = \overline{O}_{\Delta-\{\alpha\}} \qquad (\alpha \in \Delta). \tag{II.2.4}$$

These boundary hypersurfaces are embedded and the condition II.1.1 trivially fulfilled by taking $\mathcal{H}_\alpha = \overline{O}_{(\alpha)} (\alpha \in \Delta)$.

Let $M = \overline{X}^S_{\max}$. Then Proposition II.1.2 can be applied to obtain a closed manifold $\widetilde{M}$. The space $\widetilde{M}$ is a G-space containing the union of 2^r copies of X and consisting of 3^r orbits of G. By Proposition I.19.27, $\overline{X}^S_{\max}$ is a compact analytic G-space with corners. It follows then from II.1.4 that $\widetilde{M}$ is a closed real analytic manifold.

We shall see later that an easier way to show that $\overline{X}^S_{\max}$ is an analytic G-space with corners is by embedding X into the wonderful compactification of the complexification $\mathbf{X}$ of X. Together with the above discussion, this gives a more direct construction of the Oshima compactification $\overline{X}^O$.

II.2.2 We give here briefly Oshima's definition of $\overline{X}^O$, which is isomorphic to $\widetilde{M}$ (see Remark II.2.3 below). Given $t \in \mathbb{R}^{\Delta}$, we let ε_t be the signature $\varepsilon_t(\alpha) = \operatorname{sgn} t^\alpha (\alpha \in \Delta)$, and $s(t)$ will stand for $s(\varepsilon_t)$:

$$s(t) = s(\varepsilon_t) = \{\alpha \in \Delta | t^\alpha \neq 0\}.$$

Let $H^\alpha (\alpha \in \Delta)$ be the basis of $\mathfrak{a}$ dual to Δ. We define a map $a : \mathbb{R}^{\Delta} \to A$ by the rule

$$a(t) = \exp - \sum_{\alpha \in s(t)} \log |t^\alpha| \, H^\alpha. \tag{II.2.5}$$

Oshima defines a quotient

$$\overline{X}^O = G \times \mathbb{R}^{\Delta} / \sim$$

of $G \times \mathbb{R}^{\Delta}$ by the equivalence relation $\sim$:

$$(g, t) \sim (g, t')$$

if

1. $\varepsilon_t = \varepsilon_{t'}$,
2. $g \cdot a(t) \cdot Q_{s(t)} = g' \cdot a(t').Q_{s(t')}$ (Note that $s(t) = s(t')$ in view of (1)).

The G-action is defined by left translations on the first factor. Oshima shows that $\overline{X}^O$ is a compact G-space, into which $N^- \times \mathbb{R}^\Delta$ maps bijectively on an open chart. $\overline{X}^O$ consists of 3^r orbits, 2^r being copies of X. The hard point, however, is to prove that $\overline{X}^O$ is an analytic G-space. One has to show that the infinitesimal actions of $\mathfrak{g}$ on the orbits match to define analytic vector fields on $\overline{X}^O$, and then use one of the fundamental theorems of the original Lie theory. We refer to [Os1] for details.

Remark II.2.3 By Proposition I.19.27 and §II.2.1, the maximal Satake compactification $\overline{X}^S_{\max}$ can be self-glued into a closed real analytic manifold $\widetilde{M}$. Since $\overline{X}^S_{\max} = \widetilde{X}^S_{\max}$ by Proposition I.19.30 and $\overline{X}^O$ is the union of 2^r-copies of $\widetilde{X}^S_{\max}$ and the boundary faces fit together as in the self-gluing, it follows that $\overline{X}^O$ is the self-gluing $\widetilde{M}$ of 2^r-copies of $\overline{X}^S_{\max}$.

Remark II.2.4 Oshima has generalized this construction in [Os1] for Riemannian symmetric spaces to the case of semisimple symmetric spaces in [Os2]. Specifically, for any semisimple symmetric space G/H of rank r, there exists a closed real analytic manifold $\overline{G/H}^O$ containing the union of 2^r-copies of G/H as an open dense subset. The closure of each G/H in $\overline{G/H}^O$ is a real analytic manifold with corners and can be considered as the analogue of the maximal Satake compactification. This procedure can be reversed if we have such a compactification of G/H similar to the maximal Satake compactification to start with, i.e., $\overline{G/H}^O$ can be obtained from self-gluing the compactification as in the case of Riemannian symmetric spaces discussed above.

Also as in the case of the Riemannian symmetric space case, the compactification of G/H as a real analytic manifold with corners can be achieved by mapping G/H into the real locus of the wonderful compactification of $\mathbf{G}/\mathbf{H}$.

Remark II.2.5 The self-gluing procedure can be generalized to the situation in which different pieces can be different as long as their boundaries can be patched up. A typical application is to glue up the Oshima–Sekiguchi compactification from compactifications of the Riemannian symmetric space X and its associated semisimple symmetric spaces X_ε. Since this will not be pursued in this book in detail, we will briefly outline the procedure. The key problem is the marking of all boundary faces so that they can match and be glued together in one step.

In the notation of [OsS1], for each proper signature ε, let $\overline{X_\varepsilon}$ be the analogue of the maximal Satake compactification. This can either be constructed using embedding (or immersion) into the real locus of the wonderful compactification of $X_{\mathbb{C}}$, or constructed directly by adding all the boundary components of appropriate classes of parabolic subgroups. For each ε, take W_ε copies of X_ε and define a marking on the boundary faces of their union. First, we start with the corners. The Riemannian symmetric space X has only one corner C. Fix one copy of X. Then the corners of the other copies $w \cdot X$ are $w \cdot C$, where $w \in W$. For each ε, fix one X_ε and one corner C_ε. Then the W_ε-copies of X_ε are $w_1 X_\varepsilon$, $w_1 \in W_\varepsilon$. The corners of X_ε are $w_2 \cdot C_\varepsilon$, where $w_2 \in W(\varepsilon)$. If we denote the distinguished corner of $w_1 \cdot X_\varepsilon$ by $w_1 \cdot C_\varepsilon$, then the other corners of the piece $w_1 X_\varepsilon$ are $w_2 \cdot (w_1 \cdot C_\varepsilon)$, where $w_2 \in W(\varepsilon)$. Therefore, every corner of the union of the W_ε copies of X_ε is of the form $w_2(w_1 C_\varepsilon)$. By definition, every w in W can be written uniquely as $w_2 w_1$ with $w_2 \in W(\varepsilon)$, $w_1 \in W_\varepsilon$.

In summary, for every signature ε, the corners of the W_ε-copies of X_ε are canonically parametrized by elements of W. A basic point is that the corner C_ε can be identified with the corner C of X. Hence, the W-corners of the W_ε-copies of X_ε can be matched up with the W-corners of the W-copies of X. Other boundary faces of X and X_ε can be similarly marked and matched.

II.2.6 Summary and comments. In this section, we applied the general method of self-gluing in §II.1 to obtain the Oshima compactification $\overline{X}^O$ as a self-gluing of the maximal Satake compactification $\overline{X}^S_{\max}$. In applying this method, the crucial point is to show that $\overline{X}^S_{\max}$ is a compact manifold with corners. Since the closure of flats of X in $\overline{X}^S_{\max}$ is a manifold with corners, a natural approach is to extend this corner structure to the compactification $\overline{X}^S_{\max}$. This was carried out in Proposition I.19.27 by following the method in [Os1]. As will be shown below using the real locus of the wonderful compactification of the complexification $X_{\mathbb{C}}$ of X, it is easy to show that $\overline{X}^S_{\max}$ is a real analytic manifold with corners. This allows us to give a more direct and streamlined construction of the Oshima compactification.

5

Semisimple Symmetric Spaces G/H

Though Riemannian symmetric spaces are the most important class of symmetric spaces, non-Riemannian symmetric spaces are important in themselves and also occur naturally in the study of the former spaces, for example, when we study different real forms of complex semisimple Lie groups.

In §II.3, we study some general results on semisimple symmetric spaces. In §II.4, we construct real forms associated with signatures on simple roots.

II.3 Generalities on semisimple symmetric spaces

II.3.1 A (real) *semisimple symmetric pair* (G, H) or (G, σ) consists of a real semisimple group G, an involutive automorphism σ of G, and a closed subgroup H of finite index in the fixed point set G^σ of σ. We shall always assume that G is of finite index in the set of real points of a linear semisimple connected algebraic group $\mathbf{G}$ defined over $\mathbb{R}$. The quotient G/H is a semisimple symmetric space. We recall that H is always reductive.

The involution σ has a fixed point in the space of maximal compact subgroups of G, hence leaves a maximal compact subgroup of G stable. We fix one, denote it by K, and let θ or θ_K be the associated Cartan involution. It commutes with σ. We have the Cartan decompositions

$$\mathfrak{g} = \mathfrak{k} \oplus \mathfrak{p} = \mathfrak{h} \oplus \mathfrak{q}, \tag{II.3.1}$$

where $\mathfrak{q}$ is the orthogonal complement of $\mathfrak{h}$ in $\mathfrak{g}$ with respect to the Killing form. These spaces are all stable under σ and θ. The restriction of the latter to $\mathfrak{h}$ is a Cartan involution. We use below some known facts about this situation proved in [OsS1] and [OsS2]. Later on, we shall give a more extended review, with references.

II.3.2 It is known that $\mathfrak{g}$ contains Cartan subalgebras $\mathfrak{a}_\iota$ invariant under σ and θ and any two of them are conjugate under $K \cap H$ ([OsS1, 2.4], [Ros]). Then

$$\mathfrak{a}_\theta = \mathfrak{a}_\iota \cap \mathfrak{p},\ \mathfrak{a}_\sigma = \mathfrak{a}_\iota \cap \mathfrak{q} \text{ and } \mathfrak{a}_{(\sigma,\theta)} = \mathfrak{a}_\iota \cap \mathfrak{p} \cap \mathfrak{q} \tag{II.3.2}$$

are maximal abelian subspaces of $\mathfrak{p}$, $\mathfrak{q}$, and $\mathfrak{p} \cap \mathfrak{q}$ respectively. Their dimensions are the θ-rank, σ-rank, and (σ, θ)-rank of $\mathfrak{g}$ respectively. The θ-rank is also the $\mathbb{R}$-rank. For notational simplicity, we write $\mathfrak{a}_o$ for $\mathfrak{a}_{(\sigma,\theta)}$. The nonzero weights of $\mathfrak{a}_o$ in $\mathfrak{g}$ form a root system, to be denoted by $\Phi_o = \Phi(\mathfrak{g}, \mathfrak{a}_o)$ and we have the decomposition

$$\mathfrak{g} = \mathfrak{z}(\mathfrak{a}_o) \oplus \oplus_{\beta\in\Phi_o} \mathfrak{g}_\beta, \text{ where } \mathfrak{g}_\beta = \{x \in \mathfrak{g} \big| [h, x] = \beta(h) \cdot x, (h \in \mathfrak{a}_o)\}. \tag{II.3.3}$$

It is known ([Ros, Thm. 5]) that the Weyl group $W(\Phi_\sigma)$ may be identified with $\mathcal{N}_K(\mathfrak{a}_o)/\mathcal{Z}_K(\mathfrak{a}_o)$. We shall also denote it by W_o or $W(K, \mathfrak{a}_o)$. It may be identified with the image in $GL(\mathfrak{a}_o)$ of the subgroup of $\mathcal{N}_K(\mathfrak{a}_\theta)$ leaving $\mathfrak{a}_o$ stable. We shall have also to consider the subgroup of $W(K, \mathfrak{a}_o)$ to be denoted by $W(K \cap H, \mathfrak{a}_o)$:

$$W(K \cap H, \mathfrak{a}_o) = \mathcal{N}_{K\cap H}(\mathfrak{a}_o)/\mathcal{Z}_{K\cap H}(\mathfrak{a}_o). \tag{II.3.4}$$

We note also that $\mathfrak{a}_\theta$ is a Cartan subalgebra of the symmetric pair (G, K). We denote it by $\mathfrak{a}$. Then $\Phi = \Phi(\mathfrak{g}, \mathfrak{a})$ is as before the root system of the symmetric pair (G, K). We have

$$\mathfrak{h} = \mathfrak{z}(\mathfrak{a}_o)^\sigma \oplus \bigoplus_{\substack{\beta\in\Phi^+ \\ x\in\mathfrak{g}_\beta}} \langle x + \sigma x\rangle \qquad \mathfrak{q} = \mathfrak{z}(\mathfrak{a}_o)^{-\sigma} \oplus \bigoplus_{\substack{\beta>0 \\ x\in\mathfrak{g}_\beta}} \langle x - \sigma(x)\rangle. \tag{II.3.5}$$

If $\mathfrak{m}_o$ is the orthogonal complement of $\mathfrak{a}_o$ in $\mathfrak{z}(\mathfrak{a}_o)$, we have

$$\mathfrak{z}(\mathfrak{a}_\sigma) = \mathfrak{m}_o \oplus \mathfrak{a}_o \tag{II.3.6}$$

and $\mathfrak{m}_o$ is stable under θ and σ.

The product $\sigma' = \sigma \cdot \theta$ is also an involution of G. Write H' for G^σ. We have the decomposition

$$\mathfrak{g} = \mathfrak{h}' \oplus \mathfrak{p}', \mathfrak{h}' = \mathfrak{k} \cap \mathfrak{h} \oplus \mathfrak{p} \cap \mathfrak{q}, \mathfrak{p}' = \mathfrak{k} \cap \mathfrak{q} \oplus \mathfrak{p} \cap \mathfrak{h}; \tag{II.3.7}$$

$\mathfrak{a}_o$ is also a Cartan subalgebra of the Riemannian symmetric pair $(\mathfrak{h}', \mathfrak{k} \cap \mathfrak{h}')$. Note that $\mathfrak{k} \cap \mathfrak{h} = \mathfrak{k} \cap \mathfrak{h}'$, $K \cap H = K \cap H'$.

II.3.3 A semisimple Riemannian symmetric space is a product of irreducible ones and it is always possible to reduce the general case to the irreducible one. We want here to describe a similar reduction.

Let then (G, H) be a symmetric pair. The involution σ permutes the simple factors of G so that (G, H) is a product of the two following basic cases

(a) G is (almost) simple (i.e., the identity component G^o of G is simple modulo its center).
(b) $G = G' \times G'$ permutes the two simple factors G'. There exist then automorphisms μ, ν of G such that $\sigma\big((x, y)\big) = \big(\nu(y), \mu(x)\big)$. Since σ is an involution, we must have $\mu \cdot \nu = 1$. The full fixed point $H = G^\sigma$ is then the set of pairs $\big(x, \nu(x)\big)$. Let p be the map $G' \times G' \to G'$ defined by

$$p\big((x,y)\big) = x \cdot \nu(y)^{-1}. \tag{II.3.8}$$

It is an easy exercise to see that it is constant on the left H-cosets and induces an isomorphism (of manifolds) of G/H onto G'. Moreover

$$p(g \cdot x, h \cdot y) = g \cdot p\big((x,y)\big) \cdot \nu(h)^{-1} \qquad (x, y, g, h \in G'). \tag{II.3.9}$$

In particular the diagonal in G acts by twisted conjugation

$$g \circ x = g \cdot x \cdot \nu(g^{-1}) \qquad (x \in G/H : g \in G').$$

II.3.4 The four basic cases. When we go over to $\mathbf{G}$, these two cases subdivide each into two, depending on whether G or G' are absolutely simple or not, so that we have four basic cases, and any other real semisimple symmetric space is a product of those:

(a1) G in (a) is absolutely simple.
(a2) G is not absolutely simple. It is then a complex simple subgroup $\mathbf{G}'$ viewed as a real group, hence $\mathbf{G} = \mathbf{G}' \times \mathbf{G}'$, where the two factors are permuted by the complex conjugation $g \mapsto \bar{g}$ of $\mathbf{G}$ with respect to G. The latter group may be identified with the set of points $(g, \bar{g})$ $(g \in \mathbf{G}')$.
(b1) the complexification $\mathbf{G}'$ of G' in (b) is simple.
(b2) G' is a complex simple group $\mathbf{G}''$, viewed as a real group. Then

$$\mathbf{G} = \mathbf{G}'' \times \mathbf{G}'' \times \mathbf{G}'' \times \mathbf{G}''. \tag{II.3.10}$$

The complex conjugation of $\mathbf{G}$ with respect to G then permutes the first two factors and the last two, and σ permutes these two pairs.

II.4 Some real forms H_ε of H

II.4.1 Let Φ be a root system in a rational vector space V, and Δ the set of simple roots for some given ordering.

A *signature on* Φ (or an extended signature) is a map $\varepsilon : \Phi \to \{\pm 1, 0\}$ such that

$$\varepsilon(\beta) = \varepsilon(-\beta), \varepsilon(\beta + \gamma) = \varepsilon(\beta) \cdot \varepsilon(\gamma) \text{ if } \beta, \gamma, \beta + \gamma \in \Phi. \tag{II.4.1}$$

The support $s(\varepsilon)$ is again defined as the set of β for which $\varepsilon(\beta) \neq 0$ and ε is *proper* if $s(\varepsilon) = \Phi$. Let $\mathcal{E}(\Phi)$ and $\mathcal{E}^o(\Phi)$ be the sets of signatures and proper signatures respectively. If $\tilde{\varepsilon} \in \mathcal{E}(\Phi)$, then its restriction to Δ is a signature of Δ (proper if and only if $\tilde{\varepsilon}$ is so). Conversely, given $\varepsilon \in \mathcal{E}(\Delta)$, define $\tilde{\varepsilon}$ on Φ by

$$\tilde{\varepsilon}(\beta) = \tilde{\varepsilon}(-\beta) = \prod_{\alpha \in \Delta} \varepsilon(\alpha)^{|n_{\beta\alpha}|} \qquad (\text{ where } \beta = \sum_\alpha n_{\beta\alpha}\alpha), \tag{II.4.2}$$

whence natural bijections $\mathcal{E}(\Delta) \leftrightarrow \mathcal{E}(\Phi)$ and $\mathcal{E}^o(\Delta) \Leftrightarrow \mathcal{E}^o(\Phi)$. Both $\mathcal{E}^o(\Delta)$ and $\varepsilon^o(\Phi)$ are elementary abelian 2-groups, the composition being defined by the product of values, and the bijection $\mathcal{E}^o(\Delta) \leftrightarrow \mathcal{E}^o(\Phi)$ is a group isomorphism. The Weyl group W of Φ operates in a natural way on $\mathcal{E}^o(\Phi)$ by the rule

$$w(\varepsilon)(\beta) = \varepsilon(w^{-1} \cdot \beta) \qquad (\beta \in \Phi). \tag{II.4.3}$$

If $\varepsilon \in \mathcal{E}(\Delta)$ and $\bar{\varepsilon}$ is its extension to Φ, then it is immediately seen that

$$s(\tilde{\varepsilon}) = \langle s(\varepsilon)\rangle \cap \Phi. \tag{II.4.4}$$

II.4.2 Given $\varepsilon \in \mathcal{E}^o(\Delta)$, define a map $\sigma_\varepsilon : \mathfrak{g} \to \mathfrak{g}$ by the rule

$$\sigma_\varepsilon = \sigma \text{ on } \mathfrak{z}(\mathfrak{a}_o), \tag{II.4.5}$$

$$\sigma_\varepsilon(x) = \varepsilon(\beta)\sigma x \qquad (x \in \mathfrak{g}_\beta). \tag{II.4.6}$$

It is a linear bijective map of $\mathfrak{g}$ onto itself of order ≤ 2. It is readily checked to be compatible with the bracket and hence defines an involution of $\mathfrak{g}$. We let $\mathfrak{h}_\varepsilon = \mathfrak{g}^{\sigma_\varepsilon}$.

We have

$$\mathfrak{h}_\varepsilon = \mathfrak{g}^{\sigma_\varepsilon} = \mathfrak{z}(\mathfrak{a}_0) \oplus \bigoplus_{\beta>0, x\in\mathfrak{g}_\beta} \langle x + \varepsilon(\beta)\cdot\sigma x\rangle,$$

and the orthogonal complement of $\mathfrak{g}^{\sigma_\varepsilon}$ in $\mathfrak{g}$, with respect to the Killing form, is

$$\mathfrak{q}^\varepsilon = \mathfrak{z}(\mathfrak{a}_o)^{-\sigma} \oplus \bigoplus_{\beta>0, x\in\mathfrak{g}_\beta} \langle x - \varepsilon(\beta)\sigma_x\rangle.$$

We note that $\mathfrak{a}$ and $\mathfrak{a}_o$ play for the pair $(\mathfrak{g}, \mathfrak{h}_\varepsilon)$ the same role as for $(\mathfrak{g}, \mathfrak{h})$. In particular $(\mathfrak{g}, \mathfrak{h})$ and $(\mathfrak{g}, \mathfrak{h}_\varepsilon)$ have the same $\mathbb{R}$-rank, the σ-rank of $(\mathfrak{g}, \mathfrak{h})$ is equal to the σ_ε-rank of $(\mathfrak{g}, \mathfrak{h}_\varepsilon)$, and similarly the (θ, σ)-rank of $(\mathfrak{g}, \mathfrak{h})$ is the same as the $(\theta, \sigma_\varepsilon)$-rank of $(\mathfrak{g}, \mathfrak{h})$.

II.4.3 Definition of H_ε. We now want to globalize the above. The assumption that G is the full group of real points of a semisimple $\mathbb{R}$-group has not been used so far, but will be from now on.

Let $q : \mathbf{G} \to \operatorname{Ad}\mathfrak{g}_c$ be the adjoint representation of G. Let $\mathbf{A}_o$ (resp. $\mathbf{A}'_o$) be the torus in $\mathbf{G}$ (resp. $\mathbf{G}' = \operatorname{Ad}\mathfrak{g}_c$) with Lie algebra $\mathfrak{a}_{o,c}$. It is $\mathbb{R}$-split; thus, if $d = \dim \mathfrak{a}_o$ we have

$$\mathbf{A}_o = (\mathbb{C}^*)^d, \quad \mathbf{A}_o(\mathbb{R}) = \mathbb{R}^{*^d} = A_o \times (\mathbb{Z}/2\mathbb{Z})^d, \quad \mathbf{A} = e^{i\,\mathfrak{a}_o} \times A_o, \tag{II.4.7}$$

where $A_{o,u} = e^{i\,\mathfrak{a}_o}$ is the biggest compact subgroup of $\mathbf{A}_o$, and ${}_2\mathbf{A}_o = {}_2A_{o,u}$. Similarly for $\mathbf{A}'_o$.

Of course, q defines a surjective morphism of $\mathbf{A}_o$ onto $\mathbf{A}'_o$ that maps $A_{o,u}$ onto $A'_{o,u}$ and ${}_2\mathbf{A}_o$ *into* ${}_2\mathbf{A}'_o$. The kernel of ${}_2\mathbf{A}_o \to {}_2\mathbf{A}'_o$ is the intersection of ${}_2\mathbf{A}_o$ with the center $\mathcal{C}G$ of G, to be denoted by C_2.

The roots are now viewed as rational characters of $\mathbf{A}_o$ or $\mathbf{A}'_o$. Their restrictions to ${}_2\mathbf{A}_o$ or ${}_2\mathbf{A}'_o$ take the values ± 1. To $s \in {}_2\mathbf{A}_o$ we associate the map $\varepsilon(s)$ or $\varepsilon_s : \Phi \to \{\pm 1\}$ defined by

$$\varepsilon_s(\beta) = s^\beta \qquad (\beta \in \Phi_o). \tag{II.4.8}$$

It is clearly a signature and the map $s \mapsto \varepsilon_s$ is a homomorphism of ${}_2\mathbf{A}_o$ into $\mathcal{E}^o(\Phi_o)$ that is equivariant under the Weyl group of Φ_o, operating on ${}_2\mathbf{A}_o$ by inner automorphisms and on $\mathcal{E}^o(\Phi_o)$ by II.4.1(2). It will be called the *signature map*.

The elements of Δ_o form a basis of $\mathfrak{a}_o^*$, hence also define a basis of $X(\mathbf{A}'_o)$. Therefore the map $s \mapsto \varepsilon_s$ is an isomorphism of ${}_2\mathbf{A}'_o$ onto $\mathcal{E}^o(\Phi_o)$.

Fix $\varepsilon \in \mathcal{E}^o(\Phi_o)$, and let $s'_\varepsilon \in {}_2\mathbf{A}_o$ be the unique element such that $\varepsilon(s'_\varepsilon) = \varepsilon$. It also belongs to $A'_{o,u}$, hence we can find $\tilde{s}_\varepsilon \in A_{o,u}$ such that $q(\tilde{s}_\varepsilon) = s'_\varepsilon$. It is defined up to an element of $\mathcal{C}G$. Every $\beta \in \Phi_o$ takes the values ± 1 on $\tilde{s}_\varepsilon$ or s'_ε and hence $\tilde{s}_\varepsilon^2 \in \mathcal{C}G$. Moreover, the complex conjugation $g \mapsto \bar{g}$ of $\mathbf{G}$ with respect to G is the inversion on $A_{o,u}$; hence $\bar{\tilde{s}}_\varepsilon = \tilde{s}_\varepsilon^{-1}$. From this we see that $\operatorname{Int}\tilde{s}_\varepsilon$ leaves G invariant, and so does $\operatorname{Int}\tilde{s}_\varepsilon \circ \sigma$. Since $\tilde{s}_\varepsilon$ is defined modulo $\mathcal{C}G$, the transformation $\operatorname{Ad}\tilde{s}_\varepsilon$ is well-defined by ε. It is immediate that

$$\operatorname{Ad}\tilde{s}_\varepsilon \circ \sigma = \operatorname{Ad} s_\varepsilon \circ \sigma = \sigma_\varepsilon. \tag{II.4.9}$$

We then set

$$H_\varepsilon = G^{\sigma_\varepsilon}. \tag{II.4.10}$$

We claim that it is a real form of H. To see this, choose $u \in A_{o,u}$ such that $u^2 = \tilde{s}_\varepsilon$. We want to prove that $\operatorname{Ad} u(\mathfrak{h}_{\varepsilon,c}) = \mathfrak{h}_c$. First u centralizes $\mathfrak{z}(\mathfrak{a}_o)$. If now $x \in \mathfrak{g}_\beta$, then

$$\operatorname{Ad} u(x + s_\varepsilon(\beta)\cdot\sigma(x)) = u^\beta \cdot x + s_\varepsilon(\beta)u^{-\beta}\sigma(x) = u^\beta(x + s_\varepsilon)(\beta)u^{-2\beta}\cdot\sigma(x).$$

But $u^{-2\beta} = \tilde{s}_\varepsilon^\beta = \varepsilon(\beta)$ and hence

$$\operatorname{Ad} u(x + s(\varepsilon)\sigma x) \in \mathfrak{h}_c.$$

II.4.4 Real forms corresponding to different signatures may be isomorphic. In particular, let $w \in W(K \cap H, \mathfrak{a}_o)$ (see II.3.2(3)) and let n be a representative of w in $\mathcal{N}_{K\cap H}(\mathfrak{a}_o)$. It is easily checked that $\operatorname{In} n \circ \sigma_\varepsilon \circ \operatorname{In} n^{-1} = \sigma_{w(\varepsilon)}$; hence ${}^n H_\varepsilon = H_{w(\varepsilon)}$. Since H_ε contains $\mathcal{Z}_{K\cap H}(\mathfrak{a}_o)$, the conjugate ${}^n H_\varepsilon$ depends only on w, so we can write the previous equality as

$${}^w H_\varepsilon = H_{w(\varepsilon)} \qquad (w \in W)(K \cap H, \mathfrak{a}_o).$$

6

The Real Points of Complex Symmetric Spaces Defined over $\mathbb{R}$

Let (G, H) be a symmetric pair, and $\mathbf{G}, \mathbf{H}$ the complexification of G, H respectively. Then $\mathbf{X} = \mathbf{G}/\mathbf{H}$ is a complex symmetric space. It is an affine homogeneous space defined over $\mathbb{R}$. Its real locus $(\mathbf{G}/\mathbf{H})(\mathbb{R})$ contains G/H, of course, but it is in general the union of more than one but finitely many orbits of G. Our goal is to describe them in terms of quotients G/H_ε. We shall do this via Galois cohomology.

We first review the little needed here from Galois cohomology in §II.5. In §II.6, we use the results in §II.5 to determine the G-orbits in $\mathbf{G}/\mathbf{H}(\mathbb{R})$. In §II.7, we illustrate these results through the example of $\mathbf{G} = \mathrm{SL}(n, \mathbb{C})$ and $\mathbf{H} = \mathrm{SO}(n, \mathbb{C})$.

II.5 Galois Cohomology

In this section, we recall the definition and basic properties of the Galois cohomology of the group C of order 2. For more discussions about the general case, see [Ser1] [BS1].

II.5.1 We have only to consider Galois cohomology with respect to a group $C = \{1, t\}$, $t^2 = 1$, of order 2. Let L be a group on which C operates. We let ${}^t x$ be transform of x by t. By definition,

$$H^0(C; L) = L^C = L^t, \tag{II.5.1}$$

$$Z^1(C; L) = \{x \in L \mid x \cdot {}^t x = 1\}. \tag{II.5.2}$$

If $x \in Z^1$ and $x \in L$, then $x^{-1} \cdot z \cdot {}^t x \in Z^1$; hence L operates on Z^1 by twisted conjugation. Then

$$H^1(C; L) = Z^1(C; L)/H, \tag{II.5.3}$$

H acting by twisted conjugation. If L is commutative, H^1 is a commutative group. If L is not commutative, H^1 is just a set with a distinguished element, the twisted conjugacy class of the identity, the elements of which are also called coboundaries. Call that class 1.

II.5.2 If C operates trivially on L, then twisted conjugacy is ordinary conjugacy and $x \in L$ is a cocycle if and only if $x^2 = 1$. Therefore $H^1(C; L) = {}_2L/\mathrm{In}\, L$ is the set of conjugacy classes of elements of order ≤ 2.

II.5.3 Let H be a subgroup of L stable under C. Then C also operates on the space of left cosets L/H. We are interested in $(L/H)^C$. We consider the sequence

$$1 \to H^C \xrightarrow{i_*^0} L^C \xrightarrow{p_*^0} (L/H)^C \xrightarrow{\delta} H^1(C; H) \xrightarrow{i_*^1} H^1(C; L) \qquad \text{(II.5.4)}$$

where i_*^0, i_*^1 (resp. p_*^0) are the obvious maps defined by inclusion (resp. projection) and δ is to be defined below. By 1.12 in [BS1] or I, 5.3 in [Ser1], this sequence is exact (meaning that the image of a map is the kernel of the next one. It implies the following lemma.

Lemma. *The coboundary operator δ induces a bijection of the set $L^C \backslash (L/H)^C$ of L^C-orbits in $(L/H)^C$ onto* $\ker i_*^1$.

For the sake of completeness, we give the proof. We first define δ. Let $x \in (L/H)^C$. Choose $g \in L$ such that $g \cdot 0 = x$. Then

$${}^t g \cdot 0 = {}^t x = x = g \cdot 0;$$

hence $g^{-1} \cdot {}^t g \in H$. It is obviously a cocycle, the class of which, $[g^{-1} \cdot {}^t g] \in H^1(C; H)$, belongs to $\ker i_*^1$. If $g' \cdot 0 = x$, then $g' = g \cdot h$ $(h \in H)$; hence

$$g'^{-1} \cdot {}^t g' = h^{-1} \cdot g^{-1} \cdot {}^t g \cdot {}^t h$$

and therefore $[g'^{-1} \cdot {}^t g'] = [g^{-1} \cdot {}^t g]$. Their common value is $\delta(x)$ by definition. We have seen that $\delta(x) \in \ker i_x^1$. Conversely, if an element of $H^1(C; H)$ belongs to $\ker i_*^1$, it is represented by a cocycle of the form $g^{-1} \cdot {}^t g$ $(g \in L)$ and we see by the reverse computation that $x = g \cdot 0 \in (L/H^C)$. Therefore δ maps $(L/H)^C$ onto $\ker i_*^1$. There remains to see that $\delta(x) = \delta(y)$ $(x, y \in (L/H)^C)$ if and only if $y \in L^C \cdot x$.

If $y = r \cdot x$ $(r \in L^C)$ and $g \cdot 0 = x$, then $r \cdot g \cdot 0 = y$ and $g^{-1} \cdot {}^t g = (r \cdot g)^{-1} \cdot (r \cdot g)$, hence $\delta(x) = \delta(y)$. Conversely, assume that $\delta(x) = \delta(y)$. Let $g, g' \in L$ be such that $g \cdot 0 = x$, $g' \cdot 0 = y$, hence $\delta(x) = [g^{-1} \cdot {}^t g]$, $\delta(y) = [g'^{-1} \cdot {}^t g']$. There exists then $h \in H$ such that $g^{-1} \cdot {}^t g = h^{-1} \cdot g'^{-1} \cdot g' \cdot {}^t h$, whence

$$g \cdot h^{-1} \cdot g'^{-1} = {}^t(g \cdot h^{-1} \cdot g'^{-1}) \in L^C$$

and $g \cdot h^{-1} \cdot g'^{-1} \cdot y = x$.

II.5.4 We shall be interested in various special cases of the following situation. $\mathbf{L}$ is an algebraic group defined over $\mathbb{R}$, $\mathbf{H}$ a closed $\mathbb{R}$-subgroup. Then $\mathbf{L}/\mathbf{H}$ is defined over $\mathbb{R}$ and we want information on the orbits of L on $(\mathbf{L}/\mathbf{H})(\mathbb{R})$. We then let t be the complex conjugation of $\mathbf{L}$ and $C = Gal(\mathbb{C}/(\mathbb{R})$. Then $\mathbf{L}^C = L$. In this case, (4) reads

$$1 \to \mathbf{H}(\mathbb{R}) \to \mathbf{L}(\mathbb{R}) \to (\mathbf{L}/\mathbf{H})(\mathbb{R}) \to H^1(C;\mathbf{H}) \to H^1(C;\mathbf{L}) \tag{II.5.5}$$

and the lemma provides a canonical bijection

$$\mathbf{L}(\mathbb{R})\backslash(\mathbf{L}/\mathbf{H})(\mathbb{R}) = \ker(H^1(C;\mathbf{H}) \to H^1(C;\mathbf{L})). \tag{II.5.6}$$

We want to make it explicit and relate it to $\mathbb{R}$-forms of $\mathbf{H}$. Let $h \in Z^1\ (C;\mathbf{H})$ and assume it splits in $\mathbf{L}$, i.e., there exists $g \in \mathbf{L}$ such that $h = g^{-1} \cdot {}^t g$. Then

$${}^t(g \cdot 0) = {}^t g \cdot 0 = g \cdot h \cdot 0 = g \cdot 0$$

hence $g \cdot 0$ belongs to $(\mathbf{L}/\mathbf{H})(\mathbb{R})$. Its isotropy group $g \cdot \mathbf{H} \cdot g^{-1}$ is defined over $\mathbb{R}$, hence $g \cdot \mathbf{H} \cdot g^{-1} \cap L$ is a real form of $\mathbf{H}$. Conversely, if $x \in (\mathbf{L}/\mathbf{H})(\mathbb{R})$, then, as recalled in II.5.3, if we write $x = g \cdot 0$ we see that $g^{-1} \cdot {}^t g \in \mathbf{H}$ is an $\mathbf{H}$-cocycle that splits in $\mathbf{L}$.

The $\mathbb{R}$-forms of $\mathbf{H}$ correspond bijectively to the element of $H^1\ (C;\ \mathrm{Aut}\,\mathbf{L})$. Clearly, the $\mathbb{R}$-form attached to $h \in \ker\big(H^1(C;\mathbf{H}) \to H^1(C;\mathbf{L})\big)$ is the one associated with the image of h in $\mathrm{Aut}\,\mathbf{H}$.

Proposition II.5.5 *Let G be a real reductive linear group, K a maximal compact subgroup, and t an involution of G leaving K stable. Let $C = \{1, t\}$. Then the natural map $j : H^1(C;K) \to H^1(C;G)$ is bijective.*

Let θ be the Cartan involution and $G = K \cdot P$ the Cartan decomposition associated with K. Then P is stable under t and t commutes with θ.

Let $z \in Z^1(C;G)$. Then $z \cdot {}^t z = 1$. Write z as $z = k \cdot p$ $(k \in K, p \in P)$. The relation $z \cdot {}^t z = 1$ gives $k \cdot p \cdot {}^t k \cdot {}^t p = 1$, which can be written

$$k \cdot {}^t k \cdot {}^t k^{-1} \cdot p \cdot {}^t k = {}^t p^{-1}.$$

By the uniqueness of the Cartan decomposition, this yields

$$k \cdot {}^t k = 1, \quad {}^t k^{-1} \cdot p \cdot {}^t k = {}^t p^{-1}. \tag{II.5.7}$$

The first equality shows that k is a cocycle. We want to show that $j([k]) = [z]$.

Let $q = \exp(-1/2 \cdot lg\ p)$ be the unique square root of p^{-1} in P. The second relation in (II.5.7) then applies:

$${}^t k^{-1} \cdot q \cdot {}^t k = {}^t q^{-1} \text{ or } k^{-1} \cdot {}^t q \cdot k = q^{-1}, \quad {}^t q \cdot k = k \cdot q^{-1}$$

whence

$${}^t q \cdot k \cdot p \cdot q^{-1} = k \cdot q^{-1} \cdot p \cdot q^{-1} = k,$$

which proves that $[z] = j([k])$, hence j is surjective. Let now $k \in Z^1(C;K)$ and assume it splits in G, i.e., there exist $u \in K$ and $v \in P$ such that

$$k = (u.v)^{-1} \cdot {}^t(u \cdot v) = v^{-1} \cdot u^{-1} \cdot {}^t u \cdot {}^t v$$

hence

$$k \cdot {}^t v^{-1} = (u^{-1} \cdot {}^t u) \cdot ({}^t u^{-1} \cdot u \cdot v^{-1} \cdot u^{-1} \cdot {}^t u),$$

which implies

$$k = u^{-1} \cdot {}^t u, \qquad {}^t v^{-1} = {}^t u^{-1} \cdot u \cdot v^{-1} \cdot u^{-1} \cdot {}^t u,$$

and shows that k splits already in K.

II.5.6 Applications. Let K be a compact Lie group and K_c its complexification. Then $K = K_c(\mathbb{R})$ (by the algebraicity of compact linear groups, or Tannaka duality), and is a maximal compact subgroup of K_c. Therefore II.5.5 implies that

$$j : H^1(C; K) \to H^1(C; K_c) \tag{II.5.8}$$

is bijective if C is generated by the complex conjugation of K_c with respect to K (cf. [BS1, 6.8]). By II.5.2, $H^1(C; K) = {}_2K/\text{In}\, K$.

For an $\mathbb{R}$-group $\mathbf{G}$, the set of isomorphisms of $\mathbb{R}$-forms of $\mathbf{G}$ may be identified with $H^1(C; \text{Aut}\, \mathbf{G})$. Thus

$$\{\mathbb{R}\text{-forms of } K_c\} \Leftrightarrow H^1(C; \text{Aut}\, K_c) = {}_2(\text{Aut}\, K)/\text{Aut}\, K$$

(Cartan's classification).

If G is connected, let T be a maximal torus of K. Then any element of ${}_2K$ is conjugate to one of ${}_2T$ and, as is well known, ${}_2K/\text{In}\, K = {}_2T/W(K, T)$.

II.6 Orbits of G in $(\mathbf{G}/\mathbf{H})(\mathbb{R})$

We now come back to the situation of the earlier sections and want to apply (II.5.4) to our main case of interest, i.e., to determine the G-orbits in the real locus $\mathbf{G}/\mathbf{H}(\mathbb{R})$.

We have therefore to investigate

$$\ker j : H^1(C; \mathbf{H}) \to H^1(C; \mathbf{G}), \tag{II.6.1}$$

where C is generated by the complex conjugation of $\mathbf{G}$ with respect to G, and determine the orbits of G associated with its elements. They will all be of the form G/H_ε. We have first to fix some notation and recall a generalization of the Cartan decomposition $K \cdot A^+ \cdot K$. The G-orbits in $\mathbf{G}/\mathbf{H}(\mathbb{R})$ are listed in Theorem II.6.5.

II.6.1 Let $R_{\mathbb{C}/\mathbb{R}}$ denote restriction of scalars from $\mathbb{C}$ or $\mathbb{R}$. Then $R_{\mathbb{C}/\mathbb{R}}(\mathfrak{g}_c)(\mathbb{R})$ is $\mathfrak{g}_c$ viewed as a real Lie algebra, to be denoted by $\mathfrak{g}_{c,r}$ for simplicity. Similarly, if needed, we let $G_{c,r}$ stand for $R_{\mathbb{C}/\mathbb{R}} G_c(\mathbb{R})$, i.e., G_c viewed as a real group. The real Lie algebra

$$\mathfrak{g}_u = \mathfrak{k} \oplus i\mathfrak{p} \tag{II.6.2}$$

is a compact real form of $\mathfrak{g}_c$, or the Lie algebra of a maximal compact subgroup G_u of G_c. The corresponding Cartan decompositions are

$$\mathfrak{g}_{c,r} = \mathfrak{g}_u \oplus \mathfrak{p}_u \qquad (\mathfrak{p}_u = i\mathfrak{g}_u), \qquad G_{c,r} = G_u \cdot e^{\mathfrak{p}_u}. \tag{II.6.3}$$

We let θ_u be the Cartan involution of $G_{c,r}$ with respect to G_u. It is the complex conjugation of G_c with respect to G_u. Let s_c, θ_c be the automorphisms of $\mathfrak{g}_c$ obtained by extension of scalars from s and θ. The involutions s_c, θ_c, θ_u and the complex conjugation of G_c with respect to G are pairwise commuting automorphisms of $G_{c,r}$. They leave stable the decomposition (2) and the decompositions obtained by complexifications of II.3.1(1). The Cartan involution θ_u leaves H_c stable and induces on H_c, or rather $H_{c,r}$, a Cartan involution with maximal compact subgroup

$$H_u = G_u \cap H_c \text{ with Lie algebra } \mathfrak{h}_u = \mathfrak{k} \cap \mathfrak{h} \oplus i(\mathfrak{h} \cap \mathfrak{p}). \tag{II.6.4}$$

We need the following generalization of the Cartan decomposition.

Lemma II.6.2 *We have $G_u = K \cdot e^{i\mathfrak{a}_o} \cdot H_u$ and $G = K \cdot A_o \cdot H$. In particular, if $A_o = \{1\}$, then $G = K \cdot H$, then $G/H = K/(K \cap H)$, and $G/K = H/(H \cap K)$.*

Proof. The first two equalities are proved in the same way, and the second one is well known (cf., e.g., [Sch, 7.1.3]). For the sake of completeness, we prove the first one. By the Cartan decomposition for the symmetric pair (G_u, K) we have

$$G_u = K \cdot e^{i\mathfrak{p}} = K \cdot e^{i(\mathfrak{p}\cap\mathfrak{q})} \cdot e^{\mathfrak{q}\cap\mathfrak{h}}. \tag{II.6.5}$$

$K \cap H$ leaves all three factors of the right-hand side stable and we have, by the Cartan theory for the pair $(H', K \cap H')$ (see II.3.2) or, equivalently, $(H'_u, K \cap H')$,

$$e^{i(\mathfrak{p}\cap\mathfrak{q})} = \bigcup_{k \in K \cap H'} k \cdot e^{i\mathfrak{a}_o} \cdot k^{-1}; \tag{II.6.6}$$

hence, taking into account the equality $K \cap H' = K \cap H$ (II.3.2),

$$G = K \cdot e^{i(\mathfrak{p}\cap\mathfrak{q})} \cdot (K \cap H) \cdot e^{i(\mathfrak{p}\cap\mathfrak{h})} = K \cdot e^{i\mathfrak{a}_o} \cdot (K \cap H) \cdot e^{i(\mathfrak{p}\cap\mathfrak{h})} = K \cdot e^{i\mathfrak{a}_o} \cdot H_u. \tag{II.6.7}$$

Remark. The lemma and its proof are valid for $\mathfrak{g}$ reductive.

II.6.3 Group inclusions yield the following commutative diagram:

$$\begin{array}{ccc} H^1(C; H_u) & \xrightarrow{j_u} & H^1(C; G_u) \\ \downarrow & & \downarrow \\ H^1(C; \mathbf{H}) & \xrightarrow{j} & H^1(C; \mathbf{G}) \end{array} \tag{II.6.8}$$

where C is as above. By II.5.5, the two vertical arrows are isomorphisms. This reduces the determination of $\ker j$ to that of $\ker j_u$.

We use the notation and definitions of II.3.2 and II.4.3. We have $H^1(C; {}_2\mathbf{A}_o) = {}_2\mathbf{A}_o$ since ${}_2\mathbf{A}_o$ is commutative and consists of elements of order ≤ 2, and t acts trivially on it. We consider the maps

$$ {}_2\mathbf{A}_o \xrightarrow{m_1} H^1(C; K \cap H) \xrightarrow{m_2} H^1(C; H_u) \text{ and let } m = m_2 \circ m_1. \qquad \text{(II.6.9)} $$

Proposition II.6.4 *We keep the previous notation.*

(i) m_1 induces an injective map of ${}_2\mathbf{A}_o/W(K \cap H, \mathfrak{a}_o)$ into $H^1(C; K \cap H)$.
(ii) m_2 is an isomorphism of $\operatorname{Im} m_1$ *onto* $\ker j_u$.

(i) By II.6.4, $H^1(C; K \cap H)$ is equal to ${}_2(H \cap K)$ modulo conjugacy. Hence $\operatorname{Im} m_1$ is equal to ${}_2\mathbf{A}_o$ modulo conjugacy in $H \cap K$. But the latter is on ${}_2\mathbf{A}_o$ the same as the conjugacy under $W(K \cap H, \mathfrak{a}_o)$, whence (i).
(ii) (In the Riemannian case, $K \cap H$ and H_u are both equal to K, and there is nothing to prove.) We have to show that (a) m_2 maps $\operatorname{Im} m_1$ onto $\ker j_u$; (b) m_2 is injective.

Proof of (a): The map ${}_2\mathbf{A}_o \to H^1(C; G_u)$ factors through $H^1(C; {}_2\mathbf{A}_o) \to H^1(C; A_{o,u})$. However, this last group is zero: since t acts by inversion on $A_{o,u}$, every element is a cocycle. Moreover, every element $s \in A_{o,u}$ has (at least) one square root, say u, in $A_{o,u}$; therefore $s = u^2 = u \cdot {}^t(u^{-1})$ is a coboundary. This shows that $\operatorname{Im} m_2 \subset \ker j_u$.

Let now $h \in H^1(C; H_u)$ and assume that $j_u(h) = 0$. By II.6.2, there exist then $k \in K$, $z \in A_{o,u}$, and $y \in H_u$ such that

$$ h = (k \cdot z \cdot y)^{-1} \cdot {}^t(k \cdot z \cdot y) = y^{-1} \cdot z^{-1} \cdot k^{-1} \cdot k \cdot z^{-1} \cdot {}^t y = y^{-1} \cdot z^{-2} \cdot {}^t y, $$

and therefore $h = z^{-2}$ in $H^1(C; H_u)$. We have $z^{-2} \in A_{o,u} \cap H_u$. The Cartan involution of G_u with respect to H_u is the inversion on $A_{o,u}$ and the identity on H_u; hence $z^{-2} \in {}_2A_o$, which shows that $\ker j_2 \subset \operatorname{Im} m_2$ and concludes the proof of (a).

Proof of (b). Let $z \in m_1({}_2\mathbf{A}_o)$ and assume it splits in H_u, i.e., that there exists $y \in H_u$ such that $z = y^{-1} \cdot {}^t y$. By the Cartan decomposition in H_u with respect to $K \cap H$, we can write

$$ y = u \cdot v \qquad (u \in K \cap H, v \in e^{i(\mathfrak{p} \cap \mathfrak{h})}); $$

hence $z = v^{-1} \cdot u^{-1} \cdot u \cdot {}^t v = v^{-2}$. This implies that $z \in e^{i(\mathfrak{p} \cap \mathfrak{h})}$; therefore, to prove (b) it suffices to show that

$$ {}_2\mathbf{A}_o \cap e^{i(\mathfrak{p} \cap \mathfrak{h})} = \{1\}. \qquad \text{(II.6.10)} $$

We use the notation of II.3.2 and set $\mathfrak{b} = \mathfrak{a} \cap \mathfrak{h}$. Then $\mathfrak{b}$ is a Cartan subalgebra of the Riemannian symmetric pair $(\mathfrak{h}, \mathfrak{h} \cap \mathfrak{k})$. We have $\mathfrak{a} = \mathfrak{a}_o \oplus \mathfrak{b}$. Let $\mathbf{B}$ be the maximal

$\mathbb{R}$-split torus of $\mathbf{H}$ with Lie algebra $\mathfrak{b}_c$. The maximal $\mathbb{R}$-split torus $\mathbf{A}$ of $\mathbf{G}$ with Lie algebra $\mathfrak{a}_c$ is the direct product of $\mathbf{A}_o$ and $\mathbf{B}$. In particular

$$ {}_2\mathbf{A} = {}_2\mathbf{A}_o \times {}_2B. \tag{II.6.11}$$

By the Cartan theory in H_u, we have

$$ e^{i(\mathfrak{p}\cap\mathfrak{h})} = \bigcup_{k\in K\cap H} {}^kB_u. \tag{II.6.12}$$

In order to prove (II.6.9), it suffices to establish

$$ {}_2\mathbf{A}_o \cap {}^kB_u = \{1\} \text{ for every } k \in K \cap H. \tag{II.6.13}$$

Assume, contrary to (II.6.13), that there exist $k \in K \cap H$ and $z \in {}_2\mathbf{A}_o$, $z \neq 1$, such that

$$ z \in {}^kB_u, \text{ hence } z \in {}^k_2B.$$

Let $Z = \mathcal{Z}_H(z)$. It is reductive. Since $z \in K$, the group Z is stable under θ; hence the restriction of θ to Z is a Cartan involution of Z, whence the Cartan decompositions

$$ Z = (K \cap Z) \cdot e^{\mathfrak{p}\cap\mathfrak{z}}, \qquad Z_u = (K \cap Z) \cdot e^{i(\mathfrak{p}\cap\mathfrak{z})}. \tag{II.6.14}$$

The group B_u and ${}^kB_u$ are Cartan subgroups of the symmetric pair $(H_u, K \cap H)$ hence a fortiori of the symmetric pair $(Z_u, K \cap Z)$. There exists therefore $q \in K \cap Z$ such that ${}^{qk}B_u = B_u$. Since q centralizes z, it follows that $z \in B_u$, hence that ${}_2\mathbf{A}_o \cap {}_2B \neq \{1\}$, but this contradicts (II.6.11).

This proves (II.6.14), hence also (II.6.12) and (b).

Theorem II.6.5 *We keep the previous notation:*

(i) There is a natural bijection

$$ \mu : {}_2\mathbf{A}_o/W(K \cap H, \mathfrak{a}_o) = G\backslash(\mathbf{G}/\mathbf{H})(\mathbb{R}). \tag{II.6.15}$$

(ii) let $s \in {}_2\mathbf{A}_o$ and let $\varepsilon_s \in \mathcal{E}^o(\Phi_o)$ be its image under the signature map (II.4.3).

Then the orbit of G assigned to s by μ is isomorphic to G/H_{ε_s}.

Let ν be the composition of $m : {}_2\mathbf{A}_o \to H^1C; H_u)$ with the isomorphism $H^1(C; H_u) \tilde{\to} H^1(C; \mathbf{G})$ of (II.5.5). Using the diagram (II.6.8) we get a bijection

$$ {}_2\mathbf{A}_o/W(K \cap H, \mathfrak{a}_o) \leftrightarrow \ker\big(H^1(C; \mathbf{H}) \to H^1(C; \mathbf{G})\big).$$

Then μ is the composition of ν and the inverse of the bijection II.5.5.

(ii) Let $s \in {}_2\mathbf{A}_o$. Then $m(s)$ splits in $\mathbf{G}$. But in proving (a) in II.6.4(ii), we pointed out that it already splits in $A_{o,u}$, namely $s = u \cdot {}^tu^{-1}$ with $u \in A_{o,u}$ a square root of s. According to the recipe given in II.5.3, the orbit assigned to $\mu(s)$ is that of $u \cdot o$, where o is the fixed point of $\mathbf{H}$. Therefore we have to see that the intersection of G with the stability group $u^{-1} \cdot \mathbf{H} \cdot u$ of $u \cdot 0$ is H_ε, but the computation made at the end of II.4.3 shows this.

Remark. Let $\pi : \mathbf{G} \to \mathbf{G}/\mathbf{H}$ be the natural projection. Clearly, the inverse image of $\mathbf{G}/\mathbf{H}(\mathbb{R})$ is the set

$$E = \{g \in G, {}^t g \in g \cdot \mathbf{H}\},$$

and the real orbits of G are the orbits of the points $g \cdot o$, where $g \in E$.

The set E contains the finite subgroup ${}_4A_{o,u}$ of elements of order ≤ 4 in $A_{o,u'}$, which is generated by the square roots of the elements in ${}_2\mathbf{A} = A_{o,u} \cap H$. The gist of II.6.5 is that already $G \cdot {}_4A_{o,u} \cdot o$ is the union of all orbits of G in $(\mathbf{G}/\mathbf{H})(\mathbb{R})$. More precisely, $\mu(s)$ is equal to $G \cdot u_s \cdot o$, where u_s is a square root of s in ${}_4A_{o,u}$.

II.6.6 A free action of $C_2 = \mathcal{C}G \cap \mathbf{A}_o$ on G/H. To define it, it is convenient to use Cartan's way of looking at symmetric spaces as subspaces of the group (see e.g. [Bo11] IV, §3)). For simplicity, let us write g^* for ${}^\sigma(g^{-1})$. The map $g \mapsto g^*$ is an antivolution of $\mathbf{G}$,

$$g^{**} = g, (g \cdot h)^* = h^* \cdot g^*, \tag{II.6.16}$$

and $H = \{g \in \mathbf{G} | g \cdot g^* = 1\}$. Let $\varphi : \mathbf{G} \to \mathbf{G}$ be the map defined by $\varphi(g) = g \cdot g^*$. This is an $\mathbb{R}$-morphism of varieties of $\mathbf{G}$ into itself that induces an $\mathbb{R}$-isomorphism of $\mathbf{G}/\mathbf{H}$ onto a subvariety $\mathbf{M}$ of $\mathbf{G}$ that is pointwise fixed under $*$. The map φ is G-equivariant if we let $\mathbf{G}$ act on $\mathbf{G}/\mathbf{H}$ by left translations and on $\mathbf{M}$ by $g \circ m = g \cdot m \cdot g^*$. In particular, $\mathbf{H}$ acts by inner automorphisms and $\mathbf{M}$ acts on itself by the "transvections" $g \circ m = g \cdot m \cdot g$ $(g, m \in \mathbf{M})$.

Lemma. $\mathbf{M}$ *is stable under right translations (in* $\mathbf{G}$*) by* C_2*. This defines a free action by* C_2 *on* $\mathbf{M}$ *that commutes with* $\mathbf{G}$ *and leaves* $M = \mathbf{M}(\mathbb{R})$ *stable.*

The second assertion is obvious, once the first one is proved. Let $c \in C_2$. There exists $u \in A_{o,u}$ such that $u^2 = c$. We have $u^* = u$, whence $c = u \cdot u^*$ and

$$g \cdot g^* \cdot c = g \cdot c \cdot g^* = g \cdot u \cdot u^* \cdot g^* = g \cdot u \cdot (g \cdot u)^* \in M.$$

Remark. Although we shall not need it, it is worth noting that $\mathbf{M}$ is (Zariski) closed in $\mathbf{G}$. Moreover, the full fixed-point set $\mathbf{Q}$ of $x \mapsto x^*$ consists of finitely many connected components, acted upon transitively by $\mathbf{G}$ via the $*$ action, one of which is $\mathbf{M}$.

See [Ri1], Prop. 9.1. This is in fact valid whether $\mathbf{G}$ and $\mathbf{H}$ are defined over $\mathbb{R}$ or not, or more generally over an algebraically closed ground field of characteristic not 2.

In our situation, it implies that Q and M are closed in G, in the ordinary topology, and are finite unions of closed orbits of G.

II.6.7 The group C_2 is the kernel of the signature map (II.4.3). Its elements are fixed under $W(K \cap H, \mathfrak{a}_o)$. The signature map is constant on the left cosets of C_2 in

${}_2\mathbf{A}$. Each such coset provides $|C_2|$ orbits isomorphic to G/H, where ε is the image of the coset under the signature map, which are permuted in a simply transitive manner by C_2. Thus, if $C_2 \neq \{1\}$, only some of the H_ε appear as isotropy groups of *real* orbits. Let now ε not be the image of the signature map. Then H_ε, which is defined by II.4.3(10), is the isotropy group of a G-orbit isomorphic to G/H_ε, but which does not contain any real point, though it is stable under complex conjugation. To see this, let $\tilde{s}_\varepsilon$ be as defined in II.4.3. Then $q(s_\varepsilon^2) = 1$; hence $\tilde{s}_\varepsilon^2 = c \in C_2$. We have

$$\overline{\tilde{s}}_\varepsilon = \tilde{s}_\varepsilon^{-1} = \tilde{s}_\varepsilon \cdot c.$$

The isotropy group in G of $\tilde{s}_\varepsilon \cdot o$ is H_ε by the definition (II.4.10), and $\overline{\tilde{s}_\varepsilon \cdot o} = \tilde{s}_\varepsilon \cdot o \cdot c$. Thus $G \cdot \tilde{s}_\varepsilon \cdot o \cong G/H_\varepsilon$. This orbit is stable under c, and c transforms each point onto its complex conjugate.

II.6.8 As in II.4.3, let $\mathbf{G}' = \mathrm{Ad}\,\mathfrak{g}_c$ and $q : \mathbf{G} \to \mathbf{G}'$ the canonical morphism. Let us write $\mathbf{H}'$ for the fixed-point set of σ in $\mathbf{G}'$. The homomorphism $\mathbf{H} \to \mathbf{H}'$ has $\mathbf{H} \cap C\mathbf{G}$ as its kernel and $q(H)$ may have finite index in H'. The morphism q induces a morphism $q_\mathbb{R} : (\mathbf{G}/\mathbf{H})(\mathbb{R}) \to (\mathbf{G}'/\mathbf{H}')(\mathbb{R})$. The image of $q_\mathbb{R}$ is open, but $q_\mathbb{R}$ is not necessarily surjective. The image of $q_\mathbb{R}$ is isomorphic to $(\mathbf{G}/\mathbf{H})(\mathbb{R})/C_2$, where C_2 acts by II.6.7. The C_2-orbits assigned to the elements of a C_2-coset map onto the orbit isomorphic to G'/H'_ε. The orbits G/H_ε without real points consist of pairs of complex conjugate points m and $m \cdot c$ for some $c \in C_2$, which map therefore onto real points belonging to an orbit isomorphic to G'/H'_ε.

II.6.9 As a first example take $G = \mathrm{SL}_2(\mathbb{R})$. It will be convenient to let $\mathbf{L}_2$ be the subgroup of $\mathbf{GL}_2(\mathbb{C})$ consisting of matrices of determinant ± 1 (we do not know of any standard notation).

Let then $G = \mathrm{SL}_2(\mathbb{R})$, $H = K = \mathrm{SO}_2$, and let $\sigma = \theta$ be the standard Cartan involution. As usual, take as a basis of $\mathfrak{sl}_2(\mathbb{R})$

$$h = \begin{pmatrix} 1 & 0 \\ 0 & -1 \end{pmatrix}, \qquad e = \begin{pmatrix} 0 & 1 \\ 0 & 0 \end{pmatrix}, \qquad f = \begin{pmatrix} 0 & 0 \\ 1 & 0 \end{pmatrix}.$$

Then θ_ε maps (h, e, f) into $(-h, -f, -e)$ and $e - f$ spans the Lie algebra of SO_2. Let ε be the signature that is -1 on the unique simple root. Then θ_ε maps (h, e, f) onto $(-h, f, e)$. The Lie algebra of $\mathfrak{k}_\varepsilon$ is spanned by $e + f$, which generates the group of hyperbolic rotations

$$K_\varepsilon^o = \begin{pmatrix} \cosh t & \sinh t \\ \sinh t & \cosh t \end{pmatrix} \qquad (t \in \mathbb{R}).$$

The group K_ε is generated by K_ε^o and $-\mathrm{Id}$. We let of course $\mathfrak{a} = \mathfrak{a}_o$ be the Lie algebra spanned by h and $\mathbf{A}_o$ the torus of diagonal matrices (x, x^{-1}) $(x \in \mathbb{C}^*)$.

We view $\mathbf{G}/\mathbf{H}$ as the space of complex quadratic forms of determinant one, acted upon in the usual manner. In our case, $C_2 = C G = {}_2\mathbf{A}_o$. Let also $c = -\mathrm{Id}$; then II.6.5 shows the existence of two orbits of G in $(\mathbf{G}/\mathbf{H})(\mathbb{R})$, isomorphic to G/K, the spaces of positive and negative definite real quadratic forms of determinant 1, translates of one another by c.

There are no real indefinite quadratic forms of determinant one, but there are complex ones, e.g., $F = i(x^2 - y^2)$. Its stability group is K_ε. We have $\overline{F} = c \cdot F$, hence right translation by c transforms a point into its complex conjugate, as in II.6.7.

We view $\mathbf{G}'$ as the quotient $\mathbf{GL}_2(\mathbb{C})/\mathbb{C}^*$. This presentation shows that $G'(\mathbb{R})$ is the image of $L_2(\mathbb{R})$ and has two connected components, one of which is $q(G)$. Similarly, the standard maximal compact subgroup K' of G' is the image of the full orthogonal group and has two connected components.

q:$\mathbf{G}/\mathbf{H} \to \mathbf{G}'/\mathbf{H}'$ is a twofold covering and $q_\mathbb{R}$ maps $(\mathbf{G}/\mathbf{H})(\mathbb{R})$ onto one orbit of G' in $(\mathbf{G}'/\mathbf{H}')(\mathbb{R})$, isomorphic to the upper half-plane. The other orbit is G'/H'_ε, and G/H_ε is a twofold covering of it.

Note that G/K_ε may be identified with G modulo the diagonal matrices and is an open cylinder. It is a twofold covering of G'/K'_ε, which is diffeomorphic to an open Möbius band.

II.6.10 In II.6.6–II.6.8 we have carried the discussion in general, but it should be noted that C_2 is quite small in the basic cases (II.3.4). It is at most of order 2, except when $\mathbf{G}$ is simply connected of type D_{4n}, or is a product of two such groups, in which case it can be of order four.

II.7 Examples

II.7.1 In these first subsections, we consider the case $G = \mathrm{SL}_n(\mathbb{R})$, $H = K = \mathrm{SO}_n$, and $\sigma = \theta$ is the standard Cartan involution, or also where G is the group of real points of the adjoint group of $\mathbf{SL}_n(\mathbb{C})$.

G splits over $\mathbb{R}$ and $\mathfrak{a}_o$ is a Cartan subalgebra of $\mathfrak{g}$. We choose of course $\mathfrak{a}_o$ to be the algebra of real diagonal matrices with trace zero. Then $\mathbf{A}_o$ is the group of complex diagonal matrices of determinant one, and ${}_2\mathbf{A}_o$ its subgroup of matrices with entries ± 1. Let x_i be the rational character of $\mathbf{A}_o$ that associates to each element its ith coefficient. The root system $\Phi = \Phi(G, A_o)$ is of type A_{n-1}, and consists of the roots $x_i - x_j$ $(i \neq j)$. As usual, we choose as set Δ of simple roots the forms $\alpha_i = x_i - x_{i+1}$ $(1 \leq i \leq n-1)$. The Weyl group W is the group of permutations of the diagonal entries.

II.7.2 Let $\tilde{\mathcal{E}}$ be the set of proper signatures on $\{x_1, \ldots, x_n\}$ and $\mu : \tilde{\mathcal{E}} \to \mathcal{E}^o(\Phi)$ the map that associates to $\delta \in \tilde{\mathcal{E}}$ the signature defined by

$$\varepsilon(x_i - x_j) = \varepsilon(x_i) \cdot \varepsilon(x_j) \tag{II.7.1}$$

(it is immediate that it is indeed a signature). The map μ is a surjective homomorphism with kernel the two signatures that are constant on the x_i.

We also view the x_i as coordinates on $\mathbb{R}^n$ or $\mathbb{C}^n$, in which case they are of course allowed to take the value zero. To $\delta \in \tilde{\mathcal{E}}$ we associate the quadratic form $F(\delta) := \sum_1^n \delta(x_i) \cdot x_i^2$. If δ, δ' form a fiber of μ, then $\delta = \delta'$ and $F(\delta) = -F(\delta')$. If n is odd, one of them has determinant one. If n is even, both have the same determinant, which is one if and only if the number of x_i on which δ takes the value 1 (or -1) is even.

II.7.3 Recall (II.4.3) the signature map $s \mapsto \varepsilon_s$ from ${}_2\mathbf{A}_o$ to $\mathcal{E}^o(\Phi)$, where

$$\varepsilon_s(\beta) = s^\beta \qquad (\beta \in \Phi).$$

By a slight abuse of notation, we let $F(\varepsilon_s) = F(\delta)$ where $\delta \in \nu^{-1}(\varepsilon_s)$. It is defined only up to sign, which does not change its isotropy group. We claim that the latter is equal to H_{ε_s}. The computation is as in II.4.3: let $u \in A_{o,u}$ be such that $u^{-2} = s = u^2$. Let F_o be the form $\sum x_i^2$. Then $u \cdot F_o \cdot u = F(\varepsilon_s)$, and its isotropy group in $\mathbf{G}$ is $u^{-1} \cdot \mathbf{K} \cdot u$. By the computation in II.4.3, its intersection with G is K_{ε_s}.

Let n be odd. Then the signature map is a bijection. The cosets ${}_2\mathbf{A}_o/W$ or $\mathcal{E}^o(\Phi)/W$ are represented by the quadratic forms $\mathrm{SO}(p,q)$ $(0 \leq p \leq [n/2])$ and we get

$$G\backslash(\mathbf{G}/\mathbf{H})(\mathbb{R}) \simeq \coprod_{0 \leq p \leq [n/2]} G/\mathrm{SO}(p,q).$$

Let n be even. The signature map has kernel the intersection C_2 of ${}_2\mathbf{A}_o$ with the center of G, consisting of $\pm\mathrm{Id}$. The image of the signature map has index 2 in $\mathcal{E}_o(\Phi)$ and the forms associated with its image are those with an even number of positive (or negative) squares. Modulo W, they are the forms $F(p,q)$ with an even number p of positive squares. For each of them, there are two orbits of G isomorphic to $G/\mathrm{SO}(p,q)$, translates of one another by $-\mathrm{Id}$.

As in (II.6.7), one can construct orbits of G in $\mathbf{G}/\mathbf{H}$ of the form $G/\mathrm{SO}(p,q)$ with p odd consisting of pairs of complex conjugate points permuted by C_2.

II.7.4 Let $\mathbf{G}' = \mathbf{GL}_n(\mathbb{C})/\mathbb{C}^*$ be the adjoint group of $\mathbf{SL}_n(\mathbf{C})$, K' the standard maximal compact subgroup of G'. The quotient $\mathbf{G}'/\mathbf{K}'$ is the space of nondegenerate quadrics in $\mathbb{C}^n$. The group G' is the group $\mathbf{L}_n(\mathbb{R})$ of $n \times n$ real matrices with determinant ± 1 and K' is the full orthogonal group O_n of orthogonal matrices of determinant ± 1 (in both cases modulo the center if n is even).

If n is odd, the parametrization of the real orbits is the same as for SL_n. If n is even, the real orbits are now $G'/O(p,q)$ $(0 \leq p \leq n/2)$, each occurring once. The quotient G/K is a covering of order 2 of G'/K', under the free action of C_2 (II.6.6). Each pair of real orbits projects onto one orbit $G'/O(p,q)$ (p even, $p \leq n/2$). The orbits $G'/O(p,q)$ for p odd are the images of the orbits of type $G/\mathrm{SO}(p,q)$ (p odd) consisting of pairs of complex conjugate points.

7

The DeConcini–Procesi Compactification of a Complex Symmetric Space and Its Real Points

The wonderful compactification $\overline{\mathbf{X}}^W$ of a complex symmetric space $\mathbf{X} = \mathbf{G}/\mathbf{H}$ was motivated by problems in enumerative algebraic geometry and is a generalization of the variety of complete quadrics (see [DGMP]). Though its construction is similar to the Satake compactifications of Riemannian symmetric spaces, its relations to compactifications of symmetric spaces (Riemannian or semisimple) have not been studied before.

In this chapter, we establish such connections by identifying its real locus $\overline{\mathbf{X}}^W(\mathbb{R})$ in terms of compactifications of symmetric spaces when $\mathbf{X} = \mathbf{G}/\mathbf{H}$ is defined over $\mathbb{R}$. Such an identification can be used to show that the maximal Satake compactification $\overline{X}^S_{\max}$ is a real analytic manifold with corners, which can be used in §II.2 to give a simple construction of the Oshima compactification $\overline{X}^O$. It can also be used to explain the Oshima–Sekiguchi compactification $\overline{X}^{OS}$ in the next chapter.

In §II.8, we recall some general results on relations of different commutative subalgebras and associated root systems. Then we define Λ-relevant parabolic subgroups, which describe the geometry at infinity (or rather boundary components) of semisimple symmetric spaces. In §II.9, we recall the wonderful compactification $\overline{\mathbf{X}}^W$ of a complex symmetric space $\mathbf{X}$. In §II.10, we describe the real locus $\overline{\mathbf{X}}^W(\mathbb{R})$ when $\mathbf{X}$ and $\overline{\mathbf{X}}^W$ are defined over $\mathbb{R}$. As an application, we show that the maximal Satake compactification $\overline{X}^S_{\max}$ is a real analytic manifold with corners, as mentioned earlier.

II.8 Generalities on semisimple symmetric spaces

II.8.1 We now come back to §II.3, and consider again the four commutative algebras $\mathfrak{a}_\iota$, $\mathfrak{a}_\theta$, $\mathfrak{a}_\sigma$, and $\mathfrak{a}_{\sigma,\theta}$. The last one was denoted by $\mathfrak{a}_o$ above, for brevity, but now we restore the notation $\mathfrak{a}_{\sigma,\theta}$.

We have a commutative diagram

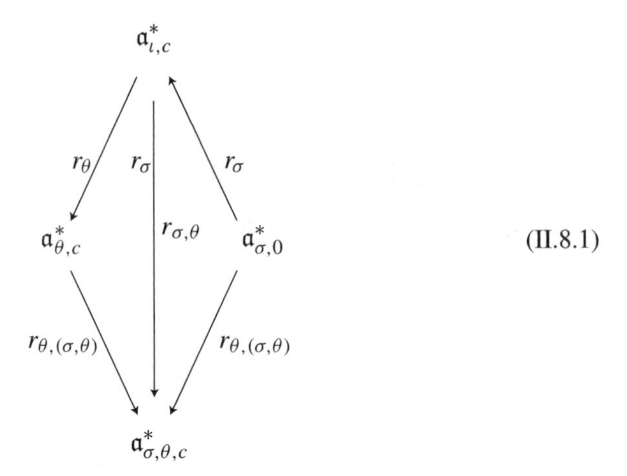

(II.8.1)

where the arrows are the restriction maps. To avoid repetitions, let us introduce the set

$$\sum = \{\iota, \theta, \sigma, (\sigma, \theta)\}. \tag{II.8.2}$$

For every $\Lambda \in \sum$, the set $\Phi_\Lambda = \Phi(\mathfrak{g}_c, \mathfrak{a}_{\Lambda,c})$ is a root system ([OsS1] 2.4, [Ros]). The root spaces are first defined on $\mathfrak{g}_c$, and we have to do so if $\Lambda = \sigma, \iota$ since in this case the roots may have complex values on $\mathfrak{a}_\Lambda$. On the other hand, they are real valued on $\mathfrak{a}_\Lambda$ for $\Lambda = \theta$, (θ, σ) and, as was always done, we consider the root spaces in $\mathfrak{g}$. Those in $\mathfrak{g}_c$ are just their complexifications. Φ is the restricted root system of the Riemannian symmetric pair (G, K) and $\Phi_{\sigma,\theta}$ is our previous Φ_o. We choose compatible orderings on these root systems and denote by Δ_Λ the set of simple roots in Φ_Λ.

The Λ-rank $r_\Lambda = rk_\Lambda(\mathbf{G})$ is the cardinality of Δ_Λ. Let Λ, Λ' be the initial and end points of an arrow in (1). By compatibility

$$\Delta_{\Lambda'} \subset r_{\Lambda,\Lambda'}(\Delta_\Lambda) \subset \Delta_{\Lambda'} \cup \{0\}. \tag{II.8.3}$$

II.8.2 Given Λ, we have to distinguish a family of standard parabolic subgroups of $\mathbf{G}$ parametrized by the subset of Δ_Λ, to be called the Λ-*relevant standard parabolic subgroups*. Let first $\mathfrak{n}_\iota$ be the standard maximal nilpotent subalgebra

$$\mathfrak{n}_\iota = \oplus_{\beta\in\Phi_\iota^+} \mathfrak{g}_{\iota,\beta}, \text{ where } \mathfrak{g}_{\iota,\beta} = \{x \in \mathfrak{g}_c, [h, x] = \beta(h)\cdot x \quad (h \in \mathfrak{a}_\Lambda)\}. \tag{II.8.4}$$

Given $I \subset \Delta_\Lambda$, let, as usual

$$\mathfrak{a}_{\Lambda,I} = \cap_{\alpha\in I} \ker \alpha. \tag{II.8.5}$$

Then the standard parabolic subalgebra $\mathfrak{p}_{\Lambda,I}$ is generated by $\mathfrak{z}(\mathfrak{a}_{\Lambda,I})$ and $\mathfrak{n}_\iota$. By definition, the standard relevant parabolic subgroup $\mathbf{P}_{\Lambda,I}$ is the normalizer in $\mathbf{G}$ of $\mathfrak{p}_{\Lambda,I}$. It is the semidirect product

$$\mathbf{P}_{\Lambda,I} = \mathbf{L}_{\Lambda,I}\cdot \mathbf{N}_{\Lambda,I}, \text{ where } \mathbf{L}_{\Lambda,I} = \mathcal{Z}_\mathbf{G}(\mathbf{A}_{\Lambda,I}), \tag{II.8.6}$$

$\mathbf{A}_{\Lambda,I}$ is the torus with Lie algebra $\mathfrak{a}_{\Lambda,I,c}$, and

$$\mathfrak{n}_{\Lambda,I} = \oplus' \, \mathfrak{g}_{\Lambda,\beta}, \tag{II.8.7}$$

where the sum runs over the $\beta \in \Phi_\Lambda^+$ that are not linear combinations of elements of I.

The Lie algebras $\mathfrak{z}(\mathfrak{a}_{\Lambda,I})$ and $\mathfrak{a}_{\Lambda,I}$ are reductive in $\mathfrak{g}$ and we can write uniquely

$$\mathfrak{z}(\mathfrak{a}_{\Lambda,I}) = \mathfrak{m}_{\Lambda,I} \oplus \mathfrak{a}_{\Lambda,I}, \tag{II.8.8}$$

where $\mathfrak{m}_{\Lambda,I}$ is the orthogonal complement of $\mathfrak{a}_{\Lambda,I}$ in $\mathfrak{z}(\mathfrak{a}_{\Lambda,I})$ with respect to the Killing form (whose restrictions to these two subalgebras are nondegenerate). Since the centralizer of a torus is always connected, we also have $\mathbf{L}_{\Lambda,I} = \mathbf{M}_{\Lambda,I} \cdot \mathbf{A}_{\Lambda,I}$ with $\mathbf{M}_{\Lambda,I} \cap \mathbf{A}_{\Lambda,I}$ finite, and

$$\mathbf{P}_{\Lambda,I} = \mathbf{M}_{\Lambda,I} \cdot \mathbf{A}_{\Lambda,I} \cdot \mathbf{N}_{\Lambda,I}. \tag{II.8.9}$$

The group $\mathbf{M}_{\Lambda,I}$ is stable under Λ. For later use, we also introduce the notation

$$\mathbf{Q}_{\Lambda,I} = \mathbf{M}^\sigma_{\Lambda,I} \cdot A_{\Lambda,I} \cdot \mathbf{N}_{\Lambda,I}, \qquad \mathbf{O}_{\Lambda,I} = \mathbf{G}/\mathbf{Q}_{\Lambda,I} \qquad (I \subset \Delta_\Lambda). \tag{II.8.10}$$

Note that

$$\mathbf{P}_{\Lambda,\Delta_\Lambda} = \mathbf{G}, \mathbf{Q}_{\Lambda,\Delta_\Lambda} = \mathbf{H} \quad \text{and} \quad \mathbf{O}_{\Lambda,\Delta_\Lambda} = \mathbf{X}. \tag{II.8.11}$$

There is a natural projection $\mathbf{G}/\mathbf{Q}_{\Lambda,I} \to \mathbf{G}/\mathbf{P}_{\Lambda,I}$ with fibers isomorphic to $\mathbf{M}_{\Lambda,I}/\mathbf{M}^\sigma_{\Lambda,I}$.

By definition, the conjugates of the $\mathbf{P}_{\Lambda,I}$ are the *Λ-relevant parabolic subgroups of* $\mathbf{G}$. The group $\mathbf{P}_{\Lambda,I}$ is of course a standard parabolic subgroup of $\mathbf{G}$ hence can also be written $\mathbf{P}_{\Lambda,I} = \mathbf{P}_{J(I)}$ with $J(I) \subset \Delta_\iota$. It is clear from the definition that

$$J(I) = r_\Lambda^{-1}(I \cup \{0\}) \cap \Delta_2. \tag{II.8.12}$$

If $\Lambda = \iota$, the Λ-relevant parabolic subgroups are all standard parabolic subgroups. If $\Lambda = \theta$, the Λ-relevant parabolic subgroups are the standard parabolic subgroups of $\mathbf{G}$ that are defined over $\mathbb{R}$, and the parabolic $\mathbb{R}$-subgroups are all Λ-relevant (recall that $\mathbf{A}_\Lambda$ is in this case a maximal $\mathbb{R}$-split torus). By 2.8 in [OsS1]

$$r_{\sigma,\theta}(\beta) = 0 \Leftrightarrow r_\theta(\beta) = 0 \text{ or } r_\sigma(\beta) = o \quad (\beta \in \Delta_t), \tag{II.8.13}$$

or equivalently,

$$r_{\sigma,\theta}^{-1}(0) = r_\theta^{-1}(0) \cup r_\sigma^{-1}(0). \tag{II.8.14}$$

This can also be expressed by the following proposition.

Proposition II.8.3 *Let $\mathbf{P}_{\iota,J}$ be a standard parabolic subgroup. Then the following conditions are equivalent*

(a) $\mathbf{P}_{\iota,J}$ is (σ,θ)-relevant; (b) $\mathbf{P}_{\iota,J}$ is both θ and σ-relevant; (c) $\mathbf{P}_{\iota,J}$ is σ-relevant and defined over $\mathbb{R}$.

II.9 The DeConcini–Procesi wonderful compactification of $\mathbf{G}/\mathbf{H}$

In this section, $\mathbf{G}$ is a connected complex semisimple group, σ an automorphism of order two of $\mathbf{G}$, and $\mathbf{H} = \mathbf{G}^\sigma$ the fixed-point set of σ. Let $\mathbf{X} = \mathbf{G}/\mathbf{H}$.

This is the same situation as before, except that $\mathbf{G}$, σ, $\mathbf{H}$ are not assumed to be defined over $\mathbb{R}$, a condition we shall return to in the next section.

II.9.1 We note first that $\mathbf{H}^o$ is reductive ([Vu1, §1]). Another argument is the following: $\mathfrak{g}$ is the direct sum of the 1 and -1 eigenspaces of σ, which are necessarily orthogonal with respect to the Killing form. Hence the restriction of the Killing form to each of them is nondegenerate. The 1-space is of course $\mathfrak{h}$. If the unipotent radical $\mathcal{R}_u\mathbf{H}^o$ of $\mathbf{H}^o$ is $\neq \{1\}$, then, putting it in triangular form (in any embedding of $\mathbf{G}$ in some $\mathbf{GL}_N$), one sees that its Lie algebra is orthogonal to $\mathfrak{h}$ with respect to the restriction of the Killing form, a contradiction.

Let $\mathbf{T}$ be an (n algebraic) torus in $\mathbf{G}$ stable under σ. The decomposition of its Lie algebra into 1 and -1 eigenspaces for σ gives rise to a decomposition

$$\mathbf{T} = \mathbf{T}_1 \cdot \mathbf{T}_{-1} \qquad (\mathbf{T}_1 \cap \mathbf{T}_{-1} \text{ a finite elementary abelian 2-group}) \tag{II.9.1}$$

such that σ is the identity on $\mathbf{T}_1$ and the inversion on $\mathbf{T}_{-1}$. Equivalently, the action of σ on $X(\mathbf{T})$ induces the identity on $X(\mathbf{T}_1)$ and $-\mathrm{Id}$ on $X(\mathbf{T}_{-1})$. The torus $\mathbf{T}_{-1}$ is the σ-anisotropic part of $\mathbf{T}$ (a terminology introduced in [Vu1]). We shall also call T_1 the σ-isotropic part of T (although this is not an accepted terminology). Note that σ-split is often used for σ-anisotropic. The torus $\mathbf{T}$ is σ-anisotropic (resp. σ-isotropic) if $\mathbf{T}_1$ (resp. $\mathbf{T}_{-1}$) is reduced to the identity.

II.9.2 The following is proved in [Vu1]: The maximal σ-anisotropic tori of $\mathbf{G}$ are conjugate under $\mathbf{H}^o$. Let $\mathbf{S}$ be one. Then any maximal torus of $\mathbf{G}$ containing $\mathbf{S}$ is stable under σ. The group $\mathcal{Z}_\mathbf{G}(\mathbf{S})$ is the almost direct product of $\mathbf{S}$ and $\mathcal{Z}_\mathbf{H}(\mathbf{S})$. The derived group of $Z_\mathbf{G}(\mathbf{S})$ is contained in $\mathbf{H}$.

Let $\mathbf{T}$ be a maximal torus of $\mathbf{G}$ containing $\mathbf{S}$, and $\Phi_\iota = \Phi(\mathbf{G}, \mathbf{T})$ the root system of $\mathbf{G}$ with respect to $\mathbf{T}$. Its Weyl group W_σ may be identified with $\mathcal{N}_\mathbf{G}(\mathbf{T})/\mathbf{T}$. The nonzero restrictions to $\mathbf{S}$ of elements of Φ also form a root system $\Phi_\sigma := \Phi(\mathbf{G}, \mathbf{S})$. Its Weyl group W_σ may be identified with $\mathcal{N}_{\mathbf{H}^o}(\mathbf{S})/\mathcal{Z}_{\mathbf{H}^o}(\mathbf{S})$ ([Ri2], §4). In the situation of II.8.2, the present Φ_σ is the same as the one there (and $\mathbf{S} = \mathbf{T}_\sigma$). Once a simple system Δ_σ of σ-roots is chosen, the σ-relevant parabolic subgroups are defined as in II.8.2 (with $\Lambda = \sigma$). The minimal ones are conjugate under $\mathbf{H}^o$ ([Vu1], Prop. 5). The standard ones are parametrized by the subsets of Δ_σ, as in II.8.2. In the next statement, we use the notation of II.8.2 for $\Lambda = \sigma$.

Theorem II.9.3 *(DeConcini–Procesi [DP1]). We assume $\mathbf{G}$ to be of adjoint type. There exists a compactification of $\mathbf{X}$, to be denoted by $\overline{\mathbf{X}}^W$, with the following properties: (i) $\overline{\mathbf{X}}^W$ is a smooth projective variety, on which $\mathbf{G}$ acts morphically. It is a disjoint union of 2^r orbits isomorphic to the $\mathbf{O}_{\sigma,J}(J \subset \Delta_\sigma)$, where $\mathbf{O}_{\sigma,J}$ is locally closed of complex codimension $|\Delta_\sigma - J|$. In particular $\mathbf{O}_{\sigma,\Delta_\sigma} = \mathbf{G}/\mathbf{H}$ is open, Zariski-dense.*

The compactification $\overline{\mathbf{X}}^W$ is the *DeConcini–Procesi or wonderful compactification of X*. More precisely, it is the minimal wonderful compactification since there are other compactifications that also deserve the title wonderful according to [DP1] and are certain blow-ups of the minimal one. We shall omit the word minimal as long as we do not have to consider others. In fact, the main statements of [DP1] are on the closures of the orbits, which we summarize in the following:

II.9.3′ Complement. *Let $\mathbf{D}_{\sigma,J}$ be the Zariski closure of $\mathbf{O}_{\sigma,J}$ and let us also write $\mathbf{D}_\sigma^J$ for $\mathbf{D}_{\sigma,\Delta_\sigma - J}$ $(J \subset \Delta_\sigma)$. Then*

$$\mathbf{D}_{\sigma,J} = \cup_{I \subset J} \mathbf{O}_{\sigma,I}. \tag{II.9.2}$$

The variety $\mathbf{D}_\sigma^J$ is smooth of codimension $|J|$. We have

$$\mathbf{D}_{\sigma,I} \cap \mathbf{D}_{\sigma,J} = \mathbf{D}_{\sigma,I\cap J} \text{ or } \mathbf{D}_\sigma^I \cap \mathbf{D}_\sigma^J = \mathbf{D}_\sigma^{I\cup J},\ (I, J \subset \Delta_o). \tag{II.9.3}$$

The $\mathbf{D}^{\{\alpha\}}$ $(\alpha \in \Lambda_\sigma)$ are smooth divisors with normal crossings.

Note that $J = \emptyset$ is always allowed:

$$\mathbf{D}_{\sigma,\emptyset} = \mathbf{D}_\sigma^{\Delta_\sigma} = \mathbf{O}_{\sigma,\emptyset} = \cap_{\alpha\in\Delta_\sigma} \mathbf{D}_\sigma^\alpha = \mathbf{G}/\mathbf{P}_\emptyset \tag{II.9.4}$$

is the only closed orbit.

Since $\mathbf{Q}_{\sigma,J} \subset \mathbf{P}_{\sigma,J}$ we have a natural $\mathbf{G}$-equivariant morphism $\mathbf{O}_{\sigma,J} \to \mathbf{G}/\mathbf{P}_{\sigma,J}$. If $I \subset J$ then $\mathbf{G}/\mathbf{P}_{\sigma,I}$ maps naturally onto $\mathbf{G}(\mathbf{P}_{\sigma,J})$.

By the theorem in [DP1, 5.2], these maps combine to a $\mathbf{G}$-equivariant fibration

$$\mathbf{D}_{\sigma,J} \to \mathbf{G}/\mathbf{P}_{\sigma,J}$$

with typical fiber the wonderful compactification of the quotient $\mathcal{D}\mathbf{L}'_{\sigma,J}$, where $\mathcal{D}\mathbf{L}'_{\sigma,J}$ is the adjoint group of the derived group of $\mathbf{L}_{\sigma,J}$.

[DP1] gives two definitions of the wonderful compactification. The first one is the closure of the $\mathbf{G}$-orbit of $\mathfrak{h}$ under the adjoint action in the Grassmannian of h-planes in $\mathfrak{g}$ ($h = \dim \mathfrak{h}$). The second one is representation-theoretic and may be viewed as a complex analogue of the original definition of the maximal Satake compactification. The first one is similar to the subalgebra compactification $\overline{X}^{sba}$ in §I.16.

Let $(\tau, \mathbf{V})$ be an irreducible faithful linear rational representation of $\mathbf{G}$. It is known that $V^{\mathbf{H}}$ is at most one-dimensional [Vu1]. The representation is called *spherical* or of *class one* if $\mathbf{V}^{\mathbf{H}}$ has dimension one. Let (τ, V) be one and assume moreover that its highest weight is *regular*, meaning that the stability group of the highest weight line is $\mathbf{P}_{\sigma,\phi}$ (see II.9.5). Then $\overline{\mathbf{X}}^W$ is, by definition, the closure of the $\mathbf{G}$-orbit of the point $[V^H]$ in the projective space $\mathbf{P}(\mathbf{V})$ of one-dimensional subspaces of V. Up to isomorphism, it is independent of the representation satisfying the previous conditions.

We refer to [DP1] for the proof. We content ourselves here with some indications to make the structure of $\overline{\mathbf{X}}^W$ plausible. This is not needed in the sequel.

II.9.4 We first intercalate some remarks on morphic actions of $\mathbb{C}^* = \mathbf{GL}_1(\mathbb{C})$ on projective spaces. We have $\mathbf{P}_1(\mathbb{C}) = \mathbb{C}^* \cup \{0\} \cup \{\infty\}$. It is a standard fact that any morphism of $\mathbb{C}^*$ into a projective (or complete) variety $\mathbf{M}$ over $\mathbb{C}$ extends to $\mathbf{P}_1(\mathbb{C})$. If this morphism is an orbit map for an algebraic action, the images of 0 and ∞ are fixed points of $\mathbb{C}^*$. We can check this directly in the only case of interest here: $\mathbf{M} = \mathbf{P}_N(\mathbb{C})$ and $\mathbb{C}^*$ acts on $\mathbf{M}$ via a linear action on the underlying vector space $W = \mathbb{C}^{N+1}$. Write W as a direct sum of one-dimensional subspaces W_i invariant under $\mathbb{C}^*$. Then $\mathbb{C}^*$ acts on W_i by means of a character $t \mapsto t^{m_i}$ $(m_i \in \mathbb{Z})$. We may assume that $m_i \leq m_j$ if $i < j$. Let now $w = (w_i) \in W$, $w \neq 0$. The point $t \cdot w$ has coordinates $(t^{m_i} \cdot w_i)$. In $\mathbf{P}_N(\mathbb{C})$, it has the same homogeneous coordinates as $(\tilde{t}^{m_i - m_1} \cdot w_i)$. Now all the exponents are ≥ 0, so if $t \to 0$, the point has the limit $0 \cdot w$ with coordinates $(0 \cdot w)_i$ equal to 1 if $m_i = m_1$ and to 0 otherwise. Similarly, if we divide by $t^{m_{N+1}}$, all exponents are ≤ 0, and if $t \to \infty$ the point $t \cdot w \cdot$ tends to the point $\infty \cdot \omega$ with homogeneous coordinates $(\infty \cdot \omega)_i$ equal to 1 if $m_i = m_{N+1}$, and to 0 otherwise.

II.9.5 Write x_o for $[V^{\mathbf{H}}]$. It is of course fixed under $\mathbf{H}$, but since $\mathbf{G}$ is of adjoint type by assumption, $\mathbf{H}$ is the full isotropy group in $\mathbf{G}$ of x_o ([DP1], (iii) of the lemma on p. 7), hence the orbit map $g \mapsto g \cdot x_o$ induces an embedding of $\mathbf{G/H}$ into $\mathbf{P(V)}$. [In the case of G/K, the group K is self-normalizing in G and this point was not an issue.] Write $\mathbf{V} = \oplus \mathbf{V}_\mu$, where μ runs through the weights of $\mathbf{S}$ in $\mathbf{V}$:

$$\mathbf{V}_\mu = \{v \in \mathbf{V}, \tau(s) \cdot v = s^\mu \cdot v \ (s \in S)\}. \tag{II.9.5}$$

Let ν be the highest weight of τ (it is noted 2λ in [DP1]). A general weight is of the form

$$\mu = \nu - \sum_{\alpha \in \Delta} c_\alpha(\mu) \cdot \alpha \qquad (c_\alpha(\mu) \in \mathbb{N}). \tag{II.9.6}$$

We view $\mathbf{S}$ as embedded in $\mathbb{C}^\Delta$ by the map $s \mapsto (s^{-\alpha_1}, \ldots, s^{-\alpha_r})$ so that s tends to a coordinate plane if some of the $s^{-\alpha_i}$ tend to infinity while the others remain constant. In studying Zariski closure of $\mathbf{G} \cdot x_o$, we may ignore the common factors s^ν and look at sequences for which some of the s^{α_i} remain constant while the others tend to infinity. This yields an isomorphism of $\mathbf{N}^- \times \mathbb{C}^\Delta$ onto an open chart in $\overline{\mathbf{X}}^W$, the intersection of which with $\mathbf{X}$ is $\mathbf{N}^- \times \mathbf{S}$. An analysis similar to that of §I.4 shows that the complement of $\mathbf{X}$ is the union of orbits $\mathbf{O}_{\sigma,J} (J \subset \Delta_\sigma)$.

II.9.6 To complete this sketch, we give some indications on the class of representations with regular highest weight. (See [DP1, §1], [Vu1, Théor. 3].)

Let $\mathbf{T}_1$ be the σ-split part of $\mathbf{T}$ (II.9.1). Thus $\mathbf{T}$ is the almost direct product of $\mathbf{T}_1$ and $\mathbf{S}$ (and $\mathbf{S}$ is the subtorus denoted by $\mathbf{T}_{-1}$ in II.9.1). The involution σ operates on $\Phi = \Phi(G, T)$, and $\mathfrak{g}_{\alpha}\sigma = \sigma(\mathfrak{g}_\alpha)$ $(\alpha \in \Phi)$. Following [DP1], we let

$$\Phi_o = \{\beta \in \Phi, s^\beta = 1 \ (s \in S)\}, \qquad \Phi_1 = \Phi - \Phi_o. \tag{II.9.7}$$

They are stable under σ. The elements of Φ_o are completely determined by their restrictions to $\mathbf{T}_1$, on which σ acts trivially; hence they are fixed under σ. In fact $\mathfrak{g}_\alpha (\alpha \in \Phi_o)$ is also pointwise fixed under σ and the $\mathfrak{g}_\alpha$ $(\alpha \in \Phi_o)$ are the roots subspaces with respect to $\mathbf{T}_1$ in $\mathcal{Z}_{\mathbf{G}}(\mathbf{S})$ ([DP1, 1.3]). Let also

$$\Delta_o = \Delta_\iota \cap \Phi_o, \Delta_1 = \Delta_\iota \cap \Phi_1. \tag{II.9.8}$$

As in II.9.1, we assume orderings in Δ_ι and Δ_σ to be compatible. Therefore $r_{\iota,\sigma}(\Delta_1) = \Delta_\sigma$. Since σ is the inversion on **S**, it is clear that if $\alpha \in \Phi_1$ is > 0, then $\sigma(\alpha) < 0$. More precisely, [DP1, 1.4] shows the existence of a permutation $\tilde{\sigma}$ of Δ_1, of order 2, such that

$$\sigma(\alpha) \equiv -\alpha_{\tilde{\sigma}(\alpha)} \tag{II.9.9}$$

modulo a negative linear integral combination of elements in Δ_o. It is then easily seen that the fibers of $r_{\iota,\sigma}$ are the orbits of $\tilde{\sigma}$ in Δ_1.

By Théor. 3 in [Vu1], a dominant weight ω in $X(\mathbf{T})$ is the highest weight of a spherical irreducible representation if and only if

$$\sigma(\omega) = -\omega \qquad \omega|\mathbf{S} \in 2X(\mathbf{S}). \tag{II.9.10}$$

Following [DP1], we give an explicit description of weights satisfying those conditions. Let $\{\omega_\alpha\}(\alpha \in \Delta_\iota)$ be the fundamental highest weights, defined as usual as the elements of the dual basis to the basis formed by the simple coroots. It is shown in [DP1, 1.4] that

$$\sigma(\omega_\alpha) = -\omega_{\tilde{\sigma}(\alpha)} \qquad (\alpha \in \Delta_1). \tag{II.9.11}$$

On the other hand, since σ acts trivially on Φ_σ, it also fixes the ω_α for $\alpha \in \Delta_o$. As a consequence, a dominant weight

$$\omega = \sum_{\alpha \in \Delta_\iota} n_\alpha \cdot \omega_\alpha \tag{II.9.12}$$

satisfies the first condition of (II.9.10) if and only

$$n_\alpha = 0 \ (\alpha \in \Delta_o), \ n_\alpha = n_{\tilde{\sigma}(\alpha)} \qquad (\alpha \in \Delta_1), \tag{II.9.13}$$

in which case it is called *special*; ω is moreover said to be *regular* if $n_\alpha \neq 0$ for all $\alpha \in \Delta_1$, or, equivalently, if $\mathbf{P}_{\iota,\phi}$ is the isotropy group of the highest weight line. The doubles of the special regular dominant weights also satisfy the second condition of (II.9.10). The corresponding representations are those used in [DP1] to construct $\overline{\mathbf{X}}^W$.

Note that the spherical representations may have a finite kernel, and so may not be linear representations of the adjoint group. However, only the associated projective representations matter in defining the wonderful compactifications, and those are of course defined on the adjoint group.

II.10 Real points of G/H

G, **H**, σ, **X** *are as in §II.9, except that they are moreover assumed to be defined over* $\mathbb{R}$.

The main goal of this section is to describe $\overline{\mathbf{X}}^W(\mathbb{R})$ as a union of orbits of G.

II.10.1 We first recall some known facts about real points of algebraic varieties defined over $\mathbb{R}$. Let $\mathbf{V}$ be a real algebraic variety and let n be its (complex) dimension. The dimension of $\mathbf{V}(\mathbb{R}) = V$, as a real algebraic variety, is at most n. It is equal to n if V is Zariski dense in $\mathbf{V}$. Assume this is the case and let $x \in \mathbf{V}$. If x is simple on $\mathbf{V}$, then it is also simple on V, and V is a manifold of dimension n around x. In particular if $\mathbf{V}$ is smooth (our case of interest), so is V. For all this, see [Wh], especially §§10, 11.

II.10.2 By II.9.3, $\overline{\mathbf{X}}^W$ is a smooth projective variety. Let n be its complex dimension. Let $\overline{X}^W = \overline{\mathbf{X}}^W(\mathbb{R})$ be the real locus of $\overline{\mathbf{X}}^W$. Then $\overline{X}^W$ contains G/H, which has real dimension n, hence is Zariski dense in $\overline{\mathbf{X}}^W$. Thus $\overline{X}^W$ is a smooth real projective variety of dimension n, and the action of G makes it an algebraic transformation group. In particular $\overline{X}^W$ is an analytic G-space. By II.9.3

$$\overline{X}^W = \coprod_{J\in\Delta_\sigma} \mathbf{O}_{\sigma,J}(\mathbb{R}), \text{ where } \mathbf{O}_{\sigma,J} = \mathbf{G}/\mathbf{Q}_{\sigma,J}. \tag{II.10.1}$$

We have to know when a term on the right-hand side is not empty. Let $x \in \mathbf{O}_{\sigma,J}(\mathbb{R})$. Its isotropy group is defined over $\mathbf{R}$, and so is the unipotent radical of the latter. But the unipotent radical of $\mathbf{Q}_{\sigma,J}$ is the same as the unipotent radical of $\mathbf{P}_{\sigma,J}$; therefore the latter is defined over $\mathbb{R}$. By II.8.3, it is therefore (σ, θ) relevant. On the other hand, if $\mathbf{P}_{\sigma,J}$ is defined over $\mathbb{R}$, so is $\mathbf{Q}_{\sigma,J}$, and $\mathbf{O}_{\sigma,J}$ has real points. The parametrization in (II.10.1) is effectively by subsets of $\Delta_{(\sigma,\theta)}$ and we have to replace $\mathfrak{a}_\sigma$ by $\mathfrak{a}_{\sigma,\theta}$. As in Chapter 6, we shall denote the index (σ, θ) by o. We have then

$$\overline{X}^W = \coprod_{J\subset\Delta_o} (\mathbf{G}/\mathbf{Q}_{o,J})(\mathbb{R}). \tag{II.10.2}$$

For $J = \Delta_\sigma$, the right-hand side is $(\mathbf{G}/\mathbf{H})(\mathbb{R})$ and the description of the G-orbits is given by II.6.5. We want to reduce the general case to that one by going over to the subgroups $\mathbf{M}_{o,J}$.

Lemma II.10.3 *Let $J \subset \Delta_o$. The orbits of G in $(\mathbf{G}/\mathbf{Q}_{o,J})(\mathbb{R})$ may be identified naturally with the orbits of $M_{o,J}$ in $(\mathbf{M}_{o,J})(\mathbb{R})$.*

Note first that we have the $\mathbb{R}$-isomorphism

$$\mathbf{P}_{o,J}/\mathbf{Q}_{o,J} = \mathbf{M}_{o,j}/\mathbf{M}^\sigma_{o,J}, \tag{II.10.3}$$

whence a bijection

$$(\mathbf{P}_{o,J}/\mathbf{Q}_{o,J})(\mathbb{R}) = (\mathbf{M}_{o,J}/\mathbf{M}^\sigma_{o,J})(\mathbb{R}), \tag{II.10.4}$$

which identifies the orbits of $M_{o,J}$ on the right-hand side with the orbits of $P_{o,J}$ on the left-hand side. There remains therefore to show the existence of a bijection

$$(\mathbf{G}/\mathbf{Q}_{o,J})(\mathbb{R}) \simeq (\mathbf{P}_{o,J}/\mathbf{Q}_{o,J})(\mathbb{R}) \tag{II.10.5}$$

that maps the orbits of G on the left-hand side onto the orbits of $P_{o,J}$ on the right-hand side. The main point to establish (II.10.5) is the following assertion:

(∗) *Let $x \in (\mathbf{G}/\mathbf{Q}_{o,J})(\mathbb{R})$. Then $G \cdot x \cap (\mathbf{P}_{o,J}/\mathbf{Q}_{o,J})(\mathbb{R})$ consists of one orbit of $P_{o,J}$ and every such orbit occurs.*

Proof. The last part is obvious. We recall that

$$(\mathbf{G}/\mathbf{P}_{o,J})(\mathbb{R}) = G/P_{o,J}, \tag{II.10.6}$$

a standard consequence of the fact that the fibration of $\mathbf{G}$ by a parabolic $\mathbb{R}$-subgroup has many local sections defined over $\mathbb{R}$, provided by the big cell of the Bruhat decomposition and its translates by G.

Let r be the projection $G/Q_{o,J} \to G/P_{o,J}$. By (II.10.6), there exists $g \in G$ such that $g \cdot 0 = r(x)$, where 0 denotes the identity coset in $G/P_{o,J}$; therefore

$$G \cdot x \cap (\mathbf{P}_{o,J}/\mathbf{Q}_{o,J})(\mathbb{R}) \neq \emptyset. \tag{II.10.7}$$

The second term is clearly a union of $P_{o,J}$-orbits.

Replacing x by $g^{-1} \cdot x$ if necessary, we may assume that $r(x) = 0$. Of course, r maps $\mathbf{P}_{o,j}/\mathbf{Q}_{o,J}$ onto 0. The second term in (II.10.7) is a union of $P_{o,J}$ orbits. Assume that $y = h \cdot x$ $(h \in G)$ also belongs to that intersection. Then

$$0 = r(x) = r(y) = h \cdot r(y) = h \cdot 0;$$

therefore h normalizes the isotropy group $\mathbf{P}_{o,J}$ of 0, hence belongs to it since a parabolic subgroup is self-normalizing. This concludes the proof of (∗) and therefore of II.10.3.

II.10.4 We want to apply II.6.5 to $M_{o,J}$, endowed with the restriction σ_J of σ. For this we have to introduce some definitions and notation. We let θ_J be the restriction of θ to $M_{o,J}$, and write $H_{o,J}$ for $M^{\sigma}_{o,J}$.

Let $\mathfrak{a}^J_o = \mathfrak{m}_{o,J} \cap \mathfrak{a}_o$. The algebra $\mathfrak{a}_o$ is the direct sum of $\mathfrak{a}_{o,J}$ and $\mathfrak{a}^J_o$, and $\mathfrak{a}^J_o$ is a maximal (σ_J, θ_J)-anisotropic subalgebra of $\mathfrak{m}_{o,J}$. The (σ_J, θ_J)-root system $\Phi_{o,J}$ of $\mathfrak{m}_{o,J}$ with respect to $\mathfrak{a}^J_o$ consists of the (σ, θ)-roots that are linear combinations of elements in J. Its Weyl group $W_{o,J}$ is generated by the reflection $s_\alpha (\alpha \in J)$. Let $\mathbf{A}^J_o$ be the (σ_J, θ_J)-anisotropic torus with Lie algebra $\mathfrak{a}^J_{o,c}$. To each element $s \in \mathbf{A}^J_o$ is attached a proper signature ε_s of J and hence a real form $H_{o,J,\varepsilon}$ of $H_{o,J}$ (II.4.3) and a real form $Q_{o,J,\varepsilon} = H_{o,J,\varepsilon} \cdot A_{o,J} \cdot N_J$ of $\mathbf{Q}_{o,J}$. II.6.5, (II.10.2) and II.10.3, imply the first three assertions of the following theorem.

Theorem II.10.5

(i) $\overline{\mathbf{X}}^W(\mathbb{R}) = \coprod_{J \subset \Delta_o} (\mathbf{G}/\mathbf{Q}_{o,J})(\mathbb{R})$.

(ii) *The orbits of G in $(\mathbf{G}/\mathbf{Q}_{o,J})(\mathbb{R})$ may be naturally identified with the orbits of $M_{o,J}$ in $(\mathbf{M}_{o,J}/\mathbf{H}_{o,J})(\mathbb{R})$.*

(iii) *There is a natural bijection of ${}_2A_o^J/W_{o,J}$ onto the orbits of G in $(\mathbf{G}/\mathbf{Q}_{o,J})(\mathbb{R})$. It assigns to ${}_2\mathbf{A}_o^J$ the orbit*

$$G \cdot u_s \cdot o = G/Q_{o,J,\varepsilon s},$$

where ε_s is the image of $s \in {}_2A_o^J$ under the signature map (II.4.3) and $u_s \in {}_4A_{o,u}^J$ is a square root of s.

(iv) *The codimension of $(\mathbf{G}/\mathbf{Q}_{o,J})(\mathbb{R})$ in $\overline{\mathbf{X}}^W(\mathbb{R})$ is equal to $|\Delta_o - J|$.*

II.10.6 There remains to prove II.10.5(iv). In the following, we also use the notation $|M|$ to denote the dimension of the manifold M. In agreement with previous conventions, we let $O_{o,J}$ stand for $(\mathbf{G}/\mathbf{Q}_{o,J})(\mathbb{R})$. We have therefore to show that codim $O_{o,J} = |\Delta_o - J|$. The product $N_\emptyset^- \cdot M_\emptyset \cdot A_\emptyset \cdot N_\emptyset$ is open in G; hence we see, using (II.3.5), that

$$|H| = |N_\emptyset| + |M_\emptyset^\sigma|, \tag{II.10.8}$$

also taking into account the obvious relation $\mathcal{Z}_G(\mathfrak{a}_o)^\sigma = M_\emptyset^\sigma$. We have

$$Q_{o,J} = M_{o,J}^\sigma \cdot A_{o,J} \cdot N_{o,J}, \tag{II.10.9}$$

hence

$$|Q_{o,J}| = |M_{o,J}^\sigma| + |J| + |N_{o,J}|. \tag{II.10.10}$$

Let

$$P_o^J = M \cap P_{o,J}. \tag{II.10.11}$$

It is a parabolic subgroup of $M_\emptyset$, which can be written

$$P_o^J = \mathcal{Z}_{M_\emptyset}(A_o^J) \cdot N_o^J \qquad (N_o^J \text{ unipotent radical of } P_o^J). \tag{II.10.12}$$

If N_o^{-J} is the unipotent radical of the parabolic subgroup of $M_\emptyset$ opposed to P_o^J, then $N_o^{-J} \cdot \mathcal{Z}_{M_\emptyset}(A_o^J) \cdot N_o^J$ is an open submanifold of $M_\emptyset$. By the same calculation as above, we see that

$$|M_{o,J}^\sigma| = |N_o^J| + |\mathcal{Z}_{M_\emptyset}(A_o^J)^\sigma|. \tag{II.10.13}$$

But $\mathcal{Z}_{M_\emptyset}(A_o^J) \cdot A_{o,J} = \mathcal{Z}_G(A_o)$, and therefore

$$|\mathcal{Z}_{M_\emptyset}(A_o^J)^\sigma| = |\mathcal{Z}_G(A_o)^\sigma| = |M_{o,\emptyset}^\sigma|. \tag{II.10.14}$$

On the other hand, $N_\emptyset$ is the semidirect product of N^J and $N_{o,J}$, so that (II.10.8), (II.10.10), (II.10.13), (II.10.14) imply

$$|Q_{o,J}| = |H| + |J|, \tag{II.10.15}$$

and our assertion follows.

We now consider the homomorphism $\sigma(\theta, \sigma)$ of II.8.1. As before, we replace (θ, σ) by ${}_o$.

Corollary II.10.7 *Let $\alpha \in \Delta_\sigma$. Then $r_{\sigma,o}^{-1}(\alpha)$ consists of one element in Δ_σ.*

Proof. Fix $\alpha \in \Delta_0$, and let $J = \Delta_o - \{\alpha\}$. Let $J' = r_{\sigma,o}^{-1}(J \cup \{0\}$. Then $\mathbf{P}_{o,J}$, viewed as a σ-relevant parabolic subgroup, can be written $\mathbf{P}_{\sigma,J'}$. Similarly, $\mathbf{Q}_{o,J} = \mathbf{Q}_{\sigma,J'}$ and we have

$$\mathbf{O}_{\sigma,J'}(\mathbb{R}) = O_{o,J}. \qquad \text{(II.10.16)}$$

By II.10.5, $\operatorname{codim}_{\mathbb{R}}(O_{o,J}) = |\Delta_o - J| = 1$.

Therefore $\operatorname{codim}_{\mathbb{C}}(\mathbf{O}_{\sigma,J'}) = 1$. But it is equal to $|\Delta_\sigma - J|$ (II.9.3). Hence the complement of J' in Δ_σ consists of one root, say α', and then $r_{\sigma,o}^{-1}(\alpha) = \alpha'$.

II.10.8 Let Δ^o be the kernel of $\sigma_{\sigma,o}$. The corollary implies that if $J \subset \Delta_o$, then $r_{\sigma,o}^{-1}(J) \cap \Delta_\sigma$ has the same cardinality as J. Thus $\Delta_\sigma = \Delta^o \cup \Delta_1$, where Δ_1 is mapped bijectively onto Δ_o.

In terms of σ-relevant parabolic subgroups, the union of the orbits of $\mathbf{G}$ in $\overline{\mathbf{X}}^W$ that contain real points can therefore be written

$$\coprod_{J',\Delta^o \subset J' \subset \Delta_\sigma} \mathbf{O}_{\sigma,J'}. \qquad \text{(II.10.17)}$$

Proposition II.10.9 *Let $\alpha \in \Delta_o$. Then $r_{\iota,o}^{-1}(\alpha) \cap \Delta_\iota$ has one or two elements. If it has two, it is an orbit of the involution $\tilde{\sigma}$ (notation of II.9.6).*

The map $r_{\iota,o}$ is the composition of $r_{\iota,\sigma}$ and $r_{\sigma,o}$. The proposition then follows from II.10.7 and the results of [DP1] recalled in II.9.6.

II.10.10 We use the notation of II.9.3′. Then

$$D_o^{\Delta_o - J} = D_{o,J} = \mathbf{D}_{o,J}(\mathbb{R}) = \mathbf{D}_{\sigma,J'}(\mathbb{R}) \qquad \text{(II.10.18)}$$

(where $J \subset \Delta_o$ and J' is as in II.10.7) is the closure of $O_{o,J}$ in $\overline{X}^W$. In view of II.9.3′ and II.10.1, we see that $D_{o,J}$ has (real) codimension $|\Delta_o - J|$ and is a smooth manifold. Furthermore, for any $I, J \subset \Delta_o$,

$$D_{o,I} \cap D_{o,J} = D_{o,I \cap J}. \qquad \text{(II.10.19)}$$

In particular the $D_o^{\{\alpha\}}$ ($\alpha \in \Delta_o$) are smooth hypersurfaces with transversal intersections. It follows that the closure of G/H in $\overline{X}^W$ is a semialgebraic G-space with corners, of rank r_o. In particular it is an analytic G-space.

II.10.11 Remarks.

(1) If we apply the remark II.6.5 to all the $\mathbf{G}/\mathbf{Q}_{o,J}$, we see that $\overline{\mathbf{X}}^W(\mathbb{R})$ is the union over $J \subset \Delta$ of the orbits $G \cdot {}_4A_{o,u}^J \cdot o$.

(2) It is worth noting explicitly what happens if J is the empty set. It is implied by the above that $(\mathbf{M}_{o,\emptyset}/\mathbf{H}_{o,\emptyset})(\mathbb{R})$ is compact, although, unless Δ^o is empty, $\mathbf{M}_{o,\emptyset}/\mathbf{H}_{o,\emptyset}$ is not a projective variety. However, $(M_{o,\emptyset}, H_{o,\emptyset})$ is of $(\theta_\emptyset, \sigma_\emptyset)$-rank zero and so are the pairs $(M_{o,\theta}, H_{o,\theta,\varepsilon})$ and therefore the orbits of $M_{o,\emptyset}$ in $(\mathbf{M}_{o,\emptyset}, \mathbf{H}_{o,\emptyset})(\mathbb{R})$ are indeed compact, and equal to orbits of the standard maximal compact subgroup of $M_{o,\emptyset}$, by II.6.2.

Remark II.10.12 As pointed out in II.10.2, $\overline{\mathbf{X}}^W(\mathbb{R})$ is a smooth real projective variety. The symmetric space X is clearly contained in $\overline{\mathbf{X}}^W(\mathbb{R})$ and its closure in the regular topology is a compactification of X. Since $\overline{\mathbf{X}}^W$ is the closure of the $\mathbf{G}$-orbit of $\mathfrak{k}$ in $Gr(\mathfrak{g}, h)$ (see the comments after II.9.3′), by Proposition I.17.9, the closure of X in $\overline{\mathbf{X}}^W(\mathbb{R})$ is isomorphic to $\overline{X}^S_{\max}$. By Theorem II.9.3 (II.9.3′) and Theorem II.10.5, the closure of X in $\overline{\mathbf{X}}^W(\mathbb{R})$ is a real analytic manifold with corners. This implies that $\overline{X}^S_{\max}$ is a real analytic manifold with corners and hence can be self-glued into a closed real analytic manifold isomorphic to $\overline{X}^O$, as pointed out near the end of §I.19 and §II.2.1.

II.11 A characterization of the involutions σ_ε

This section answers a rather natural question, but its main result (II.11.2) will not be used later. In the proof we need a theorem of [ABS]. We begin by reviewing just what we need from [ABS], where the results are proved in greater generality.

II.11.1 As usual, $\mathbf{G}$ is a connected complex semisimple group, $\mathbf{A}$ a maximal torus, $\Phi = \Phi(\mathbf{G}, \mathbf{A})$, the system of roots of $\mathbf{G}$ with respect to $\mathbf{A}$, and Δ the set of simple roots for a given ordering on Φ. For $J \subset \Delta$, the standard parabolic subgroup $\mathbf{P}_J$ is the semidirect product of the Levi subgroup $\mathcal{Z}_{\mathbf{G}}(\mathbf{A}_J)$ and its unipotent radical $\mathbf{N}_J$. The roots of $\mathbf{G}$ with respect to $\mathbf{A}_J$ are the restrictions to $\mathbf{A}_J$ of roots that are not linear combinations of elements in J. The positive ones form $\Phi(\mathbf{P}_J, \mathbf{A}_J)$ and are positive linear combinations of elements in $J' = \Delta - J$, or rather of their restrictions to $\mathbf{A}_J$.

Any root α can be written uniquely as $\alpha = \alpha_J + \alpha_{J'}$, where α_J (resp. $\alpha_{J'}$) is a linear combination of elements in J (resp. J') (with integral coefficients of the same sign). The element $\alpha_{J'}$ is called in [ABS] the shape or J-shape of α. It will also be denoted by $sh_J(\alpha)$. By restriction to $\mathbf{A}_J$ the (nonzero) positive shapes may be identified with the elements of $\Phi(\mathbf{P}_J, \mathbf{A}_J)$. For $\alpha \in \Phi(\mathbf{G}, \mathbf{A}_J)$ we obviously have

$$\mathfrak{g}_{\iota,\alpha} = \oplus_{\beta\in\Phi, sh_J(\beta)=\alpha} \quad \mathfrak{g}_{\iota,\beta}. \tag{II.11.1}$$

This space is denoted by V_S in [ABS], where S stands for the shape α. The group $\mathcal{Z}_G(\mathbf{A}_J)$, operating by the adjoint representation, leaves invariant each such space and we need to know that

(+) *The space* $\mathfrak{g}_{\iota,\alpha}$ $(\alpha \in \Phi(\mathbf{G}, \mathbf{A}_J))$ *is an irreducible* $\mathcal{Z}_{\mathbf{G}}(\mathbf{A}_J)$*-module.*

This is contained in Theorem 2 of [ABS].

We now come back to the situation of Chapter 5 and let $G, \sigma, \theta, \mathfrak{a}_o$ be as before. We want to prove the folowing result:

Proposition II.11.2 *Let τ be an involution of G that commutes with σ and let θ be a Cartan involution commuting with σ and τ (it exists, see below). Assume that $\mathfrak{a}_o = \mathfrak{a}_{\theta,\tau}$ and that $\sigma = \tau$ on $\mathcal{Z}_G(A_o)$. Then there exists $\varepsilon \in \mathcal{E}^o(\Phi_o)$ such that $\tau = \sigma_\varepsilon$.*

It is clear that the conditions imposed on τ are necessary. So we are proving that they are sufficient. We divide the proof into five steps.

(a) σ and τ generate a group of order 4 (we assume $\sigma \neq \tau$) in the automorphism group of G. By Cartan's fixed-point theorem, it has a fixed point on the space of maximal compact subgroups of G, and this provides a θ commuting with σ and τ. The product $\delta = \sigma \cdot \tau$ is the identity on $\mathcal{Z}_{\mathbf{G}}(A_o)$. In particular, it leaves each root space $\mathfrak{g}_{o,\beta}$ stable. We claim that it suffices to prove that

$$(*) \qquad \text{The restriction of } \delta \text{ to } \mathfrak{g}_{o,\beta} \text{ is equal to } \pm\mathrm{Id}. \qquad (\beta \in \Phi_o).$$

Indeed, let us denote by $c(\beta)\cdot\mathrm{Id}$ the restriction of δ to $\mathfrak{g}_{o,\beta}$. Since $[\mathfrak{g}_{o,\beta}, \mathfrak{g}_{o,-\beta}] \subset \mathfrak{z}(\mathfrak{a}_o)$ and is not zero, $[\mathfrak{g}_{o,\beta}, \mathfrak{g}_{o,\gamma}] \neq 0$ and belongs to $\mathfrak{g}_{o,\beta,\gamma}$ if β, γ, and $\beta + \gamma$ are σ-roots, we see t $\beta \mapsto c(\beta)$ is a signature on Φ_o, and we have $\tau = \sigma_\varepsilon$ by definition.

(b) Let J be the kernel of the map $r_{\iota,o} : \Delta_\iota \to \Delta_o$. Then $\mathbf{P}_{o,\emptyset}$ can be written $\mathbf{P}_{\iota,J}$. The torus $\mathbf{A}_{\iota,J}$ is a maximal torus of the radical of $\mathbf{P}_{\iota,J}$ and we have

$$\mathcal{Z}_{\mathbf{G}}(\mathbf{A}_o) = \mathcal{Z}_{\mathbf{G}}(\mathbf{A}_{\iota,I}). \tag{II.11.2}$$

We shall apply II.11.1 to $\mathbf{P}_{\iota,J}$. The J-shapes are therefore the elements of $\Phi(G, A_{\iota,J})$. We let

$$\nu : X(\mathbf{A}_{\iota,J}) \to X(\mathbf{A}_o) \tag{II.11.3}$$

be the restriction map. It sends $\Phi(\mathbf{P}_{\iota,J}, \mathbf{A}_{\iota,J})$ onto $\Phi(\mathbf{P}_{o,\emptyset}, \mathbf{A}_o)$. For $\gamma \in \Phi(\mathbf{G}, \mathbf{A}_{\iota,J})$, let us denote by $\mathfrak{u}_\gamma$ the subspace of elements of J-shape γ, i.e.,

$$\mathfrak{u}_\gamma = \oplus_{\beta\in\Phi_\iota, sh_J(\beta)=\gamma} \qquad \mathfrak{g}_{\iota,\beta}. \tag{II.11.4}$$

By (+), the representation of $\mathcal{Z}_{\mathbf{G}}(\mathbf{A}_{\iota,J})$ in $\mathfrak{u}_\gamma$ is irreducible. Since δ fixes elementwise $\mathcal{Z}_G(\mathbf{A}_o)$, which is equal to $\mathcal{Z}_{\mathbf{G}}(A_{\iota,J})$ by (1), and is of order 2, there exists a constant $c(\gamma) = \pm 1$ such that the restriction of δ to $\mathfrak{u}_\gamma$ is equal to $c(\gamma)\cdot\mathrm{Id}$. We note that

$$c(\gamma) \cdot c(-\gamma) = 1, \tag{II.11.5}$$

$$c(\gamma + \gamma') = c(\gamma) \cdot c(\gamma) \quad (\text{if } \gamma, \gamma', \text{ and } \gamma + \gamma' \text{ belong to } \Phi(\mathbf{P}_{\iota,J}\mathbf{A}_{\iota,J}).) \tag{II.11.6}$$

This follows from the fact that $[\mathfrak{u}(\gamma), \mathfrak{u}(-\gamma)]$ and $[u(\gamma), u(\gamma')]$ are not zero and belong to $\mathfrak{z}(\mathfrak{a}_o)$ and $\mathfrak{u}_{\gamma+\gamma'}$ respectively. By definition,

$$\mathfrak{g}_{o,\alpha} = \oplus_{\gamma, \nu(\gamma)=\alpha} \quad \mathfrak{u}_\gamma. \tag{II.11.7}$$

The right-hand side may have several summands, since ν restricted to $\Delta(\mathbf{P}_{\iota,J}, \mathbf{A}_{\iota,J})$ is not necessarily injective. Therefore to deduce $(*)$ from $(+)$, we have to prove

$$c(\gamma) = c(\gamma') \text{ if } \nu(\gamma) = \nu(\gamma'). \tag{II.11.8}$$

(c) We first consider the case that $\mathfrak{a}_o$ has dimension one, hence where $\Delta_o = \{\alpha\}$ is a singleton. This subdivides into two cases, in view of II.10.9.

  (i) There exists exactly one element $\alpha' \in \Delta_\iota$ such that $r_{\iota,o}(\alpha') = \alpha$. Then $\mathbf{P}_{o,\emptyset}$ is proper maximal, equal to $\mathbf{P}_{\iota,J}$ $(J = \Delta_\iota - \{\alpha'\})$.
  In the highest root of the simple summand of $\mathfrak{g}$ of which γ is a root, the latter has coefficient 1 if Φ_o is reduced, and 2 if Φ_o is not reduced. In this case ν is bijective and $\mathfrak{g}_{0,\alpha} = \mathfrak{u}_\gamma$, $\mathfrak{g}_{o,2\alpha} = \mathfrak{u}_{2\gamma}$.

  (ii) There are two distinct simple roots γ, γ' that map onto α. Then

$$\mathfrak{g}_{o,\alpha} = \mathfrak{u}_\gamma \oplus \mathfrak{u}'_\gamma. \tag{II.11.9}$$

As recalled in II.9.6, it is shown in [DP1] that $\sigma(\gamma) = -\gamma'$, modulo a linear combination of elements in a subset Δ_1 of Δ_2, mapped to 0 under $r_{\iota,\sigma}$, hence also under $r_{\iota,o}$, which belongs therefore to J. As a consequence,

$$\sigma(\mathfrak{u}_\gamma) = \mathfrak{u}_{-\gamma'}, \qquad \sigma(\mathfrak{u}_{\gamma'}) = \mathfrak{u}_{-\gamma}.$$

Since σ normalizes $\mathcal{Z}_{\mathbf{G}}(\mathcal{A}_o)$, it follows that

$$c(\gamma) = c(-\gamma'), \qquad c(\gamma') = c(-\gamma),$$

so that (II.11.8) follows in this case from (II.11.5). Therefore $(*)$ holds for α. This concludes the proof if Φ_o is reduced. Assume it is not. Then either γ, γ' belongs to the same component of the Dynkin diagram of $\mathfrak{g}$, has coefficient one in its highest root, and

$$\mathfrak{g}_{o,2\alpha} = \mathfrak{u}_{\gamma+\gamma'}$$

is irreducible or they belong to different components. At least one of them has coefficient 2 in the relevant highest root.
If both have coefficient 2 in their respective highest roots, then

$$\mathfrak{g}_{o,\alpha} = \mathfrak{u}_{2\gamma} \oplus \mathfrak{u}_{2\gamma'}$$

and we see that $c(2\gamma) = c(2\gamma')$ as before (or use the above and (II.11.6)). If only one of them, say γ, has coefficient 2, then $\mathfrak{g}_{o,\alpha} = \mathfrak{u}_{2\gamma}$ is irreducible and we apply $(+)$.

(d) We now assume that $\dim \mathfrak{a}_o > 1$. Fix $\alpha \in \Delta_o$. We want to show that $(*)$ holds for α. This is a simple reduction to the previous case. Let $I = \{\alpha\}$. Consider the parabolic subgroups $\mathbf{P}_{o,I} = \mathbf{M}_{o,I} \cdot \mathbf{A}_{o,I} \cdot \mathbf{N}_{o,I}$. The subspace $\mathfrak{g}_{o,\alpha}$ belongs to $\mathfrak{m}_{o,I}$. The group $\mathbf{M}_{o,I}$ has (θ_I, σ_I)-rank one with maximal (θ_I, σ_I)-anisotropic torus $\mathbf{A}_o^I$ and simple root α (see II.10.4). Moreover, $\mathcal{Z}_{\mathbf{M}_{o,I}}(\mathbf{A}_o^I) \cdot \mathbf{A}_I = \mathcal{Z}_{\mathbf{G}}(\mathbf{A}_o)$. Hence (c), applied to $\mathbf{M}_{o,I}$ with respect to σ_I, shows that $(*)$ is true for α.

(e) We have now shown the existence of a constant $\varepsilon(\alpha) = \pm 1$ $(\alpha \in \Delta_o)$ such that the restriction of δ to $\mathfrak{g}_{o,\alpha}$ is equal to $\varepsilon(\alpha)\cdot\mathrm{Id}$. The assignment $\alpha \mapsto \varepsilon(\alpha)$ is a signature on Δ_o. It extends uniquely to a signature on Φ_o. We want to prove that the restriction of δ to $\mathfrak{g}_{o,\alpha}$ is equal to $\varepsilon(\alpha)\cdot\mathrm{Id}$ for all $\alpha \in \Phi_o$. From (b) we see that it is also true for $\alpha \in -\Delta_o$ and that it suffices to prove it for $\alpha > 0$. In view of (4), we are reduced to showing that if $\gamma \in \Phi(\mathbf{P}_{\iota,J}, \mathbf{A}_{\iota,J})$ is a J-shape, then

$$c(\gamma) = \varepsilon\big(\nu(\gamma)\big). \tag{II.11.10}$$

As usual the degree $d(\alpha)$ of α is the sum of the coefficients of the simple roots when α is written as a sum of simple roots. We want to prove (II.11.10) by induction on the degree of α. If it is one, this was shown in (c) and (d). So we let $d(\alpha) > 1$ and assume (II.11.10) to hold for all $\alpha' \in \Phi_o$ of degrees $< d(\alpha)$.
Let $\gamma \in \Phi(\mathbf{P}'_{\iota,J}\mathbf{A}_{\iota,J})$ be a shape such that $\nu(\gamma) = \alpha$. There exists $\beta \in \Phi_\iota^+$ that restricts to γ on $\mathbf{A}_{o,J}$ and therefore to α on A_o. Choose one of smallest possible degree. It is of course a sum of simple roots. There is at least one, say β_1, that has a strictly positive scalar product with β, (since the scalar product of β with itself is > 0). Then $\beta = \beta_1 + \beta_2$, where β_2 is also a root. We have $r_{\iota,o}(\beta_1) \neq 0$, because otherwise β_2 would be a root restricting to γ with degree $< d(\beta)$. Let $\alpha_1 = r_{\iota,o}(\beta_1)$. It belongs to Δ_0. We have $\gamma = \gamma_1 + \gamma_2$ where γ_1 and γ_2 are the shapes of β_1 and β_2, whence also $\alpha = \alpha_1 + \alpha_2$, where $\alpha_2 = \nu(\gamma_2)$ has degree $< d(\alpha)$. By definition, $\varepsilon(\alpha) = \varepsilon(\alpha_1) \cdot \varepsilon(\alpha_2)$; by (5), $c(\gamma) = c(\gamma_1) \cdot c(\gamma_2)$. By the induction assumption

$$\varepsilon(\alpha_1) = c(\gamma_1), \;\; \varepsilon(\alpha_2) = c(\gamma_2),$$

whence $\varepsilon(\alpha) = c(\gamma)$.

8

The Oshima–Sekiguchi Compactification of G/K and Comparison with $\overline{\mathbf{G}/\mathbf{H}}^W(\mathbb{R})$

This chapter has two main purposes: first to review the construction in [OsS1] of a smooth compactification of G/K, different from the Oshima compactification $\overline{X}^O$ (see §II.2.2), and second to compare it with the real locus of the wonderful compactification of $\mathbf{G}'/\mathbf{K}'$, where G' is the adjoint group of G and $\mathbf{G}'$ the complexification of G', and K' is a maximal compact subgroup of G' and $\mathbf{K}'$ the complexification of K'. In order to be consistent with the notation of the previous sections, we shall have to deviate in part from the conventions of [OsS1], but we shall try to relate the two.

After discussing some preliminaries on semisimple symmetric spaces such as the Weyl groups in §II.12, we introduce the Oshima–Sekiguchi compactification in §II.13. Its main properties are listed in §II.15. To compare with the real locus $\overline{\mathbf{X}}^W(\mathbb{R})$, an action of a Weyl group W' on $\overline{X}^{OS}$ is defined in §II.13.6. In §II.14, a different definition of $\overline{\mathbf{X}}^W$ due to Springer is introduced in order to facilitate the comparison with $\overline{X}^{OS}$. Then Theorem II.4.5 shows that $\overline{\mathbf{X}}^W(\mathbb{R})$ is a finite quotient of $\overline{X}^{OS}$.

II.12 Preliminaries on semisimple symmetric spaces

II.12.1 The starting point of [OsS1] is a connected semisimple real Lie group with finite center. We shall limit ourselves to the case that the group is linear but, on the other hand, is the full group of real points of the connected complex semisimple Lie group $\mathbf{G}$. It may therefore have several connected components (only one if $\mathbf{G}$ is simply connected). We note also that the construction of [OsS1] does not change if G is replaced by a locally isomorphic group (still with finite center). In particular it may be assumed to be of adjoint type (see 1.4 in [OsS1]).

We use the usual notation for Riemann symmetric spaces. In particular $\mathfrak{g} = \mathfrak{k} \oplus \mathfrak{p}$ is the Cartan decomposition of $\mathfrak{g}$ with respect to $\mathfrak{k}$ and $\mathfrak{a}$ is a maximal abelian subalgebra of $\mathfrak{p}$. In the general notation of II.8.1, it would be $\mathfrak{a}_\theta$ and would be equal to $\mathfrak{a}_{\theta,\sigma}$. Let $\Phi = \Phi(\mathfrak{g}, \mathfrak{a})$ and Δ the set of simple roots for a given ordering. Let $P_\emptyset = M_\emptyset \cdot A \cdot N_\emptyset$ be the standard minimal parabolic subgroup; hence

$$M_\emptyset = K \cap \mathcal{Z}_G(A) = \mathcal{Z}_K(A) \text{ and } \mathcal{Z}_G(A) = M_\emptyset \times A.$$

We claim that $M_\emptyset$ meets every connected component of K, hence of G.

Let $W = W(\mathfrak{g}, \mathfrak{a})$ be the Weyl group of G. By definition it is equal to $\mathcal{N}_K(\mathfrak{a})/\mathcal{Z}_K(\mathfrak{a})$. It is well known that it can also be written $N_G(A)/Z_G(A)$, but it is also equal to $\mathcal{N}_G(A)/\mathcal{Z}_G(A)$. In fact, the complexification $\mathbf{A}$ of A is a maximal $\mathbb{R}$-split torus of $\mathbf{G}$, so that Φ can also be viewed as the system of $\mathbb{R}$-roots of $\mathbf{G}$, the Weyl group of which is defined in the theory of algebraic groups as $\mathcal{N}_G(\mathbf{A})/\mathcal{Z}_G(\mathbf{A})$ and this is clearly equal to $\mathcal{N}_G(A)/\mathcal{Z}_G(A)$. Since the Cartan subalgebras of the symmetric pair $(\mathfrak{g}, \mathfrak{k})$ are conjugate under K^o, we see that any connected component of K contains an element centralizing $\mathfrak{a}$, as claimed above.

II.12.2 We specialize the considerations of §4 to the case $\sigma = \theta$. Given a proper signature ε on Φ, we have associated to it an involution θ_ε of G and defined the $\mathbb{R}$-form K_ε of K as G^θ (II.4.2, II.4.3). Besides, we will have to consider a subgroup of finite index of K_ε used in [OsS1], to be denoted here by K^*_ε.

θ_ε commutes with θ, as follows from its definition, hence θ leaves K_ε stable and induces on it a Cartan involution. Let

$$L_\varepsilon = K \cap K_\varepsilon \tag{II.12.1}$$

be its fixed point set. It is a maximal compact subgroup of K_ε and we have the Cartan decompositions

$$K_\varepsilon = L_\varepsilon \cdot \exp \mathfrak{p}_\varepsilon, \text{ where } \mathfrak{p}_\varepsilon = \mathfrak{p} \cap \mathfrak{k}_\varepsilon. \tag{II.12.2}$$

The Cartan algebra $\mathfrak{a}$ of (G, K) is contained in $\mathfrak{l}_\varepsilon$ and $A \cdot M = \mathcal{Z}_G A \subset L_\varepsilon$. Let U_ε be the subgroup of W defined by

$$U_\varepsilon = \mathcal{N} A \cap L_\varepsilon / \mathcal{Z} A. \tag{II.12.3}$$

It is the Weyl group of the symmetric pair $(K_\varepsilon, L_\varepsilon)$ with respect to A.

It is clear from the definitions that ε and K_ε determine one another. Therefore we also have

$$U_\varepsilon = \{w \in W \mid w(\varepsilon) = \varepsilon\}. \tag{II.12.4}$$

Let

$$K^*_\varepsilon = K^o_\varepsilon \cdot M. \tag{II.12.5}$$

This is the group denoted by K_ε in [OsS1]. It is normal, of finite index in our K_ε.

The last remark in II.12.1 shows that

$$G/K^*_\varepsilon = G^o/(K^*_\varepsilon \cap G^o). \tag{II.12.6}$$

Let

$$\Phi_\varepsilon = \{\alpha \in \Phi, \varepsilon(\alpha) = 1\}. \tag{II.12.7}$$

It is a root system in the subspace of $\mathfrak{a}^*$ it generates, closed in Φ (i.e., $\alpha, \beta \in \Phi_\varepsilon, \alpha + \beta \in \Phi \Rightarrow \alpha + \beta \in \Phi_\varepsilon$). Let W_ε be the Weyl group of Φ_ε. It follows from 1.6, 1.7 in [OsS1] that Φ_ε is the restricted root system of the Riemannian symmetric pair $(K_\varepsilon^*, K_\varepsilon^* \cap K)$, hence

$$W_\varepsilon = (K_\varepsilon^*) \cap \mathcal{N}(A)/\mathcal{Z}A. \tag{II.12.8}$$

The group K_ε^* is normal in K_ε, and

$$K_\varepsilon/K_\varepsilon^* = U_\varepsilon/W_\varepsilon. \tag{II.12.9}$$

Therefore $U_\varepsilon/W_\varepsilon$ acts freely via right translations on G/K_ε^* and the quotient is G/K_ε.

II.12.3 In II.12.2, ε was assumed to be proper. We now have to extend these considerations to nonproper signatures.

Fix $J \subset \Delta$. We shall now consider signatures with support $J_\varepsilon = J$. In §II.4, II.6 we have introduced the parabolic $\mathbb{R}$-subgroup $\mathbf{P}_J = \mathbf{M}_J \cdot \mathbf{A}_J \cdot \mathbf{N}_J$, and its subgroup $\mathbf{Q}_J = \mathbf{K}_J \cdot \mathbf{A}_J \cdot \mathbf{N}_J$, where $\mathbf{K}_J = \mathbf{M}_J^\theta$ is the fixed point of the restriction of θ to $\mathbf{M}_J$. Thus K_J is a maximal compact subgroup of M_J. To $\varepsilon \in \mathcal{E}^o(J)$ there is associated a real form $K_{J,\varepsilon}$ of $\mathbf{K}_J$ and the subgroup $Q_{J,\varepsilon} = K_{J,\varepsilon} \cdot A_J \cdot N_J$, where K_{J_ε} is the fixed point set of θ_{J_ε} on M_J.

We have also defined the root system Φ_J, consisting of the roots that are linear combinations of elements in J and its Weyl group W_J. As above, we let

$$\Phi_{J,\varepsilon} = \{\alpha \in \Phi_J, \varepsilon(\alpha) = 1\} \tag{II.12.10}$$

and let $W_{J,\varepsilon}$ be its Weyl group.

The counterparts in [OsS1] of our groups $K_{J,\varepsilon}$, $Q_{J,\varepsilon}$ are denoted here by $K_{J,\varepsilon}^*$ and $Q_{J,\varepsilon}^* = K_{J,\varepsilon}^* A_J \cdot N_J$. By definition, $K_{J,\varepsilon}^*$ is generated by $(K_{J,\varepsilon})^o$ and the centralizer of A in K_J or, what is the same, of A^J in K_J. By 1.6, 1.7 of [OsS1] applied to M_J we have

$$W_{J,\varepsilon} = \big(K_{J,\varepsilon}^* \cap \mathcal{N}(A)\big)/\mathcal{Z}A. \tag{II.12.11}$$

The group $K_{J,\varepsilon}^*$ is normal in $K_{J,\varepsilon}$ and we have, as before,

$$K_{J,\varepsilon}/K_{J,\varepsilon}^* = U_{J,\varepsilon}/W_{J,\varepsilon}, \tag{II.12.12}$$

where $U_{J,\varepsilon}$ is the stability group of ε in W_J.

As in II.12.2, it follows that $U_{J,\varepsilon}/W_{J,\varepsilon}$ operates freely on $K_{J,\varepsilon}^*$ by right translations and the quotient is $K_{J,\varepsilon}$. Similarly, since W_J leaves A_J and N_J stable, right translations also induce a free action of $U_{J,\varepsilon}/W_{J,\varepsilon}$ on $G/Q_{J,\varepsilon}^*$ and the quotient is $G/Q_{J,\varepsilon}$.

As in [OsS1], $W(\varepsilon)$ denotes the canonical set of representatives of $W_{J,\varepsilon}\backslash W_{J,\varepsilon}$ defined by

$$W_J(\varepsilon) = \{w \in W_{J_\varepsilon}, \Phi_\varepsilon \cap w\Phi^+ = \Phi_\varepsilon \cap \Phi^+\}.$$

[It is indeed a set of representatives since $W_{J,\varepsilon}$ is simply transitive on the Weyl chambers of $\Phi_{J,\varepsilon}$.]

Remark. In the above, J is implied by ε so that it could be omitted from the notation and we could use just ε as index, whether it is proper or not. This is the point of view of [OsS1]. Although it leads to a clumsier notation, we prefer to make J explicit, except when ε is proper, i.e., $J = \Delta$.

II.12.4 In case Φ is not reduced, [OsS1] introduces some subgroups or subsets of the above. Let

$$\Phi' = \{\alpha \in \Phi, 2\alpha \notin \Phi, \alpha/2 \notin \Phi), \qquad W' = \langle s_\alpha, \alpha \in \Phi)\rangle.$$

The intersection with Φ' or W' of the previously defined objects will be indicated by a superscript $'$, for instance $W'_\varepsilon = W_\varepsilon \cap W'$, $W'_{J_\varepsilon} = W_{J_\varepsilon} \cap W'$, etc.

If Φ is reduced, then $\Phi = \Phi'$ and $W = W'$. If not, and if Φ is irreducible, then Φ is of type BC_r (in the notation of [Bu2]), and Φ' is a root system of type D_r, with Weyl group W' of index 2 in W. (But Φ' is not closed in Φ.)

By 2.5(iii) in [OsS1], $W(\varepsilon)$ is also a set of representatives for $W'_\varepsilon\backslash W'_{J_\varepsilon}$, i.e.,

$$W'_{J_\varepsilon} = W'_\varepsilon \cdot W'_{J_\varepsilon}(\varepsilon), \quad W_{J_\varepsilon} = W_\varepsilon \cdot W_{J_\varepsilon}(\varepsilon). \tag{II.12.13}$$

We summarize the proof. It suffices to consider the case of a simple nonreduced root system of type BC_r. In this case the reflection s_i to $x_i = 0 (1 \le i \le r)$ is a representative of W/W'. Note that if $\varepsilon(x_r) \neq 0$, then $\varepsilon(x_{2r}) = 1$, hence $s_r \in W_{J,\varepsilon}$, and a fortiori $s_r \in W_J$. Thus $W'_{J,\varepsilon}$ (resp. W'_J) is of index 2 in $W_{J,\varepsilon}$ (resp. W_J) and our assertion is proved. If now $\varepsilon(x_r) = 0$, then $W_{J,\varepsilon} = W'_{J,\varepsilon}$ and the claim is obvious.

In analogy with (II.12.13), we want to prove

$$K_{J,\varepsilon}/K^*_{J,\varepsilon} = U'_{J,\varepsilon}/W'_{J,\varepsilon}, \qquad \text{where } U'_{J,\varepsilon} = W'_J \cap U_{J,\varepsilon}. \tag{II.12.14}$$

The argument is the same as above: assume that $\varepsilon(x_r) \neq 0$. Then $s_r \in W_{J,\varepsilon}$; hence a fortiori $s_p \in U_{J,\varepsilon}$, and again $U'_{J,\varepsilon}$ (resp. $W'_{J,\varepsilon}$) is of index 2 in $U_{J,\varepsilon}$ (resp $W_{J,\varepsilon}$). If $\varepsilon(x_r) = 0$ then $W_J = W'_J$; hence $U'_{J,\varepsilon} = U'_{J,\varepsilon}$ and $W_{J,\varepsilon} = W'_{J,\varepsilon}$. Therefore, it is again true that the quotient of $K^*_{J,\varepsilon}$ (resp. $G/Q^*_{J,\varepsilon}$) by right translations of $U'_{J,\varepsilon}/W'_{J,\varepsilon}$ is equal to $K_{J,\varepsilon}$ (resp. $G/Q_{J,\varepsilon}$).

We shall later need the following lemma, where we keep the previous notation.

II.12.5 Lemma. *Let $J \subset \Delta$. Every element of W_J not contained in W'_J leaves invariant every signature in $\mathcal{E}^o(\Phi_J)$. In particular $\mathcal{E}^o(\Phi_J)/W_J = \mathcal{E}^o(\Phi_J)/W'_J$.*

Proof. It suffices to consider the case that Φ_J is of type BC_m ($m = |J|$). Let x_i be coordinates in $\mathbb{R}^m$. Then the roots are

$$\pm x_i \pm x_j \qquad (1 \leq i < j \leq m), \text{ and } \pm x_i, \pm 2x_i (1 \leq i \leq m).$$

Let $\varepsilon \in \mathcal{E}^o(J)$. Then $\varepsilon(2x_j) = \varepsilon(x_j)^2 = 1$, and therefore, if $i \neq j$,

$$\varepsilon(x_i + x_j) = \varepsilon(x_i - x_j + 2x_j) = \varepsilon(x_i - x_j).$$

The elements of W_J not contained in W'_J are the reflections s_i to the hyperplanes $x_i = 0$. The map s_i transforms x_i to $-x_i$ and leaves x_j fixed for $j \neq i$. Let $\alpha = x_a \mp x_b$ $(a \neq b)$. Then it is fixed by s_i if $a, b \neq i$. If one of them, say b, is equal to i, then $s_i(x_a \pm x_b) = x_a \mp x_b$, and the lemma follows from the previous remarks.

II.13 The Oshima–Sekiguchi compactification of G/K

II.13.1 It will be convenient to introduce one more notation, borrowed from [OsS1]. Let $t \in \mathbb{R}^\Delta$. There is associated to it a signature $\varepsilon_t \in \mathcal{E}(\Phi)$, a support J_t of t or ε_t in Δ, a standard parabolic subgroup P_{J_t}, and then a subgroup Q^*_J. All these data are determined by t and so can also be indexed simply by t. In particular, we shall also write Q^*_t for $Q^*_{J,\varepsilon}$ where $J = J_t$, and ε is identified with its restriction to J_t, and ϕ_t for ϕ_{ε_t}, etc.

Let $\{H_\alpha\}$ $(\alpha \in \Delta)$ be the basis of $\mathfrak{a}$ dual to Δ. For $J \subset \Delta$, let

$$\mathfrak{a}^J = \langle H_\alpha, \alpha \in J \rangle,\ A^J = \exp \mathfrak{a}^J. \tag{II.13.1}$$

Then A is the direct product of A^J and A_J. In analogy with [OsS1] we define a map $a : \mathbb{R}^\Delta \to A$ by the rule

$$a(t) = \exp -\frac{1}{2} \sum_{\alpha \in J_t} \log |t^\alpha|\ H_\alpha \qquad (t \in R), \tag{II.13.2}$$

or equivalently,

$$a(t)^\alpha = |t^\alpha|^{-1/2}\ (\alpha \in J_t),\ a(t)^\alpha = 1\ (\alpha \notin J_t). \tag{II.13.3}$$

For $J \subset \Delta, \varepsilon \subset \mathcal{E}^o(J)$, the map a is an isomorphism of $\mathbb{R}^{J,\varepsilon}$ onto A^J.

Remark. This notation does not agree with that used in I for A^J. Call here $\tilde{A}^J$ the A^J as defined there using a basis of $\mathfrak{a}$ formed by the simple coroots. We also have $A = \tilde{A}^J \cdot A_J$ and so $\tilde{A}^J$ and A^J are isomorphic mod A_J, and two equivalent points act in the same way on M_J. In particular the root spaces of A^J are the same as the weight spaces of $\tilde{A}^J$.

II.13.2 Let $\tilde{X} = G \times \mathbb{R}^\Delta \times W'$. [OsS1] introduces the following relation $\sim$ on $\tilde{X}$:

$$(g, t, w) \sim (g', t', \omega')$$

if

(i) $w(\varepsilon_t) = w'(\varepsilon_{t'})$,
(ii) $w^{-1} \cdot w' \in W(\varepsilon_t)$,
(iii) $g \cdot a(t) \cdot Q_t^* \cdot w^{-1} = g' \cdot a(t') \cdot Q_{t'}^* \cdot w'^{-1}$.

The Oshima–Sekiguchi compactification, to be denoted here by $\overline{X}^{OS}$, is by definition the quotient $\tilde{X}/\sim$, the action of G being defined by left translations on the first factor of $\tilde{X}$.

II.13.3 We have first to see that $\sim$ is an equivalence relation and that left translations by G on $\tilde{X}$ are compatible with it.

From (i) and the definitions, we get

$$w \cdot \Phi_t = w' \cdot \Phi_{t'} \; w \cdot \Phi_{J_t} = w' \cdot \Phi_{J_{t'}}, \tag{II.13.4}$$

hence also

$$^{w}W_{J_t} = {}^{w'}W_{Jt'} \cdot {}^{w}W_t = ^{w'} Wt'. \tag{II.13.5}$$

By (ii), $w^{-1} \cdot w' \in W(t)$, hence $w'^{-1} \cdot w \subset W_{J_t}$, and we get from (II.13.3)

$$W_{J_t} = W_{J_{t'}} \text{ hence } \Phi_{J_t} = \Phi_{J_{t'}}, \quad J_t = J_{t'}. \tag{II.13.6}$$

By (ii),

$$\Phi_t \cap w^{-1} \cdot w'\Phi^+ = \Phi_t \cap \Phi^+. \tag{II.13.7}$$

Together with (II.13.4) this implies

$$\begin{aligned} &w(\Phi_t \cap \Phi^+) = w(\Phi_t \cap w'\Phi^+) = w'\Phi_{t'} \cap w'\phi^+, \\ &w(\Phi_t \cap \Phi^+) = w'(\Phi_{t'} \cap \Phi^+); \end{aligned} \tag{II.13.8}$$

therefore

$$\Phi_{t'} \cap \Phi^+ = w'^{-1} \cdot w(\Phi_t \cap \Phi^+) = (\Phi_{t'} \cap w'^{-1} \cdot w\phi^+), \tag{II.13.9}$$

which shows that $w'^{-1} \cdot w \in W(\varepsilon_t)$, hence that $\sim$ is symmetric.

The proof that it is transitive is left to the reader.

II.13.4 For $w \in W$, we fix a representative $\bar{w}$ of w in $\mathcal{N}A$. Note however that a conjugate $^{\bar{w}}L$ or a translate $\bar{w} \cdot L$ by $\bar{w}$ of a subgroup L containing M depends only on w and we shall also denote by ^{w}L or $w \cdot L$. Similarly, W is viewed as usual as a group of automorphisms of A.

The subgroup A_{J_t} (resp. $A_{J_{t'}}$) is the intersection of kernels of the elements in Φ_{J_t} (resp. $\Phi_{J_{t'}}$). Therefore (II.13.6) implies

$$A_{J_t} = A_{J_{t'}}, \quad N_{J_t} = N_{J_{t'}}. \tag{II.13.10}$$

We claim that

$$^{\bar{w}^{-1}\cdot\bar{w}'}Q_{t'}^* = Q_t^*. \tag{II.13.11}$$

Assume for the moment this to be true. Then (iii) can be written

$$g \cdot a(t) \cdot \bar{w}^{-1} \cdot \bar{w}' \cdot Q_t^* = g' \cdot a(t') \cdot Q_t^* \ \text{ in } G/Q_t^*, \tag{II.13.12}$$

which makes it clear that $\sim$ is compatible with left translations on G.

We still have to prove (II.13.11). Since $w^{-1} \cdot w' \in W_{J_t} = W_{J_{t'}}$, it normalizes A_{J_t} and N_{J_t}, as well as $M_{J_t}^* = M_{J_{t'}}^*$. There remains to see that $\bar{w}^{-1} \cdot \bar{w}'$ transforms $M_{J_{t'},\varepsilon'}^*$ onto $M_{J_t,\varepsilon}^*$. By definition, these groups are generated by M and their identity components. It suffices therefore to prove

$$\mathrm{Ad}\,(\bar{w}^{-1} \cdot \bar{w}')\mathfrak{m}_{J_{t'\varepsilon'}} = \mathfrak{m}_{J_t,\varepsilon}. \tag{II.13.13}$$

This is a straightforward computation carried out in [OsS1], p. 15 (in the case that ε_t and $\varepsilon_{t'}$ are proper): by definition

$$\mathfrak{m}_{J_{t'},\varepsilon'} = \mathfrak{m} \oplus \oplus_{\alpha\in\Phi_{J_{t'}}} \langle x + \varepsilon(\alpha)\theta x\rangle_{x\in\mathfrak{g}_\alpha};$$

hence

$$\mathrm{Ad}\,(\bar{w}^{-1} \cdot \bar{w}')(\mathfrak{m}_{J_{t'},\varepsilon'}) = \mathfrak{m} \oplus \oplus_{\alpha\in\Phi_{J_t}} \langle\ \mathrm{Ad}\,(\bar{w}^{-1} \cdot w')x + \varepsilon(\alpha)\theta(\mathrm{Ad}\,\bar{w}^{-1} \cdot \bar{w}'x\rangle_{x\in\mathfrak{g}_\alpha}.$$

We have

$$\mathrm{Ad}\,(\bar{w}^{-1} \cdot \bar{w}')x \in \mathfrak{g}_{(w^{-1}\cdot w')\alpha}\ \varepsilon_{t'}(\alpha) = \varepsilon_t(\bar{w}^{-1} \cdot \bar{w}' \cdot \alpha),$$

(II.13.11) now follows immediately.

Remark. We see from (II.13.12) that the equivalence classes are the left cosets of the various groups Q_t^*, and that two cosets Q_t^* and $h \cdot Q_t^*$ represent the same equivalence class if and only if $h \in W(\varepsilon)$. This explains (II.13.4) in II.13.5.

II.13.5 Main Properties of $\overline{X}^{OS}$.

They are stated in Theorem 2.7 of [OsS1] and proved there on pp.16–19. We review here the main ones. Let $\pi : \tilde{X} \to \overline{X}^{OS}$ be the canonical projection.

(a) $\overline{X}^{OS}$ is a connected compact Hausdorff analytic G-space. If $x = (g, t, w) \in \tilde{X}$, the orbit of $\pi(x)$ is isomorphic to G/Q_t^*. It is open if and only if ε_t is proper, closed if and only if $t = 0$, in which case the orbit is isomorphic to $G/P_\emptyset$.
(b) We shall again use the labeling of orbit types by $\varepsilon \in \mathcal{E}(\Delta)$ and the support $J = J_\varepsilon$ of ε. i.e., $O_{J,\varepsilon}^* = G/Q_{J,\varepsilon}^*$. A given orbit $O_{J,\varepsilon}^*$ may occur several times. More precisely, since $[W' : W_\varepsilon'] = W(\varepsilon)$, the number of orbits of type $O_{J,\varepsilon}$ is

$$[W' : W_J'] \cdot |W_\varepsilon'| = |W'| \cdot |W(\varepsilon)|^{-1}. \tag{II.13.14}$$

If ε is proper, then $J = \Delta$ and $W_J = W'$; hence

$$G/K_\varepsilon^* \text{ occurs } |W'_\varepsilon| \text{ times.} \tag{II.13.15}$$

In particular, if $\mathcal{E} = \mathcal{E}_1$ is the signature $\mathcal{E}(\alpha) = 1$ for all α, then $K_\varepsilon^* = K$; hence

$$\text{there are } |W'| \text{ copies of } G/K. \tag{II.13.16}$$

If ε is the zero signature, then $W'_J = W'_\varepsilon = W(\varepsilon) = \{1\}$ and

$$\text{there are } |W'| \text{ copies of } G/P_\emptyset. \tag{II.13.17}$$

It is also shown that every orbit G/K has one orbit $G/P_\emptyset$ in its closure. More generally G/K_ε^* has $[W' : W'_\varepsilon]$ orbits $G/P_\emptyset$ in its closure.

We shall establish these properties later (II.14.8), by arguments different from those of [OsS1], except for the connectedness of $\overline{X}^{OS}$.

II.13.6 A free action of W' on $\overline{X}^{OS}$.

First we define a right action of W' on $\tilde{X}$ by translations on the last factor:

$$(g, t, w) \circ u = (g, t, u^{-1} \cdot w) \qquad (g \in G, t \in \mathbb{R}^\Delta, u, w \in W'). \tag{II.13.18}$$

It is obviously a right action:

$$(\circ\, u) \circ (\circ\, v) = \circ(u \cdot v) \qquad (u, v \in W'),$$

which is free and which commutes with G. We leave it to the reader to check that it is compatible with $\sim$, whence an action on $\overline{X}^{OS}$, to be also denoted by $\circ\, u$, which commutes with G. We want to prove that it is *free*. Let then $u \in W'$ and assume that $(g.t.w)$ is equivalent to $(g, t, u^{-1} \cdot w)$. We have to show that $u = 1$. By (i) in II.13.2,

$$w(\varepsilon) = u^{-1} \cdot w(\varepsilon), \text{ hence } {}^{u^{-1} \cdot w}U_\varepsilon = {}^{w}U_\varepsilon, \tag{II.13.19}$$

where ε stands for ε_t. This implies

$${}^{u^{-1} \cdot w}Q_t^* = {}^{w}Q_t^*, \; {}^{u^{-1} \cdot w}W_\varepsilon = {}^{w}W_\varepsilon. \tag{II.13.20}$$

By (ii) in II.13.2 we have $w^{-1} \cdot u^{-1} \cdot w \in W'(\varepsilon)$, hence

$$u^{-1} \in {}^{w}W'(\varepsilon). \tag{II.13.21}$$

By (iii) of II.13.2, or rather the equivalent statement (II.13.12), we have

$$g \cdot a(t) \cdot w^{-1} \cdot u^{-1} \cdot w \cdot Q_t^* = g \cdot a(t) \cdot Q_t^*,$$

hence, by (II.12.12)

$$u^{-1} \in {}^{w}W'_\varepsilon. \tag{II.13.22}$$

Since $W'_\varepsilon \cap W'(\varepsilon) = \{1\}$, our assertion now follows from (I.13.21) and (I.13.22).

II.14 Comparison with $\overline{\mathbf{G}/\mathbf{H}}^W(\mathbb{R})$

II.14.1 We first recall a definition of the wonderful compactification $\overline{\mathbf{X}}^W$ of the symmetric variety $\mathbf{X}$ analogous to that of $\overline{X}^O$, due to Springer [Sp1]. There, a general involution is considered, but we limit ourselves to the case of a Cartan involution. $\mathbf{G}$ is assumed to be adjoint. Let $\mathbb{C}^\Delta$ be the affine space with coordinates in Δ. As usual, the value of α on $c \in \mathbb{C}^\Delta$ is denoted by c^α. The coordinate ring of $\mathbb{C}^\Delta$ is of course $\mathbb{C}[\Delta]$. The coordinate ring of $\mathbf{A}$ is the polynomial algebra in α and $\alpha^{-1} (\alpha \in \Delta)$. The involution θ is the inversion on $\mathbf{A}$ and its fixed points form the subgroup ${}_2\mathbf{A} = \mathbf{A} \cap \mathbf{K}$ of elements of order ≤ 2. Let $\bar{\mathbf{A}} = \mathbf{A}/{}_2\mathbf{A}$. Its coordinate ring is the polynomial algebra in $\alpha^2, \alpha^{-2} (\alpha \in \Delta)$. Given $J \subset \Delta$, we let

$$
{}^o\mathbb{C}^J = \{c \in \mathbb{C}^\Delta, c^\alpha \neq 0 \text{ if and only if } \alpha \in J\}. \tag{II.14.1}
$$

Therefore

$$
\mathbb{C}^\Delta = \coprod_{J \subset \Delta} {}^o\mathbb{C}^J. \tag{II.14.2}
$$

and given $c \in \mathbb{C}^\Delta$, there is a unique subset of Δ, to be denoted by J_c, such that $c \in {}^o\mathbb{C}^J$. In analogy with II.13.1, we define a map $\tilde{a} : \mathbb{C}^\Delta \to \bar{\mathbf{A}}$ by the rule

$$
\tilde{a}(c)^{2\alpha} = c^\alpha \ (\alpha \in J_c) \quad \tilde{a}(c)^{2\alpha} = 1 \ (\alpha \notin J_c). \tag{II.14.3}
$$

Thus $\tilde{a}({}^o\mathbb{C}^J) = \bar{\mathbf{A}}^J = \{t \in \bar{\mathbf{A}}, t^{2\beta} = 1 \text{ if } \beta \notin J\}$. In particular, $\bar{\mathbf{A}} = \bar{\mathbf{A}}_J \cdot \bar{\mathbf{A}}^J$.

II.14.2 Springer defines on $\tilde{\mathbf{Y}} = \mathbf{G} \times \mathbb{C}^\Delta$ an equivalence relation $\sim$ by the rule $(g, c) \sim (g', c')$ if and only if

(i) $J_c = J_{c'}$,
(ii) $g \cdot \tilde{a}(c) \cdot \mathbf{Q}_{J_c} = g' \cdot \tilde{a}(c') \cdot \mathbf{Q}_{J_c}$.

The equivalence classes are therefore the left cosets of the groups $\mathbf{Q}_J$ $(J \subset \Delta)$, and the orbits of $\mathbf{G}$ are the quotients $\mathbf{O}_J$, as in $\overline{\mathbf{X}}^W$. Let $\mathbf{N}^-$ be the unipotent radical of $\theta(\mathbf{P}_\emptyset)$. It is obvious that

$$
\{(u, c) \sim (u', c'), u, u' \in \mathbf{N}^-, c, c' \in \mathbb{C}^\Delta\} \Leftrightarrow u = u', c = c'. \tag{II.14.4}
$$

The projection $\tilde{\pi} : \tilde{\mathbf{Y}} \to \mathbf{Y} = \tilde{\mathbf{Y}}/\sim$ maps $\mathbf{N}^- \times \mathbb{C}^\Delta$ bijectively onto a subset U_o of $\mathbf{Y}$. It admits of course a structure of an affine variety. The assertion is that its translates by $\mathbf{G}$ are open charts of a structure of a smooth projective variety naturally isomorphic to $\overline{\mathbf{X}}^W$. The set U_o is the disjoint union of the $\mathbf{N}^- \times \bar{\mathbf{A}}^J$. In particular its intersection with $\mathbf{G}/\mathbf{K}$ is $\mathbf{N}^- \times \bar{\mathbf{A}}$. More generally, the intersection with $\mathbf{O}_J$ is the image of $\mathbf{N}^- \times {}^o\mathbb{C}^J$ and is isomorphic to $\mathbf{N}^- \times \mathbf{A}^J$.

II.14.3 Let $j : \mathbb{R}^\Delta \to \mathbb{C}^\Delta$ be defined by

$$j(t)^\alpha = t^\alpha \qquad (\alpha \in \Delta). \tag{II.14.5}$$

It identifies the left-hand side with the real points of the right-hand side. It maps $\mathbb{R}^{J,\varepsilon}(\varepsilon \in \mathcal{E}^o(J))$ into ${}^o\mathbb{C}^J$; the union over ε of the images is the set of real points of ${}^o\mathbb{C}^J$. We also assume the group G involved in the definition of $\overline{X}^{OS}$ to be adjoint. Let

$$\mu : \tilde{X} = G \times \mathbb{R}^\Delta \times W' \to \tilde{\mathbf{Y}} = \mathbf{G} \times \mathbb{C}^\Delta \tag{II.14.6}$$

be defined by

$$\mu(g, t, w) = \big(g, j(t)\big). \tag{II.14.7}$$

It commutes with left translations by G and with W' acting on the right as in II.13.6. Our goal is to prove that it is compatible with the given equivalence relations on both sides and induces a continuous bijection of $\overline{X}^{OS}/W'$ onto $\overline{\mathbf{X}}^W(\mathbb{R})$. The next subsection is denoted to some preparation.

II.14.4 We can arrange that the inclusion of G in $\mathbf{G}$ maps A isomorphically onto $\bar{A}^o$. Let us denote by $\bar{j}$ this isomorphism. It also identifies A^J to $(\bar{A}^J)^o$ $(J \subset \Delta)$.

Fix $t \in \mathbb{R}^\Delta$. It defines a signature $\varepsilon = \varepsilon_t$ and a support $J = J_t$. The signature also defines an element $s_t \in {}_2\bar{\mathbf{A}}$ and we let $u_t \in {}_4\bar{\mathbf{A}}_u$ be a square root of s_t. We claim that

$$\tilde{a}\big(j(t)\big) = \bar{j}\big(a(t)\big) \cdot u_t. \tag{II.14.8}$$

This is a straightforward computation. By definition,

$$a(t)^{-\alpha} = |t^\alpha|^{1/2} (\alpha \in J_t),\ a(t)^\alpha = 1\ (\alpha \in J_t),$$

hence also

$$a(t)^{-2\alpha} = \begin{cases} |t^\alpha| & \alpha \in J_t, \\ 1 & \alpha \notin J_t, \end{cases}$$

$$\tilde{a}\big(j(t)\big)^{-2\alpha} = \begin{cases} j(t)^\alpha & \alpha \in J_t, \\ 1 & \alpha \notin J_t. \end{cases}$$

On the other hand,

$$\big(\bar{j}\big(a(t)\big) \cdot u_t\big)^{-2\alpha} = \bar{j}\big(a(t)\big)^{-2\alpha} \cdot u_t^{-2\alpha},$$

$$\big(j(t)\big)^\alpha = \begin{cases} t^\alpha = & \varepsilon(\alpha') \cdot |t^\alpha|\ (\alpha \in J_t), \\ 1 & \alpha \notin J_t, \end{cases}$$

$$(u_t)^{-2\alpha} = s_t^{-\alpha} = \begin{cases} \varepsilon(\alpha) & \alpha \in J_t, \\ 1 & \alpha \notin J_t, \end{cases}$$

so that

$$\big(\bar{j}(a(t)) \cdot u_\varepsilon\big)^{-2\alpha} = t^\alpha = \tilde{a}\big(j(t)\big)^{-2\alpha} \quad (\alpha \in J_t),$$
$$\big(\bar{j}(a(t)) \cdot u_\varepsilon\big)^{-2\alpha} = \tilde{a}\big(j(t)^{-2\alpha} = 1 \quad (\alpha \notin J_t).$$

Note that the right hand side of (II.14.8) belongs to the real points of $\mathbf{G}/\mathbf{O}_J$, by the remark 1 in II.10.11.

Let us denote in the following proof by $\sim_{OS}$ and $\sim_S$ the equivalence relations used in (II.13.2) and (II.14.2).

II.14.5 Theorem. *The map $\mu : \tilde{X} \to \tilde{\mathbf{Y}}$ defined in II.14.3 is compatible with $\sim_{OS}$ and $\sim_S$. It induces a continuous G-equivariant bijection of $\overline{X}^{OS}/W'$ onto $\overline{\mathbf{X}}^W(\mathbb{R})$.*

Proof. We have to show that

$$(g, t, w) \sim_{OS} (g', t', w') \Rightarrow \big(g, g(t)\big) \sim_S \big(g', j(t')\big).$$

By (II.13.10), $J_t = J_{t'}, N_{J_t} = N_{J_{t'}}$. This implies (see II.14.3) that $J_{j(t)} = J_{j(t')}$, whence the first condition of $\sim_S$.

By (II.13.12),

$$g \cdot a(t) \cdot \bar{w}^{-1} \cdot \bar{w}' Q_t^* = g' \cdot a(t') \cdot Q_t^*$$

which we can also write

$$g \cdot \bar{j}\big(a(t)\big) \cdot \bar{w}^{-1} \cdot \bar{w}' Q_t^* = g' \bar{j}\big(a(t')\big) \cdot Q_t^*. \tag{II.14.9}$$

Multiply both sides by u_t on the right. Note that $u_\varepsilon \cdot Q_{J,\varepsilon} \cdot u_\varepsilon^{-1} \subset \mathbf{Q}_J$. Moreover, $\bar{w}^{-1} \cdot \bar{w}' \in W_J \subset \mathbf{Q}_J$. Therefore, (II.14.8) shows that the second condition of $\sim_S$ is fulfilled.

As a consequence, μ defines a G-equivariant continuous map

$$\bar{\mu} : \overline{X}^{OS}/W' \to \overline{\mathbf{X}}^W(\mathbb{R}).$$

The G-orbits in $\overline{X}^{OS}$ are the quotients $G/Q^*_{J,\varepsilon'}\big(\varepsilon \in \mathcal{E}^o(J)\big)$, which are finite coverings of our $G/Q_{J,\varepsilon} = O_{J,\varepsilon}$. There are $[W' : W'_J] \cdot |W'_{J,\varepsilon}|$ copies of $G/Q_{J,\varepsilon}$. If we divide out by representatives of $[W' : W'_J] \cdot W'_{J,\varepsilon}$, they collapse to one copy of $O_{J,\varepsilon}$. If we divide further by $U_{J,\varepsilon}/W_{J,\varepsilon}$ (see II.12.4), then we get one copy of $O_{J,\varepsilon}$. Translation by an element of W'_J maps $O_{J,\varepsilon}$ isomorphically onto $O_{J,w(\varepsilon)}$ (see II.4.4). Altogether, the orbits in $\overline{X}^{OS}/W'$ are in a natural bijection with $\prod_{J \subset \Delta} \mathcal{E}^o(\Phi_J)/W'_J$, and we can replace W'_J by W_J in view of II.12.5. As we saw in II.6, this last quotient may also be identified with ${}_2\mathbf{A}^J/W_J$. As a consequence the G-orbits in $\overline{X}^{OS}/W'$ are parametrized in the same way as its orbits in $\overline{X}^W(\mathbb{R})$ (see II.6.5, II.10.11). There remains to see that this bijection is induced by $\bar{\mu}$. This follows from (II.14.8): if we let $\varepsilon = \varepsilon_t$ and $J = J_t$, this shows that the orbit $O_{J,\varepsilon}$ goes over to the orbit of $u_\varepsilon \cdot o$ in $(\mathbf{G}/\mathbf{Q}_J)(\mathbb{R})$.

II.14.6 For general topology, we refer to [Bu1]. In particular "compact" implies Hausdorff separation, and "quasi-compact" means that the Borel–Lebesgue cover axiom is satisfied: every open cover contains a finite one. These conventions are different from those of [OsS1], where compact (resp. compact and Hausdorff) stands for quasi-compact (resp. compact) of [Bu1].

II.14.7 The map $\bar{\mu}$ is continuous, bijective. We claim that it is a homeomorphism. First we shall show that it is open. We have a commutative diagram

$$\begin{array}{ccc} G \times \mathbb{R}^{\Delta} \times W' & \xrightarrow{\mu} & \mathbf{G} \times \mathbb{C}^{\Delta} \\ \downarrow{\scriptstyle \pi_1} & & \\ \overline{X}^{OS} & & \downarrow{\scriptstyle \pi_3} \\ \downarrow{\scriptstyle \pi_2} & & \\ \overline{X}^{OS}/W' & \xrightarrow{\bar{\mu}} & \overline{\mathbf{X}}^{W} \end{array} \tag{II.14.10}$$

where π_i $(i = 1, 2, 3)$ is the natural projection. We have seen in II.14.2 that π_3 maps $\mathbf{N}^- \times \mathbb{C}^{\Delta}$ bijectively onto an open chart, denoted by U_0', of $\overline{\mathbf{X}}^W$. It translates by g from an open cover of $\overline{\mathbf{X}}^W$. This implies that $\pi_3(N^- \times \mathbb{R}^{\Delta})$ is an open chart on $\overline{\mathbf{X}}^W(\mathbb{R})$. Since u_ε normalizes $\mathbf{N}^-$, it also follows that the translates $u_\varepsilon(N^- \times R^{\Delta}) = N^- \times u_\varepsilon \cdot o$ are open charts on $\overline{\mathbf{X}}^W(\mathbb{R})$, which then form an open cover by real affine sets.

μ maps $(N^- \times \mathbb{R}^{\Delta} \times w)$ homeomorphically onto $N^- \times \mathbb{R}^{\Delta}$. The commutativity of (II.14.10) implies that given $g \in G$, $w \in W$, π_1 induces a homeomorphism of $(g \cdot N^- \times \mathbb{R}^{\Delta} \times w)$ onto an open subset U_g^w of $\overline{X}^{OS}$, as is proved directly in [OsS1], see 2.6 there.

The equivalence relation defined by the action of a finite group is obviously both open and closed, hence π_2 is open and closed. In particular $\pi_2(U_g^w)$ is an open set mapped homeomorphically onto an open subset of $\overline{\mathbf{X}}^W(\mathbb{R})$. These sets form an open cover, hence $\bar{\mu}$ is open. Since it is continuous and bijective, it is a homeomorphism, as claimed, and $\overline{X}^{OS}/W'$ is therefore compact.

II.14.8 We want to prove that $\overline{X}^{OS}$ itself is *compact*. First we show that it is Hausdorff. Let $x, y \in \overline{X}^{OS}$ be distinct. If $\pi_2(x) \neq \pi_2(y)$ they can be separated (since $\overline{X}^{OS}/W'$ is Hausdorff, being homeomorphic to $\overline{\mathbf{X}}^W(\mathbb{R})$), hence so can x and y. Assume now that x and $y = x \circ w$ $(w \in W')$ are on the same W'-orbit. We have already seen that x has an open neighborhood U on which π_2 is injective. But then $U \circ w$ is an open neighborhood of y that has no point in common with U, hence x and y are also separated. This also implies that W' operates properly on $\overline{X}^{OS}$ ([Bu1], III, §4, n° 1, Prop. 2), hence that π_2 is *proper*. Then $\overline{X}^{OS} = \pi_2^{-1}(\overline{X}^{OS}/W')$ is compact, since $\overline{X}^{OS}/W'$ is so.

As a consequence, $\overline{X}^{OS}$ may be viewed as a regular finite covering of $\overline{\mathbf{X}}^W(\mathbb{R})$, with projection map $\bar{\mu} \circ \pi_2$. The space $\overline{\mathbf{X}}^W(\mathbb{R})$ is a smooth real projective variety on which G acts as an algebraic transformation group. In particular $\overline{\mathbf{X}}^W(\mathbb{R})$ is a smooth analytic G-space. The covering map $\bar{\mu} \circ \pi_2$ then allows one to lift that structure to $\overline{X}^{OS}$ and shows that it is a smooth compact analytic G-space too.

There remains to see that $\overline{X}^{OS}$ is connected. This is the first step of the proof of Theorem 2.7 in [OsS1] (see the first seven lines of the proof of Theorem 2.7, p. 19). For the sake of completeness, we repeat the argument.

Let $\{\beta_1, \dots, \beta_{r'}\}$ be a simple set of roots for Φ'. For $i \in [1, r']$ we let $w_i = s_{\beta_i}$ and fix a signature ε_i, which is -1 on β_i. We have $\Phi_{w_i} \cap W' = \{\beta_i\}$ ([Bu2], VI, §1, n° 6, Cor. 1, p. 157), whence $w_i \in W(\varepsilon_i)$ (cf. [OsS1], Lemma 2.5). As a consequence,

$$\big(1, \varepsilon_i(\alpha_1), \dots, \varepsilon_i(\alpha_r), w\big) \sim \big(\bar{w}, (\varepsilon_i(w_i(\alpha_1), \dots, \varepsilon_i(w_i(\alpha_r), ww_i)$$

for every $w \in W'$ and therefore

$$\pi(G \times \mathbb{R}^\Delta \times w) \cap \pi(G \times \mathbb{R}^\Delta \times ww_i) \neq \emptyset.$$

Since $(G \times \mathbb{R}^\Delta \times x \backslash w)$ is connected and W' is generated by the w_i, the claim follows.

II.14.9 Example. Let $G = \mathrm{SL}_2(\mathbb{R})$. We refer to II.6.9 for some notation. In particular, for the nontrivial signature, $K^*_\varepsilon = G/\mathrm{SO}(1,1)$.

The compactification $\overline{X}^{OS}$ may be identified with the 2-sphere with five orbits: Two copies of G/K, one of $G/\mathrm{SO}(1,1)$, and two copies of $G/P_\emptyset$, which separate the three open orbits (see [Os3]). It may be interpreted in the following manner: identify $\mathbb{R}^3 - \{0\}$ with the space of nonzero binary symmetric forms. The quotient by $\mathbb{R}^*_{>0}$, operating by dilation, is the 2-sphere, and the five orbits correspond to the positive definite forms, the positive forms of rank one, the indefinite forms, the negative forms of rank 1, and the negative definite forms. The Weyl group is of order 2. The nontrivial element w is the antipodal map. It identifies the two orbits isomorphic to G/K, and the two lines $G/P_\emptyset$. It acts freely on $G/\mathrm{SO}(1,1)$, which is an open cylinder, and the quotient is $G/\mathcal{N}(K^*_\varepsilon) = G/K_\varepsilon$, which is a Möbius band. The quotient is $P_2(\mathbb{R})$. It is the union of three orbits: the definite, indefinite, and degenerate forms.

On the other hand, $\overline{\mathbf{X}}^W$ is $P_2(\mathbb{C})$, with two orbits: the nondegenerate conics and the degenerate ones, which form $P_1(\mathbb{C})$. The real locus $\overline{\mathbf{X}}^W(\mathbb{R})$ is $P_2(\mathbb{R})$. The group G has one orbit $G/P_\emptyset$ in the space of degenerate conics, and two in the space of nondegenerate conics corresponding to definite and indefinite forms.

Remark II.14.10 By Theorem 2.10.5, different semisimple symmetric spaces X_ε (or rather their finite quotients) appear together with X in the real locus $\overline{\mathbf{X}}^W(\mathbb{R})$. The closure of X in $\overline{\mathbf{X}}^W(\mathbb{R})$ is the maximal Satake compactification. The closure of the finite quotient of X_ε in $\overline{\mathbf{X}}^W(\mathbb{R})$ can be used to define a compactification of X_ε, which is a manifold with corners and should be an analogue of the maximal Satake compactification. In fact, since the finite group W' acts freely on $\overline{X}^{OS}$ (II.13.6), the local structure of boundary points in the closure of the finite quotient of X_ε can be used to compactify X_ε into a real analytic manifold with corners. This should be the analogue of the maximal Satake compactification for X_ε. Theorem II.14.5 shows that this compactification of X_ε is isomorphic to the closure of X_ε in $\overline{X}^{OS}$, and also gives a natural explanation why different pieces of X and X_ε in the Oshima–Sekiguchi compactification $\overline{X}^{OS}$ can be glued together.

Part III

Compactifications of Locally Symmetric Spaces

In this part, we study compactifications of locally symmetric spaces. Locally symmetric spaces arise naturally from many different areas in mathematics. For example, many interesting moduli spaces in algebraic geometry such as the moduli space of principally polarized abelian varieties or of other polarizations and additional endomorphism structures are locally symmetric spaces and are often noncompact. An important problem is to compactify them. For example, the Baily–Borel compactification $\overline{\Gamma\backslash X}^{BB}$, the Borel–Serre compactification $\overline{\Gamma\backslash X}^{BS}$, the Satake compactifications $\overline{\Gamma\backslash X}^{S}$, the toroidal compactifications $\overline{\Gamma\backslash X}_{\Sigma}^{tor}$, have been constructed for many different purposes.

In Chapter 9, we recall the motivations and the original construction of many known compactifications. In Chapter 10, we discuss a uniform approach to compactifications of locally symmetric spaces and apply it to reconstruct the known compactifications. This uniform construction is very similar to and motivates the uniform method to compactify symmetric spaces in §I.8. In Chapter 11, we study relations between and more refined properties of these compactifications of locally symmetric spaces. For example, $\overline{\Gamma\backslash X}^{BS}$ is a real analytic manifold with corners and can be self-glued into a closed real analytic manifold $\overline{\Gamma\backslash X}^{BSO}$. In Chapter 12, we study another type of compactifications of locally symmetric spaces by embedding them into compact spaces such as the space of closed subgroups in G and the space of flags in $\mathbb{R}^n$ and flag lattices. This approach is similar to the compactifications of symmetric spaces X such as the Satake and Furstenberg compactifications, and more directly related to the subgroup compactification $\overline{X}^{sb}$ in §I.17. In Chapter 13, we study metric properties of compactifications of locally symmetric spaces. Then we explain applications to a conjecture of Siegel on comparison of two metrics on Siegel sets and a result on extension of holomorphic maps from a punctured disk to a Hermitian locally symmetric space to its Baily–Borel compactification. Finally, we relate the boundary components of the reductive Borel–Serre compactification $\overline{\Gamma\backslash X}^{RBS}$ and the geodesic compactification $\Gamma\backslash X \cup \Gamma\backslash X(\infty)$ to the continuous spectrum of $\Gamma\backslash X$, and a *Poisson relation* relating normalized lengths of geodesics that go to infinity, the sojourn times of scattering geodesics, and the spectral measure of the continuous spectrum.

9

Classical Compactifications of Locally Symmetric Spaces

Motivated by problems in automorphic forms, algebraic geometry, and topology, many compactifications of locally symmetric spaces have been constructed. In this chapter, we recall these compactifications and their motivations. For example, the Satake compactifications and the Baily–Borel compactification were motivated by automorphic forms, the Borel–Serre compactification $\overline{\Gamma\backslash X}^{BS}$ by topology, the toroidal compactifications by algebraic geometry, and the reductive Borel–Serre compactification $\overline{\Gamma\backslash X}^{RBS}$ by L^2-analysis.

In §III.1, we recall some of the basics of linear algebraic groups defined over $\mathbb{Q}$ and rational parabolic subgroups, in particular, the Langlands decomposition and the horospherical decomposition. In §III.2, we recall arithmetic subgroups of linear algebraic groups and give several important examples, and summarize reduction theories for arithmetic groups: both the classical and the precise ones. In §III.3, we recall the Satake compactifications. The Satake compactifications initiated the modern study of compactifications of locally symmetric spaces and give a general method of passing from (partial) compactifications of symmetric spaces to compactifications of locally symmetric spaces. In §III.4, we recall the Baily–Borel compactification. Unlike the Satake compactifications, which are only topological compactifications, the Baily–Borel compactification is a normal projective variety defined over a number field, and hence plays an important role in the Langlands program. In §III.5, we recall the Borel–Serre compactification, which plays an important role in understanding topological properties of arithmetic groups, for example, the cohomology of arithmetic groups. In §III.6, we recall the reductive Borel–Serre compactification, a variant of the Borel–Serre compactification, which is useful in the study of L^2-cohomology and other cohomology groups. In §III.7, we briefly recall the toroidal compactifications by emphasizing the role of torus embeddings and explaining how such tori arise, and discuss an alternative approach using the uniform method formulated in §III.8 below.

III.1 Rational parabolic subgroups

In this section, we recall basic facts about linear algebraic groups defined over $\mathbb{Q}$, rational parabolic subgroups of reductive linear algebraic groups defined over $\mathbb{Q}$, and their associated Langlands, horospherical decompositions. Results are similar to those for real parabolic subgroups in §I.1. The basic reference of this section is [Bo9].

This section is organized as follows. After recalling the definition of linear algebraic groups in III.1.1, we study several examples of tori in III.1.2, which show that the various ranks with respect to different fields could be different. Then we define the class of semisimple and reductive linear algebraic groups in III.1.4, and study the structure of rational parabolic subgroups in III.1.6 and III.1.7. To describe the geometry at infinity of locally symmetric spaces using rational parabolic subgroups, we introduce the rational Langlands decomposition in III.1.9 and compare it in III.1.12 with the real Langlands decomposition discussed in §I.1. The relative Langlands decomposition is defined in III.1.15. Finally, we define Siegel sets associated with rational parabolic subgroups in III.1.17.

III.1.1 Let $\mathbf{G}$ be a linear algebraic group, i.e., a Zariski-closed subgroup of some general linear group $GL(n, \mathbb{C})$:

$$\mathbf{G} = \{g = (g_{ij}) \in GL(n, \mathbb{C}) \mid P_a(g_{ij}) = 0, a \in A\},$$

where each P_a is a polynomial in g_{ij}, A a parameter space. If the polynomials P_a have $\mathbb{Q}$ coefficients, $\mathbf{G}$ is called a linear algebraic group defined over $\mathbb{Q}$.

Linear algebraic groups often occur as the *automorphism group* of some structures such as determinant and quadratic forms. For example,

$$\mathrm{SL}(n, \mathbb{C}) = \{g \in GL(n, \mathbb{C}) \mid \det g = 1\};$$

$$Sp(n, \mathbb{C}) = \{g \in GL(2n, \mathbb{C}) \mid \det g = 1, \omega(gX, gY) = \omega(X, Y), X, Y \in \mathbb{C}^{2n}\},$$

where

$$\omega(X, Y) = x_1 y_{2n} + x_2 y_{2n-1} + \cdots + x_n y_{n+1} - x_{n+1} y_n - \cdots - x_{2n} y_1$$

is a skew-symmetric form; and

$$\mathrm{SO}(2n, \mathbb{C}) = \{g \in GL(2n, \mathbb{C}) \mid \det g = 1, \langle gX, gY \rangle = \langle X, Y \rangle, X, Y \in \mathbb{C}^{2n}\},$$

where

$$\langle X, Y \rangle = x_1 y_{2n} + \cdots + x_{2n} y_1$$

a symmetric quadratic form.

III.1.2 A linear algebraic group $\mathbf{T}$ is called an *algebraic torus* if it is isomorphic to a product of $\mathbb{C}^* = GL(1, \mathbb{C})$. If the isomorphism is defined over $\mathbb{Q}$ (resp. $\mathbb{R}$), the torus T is said to *split over* $\mathbb{Q}$ (resp. $\mathbb{R}$). By definition, it always splits over $\mathbb{C}$. We discuss several examples of tori to show that these splittings are different.

First, consider the algebraic group

$$\mathbf{T}_1 = \left\{ g \in \mathrm{SL}(2, \mathbb{C}) \mid {}^t g \begin{pmatrix} 0 & 1 \\ 1 & 0 \end{pmatrix} g = \begin{pmatrix} 0 & 1 \\ 1 & 0 \end{pmatrix} \right\}.$$

It can be checked easily that if $g = \begin{pmatrix} a & b \\ c & d \end{pmatrix} \in T_1$, then $b = c = 0$, $d = a^{-1}$, and hence $g = \begin{pmatrix} a & 0 \\ 0 & a^{-1} \end{pmatrix}$. This implies that T_1 is isomorphic to $GL(1, \mathbb{C})$ over $\mathbb{Q}$ under the map $g \mapsto a$.

On the other hand, the algebraic group

$$\mathbf{T}_2 = \left\{ g \in \mathrm{SL}(2, \mathbb{C}) \mid {}^t g \begin{pmatrix} 1 & 0 \\ 0 & 1 \end{pmatrix} g = \begin{pmatrix} 1 & 0 \\ 0 & 1 \end{pmatrix} \right\}$$

is also a torus defined over $\mathbb{Q}$ but does not split over $\mathbb{Q}$ or $\mathbb{R}$. In fact, the real locus

$$\mathbf{T}_2(\mathbb{R}) = \mathrm{SO}(2, \mathbb{R})$$

is compact. To see that $\mathbf{T}_2$ is a torus, we note that $\mathbf{T}_2$ preserves the quadratic form $\langle X, X \rangle = x_1^2 + x_2^2$, while $\mathbf{T}_1$ preserves the quadratic form $2x_1x_2$, and these two forms are equivalent over $\mathbb{C}$, i.e., the quadratic form $x_1^2 + x_2^2$ splits over $\mathbb{C}$ as $(x_1 + ix_2)(x_1 - ix_2)$.

Consider another algebraic group defined over $\mathbb{Q}$,

$$\mathbf{T}_3 = \left\{ g \in \mathrm{SL}(2, \mathbb{C}) \mid {}^t g \begin{pmatrix} 1 & 0 \\ 0 & -2 \end{pmatrix} g = \begin{pmatrix} 1 & 0 \\ 0 & -2 \end{pmatrix} \right\}.$$

Since the quadratic form preserved by $\mathbf{T}_3$ is $x_1^2 - 2x_2^2$, which splits over $\mathbb{R}$ but not over $\mathbb{Q}$, it follows that $\mathbf{T}_3$ splits over $\mathbb{R}$ but not over $\mathbb{Q}$.

III.1.3 A linear algebraic group $\mathbf{G}$ is called *unipotent* if every element g of $\mathbf{G}$ is unipotent, i.e., $(g - I)^k = 0$ for some integer k.

Clearly, the subgroup of $\mathrm{SL}(n, \mathbb{C})$ consisting of upper triangular matrices with ones on the diagonal is unipotent. The converse is also true, i.e., any connected unipotent algebraic group is isomorphic to a group of upper triangular matrices with ones on the diagonal.

A linear algebraic group is called *solvable* if it is solvable as an abstract group, i.e., the derived series terminates, $G = G^{(0)} \supset G^{(1)} \supset \cdots G^{(l)} = \{e\}$ for some l, where $G^{(i)} = [G^{(i-1)}, G^{(i-1)}]$.

It can be checked easily that the subgroup of $GL(n, \mathbb{C})$ of upper triangular matrices is solvable. Hence, the above discussions show that a unipotent group is always solvable. On the other hand, for a solvable algebraic group $\mathbf{G}$ defined over $\mathbb{Q}$, let $\mathbf{U}$ be its normal subgroup consisting of all the unipotent elements. Then there exists a maximal torus $\mathbf{T}$ defined over $\mathbb{Q}$ such that $\mathbf{G}$ is the semidirect product of $\mathbf{T}$ and $\mathbf{U}$.

III.1.4 The *radical* $\mathbf{R}(\mathbf{G})$ of an algebraic group $\mathbf{G}$ is the maximal connected normal solvable subgroup of $\mathbf{G}$, and the *unipotent radical* $\mathbf{R}_\mathbf{U}(\mathbf{G})$ is the maximal connected unipotent normal subgroup of $\mathbf{G}$. If $\mathbf{G}$ is defined over $\mathbb{Q}$, then $\mathbf{R}(\mathbf{G})$, $\mathbf{R}_\mathbf{U}(\mathbf{G})$ are also defined over $\mathbb{Q}$. A linear algebraic group $\mathbf{G}$ is called *semisimple* if the radical $\mathbf{R}(\mathbf{G})$ is equal to $\{e\}$, and *reductive* if the unipotent radical $\mathbf{R}_\mathbf{U}(\mathbf{G})$ equals $\{e\}$.

Clearly, $\mathbf{G}/\mathbf{R}(\mathbf{G})$ is semisimple and $\mathbf{G}/\mathbf{R}_\mathbf{U}(\mathbf{G})$ is reductive. It is known that if $\mathbf{G}$ is defined over $\mathbb{Q}$, there exists a maximal reductive group $\mathbf{H}$ defined over $\mathbb{Q}$ such that

$$\mathbf{G} = \mathbf{H} \cdot \mathbf{R}_\mathbf{U}(\mathbf{G}) = \mathbf{R}_\mathbf{U}(\mathbf{G}) \cdot \mathbf{H}.$$

This is the so-called *Levi decomposition*, and $\mathbf{H}$ is called a Levi subgroup.

Though we are mainly interested in semisimple linear algebraic groups, reductive groups occur naturally when we consider parabolic subgroups and boundary components of compactifications of locally symmetric spaces. If $\mathbf{G}$ is a connected reductive algebraic group, then the derived subgroup $\mathbf{G}' = \mathbf{G}^{(1)} = [\mathbf{G}, \mathbf{G}]$ is semisimple, and there exists a central torus $\mathbf{T}$ such that $\mathbf{G} = \mathbf{T} \cdot \mathbf{G}'$.

III.1.5 For an algebraic group defined over $\mathbb{Q}$, an important notion is its $\mathbb{Q}$-rank, which plays a fundamental role in the geometry at infinity of locally symmetric spaces and reduction theories of arithmetic subgroups.

Let $\mathbf{G}$ be a connected linear algebraic group. Then all the maximal tori of $\mathbf{G}$ are conjugate, and this common dimension is called the absolute (or $\mathbb{C}$) rank of $\mathbf{G}$, denoted by $rk_\mathbb{C}(\mathbf{G})$. If $\mathbf{G}$ is defined over $\mathbb{Q}$, then all the maximal $\mathbb{Q}$-split tori of $\mathbf{G}$ are conjugate over $\mathbb{Q}$, i.e., by elements of $\mathbf{G}(\mathbb{Q})$, and the common dimension is called the *$\mathbb{Q}$-rank* of $\mathbf{G}$, denoted by $rk_\mathbb{Q}(\mathbf{G})$. If the $\mathbb{Q}$-rank of $\mathbf{G}$ is equal to 0, $\mathbf{G}$ is called an *anisotropic group* over $\mathbb{Q}$.

Similarly, the common dimension of maximal $\mathbb{R}$-split tori is called the *$\mathbb{R}$-rank* of $\mathbf{G}$, denoted by $rk_\mathbb{R}(\mathbf{G})$. The examples of tori in §III.1.2 considered as reductive algebraic groups over $\mathbb{Q}$ show that these ranks of the same group are in general not equal to each other.

If $\mathbf{G} = \mathrm{SL}(n)$, then $rk_\mathbb{Q}(\mathbf{G}) = rk_\mathbb{R}(\mathbf{G}) = rk_\mathbb{C}(\mathbf{G}) = n - 1$, and hence $\mathrm{SL}(n)$ is an example of groups split over $\mathbb{Q}$. In general, $\mathbf{G}$ is called split over $\mathbb{Q}$ if the $\mathbb{Q}$-rank of $\mathbf{G}$ is equal to the $\mathbb{C}$-rank of $\mathbf{G}$.

If F is a nondegenerate quadratic form on a $\mathbb{Q}$-vector space V with coefficients in $\mathbb{Q}$, then $\mathbf{G} = O(F)$, the orthogonal group of F, is a linear algebraic group defined over $\mathbb{Q}$. It can be shown (see [Bo3]) that $rk_\mathbb{Q}(\mathbf{G}) > 0$ if and only if F represents 0 over $\mathbb{Q}$, i.e., $F = 0$ has a nontrivial solution over $\mathbb{Q}$.

III.1.6 Let $\mathbf{G}$ be a connected linear algebraic group. Then a closed subgroup $\mathbf{P}$ of $\mathbf{G}$ is called a *parabolic subgroup* if $\mathbf{G}/\mathbf{P}$ a projective variety, which is equivalent to that $\mathbf{P}$ contains a maximal connected solvable subgroup, i.e., a *Borel subgroup*. These conditions are also equivalent to that $\mathbf{G}/\mathbf{P}$ is compact, as pointed out in §I.1.

III.1.7 Assume for the rest of this section that $\mathbf{G}$ is a connected reductive linear algebraic group defined over $\mathbb{Q}$ such that its center is an anisotropic subgroup over $\mathbb{Q}$, i.e., of $\mathbb{Q}$-rank 0. This condition is clearly satisfied if $\mathbf{G}$ is semisimple.

The real locus $G = \mathbf{G}(\mathbb{R})$ is a reductive Lie group with finitely many connected components and satisfies the condition (*) in §I.1.1.

If a parabolic subgroup $\mathbf{P}$ is defined over $\mathbb{Q}$, it is called a *rational parabolic subgroup* or *$\mathbb{Q}$-parabolic subgroup*, and its real locus $P = \mathbf{P}(\mathbb{R})$ is a parabolic subgroup of G in the sense of §I.1.

It is known that minimal rational parabolic subgroups of $\mathbf{G}$ are conjugate over $\mathbb{Q}$, and rational parabolic subgroups containing a minimal rational parabolic subgroup correspond to subsets of simple roots as in the real case in §I.1.

Specifically, let $\mathbf{S}$ be a maximal $\mathbb{Q}$-split torus in $\mathbf{G}$. Then the adjoint action of $\mathbf{S}$ on the Lie algebra $\mathfrak{g}$ of $\mathbf{G}$ gives a root space decomposition:

$$\mathfrak{g} = \mathfrak{g}_0 + \sum_{\alpha \in \Phi(\mathbf{G},\mathbf{S})} \mathfrak{g}_\alpha,$$

where

$$\mathfrak{g}_\alpha = \{X \in \mathfrak{g} \mid Ad(s)X = s^\alpha X, s \in S\},$$

and $\Phi(\mathbf{G}, \mathbf{S})$ consists of those nontrivial characters α such that $\mathfrak{g}_\alpha \neq 0$. These characters will also be viewed as linear functionals on the Lie algebra $\mathfrak{s}$ of $\mathbf{S}$. It is known that $\Phi(\mathbf{G}, \mathbf{S})$ is a root system and the Weyl group is isomorphic to

$${}_{\mathbb{Q}}W(\mathbf{G}, \mathbf{S}) = \mathcal{N}(\mathbf{S})/\mathcal{Z}(\mathbf{S}),$$

where $\mathcal{N}(\mathbf{S})$ is the normalizer of $\mathbf{S}$ in $\mathbf{G}$, and $\mathcal{Z}(\mathbf{S})$ the centralizer of $\mathbf{S}$ in $\mathbf{G}$.

Fix an order on $\Phi(\mathbf{G}, \mathbf{S})$, and denote the corresponding set of positive roots by $\Phi^+(\mathbf{G}, \mathbf{S})$. Define

$$\mathfrak{n} = \sum_{\alpha \in \Phi^+(\mathbf{G},\mathbf{S})} \mathfrak{g}_\alpha,$$

a subalgebra of $\mathfrak{g}$. Let $\mathbf{N}$ be the corresponding subgroup of $\mathfrak{n}$. Then $\mathbf{N}$ is normalized by $\mathcal{Z}(\mathbf{S})$, and $\mathbf{P} = \mathbf{N}\mathcal{Z}(\mathbf{S})$ is a minimal rational parabolic subgroup. Every minimal rational parabolic subgroup containing $\mathbf{S}$ is of this form with respect to some order on $\Phi(\mathbf{G}, \mathbf{S})$.

The set of simple roots in $\Phi^+(\mathbf{G}, \mathbf{S})$ is denoted by $\Delta(\mathbf{G}, \mathbf{S})$. Then the rational parabolic subgroups $\mathbf{Q}$ containing the minimal parabolic subgroup $\mathbf{P}$ correspond to proper subsets $\Delta(\mathbf{G}, \mathbf{S})$ as in §I.1, i.e., $\mathbf{Q} = \mathbf{P}_I$, where $I \subset \Delta(\mathbf{G}, \mathbf{S})$. These groups $\mathbf{P}_I$ are called the *standard rational parabolic subgroups*.

Explicitly, for any proper subset $I \subset \Delta(\mathbf{G}, \mathbf{S})$, let $\mathbf{S}_I$ be the identity component of the subgroup $\{g \in \mathbf{S} \mid g^\alpha = 1, \alpha \in I\}$ of $\mathbf{S}$. Then $\mathbf{S}_I$ is a $\mathbb{Q}$-split torus, and $\mathbf{P}_I = \mathbf{N}\mathcal{Z}(\mathbf{S}_I)$ is a rational parabolic subgroup containing $\mathbf{P}$; and any rational parabolic subgroup containing $\mathbf{P}$ is of this form for a unique I.

Any rational parabolic subgroup is conjugate by some element of $\mathbf{G}(\mathbb{Q})$ to a unique standard rational parabolic subgroup $\mathbf{P}_I$.

III.1.8 We recall the spherical *Tits building* $\Delta_{\mathbb{Q}}(\mathbf{G})$ associated with $\mathbf{G}$ over $\mathbb{Q}$ [Ti1, Theorem 5.2] [Ti2]. It is similar to the Tits building $\Delta(G)$ of a semisimple Lie group G in §I.2. Simplexes of $\Delta_{\mathbb{Q}}(\mathbf{G})$ correspond bijectively to proper rational parabolic subgroups of $\mathbf{G}$. Each proper maximal rational parabolic subgroup $\mathbf{Q}$ corresponds to a vertex of $\Delta_{\mathbb{Q}}(\mathbf{G})$, denoted by $\mathbf{Q}$. Let $\mathbf{Q}_0, \dots, \mathbf{Q}_k$ be different maximal rational parabolic subgroups. Then they form the vertices of a k-simplex if and only if $\mathbf{Q}_0 \cap \cdots \cap \mathbf{Q}_k$ is a rational parabolic subgroup, and this simplex corresponds to the parabolic subgroup $\mathbf{Q}_0 \cap \cdots \cap \mathbf{Q}_k$.

If $\mathbf{G}$ has $\mathbb{Q}$-rank one, $\Delta_{\mathbb{Q}}(\mathbf{G})$ is a countable collection of points. Otherwise, $\Delta_{\mathbb{Q}}(\mathbf{G})$ is a connected infinite simplicial complex of dimension $rk_{\mathbb{Q}}(\mathbf{G}) - 1$. For any maximal $\mathbb{Q}$-split torus $\mathbf{S}$, all the rational parabolic subgroups containing $\mathbf{S}$ form an apartment in this building. This subcomplex gives a simplicial triangulation of the sphere of dimension $r_{\mathbb{Q}}(\mathbf{G}) - 1$. This is the reason why $\Delta_{\mathbb{Q}}(\mathbf{G})$ is called a spherical building.

The rational points $\mathbf{G}(\mathbb{Q})$ of $\mathbf{G}$ act on the set of rational parabolic subgroups by conjugation and hence on $\Delta_{\mathbb{Q}}(\mathbf{G})$: For any $g \in \mathbf{G}(\mathbb{Q})$ and any rational parabolic subgroup $\mathbf{P}$, the simplex of $\mathbf{P}$ is mapped to the simplex of $g\mathbf{P}g^{-1}$.

III.1.9 For any rational parabolic subgroup $\mathbf{P}$ of $\mathbf{G}$, let $\mathbf{N_P}$ be the unipotent radical of $\mathbf{P}$, and $\mathbf{L_P} = \mathbf{N_P}\backslash\mathbf{P}$ be the *Levi quotient* of $\mathbf{P}$. Then both $\mathbf{N_P}$ and $\mathbf{L_P}$ are rational algebraic groups. Let $N_P = \mathbf{N_P}(\mathbb{R})$, $P = \mathbf{P}(\mathbb{R})$, $L_P = \mathbf{L_P}(\mathbb{R})$ be their real loci. Let $\mathbf{S_P}$ be the split center of $\mathbf{L_P}$ over $\mathbb{Q}$, and $A_\mathbf{P}$ the connected component of the identity in $\mathbf{S_P}(\mathbb{R})$. Let

$$\mathbf{M_P} = \cap_{\chi \in X(\mathbf{L_P})} \mathrm{Ker}\chi^2.$$

Then $\mathbf{M_P}$ is a reductive algebraic group defined over $\mathbb{Q}$ whose center is anisotropic over $\mathbb{Q}$. Let $M_\mathbf{P} = \mathbf{M_P}(\mathbb{R})$. Then L_P admits a decomposition

$$L_P = A_\mathbf{P} M_\mathbf{P} \cong A_\mathbf{P} \times M_\mathbf{P}. \tag{III.1.1}$$

To obtain the rational Langlands decomposition of P, we need to lift L_P and its subgroups into P. Let X be the symmetric space of maximal compact subgroups of $G = \mathbf{G}(\mathbb{R})$. Let K be a maximal compact subgroup of G. Then $X \cong G/K$. Let $x_0 \in X$ be the basepoint corresponding to K. The Cartan involution θ of G associated with K extends to an involution of $\mathbf{G}$. It is shown in [BS2, 1.9] (see also [GHM, pp. 149–151]) that there exists a unique Levi subgroup $\mathbf{L}_{P,x_0}$ of $\mathbf{G}$ that is stable under the extended Cartan involution. The canonical projection $\pi_P : \mathbf{L}_{\mathbf{P},x_0} \to \mathbf{P}/\mathbf{N_P}$ yields an isomorphism of $\mathbf{L}_{\mathbf{P},x_0}$ onto $\mathbf{L_P}$. We let i_{x_0} be the inverse to the restriction of π_P to L_{P,x_0}. In particular, it is an isomorphism of L_P onto L_{P,x_0}. We let $A_{\mathbf{P},x_0}$ and $M_{\mathbf{P},x_0}$ denote the images of $A_\mathbf{P}$ and $M_\mathbf{P}$ under i_{x_0}.

Note that though $\mathbf{L_P}$, $\mathbf{M_P}$, $\mathbf{S_P}$ are algebraic groups defined over $\mathbb{Q}$, their lifts $\mathbf{L}_{\mathbf{P},x_0}$, $\mathbf{M}_{P,x_0}$, $\mathbf{S}_{P,x_0}$ are not necessarily defined over $\mathbb{Q}$. Of course, they are defined over $\mathbb{R}$.

Lemma III.1.10 *For any other basepoint $x_1 = px_0 = pKp^{-1} \in X$, where $p \in P$, the Levi subgroup $\mathbf{L}_{P,x_1}$ associated with the basepoint x_1 is $p\mathbf{L}_{P,x_0}p^{-1}$, and*

$A_{\mathbf{P},x_1} = pA_{\mathbf{P},x_0}p^{-1}$, $M_{\mathbf{P},x_1} = pM_{\mathbf{P},x_0}p^{-1}$. *If n is the N_P-component of p in $P = N_P A_{\mathbf{P},x_0} M_{\mathbf{P},x_0}$, then* $A_{\mathbf{P},x_1} = nA_{\mathbf{P},x_0}n^{-1}$, $M_{\mathbf{P},x_1} = nM_{\mathbf{P},x_0}n^{-1}$.

Proof. Let θ be the Cartan involution for the basepoint x_0. Then the Cartan involution for x_1 is given by $\mathrm{Int}(p)\circ\theta\circ\mathrm{Int}(p)^{-1}$. Since $p \in P$, $pL_{P,x_0}p^{-1}$ belongs to P and is invariant under $\mathrm{Int}(p)\circ\theta\circ\mathrm{Int}(p)^{-1}$. This implies that $p\mathbf{L}_{P,x_0}p^{-1}$ is the lift associated with x_1. The rest is clear.

Proposition III.1.11 *For any rational parabolic subgroup* $\mathbf{P}$*, there exists a basepoint $x_1 \in X$ and a lift map i_{x_1} such that i_{x_1} is rational in the sense that the images $i_{x_1}(\mathbf{L_P})$, $i_{x_1}(\mathbf{M_P})$, $i_{x_1}(\mathbf{S_P})$ are algebraic subgroups defined over $\mathbb{Q}$, and the lift i_{x_1} is a morphism defined over $\mathbb{Q}$.*

Proof. Since $\mathbf{P}$ is defined over $\mathbb{Q}$, there is a Levi subgroup $\mathbf{L}'_{\mathbf{P}}$ defined over $\mathbb{Q}$. Since all the Levi subgroups of $\mathbf{P}$ are conjugate under N_P, there exists $n \in N_P$ such that $\mathbf{L}'_{\mathbf{P}} = ni_{x_0}(\mathbf{L_P})n^{-1}$. Let $x_1 = nx_0$. Then the proof of the above lemma shows that $i_{x_1}(\mathbf{L_P}) = \mathbf{L}'_{\mathbf{P}}$.

The lift $i_{x_0}(L_P)$ splits the exact sequence

$$\{e\} \to N_P \to P \to L_P \to \{e\}, \tag{III.1.2}$$

and gives rise to the *rational Langlands decomposition* of P:

$$P = N_P A_{\mathbf{P},x_0} M_{\mathbf{P},x_0} \cong N_P \times A_{\mathbf{P},x_0} \times M_{\mathbf{P},x_0}, \tag{III.1.3}$$

i.e., for any $g \in P$,

$$g = n(g)a(g)m(g),$$

where $n(g) \in N_P$, $a(g) \in A_{\mathbf{P},x_0}$, $m(g) \in M_{\mathbf{P},x_0}$ are uniquely determined by g, and the map $g \to (n(g), a(g), m(g))$ gives a real analytic diffeomorphism between P and $N_P \times A_{\mathbf{P},x_0} \times M_{\mathbf{P},x_0}$.

The map $P \to N_P \times A_{\mathbf{P},x_0} \times M_{\mathbf{P},x_0}$ is equivariant with respect to the P-action defined on the left by

$$n_0a_0m_0(n, a, m) = (n_0\, {}^{a_0m_0}n, a_0a, m_0m)$$

for $p = n_0a_0m_0 \in P$. Since $G = PK$, the subgroup P acts transitively on $X = G/K$, and the Langlands decomposition of P gives the following *rational horospherical decomposition* of X:

$$X = N_P \times A_{\mathbf{P},x_0} \times X_{\mathbf{P},x_0}, \tag{III.1.4}$$

where

$$X_{\mathbf{P},x_0} = M_{\mathbf{P},x_0}/K \cap M_{\mathbf{P},x_0} \tag{III.1.5}$$

is called the *boundary symmetric space* associated with **P**. The Langlands decomposition of P also induces the following *rational horospherical decomposition* of G:

$$G = N_P A_{\mathbf{P},x_0} M_{\mathbf{P},x_0} K = N_P \times A_{\mathbf{P},x_0} \times M_{\mathbf{P},x_0} K, \qquad \text{(III.1.6)}$$

i.e., any element $g \in G$ can be written uniquely in the form $g = n(g)a(g)m(g)$, where $n(g) \in N_P$, $a(g) \in A_{\mathbf{P},x_0}$, $m(g) \in M_{\mathbf{P},x_0} K$, and

$$G \to N_P \times A_{\mathbf{P},x_0} \times M_{\mathbf{P},x_0} K, \quad g \mapsto (n(g),\ a(g),\ m(g)) \qquad \text{(III.1.7)}$$

gives a real analytic diffeomorphism.

To indicate the dependence on the basepoint x_0 of the above horospherical decompositions of X and G, these maps are also denoted by

$$\nu_{x_0} : N_P \times A_{\mathbf{P},x_0} \times X_{\mathbf{P},x_0} \to X,$$
$$\nu_{x_0} : N_P \times A_{P,x_0} \times M_{\mathbf{P},x_0} K \to G.$$

In the following, $\nu_{x_0}(n, a, m)$ is also denoted by (n, a, m) or nam for simplicity.

Remark III.1.12 The real locus P of a rational parabolic subgroup **P** is a parabolic subgroup of G. Hence by §I.1, P also admits the (real) Langlands decomposition with respect to the basepoint $x_0 = K \in G/K = X$,

$$P = N_P A_P M_P,$$

and the induced horospherical decomposition

$$X \cong N_P \times A_P \times X_P,$$

where $X_P = M_P/K \cap M_P$. Note that A_P and M_P are stable under the Cartan decomposition θ on G associated with K. Hence they are the Langlands decomposition with respect to the basepoint $x_0 = K$. These decompositions are different from the above rational decompositions in equations (III.1.3, III.1.4). In fact, $A_{\mathbf{P},x_0} \subseteq A_P$, but the equality $A_{\mathbf{P},x_0} = A_P$ does not hold when the $\mathbb{R}$-rank of **P** is strictly greater than the $\mathbb{Q}$-rank of **P**. The subgroup $A_{\mathbf{P},x_0}$ is called the rational split component with respect to the basepoint x_0. (Note that if P is a minimal real parabolic subgroup, then $\mathfrak{a}_P$ is a maximal abelian subspace of $\mathfrak{p}$, where $\mathfrak{g} = \mathfrak{k} \oplus \mathfrak{p}$ is the Cartan decomposition of $\mathfrak{g}$ determined by K, which corresponds to the basepoint x_0.)

To distinguish it from the real split component A_P, we have used the subscript **P** in $A_{\mathbf{P}}$ instead of P. Let $\mathfrak{a}_P$, $\mathfrak{a}_{\mathbf{P}}$ be the Lie algebras of A_P and $A_{\mathbf{P},x_0}$ respectively, and let $\mathfrak{a}_{\mathbf{P}}^{\perp} \subset \mathfrak{a}_P$ be the orthogonal complement. Then

$$X_{\mathbf{P}} \cong X_P \times \exp \mathfrak{a}_{\mathbf{P}}^{\perp}, \quad A_P = A_{\mathbf{P},x_0} \times \exp \mathfrak{a}_{\mathbf{P}}^{\perp}. \qquad \text{(III.1.8)}$$

Therefore, $X_{\mathbf{P}}$ is different from X_P in general. This explains the reason for the subscript **P** in $X_{\mathbf{P}}$.

III.1.13 In the following, the reference to the basepoint x_0 in various subscripts will be omitted unless needed.

The unipotent subgroup N_P is normal in P, and $A_{\mathbf{P}}$ acts on it by conjugation and on its Lie algebra $\mathfrak{n}_P$ by the adjoint representation. We let $\Phi(P, A_{\mathbf{P}})$ be the set of characters of $A_{\mathbf{P}}$ in $\mathfrak{n}_P$, the "roots of P with respect to $A_{\mathbf{P}}$." The value of $\alpha \in \Phi(P, A_{\mathbf{P}})$ on $a \in A_{\mathbf{P}}$ is denoted by a^α. The differential $d\alpha$ of α, also denoted by α below, is a weight of $\mathfrak{a}_{\mathbf{P}}$ in $\mathfrak{n}_P$, and we have

$$a^\alpha = \exp d\alpha(\log a).$$

There is a unique subset $\Delta(P, A_{\mathbf{P}})$ of $\Phi(P, A_{\mathbf{P}})$, consisting of $\dim A_{\mathbf{P}}$ linearly independent roots, such that any element of $\Phi(P, A_{\mathbf{P}})$ is a linear combination with positive integral coefficients of elements of $\Delta(P, A_{\mathbf{P}})$, to be called the simple roots of P with respect to $A_{\mathbf{P}}$.

Remark III.1.14 We recall how $\Phi(P, A_{\mathbf{P}})$ and $\Delta(P, A_{\mathbf{P}})$ are related to $\mathbb{Q}$-roots. Fix a minimal parabolic $\mathbb{Q}$-subgroup $\mathbf{P}_0$ and a maximal $\mathbb{Q}$-split torus $\mathbf{S}$ of $\mathbf{P}_0$. Let $\Phi(\mathbf{S}, \mathbf{G})$ the set of roots of $\mathbf{G}$ with respect to $\mathbf{S}$ (the $\mathbb{Q}$-roots) and $\Delta(\mathbf{S}, \mathbf{G})$ be the set of $\mathbb{Q}$-simple roots for the ordering of Φ defined by $\mathbf{P}_0$ (see III.1.7). There is a unique subset $I \subset \Delta(S, G)$ such that $\mathbf{P}$ is conjugate to the standard parabolic $\mathbb{Q}$-subgroup $\mathbf{P}_{0,I}$, by a conjugation that brings the Zariski-closure of $\mathbf{S}_{\mathbf{P}}$ of $A_{\mathbf{P}}$ onto $\mathbf{S}_I = (\cap_{\alpha \in I} \ker \alpha)^o$. Then, up to conjugation, the elements of $\Phi(P, A_{\mathbf{P}})$ are the nonzero restrictions of the elements in $\Phi^+(\mathbf{S}, \mathbf{G})$ and $\Delta(P, A_{\mathbf{P}})$ is the set of restrictions of $\Delta(\mathbf{S}, \mathbf{G}) - I$.

III.1.15 For any rational parabolic subgroup $\mathbf{P}$ that is not necessarily minimal, the set of rational parabolic subgroups $\mathbf{Q}$ containing $\mathbf{P}$ also corresponds bijectively to the collection of proper subsets of $\Delta(P, A_{\mathbf{P}})$ as in the case of minimal rational parabolic subgroups in III.1.7.

Specifically, for any proper subset $I \subset \Delta(P, A_{\mathbf{P}})$, define

$$A_{\mathbf{P},I} = \{a \in A_{\mathbf{P}} \mid a^\alpha = 1, \quad \alpha \in I\}.$$

Then there exists a unique rational parabolic subgroup $\mathbf{Q}$ containing $\mathbf{P}$ such that

$$A_{\mathbf{Q}} = A_{\mathbf{P},I}.$$

Denote this parabolic subgroup by $\mathbf{P}_I$. Any rational parabolic subgroup containing $\mathbf{P}$ is of this form.

III.1.16 For any rational parabolic subgroup $\mathbf{Q}$ containing $\mathbf{P}$, there is a *relative Langlands decomposition* of $\mathbf{Q}$ with respect to $\mathbf{P}$ and the related horospherical decomposition. Let $\mathbf{Q} = \mathbf{P}_I$. Then

$$A_{\mathbf{Q}} = A_{\mathbf{P}_I} = \{e^H \in A_{\mathbf{P}} \mid \alpha(H) = 0, \alpha \in I\}.$$

Define

$$\mathfrak{a}_{\mathbf{P}}^{\mathbf{Q}} = \mathfrak{a}_{\mathbf{P}}^{I} = \{H \in \mathfrak{a}_{\mathbf{P}} \mid H \perp \mathfrak{a}_{P_I}\}$$

and

$$A_{\mathbf{P}}^{\mathbf{Q}} = A_{\mathbf{P}}^{I} = \exp \mathfrak{a}_{\mathbf{P}}^{I}.$$

Then

$$A_{\mathbf{P}} = A_{\mathbf{Q}} A_{\mathbf{P}}^{\mathbf{Q}} \cong A_{\mathbf{Q}} \times A_{\mathbf{P}}^{\mathbf{Q}}. \tag{III.1.9}$$

There is a related but different decomposition. Define

$$\begin{aligned}\mathfrak{a}_{\mathbf{P},\mathbf{Q}} &= \mathfrak{a}_{\mathbf{P},\mathbf{P}_I} = \{e^H \mid \alpha(H) = 0, \quad \alpha \in \Delta - I\},\\ A_{\mathbf{P},\mathbf{Q}} &= A_{\mathbf{P},\mathbf{P}_I} = \exp \mathfrak{a}_{\mathbf{P},\mathbf{P}_I}.\end{aligned} \tag{III.1.10}$$

Then

$$A_{\mathbf{P}} = A_{\mathbf{Q}} A_{\mathbf{P},\mathbf{Q}} = A_{\mathbf{Q}} \times A_{\mathbf{P},\mathbf{Q}}. \tag{III.1.11}$$

Combined with the horospherical decomposition of X, $X = N_P \times A_{\mathbf{P}} \times X_{\mathbf{P}}$, we get another decomposition of X:

$$X = N_P \times X_{\mathbf{P}} \times A_{\mathbf{P},\mathbf{Q}} \times A_{\mathbf{Q}}. \tag{III.1.12}$$

As in equation (I.1.21), $\mathbf{P}$ determines a unique rational parabolic subgroup $\mathbf{P}'$ of $\mathbf{M}_{\mathbf{Q}}$ (see [HC, Lemma 2]).

Identify $M_{\mathbf{Q}}$ with a subgroup of Q under the canonical lift i_{x_0}. Then $M_{\mathbf{Q}} \cap P$ is the lift of P' under i_{x_0}. Similarly, under the lift i_{x_0}, $A_{\mathbf{P}'}$ and $N_{P'}$ can be identified with subgroups of P. Then $\mathbf{P}'$ satisfies the following properties:

$$M_{\mathbf{P}'} = M_{\mathbf{P}}, \quad A_{\mathbf{P}} = A_{\mathbf{Q}} A_{\mathbf{P}'} = A_{\mathbf{Q}} \times A_{\mathbf{P}'}, \quad N_P = N_Q N_{P'} = N_Q \rtimes N_{P'}. \tag{III.1.13}$$

The parabolic subgroup P' induces a Langlands decomposition of M_Q,

$$M_Q = N_{P'} \times A_{P'} \times (M_{P'} K_Q), \tag{III.1.14}$$

and the horospherical decomposition of $X_{\mathbf{Q}}$,

$$X_{\mathbf{Q}} = N_{P'} \times A_{P'} \times X_{\mathbf{P}'} = N_{P'} \times A_{P'} \times X_{\mathbf{P}}. \tag{III.1.15}$$

III.1.17 For any $t > 0$, define

$$A_{\mathbf{P},t} = \{a \in A_{\mathbf{P}} \mid a^{\alpha} > t, \ \alpha \in \Delta(P, A_{\mathbf{P}})\}. \tag{III.1.16}$$

For any bounded sets $U \subset N_P, \ V \subset X_{\mathbf{P}}$, the subset

$$S_{\mathbf{P},U,t,V} = U \times A_{\mathbf{P},t} \times V \subset N_P \times A_{\mathbf{P}} \times X_{\mathbf{P}} = X \tag{III.1.17}$$

is called a *Siegel set* in X associated to $\mathbf{P}$. For the improper parabolic subgroup $\mathbf{P} = \mathbf{G}$, the Siegel sets are bounded sets.

Due to the difference between the rational and real Langlands decompositions of P and the induced horospherical decompositions pointed out in Remark III.1.12, the Siegel sets defined here are different from the Siegel sets associated to P in §I.9.1 if $A_{\mathbf{P}} \neq A_P$.

To construct compactifications of arithmetic quotients of G later, we also need Siegel sets in G.

For any bounded sets $U \subset N_P$, $W \subset M_P K$, the subset

$$U \times A_{\mathbf{P},t} \times W \subset N_P \times A_{\mathbf{P}} \times (M_{\mathbf{P}} K) = G \qquad \text{(III.1.18)}$$

is called a *Siegel set* in G associated to $\mathbf{P}$ and K, or the basepoint x_0.

Clearly, any Siegel set $U \times A_{\mathbf{P},t} \times V$ in X lifts to a Siegel set $U \times A_{\mathbf{P},t} \times VK$ in G; and any Siegel set $U \times A_{\mathbf{P},t} \times W$ in G with a right K-invariant W gives a Siegel set $U \times A_{\mathbf{P},t} \times W/K$ in X.

In general, a Siegel set in G is not necessarily a lift of a Siegel set in X.

III.1.18 Summary and comments. We recalled basic facts about linear algebraic groups. We gave several examples of tori defined over $\mathbb{Q}$ to show that the rank of an algebraic group depends on the field in question. The Langlands decomposition associated with a parabolic subgroup also depends on the field of definition of the parabolic subgroup.

For a more systematical summary of algebraic groups, see [Bo14]. For thorough discussions, see [Bo9] and [Sp4].

III.2 Arithmetic subgroups and reduction theories

In this section we recall arithmetic groups and the reduction theories for them. The basic references of this section are [Bo3] [Bo4] [Bo5] [PR] [OW1] [Sap1].

This section is organized as follows. In Definition III.2.1, we introduce arithmetic subgroups. In Proposition III.2.2, we show that arithmetic subgroups are independent of the choice of a $\mathbb{Q}$-basis of the vector space. Then we discuss some important examples of arithmetic subgroups: the Hilbert modular groups in III.2.7, arithmetic Fuchsian groups in III.2.8, the Bianchi groups in III.2.9, and the Picard modular groups in III.2.10. Then we introduce fundamental domains in Definition III.2.13, fundamental sets in III.2.14. In Proposition III.2.15, we state conditions that are equivalent to that the quotient $\Gamma\backslash G$ is compact. For example, when $\mathbf{G}$ is semisimple, one condition is that the $\mathbb{Q}$-rank of $\mathbf{G}$ is equal to 0. When the $\mathbb{Q}$-rank of $\mathbf{G}$ is positive, we state the classical reduction theory in Proposition III.2.16, which implies finite generation of Γ in Corollary III.2.17. Separation of Siegel sets is given in Proposition III.2.19. The precise reduction theory gives a fundamental domain and is stated in Proposition III.2.21.

Let $\mathbf{G} \subset GL(n, \mathbb{C})$ be a linear algebraic group defined over $\mathbb{Q}$, not necessarily reductive. Let $\mathbf{G}(\mathbb{Q}) \subset GL(n, \mathbb{Q})$ be the set of its rational points, and $\mathbf{G}(\mathbb{Z}) \subset$

$GL(n, \mathbb{Z})$ the set of its elements with integral entries, which can be identified with the stabilizer of the standard lattice $\mathbb{Z}^n$ in $\mathbb{R}^n$.

Definition III.2.1 *A subgroup $\Gamma \subset \mathbf{G}(\mathbb{Q})$ is called an arithmetic subgroup if it is commensurable with $\mathbf{G}(\mathbb{Z})$, i.e., $\Gamma \cap \mathbf{G}(\mathbb{Z})$ has finite index in both Γ and $\mathbf{G}(\mathbb{Z})$.*

As an abstract affine algebraic group defined over $\mathbb{Q}$, $\mathbf{G}$ admits different embeddings into $GL(n', \mathbb{C})$, where n' might be different from n. The above definition depends on the embedding $\mathbf{G} \subset GL(n, \mathbb{C})$ and the integral subgroup $GL(n, \mathbb{Z})$. If we choose a different embedding, for example using a basis of $\mathbb{R}^n$ over $\mathbb{Q}$ different from the standard basis $e_1 = (1, 0, \dots, 0), \dots, e_n = (0, \dots, 0, 1)$, then we will get a different integral subgroup $\mathbf{G}(\mathbb{Z})$ of $GL(n, \mathbb{C})$ defined with respect to this basis.

It turns out that these different embeddings $\mathbf{G} \subset GL(n', \mathbb{C})$ and different choices of integral structures lead to the same class of arithmetic groups (see [PR] [Ji6]).

Proposition III.2.2 *Let $\mathbf{G}, \mathbf{G}'$ be two linear algebraic groups defined over $\mathbb{Q}$, and $\varphi : \mathbf{G} \to \mathbf{G}'$ an isomorphism defined over $\mathbb{Q}$. Then $\varphi(\mathbf{G}(\mathbb{Z}))$ is commensurable with $\mathbf{G}'(\mathbb{Z})$.*

Corollary III.2.3 *If Γ is an arithmetic subgroup of $\mathbf{G}$, then for any $g \in \mathbf{G}(\mathbb{Q})$, $g\Gamma g^{-1}$ is also an arithmetic subgroup of $\mathbf{G}$.*

Remark III.2.4 To discuss some important examples of arithmetic subgroups such as the Hilbert modular groups, the Bianchi groups, and the Picard modular groups below, we need a slightly more general set-up for arithmetic groups. Let F be a number field, i.e., a finite extension of $\mathbb{Q}$, and $\mathcal{O}_F$ its ring of integers. Let $\mathbf{G} \subset GL(n, \mathbb{C})$ be a linear algebraic group defined over F. A subgroup Γ of $\mathbf{G}(F)$ is called *arithmetic* if it is commensurable with $G(\mathcal{O}_F) = \mathbf{G} \cap GL(n, \mathcal{O}_F)$. It turns out that such an arithmetic subgroup is also an arithmetic subgroup according to the previous definition and hence we do not get more arithmetic subgroups by considering general number fields. In fact, by the functor of restriction of scalars (see [PR] [Ji6]), there is an algebraic group $Res_{F/\mathbb{Q}}\mathbf{G}$ defined over $\mathbb{Q}$ such that $Res_{K/\mathbb{Q}}\mathbf{G}(\mathbb{Q}) = \mathbf{G}(F)$, and $\mathbf{G}(\mathcal{O}_F)$ is commensurable with $Res_{K/\mathbb{Q}}\mathbf{G}(\mathbb{Z})$ under this identification.

On the other hand, it is often convenient to use some naturally occurring number fields to define some arithmetic groups.

III.2.5 As in §III.1, we assume that $\mathbf{G}$ is a reductive linear algebraic group defined over $\mathbb{Q}$ such that its center is an anisotropic subgroup defined over $\mathbb{Q}$, and its real locus $G = \mathbf{G}(\mathbb{R})$ is a reductive Lie group with finitely many connected components. Let $K \subset G$ be a maximal compact subgroup, and $X = G/K$ the associated symmetric space, which is the product of a symmetric space of noncompact type and a possible Euclidean factor.

Let $\Gamma \subset \mathbf{G}(\mathbb{Q})$ be an arithmetic subgroup. By a known result of Selberg (see [Sel3], and also [Bo4] [Ji6]), Γ admits torsion-free subgroups of finite index. Hence, we can assume that Γ is torsion-free if necessary by passing to a subgroup of finite index.

Since Γ is a discrete subgroup of G, Γ acts properly on X, and the quotient

$$\Gamma\backslash X = \Gamma\backslash G/K$$

is called a *locally symmetric space*, which is smooth if Γ is torsion-free and has finite-quotient singularities in general, by the above result of Selberg.

To understand the geometry of $\Gamma\backslash X$, we need the reduction theories for Γ. First, we recall several important examples of arithmetic subgroups and their associated locally symmetric spaces.

III.2.6 The best-known and simplest arithmetic subgroups are

$$\mathrm{SL}(n, \mathbb{Z}) \subset \mathrm{SL}(n, \mathbb{R}),$$

and its congruence subgroups, where a subgroup of $\mathrm{SL}(n, \mathbb{Z})$ is called a *congruence subgroup* if it contains a *principal congruence subgroup* $\Gamma(N)$,

$$\Gamma(N) = \{g \in \mathrm{SL}(n, \mathbb{Z}) \mid g \equiv Id \mod N\}.$$

The group $\mathrm{SL}(n, \mathbb{Z})$ is not torsion-free, but for any $N \geq 3$, $\Gamma(N)$ is torsion-free (see [Br2] for example). The quotient $\mathrm{SL}(n, \mathbb{Z})\backslash\mathrm{SL}(n, \mathbb{R})$ can be identified with the *moduli space of unimodular lattices* in $\mathbb{R}^n$. Since the lattices in $\mathbb{R}^n$ can degenerate, for example, the minimum norm of lattice vectors can converge to 0, $\mathrm{SL}(n, \mathbb{Z})\backslash\mathrm{SL}(n, \mathbb{R})$ is noncompact, i.e., $\mathrm{SL}(n, \mathbb{Z})$ is not a uniform discrete subgroup (or lattice).

III.2.7 Hilbert modular groups. Let F be a real quadratic field, $F = \mathbb{Q}(\sqrt{d})$, with d is a square-free positive integer. Then F has two real embeddings and no complex embedding. The group SL(2) is defined over $\mathbb{Q}$ and hence also over F. The group $Res_{F/\mathbb{Q}}\mathrm{SL}(2)$ is defined over $\mathbb{Q}$ and is of $\mathbb{Q}$-rank 1, and

$$Res_{F/\mathbb{Q}}\mathrm{SL}(2, \mathbb{R}) = \mathrm{SL}(2, \mathbb{R}) \times \mathrm{SL}(2, \mathbb{R}).$$

The arithmetic group $\Gamma = \mathrm{SL}(2, \mathcal{O}_F)$ embeds into $\mathrm{SL}(2, \mathbb{R}) \times \mathrm{SL}(2, \mathbb{R})$ as a discrete subgroup, called the *(principal) Hilbert modular group*. Let $\mathbf{H}$ be the upper half-plane with the Poincaré metric. Then the Hilbert modular group Γ acts on the product $\mathbf{H} \times \mathbf{H}$ properly and the quotient $\Gamma\backslash\mathbf{H} \times \mathbf{H}$ has finite volume, called the *Hilbert modular surface* associated with F. The geometry of the Hilbert modular surface is closely related to the properties of the field F. For example, the number of ends of $\Gamma\backslash\mathbf{H} \times \mathbf{H}$ is equal to the class number of F (see [Fr, 3.5]).

The $\mathbb{Q}$-rank of $Res_{F/\mathbb{Q}}\mathrm{SL}(2, \mathbb{R})$ is equal to 1, but the $\mathbb{R}$-rank is equal to 2 and hence strictly greater than 1. Let $\mathbf{P}$ be a minimal rational parabolic subgroup. Then

$$\dim A_P = 2, \quad \dim A_{\mathbf{P}} = 1,$$

and hence the real boundary space X_P consists of one point, and the rational boundary symmetric space satisfies

$$X_{\mathbf{P}} \cong \mathbb{R}.$$

More generally, we can consider a *totally real number field* F of degree d over $\mathbb{Q}$, i.e., F admits no complex embedding, $s = d$, $t = 0$. Then $\mathrm{SL}(2, \mathcal{O}_F)$ is a discrete subgroup of $\mathrm{SL}(2, \mathbb{R})^d$ and defines the *Hilbert modular variety* $\Gamma \backslash \mathbf{H}^d$. If $\mathbf{P}$ is a minimal rational parabolic subgroup then X_P consists of a point, and

$$X_{\mathbf{P}} \cong \mathbb{R}^{d-1}.$$

III.2.8 Arithmetic Fuchsian groups. It is known that the subgroup $\mathrm{SL}(2, \mathbb{Z})$ and its subgroups of finite index are not uniform lattices, i.e., $\mathrm{SL}(2, \mathbb{Z}) \backslash \mathbf{H}$ is noncompact. To obtain uniform arithmetic subgroups of $\mathrm{SL}(2, \mathbb{R})$, we can use *quaternion algebras* over $\mathbb{Q}$.

For any two nonzero elements $a, b \in \mathbb{Q}$, there is a *quaternion algebra* $\mathbb{H}(a, b)$ defined as the 4-dimensional vector space over $\mathbb{Q}$ with a basis $1, i, j, k$ such that it is an algebra over $\mathbb{Q}$ with multiplication determined by

$$i^2 = a, \quad j^2 = b, \quad ij = -ji = k.$$

When $a = b = -1$, we get the usual quaternion algebra.

The algebra $\mathbb{H}(a, b)$ can be embedded into $M_{2\times 2}(\mathbb{Q}(\sqrt{a}))$ by

$$\rho : x = x_0 + ix_1 + jx_2 + kx_3 \mapsto \begin{pmatrix} x_0 + x_1\sqrt{a} & x_2 + x_3\sqrt{a} \\ b(x_2 - x_3\sqrt{a}) & x_0 - x_1\sqrt{a} \end{pmatrix}. \tag{III.2.19}$$

It is known that $\mathbb{H}(a, b)$ is either a *division algebra*, i.e., every nonzero element is divisible, or it is isomorphic to $M_{2\times 2}(\mathbb{Q})$ over $\mathbb{Q}$, which is clearly not a division algebra.

Assume that a, b are positive integers, and $\mathbb{H}(a, b)$ is a division algebra. Define a norm on $\mathbb{H}(a, b)$ by

$$Nr(x_0 + ix_1 + jx_2 + kx_3) = x_0^2 - ax_1^2 - bx_2^2 + abx_3^2.$$

Define

$$\Gamma = \{x = x_0 + ix_1 + jx_2 + kx_3 \mid x_0, x_1, x_2, x_3 \in \mathbb{Z}, Nr(x) = 1\},$$

the norm-1 subgroup of the order

$$\mathcal{O} = \{x = x_0 + ix_1 + jx_2 + kx_3 \mid x_0, x_1, x_2, x_3 \in \mathbb{Z}\}$$

in $\mathbb{H}(a, b)$. Since $a > 0$, a is the square of a real number, and hence

$$\mathbb{H}(a, b) \otimes \mathbb{R} = M_{2\times 2}(\mathbb{R}).$$

In other words, $\mathbb{H}(a, b)$ gives a rational structure on $M_{2\times 2}(\mathbb{R})$ different from the standard one $M_{2\times 2}(\mathbb{Q})$. Since

$$\det \begin{pmatrix} x_0 + x_1\sqrt{a} & x_2 + x_3\sqrt{a} \\ b(x_2 - x_3\sqrt{a}) & x_0 - x_1\sqrt{a} \end{pmatrix} = Nr(h),$$

the image $\rho(\Gamma)$ under the embedding ρ in equation (III.2.19) belongs to $\mathrm{SL}(2, \mathbb{R})$.

It is known that $\Gamma \cong \rho(\Gamma)$ is a discrete subgroup of $\mathrm{SL}(2, \mathbb{R})$ with compact quotient $\Gamma\backslash\mathrm{SL}(2, \mathbb{R})$ (see [Kat] [Ji6]).

III.2.9 Bianchi groups. Let $F = \mathbb{Q}(\sqrt{-d})$ be an imaginary quadratic field, where d is a positive square-free integer. Then $Res_{F/\mathbb{Q}}\mathrm{SL}(2)$ is defined over $\mathbb{Q}$ of $\mathbb{Q}$-rank 1 and

$$Res_{F/\mathbb{Q}}(\mathbb{R}) = \mathrm{SL}(2, \mathbb{C}).$$

The arithmetic subgroup $\mathrm{SL}(2, \mathcal{O}_F)$ is a discrete subgroup of $\mathrm{SL}(2, \mathbb{C})$ and is called the *Bianchi group* associated with the field F. The symmetric space $X = G/K$ for $G = \mathrm{SL}(2, \mathbb{C})$ is the real hyperbolic space $\mathbf{H}^3$ of dimension 3, i.e., the simply connected Riemannian manifold with constant curvature equal to -1. As discussed in I.4.2, $\mathbf{H}^3$ can be realized as

$$\mathbf{H}^3 = \{(x, y, t) \mid x, y \in \mathbb{R}, t > 0\}, \quad ds^2 = \frac{dx^2 + dy^2 + dt^2}{t^2}.$$

Let $\mathcal{O}_F$ be the ring of integers in F. The quotient $\mathrm{SL}(2, \mathcal{O}_F)\backslash\mathbf{H}^3$ is a typical noncompact arithmetic 3-dimensional *hyperbolic manifold* of finite volume and has been extensively studied in topology (see [EGM] and [MR]). There are also cocompact arithmetic subgroups of $\mathrm{SL}(2, \mathbb{C})$ constructed via *quaternion algebras* over F (see [MR]).

III.2.10 Picard modular groups. Let $\langle \cdot, \cdot \rangle$ be the Hermitian form on $\mathbb{C}^3$ defined by

$$\langle z, w \rangle = \bar{z}_1 w_1 + \bar{z}_2 w_2 - \bar{z}_3 w_3, \quad z = (z_1, z_2, z_3), \quad w = (w_1, w_2, w_3).$$

Let $SU(2, 1)$ be the associated special unitary group

$$SU(2, 1) = \{g \in \mathrm{SL}(3, \mathbb{C}) \mid \langle gz, gw \rangle = \langle z, w \rangle\}.$$

Clearly $SU(2, 1)$ is defined over $\mathbb{Q}$ and hence also defined over any *imaginary quadratic field* $F = \mathbb{Q}(\sqrt{-d})$, where d is a square-free positive integer. $Res_{F/\mathbb{Q}}SU(2, 1)$ is defined over $\mathbb{Q}$ and is of $\mathbb{Q}$-rank 1, and

$$Res_{F/\mathbb{Q}}SU(2, 1)(\mathbb{R}) = SU(2, 1; \mathbb{C}),$$

which is often denoted by $SU(2, 1)$ as above. The arithmetic subgroup $SU(2, 1; \mathcal{O}_F)$ is a discrete subgroup of $SU(2, 1)$ and is called the *Picard modular group* associated with F, where $\mathcal{O}_F$ is the ring of integers in F as above. The symmetric space $X = G/K$ for $G = SU(2, 1)$ is the unit ball in $\mathbb{C}^2$,

$$SU(2, 1)/S(U(2) \times U(1)) \cong B^2_{\mathbb{C}} = \{(z_1, z_2) \in \mathbb{C}^2 \mid |z_1|^2 + |z_2|^2 < 1\},$$

and the quotient $SU(2,1;\mathcal{O}_F)\backslash B^2_{\mathbb{C}}$ is called the *Picard modular surface* associated with the field F.

III.2.11 Siegel modular groups. Except for the example of $\mathrm{SL}(n,\mathbb{Z})$, all other examples of arithmetic groups mentioned above are of $\mathbb{Q}$-rank 1. The Siegel modular group is one of the most important arithmetic groups of higher $\mathbb{Q}$-rank.

For any $n \geq 1$, define

$$Sp(n,\mathbb{Z}) = \left\{ \gamma = \begin{pmatrix} A & B \\ C & D \end{pmatrix} \mid A, B, C, D \in M_n(\mathbb{Z}), {}^t\gamma J_n \gamma = J_n \right\}, \qquad \text{(III.2.20)}$$

where

$$J_n = \begin{pmatrix} 0 & I_n \\ -I_n & 0 \end{pmatrix}.$$

Let

$$\mathbf{H}_n = \{Z \in M_n(\mathbb{C}) \mid tZ = Z, \operatorname{Im} Z > 0\}$$

be the *Siegel upper half-space* of degree n. When $n = 1$, it reduces to the usual upper half-plane $\mathbf{H} = \mathbf{H}_1$. Then $Sp(n,\mathbb{Z})$ acts on $\mathbf{H}_n$ by

$$\gamma \cdot Z = (AZ + B)(CZ + D)^{-1},$$

where γ is given in equation (III.2.20).

The quotient $Sp(n,\mathbb{Z})\backslash\mathbf{H}_n$ is the moduli space of principally polarized abelian varieties of dimension n (see [Mum2, Theorem 4.7]). When $n = 1$, abelian varieties become elliptic curves and each elliptic curve has a canonical principal polarization; furthermore, $Sp(1,\mathbb{Z}) = \mathrm{SL}(2,\mathbb{Z})$.

III.2.12 The above short list gives some typical examples of arithmetic subgroups. To study the quotient $\Gamma\backslash X$, we introduce the notions of fundamental domains and fundamental sets.

Definition III.2.13 A fundamental domain for the arithmetic subgroup Γ acting on X is an open subset $\Omega \subset X$ such that

1. Each coset $\Gamma \cdot x$ contains at least one point in the closure $\overline{\Omega}$, i.e., $X = \Gamma\overline{\Omega}$.
2. No two interior points of Ω lie in one Γ-orbit, i.e., the translates $\gamma\Omega$ for $\gamma \in \Gamma$ are disjoint open subsets.

By definition, Ω is mapped injectively into $\Gamma\backslash X$, and its closure $\overline{\Omega}$ is mapped surjectively onto $\Gamma\backslash X$.

A general geometric method to construct fundamental domains for a discrete group acting isometrically and properly on a metric space uses the *Dirichlet domain*. Specifically, let $x_1 \in X$ be any basepoint that is not a fixed point of Γ. Let d be the distance function. Then

$$D(\Gamma, x_1) = \{x \in X \mid d(x, x_1) < d(\gamma x, x_1), \gamma \in \Gamma, \gamma \neq e\}$$

is the Dirichlet domain for Γ with center x_1. It can be seen that $D(\Gamma, x_1)$ is a fundamental domain for Γ acting on X. In fact, for any $x \in X$, since Γ acts properly, $\inf\{d(\gamma x, x_1) \mid \gamma \in \Gamma\}$ is achieved, say by γ_0. Then $\gamma_0 x \in \overline{D(\Gamma, x_1)}$.

If X is not a symmetric space of constant curvature and $\Gamma\backslash X$ is noncompact, the shape of $D(\Gamma, x_1)$ near infinity is complicated and its relation to parabolic subgroups is not clear. For this reason, we introduce the notion of fundamental sets.

Definition III.2.14 A subset $\mathcal{S}$ of X is called a fundamental set if

1. $\Gamma\mathcal{S} = X$.
2. For any $g \in \mathbf{G}(\mathbb{Q})$, the set $\{\gamma \in \Gamma \mid g\mathcal{S} \cap \gamma\mathcal{S} \neq \emptyset\}$ is finite.

To define the fundamental sets of a fixed arithmetic group Γ, we can replace (2) above by a weaker condition:

(2′) The set $\{\gamma \in \Gamma \mid \mathcal{S} \cap \gamma\mathcal{S} \neq \emptyset\}$ is finite.

But we need condition (2) to relate fundamental sets of different arithmetic subgroups and different algebraic groups, for example, to derive fundamental sets of general Γ from the special case $\mathbf{G} = \mathrm{SL}(n)$, $\Gamma = \mathrm{SL}(n, \mathbb{Z})$.

This condition (2) is called the *Siegel finiteness property* and also plays an important role in defining the topology of compactifications of $\Gamma\backslash X$ and showing that it is Hausdorff.

The first result concerning the quotient $\Gamma\backslash X$ is the following compactness criterion, which was conjectured by Godement and proved by Borel and Harish-Chandra [BHC], and Mostow and Tamagawa [MT].

Proposition III.2.15 *Let $\mathbf{G}$ be a connected reductive linear algebraic group defined over $\mathbb{Q}$ whose center is an anisotropic subgroup defined over $\mathbb{Q}$, and $\Gamma \subset \mathbf{G}(\mathbb{Q})$ an arithmetic subgroup. Then the following conditions are equivalent:*

1. *The locally symmetric space $\Gamma\backslash X$ is compact.*
2. *$\mathbf{G}(\mathbb{Q})$ does not contain any nontrivial unipotent element.*
3. *The $\mathbb{Q}$-rank of $\mathbf{G}$ is equal to 0.*

For the rest of this section, we assume that $\mathbf{G}$ is a connected reductive linear algebraic group defined over $\mathbb{Q}$ whose $\mathbb{Q}$-rank is positive and whose center is an anisotropic subgroup. The positivity of the $\mathbb{Q}$-rank of $\mathbf{G}$ implies that there are proper rational parabolic subgroups of $\mathbf{G}$. The reduction theory gives fundamental sets for Γ in terms of Siegel sets of rational parabolic subgroups.

Proposition III.2.16 *Let $\mathbf{G}$ be a reductive linear algebraic group defined over $\mathbb{Q}$, and Γ an arithmetic subgroup. If $\mathbf{P}$ is a minimal rational parabolic subgroup of $\mathbf{G}$, then $\Gamma\backslash\mathbf{G}(\mathbb{Q})/\mathbf{P}(\mathbb{Q})$ is finite, i.e., there are only finitely many Γ-conjugacy classes of minimal rational parabolic subgroups. Furthermore, there exists a Siegel set $\mathcal{S} = U \times A_{\mathbf{P},t} \times V$ associated with $\mathbf{P}$ and a finite subset $C \subset \mathbf{G}(\mathbb{Q})$ such that $C\mathcal{S}$ is a fundamental set for Γ.*

Together with the Siegel finiteness of the Siegel sets, the above proposition immediately implies the following results.

Corollary III.2.17 *Any arithmetic subgroup Γ is finitely generated.*

Corollary III.2.18 *Under the further assumption that the center of $\mathbf{G}$ is an anisotropic subgroup, i.e., of $\mathbb{Q}$-rank equal to 0, the volume of $\Gamma\backslash G$ is finite. In particular, if $\mathbf{G}$ is semisimple, $\Gamma\backslash G$ has finite volume.*

For the proof of these results, see [Bo4] and [PR]. To relate the geometry of $\Gamma\backslash X$ at infinity to all rational parabolic subgroups rather than only the minimal ones, we need more refined versions of the reduction theory and some separation properties of Siegel sets.

Proposition III.2.19

1. *There are only finitely many Γ-conjugacy classes of rational parabolic subgroups. Let $\mathbf{P}_1, \dots, \mathbf{P}_k$ be a set of representatives of the Γ-conjugacy classes of rational parabolic subgroups. There exist Siegel sets $U_i \times A_{P_i,t_i} \times W_i$ in G associated with $\mathbf{P}_i$ ($1 \le i \le k$) whose images in $\Gamma\backslash G$ cover the whole space.*
2. *For any two rational parabolic subgroups $\mathbf{P}_i$, $i = 1, 2$, and Siegel sets $U_i \times A_{\mathbf{P}_i,t_i} \times W_i$ associated with $\mathbf{P}_i$, the set*
$$\{\gamma \in \Gamma \mid \gamma(U_1 \times A_{\mathbf{P}_1,t_1} \times W_1) \cap U_2 \times A_{\mathbf{P}_2,t_2} \times W_2\} \neq \emptyset$$
is finite.
3. *Suppose that $\mathbf{P}_1$ is not Γ-conjugate to $\mathbf{P}_2$. Fix U_i, W_i, $i = 1, 2$. Then*
$$\gamma(U_1 \times A_{\mathbf{P}_1,t_1} \times W_1) \cap U_2 \times A_{\mathbf{P}_2,t_2} \times W_2 = \emptyset$$
for all $\gamma \in \Gamma$ if $t_1, t_2 \gg 0$.
4. *For any fixed U, W, when $t \gg 0$,*
$$\gamma(U \times A_{\mathbf{P},t} \times W) \cap U \times A_{\mathbf{P},t} \times W = \emptyset$$
for all $\gamma \in \Gamma - \Gamma_P$.
5. *For any two different parabolic subgroups $\mathbf{P}_1, \mathbf{P}_2$, when $t_1, t_2 \gg 0$,*
$$U_1 \times A_{\mathbf{P}_1,t_1} \times W_1 \cap U_2 \times A_{\mathbf{P}_2,t_2} \times W_2 = \emptyset.$$
6. *The analogous properties in (1)–(5) hold for Siegel sets $U_i \times A_{\mathbf{P}_i,t} \times V_i$ in X.*

These results are not stated in exactly the same form in [Bo4] but parts (1) to (4) follow from Theorem 15.5, Proposition 15.6, and Proposition 12.6 there, and part (5) follows from part (3) and the fact that for any two different parabolic subgroups $\mathbf{P}_1, \mathbf{P}_2$, there exists an arithmetic subgroup Γ such that $\mathbf{P}_1$ is not Γ-conjugate to $\mathbf{P}_2$. These results, except for part (5), are also stated in [OW, Theorem 2.11] for slightly more general discrete subgroups Γ.

III.2.20 The reduction theory above gives only a fundamental set for Γ. There is a refined version that allows one to get a fundamental domain, or an exact fundamental set without overlap under Γ-translates. Such a theory is called the *precise reduction theory*. To state this theory, we need to introduce a variant of Siegel sets. For any $T \in A_{\mathbf{P}}$, define

$$A_{\mathbf{P},T} = \{a \in A_{\mathbf{P}} \mid a^{\alpha} > T^{\alpha} \text{ for all } \alpha \in \Phi(P, A_{\mathbf{P}})\}. \tag{III.2.21}$$

When $T = e$, the identify element, $A_{\mathbf{P},e}$ is the positive chamber $A_{\mathbf{P}}^{+}$ associated with $\mathbf{P}$; in general, $A_{\mathbf{P},T}$ is the translate of the positive chamber $A_{\mathbf{P}}^{+}$ under T. (Note that this shifted chamber is similar to the shifted chamber $A_{P,T}$ in the real split component A_P in §I.9.6.) Then we can define *Siegel sets*

$$U \times A_{\mathbf{P},T} \times V$$

in X and $U \times A_{\mathbf{P},T} \times W$ in G.

Proposition III.2.21 *Let $\mathbf{P}_1 \dots, \mathbf{P}_k$ be representatives of Γ-conjugacy classes of all proper rational parabolic subgroups of $\mathbf{G}$. There exist a bounded set Ω_0 in X and Siegel sets $U_i \times A_{\mathbf{P}_i,T_i} \times V_i$, $i = 1, \dots, k$, such that*

1. *each $U_i \times A_{\mathbf{P}_i,T_i} \times V_i$ is mapped injectively into $\Gamma\backslash X$ under the projection $\pi : X \to \Gamma\backslash X$,*
2. *the image of $U_i \times V_i$ in $\Gamma \cap P_i\backslash N_{P_i} M_{\mathbf{P}_i}/(K \cap M_{\mathbf{P}_i}) \cong \Gamma \cap P_i\backslash N_{P_i} \times X_{\mathbf{P}_i}$ is compact,*
3. *and $\Gamma\backslash X$ admits the following disjoint decomposition*

$$\Gamma\backslash X = \Omega_0 \cup \coprod_{i=1}^{k} \pi(U_i \times A_{\mathbf{P}_i,T} \times V_i). \tag{III.2.22}$$

In the above, we have used the fact that $\Gamma \cap P_i$ is contained in $N_{P_i} M_{\mathbf{P}_i}$. The elements of $T_i \in A_{\mathbf{P}_i}$ are related and should be chosen in a compatible way. For more detailed discussions of the precise reduction theory, see [OW, Theorem 3.4], [Sap1] and also [JM, §4.5].

Remark III.2.22 The formulation in [Sap1] is different and given by a Γ-equivariant decomposition of X, called Γ-*equivariant tiling*. Briefly, the disjoint decomposition in equation (III.2.22) can be lifted to a Γ-equivariant disjoint decomposition of X. The connected components of this decomposition, i.e., the connected components of $\pi^{-1}(\pi(U_i \times A_{\mathbf{P}_i,T_i} \times V_i))$, are called *tiles* in [Sap1], and they are parametrized by rational parabolic subgroups. Of course, we can reverse the process. Starting from a Γ-equivariant tiling, take the tiles of the representatives $\mathbf{P}_1, \dots, \mathbf{P}_k$ together with the central tile associated with $\mathbf{G}$. For each tile, take a fundamental domain given by a Siegel set $U_i \times A_{\mathbf{P}_i,T} \times V_i$ for the discrete subgroup $\Gamma \cap P_i$. They give the disjoint decomposition of $\Gamma\backslash X$ as in equation (III.2.22).

III.2.23 Summary and comments. Two versions of reduction theory for arithmetic subgroups of linear reductive algebraic groups are stated. When X is a linear symmetric space, i.e., it is either a symmetric cone, for example $X = GL(n, \mathbb{R})/O(n)$, or a homothety section of a symmetric cone, for example $X = \mathrm{SL}(n, \mathbb{R})/\mathrm{SO}(n)$, there is another version of reduction theory. In this case, the reduction theory is closely related to the geometry of numbers, and a fundamental domain is given by a union of polyhedral cones or their homothety sections. See [AMRT] [As1]–[As4] and the references there. For the geometry of numbers, see [Cass1-2] [Con] [CoS] [Gru] [GruL] [MM1-2] [MR] [Ma] [PR] [So1] [Si3].

III.3 Satake compactifications of locally symmetric spaces

In this section, we recall the procedure in [Sat2] of constructing the Satake compactifications of $\Gamma\backslash X$ from the Satake compactifications of X. The basic idea is to obtain a *partial compactification* of X from a compactification $\overline{X}$ of X by adding only *rational boundary components* of $\overline{X}$ and to endow the partial compactification with the *Satake topology* on which the arithmetic subgroup Γ acts continuously with a compact Hausdorff quotient. In this construction, the notion of rational boundary components is important and depends on the closure or compactifications of fundamental sets.

This section is organized as follows. First, we recall a general result on defining the Satake topology on a Γ-space that contains X if a fundamental set has a Hausdorff compactification in Proposition III.3.2. Then we introduce three different notions of rational boundary components: Siegel rational, weakly rational, and rational boundary components, and the notion of (geometrically) rational compactifications of X in III.3.3, III.3.6–III.3.8. After showing that rational compactifications of X induce Hausdorff compactifications of locally symmetric spaces in III.3.5 and III.3.9, we discuss several examples of rational compactifications of symmetric spaces in III.3.10, III.3.13, and III.3.14. Finally, we comment informally on how compactifications of fundamental sets could lead directly to compactifications of locally symmetric spaces and difficulties of this approach in III.3.17.

In this section, we assume that $\mathbf{G}$ is a connected semisimple linear algebraic group defined over $\mathbb{Q}$, $X = G/K$ the associated symmetric space of noncompact type, and $\Gamma \subset \mathbf{G}(\mathbb{Q})$ an arithmetic subgroup.

III.3.1 In [Sat2], Satake gave a general procedure of compactifying $\Gamma\backslash X$ using a compactification of a *fundamental domain*. In fact, the compactification of the fundamental domain is used to construct a Γ-equivariant partial compactification of X. Since compactifications of the symmetric space X induce compactifications of the fundamental domain, this connects the compactifications of the symmetric spaces X and locally symmetric spaces $\Gamma\backslash X$.

Proposition III.3.2 *Let Γ be an arithmetic subgroup acting on the symmetric space X as above. Let*

$$X^* = X \cup \partial^* X$$

be a Γ*-space without topology containing* X *and extending the canonical action of* Γ *on* X. *Suppose there exists a subset* $\overline{\Sigma}$ *of* X^* *satisfying the following conditions:*

1. $X^* = \Gamma\overline{\Sigma}$.
2. *The space* $\overline{\Sigma}$ *admits a compact Hausdorff topology such that its induced subset topology on* $\overline{\Sigma} \cap X$ *gives the induced topology from* X, *and the* Γ*-action on* $\overline{\Sigma}$ *is continuous in the following sense: for any* $x \in \overline{\Sigma}$ *and* $\gamma \in \Gamma$, *(a) if* $\gamma x \in \overline{\Sigma}$, *then for any neighborhood* U' *of* γx *in* $\overline{\Sigma}$, *there exists a neighborhood* U *of* x *in* $\overline{\Sigma}$ *such that* $\gamma U \cap \overline{\Sigma} \subset U'$; *(b) if* $\gamma x \notin \overline{\Sigma}$, *then there exists a neighborhood* U *of* x *in* $\overline{\Sigma}$ *such that* $\gamma U \cap \overline{\Sigma} = \emptyset$.
3. *There exist finitely many* γ_i, $i \in I$, *such that if* $\gamma\overline{\Sigma} \cap \overline{\Sigma} \neq \emptyset$ *for some* $\gamma \in \Gamma$, *then for some* γ_i,

$$\gamma|_{\overline{\Sigma}\cap\gamma^{-1}\overline{\Sigma}} = \gamma_i|_{\overline{\Sigma}\cap\gamma^{-1}\overline{\Sigma}}. \tag{III.3.23}$$

Then there exists one and only one topology on X^*, *called the Satake topology, satisfying the following conditions:*

1. *it induces the original topology on* $\overline{\Sigma}$ *and on* X,
2. *the* Γ*-action on* X^* *is continuous,*
3. *for every point* $x \in X^*$, *there exists a fundamental system of neighborhoods* $\{U\}$ *of* x *such that*

$$\gamma U = U, \quad \gamma \in \Gamma_x; \quad \gamma U \cap U = \emptyset, \quad \gamma \notin \Gamma_x,$$

where $\Gamma_x = \{\gamma \in \Gamma \mid \gamma x = x\}$ *is the stabilizer of* x *in* Γ,

4. *if* $x, x' \in X^*$ *are not in one* Γ*-orbit, then there exist neighborhoods* U *of* x *and* U' *of* x' *such that*

$$\Gamma U \cap U' = \emptyset.$$

The quotient $\Gamma\backslash X^*$ *is a compact Hausdorff space containing* $\Gamma\backslash X$, *and it induces the canonical topology on the latter. If* $\overline{\Sigma} \cap X$ *is open and dense in* $\overline{\Sigma}$, *then* $\Gamma\backslash X^*$ *contains* $\Gamma\backslash X$ *as a dense, open subset and is hence a compactification of* $\Gamma\backslash X$.

The basic idea of the proof is as follows. Since $X^* = \Gamma\overline{\Sigma}$, we define neighborhoods of points in X^* by saturating neighborhoods of points in $\overline{\Sigma}$ under the action of the stabilizer Γ_x. The finiteness in equation (III.3.23) allows us to show that the topology is Hausdorff. The compactness follows from the fact that $\overline{\Sigma}$ is compact and mapped surjectively to $\Gamma\backslash X^*$. Details of the proof are given in [Sat2, pp. 561–563].

III.3.3 It is reasonable to expect that in the above approach, $\Sigma = \overline{\Sigma} \cap X$ is a fundamental set for the Γ-action on X. To apply this approach, we need a choice of the boundary $\partial^* X$ of X and a suitable compactification of a fundamental set Σ compatible with the structure of $X^* = X \cup \partial^* X$. This is achieved by the so-called *rational boundary components* of compactifications of X.

Let $\overline{X}^S_\tau$ be the Satake compactification of X associated with a faithful projective representation τ of the adjoint group $G/\mathcal{Z}(G)$ of G in §I.4. let μ_τ be the highest weight of τ with respect to a fixed order on $\Phi(\mathfrak{g}, \mathfrak{a})$. Recall from Proposition I.4.38 and discussions before it that for each μ_τ-saturated parabolic subgroup Q, its boundary component is given by

$$e(Q) = X_{P(Q)},$$

where $P(Q)$ is a μ_τ-reduction of Q. Then

$$\overline{X}^S_\tau = X \cup \coprod_{\mu_\tau\text{-saturated } Q} e(Q) = X \cup \coprod_{\mu_\tau\text{-saturated } Q} X_{P(Q)}.$$

Let Σ be a fundamental set for Γ constructed from Siegel sets of rational parabolic subgroups in Propositions III.2.16 and III.2.19. Let $\overline{\Sigma}$ be the closure of Σ in $\overline{X}^S_\tau$. Clearly, the Γ-action on $\overline{\Sigma}$ in the sense of Proposition III.3.2 is continuous.

Assume that $\overline{\Sigma}$ satisfies the following conditions:

III.3.4 Condition.

1. *There exist only finitely many boundary components $X_{P(Q)}$ of $\overline{X}^S_\tau$ meeting $\overline{\Sigma}$. Let $X_{P(Q_1)}, \dots, X_{P(Q_m)}$ be a maximal set of Γ-nonconjugate boundary components among them. For each i, let $I_i \subset \Gamma$ be the finite subset such that when $i = 1, \dots, m$ and $\gamma \in I_i$, $X_{P({}^\gamma Q_i)}$, exhaust all the boundary components in the orbit $\Gamma X_{P(Q_i)}$ that meet $\overline{\Sigma}$.*
2. *For each i, the action of the subgroup $\Gamma \cap Q_i$ on $X_{P(Q_i)}$ induces a discrete group $\Gamma_{X_{P(Q_i)}}$ in the automorphism of $X_{P(Q_i)}$, and the subset*

$$\cup_{\gamma \in I_i} \overline{\Sigma} \cap \gamma^{-1} X_{P({}^\gamma Q_i)} \subset X_{P(Q_i)}$$

is a fundamental set for the $\Gamma_{X_{P(Q_i)}}$-action on $X_{P(Q_i)}$.

The first condition (1) is automatically satisfied (see the proof of Proposition III.3.9 below), and the nontrivial part is condition (2). Under the above assumption, put

$$\partial^* X = \cup_{i=1}^m \Gamma X_{P(Q_i)}. \tag{III.3.24}$$

Proposition III.3.5 *Assume that the compactification $\overline{X}^S_\tau$ satisfies Condition III.3.4 above. For the choice of $\partial^* X$ in equation (III.3.24) and $\overline{\Sigma}$ above, the conditions in Proposition III.3.2 are satisfied, and hence the quotient*

$$\Gamma\backslash X \cup \partial^* X = \Gamma\backslash X \cup \coprod_{i=1}^k \Gamma_{X_{P(Q_i)}}\backslash X_{P(Q_i)}$$

defines a compactification of $\Gamma\backslash X$.

The basic point is to check that the finiteness condition in equation (III.3.23) in Proposition III.3.2 is satisfied. This follows from (2) in Condition III.3.4. See [Sat2, p. 562] for details.

Definition III.3.6 A boundary component $X_{P(Q)}$ of $\overline{X}^S_\tau$ is called *Siegel rational* if $\Gamma X_{P(Q)}$ meets the closure of some Siegel sets of rational parabolic subgroups of $\mathbf{G}$, or equivalently the closure $\overline{\Sigma}$ of a fundamental set Σ in $\overline{X}^S_\tau$ given in the above proposition.

Since all arithmetic subgroups of $\mathbf{G}(\mathbb{Q})$ are commensurable with each other, the above definition does not depend on the choice of Γ, and the Siegel rational boundary components depend only on the rational structure of G given by $\mathbf{G}$ and the compactification $\overline{X}^S_\tau$.

Therefore, the boundary $\partial^* X$ is the union of Siegel rational boundary components of $\overline{X}^S_\tau$. Certainly it is natural to expect that Siegel rational boundary components are given by those for which Q is the real locus of rational parabolic subgroups. To be precise, we need the following notion.

Definition III.3.7

1. A boundary component $X_{P(Q)}$ of $\overline{X}^S_\tau$ is called *weakly rational* if its stabilizer (or normalizer) $\mathcal{N}(X_{P(Q)}) = Q$ is the real locus of a rational parabolic subgroup $\mathbf{Q}$ of $\mathbf{G}$, which is equivalent to that $\Gamma \cap N_Q$ is a cocompact lattice in N_Q;
2. A boundary component $X_{P(Q)}$ of $\overline{X}^S_\tau$ is called *rational* if it is weakly rational and the centralizer $\mathcal{Z}(X_{P(Q)})$ contains a cocompact closed subgroup Z that is a normal subgroup of Q and is the real locus of an algebraic group $\mathbf{Z}$ defined over $\mathbb{Q}$.

The second condition for the rational boundary components implies that the $\Gamma \cap Q$-action on $X_{P(Q)}$ induces a discrete subgroup $\Gamma_{X_{P(Q)}}$, or equivalently the image of $\Gamma \cap Q$ in $Q/\mathcal{Z}(X_{P(Q)})$ is discrete, since the action of Q on $X_{P(Q)}$ factors through

$$Q \to Q/Z \to Q/\mathcal{Z}(X_{P(Q)}).$$

Definition III.3.8 A Satake compactification $\overline{X}^S_\tau$ is called *geometrically rational*, or simply *rational*, if every Siegel rational boundary component is rational.

Proposition III.3.9 *If a Satake compactification $\overline{X}^S_\tau$ is geometrically rational, then the construction in Proposition III.3.5 gives a Hausdorff compactification of $\Gamma\backslash X$,*

$$\overline{\Gamma\backslash X}^S_\tau = \Gamma\backslash X \cup \partial^* X = \Gamma\backslash X \cup \coprod_{i=1}^{m} \Gamma_{X_{P(Q_i)}}\backslash X_{P(Q_i)},$$

where $\mathbf{Q}_1, \dots, \mathbf{Q}_m$ are representatives of Γ-conjugacy classes of μ_τ-saturated rational parabolic subgroups of $\mathbf{G}$.

Proof. We need to check that the conditions in III.3.4 are satisfied. Let **P** be a minimal rational parabolic subgroup of **G**, and let P_0 be a minimal real parabolic subgroup of G contained in the real locus $P = \mathbf{P}(\mathbb{R})$. Then a Siegel set $U \times A_{\mathbf{P},t} \times V$ for the rational parabolic subgroup **P** is contained in a Siegel set $U_0 \times A_{P_0,t_0} \times V_0$ of the real parabolic subgroup P_0 for suitable U_0, V_0, t_0. Clearly, only the boundary components of the standard parabolic subgroups containing P_0 can meet the closure of $U_0 \times A_{P_0,t_0} \times V_0$, and hence of the closure of $U \times A_{\mathbf{P},t} \times V$. Since a fundamental set Σ consists of finitely many Siegel sets, the condition (1) in III.3.4 is satisfied. By the assumption, these finitely many rational boundary components are geometrically rational. Then the condition (2) follows from the observation that for any such boundary component $X_P(Q)$, the intersection with the closure of $U_0 \times A_{P_0,t_0} \times V_0$ is contained in a Siegel set.

III.3.10 To show that a compactification $\overline{X}^S_\tau$ is geometrically rational, we need to carry out two steps:

1. Understand the closure of Siegel sets in $\overline{X}^S_\tau$ and show that Siegel rational boundary components are weakly rational.
2. Show that weakly rational boundary components are rational.

To carry out step (1), we need to understand how the rational split component $A_{\mathbf{P}}$ is embedded into the real split component A_P of a rational parabolic subgroup **P**, since the convergence of interior points to boundary points in the Satake compactification $\overline{X}^S_\tau$ is described in terms of the real split component A_P. To carry out step (2), we need to determine the normalizer and the centralizer of each boundary component.

In all the known examples of Satake compactifications, weakly rational boundary components coincide with rational boundary components.

Lemma III.3.11 *In the maximal Satake compactification* $\overline{X}^S_{\max}$, *any weakly rational boundary component is rational.*

Proof. In this case, the centralizer of the boundary component X_P contains the normal cocompact subgroup $N_P A_P$, which is the radical of P and is hence rational, i.e., the real locus of an algebraic group defined over $\mathbb{Q}$ if P is the real locus of a rational parabolic subgroup **P**, i.e., X_P is weakly rational.

It is not known whether this is true in general, or equivalently, whether the second condition in rational boundary components is necessary.

III.3.12 We consider some examples of geometrically rational Satake compactifications. The first example is given by $\mathbf{G} = \mathrm{SL}(n)$ and the standard representation. In this case, $G = \mathrm{SL}(n, \mathbb{R})$, and

$$X = \mathrm{SL}(n, \mathbb{R})/\mathrm{SO}(n)$$

is the space of positive definite (real) symmetric matrices (or quadratic forms) of determinant one. Denote this symmetric space by X_n. Let $\tau : \mathrm{SL}(n, \mathbb{R}) \to \mathrm{SL}(n, \mathbb{R}) \subset \mathrm{SL}(n, \mathbb{C})$ be the standard representation. Then the Satake compactification $\overline{(X_n)}_\tau^S = \overline{X}_\tau^S$ is similar to the standard compactification $\overline{\mathcal{P}_n}^S$ studied in §I.4, and denoted by $\overline{X_n}^S$.

For any $k \leq n$, X_k is contained in $\overline{X_n}^S$ as a boundary component as follows:

$$A \in X_k \mapsto \left[\begin{pmatrix} A & 0 \\ 0 & 0 \end{pmatrix}\right] \in P(\mathcal{H}_n).$$

By computations similar to those in §I.4.1, we can show that the *Siegel rational boundary components* of $\overline{X_n}^S$ are $\mathrm{SL}(n, \mathbb{Q})$-conjugates of X_k, $k = 1, \dots, n-1$.

Their centralizers and normalizers can be determined as in Propositions I.4.6 and I.4.7 and can be seen to be the real locus of algebraic groups defined over $\mathbb{Q}$, and hence the Siegel rational boundary components are rational. Therefore, the standard compactification $\overline{X_n}^S$ is geometrically rational and induces the standard Satake compactification $\overline{\Gamma\backslash X_n}^S$ of $\Gamma\backslash X_n$. See [GT2] for interpretations of this compactification.

A generalization of this example is given by the following result.

Proposition III.3.13 *If the representation τ of G is defined over $\mathbb{Q}$, then the Satake compactification $\overline{X}_\tau^S$ is geometrically rational, and hence induces a compactification $\overline{\Gamma\backslash X}_\tau^S$ of $\Gamma\backslash X$.*

The original proof was given in [Bo13], and an alternative proof is given in [Sap2].

When $\mathbf{G}$ splits over $\mathbb{Q}$, i.e., the $\mathbb{Q}$-rank of $\mathbf{G}$ is equal to its $\mathbb{C}$-rank, every Satake compactification $\overline{X}_\tau^S$ is G-isomorphic to a Satake compactification associated with a representation defined over $\mathbb{Q}$ and is hence geometrically rational.

Proposition III.3.14 *Let $\mathbf{P}$ be a minimal rational parabolic subgroup of $\mathbf{G}$. Suppose no root of $\Phi(P, A_P)$ restricts trivially on $A_{\mathbf{P}}$. Then the maximal Satake $\overline{X}_{\max}^S$ is geometrically rational, and the associated compactification of $\Gamma\backslash X$ is also called the maximal Satake compactification and denoted by $\overline{\Gamma\backslash X}_{\max}^S$, which admits the following disjoint decomposition:*

$$\overline{\Gamma\backslash X}_{\max}^S = \Gamma\backslash X \cup \coprod_{\mathbf{P}} \Gamma_{X_P}\backslash X_P.$$

(It should be emphasized that X_P is the real boundary symmetric space.)

Proof. As pointed out earlier in Lemma III.3.11, it suffices to show that every Siegel rational boundary component is weakly rational. Let $\mathbf{P}$ be a minimal rational parabolic subgroup of $\mathbf{G}$, and $U \times A_{\mathbf{P},t} \times V$ a Siegel set. It suffices to show that if a boundary component X_Q meets the closure of $U \times A_{\mathbf{P},t} \times V$ in $\overline{X}_{\max}^S$, then Q is the real locus of a rational parabolic subgroup $\mathbf{Q}$.

Let $y_j = (n_j, e^{H_j}, m_j) \in U \times A_{\mathbf{P},t} \times V$ converging to $m_\infty \in X_Q$. By passing to a subsequence, we can assume

1. $n_j \to n_\infty$ in N_Q, and $m_j \to m'_\infty$ in $X_{\mathbf{P}}$,
2. and there exists a subset $I \subset \Delta(P, A_{\mathbf{P}})$ such that for $\alpha \in \Delta - I, \alpha(H_j) \to +\infty$; and for $\alpha \in I, \alpha(H_j)$ converges to a finite number.

We claim that $Q = \mathbf{P}_I(\mathbb{R})$. In fact, $A_{\mathbf{P}_I} \subset A_{P_I}$. By assumption, none of the real roots in $\Phi(P, A_P)$ restricts to zero on the rational split component $A_{\mathbf{P}}$. This implies that

$$A_{\mathbf{P}}^+ \subset A_P^+, \quad A_{\mathbf{P}_I}^+ \subset A_{\mathbf{P}_I(\mathbb{R})}^+.$$

Since e^{H_j} goes to infinity and away from the walls of $A_{\mathbf{P}_I}^+$, it also goes to infinity of $A_{P_I}^+$ and away from its walls. This implies that y_j converges to the image of $m'_j \in X_{\mathbf{P}_I}$ in $X_{\mathbf{P}_I(\mathbb{R})}$ in $\overline{X}^S_{\max}$. Since the boundary components of $\overline{X}^S_{\max}$ are disjoint, $X_Q = X_{\mathbf{P}_I(\mathbb{R})}$, and hence $Q = \mathbf{P}_I(\mathbb{R})$.

III.3.15 The condition in the above proposition is satisfied when $\mathbf{G}$ is *quasi-split* over $\mathbb{Q}$, i.e., when $\mathbf{G}$ has a Borel subgroup defined over $\mathbb{Q}$. In this case, there exist generic representations of G that are defined over $\mathbb{Q}$, and the proposition follows from the earlier one as well.

There are some important examples of geometrically rational Satake compactifications that are not defined by rational representations. One such example is the Baily–Borel compactification of Hermitian symmetric spaces. In fact, by §I.5, the Baily–Borel compactification $\overline{X}^{BB}$ is isomorphic to a minimal Satake compactification. It will be shown in the next section that $\overline{X}^{BB}$ is a geometrically rational compactification.

III.3.16 The construction of compactifications of locally symmetric spaces in this section can be summarized as follows:

1. Choose the collection of rational boundary components of $\overline{X}^S_\tau$.
2. Attach these rational boundary components at infinity to form a partial compactification $X \cup \partial^* X$ that has a Satake topology, so that its quotient $\Gamma\backslash X \cup \partial^* X$ is a compact Hausdorff space.

Step (1) depends on the G-action on the compactification $\overline{X}^S_\tau$, and the attaching of the rational boundary components and the Satake topology in step (2) depends on the topology of the compactification $\overline{X}^S_\tau$ as well. Since the G-action and the compactification $\overline{X}^S_\tau$ are real structures, the passage to the rational structures causes difficulties. In this sense, this process is not natural. It would be desirable to carry out all these steps using only the rational structures of $\mathbf{G}$ and X. In §III.11, we will discuss such an alternative construction of the maximal Satake compactification.

Remark III.3.17 In the above passage from a compactification of the symmetric space X to a compactification of the locally symmetric space $\Gamma\backslash X$, the fundamental set Σ of Γ in X played an essential role. Here we explain this approach in a slightly different, informal way to emphasize the difference between compactifications of a fundamental set or domain (but without using compactifications of X) and compactifications of $\Gamma\backslash X$ and to point out the difficulties in passing from the former to the latter.

By definition, $X = \Gamma\Sigma$, and hence Σ projects surjectively to $\Gamma\backslash X$. If Σ admits a Hausdorff compactification $\overline{\Sigma}$ that is equivariant in a suitable sense, then it is reasonable to expect that $\overline{\Sigma}$ is mapped onto a compactification of $\Gamma\backslash X$. As will be seen, the problem is to show the Hausdorff property.

For simplicity, we assume that Σ is a closed subset of X. The equivariant condition on $\overline{\Sigma}$ can be formulated as follows: For any sequence x_j in Σ, let

$$\Gamma' = \{\gamma \in \Gamma \mid \gamma x_j \in \Sigma \text{ eventually}\},$$

which is a finite set. Then we require that x_j converge in $\overline{\Sigma}$ if and only if for every $\gamma \in \Gamma'$, γx_j converges in $\overline{\Sigma}$. If $\overline{X}$ is a G-compactification of X, then the closure of Σ in $\overline{X}$ is clearly such an equivariant compactification $\overline{\Sigma}$.

To obtain a compactification of $\Gamma\backslash X$ as a quotient of $\overline{\Sigma}$, we need an equivalence on $\overline{\Sigma}$. Two points x_∞, x'_∞ in $\overline{\Sigma}$ are defined to be related if there exists a sequence x_j in Σ such that $x_j \to x_\infty$ and an element $\gamma \in \Gamma$ with $\gamma x_j \in \Sigma$ eventually and $\gamma x_j \to x'_\infty$. If this were an equivalence relation, the finiteness of the subset Γ' of Γ implies that the quotient of $\overline{\Sigma}$ is a compact Hausdorff space containing X as an open dense subset. But this relation is not necessarily an equivalence relation. In fact, it is not clear that it is transitive. We need to extend it to an equivalence. Once extended, it is not clear anymore that the induced quotient of $\overline{\Sigma}$ is Hausdorff.

In a certain sense, Proposition III.3.2 gives the precise conditions on $\overline{\Sigma}$ if Γ-translates of $\overline{\Sigma}$ can be defined. If X has a compactification $\overline{X}$ and $\overline{\Sigma}$ is the closure of Σ in $\overline{X}$, then we can certainly define Γ-translates of $\overline{\Sigma}$ via the Γ-action on $\overline{X}$.

III.3.18 Summary and comments. We followed the method in [Sat2] to construct compactifications of $\Gamma\backslash X$ from certain compactifications of X. The key point is a suitable notion of rational boundary components. In the case of $\mathbf{G} = \mathrm{SL}(2)$ and $X = \mathbf{H}$, the definition is simple. But it is not so simple already in the case of the Hilbert modular group $\Gamma = \mathrm{SL}(2, \mathcal{O}_F)$. In this case, $X = \mathbf{H}\times\mathbf{H}$. Let $i\infty$ be the distinguished boundary point in the boundary $\mathbb{R}\cup\{i\infty\}$ of $\mathbf{H}$. Then the point $(i\infty, i\infty)$ is a rational boundary point for the Hilbert modular group Γ. The other rational boundary points are points in the orbit $\mathrm{SL}(2, \mathcal{O}_F) \cdot (i\infty, i\infty)$, most of which do not belong to $\mathbb{Q} \cup \{i\infty\} \times \mathbb{Q} \cup \{i\infty\}$, the product of the rational points of $\mathbf{H}$.

III.4 Baily–Borel compactification

In this section, we discuss the Baily–Borel compactification $\overline{\Gamma\backslash X}^{BB}$ of Hermitian locally symmetric spaces. As a topological compactification of $\Gamma\backslash X$, it is a Satake compactification. The Satake compactifications are only topological spaces, but the

Baily–Borel compactification is a normal projective variety. Due to this property, the Baily–Borel compactification is one of the most important among the finitely many Satake compactifications. In this section, we will mainly discuss the structures of the Baily–Borel compactification as a topological and analytic compactification.

To construct $\overline{\Gamma\backslash X}^{BB}$ as a topological space, we apply the general procedure in the previous section. To do this, we need to show that the Baily–Borel compactification $\overline{X}^{BB}$ is (geometrically) rational. The first part of this section is concerned with this. In the second part, we put the structure of a normal analytic space on $\overline{\Gamma\backslash X}^{BB}$. Finally we briefly comment on more refined structures such as the projective embedding of $\overline{\Gamma\backslash X}^{BB}$, its models over number fields, and applications to L^2-cohomology.

More precisely, this section is organized as follows. In III.4.1, we discussed the original motivation of the Baily–Borel compactification. To determine the Siegel rational boundary components, we need to understand the root structure (III.4.3, III.4.4). The Siegel rational boundary components are given in III.4.5, which are shown to be rational in III.4.6. Hence $\overline{X}^{BB}$ is shown to be geometrically rational in III.4.7. A general procedure of putting a sheaf of analytic functions on a union of normal analytic spaces is given in III.4.9, which is applied in III.4.10 to show that $\overline{\Gamma\backslash X}^{BB}$ is a normal analytic space (III.4.12). In general, the compactification $\overline{\Gamma\backslash X}^{BB}$ is singular (III.4.14).

III.4.1 We start with the original motivation of the Baily–Borel compactification. For the rest of this section, X is a Hermitian symmetric space, and $\Gamma\backslash X$ is a noncompact Hermitian locally symmetric space.

Let $\mathcal{M}(\Gamma\backslash X)$ be the field of meromorphic functions on $\Gamma\backslash X$. If f, g are two holomorphic modular forms on X with respect to Γ of the same weight, then f/g belongs to $\mathcal{M}(\Gamma\backslash X)$. Siegel raised the question whether the *transcendental degree* of $\mathcal{M}(\Gamma\backslash X)$ is equal to $\dim_{\mathbb{C}} \Gamma\backslash X$. This question is related to the growth of the dimension of the space of modular forms when the weight goes to infinity. If $\Gamma\backslash X$ admits a compactification $\overline{\Gamma\backslash X}$ that is a normal projective variety such that the boundary $\overline{\Gamma\backslash X} - \Gamma\backslash X$ is of codimension at least 2, then by the Riemannian extension theorem, every function in $\mathcal{M}(\Gamma\backslash X)$ extends to a rational function on $\overline{\Gamma\backslash X}$. Since the field of rational functions of the projective variety $\overline{\Gamma\backslash X}$ clearly has transcendental degree equal to $\dim_{\mathbb{C}} \Gamma\backslash X$, it answers the Siegel question positively.

When X is the Siegel upper space $\mathbf{H}_n$ in §III.2.11 and Γ the Siegel modular group $Sp(n, \mathbb{Z})$, such a compactification was first constructed in [Sat3] as an analytic space. It was proved later by Baily in [Ba1] that it is a normal projective variety. When X is a general Hermitian symmetric space of noncompact type, such a compactification of $\Gamma\backslash X$ was constructed in [BB1] (when X is a classical domain, a related compactification was also constructed in [PS] using different methods). The compactification is usually called the Baily–Borel compactification in the literature.

III.4.2 As proved in §I.5, the Baily–Borel compactification $\overline{X}^{BB}$ is isomorphic to a minimal Satake compactification as a G-space. To determine the Siegel rational boundary components, we need to compare the rational roots with the real roots, since the Siegel sets are defined in terms of rational roots, while the convergence of a sequence of points in $\overline{X}^{BB}$ is determined in terms of the real roots.

Recall that $X = G/K$, where G is the real locus of a semisimple linear algebraic group $\mathbf{G}$ defined over $\mathbb{Q}$. Without loss of generality, we can assume that the center of $\mathbf{G}$ is trivial. Then $\mathbf{G}$ is the direct product of its normal $\mathbb{Q}$-simple subgroups $\mathbf{G}_i$, $1 \leq i \leq m$. Let

$$\Gamma_i = \Gamma \cap \mathbf{G}_i(\mathbb{R}), \quad X_i = \mathbf{G}_i(\mathbb{R})/K \cap \mathbf{G}_i(\mathbb{R}).$$

Then the subgroup

$$\Gamma' = \Gamma_1 \cdots \Gamma_m \cong \Gamma_1 \times \cdots \times \Gamma_m$$

is commensurable with Γ, and X admits a product decomposition $X = X_1 \times \cdots \times X_m$, and hence

$$\Gamma' \backslash X = \Gamma_1 \backslash X_1 \times \cdots \times \Gamma_m \backslash X_m. \tag{III.4.25}$$

Since the collection of rational boundary components and the Satake topology on the partial compactification $X \cup \partial^* X$ in the previous section depend only on the commensurable class of the arithmetic subgroups, we can assume that $\Gamma = \Gamma'$. By equation (III.4.25), it suffices to consider the case that $\mathbf{G}$ is $\mathbb{Q}$-simple, which will be assumed for the rest of this section.

Proposition III.4.3 *Under the above assumption on G and X, there exist a totally real number field k and a connected, absolutely simple linear algebraic group $\mathbf{G}'$ defined over k such that*

$$\mathbf{G} = Res_{k/\mathbb{Q}} \mathbf{G}'.$$

Under the additional assumption that $\Gamma \backslash X$ is noncompact, G has no normal compact subgroup of positive dimension.

A special case of this proposition is that of Hilbert modular surfaces. See [BB1, p. 469] for a proof.

Let $\nu_1, \ldots, \nu_p$ be all the normalized archimedean valuations of k, and $\sigma_1, \ldots, \sigma_p$ the corresponding embeddings of k into $\mathbb{R}$ (note that k is totally real and hence every embedding is real). Let k_{ν_i} be the completion of k with respect to ν_i. Then $k_{\nu_i} \cong \mathbb{R}$ and is equal to the closure of $\sigma_i(k)$ in $\mathbb{R}$.

Under the embedding $\sigma_i : k \to \mathbb{R}$, $\mathbf{G}$ can be looked upon as an algebraic group defined over $\sigma(k)$, which is denoted by ${}^{\sigma_i}\mathbf{G}'$. Let ${}^{\sigma_i}\mathbf{G}'(\mathbb{R})$ be the real locus of the linear algebraic group ${}^{\sigma_i}\mathbf{G}'$.

Then

$$G = \mathbf{G}(\mathbb{R}) = \prod_{i=1}^{p} {}^{\sigma_i}\mathbf{G}'(\mathbb{R}),$$

and hence

$$X = \prod_{i=1}^{p} X_i,$$

where $X_i = {}^{\sigma_i}\mathbf{G}'(\mathbb{R})/K_i$, and $K_i = K \cap {}^{\sigma_i}\mathbf{G}'(\mathbb{R})$ is a maximal compact subgroup of ${}^{\sigma_i}\mathbf{G}'(\mathbb{R})$.

Let $\mathbf{S}'$ be a maximal k-split torus of $\mathbf{G}'$. Under the embedding $\sigma_i : k \to \mathbb{R}$, $\mathbf{S}'$ is mapped isomorphically to a maximal $\sigma_i(k)$-split torus ${}^{\sigma_i}\mathbf{S}'$ of the algebraic group ${}^{\sigma_i}\mathbf{G}'$ defined over $\sigma_i(k)$. The maximal $\mathbb{Q}$-split subtorus $\mathbf{S}$ of $Res_{k/\mathbb{Q}}\mathbf{S}'$ is canonically isomorphic to $\mathbf{S}'$ and diagonally embedded in $Res_{k/\mathbb{Q}}\mathbf{S}'$, and gives a maximal $\mathbb{Q}$-split torus of $\mathbf{G}$.

In each group ${}^{\sigma_i}\mathbf{G}'$, we choose a maximal $\mathbb{R}$-split torus $\mathbf{T}_{\sigma_i} \supseteq {}^{\sigma}\mathbf{S}'$, which is contained in a maximal torus defined over $\sigma_i(k)$. Then

$$\mathbf{T}_{\sigma_1}(\mathbb{R})^0 \times \cdots \times \mathbf{T}_{\sigma_p}(\mathbb{R})^0$$

is the $\mathbb{R}$-split component A_P of a minimal real parabolic subgroup P of G. Since the identity component of the real locus $\mathbf{S}(\mathbb{R})$ is the $\mathbb{Q}$-split component $A_{\mathbf{P}_0}$ of the real locus of a minimal rational parabolic subgroup $\mathbf{P}_0$ of $\mathbf{G}$, we need to study the restriction of the roots in $\Phi({}^{\sigma_i}G', T^0_{\sigma_i})$ to ${}^{\sigma_i}S'^{,0}$, where ${}^{\sigma_i}G' = {}^{\sigma_i}\mathbf{G}'(\mathbb{R})$, and $T^0_{\sigma_i}$ is the identity component of $\mathbf{T}_{\sigma_i}(\mathbb{R})$, and ${}^{\sigma_i}S'^{,0}$ is the identity component of ${}^{\sigma_i}\mathbf{S}'(\mathbb{R})$.

For each i, X_i is an irreducible Hermitian symmetric space, and hence the root system $\Phi({}^{\sigma_i}G', T^0_{\sigma_i})$ is either of type B_t or of BC_t, by Proposition I.5.18 in §I.5. Let

$$r : \Phi({}^{\sigma_i}G', T^0_{\sigma_i}) \to \{0\} \cup \Phi({}^{\sigma}G', {}^{\sigma_i}S'^{,0})$$

be the restriction map. For each ordering on the latter, we can choose a compatible ordering such that for every $\alpha \in \Phi^+({}^{\sigma_i}G', T^0_{\sigma_i})$, if $r(\alpha) \neq 0$, then $r(\alpha) > 0$.

Let $\Delta({}^{\sigma_i}G', T^0_{\sigma_i})$, $\Delta({}^{\sigma}G', {}^{\sigma_i}S'^{,0})$ be the simple roots for the compatible ordering.

Proposition III.4.4 *The root system $\Phi({}^{\sigma}G', {}^{\sigma_i}S'^{,0})$ is of type BC_s if either $\Phi({}^{\sigma_i}G', T^0_{\sigma_i})$ is of type BC_t or if $\Phi({}^{\sigma_i}G', T^0_{\sigma_i})$ is of type C_t and $r(\alpha_t) = 0$, and is of type C_s otherwise. Each simple β in $\Delta({}^{\sigma}G', {}^{\sigma_i}S'^{,0})$ is the restriction of a unique simple root α in $\Delta({}^{\sigma_i}G', T^0_{\sigma_i})$.*

A simple root in $\Delta({}^{\sigma_i}G', T^0_{\sigma_i})$ that does not restrict trivially to ${}^{\sigma_i}S'^{,0}$ is called *critical*.

Let $\beta_1, \dots, \beta_s$ be the set of simple roots in $\Delta(G, S^0)$, where $S^0 = \mathbf{S}(\mathbb{R})^0$. Since $\mathbf{S} \cong {}^{\sigma_i}\mathbf{S}'$, under the map $S \to {}^{\sigma_i}S'$, we choose the ordering on $\Phi({}^{\sigma_i}G', {}^{\sigma_i}S'^{,0})$ so that $\beta_1, \dots, \beta_s$ are mapped to simple roots in $\Delta({}^{\sigma}G', {}^{\sigma_i}S'^{,0})$. Let $\alpha_{i,1}, \dots, \alpha_{i,t}$ be the simple roots in $\Delta({}^{\sigma_i}G', T^0_{\sigma_i})$. For each β_j, let $c(\sigma_i, j)$ be the unique integer such that $\alpha_{c(\sigma_i,j)}$ restricts to β_j. We can choose the numbering such that for each σ_i, $c(\sigma_i, j)$ is an increasing function in j. This numbering is called the *canonical numbering*.

Let ${}^{\sigma_i}P$ be the minimal real parabolic subgroup of ${}^{\sigma_i}G'$ corresponding to the simple roots $\alpha_{i,1}, \dots, \alpha_{i,t}$. For each j, let

$$X_{\sigma_i,j} = X_{{}^{\sigma_i}P_I}, \quad \text{where } I = \{\alpha_{j+1}, \dots, \alpha_t\}.$$

Let $\mathbf{P}_0$ be the minimal rational parabolic subgroup of $\mathbf{G}$ corresponding to the simple roots $\beta_1, \dots, \beta_s$. For each $b \in \{1, \dots, s\}$, then

$$X_{P_{0,\{\beta_{b+1},\dots,\beta_s\}}} = \prod_{i=1}^{p} X_{\sigma_i,j}$$

is the boundary symmetric space of the real locus of the standard rational parabolic subgroup $\mathbf{P}_{0,\{\beta_{b+1},\dots,\beta_s\}}$. Since each $X_{\sigma_i,j}$ is a Hermitian symmetric space, $X_{P_{0,\{\beta_{b+1},\dots,\beta_s\}}}$ is a Hermitian symmetric space. In fact, it is a boundary component of $\overline{X}^{BB}$ and its normalizer is the real locus of the maximal rational parabolic subgroup $\mathbf{P}_{0,\Delta-\{\beta_b\}}$, and hence is a weakly rational boundary component. The conjugates under $\mathbf{G}(\mathbb{Q})$ of these weakly rational standard boundary components give all the weakly rational boundary components.

Proposition III.4.5 *Every Siegel rational boundary component of $\overline{X}^{BB}$ is a $\mathbf{G}(\mathbb{Q})$-conjugate of the standard boundary components and hence weakly rational.*

The idea of the proof is as follows.

$$\overline{X}^{BB} = \overline{X_1}^{BB} \times \cdots \times \overline{X_k}^{BB}.$$

Recall from Part I, Proposition I.5.18, that the roots in $\Phi({}^{\sigma_i}G', T^0_{\sigma_i})$ are given by

$$\pm \frac{\gamma_{\sigma_i,m} \pm \gamma_{\sigma_i,n}}{2}.$$

The convergence of a sequence $e^{H_j} x_0 \in {}^{\sigma_i}S'^{,0} x_0$ in $\overline{X_i}^{BB}$ is determined by the limits of $\gamma_{\sigma_i,m}(H_j)$. Note that $A_{\mathbf{P}_0}$ is embedded diagonally into the product $T^0_{\sigma_1} \times \cdots \times T^0_{\sigma_k}$, $a \mapsto (a_{\sigma_1}, \dots, a_{\sigma_p})$, and the image satisfies the equations

$$\gamma_{\sigma_i,m}(\log a_{\sigma_i}) = \gamma_{\sigma_i,m+1}(\log a_{\sigma_i})$$

when m is not a critical index, where $\gamma_{\sigma_i,t+1} = 0$. Using this, the above proposition follows easily. See [BB1, p. 480] for details.

Proposition III.4.6 *Every weakly rational boundary component of the Baily–Borel compactification is rational.*

The idea of the proof is to compute explicitly the centralizer of the standard boundary component and show that it contains a cocompact normal subgroup defined over $\mathbb{Q}$. For details, see [BB1, Theorem 3.7].

By combining the above results and the general procedure in §III.3, we obtain the following result.

Proposition III.4.7 *The Baily–Borel compactification $\overline{X}^{BB}$ is geometrically rational, and hence induces a Hausdorff compactification of $\Gamma\backslash X$, called the Baily–Borel compactification and denoted by $\overline{\Gamma\backslash X}^{BB}$.*

Let $\mathbf{P}_1, \dots, \mathbf{P}_m$ be a set of representatives of Γ-conjugacy classes of maximal rational parabolic subgroups of $\mathbf{G}$. Let $X_{P_1,h}, \dots, X_{P_m,h}$ be the boundary components in $\overline{X}^{BB}$ associated with them. As pointed out earlier, each $X_{P_i,h}$ is a Hermitian symmetric space. Let $\Gamma_{X_{P_i,h}}$ be the induced arithmetic subgroup acting on $X_{P_i,h}$. Then

$$\overline{\Gamma\backslash X}^{BB} = \Gamma\backslash X \cup \coprod_{i=1}^{m} \Gamma_{X_{P_i,h}}\backslash X_{P_i,h}. \tag{III.4.26}$$

In the above decomposition, each piece is a normal analytic space. It is natural to expect that $\overline{\Gamma\backslash X}^{BB}$ is a compact normal analytic space.

For this purpose, we need a general criterion on how to patch up finitely many analytic spaces into an analytic space.

III.4.8 Let V be a second-countable compact Hausdorff space. Suppose V is the disjoint union of finitely many subspaces

$$V = V_0 \coprod V_1 \coprod \cdots \coprod V_m, \tag{III.4.27}$$

where each V_i is an irreducible normal analytic space.

Define a sheaf of $\mathcal{A}$-functions on V as follows. For any open subset $U \subset V$, a complex-valued continuous function on U is an $\mathcal{A}$-function if its restriction to each $U \cap V_i$, $0 \le i \le m$, is analytic. In other words, it is patched up continuously from analytic functions on the subsets $U \cap V_i$.

Proposition III.4.9 *Assume that V and the sheaf of $\mathcal{A}$-functions satisfy the following conditions:*

1. *For each positive integer d, the union $V_{(d)}$ of V_i with $\dim_{\mathbb{C}} V_i \le d$ is closed; for any $i > 0$, $\dim_{\mathbb{C}} V_i < \dim_{\mathbb{C}} V_0$; and V_0 is open and dense in V.*
2. *Each point $v \in V$ has a fundamental set of open neighborhoods $\{U_j\}$ such that $U_j \cap V_0$ is connected for every j.*
3. *The restrictions to V_i of local $\mathcal{A}$-functions define the structure sheaf of V_i.*
4. *Each point $v \in V$ has a neighborhood U_v whose points are separated by the $\mathcal{A}$-functions defined on U.*

Then V with the sheaf of $\mathcal{A}$-functions is an irreducible normal analytic space and for each $d \le \dim_{\mathbb{C}} V_0$, the union $V_{(d)}$ defined earlier is an analytic subspace of V with dimension equal to $\max\{\dim_{\mathbb{C}} V_i \mid V_i \subset V_{(d)}\}$.

We will apply this criterion to $\overline{\Gamma\backslash X}^{BB}$ and prove the following result.

Proposition III.4.10 *The compactification $\overline{\Gamma\backslash X}^{BB}$ is a normal analytic space that induces the natural analytic structure on $\Gamma\backslash X$ and the boundary components.*

We will briefly outline the proof. Clearly, the decomposition in equation (III.4.26) satisfies the conditions in equation (III.4.27).

Since $\Gamma\backslash X$ is open and dense in $\overline{\Gamma\backslash X}^{BB}$, the condition (1) in Proposition III.4.9 is satisfied with $V_0 = \Gamma\backslash X$. A basis of neighborhoods of boundary points in $\overline{\Gamma\backslash X}^{BB}$ is given by the closure of the image of truncated Siegel sets, and suitable ones are shown to be connected in [BB1, Proposition 4.15]. Hence (2) is connected.

The idea to check the condition (3) is to push analytic functions on the boundary into the interior. Roughly, if for any pair of boundary components V_i, V_j with $V_i \subset \overline{V_j}$ there is a (local) holomorphic map from a neighborhood of V_i into $\overline{V_j}$ such that these maps are compatible for all such pairs, then any analytic functions near a point v in V_i can be extended to an $\mathcal{A}$-function on a neighborhood of v in V. Since the problem is local, for simplicity, we consider the lift to the universal covering. By the horopsherical decomposition of X,

$$X = N_{P_i} \times A_{P_i} \times X_{P_i,l} \times X_{P_i,h},$$

where $X_{P_i,l}$, $X_{P_i,h}$ are two factors of X_{P_i}, $X_{P_i} = X_{P_i,l} \times X_{P_i,h}$, we have the projection map

$$X \to X_{P_j,h}.$$

Though the horospherical decomposition of X is not holomorphic, this projection map is holomorphic. The relative horospherical decomposition shows that if $X_{P_i,h}$ is contained in the closure of $X_{P_j,h}$, there is an analogous holomorphic map from $X_{P_j,h}$ to $X_{P_i,h}$. These maps are compatible in the sense that the map $X \to X_{P_i,h}$ is the composition of the two maps $X \to X_{P_j,h}$ and $X_{P_j,h} \to X_{P_i,h}$.

Condition (4) is the most difficult to check. Clearly, the analytic functions obtained by pushing in from the boundary in the previous paragraph do not separate points in the neighborhoods in V. To obtain the desired functions, we have to realize X as a Siegel domain of the third kind over $X_{P_i,h}$, and define the *Poincaré–Eisenstein series* adapted to this realization. Briefly, a *Siegel domain of the third kind* is a holomorphic family of Siegel domains of the second kind over $X_{P_i,h}$, where a *Siegel domain of the second kind* is a holomorphic family of tube domains, i.e., *Siegel domains of the first kind*, over a complex vector space. In this realization, the base $X_{P_i,h}$ is realized as a bounded symmetric domain, and hence it supports (holomorphic) polynomial functions that separate the points on $X_{P_i,h}$. The construction of the Poincaré–Eisenstein series allows one to lift these functions to automorphic forms on X with respect to Γ. Once their boundary behaviors near the rational boundary components are understood, it can be shown that they give rise to $\mathcal{A}$-functions that separate points in neighborhoods of boundary points in $\overline{\Gamma\backslash X}^{BB}$. For details, see [BB1, 8.6, 8.8, 8.9]. Specifically, The Poincaré–Eisenstein series are defined in [BB1, 7.1, equation 1], their boundary behaviors are given in [BB1, 8.6]; separation of points in [BB1, 8.8, 8.9], and the condition (4) is checked in [BB1, 10.4].

Remarks III.4.11

(1) In [BB1, 10.3, 1.7], the projection map $X \to X_{P_i,h}$ is defined in terms of the realization of X as a Siegel domain of the third kind over $X_{P_i,h}$. To see that it agrees with the definition given here, we check that they agree on a flat (or a polydisk) passing through the basepoint x_0. Then apply the polydisk theorem to globalize the map.

(2) The construction of $\overline{\Gamma\backslash X}^{BB}$ in [PS] is slightly different. Instead of the Satake topology on the partial compactification

$$X \coprod \cup_Q X_{Q,h},$$

where Q runs over the real locus of maximal rational parabolic subgroups, a *cylindrical topology* based on the realization of X as Siegel domains of the third kind is used. For the equivalence of these two topologies on $\overline{\Gamma\backslash X}$, see [Ki] and [KK1].

Proposition III.4.12 *The space $\overline{\Gamma\backslash X}^{BB}$ is a normal projective variety that induces the analytic structure above.*

In fact, using suitable Poincaré series, which correspond to cusp forms, and Poincaré–Eisenstein series, we can embed $\overline{\Gamma\backslash X}^{BB}$ as a normal projective subvariety of a projective space. See [BB1, 10.10, 10.11] for details.

A stronger result that $\overline{\Gamma\backslash X}^{BB}$ is defined over a number field, or even over its ring of integers, also holds. See [Mi1] [Mi2] [Ch1] [Ch2] [FC] for precise statements and applications to number theory.

III.4.13 When $\Gamma\backslash X$ is a Riemann surface, $\overline{\Gamma\backslash X}^{BB}$ is obtained by adding a point to each cusp neighborhood, and is a compact smooth Riemann surface. In general, $\overline{\Gamma\backslash X}^{BB}$ is a singular variety. In fact, in the example of Hilbert modular surfaces, $\overline{\Gamma\backslash X}^{BB}$ is also obtained by adding a point to each end and the link of such a point is a $\Gamma_{N_P}\backslash N_P$-bundle over $\Gamma_{M_\mathbf{P}}\backslash X_\mathbf{P}$, which is clearly not a sphere; and hence $\overline{\Gamma\backslash X}^{BB}$ is not smooth. It is reasonable to expect that for general $\Gamma\backslash X$, the links of boundary points in $\overline{\Gamma\backslash X}^{BB}$ are complicated.

Proposition III.4.14 *The compactification $\overline{\Gamma\backslash X}^{BB}$ is a singular variety when $\Gamma\backslash X$ is not a product of Riemann surfaces, and the singularities consist of the boundary $\overline{\Gamma\backslash X}^{BB} - \Gamma\backslash X$.*

This result was basically proved in [Ig3], and the idea is to construct a holomorphic form on $\Gamma\backslash X$ that does not extend to a holomorphic form on a smooth compactification. On the other hand, it extends holomorphically to $\overline{\Gamma\backslash X}^{BB}$. This implies that $\overline{\Gamma\backslash X}^{BB}$ has no finite cover that is a smooth variety, and hence it is singular.

III.4.15 The singular projective variety $\overline{\Gamma\backslash X}^{BB}$ has the natural (middle perversity) *intersection cohomology groups*, which have the important property of Poincaré duality. An important conjecture of Zucker states that this intersection cohomology is naturally isomorphic to the L^2-*cohomology* of $\Gamma\backslash X$. This conjecture has been proved independently by Saper-Stern [SaS] and Looijenga [Lo6]. See [Zu4] for a survey of this conjecture.

III.4.16 Summary and comments. We showed that $\overline{X}^{BB}$ is geometrically rational and hence $\overline{\Gamma\backslash X}^{BB}$ is defined. Then we outlined the steps of showing that $\overline{\Gamma\backslash X}^{BB}$ is a normal projective variety. In fact, it is defined over a number field (see [Mi1] [Mi2]). This property is important in applications to number theory and the Langlands program.

III.5 Borel–Serre compactification

In this section, we assume that $\mathbf{G}$ is a connected reductive linear algebraic group defined over $\mathbb{Q}$ whose center is anisotropic over $\mathbb{Q}$. This condition is satisfied if $\mathbf{G}$ is semisimple. We recall the Borel–Serre compactification $\overline{\Gamma\backslash X}^{BS}$ of a locally symmetric space [BS2], which has important applications to the cohomology of arithmetic groups and automorphic forms. A variant of the construction of this compactification gives the uniform method for compactifying symmetric spaces in §I.8 and locally symmetric spaces in §III.8.

More specifically, this section is organized as follows. We first motivate the Borel–Serre compactification $\overline{\Gamma\backslash X}^{BS}$ by the problem of finding a finite classifying space of Γ in topology (III.5.1). Its applications to cohomology groups of Γ are briefly discussed in III.5.2. The geodesic action is given in III.5.3 and used in III.5.4 to construct a real analytic corner $X(\mathbf{P})$ for each rational parabolic subgroup $\mathbf{P}$, which is a partial compactification of X along the direction of $\mathbf{P}$. Then we show that the real analytic structures of these corners are compatible (III.5.5–III.5.7), and hence they glue into a real analytic manifold with corners, the partial Borel–Serre compactification $\overline{X}^{BS}$ (III.5.11),[1] whose quotient by Γ gives $\overline{\Gamma\backslash X}^{BS} = \Gamma\backslash_{\mathbb{Q}}\overline{X}^{BS}$ (III.5.14). The construction is illustrated in III.5.15 through the example $\mathbf{G} = \mathrm{SL}(2)$.

III.5.1 Let Γ be an arithmetic subgroup as above. Recall that a topological space B is a $K(\Gamma, 1)$-space if

$$\pi_1(B) = \Gamma, \quad \pi_i(B) = \{1\}, \quad i \geq 2.$$

An important problem in topology to find a finite $K(\Gamma, 1)$-space, i.e., a $K(\Gamma, 1)$-space that is homotopic to a finite CW-complex.

It is known that a necessary condition for the existence of a finite $K(\Gamma, 1)$-space is that Γ is torsion-free (see [Br2]).

Assume that Γ is a torsion-free arithmetic subgroup. Note that X is simply connected and nonpositively curved, and hence contractible. Since Γ acts fixed-point

---

[1] In §III.9 below, the partial Borel–Serre compactification is denoted by ${}_{\mathbb{Q}}\overline{X}^{BS}$. The reason is that $\overline{X}^{BS}$ is basically the notation in [BS2], and we want to use ${}_{\mathbb{Q}}\overline{X}^{BS}$ below to show that it is constructed differently. For example, the topologies on them look different.

freely on X, $\Gamma\backslash X$ is a $K(\Gamma, 1)$-space. If $\Gamma\backslash X$ is compact, then it is a closed smooth manifold, which has a finite triangulation and hence is a finite $K(\Gamma, 1)$-space. Otherwise, we need a compact space that supports the structure of a finite CW-complex and is homotopic to $\Gamma\backslash X$. One natural method is to construct a compactification $\overline{\Gamma\backslash X}$ of $\Gamma\backslash X$ that is homotopic to the interior $\Gamma\backslash X$ and is a manifold with boundary (or corners).

The Satake compactifications $\overline{\Gamma\backslash X}^S$ are not homotopic to the interior. In fact, when X is the upper half-plane $\mathbf{H}$, $\Gamma\backslash\mathbf{H}$ is a Riemann surface with finitely many cusp neighborhoods. There exists only one Satake compactification $\overline{\Gamma\backslash\mathbf{H}}^S$, which is obtained by adding one point to each cusp. Clearly, the inclusion $\Gamma\backslash\mathbf{H} \to \overline{\Gamma\backslash\mathbf{H}}^S$ is not a homotopy equivalence, since the induced map $\Gamma = \pi_1(\Gamma\backslash\mathbf{H}) \to \pi_1(\overline{\Gamma\backslash\mathbf{H}}^S)$ has a large kernel containing subgroups generated by loops around the cusps, i.e., the subgroups $\Gamma \cap N_{P_i}$, where $\mathbf{P}_1, \dots, \mathbf{P}_m$ are the set of representatives of Γ-conjugacy classes of rational parabolic subgroups of $\mathbf{G}$ corresponding to the cusps of $\Gamma\backslash\mathbf{H}$.

In this case, if we compactify each cusp end of $\Gamma\backslash\mathbf{H}$ by adding a circle, which is the Borel–Serre compactification $\overline{\Gamma\backslash\mathbf{H}}^{BS}$, then the inclusion $\Gamma\backslash\mathbf{H} \to \overline{\Gamma\backslash\mathbf{H}}^{BS}$ is a homotopic equivalence. Since $\overline{\Gamma\backslash\mathbf{H}}^{BS}$ is a compact manifold with boundary, it has a finite triangulation, and hence is a desired finite $K(\Gamma, 1)$-space.

In general, the Borel–Serre compactification $\Gamma\backslash\overline{X}^{BS}$ is a compact manifold with corners, which is homotopic to $\Gamma\backslash X$ and also admits a finite triangulation, and hence is a finite $K(\Gamma, 1)$-space when Γ is torsion-free. The easiest way to see how the corner structures arise is to consider $\Gamma\backslash X = \Gamma_1\backslash\mathbf{H} \times \Gamma_2\backslash\mathbf{H}$. Then $\overline{\Gamma\backslash X}^{BS} = \overline{\Gamma_1\backslash\mathbf{H}}^{BS} \times \overline{\Gamma_2\backslash\mathbf{H}}^{BS}$ is a product of two manifolds with boundary, and hence a manifold with corners of codimension 2.

III.5.2 The homotopic equivalence between $\Gamma\backslash X$ and $\Gamma\backslash\overline{X}^{BS}$ allows us to identify $H^*(\Gamma)$ with $H^*(\Gamma\backslash\overline{X}^{BS})$ when Γ is torsion free, and this identification allows us to use the compact space to study $H^*(\Gamma)$.

For example, we have the long exact sequence

$$\to H^*(\Gamma\backslash\overline{X}^{BS}, \partial\Gamma\backslash\overline{X}^{BS}) \xrightarrow{\alpha} H^*(\Gamma\backslash\overline{X}^{BS}) \xrightarrow{\beta} H^*(\partial\Gamma\backslash\overline{X}^{BS}) \to$$

for the pair $\Gamma\backslash\overline{X}^{BS}$, $\partial\Gamma\backslash\overline{X}^{BS} = \Gamma\backslash\overline{X}^{BS} - \Gamma\backslash X$. The image of α in $H^*(\Gamma\backslash\overline{X}^{BS})$ is called the *interior cohomology*, and the image of β is called the *boundary cohomology*. The cohomology $H^*(\Gamma\backslash X, \mathbb{C})$ can be studied using automorphic forms, and the division of $H^*(\Gamma\backslash\overline{X}^{BS})$ into the interior and boundary cohomologies corresponds to different behaviors of automorphic forms at infinity, i.e., near the boundary of $\Gamma\backslash\overline{X}^{BS}$. It is through the above identifications that one can use the theory of automorphic forms and automorphic representation to study $H^*(\Gamma, \mathbb{C})$. See [BW] [Shw] [JS] for details.

III.5.3 A basic step in the construction of $\Gamma\backslash\overline{X}^{BS}$ is the *geodesic action* on X associated with every rational parabolic subgroup $\mathbf{P}$.

Let

$$X = N_P \times A_{\mathbf{P},x_0} \times X_{\mathbf{P}} \tag{III.5.28}$$

be the horospherical decomposition of X in equation (III.1.4) in §III.1 with respect to the basepoint $x_0 = K \in X = G/K$. Define the geodesic action of $A_{\mathbf{P}}$ on X as follows. Identify $A_{\mathbf{P}}$ with $A_{\mathbf{P},x_0}$ under the lift i_{x_0} in §III.1.9. For any $b \in A_{\mathbf{P},x_0}$, and $(n, a, z) \in N_P \times A_{\mathbf{P},x_0} \times X_{\mathbf{P}}$,

$$b \cdot (n, a, z) = (n, ba, z). \tag{III.5.29}$$

Clearly, this action is equivariant with respect to the P-action on X: for any $p \in P$ and $x = (n, a, z) \in X$,

$$p(b \cdot x) = b \cdot px.$$

In fact, write $p = n'a'm'$, where $n' \in N_P$, $a' \in A_{\mathbf{P},x_0}$, $z \in M_{\mathbf{P},x_0}$. Then

$$p(b \cdot x) = n'a'm'(n, ba, z) = (n'\,{}^{a'm'}n, a'ba, m'z),$$

$$b \cdot px = b \cdot (n'\,{}^{a'm'}n, a'a, m'z) = (n'\,{}^{a'm'}n, ba'a, m'z) = (n'\,{}^{a'm'}n, a'ba, m'z).$$

It should be emphasized that $A_{\mathbf{P}}$ could be the split component of several rational parabolic subgroups, but the geodesic action depends crucially on the choice of the parabolic subgroup $\mathbf{P}$. For example, when $\mathbf{G} = \mathrm{SL}(2)$, $X = \mathbf{H}$, let $\mathbf{P}$ be the parabolic subgroup of upper triangular matrices. Then $A_{\mathbf{P}}$ is the subgroup of diagonal matrices. The orbits of the geodesic action of $A_{\mathbf{P}}$ are vertical lines. On the other hand, let $\mathbf{P}^-$ be the opposite parabolic subgroup of the lower triangular matrices. Then $A_{\mathbf{P}^-} = A_{\mathbf{P}}$, and the orbits of the geodesic action of $A_{\mathbf{P}^-}$ are half-circles with one endpoint at 0 in the boundary of $\mathbf{H}$.

On the other hand, it can be shown that this geodesic action is independent of the choice of the basepoint. Hence, the basepoint x_0 in the subscript will be dropped.

III.5.4 Let $\alpha_1, \dots, \alpha_r$ be the simple roots in $\Delta(P, A_{\mathbf{P}})$. Then

$$A_{\mathbf{P}} \cong (\mathbb{R}_+)^r, \quad a \mapsto (a^{-\alpha_1}, \dots, a^{-\alpha_r}).$$

The closure of $A_{\mathbf{P}}$ in $\mathbb{R}^r$ is equal to the corner $\mathbb{R}^r_{\geq 0}$. Denote it by $\overline{A_{\mathbf{P}}}$.

The multiplication of $A_{\mathbf{P}}$ on $A_{\mathbf{P}}$ extends to an action on $\mathbb{R}^r$ by

$$a \cdot (t_1, \dots, t_r) = (a^{-\alpha_1} t_1, \dots, a^{-\alpha_r} t_r).$$

Clearly, A_P preserves the closure $\overline{A_{\mathbf{P}}}$ and hence acts on it.

By the horospherical decomposition of X in equation (III.5.28), X is a principal $A_{\mathbf{P}}$-bundle over $N_P \times X_{\mathbf{P}} \cong X/A_{\mathbf{P}}$ under the geodesic action of $A_{\mathbf{P}}$. Define the *corner* $X(\mathbf{P})$ associated with $\mathbf{P}$ by

$$X(\mathbf{P}) = X \times_{A_{\mathbf{P}}} \overline{A_{\mathbf{P}}}. \tag{III.5.30}$$

Clearly, the corner $X(\mathbf{P})$ can be identified as follows:

$$X(\mathbf{P}) = N_P \times X_{\mathbf{P}} \times \overline{A_{\mathbf{P}}}. \tag{III.5.31}$$

Since N_P and $X_{\mathbf{P}}$ are real analytic manifolds, and $\overline{A_{\mathbf{P}}}$ is a real analytic corner, $X(\mathbf{P})$ has a structure of a real analytic manifold with corners. Clearly, $X(\mathbf{P})$ contains X as an open dense subset. Since the horospherical decomposition of $X = N_P \times A_{\mathbf{P}} \times X_{\mathbf{P}}$ is real analytic, the real analytic structure of $X(\mathbf{P})$ restricts to the canonical one on X.

An analytic submanifold S of X is called an *analytic cross-section* of the $A_{\mathbf{P}}$-principal bundle X if the map

$$S \times A_{\mathbf{P}} \to X, \quad (z, a) \mapsto a \cdot z,$$

is an analytic diffeomorphism. Clearly, $N_P M_{\mathbf{P}} x_0 \cong N_P \times X_{\mathbf{P}}$ is an analytic section, called the canonical cross-section. But other cross-sections occur naturally when corners of different rational parabolic subgroups are compared.

For any such analytic cross-section S, the corner $X(\mathbf{P})$ can also be identified with

$$X(\mathbf{P}) = S \times \overline{A_{\mathbf{P}}}. \tag{III.5.32}$$

This possibly gives another different structure of a real analytic manifold with corners on $X(\mathbf{P})$, which also restricts to the canonical real analytic structure on X.

Proposition III.5.5 *Any two analytic cross-sections S_1, S_2 induce, by Equation (III.5.32), the same real analytic structure on $X(\mathbf{P})$. Hence, the real analytic structure on $X(\mathbf{P})$ is canonical in the sense that it only depends on the geodesic action of $A_{\mathbf{P}}$ on X.*

Proof. Let

$$X \to S_i \times A_{\mathbf{P}}, \quad x \mapsto (s_i(x), a(x))$$

be the coordinates in the trivialization induced by the section S_i, $i = 1, 2$. For any $s_1 \in S_1$, there exists a unique point $s_2(s_1) \in S_2$ and $a_2(s_1) \in A_{\mathbf{P}}$ such that

$$s_1 = s_2(s_1)a_2(s_1) = (s_2(s_1), a_2(s_1)).$$

Since S_1 is a real analytic submanifold, $s_2(s_1), a_2(s_1)$ are real analytic in s_1.

Then the coordinates of $x \in X$ with respect to the trivializations induced by S_1, S_2 are related by

$$s_2(x) = s_2(s_1(x)), \quad a_2(x) = a_2(s_1(x))a_1(x).$$

In fact,

$$x = s_1(x)a_1(x) = s_2(s_1(x))a_2(s_1(x))a_1(x) = (s_2(s_1(x)), a_2(s_1(x))a_1(x)).$$

Hence, the transition function from $X = S_1 \times A_{\mathbf{P}}$ to $X = S_2 \times A_{\mathbf{P}}$ is given by

$$(s_1, a_1) \to (s_2(s_1), a_2(s_1)a_1).$$

Under the identifications $X(\mathbf{P}) = S_i \times \overline{A_{\mathbf{P}}} \cong S_1 \times (\mathbb{R}_{\geq 0})^r$, $i = 1, 2$, the transition function is given by

$$(s_1; t_1, \dots, t_r) \to (s_2(s_1); a_2(s_1)^{-\alpha_1} t_1, \dots, a_2(s)^{-\alpha_r} t_r).$$

Since $s_2(s_1), a_2(s_1)$ are real analytic functions in s_1, the above transition function from the coordinates with respect to S_1 to those with respect to S_2 is real analytic. By swapping S_1 and S_2, we can show that the other transition function is also real analytic. This proves that two trivializations give the same real analytic structure.

Let o_P be the (corner) point in $\overline{A_{\mathbf{P}}}$ corresponding to the origin in $\mathbb{R}^r$ under the embedding $\overline{A_{\mathbf{P}}} \to \mathbb{R}^r$. Then the corner

$$X \times_{A_{\mathbf{P}}} \{o_P\} \subset X(\mathbf{P})$$

can be identified with $X/A_{\mathbf{P}} \cong N_P \times X_{\mathbf{P}}$. Define

$$e(\mathbf{P}) = N_P \times X_{\mathbf{P}} \tag{III.5.33}$$

and call it the *Borel–Serre boundary component* of $\mathbf{P}$. Then $e(\mathbf{P})$ can be identified with the corner of $X(\mathbf{P})$.

Proposition III.5.6 *For any parabolic subgroup* $\mathbf{P}$, *the corner* $X(\mathbf{P})$ *can be canonically decomposed as follows:*

$$X(\mathbf{P}) = X \cup \coprod_{\mathbf{Q} \supseteq \mathbf{P}} e(\mathbf{Q}).$$

Proof. For any subset $I \subset \{\alpha_1, \dots, \alpha_r\}$, let $o_I = (t_1, \dots, t_r)$ be the point in $\overline{A_{\mathbf{P}}} = (\mathbb{R}_{\geq 0})^r$ with coordinates: $t_i = 1$ for $\alpha_i \in I$, and $t_i = 0$ for $\alpha_i \notin I$. Then the $A_{\mathbf{P}}$-orbit through o_I can be identified with $A_{\mathbf{P}}/A_{\mathbf{P}_I}$. In fact, since the simple roots in $\Phi(P_I, A_{\mathbf{P}_I})$ are the restriction of $\{\alpha_1, \dots, \alpha_r\} - I$, $A_{\mathbf{P}_I}$ is the stabilizer o_I. Then

$$X \times_{A_{\mathbf{P}}} \{o_I\} \cong X/A_{\mathbf{P}_I} \cong e(\mathbf{P}_I).$$

Since every face of $\overline{A_{\mathbf{P}}}$ is of the form $A_{\mathbf{P}} \cdot o_I$, the disjoint decomposition of $X(\mathbf{P})$ follows.

Proposition III.5.7 *For any pair of rational parabolic subgroups* $\mathbf{P}_1$, $\mathbf{P}_2$ *with* $\mathbf{P}_1 \subset \mathbf{P}_2$, *the identity map on* X *extends to an embedding of* $X(\mathbf{P}_2)$ *into* $X(\mathbf{P}_1)$ *as real analytic manifolds with corners, and the image is an open submanifold.*

Proof. Since $\mathbf{P}_1 \subset \mathbf{P}_2$, by equation (III.1.11),

$$A_{\mathbf{P}_1} = A_{\mathbf{P}_1,\mathbf{P}_2} A_{\mathbf{P}_2} = A_{\mathbf{P}_1,\mathbf{P}_2} \times A_{\mathbf{P}_2}.$$

It is clear from the definition that $A_{\mathbf{P}_1,\mathbf{P}_2} \times \overline{A_{\mathbf{P}_2}}$ is contained in $\overline{A_{\mathbf{P}_1}}$ as a face, and is stable under the action of $A_{\mathbf{P}_1}$. Hence

$$X(\mathbf{P}_1) = X \times_{A_{\mathbf{P}_1}} \overline{A_{\mathbf{P}_1}} \supset X \times_{A_{\mathbf{P}_1}} A_{\mathbf{P}_1,\mathbf{P}_2} \times \overline{A_{\mathbf{P}_2}} = X \times_{A_{\mathbf{P}_2}} \overline{A_{\mathbf{P}_2}} = X(\mathbf{P}_2).$$

To show that the analytic structures are compatible, we note that equation (III.1.12) implies that $N_{P_1} \times X_{\mathbf{P}_1} \times A_{\mathbf{P}_1,\mathbf{P}_2}$ is also an analytic cross-section of the $A_{\mathbf{P}_2}$-action on X, and hence by Proposition III.5.5,

$$X(\mathbf{P}_2) = N_{P_1} \times X_{\mathbf{P}_1} \times A_{\mathbf{P}_1,\mathbf{P}_2} \times \overline{A_{\mathbf{P}_2}},$$

as analytic manifolds. Since the decomposition $A_{\mathbf{P}_1} = A_{\mathbf{P}_1,\mathbf{P}_2} \times A_{\mathbf{P}_2}$ corresponds to the standard coordinate decomposition of $\mathbb{R}^r$ under the embedding $A_{\mathbf{P}_1} \to \mathbb{R}^r$ above, $N_{P_1} \times X_{\mathbf{P}_1} \times A_{\mathbf{P}_1,\mathbf{P}_2} \times \overline{A_{\mathbf{P}_2}}$ is an open analytic submanifold of $N_{P_1} \times X_{\mathbf{P}_1} \times \overline{A_{\mathbf{P}_1}}$, which is equal to $X(\mathbf{P}_1)$ as analytic manifolds with corners. This proves that $X(\mathbf{P}_2)$ is an analytic manifold in $X(\mathbf{P}_1)$.

III.5.8 Now we are ready to define the *Borel–Serre partial compactification* $\overline{X}^{BS}$. Define

$$\overline{X}^{BS} = X \cup \coprod_{\mathbf{P}} X(\mathbf{P}) / \sim, \tag{III.5.34}$$

where the equivalence relation $\sim$ is defined as follows. For any pair of rational parabolic subgroups $\mathbf{P}$ and $\mathbf{Q}$, let $\mathbf{R}$ be the smallest rational parabolic subgroup containing both $\mathbf{P}$ and $\mathbf{Q}$. Such an $\mathbf{R}$ exists and could be equal to the improper parabolic subgroup $\mathbf{G}$. Let $i_1 : X(\mathbf{R}) \to X(\mathbf{P})$ and $i_2 : X(\mathbf{R}) \to X(\mathbf{Q})$ be the natural embedding in Proposition III.5.7. Then for any $x \in X(\mathbf{R})$, points $i_1(x) \in X(\mathbf{P})$ and $i_2(x) \in X(\mathbf{Q})$ are defined to be equivalent. It defines an equivalence relation.

Clearly, we need only to check the transitivity. Suppose $x_1 \in X(\mathbf{P}_1), x_2 \in X(\mathbf{P}_2), x_3 \in X(\mathbf{P}_3)$, and $x_1 \sim x_2$, $x_2 \sim x_3$. Then there exist rational parabolic subgroups $\mathbf{Q}_1, \mathbf{Q}_2$ and points $y_1 \in X(\mathbf{Q}_1)$, $y_2 \in X(\mathbf{Q}_2)$ such that $X(\mathbf{Q}_1) \subseteq X(\mathbf{P}_1)$, $X(\mathbf{Q}_1) \subseteq X(\mathbf{P}_2)$, and y_1 is mapped to x_1, x_2, and y_2 is similarly mapped to x_2, x_3 in the inclusions $X(\mathbf{Q}_2) \subseteq X(\mathbf{P}_2)$, $X(\mathbf{Q}_2) \subseteq X(\mathbf{P}_3)$. Let $\mathbf{R}$ be the largest rational parabolic subgroup containing $\mathbf{Q}_1, \mathbf{Q}_2$. When considered as subsets of $X(\mathbf{P}_2)$, $X(\mathbf{Q}_1) \cap X(\mathbf{Q}_2) = X(\mathbf{R})$, and $y_1 = y_2$ since they are both mapped to x_2, and hence $y_1 = y_2 \in X(\mathbf{R})$. Denote this point in $X(\mathbf{R})$ by y. Then it is mapped to x_1 through the composition $X(\mathbf{R}) \subseteq X(\mathbf{Q}_1) \subseteq X(\mathbf{P}_1)$, and to x_3 through the composition $X(\mathbf{R}) \subseteq X(\mathbf{Q}_2) \subseteq X(\mathbf{P}_3)$. Hence $x_1 \sim x_3$.

Using the decomposition in Proposition III.5.6, we obtain from the proof of Proposition III.5.7 that each point in $e(\mathbf{P})$ determines an equivalence class and each equivalence class is of this form for some $\mathbf{P}$. Hence as a set,

$$\overline{X}^{BS} = X \cup \coprod_{\mathbf{P}} e(\mathbf{P}). \tag{III.5.35}$$

Since the natural topologies of the corners $X(\mathbf{P})$ are compatible on the intersection, they define a *sum topology* on $X \cup \coprod_{\mathbf{P}} X(\mathbf{P})$. Let the topology of $\overline{X}^{BS}$ be the quotient topology of the sum topology.

Proposition III.5.9 *For any rational parabolic subgroup* $\mathbf{P}$*, the natural projection* $X(\mathbf{P}) \to \overline{X}^{BS}$ *is an embedding onto an open subset, and the topology on* $\overline{X}^{BS}$ *is Hausdorff.*

Proof. We will prove only the first statement. Since we will give an alternative, simpler construction of $\overline{X}^{BS}$ in §III.9 below, we refer the reader to [BS2, Theorem 7.8] for the original proof (see also Remark III.5.10 below).

For any rational parabolic subgroup $\mathbf{P}$, there are only finitely many rational parabolic subgroups $\mathbf{Q}$ containing $\mathbf{P}$. Clearly, for any such $\mathbf{Q}$ and the embedding $X(\mathbf{Q}) \subseteq X(\mathbf{P})$, no two points in $X(\mathbf{P})$ can be identified. This shows that the map $X(\mathbf{P}) \to \overline{X}^{BS}$ is injective. It can also be shown that it is an open map, and hence it is a homeomorphism.

Remark III.5.10 The proof of this proposition in [BS2, Theorem 7.8] is inductive and depends on spaces of S-type, a generalization of symmetric spaces, and compactifications of spaces of S-type. In fact, the proof involves reduction to the compactifications of the boundary components $e(\mathbf{P})$. Clearly

$$e(\mathbf{P}) = P/N_P A_{\mathbf{P}} M_{\mathbf{P}}.$$

So $e(\mathbf{P})$ is a homogeneous space of a nonreductive group with noncompact stabilizer, and hence is not a symmetric space. The spaces of S-type are introduced right from the beginning of [BS2] and the geodesic action on them is used repeatedly in various proofs. Since we will not introduce such spaces of S-type, and alternative proofs of the structures of $\overline{X}^{BS}$ will be given in §III.9 below, we will state several results from [BS2] without proof as in the above Proposition III.5.9.

Proposition III.5.11 *The space* $\overline{X}^{BS}$ *has a natural structure of analytic manifolds with corners.*

Proof. By Proposition III.5.9, for any rational parabolic subgroup $\mathbf{P}$, the corner $X(\mathbf{P})$ is embedded into $\overline{X}^{BS}$ as an open subset. Clearly, they cover $\overline{X}^{BS}$ and form coordinate charts. By Proposition III.5.7, the analytic structures of the corners $X(\mathbf{P})$ are compatible, and hence $\overline{X}^{BS}$ has a natural structure of a real analytic manifold with corners induced from the analytic structures of the corners $X(\mathbf{P})$.

Proposition III.5.12 *For any two rational parabolic subgroups* $\mathbf{P}_1, \mathbf{P}_2$*,* $\mathbf{P}_1 \subseteq \mathbf{P}_2$ *if and only if* $e(\mathbf{P}_1) \subseteq \overline{e(\mathbf{P}_2)}$*, where* $\overline{e(\mathbf{P}_2)}$ *is the closure of* $e(\mathbf{P}_2)$ *in* $\overline{X}^{BS}$*.*

The proof uses the Borel–Serre compactification of $e(\mathbf{P}_2)$ as a space of S-type and its identification with the closure $\overline{e(\mathbf{P}_2)}$. See [BS2, 7.5] for details.

Proposition III.5.13 *The natural action of the group of rational points $\mathbf{G}(\mathbb{Q})$ on X extends to $\overline{X}^{BS}$. The extended action preserves the structure of real analytic manifold with corners and permutes the faces by*

$$g \cdot e(\mathbf{P}) = e({}^{g}\mathbf{P}),$$

where $g \in \mathbf{G}(\mathbb{Q})$, and hence the stabilizer of $e(\mathbf{P})$ in Γ is equal to $\Gamma \cap P$.

Proof. The basic point is to prove that for any element $g \in \mathbf{G}(\mathbb{Q})$, the action of g on X extends to a map $g : X(\mathbf{P}) \to X(\mathbf{P}^{g})$ and show that this map is real analytic. Specifically, write $g = km_0 a_0 n_0$, where $k \in K$ and $n_0 \in N_P$, $a_0 \in A_{\mathbf{P}}$ and $m_0 \in M_{\mathbf{P}}$. For any point $x = (n, a, mK_{\mathbf{P}}) \in N_{\mathbf{P}} \times A_{\mathbf{P}} \times X_{\mathbf{P}} = X$, where $K_{\mathbf{P}} = K \cap M_{\mathbf{P}}$,

$$m_0 a_0 m_0 x = ({}^{a_0 m_0}(n_0 n), a_0 a, m_0 m K_{\mathbf{P}}).$$

From the analytic structure of $X(\mathbf{P})$ given by $X(\mathbf{P}) = N_P \times X_{\mathbf{P}} \times \overline{A_{\mathbf{P}}}$, it is clear that this action $x \mapsto n_0 a_0 m_0 x$ extends to an analytic diffeomorphism of $X(\mathbf{P})$ to itself. Since ${}^{k}\mathbf{P} = {}^{g}\mathbf{P}$, the action by k transfers the horospherical coordinates of (n, a, m) with respect to $\mathbf{P}$ to those of ${}^{k}\mathbf{P} = {}^{g}\mathbf{P}$, i.e.,

$$gx = ({}^{k a_0 m_0}(n_0 n), {}^{k}a_0 a, {}^{k}m_0 m^{k} K_{\mathbf{P}}) \in N_{{}^{g}P} \times X_{{}^{g}\mathbf{P}} \times A_{{}^{k}\mathbf{P}}.$$

Though the components k, m_0 are not uniquely determined by g, km_0 is, and hence the action is well-defined. Clearly, the action $x \mapsto kx$ extends to an analytic diffeomorphism between $X(\mathbf{P})$ and $X({}^{g}\mathbf{P})$. By combining these two analytic diffeomorphisms, we obtain that the action $x \mapsto gx$ extends to an analytic diffeomorphism between the corners $X(\mathbf{P}) \to X(\mathbf{P}^{g})$ and hence to $\overline{X}^{BS}$. Under this map, clearly, the face $e(\mathbf{P})$ is mapped to $e({}^{g}\mathbf{P})$. The last sentence follows from the fact that P is equal to its normalizer.

Proposition III.5.14 *Assume that the $\mathbb{Q}$-rank r of $\mathbf{G}$ is positive. Let Γ be an arithmetic subgroup of $\mathbf{G}(\mathbb{Q})$ above. Then Γ acts properly with a compact Hausdorff space $\Gamma\backslash\overline{X}^{BS}$. If Γ is torsion-free, $\Gamma\backslash\overline{X}^{BS}$ is a compact real analytic manifold with corners, called the Borel–Serre compactification of $\Gamma\backslash X$, also denoted by $\overline{\Gamma\backslash X}^{BS}$. The highest codimension of the corners of $\overline{\Gamma\backslash X}^{BS}$ is equal to the $\mathbb{Q}$-rank r.*

The idea of the proof is to show that the closure of a Siegel set in $\overline{X}^{BS}$ is compact. Then by the reduction theory, a compact subset is mapped surjectively onto $\Gamma\backslash\overline{X}^{BS}$, and hence the latter is compact. See [BS2, Theorem 9.3] (or Theorem III.9.18 below).

III.5.15 We consider the simplest example, when $\mathbf{G} = \mathrm{SL}(2)$. In this case, $X = \mathbf{H}$, the upper half-plane. For any (proper) rational parabolic subgroup $\mathbf{P}$, the boundary component is given by

$$e(\mathbf{P}) = N_P \cong \mathbb{R}.$$

The natural boundary of $\mathbf{H}$ in $\mathbb{C} \cup \{\infty\}$ consists of $\mathbb{R} \cup \{\infty\}$. Then $\overline{\mathbf{H}}$ is obtained by adding one copy of $\mathbb{R}$ to every boundary point in $\mathbb{Q} \cup \{\infty\}$. This copy of $\mathbb{R}$ is the parameter space of all geodesics converging to the boundary point. The space $\Gamma\backslash\mathbf{H}$ has finitely many cusp neighborhoods, and $\overline{\Gamma\backslash\mathbf{H}}^{BS}$ is obtained by adding a circle at infinity of each cusp neighborhood, which is the quotient $\Gamma \cap N_P\backslash N_P$. Clearly, $\overline{\Gamma\backslash\mathbf{H}}^{BS}$ is a manifold with boundary and homeomorphic to $\Gamma\backslash\mathbf{H}^{BS}$.

III.5.16 By Proposition III.5.12, the boundary $\partial\overline{X}^{BS} = \overline{X}^{BS} - X$ is a cell complex dual to the rational Tits building $\Delta_{\mathbb{Q}}(\mathbf{G})$ of $\mathbf{G}$. In fact, it has the same homotopy type as the Tits building [BS2, 8.4.2]. This relation to the Tits building has important applications to the cohomology groups of Γ in [BS2, §11], for example, the cohomology dimension of Γ is equal to $\dim \Gamma\backslash X - rk_{\mathbb{Q}}(\mathbf{G})$.

III.5.17 Summary and comments. We recalled the construction in [BS2] of the Borel–Serre compactification $\overline{\Gamma\backslash X}^{BS}$. Contrary to the Baily–Borel compactification $\overline{\Gamma\backslash X}^{BB}$, this compactification is large and no topology of $\Gamma\backslash X$ is collapsed at infinity; hence the inclusion $\Gamma\backslash X \subset \overline{\Gamma\backslash X}^{BS}$ is a homotopy equivalence. This is important to applications in topology. Another way to obtain a compact manifold homeomorphic to $\Gamma\backslash X$ is to remove a suitable neighborhood of the infinity of $\Gamma\backslash X$ to obtain a compact submanifold in $\Gamma\backslash X$.

When the $\mathbb{Q}$-rank of $\mathbf{G}$ is equal to 1, for example, $\mathbf{G} = \mathrm{SL}(2)$, each end of $\Gamma\backslash X$ is a topological cylinder. By cutting off each end suitably, we obtain a compact submanifold with boundary that is clearly homeomorphic to $\Gamma\backslash X$. When the rank is greater than 1, we obtain a manifold with corners in general. See [Sap1] [Leu2] [Leu5] [Ra2] [Gra1] [Gra2] for the constructions and applications.

III.6 Reductive Borel–Serre compactification

In this section, we study the reductive Borel–Serre compactification $\overline{\Gamma\backslash X}^{RBS}$, which is closely related to the Borel–Serre compactification $\overline{\Gamma\backslash X}^{BS}$ and is motivated by problems in analysis [Zu1] and has found natural, important applications in cohomology groups of arithmetic groups in [GHM] and [Zu3].

This section is organized as follows. First, we explain in III.6.1 that the compactification $\overline{\Gamma\backslash X}^{BS}$ is too large to support partitions of unity, which are needed to show that the L^2-complex of sheaves of differential forms is fine. Then we explain how to blow down the boundary of $\overline{\Gamma\backslash X}^{BS}$ to obtain the smaller compactification $\overline{\Gamma\backslash X}^{RBS}$ (III.6.2, III.6.4). We show by examples that $\overline{\Gamma\backslash X}^{RBS}$ is different from the Satake compactifications (III.6.5), though it dominates $\overline{\Gamma\backslash X}^{S}_{\max}$ (III.6.6). Finally we mention several applications of $\overline{\Gamma\backslash X}^{RBS}$ to cohomology groups of $\Gamma\backslash X$ (III.6.6, III.6.7).

III.6.1 On any compactification $\overline{\Gamma\backslash X}$ of $\Gamma\backslash X$, the L^2-complex of sheaves $\mathcal{L}_{(2)}(\overline{\Gamma\backslash X})$ in [Zu1, p. 175] is defined as follows: for any open subset $U \subset \overline{\Gamma\backslash X}$,

$$\mathcal{L}^i_{(2)}(U) = L^i_{(2)}(U \cap \Gamma\backslash X),$$

where $L^i_{(2)}(U \cap \Gamma\backslash X)$ is the space of forms of degree i on $U \cap \Gamma\backslash X$, which together with their differentials are L^2-integrable.

To study the induced cohomology groups, it is important to know when these sheaves are fine. It is well known (see [GH, pp. 41–44]) that for a compact manifold M, the complex of sheaves of differential forms M is fine, which follows from the fact that for any locally finite covering of M, there is a *partition of unity* of the sheaves subordinate to it; and this fact plays an important role in proving the De Rham theorem.

The L^2-complex of sheaves on $\Gamma\backslash\overline{X}^{BS}$ does not admit such a partition of unity. The reason is that for any rational parabolic subgroup $\mathbf{P}$, in the horospherical decomposition $X = N_P \times A_\mathbf{P} \times X_\mathbf{P}$, the differential in the N_P variable at the height $a \in A_\mathbf{P}$ is not bounded as $a \to +\infty$ in $A_\mathbf{P}$ in the sense $\alpha(\log a) \to +\infty$ for all $\alpha \in \Phi(P, A_\mathbf{P})$.

Specifically, consider the example of $X = \mathbf{H}$, $\mathbf{G} = \mathrm{SL}(2)$, and $\mathbf{P}$ the parabolic subgroup of upper triangular matrices. Then the N_P orbits correspond to the y-coordinates in $\mathbf{H}$. Since the metric of $\mathbf{H}$ is given by

$$ds^2 = \frac{dx^2 + dy^2}{y^2},$$

the norm square of the differential dx is equal to y^2. Hence, dx is not L^2-integrable on a Siegel set associated with $\mathbf{P}$. For any locally finite covering $\{U_i\}$ of $\overline{\Gamma\backslash\mathbf{H}}^{BS}$, let $\{\varphi_i\}$ be a partition of unity subordinate to it. Choose a boundary point $b \in \partial\overline{\Gamma\backslash\mathbf{H}}^{BS}$ such that there exists a function φ_i that satisfies two conditions: (1) $b \in \mathrm{Supp}(\varphi_i)$, (2) $\frac{\partial}{\partial x}\varphi_i$ does not vanish at b. Such a boundary point exists when the boundary component $\Gamma_{N_P}\backslash N_P$, a circle, is not contained in a single U_i. Then the above discussions show that $d\varphi_i$ is not square integrable, and hence the L^2-complexes of sheaves on $\overline{\Gamma\backslash\mathbf{H}}^{BS}$ do not admit a partition of unity. Note that the defining function of the boundary circle is given by $r = e^{-y}$. It can be seen that dr is bounded and hence square integrable.

On the other hand, if we collapse each boundary circle to a point, each sufficiently small cusp neighborhood is covered by a single open set U_i in any locally finite covering $\{U_i\}$, and we can choose the corresponding function φ_i to be N_P-invariant, i.e., independent of x when the cusp is mapped to the standard cusp at infinity. Then the above difficulty can be avoided.

The compactification of $\Gamma\backslash\mathbf{H}$ obtained from $\overline{\Gamma\backslash\mathbf{H}}^{BS}$ by collapsing the boundary circles to points is the reductive Borel–Serre compactification $\overline{\Gamma\backslash\mathbf{H}}^{RBS}$.

In general, we can also blow down boundary nilmanifolds, i.e., manifolds diffeomorphic to quotients of N_P, in the boundary $\Gamma\backslash\overline{X}^{BS}$ to obtain the reductive Borel–

Serre compactification $\overline{\Gamma\backslash X}^{RBS}$. Then the L^2-complex of sheaves on $\overline{\Gamma\backslash X}^{RBS}$ is fine [Zu1, Proposition 4.4],

III.6.2 We now recall the construction of the reductive Borel–Serre compactification $\overline{\Gamma\backslash X}^{RBS}$ in [Zu1, p. 190].

For each rational parabolic subgroup $\mathbf{P}$, let $\Gamma_P = \Gamma \cap P$, $\Gamma_{N_P} = \Gamma \cap N_P$. Then Γ_P is an arithmetic subgroup of $\mathbf{P}$, and Γ_{N_P} is an arithmetic subgroup of $\mathbf{N_P}$ and hence $\Gamma_{N_P}\backslash N_P$ is compact. By [BS2, Proposition 1.2], the image of $\Gamma_{\mathbf{P}}$ in L_P under the projection $P \to N_P\backslash P$ is contained in $M_{\mathbf{P}}$ and is an arithmetic subgroup of $\mathbf{M_P}$, to be denoted by $\Gamma_{M_{\mathbf{P}}}$. By definition, we have an exact sequence

$$0 \to \Gamma_{N_P} \to \Gamma_P \to \Gamma_{M_{\mathbf{P}}} \to 0. \tag{III.6.36}$$

To understand the action of Γ_P on X in terms of the horospherical decomposition and the structure of the boundary components of $\Gamma\backslash\overline{X}^{BS}$, we lift $\Gamma_{M_{\mathbf{P}}}$ into P by the canonical lift i_{x_0} associated with the basepoint x_0 in §III.1.9. The subgroup $i_{x_0}(\Gamma_{M_{\mathbf{P}}})$ in $M_{\mathbf{P},x_0}$ is denoted by $\Gamma_{M_{\mathbf{P},x_0}}$, where the subscript x_0 will be dropped once the basepoint is fixed and the dependence on it is clear. It should be emphasized that this lift i_{x_0} does not split the above exact sequence. Since the symmetric space associated with $M_{\mathbf{P}}$ is the boundary symmetric space $X_{\mathbf{P}}$, $\Gamma_{M_{\mathbf{P}}}$ acts properly on $X_{\mathbf{P}}$, and the quotient

$$\Gamma_{M_{\mathbf{P}}}\backslash X_{\mathbf{P}}$$

is a locally symmetric space, called the *boundary locally symmetric space* for $\mathbf{P}$. Because of this, $\Gamma_{M_{\mathbf{P}}}$ is also denoted by $\Gamma_{X_{\mathbf{P}}}$, similar to the notation Γ_{X_P} in §III.3.

The group $\Gamma_{M_{\mathbf{P}}}$ can also be defined as the image of Γ_P under the Langlands projection

$$P = N_P A_{\mathbf{P}} M_{\mathbf{P}} \to M_{\mathbf{P}}, \quad \Gamma_{M_{\mathbf{P}}} = \pi(\Gamma_P) \tag{III.6.37}$$

(see [BJ2, Proposition 2.6]).

III.6.3 Now we analyze the structure of the boundary components of $\overline{\Gamma\backslash X}^{BS}$. Since N_P is a normal subgroup of $N_P M_{\mathbf{P}}$, using the exact sequence in equation (III.6.36), we obtain that $\Gamma_P\backslash N_P M_{\mathbf{P}}$ is a fiber bundle over $\Gamma_{M_{\mathbf{P}}}\backslash M_{\mathbf{P}}$ with fiber a nilmanifold $\Gamma_{N_P}\backslash N_P$, and hence $\Gamma_P\backslash e(\mathbf{P})$ is also a fiber bundle over $\Gamma_{M_{\mathbf{P}}}\backslash X_{\mathbf{P}}$ with the fiber equal to the nilmanifold.

By Proposition III.5.13, the stabilizer in Γ of the boundary component $e(\mathbf{P})$ in $\overline{X}^{BS}$ is equal to Γ_P. Hence $\Gamma_{\mathbf{P}}\backslash e(\mathbf{P})$ belongs to the boundary of $\Gamma\backslash\overline{X}^{BS}$, called the boundary component of $\overline{\Gamma\backslash X}^{BS}$ associated to $\mathbf{P}$, or rather the Γ-conjugacy class of $\mathbf{P}$.

Let $\mathbf{P}_1, \dots, \mathbf{P}_m$ be a set of representatives of Γ-conjugacy classes of rational parabolic subgroups of $\mathbf{G}$. Then

$$\Gamma\backslash\overline{X}^{BS} = \Gamma\backslash X \cup \coprod_{i=1}^{m} \Gamma_{P_i}\backslash e(\mathbf{P}_i). \tag{III.6.38}$$

Define a space

$$\overline{\Gamma\backslash X}^{RBS} = \Gamma\backslash X \cup \coprod_{i=1}^{m} \Gamma_{M_{\mathbf{P}_i}}\backslash X_{\mathbf{P}_i}. \tag{III.6.39}$$

Then there is a canonical surjective map from $\Gamma\backslash\overline{X}^{BS}$ to $\overline{\Gamma\backslash X}^{RBS}$ that is equal to the identity map on $\Gamma\backslash X$, and on each boundary component $\Gamma_{P_i}\backslash e(\mathbf{P}_i)$, it is the projection to the base $\Gamma_{M_{\mathbf{P}_i}}\backslash X_{\mathbf{P}_i}$ when $\Gamma_{P_i}\backslash e(\mathbf{P}_i)$ is considered as a fiber bundle over the latter as above. In other words, $\overline{\Gamma\backslash X}^{RBS}$ is the quotient of $\Gamma\backslash\overline{X}^{BS}$ when the nilmanifolds $\Gamma_{N_{P_i}}\backslash N_{P_i}$ in the boundary of $\Gamma\backslash\overline{X}^{BS}$ are collapsed to points.

Endow the space $\overline{\Gamma\backslash X}^{RBS}$ with the quotient topology induced from $\Gamma\backslash\overline{X}^{BS}$, and call it the *reductive Borel–Serre compactification* of $\Gamma\backslash X$, in view of the following result. See §III.10 below for a construction of $\overline{\Gamma\backslash X}^{RBS}$ independent of $\overline{\Gamma\backslash X}^{BS}$.

Proposition III.6.4 *The topological space $\overline{\Gamma\backslash X}^{RBS}$ is compact, Hausdorff, and contains $\Gamma\backslash X$ as an open dense subset.*

Proof. By definition, the projection map from $\Gamma\backslash\overline{X}^{BS}$ to $\overline{\Gamma\backslash X}^{RBS}$ is continuous. Since $\Gamma\backslash\overline{X}^{BS}$ is compact, $\overline{\Gamma\backslash X}^{RBS}$ is also compact. Since the inverse image of the boundary $\overline{\Gamma\backslash X}^{RBS} - \Gamma\backslash X$ is the boundary $\Gamma\backslash\overline{X}^{BS} - \Gamma\backslash X$, and $\Gamma\backslash X$ is open and dense in $\Gamma\backslash\overline{X}^{BS}$, $\Gamma\backslash X$ is also open and dense in $\overline{\Gamma\backslash X}^{RBS}$. It remains to prove the Hausdorff property.

Let $\sim$ be the equivalence relation defining the quotient $\overline{\Gamma\backslash X}^{RBS}$. Since $\Gamma\backslash\overline{X}^{BS}$ is a manifold with corners, it is a normal space. By [Fr, p. 33], it suffices to prove that the $\sim$-saturation of closed subsets of $\Gamma\backslash\overline{X}^{BS}$ is also closed. By definition, the $\sim$-equivalence classes consist of single points in X, and of nilmanifolds $\Gamma_{N_{P_i}}\backslash N_{P_i}$ in the boundary component $\Gamma_{\mathbf{P}_i}\backslash e(\mathbf{P}_i)$. By Proposition III.5.12, for any two rational parabolic subgroups $\mathbf{P}, \mathbf{Q}$, $e(\mathbf{P}) \subseteq \overline{e(\mathbf{Q})}$ if and only if $\mathbf{Q} \supseteq \mathbf{P}$, which is in turn equivalent to $N_P \supseteq N_Q$. Intuitively, this immediately implies that the $\sim$-saturation of any closed subset in $\Gamma\backslash\overline{X}^{BS}$ is closed, since when we move to smaller boundary components in the boundary, the saturation process is increasing.

More precisely, let $C \subset \Gamma\backslash\overline{X}^{BS}$ be a closed subset. Let $C^{\sim}$ be its $\sim$-saturation. Suppose that $y_j \in C^{\sim}$ is a sequence converging to $y_\infty \in \Gamma\backslash\overline{X}^{BS}$. We need to prove that $y_\infty \in C^{\sim}$. Let $y'_j \in C$ be such that $y'_j \sim y_j$. Since $\Gamma\backslash\overline{X}^{BS}$ is compact, C is also compact. By passing to a subsequence if necessary, we assume that $y'_j \to y'_\infty$ for some $y'_\infty \in C$. Assume that $y'_\infty \in \Gamma_P\backslash e(\mathbf{P})$ for some rational parabolic subgroup $\mathbf{P}$. Since each rational parabolic subgroup $\mathbf{P}$ is contained in only finitely many rational parabolic subgroups $\mathbf{Q}$, Proposition III.5.12 implies that the boundary component

$e(\mathbf{P})$ is contained in the closure of these finitely many boundary components $e(\mathbf{Q})$ in $\overline{X}^{BS}$. By passing to a further subsequence if necessary, we can assume that there exists a rational parabolic subgroup $\mathbf{Q}$ containing $\mathbf{P}$ such that all $y'_j \in \Gamma_Q\backslash e(\mathbf{Q})$. Since $N_P \supset N_Q$, $y_j \to y_\infty$, $y'_j \to y'_\infty$, and $y'_j \sim y_j$, it follows that $y_\infty \sim y'_\infty$, and hence $y_\infty \in C^\sim$.

III.6.5 We use the example of Hilbert modular surfaces to show that the reductive Borel–Serre compactification is different from all the previous compactifications. In this case, for any rational parabolic subgroup $\mathbf{P}$, $N_P \cong \mathbb{R}^2$, $\Gamma_{N_P}\backslash N_P$ is a two-dimensional torus; $X_\mathbf{P} \cong \mathbb{R}$, since $\dim A_P = 2$, but $\dim A_\mathbf{P} = 1$, and the real boundary symmetric space X_P consists of one point, and hence $\Gamma_{M_\mathbf{P}}\backslash X_\mathbf{P}$ is a circle. Therefore, the Borel–Serre boundary component $\Gamma_P\backslash e(\mathbf{P})$ is a torus bundle over the circle $\Gamma_P\backslash X_\mathbf{P}$.

Since the $\mathbb{Q}$-rank is equal to 1, there is a one-to-one correspondence between the ends of $\Gamma\backslash X$ and the Γ-conjugacy classes of rational parabolic subgroups, and $\overline{\Gamma\backslash X}^{RBS}$ is obtained by adding a circle $\Gamma_{M_P}\backslash X_\mathbf{P}$ to each end. Hence it is different from the Baily–Borel compactification $\overline{\Gamma\backslash X}^{BB}$ and other Satake compactifications $\overline{\Gamma\backslash X}^{S}$, which all agree in this case and are obtained by adding one point to each end.

In fact, since $X = \mathbf{H} \times \mathbf{H}$, all the Satake compactifications of X are equal to $\overline{\mathbf{H}} \times \overline{\mathbf{H}}$, where $\overline{\mathbf{H}} = \mathbf{H} \cup \mathbb{R} \cup \{\infty\}$. Since it can be considered either as the Baily–Borel compactification or the maximal Satake compactification satisfying the conditions in Proposition III.3.14, it is geometrically rational, and hence $\overline{\Gamma\backslash \mathbf{H} \times \mathbf{H}}^{S}$ is defined. For any rational parabolic subgroup $\mathbf{P}$, its real locus P is of the form $P_1 \times P_2$, where P_1, P_2 are real parabolic subgroups in $\mathrm{SL}(2, \mathbb{R})$, and hence the real boundary symmetric space X_P consists of only one point, which implies that $\overline{\Gamma\backslash \mathbf{H} \times \mathbf{H}}^{S}$ is obtained by adding one point to each end.

III.6.6 In general, suppose that the maximal Satake compactification of X is geometrically rational. If the $\mathbb{Q}$-rank of $\mathbf{G}$ is strictly less than the $\mathbb{R}$-rank of $\mathbf{G}$, then the induced maximal Satake compactification $\overline{\Gamma\backslash X}^{S}_{\max}$ is different from $\overline{\Gamma\backslash X}^{RBS}$. In fact, by equation (III.1.8) in §III.1, $\mathfrak{a}_\mathbf{P}^\perp \neq 0$, and then,

$$X_\mathbf{P} = X_P \times \exp \mathfrak{a}_\mathbf{P}^\perp \neq X_P.$$

The subgroup $\Gamma_{M_\mathbf{P}}$ acting on $X_\mathbf{P}$ preserves the product, and induces the subgroup Γ_{X_P}, and hence $\Gamma_{M_\mathbf{P}}\backslash X_\mathbf{P}$ is a fiber bundle over $\Gamma_{X_P}\backslash X_P$. Denote the projection $\Gamma_{M_\mathbf{P}}\backslash X_\mathbf{P} \to \Gamma_{X_P}\backslash X_P$ by π. Let $\mathbf{P}_1, \dots, \mathbf{P}_m$ be the set of representatives of Γ-conjugacy classes of rational parabolic subgroups of $\mathbf{G}$ as above. Then

$$\overline{\Gamma\backslash X}^{S}_{\max} = \Gamma\backslash X \cup \coprod_{i=1}^{m} \Gamma_{X_{P_i}}\backslash X_{P_i}.$$

Combining the projection maps on the boundary components of $\overline{\Gamma\backslash X}^{RBS}$ with the identity map on $\Gamma\backslash X$, we obtain a surjective map

$$\overline{\Gamma\backslash X}^{RBS} \to \overline{\Gamma\backslash X}^{S}_{\max}, \tag{III.6.40}$$

which is shown to be continuous in [Zu2].

III.6.7 As mentioned earlier, the original motivation for $\overline{\Gamma\backslash X}^{RBS}$ is to study the L^2-*cohomology groups* of $\Gamma\backslash X$. Similarly, one can define the L^p-*cohomology* groups of $\Gamma\backslash X$ for all $p > 1$. In [Zu3], Zucker proved that when $p \gg 1$, the L^p-cohomology groups of $\Gamma\backslash X$ are canonically isomorphic to the usual (singular) cohomology groups of $\overline{\Gamma\backslash X}^{RBS}$. This result gives a natural explanation for $\overline{\Gamma\backslash X}^{RBS}$.

III.6.8 In [GHM], the reductive Borel–Serre compactification $\overline{\Gamma\backslash X}^{RBS}$ is used to define the *weighted cohomology groups*, which play an important role in understanding the action of the *Hecke operators* on cohomology groups of Hermitian locally symmetric spaces.

By Proposition III.4.14, $\overline{\Gamma\backslash X}^{BB}$ is a singular variety when $\Gamma\backslash X$ is not a product of Riemann surfaces. On the other hand, by Proposition III.5.14, $\Gamma\backslash\overline{X}^{BS}$ is a real analytic manifold with corners. Zucker proved in [Zu2] (see also §III.15 below) that $\Gamma\backslash\overline{X}^{BS}$ dominates $\overline{\Gamma\backslash X}^{BB}$, i.e., the identity map on $\Gamma\backslash X$ extends to a continuous map $\Gamma\backslash\overline{X}^{BS} \to \overline{\Gamma\backslash X}^{BB}$. Hence $\Gamma\backslash\overline{X}^{BS}$ is a resolution of $\Gamma\backslash\overline{X}^{BS}$ in a topological sense. The above discussions show that $\Gamma\backslash\overline{X}^{BS}$ is too large in studying the L^2-cohomology groups, and the reductive Borel–Serre compactification $\overline{\Gamma\backslash X}^{RBS}$ is an intermediate resolution. Though $\overline{\Gamma\backslash X}^{RBS}$ is still singular, the structure of its singularities, for example the *links of the singular strata*, can be described explicitly, which permits one to compute the stalk cohomology of various sheaves along the boundary of $\overline{\Gamma\backslash X}^{RBS}$. For the sheaf of L^2-differential forms, the stalk cohomology groups at the boundary points might be infinite-dimensional. This problem is solved by picking out certain summands in the stalk cohomology according to the weights of the geodesic action of the split components of rational parabolic subgroups, which explains the name of weighted cohomology. See [Go] for an introduction to the weighted cohomology groups and a survey of applications and other related topics.

III.6.9 **Summary and comments.** The reductive Borel–Serre compactification is defined as a quotient of the Borel–Serre compactification $\overline{\Gamma\backslash X}^{BS}$. The collapsing of the boundary of $\overline{\Gamma\backslash X}^{BS}$ was motivated by application to the L^2-cohomology group. This compactification is also natural for other purposes. For example, its cohomology group is naturally isomorphic to the L^p-cohomology group of $\Gamma\backslash X$ when $p \gg 0$.

III.7 Toroidal compactifications

Assume that $\Gamma\backslash X$ is a noncompact, Hermitian locally symmetric space. In this section, we briefly discuss the toroidal compactifications $\overline{\Gamma\backslash X}^{tor}_{\Sigma}$. They are defined quite differently from all the previous compactifications, and there are infinitely many of them, depending on some combinatorial data Σ. Since there are several books [AMRT] [Nam1] devoted entirely to them, we will recall only the main steps. There is an alternative construction of the toroidal compactifications using the uniform method in §III.8, which will be outlined here but treated in detail elsewhere.

In this section, we try to explain how torus embeddings arise and are used to compactify $\Gamma\backslash X$. In the construction, realization of X as Siegel domains of the third kind is crucial. We also explain how to relate this realization to the easier and more geometric horospherical decomposition of X. Finally, we outline an alternative construction of $\overline{\Gamma\backslash X}^{tor}_{\Sigma}$ using the uniform method in §III.8 to avoid partial compactifications of coverings of $\Gamma\backslash X$ in the original definition in [AMRT].

More precisely, we discuss the original motivation of finding explicit resolution of the singularities of $\overline{\Gamma\backslash X}^{BB}$ in III.7.1. In III.7.3, we explain how tori and torus bundles are related to Hermitian locally symmetric spaces. Some informal explanations are given in III.7.4. Realization of a Hermitian symmetric space X as Siegel domains of various kinds is given in III.7.6–III.7.12. The relation between the horospherical decomposition and the decomposition induced from the realization as Siegel domains is given in III.17.13. The notions of admissible polyhedral cone decompositions are introduced in III.7.16 and III.7.17. They are used to define partial compactifications of coverings of $\Gamma\backslash X$ in III.7.18. Suitable quotients of these partial compactifications can be fitted together into the toroidal compactifications $\overline{\Gamma\backslash X}^{tor}_{\Sigma}$ in III. 7. 20.

III.7.1 By Proposition III.4.14, $\overline{\Gamma\backslash X}^{BB}$ is a singular normal projective variety unless $\Gamma\backslash X$ is a product of Riemann surfaces. If Γ is torsion-free, then the singularities of $\overline{\Gamma\backslash X}^{BB}$ are contained in the boundary. By Hironaka's resolution theorem, there exists a smooth projective variety Y together with a morphism $\pi : Y \to \overline{\Gamma\backslash X}^{BB}$ that is proper and birational, and one-to-one on $\Gamma\backslash X$. Identify $\Gamma\backslash X$ with an open dense subset of Y. Then Y can be regarded as a smooth compactification of $\Gamma\backslash X$. Since $\Gamma\backslash X$ arises from the pair $(\mathbf{G}, \Gamma)$, it is a natural problem to construct explicitly such resolutions and to understand how such smooth compactifications depend on data related to the pair $(\mathbf{G}, \Gamma)$.

Such compactifications were first constructed for Hilbert modular surfaces by Hirzebruch [Hi1] and for quotients of tube domains by Satake [Sat5]. In [Ig2], Igusa constructed a partial desingularization when $\Gamma\backslash X$ is a Siegel modular variety. The general case was solved completely in [AMRT] using the theory of torus embeddings in [KKMS]. In fact, the theory of torus embeddings or toric varieties was partly motivated by this problem of obtaining smooth compactifications of $\Gamma\backslash X$, hence the name torus embeddings.

III.7.2 Let $T = (\mathbb{C}^\times)^r$ be the r-dimensional complex torus. A torus embedding is a variety Y containing some torus T such that the T-action on itself extends to Y. The variety Y is not required to be compact.

Let N be a lattice in $\mathbb{R}^r$. Then $N\backslash\mathbb{C}^r \cong (\mathbb{C}^\times)^r$, and its torus embeddings, which are normal varieties, are parametrized by *polyhedral cone decompositions* of $\mathbb{R}^n$, that are rational with respect to the integral structure given by N. For details about *torus embeddings*, see [Nam1, §6], [KKMS], [Oda] and [Ful].

An important example is $\mathbb{C}^r = \mathfrak{a}^* \otimes \mathbb{C}$, and the *root lattice*, i.e., the lattice spanned by roots $\Phi(\mathfrak{g}, \mathfrak{a})$. Then the polyhedral cone decomposition given by the Weyl chamber decomposition is rational.

III.7.3 To motivate how torus embeddings are related to compactifications of $\Gamma\backslash X$, we first consider the example of $\Gamma\backslash X = \Gamma\backslash \mathbf{H}$. Assume that the vertical strip $\{z = x + iy \in \mathbf{H} \mid -\frac{1}{2} \leq x \leq \frac{1}{2}, y > t_0\}$, $t_0 \gg 1$, is mapped to a cusp neighborhood. Under the exponential map $z \mapsto e^{2\pi i z}$, this cusp neighborhood is mapped to a punctured neighborhood of the origin in the torus $\mathbb{C}^\times$. Clearly, the torus $\mathbb{C}^\times$ admits a torus embedding $\mathbb{C}$, which gives a smooth compactification of the cusp neighborhood by adding in the origin. By compactifying all cusp neighborhoods in this way, we obtain a compactification of $\Gamma\backslash\mathbf{H}$ that is a smooth projective curve.

In this construction, the crucial step is to identify a cusp neighborhood, or rather $\Gamma_{N_P}\backslash\mathbf{H}$, with a subset of the torus $\mathbb{C}^\times$. For general $\Gamma\backslash X$, the first step is to embed suitable neighborhoods near infinity into suitable torus bundles.

III.7.4 A rather informal explanation why the torus embeddings are needed is as follows. By the reduction theory, $\Gamma\backslash X$ are images of finitely many Siegel sets $U \times A_{\mathbf{P},t} \times V$, where $U \subset N_P$, $V \subset X_{\mathbf{P}}$ can be taken to compact subsets. To compactify $\Gamma\backslash X$, it is important to compactify the split part $A_{\mathbf{P},t}$, or equivalently to partially compactify $A_{\mathbf{P}}$ in the positive direction of the chamber corresponding to $\mathbf{P}$. An obvious problem with this approach is that the Siegel sets are given by the horospherical decomposition $X = N_P \times A_{\mathbf{P}} \times X_{\mathbf{P}}$, which is not a holomorphic decomposition. In fact, none of the factors, in particular $A_{\mathbf{P}}$, is a complex manifold.

Let $\alpha_1, \dots, \alpha_r$ be the set of simple roots in $\Delta(P, A_{\mathbf{P}})$. Then we have the following identification

$$A_{\mathbf{P}} \cong (\mathbb{R}^\times)^r, \quad a \mapsto (a^{-\alpha_1}, \dots, a^{-\alpha_r}).$$

Naturally one tries to complexify $A_{\mathbf{P}}$ into

$$A_{\mathbf{P}} \otimes \mathbb{C} = (\mathbb{R}^\times)^r \otimes \mathbb{C} \cong (\mathbb{C}^\times)^r,$$

and hence the complex torus occurs naturally. Ideally, the next step is to obtain a decomposition of X that is holomorphic and for which the factor that causes the noncompactness is contained in the complex torus $A_{\mathbf{P}} \otimes \mathbb{C}$ as in the example of $\Gamma\backslash\mathbf{H}$. But this does not happen for several reasons.

One reason is that since X is simply connected, it cannot contain any torus in such a decomposition. To overcome this difficulty, the idea is to show that some partial quotients $\Gamma'\backslash X$ for suitable subgroups $\Gamma' \subset \Gamma$ are naturally embedded into some torus bundles. In the case of the example $X = \mathbf{H}$, the partial quotient $\Gamma_{N_P}\backslash\mathbf{H}$ is mapped into the unit disk in the complex torus $A_{\mathbf{P}} \otimes \mathbb{C} = \mathbb{C}$. In general, for such partial quotients $\Gamma'\backslash X$, only parts of the complex tori appear.

III.7.5 The holomorphic decomposition of X mentioned in the previous paragraph is given by the realization of X as a Siegel domains of the third kind.

Assume that $\mathbf{G}$ is $\mathbb{Q}$-simple. Recall from §III.4 that the Baily–Borel compactification $\overline{\Gamma\backslash X}^{BB}$ is constructed as follows. For any maximal rational parabolic subgroup $\mathbf{P}$ of $\mathbf{G}$, its boundary component $X_{P,h}$ is a Hermitian symmetric space of lower dimension. Then

$$\overline{\Gamma\backslash X}^{BB} = \Gamma\backslash \left(X \cup \coprod_{\mathbf{P}} X_{P,h} \right),$$

where $\mathbf{P}$ runs over all maximal parabolic subgroups of $\mathbf{G}$.

Since the toroidal compactifications $\overline{\Gamma\backslash X}_{\Sigma}^{tor}$ are blow-ups of $\overline{\Gamma\backslash X}^{BB}$, it is reasonable that we need to use some refined holomorphic decomposition of X based on each $X_{P,h}$.

In fact, for any such maximal rational parabolic subgroup $\mathbf{P}$, there is an unbounded realization of X as a Siegel domains of the third kind over $X_{P,h}$. Briefly, a Siegel domain of the first kind is a tube domain, and a Siegel domain of the second kind is a family of Siegel domains of the first kind over a complex vector space, and a Siegel domain of the third kind is a family of Siegel domain of the second kind over the bounded symmetric domain $X_{P,h}$. For convenience, we consider Siegel domains of the first and second kinds as special Siegel domains of the third kind.

III.7.6 Next we recall some of the basics of the realization of X as Siegel domains of the third kind and show how to compare it with and visualize it in relation to the more geometric horospherical decomposition.

Assume that $\mathbf{G}$ is $\mathbb{Q}$-simple. In the notation of §III.4, let $\mathbf{S}$ be a maximal $\mathbb{Q}$-split torus in $\mathbf{G}$. Let $S = \mathbf{S}(\mathbb{R})$ be the real locus and S^0 the identity component. Let $\beta_1, \dots, \beta_s$ be the ordered simple roots in $\Phi(G, S^0)$, which determine a positive chamber. Let $\mathbf{P}_0$ be the minimal rational parabolic subgroup corresponding to the positive chamber. In particular, $A_{\mathbf{P}_0} = S^0$. Then any rational parabolic subgroup containing $\mathbf{P}_0$ is of the form $\mathbf{P}_{0,I}$, for some $I \subset \{\beta_1, \dots, \beta_s\}$.

For any such I, let $\Delta_{I,h}$ be the connected component of I containing the distinguished root β_s. If I does not contain β_s, $\Delta_{I,h}$ is defined to be empty.

Identify the simple roots $\beta_1, \dots, \beta_s$ with simple roots $\Delta(\mathbf{G}, \mathbf{S})$ in $\Phi(\mathbf{G}, \mathbf{S})$ (see Remark III.1.14).

Then $\Delta_{I,h}$ spans a subroot system in $\Phi(\mathbf{G}, \mathbf{S})$, whose root spaces generate a semisimple algebraic subgroup $\mathbf{G}_{I,h}$ of $\mathbf{G}$. Let $\mathbf{G}_{I,l}$ be the normal $\mathbb{Q}$-subgroup in the Levi group $\mathbf{L}_{\mathbf{P}_I}(x_0)$ complementary to $\mathbf{G}_{I,h}$, i.e., $\mathbf{L}_{\mathbf{P}_I}(x_0) = \mathbf{G}_{I,h}\mathbf{G}_{I,l}$ is an almost direct product. Define $K_{I,h} = K \cap G_{I,h}$, and $K_{I,l} = K \cap G_{I,l}$, where $G_{I,h} = \mathbf{G}_{I,h}(\mathbb{R})$, $G_{I,l} = \mathbf{G}_{P,l}(\mathbb{R})$.

Lemma III.7.7 *The space $X_{I,h} = G_{I,h}/K_{I,h}$ is a Hermitian symmetric space of noncompact type. And $X_{I,l} = G_{I,l}/K_{I,l}A_{P_I}$ is a symmetric space of noncompact type. If $\mathbf{P}_I$ is a maximal rational parabolic subgroup, then $C_{\mathbf{P}_I} = G_{I,l}/K_{I,l}$ is a symmetric cone in the center of the nilpotent radical of P_I.*

Proof. The first statement follows from the fact that $\Delta_{I,h}$ spans a root system of type either BC or C. If $\mathbf{P}_I$ is maximal, the second and the third statements follow from [AMRT, Theorem 1, p. 227]; and the nonmaximal cases are similar.

Lemma III.7.8 *The boundary symmetric space $X_{\mathbf{P}_{0,I}}$ can be decomposed as a Riemannian product $X_{\mathbf{P}_{0,I}} = X_{I,h} \times X_{I,l}$.*

Proof. Since $M_{\mathbf{P}_{0,I}} = G_{I,h}G_{I,l}/A_{\mathbf{P}_I}$ and $G_{I,h}$ commutes with $G_{I,l}$, the lemma follows from the definition of $X_{I,h}$ and $X_{I,l}$.

Since every rational parabolic subgroup $\mathbf{Q}$ is conjugate to a standard parabolic subgroup $\mathbf{P}_I$, we also get subgroups $G_{\mathbf{Q},h}$, $G_{\mathbf{Q},l}$ of $M_{\mathbf{Q}}$, and the boundary spaces $X_{\mathbf{Q},h}$, $X_{\mathbf{Q},l}$.

Lemma III.7.9 *With the above notation, the space X has the following refined horospherical decomposition with respect to the parabolic subgroup $\mathbf{Q}$:*

$$X = N_Q \times X_{\mathbf{Q},h} \times X_{\mathbf{Q},l} \times A_{\mathbf{Q}}.$$

Proof. It follows from the horospherical decomposition $X = N_Q \times X_{\mathbf{Q}} \times A_{\mathbf{Q}}$ in equation (III.1.4) and the decomposition $X_{\mathbf{Q}} = X_{\mathbf{Q},h} \times X_{\mathbf{Q},l}$ in Lemma III.7.8.

For every maximal rational parabolic subgroup $\mathbf{Q}$, let U_Q be the center of the nilpotent radical N_Q of Q, and let $V_Q = N_Q/U_Q$. Then V_Q is a vector group, i.e., abelian and diffeomorphic to its Lie algebra $\mathfrak{v}_Q$. Since N_Q is a U_Q-principal bundle over V_Q, we get that as differential manifolds, $N_Q = U_Q \times V_Q$.

Lemma III.7.10 *The Lie algebra $\mathfrak{v}_Q$ of the quotient group $V_Q = N_Q/U_Q$ can be identified with a subspace of the algebra $\mathfrak{n}_Q$ of N_Q that is complementary to the Lie algebra $\mathfrak{u}_Q$ of U_Q. Denote this subspace by $\mathfrak{v}_Q$. The adjoint action of $G_{\mathbf{Q},l}$ on $\mathfrak{n}_Q$ preserves $\mathfrak{v}_Q$.*

Proof. By [BB1, Corollary 2.10], Q is also a maximal real parabolic subgroup. For simplicity, we assume that Q is the normalizer of a standard boundary component F_s in the notation of [AMRT, §4.1]. Then the third equation in [AMRT, p. 224] shows that the Lie algebra $\mathfrak{u}_Q$ is the direct sum of some of the root spaces that appear in $\mathfrak{n}_Q$, and hence $\mathfrak{v}_Q$ can be naturally identified with the direct sum of the other root spaces in $\mathfrak{n}_Q$, given by the second equation in [AMRT, p. 224]. This equation also shows that this complementary subspace of $\mathfrak{u}_Q$ in $\mathfrak{n}_Q$ is an abelian subalgebra. The root space decomposition of the Lie algebra $\mathfrak{g}_{Q,l}$ of $G_{Q,l}$ in [AMRT, p. 226] shows that the adjoint action of $G_{Q,l}$ on $\mathfrak{n}_Q$ leaves both subspaces $\mathfrak{u}_Q$ and $\mathfrak{v}_Q$ invariant. This completes the proof.

Remark III.7.11 This lemma shows that the adjoint action of $G_{Q,l}$ on the quotient group V_Q and its Lie algebra $\mathfrak{v}_Q$ can be studied by the restriction of the adjoint action of $G_{Q,l}$ on $\mathfrak{n}_Q$ to the invariant subspace $\mathfrak{v}_Q$. In fact, it is shown in [Ji3] that when $\Gamma_{Q,l}$ is infinite, or equivalently $\dim X_{P,l} > 0$, the adjoint (or holonomy) action of $\Gamma_{Q,l}$ on the compact nilmanifold $\Gamma_{V_Q}\backslash V_Q$ is ergodic, where $\Gamma_{V_Q} = \Gamma_{N_Q}/\Gamma_{U_Q}$, $\Gamma_{N_Q} = \Gamma \cap N_Q$, $\Gamma_{U_Q} = \Gamma \cap U_Q$. This ergodicity result plays an important role in comparing the two compactifications $\overline{\Gamma\backslash X}^{BS}$ and $\overline{\Gamma\backslash X}_{\Sigma}^{tor}$, i.e., the Harris–Zucker conjecture in §III.15.6 (or Proposition III.15.4.4).

After these preparations, we have the following realization of X as a Siegel domain of the third kind over $X_{Q,h}$.

Proposition III.7.12 [WK, Theorem 7.7] [AMRT, §3.4, pp. 238–239] [Nam1, §5] *With the above notation, there exists an injective holomorphic map $\pi : X \to X_{\mathbf{Q},h} \times \mathbb{C}^n \times (U_Q \otimes \mathbb{C})$ such that*

$$\pi(X) = \{(z, v, u_1 + iu_2) \mid z \in X_{\mathbf{Q},h}, v \in \mathbb{C}^n, u_1 \in U_Q, u_2 \in h_z(v, v) + C_Q\},$$

where $n = \frac{1}{2}\dim V_Q$ and $h_z(v, v) \in C_Q$ is a quadratic form in v depending holomorphically on z.

This realization represents X as a family of tube domains $U_Q + iC_Q$ over $X_{\mathbf{Q},h} \times \mathbb{C}^n$ and hence as a family of Siegel domains of the second kind over $X_{\mathbf{Q},h}$. Hence X has been realized as a Siegel domain of the third kind. Since the ambient space containing the image of the map has a product structure, this realization induces a decomposition of X, which is closely related to the refined horospherical decomposition in Lemma III.7.9.

Using the decomposition $N_Q = U_Q \times V_Q$ explained above, we can write the refined horospherical decomposition of X in Lemma III.7.9 as follows:

$$X = U_Q \times V_Q \times X_{\mathbf{Q},h} \times X_{\mathbf{Q},l} \times A_{\mathbf{Q}}. \tag{1}$$

Then the relation between the horospherical decomposition and the realization as a Siegel domain of the third kind is as follows.

Lemma III.7.13 *For any $x = (u, v, z, x_l, a) \in U_Q \times V_Q \times X_{\mathbf{Q},h} \times X_{\mathbf{Q},l} \times A_{\mathbb{Q}} = X$, denote the image $\pi(x)$ of x under the map π in Proposition III.7.12 by $(z', v', u_1' + iu_2') \in X_{\mathbf{Q},h} \times \mathbb{C}^n \times (U_Q \otimes \mathbb{C})$. Then $z' = z$, the map $v \to v'$ defines an $\mathbb{R}$-linear isomorphism from V_Q to $\mathbb{C}^n$, the map $u \to u_1'$ is an $\mathbb{R}$-linear transformation on U_Q, and $u_2' \in h_z(v', v') + C_Q$. Furthermore, for any $u \in U_Q$, $v \in V_Q$, and $z \in X_{\mathbf{Q},h}$, the image of $\{u\} \times \{v\} \times \{z\} \times X_{\mathbf{Q},l} \times A_{\mathbf{Q}}$ is exactly the shifted cone $u_1' + i(h_z(v', v') + C_Q)$ over the point $(z, v') \in X_{\mathbf{Q},h} \times \mathbb{C}^n$.*

Proof. The lemma follows from the discussions in [AMRT, pp. 235–238]; in particular, the linear isomorphisms $v \to v'$ and $u \to u'$ come from trivialization of the two principal bundles.

Since the horospherical decomposition of X describes the structure of geodesics in X, it is helpful in understanding its relation with the realization of X as a Siegel domain of the third kind in the above proposition by discussing how geodesics behave in the realization of X as Siegel domains.

Lemma III.7.14 *For any $u \in U_Q$, $v \in V_Q$, $z \in X_{\mathbf{Q},h}$, $x_l \in X_{\mathbf{Q},l}$, and $H \in \mathfrak{a}_Q^+$, $|H| = 1$, the curve $c(t) = (u, v, z, x_l, \exp(tH))$, $t \in \mathbb{R}$, is a geodesic in X. In the realization of X as a Siegel domain of the third kind above, $c(t)$ becomes a ray in the tube domain $U_Q + i(h_z(v', v') + C_Q)$ whose imaginary part is a ray starting from the vertex $h_z(v', v')$.*

Proof. By Lemma I.2.9, $c(t)$ is a geodesic in X. By Lemma III.7.13, the geodesic $c(t)$, $t \in \mathbb{R}$, is mapped into the cone $u_1' + i(h_z(v', v') + C_Q)$ over the point $(z, v') \in X_{\mathbf{Q},h} \times \mathbb{C}^n$. Since $X_{\mathbf{Q},h}$ is a section of the symmetric cone C_Q, any geodesic in the cone $u_1' + i(h_z(v', v') + C_Q)$ with respect to the invariant metric is a ray from the vertex.

This lemma shows that when x_l varies in $X_{\mathbf{Q},l}$, the family of parallel geodesics $(u, v, z, x_l, \exp(tH))$ in X is mapped to a family of rays in the cone $h_z(v', v') + C_Q$ issuing from the vertex.

III.7.15 As in the case of toroidal embeddings or toric varieties, the toroidal compactifications $\overline{\Gamma\backslash X}_{\Sigma}^{Tor}$ of $\Gamma\backslash X$ depend on polyhedral decompositions Σ.

For any maximal rational parabolic subgroup $\mathbf{Q}$, let $\Gamma_{Q,l}$ be the image of Γ_Q in $G_{\mathbf{Q},l}$ under the projection $Q = N_Q G_{\mathbf{Q},h} G_{\mathbf{Q},l} A_{\mathbf{Q}} \to G_{\mathbf{Q},l}$. Then $\Gamma_{\mathbf{Q},l}$ is a torsion-free lattice subgroup acting on $X_{\mathbf{Q},l}$. Denote the intersection $\Gamma_Q \cap U_Q$ by Γ_{U_Q}. Then Γ_{U_Q} is a torsion-free lattice in the vector group U_Q.

Definition III.7.16 [AMRT, pp. 117, 252] [Nam1, pp. 59–60] A $\Gamma_{Q,l}$-admissible polyhedral decomposition of C_Q is a collection Σ_Q of polyhedral cones satisfying the following conditions:

1. Each cone in Σ_Q is a strongly convex rational polyhedral cone in $\overline{C_Q} \subset U_Q$ with respect to the rational structure on U_Q induced by the lattice Γ_{U_Q}.
2. Every face of any $\sigma \in \Sigma_Q$ is also an element in Σ_Q.
3. For any $\sigma, \sigma' \in \Sigma_Q$, the intersection $\sigma \cap \sigma'$ is a face of both σ and σ'.
4. For any $\gamma \in \Gamma_{Q,l}$ and $\sigma \in \Sigma_Q$, $\gamma\sigma$ is also a cone in Σ_Q.
5. There are only finitely many classes of cones in Σ_Q modulo $\Gamma_{Q,l}$.
6. $C_Q \subset \cup_{\sigma \in \Sigma_Q} \sigma$, and hence $C_Q = \cup_{\sigma \in \Sigma_Q} C_Q \cap \sigma$.

Definition III.7.17 A Γ-admissible family of polyhedral cone decompositions $\Sigma = \{\Sigma_Q\}$ is a union of $\Gamma_{Q,l}$-admissible polyhedral cone decompositions Σ_Q of C_Q over all maximal rational parabolic subgroups satisfying the following compatibility conditions:

1. If $Q_1 = \gamma Q_2 \gamma^{-1}$, then $\gamma \Sigma_{Q_1} = \Sigma_{Q_2}$.
2. If C_{Q_1} is contained in the boundary of C_{Q_2}, then $\Sigma_{Q_1} = \{\sigma \cap \overline{C_{Q_1}} \mid \sigma \in \Sigma_{Q_2}\}$.

III.7.18 For any maximal rational parabolic subgroup $\mathbf{Q}$, Γ_{U_Q} is a lattice in U_Q, and $\Gamma_{U_Q}\backslash U_Q \otimes \mathbb{C}$ is a complex torus. Using Proposition III.7.12, identify X with the subset $\pi(X)$ in $X_{\mathbf{Q},h} \times \mathbb{C}^n \times (U_Q \otimes \mathbb{C})$. Then $\Gamma_{U_Q}\backslash X$ is contained in a bundle

$\Gamma_{U_Q}\backslash X_{\mathbf{Q},h} \times \mathbb{C}^n \times (U_Q \otimes \mathbb{C})$ over $X_{\mathbf{Q},h} \times \mathbb{C}^n$ with fiber $\Gamma_{U_Q}\backslash U_Q \otimes \mathbb{C}$, which is denoted by $\Gamma_{U_Q}\backslash B(Q)$. This is the torus bundle we mentioned earlier. It contains the partial quotient $\Gamma_{U_Q}\backslash X$ as a proper subset.

A $\Gamma_{Q,l}$-admissible polyhedral decomposition Σ_Q of $C_Q \subset U_Q$ defines a partial compactification (a toroidal embedding) $\overline{\Gamma_{U_Q}\backslash U_Q \otimes \mathbb{C}}_{\Sigma_Q}$ of every fiber $\Gamma_{U_Q}\backslash U_Q \otimes \mathbb{C}$ in $\Gamma_{U_Q}\backslash B(Q)$. Putting all these partial compactifications together, we get a partial compactification $\overline{\Gamma_{U_Q}\backslash B(Q)}_{\Sigma_Q}$ of the torus bundle $\Gamma_{U_Q}\backslash B(Q)$. The interior of the closure of $\Gamma_{U_Q}\backslash X$ in $\overline{\Gamma_{U_Q}\backslash B(Q)}_{\Sigma_Q}$ defines a partial compactification $\overline{\Gamma_{U_Q}\backslash X}_{\Sigma_Q}$ of $\Gamma_{U_Q}\backslash X$, which is a bundle over $X_{\mathbf{Q},h} \times \mathbb{C}^n$ [AMRT, pp. 249–250].

III.7.19 For every Γ-admissible family of polyhedral cone decompositions $\Sigma = \{\Sigma_Q\}$, we get a family of partially compactified spaces $\overline{\Gamma_{U_Q}\backslash X}_{\Sigma_Q}$ of coverings $\Gamma_{U_Q}\backslash X$ of $\Gamma\backslash X$.

Proposition III.7.20 [AMRT, Main Theorem I, p. 252] [Nam1, Main Theorem 7.10] *For any Γ-admissible family of polyhedral cone decompositions Σ, there exists a unique compact Hausdorff analytic compactification $\overline{\Gamma\backslash X}_{\Sigma}^{tor}$ satisfying the following conditions:*

1. *For every maximal rational parabolic subgroup $\mathbf{Q}$, the projection map $\pi_Q : \Gamma_{U_Q}\backslash X \to \Gamma\backslash X$ extends to an open holomorphic map $\pi_Q : \overline{\Gamma_{U_Q}\backslash X}_{\Sigma_Q} \to \overline{\Gamma\backslash X}_{\Sigma}^{tor}$.*
2. *The images $\pi_Q(\overline{\Gamma_{U_Q}\backslash X}_{\Sigma_Q})$ for all maximal rational parabolic subgroups $\mathbf{Q}$ cover $\overline{\Gamma\backslash X}_{\Sigma}^{tor}$.*
3. *The compactification $\overline{\Gamma\backslash X}_{\Sigma}^{tor}$ dominates $\overline{\Gamma\backslash X}^{BB}$.*

This compactification $\overline{\Gamma\backslash X}_{\Sigma}^{tor}$ is called the *toroidal compactification* of $\Gamma\backslash X$ associated with the polyhedral cone decomposition Σ. For any such Σ, there always exists a refinement Σ' of Σ such that the corresponding toroidal compactification $\overline{\Gamma\backslash X}_{\Sigma'}^{tor}$ is a smooth projective variety. This solves the problem mentioned in §III.7.1, i.e., gives explicit resolutions of the singularities of $\overline{\Gamma\backslash X}^{BB}$. Though they are not unique and hence canonical in general, they have exploited and depend on the underlying group theoretical structures of $\Gamma\backslash X$.

III.7.21 Though the above discussions only summarize the main steps in the construction of the toroidal compactifications $\overline{\Gamma\backslash X}_{\Sigma}^{tor}$, details of the many steps are difficult. The space $\overline{\Gamma\backslash X}_{\Sigma}^{tor}$ is defined as a suitable quotient of the union $\coprod_{\mathbf{Q}} \overline{\Gamma_{U_Q}\backslash X}_{\Sigma_Q}$. Since they are partial compactifications of different covering spaces $\Gamma_{U_Q}\backslash X$ of $\Gamma\backslash X$, the equivalence relation is complicated and makes substantial use of $\overline{\Gamma\backslash X}^{BB}$. In fact, there are a lot of overlaps between the images $\pi_Q(\overline{\Gamma_{U_Q}\backslash X}_{\Sigma_Q})$. On the other hand, the procedures to construct $\overline{\Gamma\backslash X}^{BS}$ and $\overline{\Gamma\backslash X}^{BB}$ are different: Start with a partial compactification of X, then take the quotient by Γ only once to get the compactification of $\Gamma\backslash X$.

It is certainly desirable to construct $\overline{\Gamma\backslash X}_{\Sigma}^{tor}$ in this way in order to understand $\overline{\Gamma\backslash X}_{\Sigma}^{tor}$ better (see Satake's long review [Sat7] of [AMRT] in *Math. Reviews*).

At first sight, such an approach seems impossible. The reason is that to get torus bundles and apply the techniques of torus embeddings, we need to divide the contractible X by some discrete subgroups such as Γ_{U_Q} to start with, and the passage to intermediate quotients $\Gamma_{U_Q}\backslash X$ seems to be necessary. It turns out that this difficulty can be overcome by observing that the torus embeddings of a torus $N\backslash\mathbb{C}^n$ can be constructed differently from the usual procedure. Specifically, for a polyhedral cone decomposition Σ of $\mathbb{R}^n$, one can construct a partial compactification $\overline{\mathbb{C}^n}_{\Sigma}$ such that the quotient $N\backslash\overline{\mathbb{C}^n}_{\Sigma}$ is the torus embedding associated with Σ. Using this observation, one can construct $\overline{\Gamma\backslash X}_{\Sigma}^{tor}$ by methods similar to the construction of $\overline{\Gamma\backslash X}^{BS}$, or by the general method in §III.8 below. Specifically, given a Γ-admissible family $\Sigma = \{\Sigma_Q\}$ of polyhedral cone decompositions, one proceeds in two steps:

1. For each maximal rational parabolic subgroup $\mathbf{Q}$, attach a boundary component $e_{\Sigma}(\mathbf{Q})$ to the infinity of X using the realization of X as a Siegel domain of the third kind to get a partial compactification $\overline{X}_{\Sigma}^{tor} = X \cup \coprod_Q e_{\Sigma}(\mathbf{Q})$.
2. Show that Γ acts continuously on $\overline{X}_{\Sigma}^{tor}$ with a compact quotient, which is equal to the toroidal compactification $\overline{\Gamma\backslash X}_{\Sigma}^{tor}$.

Details of this construction are under preparation and will appear elsewhere.

III.7.22 Summary and comments. In this section, we outlined the construction of toroidal compactifications $\overline{\Gamma\backslash X}_{\Sigma}^{tor}$ in [AMRT] and mentioned an alternative approach using the uniform method in the book. The latter construction is independent of the Baily–Borel compactification and allows one to understand the boundary structure of $\overline{\Gamma\backslash X}_{\Sigma}^{tor}$ better.

There are natural realizations of the toroidal compactifications of special locally symmetric spaces in terms of degenerations of varieties. See [Nam2] [Nam3] [Ale] [HKW] [Sha] for details. There are also other complex compactifications that lie between the Baily–Borel and the toroidal compactifications in [Lo1] [Lo2].

The toroidal compactifications have some important applications. One is to the generalization by Mumford [Mum3] of the Hirzebruch proportionality principle [Hi3]. Some toroidal compactifications also have models over number fields and their rings of integers [FC] [Ch1] [Ch2], which are important to applications in number theory.

10

Uniform Construction of Compactifications of Locally Symmetric Spaces

In the previous chapter, we have recalled most of the known compactifications of locally symmetric spaces and their motivations. In this chapter, we modify the approach in [BS2] to propose a general uniform method to compactify locally symmetric spaces and construct all the previous compactifications of $\Gamma\backslash X$ and some new compactifications of $\Gamma\backslash G$ in a uniform way. As mentioned in Part I, this method motivated the uniform approach to compactifications of symmetric spaces §I.8. Besides allowing easier comparison between different compactifications, this approach also avoids some of the steps in [BS2] such as the introduction of spaces of S-type, which are needed in the inductive proofs, and the issues around the geometric rationality of the Satake compactifications in §III.3.

In §III.8, we formulate precisely this uniform approach. In §III.9, we apply this method to give a uniform construction of the Borel–Serre compactification $\overline{\Gamma\backslash X}^{BS}$ and prove several results on $\overline{\Gamma\backslash X}^{BS}$ stated in §III.5. In §III.10, we use this uniform method to give a direct construction of the reductive Borel–Serre compactification $\overline{\Gamma\backslash X}^{RBS}$ independent of the Borel–Serre compactification $\overline{\Gamma\backslash X}^{BS}$, which was used crucially to define $\overline{\Gamma\backslash X}^{RBS}$ in §III.6. In §III.11, we construct the maximal Satake compactification $\overline{\Gamma\backslash X}^{S}_{\max}$ without using the maximal Satake compactification $\overline{X}^{S}_{\max}$ and hence avoid the difficulties about the geometric rationality of $\overline{X}^{S}_{\max}$. Therefore, the maximal Satake compactification $\overline{\Gamma\backslash X}^{S}_{\max}$ of $\Gamma\backslash X$ is always defined. In §III.12, we construct a compactification $\overline{\Gamma\backslash X}^{T}$ by using the Tits building $\Delta_{\mathbb{Q}}(\mathbf{G})$ as the boundary, which is shown to be related to certain geodesics going out to infinity and other geometric compactifications such as the geodesic compactification $\Gamma\backslash X \cup \Gamma\backslash X(\infty)$ and the Gromov compactification $\overline{\Gamma\backslash X}^{G}$ in §III.20. In §III.13 and §III.14, we apply the uniform method to construct compactifications $\overline{\Gamma\backslash G}^{BS}$ and $\overline{\Gamma\backslash G}^{RBS}$ of the homogeneous space $\Gamma\backslash G$ and related spaces $\Gamma\backslash G/H$, where H is a nonmaximal compact subgroup of G.

III.8 Formulation of the uniform construction

In this section we formulate a uniform approach to compactifications of locally symmetric spaces. It is suggested by [BS2]. The basic difference from [BS2] is that the geodesic action and the induced corners do not play a prominent role. In fact, many compactifications do not have corner structure; and the homogeneous space G does not admit canonical geodesic action, unlike the symmetric space X, and hence the geodesic action needs to be avoided for compactifications of the homogeneous space $\Gamma\backslash G$.

As discussed in §III.3, to pass from the Satake compactifications $\overline{X}^S_\tau$ to the Satake compactifications of $\Gamma\backslash X$, an important step is the question of geometric rationality of $\overline{X}^S_\tau$. The basic point of this chapter is that compactifications of $\Gamma\backslash X$ should be constructed directly, independently of compactifications of X.

III.8.1 A general uniform method in [BJ3], suggested by [BS2], to construct compactifications of $\Gamma\backslash X$ is as follows:

1. For every rational parabolic subgroup $\mathbf{P}$ of $\mathbf{G}$, define a boundary component $e(\mathbf{P})$ using the Langlands decomposition of P.
2. Form a partial compactification of X by attaching all the rational boundary components
$$_{\mathbb{Q}}\overline{X} = X \cup \coprod_{\mathbf{P}} e(\mathbf{P}),$$
using the rational horospherical coordinate decomposition of X with respect to P.
3. Show that Γ acts continuously on the partial compactification $_{\mathbb{Q}}\overline{X}$ with a compact Hausdorff quotient, which is a compactification of $\Gamma\backslash X$.

Different choices of the boundary component $e(\mathbf{P})$ lead to different compactifications, and they are often constructed from the factors of the rational horospherical decomposition of X with respect to P. The boundary components $e(\mathbf{P})$ are also glued at the infinity of X using the horospherical decomposition. If the boundary components $e(\mathbf{P})$ are sufficiently large, for example, as large as the Borel–Serre boundary components, the arithmetic subgroup Γ acts properly on $_{\mathbb{Q}}\overline{X}$. Otherwise, the action is not proper. It is reasonable to expect that infinite stabilizers, which is the case when the Γ-action is not proper, prevent any chances for the quotient to have differentiable structures.

The reduction theories for arithmetic groups in §III.2 play an important role in proving that the quotient is Hausdorff and compact. In fact, the compactness is often proved by showing that the closure of a Siegel set, and hence a fundamental set, is compact. The Hausdorff property is often proved with the help of the Siegel finiteness property (of Siegel sets). If the action is proper, its proof also uses the Siegel finiteness property.

III.8.2 In all the examples discussed in this chapter, the boundary components satisfy one and only one of the following conditions:

1. for every pair of rational parabolic subgroups $\mathbf{P}, \mathbf{Q}$, $e(\mathbf{P})$ is contained in the closure of $e(\mathbf{Q})$ if and only if $\mathbf{P} \supseteq \mathbf{Q}$.
2. for every pair of rational parabolic subgroups $\mathbf{P}, \mathbf{Q}$, $e(\mathbf{P})$ is contained in the closure of $e(\mathbf{Q})$ if and only if $\mathbf{P} \subseteq \mathbf{Q}$,

Usually, each boundary component $e(\mathbf{P})$ is a cell. In case (1), the boundary $\coprod_{\mathbf{P}} e(\mathbf{P})$ of ${}_{\mathbb{Q}}\overline{X}$ is a cell complex parametrized by the set of rational parabolic subgroups, and the incidence relation between the cells is the same as in the rational Tits building of $\mathbf{G}$.

In case (2), the boundary $\coprod_{\mathbf{P}} e(\mathbf{P})$ is a cell complex over all the rational parabolic subgroups whose incidence relation is dual to the incidence relation in the rational Tits building of $\mathbf{G}$, and is hence a cell complex dual to the Tits building.

Though we can obtain boundary components that are of mixed type as in the case of the Martin compactification of symmetric spaces, the resulting compactifications do not seem to occur naturally, and hence we will not construct them.

III.8.3 The above method by gluing on the boundary components of all rational parabolic subgroups can be used to construct the Borel–Serre compactification $\Gamma\backslash\overline{X}^{BS}$, the reductive Borel–Serre compactification $\overline{\Gamma\backslash X}^{RBS}$, and the maximal Satake compactification $\overline{\Gamma\backslash X}^{S}_{\max}$. In order to construct the nonmaximal Satake compactifications such as the Baily–Borel compactification and the toroidal compactifications, we need to modify the method as follows:

1. Choose a Γ-invariant collection $\mathcal{P}$ of rational parabolic subgroups of $\mathbf{G}$.
2. For every rational parabolic subgroups $\mathbf{P}$ in the collection $\mathcal{P}$, define a boundary component $e(\mathbf{P})$ using the Langlands decomposition of P or other decompositions and data, for example, the realization of a Hermitian symmetric space as a Siegel domain of the third kind and the Γ-admissible polyhedral cone decompositions.
3. Form a partial compactification of X by attaching all the rational boundary components
$$ {}_{\mathbb{Q}}\overline{X} = X \cup \coprod_{\mathbf{P}\in\mathcal{P}} e(\mathbf{P}), $$
using the rational horospherical coordinate decomposition of X with respect to P or other decompositions and data as above.
4. Show that Γ acts continuously on the partial compactification ${}_{\mathbb{Q}}\overline{X}$ with a compact Hausdorff quotient, which is a compactification of $\Gamma\backslash X$.

For the Baily–Borel compactification $\overline{\Gamma\backslash X}^{BB}$, the collection $\mathcal{P}_{\max}$ of maximal rational parabolic subgroups is used. For other Satake compactifications of $\Gamma\backslash X$, we need to pick this collection suitably. For the toroidal compactifications, the collection $\mathcal{P}_{\max}$ is also used.

III.8.4 There are several general features of this uniform approach.

1. Since the compactifications of $\Gamma\backslash X$ can be constructed by the same procedure by varying the choices of the boundary components, relations between them can be easily determined by comparing their boundary components.
2. The topologies, in particular neighborhoods of boundary points, of both the partial compactification $_{\mathbb{Q}}\overline{X}$ and the quotient $\Gamma\backslash_{\mathbb{Q}}\overline{X}$ can be described explicitly. Such explicit descriptions are useful for many applications (see [Zu2] [GHMN]).
3. Since the uniform construction of compactifications of symmetric spaces was motivated by the approach here, it suggests a close analogue between the compactifications of symmetric spaces and compactifications of locally symmetric spaces.
4. The combinatorial structure of the boundary of the compactifications is described by the rational Tits building.
5. Avoid the issue of geometrical rationality of compactifications of X, which has been an important step in constructing compactifications of $\Gamma\backslash X$ (see [Sat2] [BB1] [Ca2] and [Sap2]).

III.8.5 As in the case of compactifications of symmetric spaces, in applying this method to construct compactifications of $\Gamma\backslash X$, it is often more convenient to describe a topology in terms of convergent sequences. Such explicit convergent sequences allow us to compare the topologies of different compactifications more easily.

III.8.6 Summary and comments. The uniform method to construct compactifications of locally symmetric spaces is similar to the method for compactifications of symmetric spaces in §I.8, but there are several differences:

1. For symmetric spaces, we use $\mathbb{R}$-parabolic subgroups, while for locally symmetric spaces, we use $\mathbb{Q}$-parabolic subgroups.
2. For symmetric spaces, the real Langlands decomposition is used, while for locally symmetric spaces, the rational Langlands decomposition of the real locus of $\mathbb{Q}$-parabolic subgroups is used.

III.9 Uniform construction of the Borel–Serre compactification

In this section, we apply the uniform method in the previous section to give an alternative construction of $\Gamma\backslash\overline{X}^{BS}$. This construction avoids the introduction of spaces of S-type, which are used in the original construction in [BS2]. Details of the general method in §III.8 will be explained through this construction. We also show explicitly that $\overline{\Gamma\backslash X}^{BS}$ is a real analytic manifold with corners and prove several statements in §III.5.

This section is organized as follows. The Borel–Serre boundary component of $\mathbb{Q}$-parabolic subgroups is defined in III.9.1. The topology of the partial Borel–Serre compactification $_{\mathbb{Q}}\overline{X}^{BS}$ is defined in III.9.2 using convergent sequences. Explicit neighborhoods of boundary points are given in III.9.4. Corners associated to $\mathbb{Q}$-parabolic subgroups are assembled from the boundary components in Proposition

III.9.5. The Hausdorff property of the partial compactification ${}_{\mathbb{Q}}\overline{X}^{BS}$ is proved in III.9.14. The $\mathbf{G}(\mathbb{Q})$-action on X is shown to extend to a continuous action on ${}_{\mathbb{Q}}\overline{X}^{BS}$ in III.9.15. This action is shown to be real analytic with respect to the canonical real analytic structure on ${}_{\mathbb{Q}}\overline{X}^{BS}$ in III.9.16. The action on Γ on ${}_{\mathbb{Q}}\overline{X}^{BS}$ is shown to be proper in III.9.17. Finally, the compactification $\overline{\Gamma\backslash X}^{BS}$ is constructed in III.9.18. Its boundary components are listed in III.9.20.

III.9.1 For any rational parabolic subgroup $\mathbf{P}$, let

$$X = N_P \times A_{\mathbf{P}} \times X_{\mathbf{P}} \tag{III.9.1}$$

be the rational Langlands decomposition of X with respect to P and the basepoint $x_0 = K \in X = G/K$ in §III.1. Define the boundary component $e(\mathbf{P})$ by

$$e(\mathbf{P}) = N_P \times X_{\mathbf{P}}. \tag{III.9.2}$$

Since $X_{\mathbf{P}} = M_{\mathbf{P}}/K_{\mathbf{P}}$, where $K_{\mathbf{P}} = K \cap M_{\mathbf{P}}$, $e(\mathbf{P})$ can also be written as a homogeneous space of P,

$$e(\mathbf{P}) \cong P/A_{\mathbf{P}}K_{\mathbf{P}}, \quad (n, mK_{\mathbf{P}}) \mapsto nmA_{\mathbf{P}}K_{\mathbf{P}}.$$

Since N_P is a normal subgroup of P, $P/A_{\mathbf{P}}K_{\mathbf{P}}$ is an N_P-principal bundle over $P/N_P A_{\mathbf{P}}K_{\mathbf{P}}$, which can be identified with the rational boundary symmetric space $X_{\mathbf{P}}$. Hence, the boundary component $e(\mathbf{P})$ is a principal bundle over $X_{\mathbf{P}}$ with fiber N_P.

Define

$${}_{\mathbb{Q}}\overline{X}^{BS} = X \cup \coprod_{\mathbf{P}} e(\mathbf{P}) = X \cup \coprod_{\mathbf{P}} N_P \times X_{\mathbf{P}}, \tag{III.9.3}$$

where $\mathbf{P}$ runs over all rational parabolic subgroups of $\mathbf{G}$.

III.9.2 The topology of ${}_{\mathbb{Q}}\overline{X}^{BS}$ is defined as follows. Clearly, X and the boundary components $e(\mathbf{P})$ have the natural topology. We need to define convergence of sequences of interior points in X to the boundary points in ${}_{\mathbb{Q}}\overline{X}^{BS}$ and convergence of boundary points.

To describe the convergence of boundary points, we need the relative horospherical decomposition for a pair of rational parabolic subgroups $\mathbf{P}, \mathbf{Q}$ with $\mathbf{P} \subset \mathbf{Q}$. Let $\mathbf{P}'$ be the rational parabolic subgroup in $\mathbf{M_Q}$ corresponding to $\mathbf{P}$ by equation (III.1.13) in §III.1, and let

$$X_{\mathbf{Q}} = N_{P'} \times A_{\mathbf{P}'} \times X_{\mathbf{P}'}$$

be the rational horospherical decomposition of $X_{\mathbf{Q}}$ with respect to $\mathbf{P}'$ in equation (III.1.15) in §III.1. Since

$$N_P = N_Q N_{P'}, X_{\mathbf{P}'} = X_{\mathbf{P}},$$

we have a decomposition of the boundary component $e(\mathbf{Q})$ with respect to $\mathbf{P}$,

$$e(\mathbf{Q}) = N_Q \times X_{\mathbf{Q}} \cong N_P \times A_{\mathbf{P}'} \times X_{\mathbf{P}}. \tag{III.9.4}$$

1. An unbounded sequence y_j in X converges to a boundary point $(n_\infty, z_\infty) \in e(\mathbf{P})$ if and only if in terms of the rational horospherical decomposition of X with respect to $\mathbf{P}$ in equation (III.9.1), $y_j = (n_j, a_j, z_j)$, $n_j \in N_P$, $a_j \in A_{\mathbf{P}}$, $z_j \in X_{\mathbf{P}}$, and the components n_j, a_j, z_j satisfy the following conditions:
   (a) For every $\alpha \in \Phi(P, A_{\mathbf{P}})$, $(a_j)^\alpha \to +\infty$,
   (b) $n_j \to n_\infty$ in N_P,
   (c) and $z_j \to z_\infty \in X_{\mathbf{P}}$.
2. For every pair of rational parabolic subgroups $\mathbf{P}, \mathbf{Q}$ with $\mathbf{P} \subset \mathbf{Q}$, $\mathbf{P} \neq \mathbf{Q}$, $e(\mathbf{P})$ is contained in the boundary of $e(\mathbf{Q})$. Specifically, a sequence y_j in $e(\mathbf{Q})$ converges to a point $(n_\infty, z_\infty) \in e(\mathbf{P})$ if and only if the coordinates of y_j with respect to the decomposition in equation (III.9.4),

$$y = (n_j, a_j', z_j) \in N_P \times A_{\mathbf{P}'} \times X_{\mathbf{P}} = e(\mathbf{Q}),$$

   satisfy the following conditions:
   (a) For every $\alpha \in \Phi(P', A_{\mathbf{P}'})$, $a_j'^\alpha \to +\infty$,
   (b) $n_j \to n_\infty$ in N_P,
   (c) $z_j \to z_\infty$ in $X_{\mathbf{P}}$.

The above two types of convergent sequences are special, and their combinations give the general convergent sequences. It can be seen that they form a convergence class of sequences in the sense of Definition I.8.7 and hence define a topology on ${}_{\mathbb{Q}}\overline{X}^{BS}$. The space ${}_{\mathbb{Q}}\overline{X}^{BS}$ with the above topology is called the Borel–Serre partial compactification.

Remark III.9.3 There does not seem to be a fixed name for such a noncompact space ${}_{\mathbb{Q}}\overline{X}^{BS}$ containing X as a dense, open subset. The term partial compactification is used here to emphasize the fact that it is obtained by compactifying X only along certain directions, i.e., only along those of rational parabolic subgroups. This is similar to partial resolutions of singularities in algebraic geometry. Some other common names are bordifications, enlargements, and completions.

III.9.4 Neighborhoods of boundary points in ${}_{\mathbb{Q}}\overline{X}^{BS}$ can be described explicitly. For $(n, m) \in e(\mathbf{P})$, let U, V be neighborhoods of n, m in $N_P, X_{\mathbf{P}}$ respectively. For any parabolic subgroup $\mathbf{Q} = \mathbf{P}_I$ containing $\mathbf{P}$, let $\mathbf{P}_I'$ be the parabolic subgroup of $\mathbf{M}_{\mathbf{P}_I}$ corresponding to $\mathbf{P}$ in equation (III.1.13) in §III.1, and A_{P_I}' its split component. Then

$$\coprod_I U \times A_{P_I',t} \times V \tag{III.9.5}$$

is a neighborhood of (n, m) in ${}_{\mathbb{Q}}\overline{X}^{BS}$. It can be checked that they define the same topology as one above defined by the convergent sequences. For any sequence $t_j \to +\infty$ and bases of neighborhoods U_j, V_j, the above neighborhoods in equation (III.9.5) form a countable basis for the point (n, m) in ${}_{\mathbb{Q}}\overline{X}^{BS}$.

To understand more directly neighborhoods of the boundary points and to show that ${}_{\mathbb{Q}}\overline{X}^{BS}$ is a manifold with corners, we need to identify the closure of a Siegel set in ${}_{\mathbb{Q}}\overline{X}^{BS}$.

For any rational parabolic subgroup $\mathbf{P}$, let $\Delta = \{\alpha_1, \dots, \alpha_r\}$ be the set of simple roots in $\Phi(P, A_{\mathbf{P}})$. Then $A_{\mathbf{P}}$ can be identified with $\mathbb{R}^r_{>0}$ under the map

$$a \in A_{\mathbf{P}} \mapsto (a^{-\alpha_1}, \dots, a^{-\alpha_r}) \in (\mathbb{R}_{>0})^r \subset \mathbb{R}^r. \tag{III.9.6}$$

The closure of $A_{\mathbf{P}}$ in $\mathbb{R}^r$ under this embedding is denoted by $\overline{A_{\mathbf{P}}}$.

Proposition III.9.5 *The embedding $N_P \times A_{\mathbf{P}} \times X_{\mathbf{P}} \hookrightarrow X \subset {}_{\mathbb{Q}}\overline{X}^{BS}$ can be naturally extended to an embedding $N_P \times \overline{A_{\mathbf{P}}} \times X_{\mathbf{P}} \hookrightarrow {}_{\mathbb{Q}}\overline{X}^{BS}$. The image of $N_P \times \overline{A_{\mathbf{P}}} \times X_{\mathbf{P}}$ in ${}_{\mathbb{Q}}\overline{X}^{BS}$ is denoted by $X(\mathbf{P})$ and called the corner associated with $\mathbf{P}$. Furthermore, $X(\mathbf{P})$ is equal to the subset $X \cup \coprod_{\mathbf{Q} \supseteq \mathbf{P}} e(\mathbf{Q})$ in ${}_{\mathbb{Q}}\overline{X}^{BS}$.*

To prove this proposition, we need to decompose $\overline{A_{\mathbf{P}}}$ according to rational parabolic subgroups containing $\mathbf{P}$. Let $\Delta(P, A_{\mathbf{P}})$ be the set of simple roots in $\Phi(P, A_{\mathbf{P}})$. By §III.1.15, there is a one-to-one correspondence between the rational parabolic subgroups containing $\mathbf{P}$ and the subsets in $\Delta(P, A_{\mathbf{P}})$. For any $I \subset \Delta(P, A_{\mathbf{P}})$, the decomposition

$$A_{\mathbf{P}} = A_{\mathbf{P},\mathbf{P}_I} \times A_{\mathbf{P}_I}$$

in equation (III.1.11) corresponds to the standard coordinate decomposition under the identification in equation (III.9.6). Since the simple roots in $\Delta(P_I, A_{\mathbf{P}_I})$ are restrictions of the simple roots in $\Delta(P, A_{\mathbf{P}})$, it implies that the inclusion

$$A_{\mathbf{P},\mathbf{P}_I} \times A_{\mathbf{P}_I} \subset \overline{A_{\mathbf{P}}}$$

extends to an inclusion

$$A_{\mathbf{P},\mathbf{P}_I} \times \overline{A_{\mathbf{P}_I}} \subset \overline{A_{\mathbf{P}}}.$$

Let $o_{\mathbf{P}_I}$ be the origin of $\overline{A_{\mathbf{P}}}$. Then $A_{\mathbf{P},\mathbf{P}_I} \times o_{\mathbf{P}_I}$ is a face of the coordinate quadrant $\overline{A_{\mathbf{P}}} \cong (\mathbb{R}_{\geq 0})^{\Delta}$.

Lemma III.9.6 *The corner $\overline{A_{\mathbf{P}}}$ admits a disjoint decomposition*

$$\overline{A_{\mathbf{P}}} = A_{\mathbf{P}} \cup \coprod_{\mathbf{Q} \supseteq \mathbf{P}} A_{\mathbf{P},\mathbf{Q}} \times o_{\mathbf{Q}}.$$

In this decomposition, a sequence $a_j \in A_{\mathbf{P}}$ converges to $(a_\infty, o_Q) \in A_{\mathbf{P},\mathbf{Q}} \times o_{\mathbf{Q}}$ if and only if in the decomposition $a = (a'_j, a''_j) \in A_{\mathbf{P},\mathbf{Q}} \times A_{\mathbf{Q}}$, $a'_j \to a_\infty$ and $a''_j \to o_{\mathbb{Q}}$ in $\overline{A_{\mathbf{Q}}}$.

Proof. The disjoint decomposition follows from the identification $\overline{A_{\mathbf{P}}} = \mathbb{R}^{\Delta}_{\geq 0}$ and the one-to-one correspondence between proper subsets of Δ and the (proper) rational parabolic subgroups $\mathbf{Q}$ containing $\mathbf{P}$, and the second statement is also clear.

Lemma III.9.7 *For two rational parabolic subgroups $\mathbf{P}, \mathbf{Q}$, $\mathbf{P} \subset \mathbf{Q}$, as above, let $\mathbf{P}'$ be the unique rational parabolic subgroup of $\mathbf{M_Q}$ corresponding to $\mathbf{P}$ in equation (III.1.13). Then*

$$e(\mathbf{Q}) = N_Q \times X_{\mathbf{Q}} \cong N_P \times A_{\mathbf{P}'} \times X_{\mathbf{P}}$$

using the rational horopsherical decomposition of $X_{\mathbf{Q}}$ with respect to $\mathbf{P}'$. Furthermore, $e(\mathbf{Q}) \cong N_P \times A_{\mathbf{P}'} \times X_{\mathbf{P}}$ can be identified with $N_P \times A_{\mathbf{P},\mathbf{Q}} \times X_{\mathbf{P}}$ through the map

$$N_P \times A_{\mathbf{P},\mathbf{Q}} \times X_{\mathbf{P}} \to N_P \times A_{\mathbf{P}'} \times X_{\mathbf{P}} : (n, \exp H,\ m) \mapsto (n,\ \exp H_{\mathbf{P}'},\ m),$$

where $H_{\mathbf{P}'}$ is the component of H in $\mathfrak{a}_{\mathbf{P}'}$ in the decomposition $\mathfrak{a}_{\mathbf{P}} = \mathfrak{a}_{\mathbf{P}'} \oplus \mathfrak{a}_{\mathbf{Q}}$.

Proof. The first statement follows from

$$N_P = N_Q N_{P'}, \quad X_{\mathbf{P}} = X_{\mathbf{P}'}$$

and the horospherical decomposition

$$X_{\mathbf{Q}} = N_{P'} \times A_{\mathbf{P}'} \times X_{P'}.$$

Since $\mathfrak{a}_{\mathbf{P}} = \mathfrak{a}_{\mathbf{P},\mathbf{Q}} \oplus \mathfrak{a}_{\mathbf{Q}}$ and $\mathfrak{a}_{\mathbf{P}} = \mathfrak{a}_{\mathbf{P}'} \oplus \mathfrak{a}_{\mathbf{Q}}$, the map $H \in \mathfrak{a}_{\mathbf{P},\mathbf{Q}} \mapsto H_{\mathbf{P}'} \in \mathfrak{a}_{\mathbf{P}'}$ is a linear isomorphism, and the second statement follows.

Lemma III.9.8 *For any rational parabolic subgroup $\mathbf{Q}$ containing $\mathbf{P}$, under the identification $e(\mathbf{Q}) = N_P \times A_{\mathbf{P},\mathbf{Q}} \times X_{\mathbf{P}}$, a sequence of points y_j in X converges to a boundary point in $e(\mathbf{Q})$ if and only if in the decomposition*

$$X = N_P \times A_{\mathbf{P},\mathbf{Q}} \times A_{\mathbf{Q}} \times X_{\mathbf{P}},$$

the coordinates of $y_j = (n_j, a''_j, a_j, z_j)$ satisfy

1. *$n_j \to n_\infty$ in N_P,*
2. *$a''_j \to a''_\infty$ in $A_{\mathbf{P},\mathbf{Q}}$,*
3. *for all $\alpha \in \Phi(Q, A_{\mathbf{Q}})$, $a_j^\alpha \to +\infty$,*
4. *$z_j \to z_\infty$ in $X_{\mathbf{P}}$,*

and the limit of y_j is equal to $(n_\infty, a''_\infty, z_\infty) \in e(\mathbf{Q})$.

Proof. Since the convergence of points in X to limits $e(\mathbf{Q})$ is defined through the horospherical decomposition of X with respect to $\mathbf{Q}$, we need to relate the above decomposition of X to the horospherical decomposition associated with $\mathbf{Q}$:

$$X = N_Q \times A_{\mathbf{Q}} \times X_{\mathbf{Q}}.$$

Let $\mathbf{P}'$ be the rational parabolic subgroup of $\mathbf{M_Q}$ corresponding to $\mathbf{P}$ as above. Then this horospherical decomposition can be refined as follows:

$$\begin{aligned} X &= N_Q \times A_{\mathbf{Q}} \times X_{\mathbf{Q}} = N_Q \times A_{\mathbf{Q}} \times (N_{P'} \times A_{\mathbf{P}'} \times X_{\mathbf{P}}) \\ &= N_P \times A_{\mathbf{P}'} \times A_{\mathbf{Q}} \times X_{\mathbf{Q}}, \end{aligned} \tag{III.9.7}$$

where we have used the fact that $A_{\mathbf{Q}}$ commutes with $N_{P'}$ and $A_{\mathbf{P}'}$.

As in the previous lemma, for any $H \in \mathfrak{a}_{\mathbf{P},\mathbf{Q}}$, write

$$H = H_{\mathbf{P}'} + H_{\mathbf{Q}},$$

where $H_{\mathbf{P}'} \in \mathfrak{a}_{\mathbf{P}'}$ and $H_{\mathbf{Q}} \in \mathfrak{a}_{\mathbf{Q}}$. Then the map

$$N_P \times A_{\mathbf{P},\mathbf{Q}} \times A_{\mathbf{Q}} \times X_{\mathbf{P}} \to N_P \times A_{\mathbf{P}'} \times A_{\mathbf{Q}} \times X_{\mathbf{Q}} \tag{III.9.8}$$

is given by

$$(n,\ \exp H,\ \exp V,\ z) \mapsto (n,\ \exp H_{\mathbf{P}'}, \exp(H_{\mathbf{Q}} + V),\ z). \tag{III.9.9}$$

Similarly, for any $H \in \mathfrak{a}_{\mathbf{P}'}$, write $H = H_{\mathbf{P},\mathbf{Q}} + H_{\mathbf{Q}}$, where $H_{\mathbf{P},\mathbf{Q}} \in \mathfrak{a}_{\mathbf{P},\mathbf{Q}}$, $H_{\mathbf{Q}} \in \mathfrak{a}_{\mathbf{Q}}$. Then the transformation

$$N_P \times A_{\mathbf{P}'} \times A_{\mathbf{Q}} \times X_{\mathbf{Q}} \to N_P \times A_{\mathbf{P},\mathbf{Q}} \times A_{\mathbf{Q}} \times X_{\mathbf{P}} \tag{III.9.10}$$

is given by

$$(n,\ \exp H,\ \exp V,\ m) \mapsto (n,\ \exp H_{P,Q}, \exp(H_Q + V),\ m). \tag{III.9.11}$$

These two formulae of coordinate changes imply the lemma.

Lemma III.9.9 *For a pair of parabolic subgroups $\mathbf{P} \subset \mathbf{Q}$, let $I \subset \Delta(P, A_{\mathbf{P}})$ be the subset such that $\mathbf{Q} = \mathbf{P}_I$. Then under the identification*

$$e(\mathbf{Q}) \cong N_P \times A_{\mathbf{P},\mathbf{Q}} \times X_{\mathbf{P}}$$

in Lemma III.9.7, a sequence $y_j = (n_j,\ a_j,\ z_j)$ in $e(\mathbf{Q})$ converges to a point $(n_\infty,\ z_\infty) \in e(\mathbf{P}) = N_P \times X_{\mathbf{P}}$ if and only if for all $\beta \in \Delta(P, A_{\mathbf{P}}) \setminus I$, $a_j^\beta \to +\infty$, and $n_j \to n_\infty$, $z_j \to z_\infty$.

Proof. For all $H \in \mathfrak{a}_{\mathbf{Q}}$ and $\beta \in \Delta(P, A_{\mathbf{P}}) \setminus I$, $\beta(H) = 0$. This implies that for all $\beta \in \Delta(P, A_{\mathbf{P}}) \setminus I$ and $H \in \mathfrak{a}_{\mathbf{P},\mathbf{Q}}$, $\beta(H_{P'}) = \beta(H)$, where $H_{P'}$ is the component of H in $\mathfrak{a}_{\mathbf{P}'}$ under the direct sum $\mathfrak{a}_P = \mathfrak{a}_{P'} \oplus \mathfrak{a}_Q$. Since the simple roots in $\Phi(P', A_{\mathbf{P}'})$ are restrictions of $\Delta(P, A_P) \setminus I$ to $\mathfrak{a}_{P'}$, the lemma is clear.

More generally, the following lemma is true.

Lemma III.9.10 *Let $\mathbf{Q}_1, \mathbf{Q}_2$ be two rational parabolic subgroups containing $\mathbf{P}$. Suppose that $\mathbf{Q}_1 \subset \mathbf{Q}_2$. Let I_j be the subset of the simple roots in $\Delta(P, A_{\mathbf{P}})$ such that $\mathbf{Q}_i = \mathbf{P}_{I_i}$, $i = 1, 2$. Under the identifications*

$$e(\mathbf{Q}_1) \cong N_P \times A_{\mathbf{P},\mathbf{Q}_1} \times X_{\mathbf{P}},$$

$$e(\mathbf{Q}_2) \cong N_P \times A_{\mathbf{P},\mathbf{Q}_1} \times A_{\mathbf{Q}_1,\mathbf{Q}_2} \times X_{\mathbf{P}},$$

a sequence of points $y_j = (n_j, a_{P,Q_1,j}, a_{Q_1,Q_2,j}, m_j)$ in $e(\mathbf{Q}_2)$ converges in ${}_{\mathbb{Q}}\overline{X}$ to a point $(n_\infty, a_{P,Q_1,\infty}, m_\infty) \in e(\mathbf{Q}_1)$ if and only if $n_j \to n_\infty$, $a_{P,Q_1,j} \to a_{P,Q_1,\infty}$, $m_j \to m_\infty$, and for all $\alpha \in I_2 \setminus I_1$, $(a_{Q_1,Q_2,j})^\alpha \to +\infty$.

III.9.11 *Proof of Proposition III.9.5.* By Lemmas III.9.7 and III.9.8, the subset $N_P \times A_{\mathbf{P},\mathbf{Q}} \times o_{\mathbf{Q}} \times X_{\mathbf{P}}$ in $N_P \times \overline{A_{\mathbf{P}}} \times X_{\mathbf{P}}$ can be identified with $e(\mathbf{Q})$, and under this identification, convergence of sequences of interior points to points in $e(\mathbf{Q})$ in the topology of $N_P \times \overline{A_{\mathbf{P}}} \times X_{\mathbf{P}}$ is the same as in the topology of ${}_{\mathbb{Q}}\overline{X}^{BS}$. By Lemma III.9.9, under this identification, the convergence of sequences of points in $e(\mathbf{Q})$ to points in $e(\mathbf{P})$ in the topology of $N_P \times \overline{A_{\mathbf{P}}} \times X_{\mathbf{P}}$ is the same as the convergence in the topology of ${}_{\mathbb{Q}}\overline{X}^{BS}$. Similarly, by Lemma III.9.10, for any two boundary faces $e(\mathbf{Q}_1)$, $e(\mathbf{Q}_2)$ with $\mathbf{Q}_1 \subset \mathbf{Q}_2$, the convergence of sequences of points in $e(\mathbf{Q}_2)$ to points in $e(\mathbf{Q}_1)$ is the same in both topologies.

This implies that the embedding $N_P \times A_{\mathbf{P}} \times X_{\mathbf{P}} \hookrightarrow {}_{\mathbb{Q}}\overline{X}^{BS}$ can be extended to an embedding

$$N_P \times \overline{A_{\mathbf{P}}} \times X_{\mathbf{P}} \hookrightarrow {}_{\mathbb{Q}}\overline{X}^{BS},$$

and the image of $N_P \times \overline{A_{\mathbf{P}}} \times X_{\mathbf{P}}$ in ${}_{\mathbb{Q}}\overline{X}^{BS}$ is equal to $X \cup \coprod_{\mathbf{Q} \supseteq \mathbf{P}} e(\mathbf{Q})$.

Remark III.9.12 Proposition III.9.5 says that for any rational parabolic subgroup $\mathbf{P}$, all the boundary faces $e(\mathbf{Q})$ for $\mathbf{Q}$ containing $\mathbf{P}$ form a corner $X(\mathbf{P})$ in ${}_{\mathbb{Q}}\overline{X}^{BS}$. This implies that each corner $X(\mathbf{P})$ is an open subset, and the corner $X(\mathbf{Q})$ for any $\mathbf{Q} \supset \mathbf{P}$ is contained in $X(\mathbf{P})$ as an open subset. On the other hand, the boundary component $e(\mathbf{P})$ is contained in the closure of $e(\mathbf{Q})$.

Recall from equation (III.1.16) in §III.1 that for any $t > 0$, $A_{\mathbf{P},t}$ is defined by

$$A_{\mathbf{P},t} = \{a \in A_{\mathbf{P}} \mid a^\alpha > t,\ \alpha \in \Delta(P, A_{\mathbf{P}})\}.$$

Define

$$\overline{A_{\mathbf{P},t}} = \{a \in \overline{A_{\mathbf{P}}} \mid a^\alpha > t,\ \alpha \in \Delta(P, A_{\mathbf{P}})\},$$

where a^α could be equal to $+\infty$. The space $\overline{A_{\mathbf{P},t}}$ is a partial compactification of $A_{\mathbf{P},t}$ in the direction of $\mathbf{P}$, and is equal to the interior of the closure of $A_{\mathbf{P},t}$ in $\overline{A_{\mathbf{P}}}$.

Lemma III.9.13 *For any point $(n, z) \in N_P \times X_{\mathbf{P}} = e(\mathbf{P})$, a neighborhood basis of (n, z) in ${}_{\mathbb{Q}}\overline{X}$ is given by $U \times \overline{A_{\mathbf{P},t}} \times W$, where $n \in U$, $z \in W$ are bases of neighborhoods of $n \in N_P$, $z \in X_{\mathbf{P}}$ respectively, and $t > 0$.*

Proof. Let t be any sufficiently large number. For any interior sequence y_j converging to $(n, m) \in e(\mathbf{P})$, it follows from the definition that $y_j \in U \times A_{\mathbf{P},t} \times W$ eventually. For any rational parabolic subgroup $\mathbf{Q} \supset \mathbf{P}$ and any sequence y_j in $e(\mathbf{Q})$ converging to $(n, m) \in e(\mathbf{P})$, y_j belongs to $U \times A_{\mathbf{P}',t} \times W$ eventually, where $\mathbf{P}'$ is the unique parabolic subgroup of $\mathbf{M_Q}$ corresponding to $\mathbf{P}$ as in equation (III.1.13). By Lemmas III.9.7, III.9.8 and the proof of Lemma III.9.9, $U \times A_{\mathbf{P}',t} \times W$ can be identified with $U \times (A_{\mathbf{P},\mathbf{Q},t} \times \{o_{\mathbf{Q}}\}) \times W$ in $X(\mathbf{P}) = N_P \times \overline{A_{\mathbf{P}}} \times X_{\mathbf{P}}$, where

$$A_{\mathbf{P},\mathbf{Q},t} = \{a \in A_{\mathbf{P},\mathbf{Q}} \mid a^{\beta} > t,\ \beta \in \Delta(P, A_{\mathbf{P}}) \setminus I\},$$

I being the subset of simple roots $\mathbf{P}_I = \mathbf{Q}$. By Lemma III.9.7, $\overline{A_{\mathbf{P},t}} = A_{\mathbf{P},t} \cup \coprod_{\mathbf{Q}\supseteq\mathbf{P}} A_{\mathbf{P},\mathbf{Q},t}$, which implies that every sequence in ${}_{\mathbb{Q}}\overline{X}^{BS}$ converging to (n, m) belongs to $U \times \overline{A_{\mathbf{P},t}} \times W$ eventually. This shows that when $U,\ W$ shrink to $n,\ m$ respectively and $t \to +\infty$, $U \times \overline{A_{\mathbf{P},t}} \times W$ forms a basis of neighborhoods of $(n,\ m)$ in ${}_{\mathbb{Q}}\overline{X}^{BS}$ and hence completes the proof.

Proposition III.9.14 *The partial compactification ${}_{\mathbb{Q}}\overline{X}^{BS}$ is a Hausdorff space.*

Proof. It suffices to show that any two distinct boundary points $y_1,\ y_2 \in {}_{\mathbb{Q}}\overline{X}^{BS} - X$ have disjoint neighborhoods. Let $\mathbf{P}_i$ be the rational parabolic subgroup such that $y_i \in e(\mathbf{P}_i)$. By Lemma III.9.13, for any neighborhood $U_i \times W_i$ of y_i in $e(\mathbf{P}_i)$ and any $t > 0$, $U_i \times \overline{A_{\mathbf{P}_i,t}} \times W_i$ is a neighborhood of y_i in ${}_{\mathbb{Q}}\overline{X}^{BS}$. There are two cases to consider. Suppose first that $P_1 \neq P_2$. By Proposition III.2.19, when $U_i \times W_i$ are bounded and $t \gg 0$, the Siegel sets $U_i \times A_{\mathbf{P}_i,t} \times W_i$, $i = 1,\ 2$, are disjoint. Hence the sets $U_i \times \overline{A_{\mathbf{P}_i,t}} \times W_i$ are also disjoint, since $U_i \times A_{\mathbf{P}_i,t} \times W_i$ is an open dense subset of $U_i \times \overline{A_{\mathbf{P}_i,t}} \times W_i$. On the other hand, suppose that $\mathbf{P}_1 = \mathbf{P}_2$. Then $y_1,\ y_2$ are two distinct points on the same boundary face $e(\mathbf{P}_1)$ and hence have disjoint neighborhoods $U_1 \times W_1$, $U_2 \times W_2$. In particular, $U_1 \times A_{\mathbf{P}_1,t} \times W_1$ and $U_2 \times A_{\mathbf{P}_2,t} \times W_2$ are disjoint. As in the previous case, this implies that $U_i \times \overline{A_{\mathbf{P}_i,t}} \times W_i$ are also disjoint.

Proposition III.9.15 *The $\mathbf{G}(\mathbb{Q})$-action on X extends to a continuous action on ${}_{\mathbb{Q}}\overline{X}^{BS}$. In particular, the arithmetic subgroup Γ acts continuously on ${}_{\mathbb{Q}}\overline{X}^{BS}$.*

Proof. There are two steps in the proof. The first step is to extend the $\mathbf{G}(\mathbb{Q})$-action on X to ${}_{\mathbb{Q}}\overline{X}^{BS}$, and the second step is to prove that this extended action is continuous.

For any rational parabolic subgroup $\mathbf{P}$, and $g \in \mathbf{G}(\mathbb{Q})$, write $g = kp = kman$, where $k \in K$, $p \in P$, $n \in N_P$, $a \in A_{\mathbf{P}}$, $m \in M_{\mathbf{P}}$. Then ${}^k\mathbf{P} = {}^g\mathbf{P}$, and k defines a canonical identification

$$k \cdot e(\mathbf{P}) = e({}^k\mathbf{P}), \quad (n, mK_{\mathbf{P}}) \mapsto ({}^k n, {}^k m K_{{}^k\mathbf{P}}).$$

For any boundary point $(n', z') \in e(\mathbf{P})$, define

$$g \cdot (n', z') = k \cdot ({}^{am}(nn'), mz') \in e({}^{k}\mathbf{P}) = e({}^{g}\mathbf{P}).$$

Note that a, n are uniquely determined by g. Though each of k and m is not uniquely determined by g, the product km is, and hence the above action is well-defined.

To prove that this map is continuous, we first show that if a sequence y_j in X converges to $(n_\infty, z_\infty) \in e(\mathbf{P})$, then gy_j converges to $g \cdot (n_\infty, z_\infty) \in e({}^{g}\mathbf{P})$. Write $y_j = (n_j, a_j, z_j) \in N_P \times A_\mathbf{P} \times X_\mathbf{P}$. Then

$$\begin{aligned} gy_j &= kmann_j a_j z_j = {}^{kma}(nn_j) \cdot kma \cdot a_j z_j \\ &= {}^{kma}(nn_j) \cdot {}^{k}(aa_j) \cdot (k \cdot mz_j) \end{aligned} \tag{III.9.12}$$

with ${}^{kma}(nn_j) \in N_{{}^{k}P}$, ${}^{k}(aa_j) \in A_{{}^{k}\mathbf{P}}$, $k \cdot mz_j \in X_{{}^{k}P}$. Since $n_j \to n_\infty$, $z_j \to z_\infty$, and $(aa_j)^\alpha \to +\infty$ for all $\alpha \in \Phi(P, A_\mathbf{P})$, it is clear that gy_j converges to the point $g({}^{kma}(nn_\infty), k \cdot mz_\infty) = g \cdot (n_\infty, z_\infty) \in e({}^{k}\mathbf{P}) = e({}^{g}\mathbf{P})$.

Suppose that y_j is a sequence in the boundary of ${}_{\mathbb{Q}}\overline{X}^{BS}$ converging $y_\infty \in e(\mathbf{P})$ for some rational parabolic subgroup $\mathbf{P}$. By passing to a subsequence, we can assume that there exists a rational parabolic subgroup $\mathbf{Q}$ containing $\mathbf{P}$ such that $y_j \in e(\mathbf{Q})$.

Write

$$\begin{aligned} y_j &= (n_j, z_j) \in N_Q \times X_\mathbf{Q} = e(\mathbf{Q}), \\ z_j &= (n'_j, a'_j, z'_j) \in N_{P'} \times A_{\mathbf{P}'} \times X_{\mathbf{P}'}. \end{aligned}$$

By definition, the convergence of y_j means that n_j, n'_j, z'_j all converge with limits n_∞, n'_∞, and z'_∞ respectively, and for all $\alpha \in \Phi(P', A_{\mathbf{P}'})$, $(a'_j)^\alpha \to +\infty$. Then the limit y_∞ is given by

$$y_\infty = (n_\infty n'_\infty, z'_\infty).$$

For $g \in \mathbf{G}(\mathbb{Q})$ as above, write $g = kman$ where $k \in K, m \in M_\mathbf{Q}, a \in A_\mathbf{Q}, n \in N_Q$. Then

$$gy_j = g \cdot (n_j, m_j) = ({}^{kma}(nn_j), {}^{k}(mm_j) \cdot k) \in e({}^{g}\mathbf{Q}).$$

To compute the limit of gy_j in $\overline{e({}^{g}\mathbf{Q})}$, we decompose $m = k'm'a'n'$ where $k' \in K \cap M_\mathbf{Q}$, $m' \in M_{\mathbf{P}'}$, $a' \in A_{\mathbf{P}'}$, $n \in N_{P'}$. By computations similar to the above, the limit of gy_j in $\overline{e({}^{g}\mathbf{Q})}$ is equal to

$$({}^{kma}(nn_\infty)^{kk'm'a'}(n'n'_\infty), kk' \cdot (m'm'_\infty)) \in e({}^{g}\mathbf{P}).$$

By a direct computation, this limit is equal to

$$({}^{kk'm'a'}(n')^{kk'm'a'a}(nn_\infty n'_\infty), kk' \cdot (m'm'_\infty)) = ({}^{kk'm'a'a}(n'^{a}nn_\infty n'_\infty), kk' \cdot (m'm'_\infty)).$$

We claim that this limit is equal to gy_∞. In fact, from $g = kman$ and $m = k'm'a'n'$, we obtain

$$g = kk' \cdot m' \cdot a'a \cdot n'^{a} n$$

with $kk' \in K$, $m' \in M_{\mathbf{P}}$, $a'a \in A_{\mathbf{P}}$, and $n'^{a} n \in N_P$. Then the claim follows from the equality $y_\infty = (n_\infty n'_\infty, z'_\infty)$ and the definition of the $\mathbf{G}(\mathbb{Q})$-action on the boundary.

Proposition III.9.16 *The Borel–Serre partial compactification ${}_{\mathbb{Q}}\overline{X}^{BS}$ has a canonical structure of real analytic manifolds with corners, and the $\mathbf{G}(\mathbb{Q})$-action on ${}_{\mathbb{Q}}\overline{X}^{BS}$ is given by real analytic diffeomorphisms.*

Proof. This was proved in Proposition III.5.7. Here is a more explicit proof. For any rational parabolic subgroup $\mathbf{P}$, by Proposition III.9.5, the corner $X(\mathbf{P})$ in ${}_{\mathbb{Q}}\overline{X}^{BS}$ is an open dense subset, and has a canonical structure of a real analytic manifold with corners given by

$$X(\mathbf{P}) \cong N_P \times \overline{A_{\mathbf{P}}} \times X_{\mathbf{P}} \cong N_P \times (\mathbb{R}_{\geq 0})^{\Delta} \times X_{\mathbf{P}}, \tag{III.9.13}$$

where $\Delta = \Delta(P, A_{\mathbf{P}})$. These corners cover ${}_{\mathbb{Q}}\overline{X}^{BS}$, and we need to show that their analytic structures are compatible. For any two rational parabolic subgroups $\mathbf{P}_1, \mathbf{P}_2$, $X(\mathbf{P}_1) \subset X(\mathbf{P}_2)$ if and only if $\mathbf{P}_1 \supset \mathbf{P}_2$. This implies that in general,

$$X(\mathbf{P}_1) \cap X(\mathbf{P}_2) = X(\mathbf{Q}),$$

where $\mathbf{Q}$ is the least group among all the parabolic subgroups containing both $\mathbf{P}_1, \mathbf{P}_2$ and is set to be equal to $\mathbf{G}$ if there is no proper rational parabolic subgroup containing $\mathbf{P}_1, \mathbf{P}_2$. Hence to show the compatibility of the analytic structures of the corners, it suffices to show that for any pair of not necessarily proper rational parabolic subgroups $\mathbf{P}, \mathbf{Q}$, $\mathbf{P} \subset \mathbf{Q}$, the corner $X(\mathbf{Q})$ is included in $X(\mathbf{P})$ as an open analytic submanifold with corners.

When $\mathbf{Q} = \mathbf{G}$, then $X(\mathbf{Q}) = X$. Since the horospherical decomposition $X = N_P \times A_{\mathbf{P}} \times X_{\mathbf{P}}$ is real analytic, and the real analytic structure of $X(\mathbf{P})$ restricts to the analytic structure of $X = X(\mathbf{Q})$.

Assume that both $\mathbf{P}, \mathbf{Q}$ are proper rational parabolic subgroups. Under the inclusion $X(\mathbf{Q}) \subset X(\mathbf{P})$ and the identification of $X(\mathbf{P})$ in equation (III.9.13), $X(\mathbf{Q})$ can be identified with the subset

$$X(\mathbf{Q}) = N_{\mathbf{P}} \times A_{\mathbf{P},\mathbf{Q}} \times \overline{A_{\mathbf{Q}}} \times X_{\mathbf{P}} = N_{\mathbf{P}} \times A_{\mathbf{P},\mathbf{Q}} \times (\mathbb{R}_{\geq 0})^{r} \times X_{\mathbf{P}},$$

where $r = \dim A_{\mathbf{Q}}$, $\alpha_1, \dots, \alpha_r$ are simple roots in $\Delta(Q, A_{\mathbf{Q}})$. On the other hand, the canonical analytic structure of $X(\mathbf{Q})$ is given by

$$X(\mathbf{Q}) = N_Q \times \overline{A_{\mathbf{Q}}} \times X_{\mathbf{Q}} \cong N_P \times A_{\mathbf{P}'} \times (\mathbb{R}_{\geq 0})^{r} \times X_{\mathbf{P}},$$

where $\mathbf{P}'$ is the rational parabolic subgroup in $\mathbf{M}_{\mathbf{Q}}$ corresponding to $\mathbf{P}$ in equation (III.1.13), and the second identification is real analytic since it is obtained from the horospherical decomposition of $X_{\mathbf{Q}}$ with respect to $\mathbf{P}'$. By the formulas of the transformation between the two decompositions of X given in equations (III.9.9, III.9.11), it follows that these two analytic structures are compatible. In fact, the transformation

$$N_{\mathbf{P}} \times A_{\mathbf{P},\mathbf{Q}} \times (\mathbb{R}_{\geq 0})^r \times X_{\mathbf{P}} \to N_P \times A_{\mathbf{P}'} \times (\mathbb{R}_{\geq 0})^r \times X_{\mathbf{P}}$$

is given by

$$(n, \exp H, (t_1, \dots, t_r), z) \mapsto (n, \exp H_{\mathbf{P}'}, (e^{-\alpha_1(H_{\mathbf{Q}})}t_1, \dots, e^{-\alpha_r(H_{\mathbf{Q}})}t_r), z),$$

where $H = H_{\mathbf{P}'} + H_{\mathbf{Q}}$, $H_{\mathbf{P}'} \in \mathfrak{a}_{\mathbf{P}'}$, $H_{\mathbf{Q}} \in \mathfrak{a}_{\mathbf{Q}}$. Since $H_{\mathbf{P}'}$, $H_{\mathbf{Q}}$ are real analytic in $\exp H$, this transformation is real analytic. Similarly, the transformation in the other direction is also real analytic. This proves that ${}_{\mathbb{Q}}\overline{X}^{BS}$ has a canonical real analytic structure.

To show that $\mathbf{G}(\mathbb{Q})$ acts real analytically on ${}_{\mathbb{Q}}\overline{X}^{BS}$, we note that for any $g \in \mathbf{G}(\mathbb{Q})$ and any rational parabolic subgroup $\mathbf{P}$, $g \cdot X(\mathbf{P}) = X({}^g\mathbf{P})$. Write $g = kman$, where $k \in K$, $m \in M_{\mathbf{P}}$, $a \in A_{\mathbf{P}}$, and $n \in N_P$ as above. Then for $x = (n', a', z') \in N_{\mathbf{P}} \times A_{\mathbf{P}} \times X_{\mathbf{P}}$,

$$gx = k \cdot ({}^{ma}(nn'), aa', mz') = ({}^{kma}(nn'), {}^{k}(aa'), k \cdot mz') \in N_{{}^k\mathbf{P}} \times A_{{}^k\mathbf{P}} \times X_{{}^k\mathbf{P}}.$$

This implies that the induced map from the corner $X(\mathbf{P})$ to $X({}^g\mathbf{P})$ is real analytic.

Proposition III.9.17 *The arithmetic group Γ acts properly on ${}_{\mathbb{Q}}\overline{X}^{BS}$.*

Proof. Since Γ acts properly on X, there remains to show that a point z on the boundary of ${}_{\mathbb{Q}}\overline{X}^{BS}$ has an open neighborhood V such that

$$\{\gamma \in \Gamma \mid \gamma(V) \cap V \neq \emptyset\}$$

is finite. By Lemma III.9.13, we may take $V = U \times \overline{A}_{\mathbf{P},t} \times W$, where $V' = U \times A_{\mathbf{P},t} \times W$ is a Siegel set in X. In view of the finiteness property of Siegel sets in Proposition III.2.19, it suffices to show that

$$\gamma(V) \cap V \neq \emptyset \quad (\gamma \in \Gamma) \tag{III.9.14}$$

implies

$$\gamma(V') \cap V' \neq \emptyset. \tag{III.9.15}$$

Let y be a point in the set in equation (III.9.14). Since V is open in ${}_{\mathbb{Q}}\overline{X}^{BS}$ and Γ acts continuously, by the previous proposition, this intersection contains an open neighborhood of y. The relation in equation (III.9.15) now follows from the fact that V' is open dense in V.

Theorem III.9.18 *The quotient $\Gamma\backslash_{\mathbb{Q}}\overline{X}^{BS}$ is a compact Hausdorff space. Furthermore, if Γ is torsion-free, it has a canonical structure of a real analytic manifold with corners.*

Proof. Since Γ acts properly on ${}_{\mathbb{Q}}\overline{X}^{BS}$ and ${}_{\mathbb{Q}}\overline{X}^{BS}$ is Hausdorff, the quotient $\Gamma\backslash\overline{X}^{BS}$ is also Hausdorff. To prove that it is compact, we note that by the reduction theory in Proposition III.2.19, there are finitely many rational parabolic subgroups $\mathbf{P}_1, \ldots, \mathbf{P}_k$ and Siegel sets $U_1 \times A_{\mathbf{P}_1,t_1} \times W_1, \ldots, U_k \times A_{\mathbf{P}_k,t_k} \times W_k$ such that the images of these Siegel sets in $\Gamma\backslash X$ cover the whole space $\Gamma\backslash X$. Clearly we can assume that U_i, W_i are compact. Since the closure of $A_{\mathbf{P}_i,t_i}$ in $\overline{A_{P_i}}$ is compact, by Proposition III.9.5, the closure of $U_i \times A_{\mathbf{P}_i,t_i} \times W_i$ is equal to $U_i \times \overline{A_{\mathbf{P}_i,t_i}} \times W_i$ in ${}_{\mathbb{Q}}\overline{X}^{BS}$ and hence is compact. Since X is dense in ${}_{\mathbb{Q}}\overline{X}^{BS}$, the Γ-translates of these compact subsets $U_i \times \overline{A_{\mathbf{P}_i,t_i}} \times W_i$ cover ${}_{\mathbb{Q}}\overline{X}^{BS}$. Hence, the projections of $U_i \times \overline{A_{\mathbf{P}_i,t_i}} \times W_i$, $i = 1, \ldots, k$, project to compact subsets and cover $\Gamma\backslash_{\mathbb{Q}}\overline{X}^{BS}$, which implies that $\Gamma\backslash_{\mathbb{Q}}\overline{X}^{BS}$ is compact. The last statement follows from Proposition III.9.16, and the fact that the action of Γ on ${}_{\mathbb{Q}}\overline{X}^{BS}$ is fixed-point free.

Proposition III.9.19 *The compactification $\Gamma\backslash_{\mathbb{Q}}\overline{X}^{BS}$ constructed in the above theorem is isomorphic to the original compactification $\Gamma\backslash\overline{X}^{BS}$ in §III.5.*

Proof. Since

$$\overline{X}^{BS} = X \cup \coprod_{\mathbf{P}} N_P \times X_{\mathbf{P}} = {}_{\mathbb{Q}}\overline{X}^{BS},$$

they are the same as sets. We need to show that this identity map is a homeomorphism. Note that both $\Gamma\backslash_{\mathbb{Q}}\overline{X}^{BS}$ and $\Gamma\backslash\overline{X}^{BS}$ are compact and Hausdorff. By [GJT, Lemma 3.28], it suffices to show that if an unbounded sequence y_j in $\Gamma\backslash X$ converges to y_∞ in $\Gamma\backslash_{\mathbb{Q}}\overline{X}^{BS}$, then it also converges to y_∞ in $\Gamma\backslash\overline{X}^{BS}$. Suppose that $y_\infty \in \Gamma_P\backslash e(\mathbf{P})$. Let $(n_\infty, z_\infty) \in N_P \times X_{\mathbf{P}}$ be an inverse image of y_∞ in $e(\mathbf{P})$. Then there exists a lift $\tilde{y}_j$ in X such that the horospherical coordinates of $\tilde{y}_j = (n_j, a_j, z_j) \in N_P \times A_{\mathbf{P}} \times X_{\mathbf{P}}$ satisfy

1. $n_j \to n_\infty, z_j \to z_\infty$,
2. and for all $\alpha \in \Phi(P, A_{\mathbf{P}})$, $a_j^\alpha \to +\infty$.

Clearly $\tilde{y}_j$ converges to y_∞ in the corner $X(\mathbf{P}) = X \times_{A_{\mathbf{P}}} \overline{A_{\mathbf{P}}}$, and hence its projection y_j converges to y_∞ in $\Gamma\backslash\overline{X}^{BS}$.

Proposition III.9.20 *Let $\mathbf{P}_1, \ldots, \mathbf{P}_m$ be a set of representatives of Γ-conjugacy classes of rational parabolic subgroups $\mathbf{G}$. Then*

$$\overline{\Gamma\backslash X}^{BS} = \Gamma\backslash X \cup \coprod_{i=1}^{m} \Gamma_{\mathbf{P}_i}\backslash e(\mathbf{P}_i) = \Gamma\backslash X \cup \coprod_{i=1}^{m} \Gamma_{\mathbf{P}_i}\backslash N_{P_i} \times X_{\mathbf{P}_i}.$$

Proof. By the definition of the $\mathbf{G}(\mathbb{Q})$-action on ${}_{\mathbb{Q}}\overline{X}^{BS}$ in Proposition III.9.15, it is clear that for any $g \in \mathbf{G}(\mathbb{Q})$ and any rational parabolic subgroup $\mathbf{P}$, if $g \in P$, then $ge(\mathbf{P}) = e(\mathbf{P})$; otherwise, $g \notin P$, and $ge(\mathbf{P}) \cap e(\mathbf{P}) = \emptyset$. In other words, Γ_P is the stabilizer of $e(\mathbf{P})$ in Γ. Then the proposition follows immediately.

III.9.21 Summary and comments. We applied the uniform method to construct $\overline{\Gamma\backslash X}^{BS}$. In [BS2], the geodesic action and corners $X(\mathbf{P})$ are basic objects. In this approach, the boundary components are the basic ingredients, and the corners are assembled out of them; and the geodesic action is not used. We also gave an explicit proof that $\overline{\Gamma\backslash X}^{BS}$ is a real analytic manifold with corners.

III.10 Uniform construction of the reductive Borel–Serre compactification

In §III.6, the reductive Borel–Serre compactification $\overline{\Gamma\backslash X}^{RBS}$ was defined as a quotient of the Borel–Serre compactification $\overline{\Gamma\backslash X}^{BS}$. In this section, using the uniform method in §III.8, we give a direct construction of the reductive Borel–Serre compactification $\overline{\Gamma\backslash X}^{RBS}$ independently of the Borel–Serre compactification $\overline{\Gamma\backslash X}^{BS}$.

This section is organized as follows. The reductive Borel–Serre boundary component of $\mathbb{Q}$-parabolic subgroups is defined in III.10.1, and the topology of the reductive Borel–Serre partial compactification ${}_{\mathbb{Q}}\overline{X}^{RBS}$ is defined in terms of convergent sequences in III.10.2. Explicit neighborhoods of boundary points are given in terms of Siegel sets in III.10.3, and their separation property in III.10.4. The Hausdorff property of ${}_{\mathbb{Q}}\overline{X}^{RBS}$ is proved in III.10.5, and the action of $\mathbf{G}(\mathbb{Q})$ on X is extended continuously to ${}_{\mathbb{Q}}\overline{X}^{RBS}$ in III.10.6. The relation between ${}_{\mathbb{Q}}\overline{X}^{BS}$ and ${}_{\mathbb{Q}}\overline{X}^{RBS}$ is clarified in III.10.6. The compactification $\overline{\Gamma\backslash X}^{RBS}$ is constructed in III.10.9, and its boundary components are listed in III.10.10. From this, it is clear that $\overline{\Gamma\backslash X}^{RBS}$ is a quotient of $\overline{\Gamma\backslash X}^{BS}$.

III.10.1 For any rational parabolic subgroup $\mathbf{P}$, define the reductive Borel–Serre boundary component $e(\mathbf{P})$ by

$$e(\mathbf{P}) = X_{\mathbf{P}}. \tag{III.10.16}$$

Clearly, it is obtained from the Borel–Serre boundary component $N_P \times X_{\mathbf{P}}$ by collapsing the unipotent factor N_P. Note that $e(\mathbf{P})$ can also be written as a homogeneous space of P as follows:

$$e(\mathbf{P}) = P/N_P A_{\mathbf{P}} K_{\mathbf{P}},$$

where $K_{\mathbf{P}} = K \cap M_{\mathbf{P}}$.

Define

$$ _{\mathbb{Q}}\overline{X}^{RBS} = X \cup \coprod_{\mathbf{P}} e(\mathbf{P}), $$

where **P** runs over all rational parabolic subgroups of **G**. This space with the topology described next is called the reductive Borel–Serre partial compactification of X.

III.10.2 The topology of $_{\mathbb{Q}}\overline{X}^{RBS}$ is given in terms of convergent sequences as follows:

1. For any rational parabolic subgroup **P**, an unbounded sequence $y_j = (n_j, a_j, z_j) \in N_P \times A_{\mathbf{P}} \times X_{\mathbf{P}} = X$ converges to a point $z_\infty \in X_{\mathbf{P}} = e(\mathbf{P})$ if and only if the following two conditions are satisfied:
   (a) For all $\alpha \in \Phi(P, A_{\mathbf{P}})$, $\quad a_j^\alpha \to +\infty$.
   (b) $z_j \to z_\infty$ in $X_{\mathbf{P}}$.

   We note that unlike the case of $_{\mathbb{Q}}\overline{X}^{BS}$, there is no requirement on the N_P-component n_j.
2. For any two rational parabolic subgroups $\mathbf{P} \subset \mathbf{Q}$, the boundary face $e(\mathbf{P})$ is attached at infinity of $e(\mathbf{Q})$. Let $\mathbf{P}'$ be the rational parabolic subgroup of $\mathbf{M_Q}$ corresponding to **P** as in equation (III.1.13), and let

   $$ e(\mathbf{Q}) = X_{\mathbf{Q}} = N_{P'} A_{\mathbf{P}'} X_{\mathbf{P}} $$

   be the rational horospherical decomposition of $X_{\mathbf{Q}}$ with respect to $\mathbf{P}'$. In this decomposition of $e(\mathbf{Q})$, a sequence $y_j = (n_j, a_j, z_j)$ in $e(\mathbf{Q})$ converges to a point $z_\infty \in e(\mathbf{P})$ if and only if the following two conditions are satisfied:
   (a) For all $\alpha \in \Phi(P', A_{\mathbf{P}'})$, $a_j^\alpha \to +\infty$.
   (b) $z_j \to z_\infty$ in $e(\mathbf{P})$.

These are special convergent sequences, and their combinations give the general convergent sequences, which form a convergence class of sequences and hence define a topology on $_{\mathbb{Q}}\overline{X}^{RBS}$.

III.10.3 Neighborhoods of boundary points can also be given explicitly. For a rational parabolic subgroup **P** and a point $z \in e(\mathbf{P})$, let $\mathbf{P}_I$, $I \subset \Delta(P, A_{\mathbf{P}})$, be all the rational parabolic subgroups containing **P**. For each $\mathbf{P}_I$, let $\mathbf{P}'_I$ be the unique parabolic subgroup in $\mathbf{M_{P_I}}$ corresponding to **P** as above. Let W be a neighborhood of z in $e(\mathbf{P}) = X_{\mathbf{P}'_I} = X_{\mathbf{P}}$. Then $N_{P'_I} \times A_{\mathbf{P}'_I,t} \times W$ defines a subset in $e(\mathbf{P}_I)$. The union

$$ N_P \times A_{\mathbf{P},t} \times W \cup \coprod_{I \subset \Delta} N_{P'_I} \times A_{\mathbf{P}'_I,t} \times W \tag{III.10.17} $$

is a neighborhood of z in $_{\mathbb{Q}}\overline{X}^{RBS}$. For sequences $t_j \to +\infty$ and W_j shrinking to z, the above sequence of neighborhoods forms a countable basis at z.

To show that the above topology on $_{\mathbb{Q}}\overline{X}^{RBS}$ is Hausdorff, i.e., every sequence has a unique limit, we need the following separation property of generalized Siegel sets.

Proposition III.10.4

1. *For any bounded set* $W \subset X_{\mathbf{P}}$, *when* $t \gg 0$, *for any* $\gamma \in \Gamma - \Gamma_P$,

$$\gamma(N_P \times A_{\mathbf{P},t} \times W) \cap (N_P \times A_{\mathbf{P},t} \times W) = \emptyset.$$

2. *Let* $\Gamma_{M_{\mathbf{P}}}$ *be the arithmetic subgroup in* $M_{\mathbf{P}}$ *induced from* Γ *in equation (III.6.37). Suppose* W *satisfies the condition that for any nontrivial* $\gamma \in \Gamma_{M_{\mathbf{P}}}$, $\gamma W \cap W = \emptyset$. *Then for any* $\gamma \in \Gamma - \Gamma_{N_P}$,

$$\gamma(N_P \times A_{\mathbf{P},t} \times W) \cap (N_P \times A_{\mathbf{P},t} \times W) = \emptyset.$$

3. *For any two rational parabolic subgroups* $\mathbf{P}_1$, $\mathbf{P}_2$ *that are not conjugate under* Γ, *when* $t \gg 0$,

$$\gamma(N_{P_1} \times A_{\mathbf{P}_1,t} \times W_1) \cap (N_{P_2} \times A_{\mathbf{P}_2,t} \times W_2) = \emptyset$$

for all $\gamma \in \Gamma$.

Proof. These separation properties are generalizations of those stated in Proposition III.2.19, where the results are stated for Siegel sets $U \times A_{\mathbf{P},t} \times W$, where U is a bounded set instead of the whole N_P.

Since Γ_{N_P} acts cocompactly on N_P and the condition $\gamma \in \Gamma - \Gamma_P$ is preserved under multiplication by elements of Γ_{N_P}, (1) and (3) follow immediately from Proposition III.2.19.

To prove (2), we need to show that the separation holds for $\gamma \in \Gamma_P - \Gamma_{N_P}$. We note that for any $\gamma \in \Gamma_P$, $\gamma(N_P \times A_{\mathbf{P},t} \times W) = N_P \times A_{\mathbf{P},t} \times \gamma_M W$, where γ_M is the image of γ under the projection

$$\Gamma_P \subset P = N_P \times A_P \times M_{\mathbf{P}} \to \Gamma_{M_{\mathbf{P}}} \subset M_{\mathbf{P}}.$$

(See equation III.6.37.) If $\gamma \in \Gamma - \Gamma_{N_P}$, then γ_M is nontrivial, and by the assumption on W, $\gamma_M W \cap W \neq \emptyset$, and hence

$$\gamma(N_P \times A_{\mathbf{P},t} \times W) \cap N_P \times A_{\mathbf{P},t} \times W = N_P \times A_{\mathbf{P},t} \times \gamma_M W \cap N_P \times A_{\mathbf{P},t} \times W = \emptyset.$$

Proposition III.10.5 *Every convergent sequence in* ${}_{\mathbb{Q}}\overline{X}^{RBS}$ *has a unique limit, and hence the topology on* ${}_{\mathbb{Q}}\overline{X}^{RBS}$ *defined above is Hausdorff.*

Proof. Since every boundary face $e(\mathbf{P})$ is contained in the closure of only finitely many boundary faces $e(\mathbf{Q})$, it suffices to consider unbounded sequences in a fixed boundary face $e(\mathbf{Q})$. Let y_j be an unbounded sequence in $e(\mathbf{Q})$ converging to a limit $y_\infty \in e(\mathbf{P}_1)$ for a rational parabolic subgroup $\mathbf{P}_1$ contained in $\mathbf{Q}$. Suppose y_j converges to another limit $y'_\infty \in e(\mathbf{P}_2)$, where $\mathbf{P}_2$ is a rational parabolic subgroup contained in $\mathbf{Q}$. We claim that $\mathbf{P}_2 = \mathbf{P}_1$.

Denote the rational parabolic subgroups of $\mathbf{M_Q}$ corresponding to $\mathbf{P}_1$ and $\mathbf{P}_2$ by $\mathbf{P}'_1$ and $\mathbf{P}'_2$ respectively as in equation (III.1.13). By definition, for any bounded neighborhood W_1 of y_∞ in $e(\mathbf{P}_1)$ and $t > 0$, when $j \gg 0$,

$$y_j \in N_{P'_1} \times A_{\mathbf{P}'_1,t} \times W_1.$$

Similarly, for such a neighborhood W_2 of y'_∞ in $e(\mathbf{P}_2)$, when $j \gg 0$,

$$y_j \in N_{P'_2} \times A_{\mathbf{P}'_2,t} \times W_2.$$

If the claim is not true, i.e., $\mathbf{P}'_2 \neq \mathbf{P}'_1$, then Proposition III.10.4, applied to $\mathbf{M_Q}$ and the pair of parabolic subgroups $\mathbf{P}'_1, \mathbf{P}'_2$, shows that $N_{P'_1} \times A_{\mathbf{P}'_1,t} \times W_1$ is disjoint from $N_{P'_2} \times A_{\mathbf{P}'_2,t} \times W_2$. This contradiction proves the claim.

Now $y_\infty, y'_\infty \in e(\mathbf{P}_1)$. Since the coordinates of $y_j = (n_j, a_j, z_j)$ in $N_{P_1} \times A_{\mathbf{P}_1} \times X_{\mathbf{P}_1} = e(\mathbf{Q})$ are uniquely determined by y_j, $\lim_{j\to+\infty} z_j$ has a unique limit if it exists. This implies that $y_\infty = y'_\infty = \lim_{j\to+\infty} z_j$, and hence the sequence y_j has a unique limit.

Proposition III.10.6 *The* $\mathbf{G}(\mathbb{Q})$*-action on* X *extends to a continuous action on* ${}_{\mathbb{Q}}\overline{X}^{RBS}$*, in particular,* Γ *acts continuously on* ${}_{\mathbb{Q}}\overline{X}^{RBS}$.

Proof. First we define the extended action. For any $g \in \mathbf{G}(\mathbb{Q})$ and $z \in e(\mathbf{P}) = X_\mathbf{P}$, write $g = kman$, where $k \in K, m \in M_\mathbf{P}, a \in A_\mathbf{P}, n \in N_P$. Define

$$g \cdot z = k \cdot (mz) \in X_{{}^k\mathbf{P}} = e({}^g\mathbf{P}).$$

We note that km is uniquely determined by g, and this action is well-defined. The continuity of this extended action can be proved similarly as in Proposition III.9.15.

Proposition III.10.7 *The identity map on* X *extends to a continuous, surjective* $\mathbf{G}(\mathbb{Q})$*-equivariant map* ${}_{\mathbb{Q}}\overline{X}^{BS} \to {}_{\mathbb{Q}}\overline{X}^{RBS}$.

Proof. For every rational parabolic subgroup $\mathbf{P}$, define a projection

$$\pi : N_P \times X_\mathbf{P} \to X_\mathbf{P}, \quad (n, z) \mapsto z.$$

This projection is clearly P-equivariant. Extending the identity map on X by the map π on the boundary components, we get a $\mathbf{G}(\mathbb{Q})$-equivariant surjective map

$$\pi : {}_{\mathbb{Q}}\overline{X}^{BS} \to {}_{\mathbb{Q}}\overline{X}^{RBS}.$$

We claim that this map π is continuous. Let $y_j \to y_\infty$ be a convergent sequence in ${}_{\mathbb{Q}}\overline{X}^{BS}$. We need to show that $\pi(y_j) \to \pi(y_\infty)$ in ${}_{\mathbb{Q}}\overline{X}^{RBS}$. It suffices to consider two cases:

1. $y_\infty \in N_P \times X_\mathbf{P}$ for some $\mathbf{P}$, and $y_j \in X$.
2. $y_\infty \in N_P \times X_\mathbf{P}$, and $y_j \in N_Q \times X_\mathbf{Q}$, where $\mathbf{Q} \supset \mathbf{P}$.

Write $y_\infty = (n_\infty, z_\infty)$. In the first case, $y_j = (n_j, a_j, z_j) \in N_P \times A_{\mathbf{P}} \times X_{\mathbf{P}} = X$, and the coordinates satisfy the following conditions:

1. $n_j \to n_\infty$,
2. for all $\alpha \in \Phi(P, A_{\mathbf{P}})$, $a_j^\alpha \to +\infty$,
3. $z_j \to z_\infty$.

These conditions clearly imply that $y_j \to z_\infty \in X_{\mathbf{P}} = e(\mathbf{P})$ in ${}_{\mathbb{Q}}\overline{X}^{RBS}$. The second case can be proved similarly.

To prove that the quotient $\Gamma\backslash{}_{\mathbb{Q}}\overline{X}^{RBS}$ is Hausdorff, we need to identify neighborhoods of boundary points in ${}_{\mathbb{Q}}\overline{X}^{RBS}$.

Lemma III.10.8 *For every point $z \in e(\mathbf{P}) = X_{\mathbf{P}}$, a basis of neighborhood system of z in ${}_{\mathbb{Q}}\overline{X}^{RBS}$ is given by*

$$N_P \times A_{\mathbf{P},t} \times W \cup \coprod_{\mathbf{Q} \supseteq \mathbf{P}} N_{P'} \times A_{\mathbf{P}',t} \times W,$$

where $\mathbf{P}'$ is the parabolic subgroup in $\mathbf{M_Q}$ corresponding to $\mathbf{P}$ in equation (III.1.13), W is a neighborhood of z in $e(\mathbf{P}) = X_{\mathbf{P}}$, $t > 0$, and $e(\mathbf{Q}) = X_{\mathbf{Q}}$ is identified with $N_{P'} \times A_{\mathbf{P}'} \times X_{\mathbf{P}}$. Furthermore, if W is open, then

$$N_P \times A_{\mathbf{P},t} \times W \cup \coprod_{\mathbf{Q} \supseteq \mathbf{P}} N_{P'} \times A_{\mathbf{P}',t} \times W = Int(cl(N_P \times A_{\mathbf{P},t} \times W)),$$

the interior of the closure of $N_P \times A_{\mathbf{P},t} \times W$ in ${}_{\mathbb{Q}}\overline{X}^{RBS}$. In particular, $N_P \times A_{\mathbf{P},t} \times W$ is an open dense subset of the open neighborhood $Int(cl(N_P \times A_{\mathbf{P},t} \times W))$ of z in ${}_{\mathbb{Q}}\overline{X}^{RBS}$.

Proof. The first statement was mentioned earlier in equation (III.10.17). In fact, for any $t > 0$ and any neighborhood W of z, if a sequence y_j in X converges to z in ${}_{\mathbb{Q}}\overline{X}^{RBS}$, then $y_j \in N_P \times A_{\mathbf{P},t} \times W$. Similarly, for any $\mathbf{Q} \supset \mathbf{P}$, if a sequence $y_j \in e(\mathbf{Q}) = X_{\mathbf{Q}}$ converges to z in ${}_{\mathbb{Q}}\overline{X}^{RBS}$, then $y_j \in N_{P'} \times A_{\mathbf{P}',t} \times W$. This implies that any sequence y_j in $\overline{G}^{RBS}$ converging to z belongs to $N_P \times A_{P,t} \times W \cup \coprod_{Q \supseteq P} N_{P'} \times A_{P',t} \times W$ eventually.

To prove the second statement, we note that $X_{\mathbf{Q}} = N_{P'} \times A_{\mathbf{P}'} \times X_{\mathbf{P}}$ can be identified with $N_{P'} \times A_{\mathbf{P},\mathbf{Q}} \times X_{\mathbf{P}}$ as in Lemma III.9.8. Let $cl(N_P \times A_{\mathbf{P},t} \times W)$ be the closure in ${}_{\mathbb{Q}}\overline{X}^{RBS}$. Then the proof of Lemma III.9.9 shows that $cl(N_P \times A_{\mathbf{P},t} \times W) \cap e(\mathbf{Q})$ contains $N_{P'} \times A_{\mathbf{P},\mathbf{Q},t} \times W \cong N_{P'} \times A_{\mathbf{P}',t} \times W$ as a dense open set. This proves the second statement.

Theorem III.10.9 *The quotient $\Gamma\backslash{}_{\mathbb{Q}}\overline{X}^{RBS}$ is a compact Hausdorff space containing $\Gamma\backslash X$ as an open dense subset. This compactification is also denoted by $\overline{\Gamma\backslash X}^{RBS}$.*

Proof. Since Γ does not act properly on ${}_{\mathbb{Q}}\overline{X}^{RBS}$, it is not automatic that the Hausdorff topology of ${}_{\mathbb{Q}}\overline{X}^{RBS}$ induces a Hausdorff topology on the quotient $\Gamma\backslash{}_{\mathbb{Q}}\overline{X}^{RBS}$. Both the Hausdorff property and compactness of the quotient topology on $\Gamma\backslash\overline{G}^{RBS}$ follow from the reduction theory for Γ. We first prove the Hausdorff property.

Let $\phi : {}_{\mathbb{Q}}\overline{X}^{RBS} \to \Gamma\backslash{}_{\mathbb{Q}}\overline{X}^{RBS}$ be the quotient map. For a point $z \in e(\mathbf{P}) \subset {}_{\mathbb{Q}}\overline{X}^{RBS}$, let W be an open neighborhood of z in $e(\mathbf{P}) = X_{\mathbf{P}}$. Let $\mathrm{Int}(cl(N_P \times A_{\mathbf{P},t} \times W))$ be the interior of the closure $cl(N_P \times A_{\mathbf{P},t} \times W)$ in ${}_{\mathbb{Q}}\overline{X}^{RBS}$, which is an open neighborhood of z by Lemma III.10.8. We claim that the image $\phi(\mathrm{Int}(cl(N_P \times A_{\mathbf{P},t} \times W)))$ is an open neighborhood of $\phi(z)$ in $\Gamma\backslash{}_{\mathbb{Q}}\overline{X}^{RBS}$. In fact, the inverse image of $\phi(\mathrm{Int}(cl(N_P \times A_{\mathbf{P},t} \times W)))$ in ${}_{\mathbb{Q}}\overline{X}^{RBS}$ is equal to

$$\cup_{\gamma\in\Gamma}\ \gamma(\mathrm{Int}(cl(N_P \times A_{\mathbf{P},t} \times W))),$$

which is a union of open sets since the Γ-action is continuous, and hence open.

For two different boundary points of ${}_{\mathbb{Q}}\overline{X}^{RBS}$, we need to find two disjoint neighborhoods of them. For any $z \in {}_{\mathbb{Q}}\overline{X}^{RBS}$ and a neighborhood U of $\phi(z)$ in $\Gamma\backslash{}_{\mathbb{Q}}\overline{X}^{RBS}$, the inverse image $\phi^{-1}(U)$ in ${}_{\mathbb{Q}}\overline{X}^{RBS}$ is a Γ-invariant neighborhood of Γz. Therefore, it is equivalent to prove that for any two boundary points $z_1,\ z_2$ in ${}_{\mathbb{Q}}\overline{X}^{RBS}$ with $\Gamma z_1 \cap \Gamma z_2 = \emptyset$, there exist Γ-invariant neighborhoods of $\Gamma z_1,\ \Gamma z_2$ that are disjoint.

Let $\mathbf{P}_1,\ \mathbf{P}_2$ be parabolic subgroups such that $z_1 \in e(\mathbf{P}_1)$, $z_2 \in e(\mathbf{P}_2)$. There are two cases to consider depending on whether $\mathbf{P}_1$ is Γ-conjugate to $\mathbf{P}_2$ or not.

In the latter case, let W_i be a neighborhood of z_i in $e(\mathbf{P}_i)$. By the above discussion,

$$\cup_{\gamma\in\Gamma}\ \gamma\mathrm{Int}(cl(N_{P_i} \times A_{\mathbf{P}_i,t} \times W_i))$$

is a Γ-invariant neighborhood of Γz_i, $i = 1,\ 2$. We claim that when $t \gg 0$, they are disjoint. If not, there exist $\gamma_1,\ \gamma_2 \in \Gamma$ such that

$$\gamma_1\mathrm{Int}(cl(N_{P_1} \times A_{\mathbf{P}_1,t} \times W_1)) \cap \gamma_2\mathrm{Int}(cl(N_{P_2} \times A_{\mathbf{P}_2,t} \times W_2)) \neq \emptyset.$$

Let $\gamma = \gamma_2^{-1}\gamma_1$. Then

$$\gamma\mathrm{Int}(cl(N_{P_1} \times A_{\mathbf{P}_1,t} \times W_1)) \cap \mathrm{Int}(cl(N_{P_2} \times A_{\mathbf{P}_2,t} \times W_2)) \neq \emptyset.$$

By Lemma III.10.8, $\gamma\mathrm{Int}(cl(N_{P_1} \times A_{\mathbf{P}_1,t} \times W_1))$ and $\mathrm{Int}(cl(N_{P_2} \times A_{\mathbf{P}_2,t} \times W_2))$ are open in ${}_{\mathbb{Q}}\overline{X}^{RBS}$ and contain open dense subsets $\gamma(N_{P_1} \times A_{\mathbf{P}_1,t} \times W_1)$ and $N_{P_2} \times A_{\mathbf{P}_2,t} \times W_2$ respectively. It follows that the intersection

$$\gamma\mathrm{Int}(cl(N_{P_1} \times A_{\mathbf{P}_1,t} \times W_1)) \cap \mathrm{Int}(cl(N_{P_2} \times A_{\mathbf{P}_2,t} \times W_2))$$

is open, and hence

$$\gamma(N_{P_1} \times A_{\mathbf{P}_1,t} \times W_1) \cap N_{P_2} \times A_{\mathbf{P}_2,t} \times W_2 \neq \emptyset.$$

But this contradicts Proposition III.10.4.3 and hence proves the claim.

In the former case, assume that $\mathbf{P}_1 = \mathbf{P}_2$ for simplicity. Choose neighborhoods W_1, W_2 of z_1, z_2 such that for all $\gamma \in \Gamma_{M_{\mathbf{P}_1}}$, $\gamma W_1 \cap W_2 = \emptyset$, in particular, $W_1 \cap W_2 = \emptyset$. Let $t \gg 0$. If $\cup_{\gamma \in \Gamma} \gamma \mathrm{Int}(cl(N_{P_i} \times A_{\mathbf{P}_i,t} \times W_i))$, $i = 1, 2$, are not disjoint, then as in the above paragraph, there exists an element $\gamma \in \Gamma$ such that

$$\gamma(N_{P_1} \times A_{\mathbf{P}_1,t} \times W_1) \cap (N_{P_2} \times A_{\mathbf{P}_2,t} \times W_2) \neq \emptyset.$$

We claim that this contradicts Proposition III.10.4.(2). In fact, by Proposition III.10.4.(2), this is impossible if $\gamma \notin \Gamma_{N_{P_1}}$. On the other hand, if $\gamma \in \Gamma_{N_{P_1}}$,

$$\gamma(N_{P_1} \times A_{\mathbf{P}_1,t} \times W_1) = N_{P_1} \times A_{\mathbf{P}_1,t} \times W_1,$$

which is disjoint from $N_{P_2} \times A_{\mathbf{P}_2,t} \times W_2$ since $W_1 \cap W_2 = \emptyset$.

To prove the compactness of $\Gamma \backslash {}_{\mathbb{Q}}\overline{X}^{RBS}$, we note that for every rational parabolic subgroup $\mathbf{P}$ and a compact subset $U \subset N_P$, the closure of $U \times A_{\mathbf{P},t} \times X_{\mathbf{P}}$ in ${}_{\mathbb{Q}}\overline{X}^{RBS}$ is compact. This can be seen either from Lemma III.10.8 or from the fact that the closure of $U \times A_{\mathbf{P},t} \times X_{\mathbf{P}}$ in ${}_{\mathbb{Q}}\overline{X}^{BS}$ is compact and is mapped continuously onto the closure in ${}_{\mathbb{Q}}\overline{X}^{RBS}$. Then the reduction theory in Proposition III.2.19 implies that $\Gamma \backslash {}_{\mathbb{Q}}\overline{X}^{RBS}$ is covered by finitely many compact subsets and hence is compact.

Proposition III.10.10 *Let $\mathbf{P}_1, \dots, \mathbf{P}_m$ be a set of representatives of Γ-conjugacy classes of rational parabolic subgroups* $\mathbf{G}$. *Then*

$$\overline{\Gamma \backslash X}^{RBS} = \Gamma \backslash X \cup \coprod_{i=1}^{m} \Gamma_{M_{\mathbf{P}_i}} \backslash X_{\mathbf{P}_i}.$$

Proof. By the definition of the $\mathbf{G}(\mathbb{Q})$-action on ${}_{\mathbb{Q}}\overline{X}^{RBS}$ in Proposition III.10.6, it is clear that for any $g \in \mathbf{G}(\mathbb{Q})$ and any rational parabolic subgroup $\mathbf{P}$, if $g \in P$, then $gX_{\mathbf{P}} = X_{\mathbf{P}}$; otherwise, $g \notin P$ and $gX_{\mathbf{P}} \cap X_{\mathbf{P}} = \emptyset$. Furthermore, the quotient $\Gamma_P \backslash X_{\mathbf{P}}$ is equal to $\Gamma_{M_{\mathbf{P}}} \backslash X_{\mathbf{P}}$. Then the proposition follows immediately.

Proposition III.10.11 *The identity map on $\Gamma \backslash X$ extends to a continuous map $\overline{\Gamma \backslash X}^{BS} \to \overline{\Gamma \backslash X}^{RBS}$.*

Proof. This follows from Proposition III.10.7.

III.10.12 Summary and comments. The uniform construction of the reductive Borel–Serre compactification $\overline{\Gamma \backslash X}^{RBS}$ in this section is independent of the Borel–Serre compactification $\overline{\Gamma \backslash X}^{BS}$. This is different from the original definition of $\overline{\Gamma \backslash X}^{RBS}$ in [Zu1], where $\overline{\Gamma \backslash X}^{RBS}$ was defined as a quotient of $\overline{\Gamma \backslash X}^{BS}$. Since $\overline{\Gamma \backslash X}^{RBS}$ has played an important role in many problems and is more natural than

$\overline{\Gamma\backslash X}^{BS}$ in some ways, it is desirable to give an independent construction. Unlike $\overline{\Gamma\backslash X}^{BS}$, $\overline{\Gamma\backslash X}^{RBS}$ is not a manifold with corners. Hence, various notions of crumpled corners were introduced in [Zu2].

III.11 Uniform construction of the maximal Satake compactification

In this section we apply the uniform method in §III.8 to construct the maximal Satake compactification of the locally symmetric space $\Gamma\backslash X$. When the maximal Satake compactification $\overline{X}^S_{\max}$ of the symmetric space is geometrically rational, it is the same as the maximal Satake compactification $\overline{\Gamma\backslash X}^S_{\max}$ defined in §III.3. As discussed in §III.3, $\overline{X}^S_{\max}$ is not necessarily always geometrically rational and hence it may not induce a compactification of $\Gamma\backslash X$ by the procedure there. On the other hand, the construction in this section always works. This is one of the major differences between these two approaches.

In this section, we assume that $\mathbf{G}$ is semisimple and hence X is a symmetric space of noncompact type.

The boundary component of $\mathbb{Q}$-parabolic subgroups is given in III.11.1, and the topology of the partial compactification ${}_{\mathbb{Q}}\overline{X}^S_{\max}$ is given in III.11.2. The Hausdorff property is proved in III.11.6, and the continuous extension of the $\mathbf{G}(\mathbb{Q})$-action on X to the partial compactification ${}_{\mathbb{Q}}\overline{X}^S_{\max}$ is proved in III.11.7. The induced compactification $\overline{\Gamma\backslash X}^S_{\max}$ of $\Gamma\backslash X$ is given in III.11.10, which is shown to be isomorphic to the maximal Satake compactification constructed earlier if $\overline{X}^S_{\max}$ is geometrically rational. It is shown in III.11.12 that $\overline{\Gamma\backslash X}^S_{\max}$ is a quotient of $\overline{\Gamma\backslash X}^{RBS}$ (see also III.11.8).

III.11.1 For any rational parabolic subgroup $\mathbf{P}$, define its *maximal Satake boundary component* by

$$e(\mathbf{P}) = X_P,$$

where X_P is the boundary symmetric space associated with the real locus P of $\mathbf{P}$ and is hence a symmetric space of noncompact type when G is semisimple. In general, it is different from the reductive Borel–Serre boundary component $X_{\mathbf{P}}$ unless $A_P = A_{\mathbf{P}}$. In fact, by equation (III.1.8), we have

$$X_{\mathbf{P}} = X_P \times \exp \mathfrak{a}_{\mathbf{P}}^{\perp}. \tag{III.11.18}$$

Define

$${}_{\mathbb{Q}}\overline{X}^S_{\max} = X \cup \coprod_{\mathbf{P}} e(\mathbf{P}) = X \cup \coprod_{\mathbf{P}} X_{\mathbf{P}},$$

where $\mathbf{P}$ runs over all proper rational parabolic subgroupsof $\mathbf{G}$. The space ${}_{\mathbb{Q}}\overline{X}^S_{\max}$ with the topology described below is called the maximal Satake partial compactification.

III.11.2 For any rational parabolic subgroup of $\mathbf{G}$, the rational horospherical decomposition of X with respect to $\mathbf{P}$ can be refined as

$$X = N_P \times A_{\mathbf{P}} \times X_P \times \exp \mathfrak{a}_{\mathbf{P}}^{\perp}. \tag{III.11.19}$$

For any pair of rational parabolic subgroups $\mathbf{P}, \mathbf{Q}$, $\mathbf{P} \subsetneq \mathbf{Q}$, let $\mathbf{P}'$ be the rational parabolic subgroup of $\mathbf{M_Q}$ corresponding to $\mathbf{P}$ in equation (III.1.13). Let $A'_{\mathbf{P}}$ be the $\mathbb{Q}$-split component of the rational Langlands decomposition of $\mathbf{P}'(\mathbb{R})$. On the other hand, when considered as real parabolic subgroups, P corresponds to a real parabolic subgroup P'' of M_Q. In general P'' is different from the real locus $\mathbf{P}'(\mathbb{R})$. In fact, the rational and real Langlands decompositions

$$Q = N_Q A_Q M_Q, \quad Q = N_Q A_{\mathbf{Q}} M_{\mathbf{Q}}$$

are related by

$$A_Q = A_{\mathbf{Q}} \exp \mathfrak{a}_{\mathbf{Q}}^{\perp}, \quad M_{\mathbf{Q}} = M_Q \exp \mathfrak{a}_{\mathbf{Q}}^{\perp},$$

and hence

$$\mathbf{P}'(\mathbb{R}) = P'' \exp \mathfrak{a}_{\mathbf{Q}}^{\perp}.$$

This implies that

$$N_{P'} = N_{P''}, \quad A_{\mathbf{P}'} \subset A_{P''}, \quad X_{P''} = X_P.$$

Let $\mathfrak{a}_{P''}^{\mathbf{P}'}$ be the orthogonal complement of $\mathfrak{a}_{\mathbf{P}'}$ in $\mathfrak{a}_{P''}$. From the real horospherical coordinate decomposition $X_Q = N_{P''} \times A_{P''} \times X_{P''}$, it follows that X_Q admits the following decomposition:

$$X_Q = N_{P'} \times A_{\mathbf{P}'} \times X_P \times \mathfrak{a}_{P''}^{\mathbf{P}'}. \tag{III.11.20}$$

The topology on ${}_{\mathbb{Q}}\overline{X}^S_{\max}$ is given in terms of convergent sequences as follows.

1. For any rational parabolic subgroup $\mathbf{P}$, an unbounded sequence y_j in X converges to $z_\infty \in X_P = e(P)$ if and only if in the decomposition $y_j = (n_j, a_j, z_j, a_j^{\perp}) \in N_P \times A_{\mathbf{P}} \times X_P \times \exp \mathfrak{a}_P^{\perp}$, the coordinates satisfy the following conditions:
   (a) $z_j \to z_\infty$ in X_P,
   (b) for all $\alpha \in \Phi(P, A_{\mathbf{P}})$, $a_j^{\alpha} \to +\infty$.
   (Note that there is no condition on n_j and $a_j^{\perp}$.)
2. For any pair of rational parabolic subgroups $\mathbf{P}, \mathbf{Q}$, $\mathbf{P} \subset \mathbf{Q}$, a sequence of points y_j in $e(\mathbf{Q}) = X_Q$ converges to a point $z_\infty \in e(\mathbf{P}) = X_P$ if and only if the coordinates of y_j in the decomposition in equation (III.11.20), $y_j = (n_j, a'_j, z_j, a''_j)$, satisfy the conditions
   (a) $z_j \to z_\infty$ in X_P,
   (b) for all $\alpha(P', A_{\mathbf{P}'})$, $(a'_j)^{\alpha} \to +\infty$.

These are special convergent sequences, and their combinations give the general convergent sequences. It can be seen that they form a convergence class of sequences and hence define a topology on ${}_{\mathbb{Q}}\overline{X}^S_{\max}$.

III.11.3 Neighborhoods of boundary points in ${}_{\mathbb{Q}}\overline{X}^S_{\max}$ can also be given explicitly. For a rational parabolic subgroup $\mathbf{P}$, let $\mathbf{P}_I$, $I \subset \Delta(P, A_{\mathbf{P}})$, be all the rational parabolic subgroups containing $\mathbf{P}$. For each $\mathbf{P}_I$, let $\mathbf{P}'_I$ be the unique parabolic subgroup in $\mathbf{M}_{\mathbf{P}_I}$ corresponding to $\mathbf{P}$ as above. Let W be a neighborhood of z in $e(\mathbf{P}) = X_P$. Then $N_{P'_I} \times A_{\mathbf{P}'_I,t} \times W \times \exp \mathfrak{a}^{\mathbf{P}'_I}_{P''_I}$ defines a subset in $e(\mathbf{P}_I)$. The union

$$N_P \times A_{\mathbf{P},t} \times W \times \exp \mathfrak{a}^{\perp}_{\mathbf{P}} \cup \coprod_{I \subset \Delta} N_{P'_I} \times A_{\mathbf{P}'_I,t} \times W \times \exp \mathfrak{a}^{\mathbf{P}'_I}_{P''_I} \tag{III.11.21}$$

is a neighborhood of z in ${}_{\mathbb{Q}}\overline{X}^S_{\max}$. For sequences $t_j \to +\infty$ and W_j shrinking to z, the above sequence of neighborhoods forms a countable basis at z.

III.11.4 Once the partial compactification ${}_{\mathbb{Q}}\overline{X}^S_{\max}$ has been defined, the rest of the construction is similar to that of $\overline{\Gamma\backslash X}^{RBS}$.

To show that the above topology on ${}_{\mathbb{Q}}\overline{X}^S_{\max}$ is Hausdorff, i.e., every sequence has a unique limit, we need the following separation property of generalized Siegel sets, which is a generalization of Proposition III.10.4.

For any rational parabolic subgroup $\mathbf{P}$,

$$X_{\mathbf{P}} = X_P \times \exp \mathfrak{a}^{\perp}_{\mathbf{P}}.$$

The subgroup $M_{\mathbf{P}}$ acts on $X_{\mathbf{P}}$ and preserves this product. Denote the induced action of $M_{\mathbf{P}}$ on X_P by $m \cdot z$, for $m \in M_{\mathbf{P}}$, $z \in X_P$. The image of $\Gamma_{M_{\mathbf{P}}}$ in $\mathrm{Isom}(X_P)$ is denoted by Γ_{X_P}. Since $\exp \mathfrak{a}^{\perp}_{\mathbf{P}}$ is the real locus of an anisotropic torus defined over $\mathbb{Q}$, Γ_{X_P} is a discrete subgroup and acts properly on X_P. In fact, Γ_{X_P} is the same as the subgroup defined in §3 when $\overline{X}^S_{\max}$ is geometrically rational, and X_P is a rational boundary component.

Proposition III.11.5

1. *For any bounded set $W \subset X_P$, when $t \gg 0$, for any $\gamma \in \Gamma - \Gamma_P$,*

$$\gamma(N_P \times A_{\mathbf{P},t} \times W \times \exp \mathfrak{a}^{\mathbf{P}'}_{P''}) \cap (N_P \times A_{\mathbf{P},t} \times W \times \exp \mathfrak{a}^{\mathbf{P}'}_{P''}) = \emptyset.$$

2. *Suppose W satisfies the condition that for any nontrivial $\gamma \in \Gamma(X_P)$, $\gamma W \cap W = \emptyset$. Then for any $\gamma \in \Gamma - \Gamma_{N_P}$,*

$$\gamma(N_P \times A_{\mathbf{P},t} \times W \times \exp \mathfrak{a}^{\mathbf{P}'}_{P''}) \cap (N_P \times A_{\mathbf{P},t} \times W \times \exp \mathfrak{a}^{\mathbf{P}'}_{P''}) = \emptyset.$$

3. *For any two rational parabolic subgroups $\mathbf{P}_1$, $\mathbf{P}_2$ that are not conjugate under Γ, when $t \gg 0$,*

$$\gamma(N_{P_1} \times A_{\mathbf{P}_1,t} \times W_1 \times \exp \mathfrak{a}_{P_1''}^{\mathbf{P}_1'}) \cap (N_{P_2} \times A_{\mathbf{P}_2,t} \times W_2 \times \exp \mathfrak{a}_{P_2''}^{\mathbf{P}_2'}) = \emptyset$$

for all $\gamma \in \Gamma$.

Proof. The proof is the same as the proof of Proposition III.10.5 by observing that $\Gamma_{M_{\mathbf{P}}}$ induces a cocompact action on the factor $\exp \mathfrak{a}_{P''}^{\mathbf{P}'}$.

Proposition III.11.6 *Every convergent sequence in* ${}_{\mathbb{Q}}\overline{X}^{S}_{\max}$ *has a unique limit, and hence the topology on* ${}_{\mathbb{Q}}\overline{X}^{S}_{\max}$ *defined above is Hausdorff.*

Proof. The proof is similar to the proof of Proposition III.10.5 by using Proposition III.11.5 instead of Proposition III.10.4.

Proposition III.11.7 *The* $\mathbf{G}(\mathbb{Q})$*-action on* X *extends to a continuous action on* ${}_{\mathbb{Q}}\overline{X}^{S}_{\max}$, *in particular,* Γ *acts continuously on* ${}_{\mathbb{Q}}\overline{X}^{S}_{\max}$.

Proof. First we define the extended action, then show it is continuous. As noted before, for any rational parabolic subgroup $\mathbf{P}$,

$$X_{\mathbf{P}} = X_P \times \exp \mathfrak{a}_{\mathbf{P}}^{\perp},$$

and the subgroup $M_{\mathbf{P}}$ acts on $X_{\mathbf{P}}$ and preserves this product. Denote the induced action of $M_{\mathbf{P}}$ on X_P by $m \cdot z$, for $m \in M_{\mathbf{P}}$, $z \in X_P$.

For any $g \in \mathbf{G}(\mathbb{Q})$, and $z \in e(\mathbf{P}) = X_{\mathbf{P}}$, write $g = kman$, where $k \in K$, $m \in M_{\mathbf{P}}$, $a \in A_{\mathbf{P}}$, $n \in N_P$. Define

$$g \cdot z = k \cdot (mz) \in X_{{}^k P} = e({}^g\mathbf{P}).$$

We note that km is uniquely determined by g, and this action is well-defined.

The continuity of this extended action can be proved as in Proposition III.9.15.

Proposition III.11.8 *The identity map on* X *extends to a continuous, surjective* $\mathbf{G}(\mathbb{Q})$*-equivariant map* ${}_{\mathbb{Q}}\overline{X}^{RBS} \to {}_{\mathbb{Q}}\overline{X}^{S}_{\max}$.

Proof. For every rational parabolic subgroup $\mathbf{P}$, define a projection

$$\pi : X_{\mathbf{P}} = X_P \times \exp \mathfrak{a}_{\mathbf{P}}^{\perp} \to X_P, \quad (z, a^{\perp}) \mapsto z.$$

Extending the identity map on X by the map π on the boundary components, we get a $\mathbf{G}(\mathbb{Q})$-equivariant surjective map

$$\pi : {}_{\mathbb{Q}}\overline{X}^{RBS} \to {}_{\mathbb{Q}}\overline{X}^{S}_{\max}.$$

We claim that this map π is continuous. Let $y_j \to y_\infty$ be a convergent sequence in ${}_{\mathbb{Q}}\overline{X}^{RBS}$. We need to show that $\pi(y_j) \to \pi(y_\infty)$ in ${}_{\mathbb{Q}}\overline{X}^{RBS}$. It suffices to consider two cases:

1. $y_\infty \in X_P$ for some $\mathbf{P}$, and $y_j \in X$.
2. $y_\infty \in X_P$, and $y_j \in X_Q$, where $\mathbf{Q} \supset \mathbf{P}$.

Write $y_\infty = z_\infty$. In the first case, write $y_j = (n_j, a_j, z_j, a_j^\perp) \in N_P \times A_\mathbf{P} \times X_P \times \exp \mathfrak{a}_\mathbf{P}^\perp = X$, and the coordinates satisfy

1. for all $\alpha \in \Phi(P, A_\mathbf{P})$, $a_j^\alpha \to +\infty$,
2. $a_j^\perp$ converges to some $a_\infty^\perp$ in $\exp \mathfrak{a}_\mathbf{P}^\perp$, and $z_j \to z_\infty$.

They clearly imply that $y_j \to z_\infty \in X_P = e(\mathbf{P})$ in ${}_\mathbb{Q}\overline{X}^S_{\max}$. The second case can be proved similarly.

By using similar arguments to those in the proof of Theorem III.10.9, we can prove the following result.

Theorem III.11.9 *The quotient $\Gamma\backslash{}_\mathbb{Q}\overline{X}^S_{\max}$ is a compact Hausdorff space containing $\Gamma\backslash X$ as an open dense subset. This compactification is also denoted by $\overline{\Gamma\backslash X}^S_{\max}$, called the maximal Satake compactification.*

Proposition III.11.10 *Let $\overline{X}^S_\tau$ be a maximal Satake compactification. If $\overline{X}^S_\tau$ is geometrically rational, the induced compactification $\overline{\Gamma\backslash X}^S_\tau$ is isomorphic to the compactification $\Gamma\backslash{}_\mathbb{Q}\overline{X}^S_{\max}$ defined in the previous theorem.*

Proof. If $\overline{X}^S_\tau$ is geometrically rational, then

$$X \cup \partial^* X = X \cup \coprod_{\mathbf{P}} X_\mathbf{P} = {}_\mathbb{Q}\overline{X}^S_{\max}.$$

It follows that $\Gamma\backslash{}_\mathbb{Q}\overline{X}^S$ can be identified with $\overline{\Gamma\backslash X}^S_\tau$. By [GJT, Lemma 3.28], it suffices to prove that if an unbounded sequence y_j in $\Gamma\backslash X$ converges to a boundary point z_∞ in $\Gamma\backslash{}_\mathbb{Q}\overline{X}^S$, it also converges to z_∞ in $\overline{\Gamma\backslash X}^S_\tau$.

Let $\mathbf{P}$ be the rational parabolic subgroup such that z_∞ belongs to the image of X_P. Let $\tilde{z}_\infty \in X_P$ be a lift of z_∞. Then there exists a lift $\tilde{y}_j$ in X such that the coordinates of $\tilde{y}_j = (n_j, a_j, z_j, a_j^\perp) \in N_P \times A_\mathbf{P} \times X_P \times \exp \mathfrak{a}_\mathbf{P}^\perp$ satisfy the following conditions:

1. $n_j, a_j^\perp$ are bounded,
2. $z_j \to \tilde{z}_\infty$,
3. for all $\alpha \in \Phi(P, A_\mathbf{P})$, $a_j^\alpha \to +\infty$.

In particular, $\tilde{y}_j$ belongs to a Siegel set associated with $\mathbf{P}$. Since $\overline{X}^S_\tau$ is geometrically rational, it is Siegel rational, and the limit of this sequence $\tilde{y}_j$ in the closure of the Siegel set (or in $\overline{X}^S_\tau$) is equal to $\tilde{z}_\infty \in X_P$. This shows that the projection y_j of $\tilde{y}_j$ converges to z_∞ in $\overline{\Gamma\backslash X}^S_\tau$.

Proposition III.11.11 *Let $\mathbf{P}_1, \dots, \mathbf{P}_m$ be a set of Γ-conjugacy classes of rational parabolic subgroups of* $\mathbf{G}$. *Then*

$$\overline{\Gamma\backslash X}^S_{\max} = \Gamma\backslash X \cup \coprod_{i=1}^{m} \Gamma_{X_P}\backslash X_P,$$

where Γ_{X_P} is the image of Γ_{M_P} under the projection $M_{\mathbf{P}} = M_P \times \exp \mathfrak{a}_{\mathbf{P}}^{\perp} \to M_P$, and hence also the image of Γ_P under the composed projection

$$\pi : P = N_P A_{\mathbf{P}} M_{\mathbf{P}} \to M_{\mathbf{P}} \to M_P.$$

Proof. It follows from the definition of the $\mathbf{G}(\mathbb{Q})$-action in Proposition III.11.7 that for any $g \in \Gamma$ and rational parabolic subgroup $\mathbf{P}$, if $g \in P$, then $gX_P = X_P$, and the action of g on X_P factors through the projection $\pi(\gamma)$; hence, the quotient $\Gamma_P\backslash X_P$ is equal to $\Gamma_{X_P}\backslash X_P$. On the other hand, if $g \notin P$, then $gX_P \cap X_P = \emptyset$. Then the rest of the proposition is clear.

Proposition III.11.12 *The identity map on $\Gamma\backslash X$ extends to a continuous map*

$$\overline{\Gamma\backslash X}^{RBS} \to \overline{\Gamma\backslash X}^S_{\max}.$$

Proof. The result follows from Proposition III.11.8, and the map can be seen easily from the decompositions of $\overline{\Gamma\backslash X}^{RBS}$ and $\overline{\Gamma\backslash X}^S_{\max}$ in Propositions III.10.10 and III.11.11.

III.11.13 Summary and comments. We applied the uniform method to construct the maximal Satake compactification of $\Gamma\backslash X$. In contrast to the original construction in [Sat2], the issue of the geometrically rationality of $\overline{X}^S_{\max}$ is avoided.

III.12 Tits compactification

In this section, we use the uniform method in §III.8 to construct a compactification $\overline{\Gamma\backslash X}^T$, called the Tits compactification in [JM]. The reason why it is called the Tits compactification is that its boundary is the quotient $\Gamma\backslash\Delta_{\mathbb{Q}}(\mathbf{G})$ of the Tits building $\Delta_{\mathbb{Q}}(\mathbf{G})$ of $\mathbf{G}$. The basic reference of this section is [JM, §8].

The Tits compactification is complementary to the Borel–Serre compactification $\overline{\Gamma\backslash X}^{BS}$. As shown in Proposition III.11.12, $\overline{\Gamma\backslash X}^{RBS}$ dominates $\overline{\Gamma\backslash X}^S_{\max}$, and hence both the reductive Borel–Serre compactification $\overline{\Gamma\backslash X}^{RBS}$ and the maximal Satake compactification $\overline{\Gamma\backslash X}^S_{\max}$ are quotients of the Borel–Serre compactification $\overline{\Gamma\backslash X}^{BS}$. On the other hand, if the $\mathbb{Q}$-rank is greater than or equal to 2, $\overline{\Gamma\backslash X}^T$ is not a quotient of $\overline{\Gamma\backslash X}^{BS}$. In fact, in this case, the *greatest common quotient* of $\overline{\Gamma\backslash X}^T$ and $\overline{\Gamma\backslash X}^{BS}$

is the one-point compactification. In the case of $\mathbb{Q}$-rank 1, the greatest common quotient of $\overline{\Gamma\backslash X}^T$ and $\overline{\Gamma\backslash X}^{BS}$ is the *end compactification*, which is obtained by adding one point to each end.

The Tits compactification $\overline{\Gamma\backslash X}^T$ is isomorphic to the geodesic compactification $\Gamma\backslash X \cup \Gamma\backslash X(\infty)$ in §III.20, a natural generalization of the geodesic compactification $X \cup X(\infty)$ of the symmetric space to the nonsimply connected case, and is useful in understanding the geometry of $\Gamma\backslash X$ at infinity and analysis on $\Gamma\backslash X$.

This section is organized as follows. A concrete realization of the Tits building $\Delta_{\mathbb{Q}}(\mathbf{G})$ of $\mathbf{G}$ is given in III.12.1, which is used to give a topology on the boundary of the partial compactification ${}_{\mathbb{Q}}\overline{X}^T$ of X in III.12.2, where the boundary component of $\mathbb{Q}$-parabolic subgroups is also defined. The topology of ${}_{\mathbb{Q}}\overline{X}^T$ is defined in III.12.3 in terms of convergent sequences. Its Hausdorff property is proved in III.12.4, and the continuous extension of $\mathbf{G}(\mathbb{Q})$ in III.12.5. The Tits compactification $\overline{\Gamma\backslash X}^T$ is constructed in III.12.6. Unlike the Satake compactifications and the reductive Borel–Serre compactification, the compactification $\overline{\Gamma\backslash X}^T$ is not a quotient of $\overline{\Gamma\backslash X}^{BS}$ in general.

III.12.1 For any rational parabolic subgroup $\mathbf{P}$, define the positive chamber

$$\mathfrak{a}_{\mathbf{P}}^+ = \{H \in \mathfrak{a}_{\mathbf{P}} \mid \alpha(H) > 0,\ \alpha \in \Phi(P, A_{\mathbf{P}})\},$$

and

$$\mathfrak{a}_{\mathbf{P}}^+(\infty) = \{H \in \mathfrak{a}_{\mathbf{P}}^+ \mid \langle H, H\rangle = 1\}, \tag{III.12.22}$$

where $\langle\cdot,\cdot\rangle$ is the Killing form, and

$$\overline{\mathfrak{a}_{\mathbf{P}}^+}(\infty) = \{H \in \mathfrak{a}_{\mathbf{P}} \mid \alpha(H) \geq 0,\ \alpha \in \Phi(P, A_{\mathbf{P}})\},$$

a closed simplex.

Clearly, for any pair of rational parabolic subgroups $\mathbf{P}, \mathbf{Q}$, $\mathbf{P} \subset \mathbf{Q}$ if and only if $\mathfrak{a}_{\mathbf{Q}}^+(\infty)$ is a face of $\overline{\mathfrak{a}_{\mathbf{P}}^+}(\infty)$. When $\mathbf{P}$ is a maximal rational parabolic subgroup, $\overline{\mathfrak{a}_{\mathbf{P}}^+}(\infty)$ consists of one point. This implies that $\overline{\mathfrak{a}_{\mathbf{P}}^+}(\infty)$ is a realization of the simplex corresponding to $\mathbf{P}$ in the rational Tits building $\Delta_{\mathbb{Q}}(\mathbf{G})$ in §III.1.

Define a complex

$$\Delta_{\mathbb{Q}}(X) = \cup_{\mathbf{P}}\overline{\mathfrak{a}_{\mathbf{P}}^+}(\infty)/\sim,$$

where $\mathbf{P}$ runs over all the rational parabolic subgroups of $\mathbf{G}$, and the equivalence relation is given by the inclusion above $\overline{\mathfrak{a}_{\mathbf{Q}}^+}(\infty) \subset \overline{\mathfrak{a}_{\mathbf{P}}^+}(\infty)$ for any pair of rational parabolic subgroups $\mathbf{Q} \supset \mathbf{P}$. As commented earlier, this simplicial complex is a realization of the spherical Tits building for $\mathbf{G}$:

$$\Delta_{\mathbb{Q}}(\mathbf{G}) \cong \Delta_{\mathbb{Q}}(X) = \cup_{\mathbf{P}}\overline{\mathfrak{a}_{\mathbf{P}}^+}(\infty)/\sim. \tag{III.12.23}$$

Since the inclusion $\overline{\mathfrak{a}^+_{\mathbf{Q}}(\infty)} \subset \overline{\mathfrak{a}^+_{\mathbf{P}}(\infty)}$ is an embedding, the quotient $\Delta_{\mathbb{Q}}(X)$ has a well-defined quotient topology. As a set $\Delta_{\mathbb{Q}}(X)$ is a disjoint union of the open simplexes

$$\Delta_{\mathbb{Q}}(X) = \coprod_{\mathbf{P}} \mathfrak{a}^+_{\mathbf{P}}(\infty).$$

III.12.2 For any rational parabolic subgroup $\mathbf{P}$, define its Tits boundary component $e(\mathbf{P})$ by

$$e(\mathbf{P}) = \mathfrak{a}^+_{\mathbf{P}}(\infty), \tag{III.12.24}$$

an open simplex. Define

$${}_{\mathbb{Q}}\overline{X}^T = X \cup \coprod_{\mathbf{P}} \mathfrak{a}^+_{\mathbf{P}}(\infty), \tag{III.12.25}$$

which can be identified with

$$X \cup \Delta_{\mathbb{Q}}(X) = X \cup \Delta_{\mathbb{Q}}(\mathbf{G}).$$

The space ${}_{\mathbb{Q}}\overline{X}^T$ with the topology described below is called the *Tits partial compactification*, since the boundary is the Tits building $\Delta_{\mathbb{Q}}(\mathbf{G})$.

III.12.3 The topology on ${}_{\mathbb{Q}}\overline{X}^T$ is described in terms of convergent sequences as follows:

1. For any rational parabolic subgroup $\mathbf{P}$, a unbounded sequence y_j in X converges to a point $H_\infty \in \mathfrak{a}^+_{\mathbf{P}}(\infty)$ if and only if in the horospherical coordinate decomposition $y_j = (n_j, \exp H_j, m_j K_{\mathbf{P}})$, where $n_j \in N_P$, $H_j \in \mathfrak{a}_{\mathbf{P}}$, $m_j \in M_{\mathbf{P}}$, the coordinates satisfy
   (a) $H_j/\|H_j\| \to H_\infty$ in $\mathfrak{a}_{\mathbf{P}}$,
   (b) $d(n_j m_j x_0, x_0)/\|H_j\| \to 0$.
2. For any sequence y_j in the boundary $\coprod_{\mathbf{P}} \mathfrak{a}^+_{\mathbf{P}}(\infty) = \Delta_{\mathbb{Q}}(X)$, it converges to a point $y_\infty \in \mathfrak{a}^+_P(\infty)$ if and only if it converges to y_∞ with respect to the quotient topology in equation (III.12.23).

These are special convergent sequences, and their combinations give general convergent sequences. It can be shown that they form a convergence class of sequences (see [JG, Lemma 8.4]) and hence defines a topology on ${}_{\mathbb{Q}}\overline{X}^T$.

Proposition III.12.4 *The Tits partial compactification* ${}_{\mathbb{Q}}\overline{X}^T$ *is a Hausdorff space.*

Proof. We claim that for any unbounded sequence $\{y_n\}$ in X, there is at most one rational parabolic subgroup $\mathbf{P}$ such that y_j converges to a point in $\mathfrak{a}^+_{\mathbf{P}}(\infty)$.

To prove this claim, we recall the geodesic compactification $X \cup X(\infty)$ in §I.2. The boundary $X(\infty)$ has a simplicial structure $\Delta(G)$, called the spherical

Tits building of X (see Propositions I.2.19 and I.2.16). The rational Tits building $\Delta_{\mathbb{Q}}(\mathbf{G}) = \Delta_{\mathbb{Q}}(X)$ is embedded in $\Delta(G)$ as follows. For each real parabolic subgroup P of G, let A_P be the maximal real split torus in P. Then

$$\mathfrak{a}_P^+(\infty) = \{H \in \mathfrak{a}_P \mid \alpha(H) > 0, \alpha \in \Phi(P, A_P)\}$$

can be identified with a subset of $X(\infty)$, and $X(\infty) = \coprod_P A_P^+(\infty)$. For any rational parabolic subgroup $\mathbf{P}$ of $\mathbf{G}$, $P = \mathbf{P}(\mathbb{R})$ is a real parabolic subgroup of G. The maximal real split torus A_P of P contains the maximal rational split torus $A_{\mathbf{P}}$, and $\mathfrak{a}_{\mathbf{P}}^+ \subseteq \overline{\mathfrak{a}_P^+}$. Therefore, $\mathfrak{a}_{\mathbf{P}}^+(\infty) \subseteq \overline{\mathfrak{a}_P^+(\infty)} \subset X(\infty)$.

We can check easily that if a sequence y_j converges to $H_\infty \in \mathfrak{a}_{\mathbf{P}}^+(\infty)$ in ${}_{\mathbb{Q}}\overline{X}^T$, then it converges to H_∞ in the geodesic compactification $X \cup X(\infty)$. Since the compactification $X \cup X(\infty)$ is Hausdorff and $\mathfrak{a}_{\mathbf{P}'}^+(\infty) \cap \mathfrak{a}_{\mathbf{P}}^+(\infty) = \emptyset$ for two different rational parabolic subgroups $\mathbf{P}'$ and $\mathbf{P}$, the claim is proved.

Using the claim and the Hausdorff property of X and $\Delta_{\mathbb{Q}}(X)$, we can prove easily that every convergent sequence in ${}_{\mathbb{Q}}\overline{X}^T$ has a unique limit.

Proposition III.12.5 *The* $\mathbf{G}(\mathbb{Q})$*-action on* X *extends to a continuous action on* ${}_{\mathbb{Q}}\overline{X}^T$.

Proof. First we define the action on ${}_{\mathbb{Q}}\overline{X}^T$. Then show that it is continuous. For any rational parabolic subgroup $\mathbf{P}$, $g \in \mathbf{G}(\mathbb{Q})$, write $g = kp$, where $k \in K$ and $p \in P$. Then $Ad(k)$ gives a canonical identification between $\mathfrak{a}_{\mathbf{P}}^+(\infty)$ and $\mathfrak{a}_{{}^k\mathbf{P}}^+(\infty) = \mathfrak{a}_{{}^g\mathbf{P}}^+$. For any $H \in \mathfrak{a}_{\mathbf{P}}^+(\infty)$, define the action by

$$g \cdot H = Ad(k)H \in \mathfrak{a}_{{}^k\mathbf{P}}^+ = \mathfrak{a}_{{}^g\mathbf{P}}^+.$$

Note that k is determined by g up to an element in $K_{\mathbf{P}}$, and hence this action is well-defined.

Since $\mathbf{G}(\mathbb{Q})$ acts simplicially on the Tits building $\Delta_{\mathbb{Q}}(\mathbf{G})$ and hence by homeomorphisms, it suffices to prove that if y_j is an unbounded sequence in X converging to $H_\infty \in \mathfrak{a}_{\mathbf{P}}^+(\infty)$ for some $\mathbf{P}$, then gy_j converges to $g \cdot H_\infty$ for any $g \in \mathbf{G}(\mathbb{Q})$. Write $y_j = (n_j, \exp H_j, m_j K_{\mathbf{P}}) \in N_P \times A_{\mathbf{P}} \times X_{\mathbf{P}}$, and $g = kp$, $p = man$, where $k \in K, m \in M_{\mathbf{P}}, a \in A_{\mathbf{P}}, n \in N_{\mathbf{P}}$. Then

$$py_j = ({}^{ma}(nn_j), a \exp H_j, mm_j K_{\mathbf{P}}).$$

It can be checked easily that py_j converges to H_∞ in ${}_{\mathbb{Q}}\overline{X}^T$. Since the conjugation by k transfers the horospherical decomposition with respect to $\mathbf{P}$ to that of ${}^k\mathbf{P}$, it follows that $gy_j = k(py_j)$ converges to $g \cdot H_\infty = k \cdot H_\infty$ in ${}_{\mathbb{Q}}\overline{X}^T$.

Proposition III.12.6 *The arithmetic subgroup* Γ *acts continuously on* ${}_{\mathbb{Q}}\overline{X}^T$ *with a Hausdorff compact quotient, which contains* $\Gamma\backslash X$ *as an open dense subspace. The compactification* $\Gamma\backslash{}_{\mathbb{Q}}\overline{X}^T$ *is called the Tits compactification and is also denoted by* $\overline{\Gamma\backslash X}^T$.

Proof. By Proposition III.3.2, it suffices to construct a subset $\overline{\Sigma} \subset {}_{\mathbb{Q}}\overline{X}^T$ satisfying the conditions there.

Let $\mathbf{P}$ be a minimal rational parabolic subgroup of $\mathbf{G}$, and let Σ be the finite union $C\mathcal{S}$ of the Siegel sets in Proposition III.2.16, where $\mathcal{S} = U \times A_{\mathbf{P},t} \times V$. We can assume that $U \subset N_P$, $V \subset X_{\mathbf{P}}$ are compact. Let $\overline{\Sigma}$ be the closure of Σ in the partial compactification ${}_{\mathbb{Q}}\overline{X}^T$. We claim that $\overline{\Sigma}$ satisfies all three conditions in Proposition III.3.2.

Let $\overline{\mathcal{S}}$ be the closure of $\mathcal{S}$ in $X \cup \Delta_{\mathbb{Q}}(X)$. Then $\overline{\mathcal{S}} \supset \overline{\mathfrak{a}_{\mathbf{P}}^+(\infty)}$, because the sequence $\exp(nH), n \geq 1$, converges to $H \in \mathfrak{a}_{\mathbf{P}}^+(\infty)$. For any element $g \in \mathbf{G}(\mathbb{Q})$, by the proof of the previous lemma, there exists a Siegel domain $\mathcal{S}'$ associated with the minimal rational parabolic subgroup $g\mathbf{P}g^{-1}$ such that $g\mathcal{S} \supset \mathcal{S}'$, and hence $g\overline{\mathcal{S}} \supset \overline{\mathfrak{a}^+{}_{g\mathbf{P}g^{-1}}(\infty)}$. Since any minimal rational parabolic subgroup is Γ conjugate to one of the groups $g\mathbf{P}g^{-1}$, $g \in C$, it follows that $\Gamma\overline{\Sigma} = X \cup \Delta_{\mathbb{Q}}(X)$, and hence condition (1) is satisfied.

To show that $\overline{\Sigma}$ is compact, it suffices to show that $\overline{\mathcal{S}}$ is compact. From (3) of the definition of the topology in (III.12.3), it follows that $\overline{\mathcal{S}} = U \times \overline{A_{\mathbf{P},t}} \times V \cup \overline{\mathfrak{a}_{\mathbf{P}}^+(\infty)}$, where $\overline{A_{\mathbf{P},t}}$ is the closure of $A_{\mathbf{P},t}$ in $A_{\mathbf{P}}$. Since U, V are compact, $\overline{\mathcal{S}}$ is compact.

To check the condition (3), we note that if for some $\gamma \in \Gamma$, $\gamma\overline{\Sigma} \cap \overline{\Sigma} \neq \emptyset$, then for some $g_1, g_2 \in C$, $\gamma g_1\overline{\mathcal{S}} \cap g_2\overline{\mathcal{S}} \neq \emptyset$. Thus it suffices to show that there exist finitely many $\gamma_i \in \Gamma$ such that if $\gamma \in \Gamma$ and $\gamma g_1\overline{\mathcal{S}} \cap g_2\overline{\mathcal{S}} \neq \emptyset$, then $\gamma|_{g_1\overline{\mathcal{S}}\cap\gamma^{-1}g_2\overline{\mathcal{S}}} = \gamma_i|_{g_1\overline{\mathcal{S}}\cap\gamma^{-1}g_2\overline{\mathcal{S}}}$ for some γ_i. Assume $\gamma g_1\overline{\mathcal{S}} \cap g_2\overline{\mathcal{S}} \neq \emptyset$. If $\gamma g_1\mathcal{S} \cap g_2\mathcal{S} \neq \emptyset$, then by Proposition III.2.16, there are only finitely many such γ in Γ. Otherwise, by the previous paragraph, $\gamma g_1\overline{\mathcal{S}} \cap g_2\overline{\mathcal{S}} = \overline{\mathfrak{a}^+{}_{\mathbf{P}'_I}(\infty)}$ for some rational parabolic subgroup $\mathbf{P}'_I$ containing the minimal rational parabolic subgroup $\mathbf{P}' = g_2\mathbf{P}g_2^{-1}$. Since

$$\gamma g_1 g_2^{-1}\overline{\mathfrak{a}^+{}_{\mathbf{P}'}(\infty)} \cap \overline{\mathfrak{a}^+{}_{\mathbf{P}'}(\infty)} = (\gamma g_1 g_2^{-1})g_2\overline{\mathcal{S}} \cap g_2\overline{\mathcal{S}} \cap \Delta_{\mathbb{Q}}(X),$$

and $(\gamma g_1 g_2^{-1})g_2\overline{\mathcal{S}} \cap g_2\overline{\mathcal{S}} = \overline{\mathfrak{a}^+{}_{\mathbf{P}'_I}(\infty)}$, it follows that $\gamma g_1 g_2^{-1}\overline{\mathfrak{a}^+{}_{\mathbf{P}'}(\infty)} \cap \overline{\mathfrak{a}^+{}_{\mathbf{P}'}(\infty)} = \overline{\mathfrak{a}^+{}_{\mathbf{P}'_I}(\infty)}$, and hence $\gamma g_1 g_2^{-1}$ leaves $\overline{\mathfrak{a}^+{}_{\mathbf{P}'_I}(\infty)}$ invariant, which in turn implies that $\gamma g_1 g_2^{-1} \in \mathbf{P}'_I$. By the definition of the $\mathbf{G}(\mathbb{Q})$-action on $\Delta_{\mathbb{Q}}(X)$, $\gamma g_1 g_2^{-1}$ acts as the identity on $\mathfrak{a}_{\mathbf{P}'_I}^+(\infty)$, and hence γ acts as $g_2 g_1^{-1}$ on $\gamma^{-1}\mathfrak{a}_{\mathbf{P}'_I}^+(\infty) = g_1\overline{\mathcal{S}} \cap \gamma^{-1}g_2\overline{\mathcal{S}}$. Therefore the condition (3) is satisfied, and hence $\Gamma\backslash X \cup \Delta_{\mathbb{Q}}(X)$ is compact and Hausdorff.

As in §I.16, given any two compactifications $\overline{\Gamma\backslash X}^1$, $\overline{\Gamma\backslash X}^2$ of $\Gamma\backslash X$, there is a unique greatest common quotient (GCQ) $\overline{\Gamma\backslash X}^1 \wedge \overline{\Gamma\backslash X}^2$.

Proposition III.12.7 *If the $\mathbb{Q}$-rank of $\mathbf{G}$ is greater than or equal to 2, then the GCQ $\overline{\Gamma\backslash X}^T \wedge \overline{\Gamma\backslash X}^{BS}$ is equal to the one-point compactification; if the $\mathbb{Q}$-rank of $\mathbf{G}$ is equal to* 1, *$\overline{\Gamma\backslash X}^T$ is the end compactification, which is obtained by adding one-point to each end, and $\overline{\Gamma\backslash X}^{BS}$ dominates $\overline{\Gamma\backslash X}^T$, i.e., the identity map on $\Gamma\backslash X$ extends to a continuous map.*

Proof. Assume that the $\mathbb{Q}$-rank of $\mathbf{G}$ is greater than or equal to 2, then the Tits building $\Delta_{\mathbb{Q}}(\mathbf{G})$ is connected, and the infinity of $\Gamma\backslash X$ is also connected, i.e., $\Gamma\backslash X$ has only one end. To show that $\overline{\Gamma\backslash X}^T \wedge \overline{\Gamma\backslash X}^{BS}$ is the one-point compactification, it suffices to show that for any rational parabolic subgroup $\mathbf{P}$ and any two boundary points $H_1, H_2 \in \mathfrak{a}_{\mathbf{P}}^+(\infty)$, there exist two sequences $y_{j,1}, y_{j,2}$ such that $y_{j,1} \to H_1$, $y_{j,2} \to H_2$ in $\overline{\Gamma\backslash X}^T$, but $y_{j,1}, y_{j,2}$ converge to the same boundary point in $\overline{\Gamma\backslash X}^{BS}$. In fact, take $y_{j,1}$ to be the image of $e^{jH_1}x_0$, and $y_{j,2}$ the image of $e^{jH_2}x_0$. They satisfy the above properties.

If the $\mathbb{Q}$-rank of $\mathbf{G}$ is equal to 1, every rational parabolic subgroup $\mathbf{P}$ is minimal and $\dim A_{\mathbf{P}} = 1$. Hence $\Gamma\backslash X$ has finitely many ends corresponding to Γ-conjugacy classes of rational parabolic subgroups, and $\overline{\Gamma\backslash X}^T$ is the end compactification. Clearly $\overline{\Gamma\backslash X}^{BS}$ dominates $\overline{\Gamma\backslash X}^T$ and the GCQ is the end compactification.

III.12.8 Summary and comments. Boundaries of compactifications of locally symmetric spaces $\Gamma\backslash X$ are often related to the spherical Tits building $\Delta_{\mathbb{Q}}(\mathbf{G})$. The compactification $\overline{\Gamma\backslash X}^T$ is special in the sense that its boundary is exactly the quotient of $\Delta_{\mathbb{Q}}(\mathbf{G})$ by Γ, or equivalently, the boundary of the partial compactification of X is equal to $\Delta_{\mathbb{Q}}(\mathbf{G})$. Later, in §III.20, the compactification $\overline{\Gamma\backslash X}^T$ and in particular its boundary will be realized naturally in terms of certain geodesics in $\Gamma\backslash X$, so-called EDM-geodesics, which go to infinity.

III.13 Borel–Serre compactification of homogeneous spaces $\Gamma\backslash G$

In the earlier sections, we have recalled and constructed many compactifications of locally symmetric spaces $\Gamma\backslash X$. In this section, we construct the Borel–Serre compactification $\overline{\Gamma\backslash G}^{BS}$ of $\Gamma\backslash G$ corresponding to $\overline{\Gamma\backslash X}^{BS}$ in the sense that the quotient of $\overline{\Gamma\backslash G}^{BS}$ by K on the right gives $\overline{\Gamma\backslash X}^{BS}$. In the next section, the reductive Borel–Serre compactification $\overline{\Gamma\backslash G}^{RBS}$ corresponding to $\overline{\Gamma\backslash X}^{RBS}$ will be constructed.

This section is organized as follows. Some motivations of studying the homogeneous space $\Gamma\backslash X$ and its compactifications are discussed in III.13.1. The construction is very similar to $\overline{\Gamma\backslash X}^{BS}$ in §III.9. Specifically, the boundary components are defined in III.13.2, and the topology of the partial compactification in III.13.3. The compactification $\overline{\Gamma\backslash G}^{BS}$ is defined in III.13.7. An application to the extension of homogeneous bundles on $\Gamma\backslash X$ to $\overline{\Gamma\backslash X}^{BS}$ is discussed in III.13.9, using the fact that the right K-action on $\Gamma\backslash G$ extends to $\overline{\Gamma\backslash G}^{BS}$. On the other hand, the right G-action on $\Gamma\backslash G$ does not extend to $\overline{\Gamma\backslash G}^{BS}$ (III.13.10).

III.13.1 There are several reasons to study $\Gamma\backslash G$ and its compactifications. First, $\Gamma\backslash G$ is a homogeneous space, while $\Gamma\backslash X$ is not. This fact is quite important. In fact, understanding the regular representation of G on $L^2(\Gamma\backslash G)$ is a central problem in automorphic representations.

The locally symmetric space $\Gamma\backslash X$ is an important quotient of $\Gamma\backslash G$, but there are natural homogeneous bundles over $\Gamma\backslash X$ and it is fruitful to study them as well, for example, in decomposing the *regular representation* of G in $L^2(\Gamma\backslash G)$ (see [Ji2]).

In fact, for any finite-dimensional representation $\sigma : K \to GL(n, \mathbb{C})$, we can define a bundle E_σ over $\Gamma\backslash X$ by

$$E_\sigma = \Gamma\backslash G \otimes_K \mathbb{C}^n.$$

For some applications, it is useful to extend this bundle to compactifications of $\Gamma\backslash X$ (see [Zu3]). If $\Gamma\backslash G$ has a compactification $\overline{\Gamma\backslash G}$ such that the right K-multiplication on $\Gamma\backslash G$ extends, then $\overline{\Gamma\backslash G} \otimes_K \mathbb{C}^n$ gives a bundle over $\overline{\Gamma\backslash G}/K$, which is a compactification of $\Gamma\backslash X$. Hence the homogeneous bundle E_σ on $\Gamma\backslash X$ can be extended to the compactification of $\Gamma\backslash X$.

Another motivation comes from the theory of variation of Hodge structures. The target of the period map for variation of Hodge structures is of the form $\Gamma\backslash G/H$, where H is a compact subgroup, which is not necessarily a maximal compact subgroup and hence contained in K (or some conjugate of K). Compactifications of such period manifolds were sought after in [Gri]. Clearly compactifications of $\Gamma\backslash G$ admitting a right K-action give compactifications of $\Gamma\backslash G/H$. In §III.17–§III.19, we will see that compactifications of $\Gamma\backslash G$ can be constructed by embedding $\Gamma\backslash G$ into compact G-spaces and taking the closure, which leads to new, natural constructions of compactifications of $\Gamma\backslash X$.

III.13.2 For each rational parabolic subgroup $\mathbf{P}$, the rational Langlands decomposition of $P = N_P \mathbf{A_P} \mathbf{M_P}$ induces the horospherical decomposition

$$G = N_P \mathbf{A_P} \mathbf{M_P} K = N_P \times \mathbf{A_P} \times (\mathbf{M_P} K) \tag{III.13.26}$$

as in §III.1.

By replacing the horospherical decomposition of X, $X = N_P \times \mathbf{A_P} \times X_\mathbf{P}$, by the above decomposition of G, the construction of the Borel–Serre compactification of $\Gamma\backslash G$ is similar to the construction of $\Gamma\backslash {}_\mathbb{Q}\overline{X}^{BS}$ in §III.9. In the following, we outline the main steps and basic results and refer to [BJ3, §3] for details of the proofs.

Define its Borel–Serre compactification $e(\mathbf{P})$ by

$$e(\mathbf{P}) = N_P \times (\mathbf{M_P} K).$$

Define the Borel–Serre partial compactification ${}_\mathbb{Q}\overline{G}^{BS}$ to be

$${}_\mathbb{Q}\overline{G}^{BS} = G \cup \coprod_{\mathbf{P}} e(\mathbf{P}) = G \cup \coprod_{\mathbf{P}} \mathbf{M_P} K, \tag{III.13.27}$$

where $\mathbf{P}$ runs over all rational parabolic subgroups of $\mathbf{G}$, with a topology described below.

III.13.3 The topology of ${}_\mathbb{Q}\overline{G}^{BS}$ is described in terms of convergent sequences as follows:

1. For any rational parabolic subgroup **P**, an unbounded sequence y_j in G converges to a point $(n_\infty, m_\infty) \in e(\mathbf{P})$ if and only if in terms of the above horospherical decomposition of G, $y_j = (n_j, a_j, m_j)$, $n_j \in N_P$, $a_j \in A_{\mathbf{P}}$, $m_j \in M_{\mathbf{P}}K$, the components n_j, a_j, m_j satisfy the following conditions:
   (a) For any $\alpha \in \Phi(P, A_{\mathbf{P}})$, $(a_j)^\alpha \to +\infty$,
   (b) $n_j \to n_\infty$ in N_P, and $m_j \to m_\infty \in M_{\mathbf{P}}K$.
2. For two rational parabolic subgroups $\mathbf{P} \subset \mathbf{Q}$, $\mathbf{P} \neq \mathbf{Q}$, let $\mathbf{P}'$ be the rational parabolic subgroup in $\mathbf{M_Q}$ corresponding to **P** as in equation (III.1.13). The parabolic subgroup $\mathbf{P}'$ induces a Langlands decomposition of $M_{\mathbf{Q}}$,

$$M_{\mathbf{Q}} = N_{P'} \times A_{\mathbf{P}'} \times (M_{\mathbf{P}'}K_{\mathbf{Q}}),$$

   and hence a decomposition of $e(\mathbf{Q})$,

$$\begin{aligned} e(\mathbf{Q}) = N_Q \times (M_{\mathbf{Q}}K) &= (N_Q N_{\mathbf{P}'}) \times A_{\mathbf{P}'} \times (M_{\mathbf{P}}K) \\ &= N_P \times A_{\mathbf{P}'} \times (M_{\mathbf{P}}K). \end{aligned} \tag{III.13.28}$$

   Then a sequence y_j in $e(\mathbf{Q})$ converges to a point $(n_\infty, m_\infty) \in e(\mathbf{P})$ if and only if in the decomposition

$$y_j = (n_j, a'_j, m_j) \in N_P \times A_{\mathbf{P}'} \times (M_{\mathbf{P}}K) = e(\mathbf{Q}),$$

   the coordinates satisfy the following conditions:
   (a) $n_j \to n_\infty$ in N_P, $m_j \to m_\infty$ in $M_{\mathbf{P}}$,
   (b) for all roots $\alpha \in \Phi(P', A_{\mathbf{P}'})$, $(a'_j)^\alpha \to +\infty$.

These are special convergent sequences, and their combinations give the general convergent sequences. It can be seen that they form a convergence class of sequences in the sense of §I.8, and hence define a topology on ${}_{\mathbb{Q}}\overline{G}^{BS}$. Neighborhoods of boundary points can be given explicitly.

For any rational parabolic subgroup **P**, let $\overline{A_{\mathbf{P}}}$ be the partial compactification defined in equation (III.9.6).

By arguments similar to the proof of Proposition III.9.5, we can prove the existence of corners in ${}_{\mathbb{Q}}\overline{G}^{BS}$.

Proposition III.13.4 *The embedding $N_P \times A_{\mathbf{P}} \times M_{\mathbf{P}}K \hookrightarrow G \subset {}_{\mathbb{Q}}\overline{G}^{BS}$ can be naturally extended to an embedding $N_P \times \overline{A_{\mathbf{P}}} \times M_{\mathbf{P}}K \hookrightarrow {}_{\mathbb{Q}}\overline{G}^{BS}$. The image of $N_P \times \overline{A_{\mathbf{P}}} \times M_{\mathbf{P}}K$ in ${}_{\mathbb{Q}}\overline{G}^{BS}$ is denoted by $G(\mathbf{P})$ and called the corner associated with* **P**. *Furthermore, $G(\mathbf{P})$ is equal to $G \cup \coprod_{\mathbf{Q}\supseteq\mathbf{P}} e(\mathbf{Q})$.*

These corners allow us to write explicitly neighborhoods of boundary points. By the reduction theory in Proposition III.2.19, we can prove as in Proposition III.9.15 the following result.

Proposition III.13.5 *The partial compactification ${}_{\mathbb{Q}}\overline{G}^{BS}$ is a Hausdorff space.*

Proposition III.13.6 *The left multiplication of $\mathbf{G}(\mathbb{Q})$-action on G extends to a continuous action on ${}_{\mathbb{Q}}\overline{G}^{BS}$.*

Proof. For any rational parabolic subgroup $\mathbf{P}$ and $g \in \mathbf{G}(\mathbb{Q})$, write $g = kman$, where $k \in K$, $m \in M_{\mathbf{P}}$, $a \in A_{\mathbf{P}}$, $n \in N_P$. For any boundary point $(n', m') \in N_P \times X_{\mathbf{P}}$, define the action

$$g \cdot (n', m') = ({}^{kma}(nn'), {}^{k}(mm')k) \in N_{{}^kP} \times M_{{}^k\mathbf{P}} = e({}^g\mathbf{P}).$$

Since km is uniquely determined by g, this action is well-defined. This define an action of $\mathbf{G}(\mathbb{Q})$ on ${}_{\mathbb{Q}}\overline{G}^{BS}$. By arguments similar to the proof of Proposition III.9.15, we can prove that this extended action is continuous.

By the reduction theory and the explicit description of neighborhoods of boundary points, we can prove as in Proposition III.9.17 the following result.

Proposition III.13.7 *The arithmetic subgroup Γ acts properly on ${}_{\mathbb{Q}}\overline{G}^{BS}$ with a compact, Hausdorff quotient $\Gamma\backslash{}_{\mathbb{Q}}\overline{G}^{BS}$, which is a Hausdorff compactification of $\Gamma\backslash G$ and also denoted by $\overline{\Gamma\backslash G}^{BS}$.*

Proposition III.13.8 *The right K-multiplication on G extends to a continuous action on ${}_{\mathbb{Q}}\overline{G}^{BS}$. Similarly, the right K-multiplication on $\Gamma\backslash G$ extends to a continuous action on $\overline{\Gamma\backslash G}^{BS}$. The quotient $\overline{\Gamma\backslash G}^{BS}/K$ is isomorphic to $\overline{\Gamma\backslash X}^{BS}$.*

Proof. Since the right multiplication by elements in K preserves the conditions on convergent sequences in ${}_{\mathbb{Q}}\overline{G}^{BS}$, it is clear that it extends to a continuous action on ${}_{\mathbb{Q}}\overline{G}^{BS}$. Since the left action of Γ on ${}_{\mathbb{Q}}\overline{G}^{BS}$ commutes with the right K-multiplication, the right K-action also extends continuously to $\overline{\Gamma\backslash G}^{BS}$. The third second statement is clear from the definitions of the boundary components of ${}_{\mathbb{Q}}\overline{X}^{BS}$, ${}_{\mathbb{Q}}\overline{G}^{BS}$.

Corollary III.13.9 *Any homogeneous bundle $E_\sigma = \Gamma\backslash G \otimes_K \mathbb{C}^n$ on $\Gamma\backslash X$ extends to a bundle $\overline{\Gamma\backslash G}^{BS} \otimes_K \mathbb{C}^n$ over $\overline{\Gamma\backslash X}^{BS}$, where $\sigma : K \to GL(n, \mathbb{C})$ is a finite-dimensional representation of K.*

Proposition III.13.10 *The right G-action on G does not extend to the Borel–Serre partial compactification ${}_{\mathbb{Q}}\overline{G}^{BS}$. Similarly, the right G-action does not extend to $\Gamma\backslash{}_{\mathbb{Q}}\overline{G}^{BS}$.*

Proof. It suffices to exhibit a convergent sequence y_j in ${}_{\mathbb{Q}}\overline{G}^{BS}$ and an element $g \in G$ such that $y_j g$ are not convergent in ${}_{\mathbb{Q}}\overline{G}^{BS}$.

Let $\mathbf{P}$ be a rational parabolic subgroup. Choose $H \in \mathfrak{a}_P$ such that for all $\alpha \in \Phi(P, A_{\mathbf{P}})$, $\alpha(H) > 0$. Let $y_j = \exp t_j H$ for a sequence $t_j \to +\infty$. Clearly, y_j is convergent in $\overline{G}^{BS}$. Let $g = n \in N_P$, $n \neq e$. We claim that $y_j g$ is not convergent in ${}_{\mathbb{Q}}\overline{G}^{BS}$, and its image in $\Gamma\backslash{}_{\mathbb{Q}}\overline{G}^{BS}$ does not converge either for suitably chosen t_j.

In fact,

$$
\begin{aligned}
y_j g = (\exp t_j H)n &= Ad(\exp t_j H)(n) \cdot \exp t_j H \\
&= (Ad(\exp t_j H)(n),\ \exp t_j H,\ 1) \in N_P \times A_{\mathbf{P}} \times (M_{\mathbf{P}} K).
\end{aligned}
$$

The component $Ad(\exp t_j H)(n)$ in N_P is not bounded, and hence the sequence $y_j g$ does not converge to any point in ${}_{\mathbb{Q}}\overline{G}^{BS}$. When t_j is suitably chosen, the image of this unbounded sequence $Ad(\exp t_j H)(n)$ in $\Gamma_{N_P}\backslash N_P$ does not converge either. In fact, when $t \to +\infty$, the image of $Ad(\exp tH)(n)$ in $\Gamma_{N_P}\backslash N_P$ traces out a nonconstant continuous path, wrapping around the "cusp" of P, and hence we can pick a sequence t_j such that the image $Ad(\exp t_j H)(n)$ in $\Gamma_{N_P}\backslash N_P$ does not converge. Then the image of $y_j g$ in $\Gamma\backslash{}_{\mathbb{Q}}\overline{G}^{BS}$ does not converge to any point either.

Proposition III.13.11 *The partial compactification ${}_{\mathbb{Q}}\overline{G}^{BS}$ has a canonical structure of a real analytic manifold with corners. The Γ-action on ${}_{\mathbb{Q}}\overline{G}^{BS}$ is real analytic with respect to this structure, and hence $\overline{\Gamma\backslash G}^{BS}$ is a compact real analytic manifold with corners.*

Proof. For each rational parabolic subgroup P, the corner $G(\mathbf{P})$ has a canonical analytic structure under the identification

$$G(\mathbf{P}) = N_P \times \overline{A_{\mathbf{P}}} \times X_{\mathbf{P}}.$$

As in Proposition III.9.16, we can show that the real analytic structures of these corners are compatible. They form an open covering of ${}_{\mathbb{Q}}\overline{G}^{BS}$ and define a real analytic structure on it. By the method in Proposition III.9.16 again, we can prove that Γ acts real analytically on ${}_{\mathbb{Q}}\overline{G}^{BS}$.

III.13.12 Summary and comments. Though the space $\Gamma\backslash G$ is homogeneous and important for many purposes, its compactifications have not been studied much. The construction of $\overline{\Gamma\backslash G}^{BS}$ is almost identical to the compactification $\overline{\Gamma\backslash X}^{BS}$ in §III.9. On the other hand, unlike the symmetric space X, the Lie group G does not have the canonical geodesic action of $A_{\mathbf{P}}$ for each parabolic subgroup $\mathbf{P}$. So the method in [BS2] can not be applied directly to construct $\overline{\Gamma\backslash G}^{BS}$.

III.14 Reductive Borel–Serre compactification of homogeneous spaces $\Gamma\backslash G$

In this section, we follow the uniform method in §III.8 to define the reductive compactification $\overline{\Gamma\backslash G}^{RBS}$. An important difference from the Borel–Serre compactification $\overline{\Gamma\backslash G}^{BS}$ is that the right G-multiplication on $\Gamma\backslash G$ extends to a continuous action on $\overline{\Gamma\backslash G}^{RBS}$, which is not the case for $\overline{\Gamma\backslash X}^{BS}$ (see Proposition III.13.10).

The rest of the construction is similar to that of $\overline{\Gamma\backslash X}^{RBS}$. Specifically, the boundary component of $\mathbb{Q}$-parabolic subgroups are defined in III.14.1, and the topology of the partial compactification is described in terms of convergent sequences in III.14.2. The Hausdorff property is given in III.14.3, and the extension of the $\mathbf{G}(\mathbb{Q})$-action is given in III.14.4. The compactification $\overline{\Gamma\backslash G}^{RBS}$ is constructed in III.14.5. The right G-multiplication on $\Gamma\backslash G$ is shown to extend continuously to the compactification in III.14.6. Extension of homogeneous vector bundles on $\Gamma\backslash X$ to $\overline{\Gamma\backslash X}^{RBS}$ is proved in III.14.7. The G-orbits of the right action on $\overline{\Gamma\backslash G}^{RBS}$ are given in III.14.8. For details of some proofs, see [BJ3].

An important difference from $\overline{\Gamma\backslash G}^{BS}$ is that the right G-action on $\Gamma\backslash G$ extends continuously to $\overline{\Gamma\backslash G}^{RBS}$.

III.14.1 For any rational parabolic subgroup $\mathbf{P}$, define its boundary face $e(\mathbf{P})$ by

$$e(\mathbf{P}) = M_{\mathbf{P}}K \cong N_P A_{\mathbf{P}}\backslash G.$$

Notice that $e(\mathbf{P})$ is obtained from the Borel–Serre boundary component $N_P \times (M_{\mathbf{P}}K)$ by collapsing the unipotent radical N_P, and is hence called the reductive Borel–Serre boundary component. The identification with $N_P A_{\mathbf{P}}\backslash G$ shows that it is a homogeneous space of G.

Define the *reductive Borel–Serre partial compactification* ${}_{\mathbb{Q}}\overline{G}^{RBS}$ to be

$${}_{\mathbb{Q}}\overline{G}^{RBS} = G \cup \coprod_{\mathbf{P}} e(\mathbf{P}) = G \cup \coprod_{\mathbf{P}} M_{\mathbf{P}}K$$

with a topology described below, where $\mathbf{P}$ runs over all rational parabolic subgroups.

III.14.2 The topology on ${}_{\mathbb{Q}}\overline{G}^{RBS}$ is described in terms of convergent sequences as follows:

1. For any rational parabolic subgroup $\mathbf{P}$, an unbounded sequence $y_j = (n_j, a_j, m_j) \in N_P \times A_{\mathbf{P}} \times (M_{\mathbf{P}}K) = G$ converges to a point $m_\infty \in e(\mathbf{P})$ if and only if the following two conditions are satisfied:
   (a) For all $\alpha \in \Phi(P, A_{\mathbf{P}})$, $\quad a_j^\alpha \to +\infty$.
   (b) $m_j \to m_\infty$ in $M_{\mathbf{P}}K$.
2. For any two rational parabolic subgroups $\mathbf{P} \subseteq \mathbf{Q}$, $\mathbf{P} \neq \mathbf{Q}$, let $\mathbf{P}'$ be the rational parabolic subgroup of $\mathbf{M_Q}$ corresponding to $\mathbf{P}$ as in equation (III.1.13). The group $\mathbf{P}'$ gives a Langlands decomposition $M_{\mathbf{Q}} = N_{P'}A_{\mathbf{P}'}(M_{\mathbf{P}'}K_{\mathbf{Q}})$ and hence a decomposition of $M_Q K = N_{P'}A_{\mathbf{P}'}M_{\mathbf{P}'}K$, i.e.,

$$e(\mathbf{Q}) = N_{P'} \times A_{\mathbf{P}'} \times e(\mathbf{P}).$$

   In this decomposition of $e(\mathbf{Q})$, a sequence $y_j = (n_j, a_j, m_j)$ in $e(\mathbf{Q})$ converges to a point $m_\infty \in e(\mathbf{P})$ if and only if the following two conditions are satisfied:

(a) For all $\alpha \in \Phi(P', A_{\mathbf{P}'})$, $a_j^{\alpha} \to +\infty$.
(b) $m_j \to m_\infty$ in $e(\mathbf{P})$.

These are special convergent sequences, and their combinations give the general convergent sequences. It can be shown that they form a convergence class of sequences and hence define a topology on ${}_{\mathbb{Q}}\overline{G}^{RBS}$.

As in the case of ${}_{\mathbb{Q}}\overline{X}^{RBS}$, neighborhoods of boundary points can be given explicitly.

Using an explicit description of neighborhoods of boundary points in terms of generalized Siegel sets, we can prove the following result, as in Proposition III.10.5.

Proposition III.14.3 *Every convergent sequence in ${}_{\mathbb{Q}}\overline{G}^{RBS}$ has a unique limit, and hence the topology on ${}_{\mathbb{Q}}\overline{G}^{RBS}$ defined above is Hausdorff.*

Proposition III.14.4 *The left $\mathbf{G}(\mathbb{Q})$-multiplication on G extends to a continuous action on ${}_{\mathbb{Q}}\overline{G}^{RBS}$. In particular, Γ acts continuously on ${}_{\mathbb{Q}}\overline{G}^{RBS}$ on the left.*

Proof. For any rational parabolic subgroup P and $g \in \mathbf{G}(\mathbb{Q})$, write $g = kman$, where $k \in K, m \in M_{\mathbf{P}}, a \in A_{\mathbf{P}}, n \in N_P$. For any $m' \in M_{\mathbf{P}}K = e(\mathbf{P})$, define the action

$$g \cdot m' = {}^k(mm') \cdot k \in M_{{}^k\mathbf{P}}K = e({}^k\mathbf{P}).$$

Since km is uniquely determined by g, this action is well-defined. The continuity of the action is proved as in Proposition III.10.6.

By using the reduction theory and explicit description of neighborhoods of the boundary points, we can prove as in Theorem III.10.9 the following result.

Proposition III.14.5 *The quotient $\Gamma\backslash{}_{\mathbb{Q}}\overline{G}^{RBS}$ is a compact Hausdorff space containing $\Gamma\backslash G$ as an open dense subset. This compactification is also denoted by $\overline{\Gamma\backslash G}^{RBS}$, called the reductive Borel–Serre compactification.*

Proposition III.14.6 *The right G-multiplication on G extends to a continuous action on ${}_{\mathbb{Q}}\overline{G}^{RBS}$, and hence the right G-multiplication on $\Gamma\backslash G$ extends to a continuous G-action on $\Gamma\backslash{}_{\mathbb{Q}}\overline{G}^{RBS}$.*

Proof. First we define the action, then show that it is continuous. For any boundary point $m \in e(\mathbf{P}) = M_{\mathbf{P}}K$, and an element $g \in G$, write $mg = (n', a', m') \in N_P \times A_P \times M_{\mathbf{P}}K$. Then define

$$m \cdot g = m' \in e(\mathbf{P}).$$

Combined with the right multiplication on G, this gives a right action of G on ${}_{\mathbb{Q}}\overline{G}^{RBS}$. When $e(\mathbf{P}) = M_{\mathbf{P}}K$ is identified with $N_P A_P\backslash G$, this action of G is given by the right multiplication. Then it is clear that each boundary face $e(\mathbf{P})$ is preserved by the G-action and acted upon transitively by G, and the decomposition

$$ {}_{\mathbb{Q}}\overline{G}^{RBS} = G \cup \coprod_{P} e(\mathbf{P}) $$

is the orbit decomposition of ${}_{\mathbb{Q}}\overline{G}^{RBS}$ under this G-action.

To show that this extended action is continuous, let y_j be any sequence converging to a point $m_\infty \in e(\mathbf{P})$. Write $y_j = (n_j, a_j, m_j) \in N_P \times A_{\mathbf{P}} \times (M_{\mathbf{P}}K)$. Then $m_j \to m_\infty$ in $M_{\mathbf{P}}K$, and for all $\alpha \in \Phi(P, A_{\mathbf{P}})$, $(a_j)^\alpha \to +\infty$, but there is no condition on n_j.

Write

$$ m_j g = n'_j a'_j m'_j = (n'_j, a'_j, m'_j) \in N_P \times A_{\mathbf{P}} \times (M_{\mathbf{P}}K). $$

Since $m_j g \to m_\infty g$, the components n'_j, a'_j, m'_j all converge. Let $m'_\infty = \lim_{j\to\infty} m'_j$. Now

$$ \begin{aligned} y_j g &= n_j a_j m_j g = n_j a_j n'_j a'_j m'_j \\ &= n_j a_j n'_j a_j^{-1} a_j a'_j m'_j \\ &= (n_j a_j n'_j a_j^{-1}, a_j a'_j, m'_j). \end{aligned} $$

For all $\alpha \in \Phi(P, A_{\mathbf{P}})$, $(a_j a'_j)^\alpha = (a_j)^\alpha (a'_j)^\alpha \to +\infty$, since a'_j is bounded. This implies that $y_j g$ converges in ${}_{\mathbb{Q}}\overline{G}^{RBS}$ to $m'_\infty \in e(\mathbf{P})$. We note that the N_P-component $n_j a_j n'_j a_j^{-1}$ is unbounded in general (see the proof of Proposition III.13.10), but this does not affect the convergence in ${}_{\mathbb{Q}}\overline{G}^{RBS}$, since there is no condition on the N_P-component for the convergent sequences.

We note that the limit $m'_\infty = \lim_{j\to+\infty} m'_j$ is equal to the $M_{\mathbf{P}}K$ component of $m_\infty g$ in the decomposition $G = N_P \times A_P \times M_{\mathbf{P}}K$, which is equal to $m_\infty \cdot g$.

We can show similarly that the same conclusion holds when y_j is a sequence of points in $e(\mathbf{Q})$ converging to $m_\infty \in e(\mathbf{P})$ in ${}_{\mathbb{Q}}\overline{G}^{RBS}$, where $\mathbf{Q} \supset \mathbf{P}$.

Corollary III.14.7 *The right K-multiplication on G, $\Gamma\backslash G$ extends to ${}_{\mathbb{Q}}\overline{G}^{RBS}$ and $\overline{\Gamma\backslash G}^{RBS}$ respectively. The quotient $\overline{\Gamma\backslash G}^{RBS}/K$ is isomorphic to $\overline{\Gamma\backslash X}^{RBS}$, and hence every homogeneous bundle $E_\sigma = \Gamma\backslash G \otimes_K \mathbb{C}^n$ extends to a bundle $\overline{\Gamma\backslash G}^{RBS} \otimes_K \mathbb{C}^n$ over $\overline{\Gamma\backslash X}^{BS}$.*

Proof. The first statement is a special case of the above proposition. The second statement follows from the definitions of the boundary components and the topologies on ${}_{\mathbb{Q}}\overline{X}^{RBS}$, ${}_{\mathbb{Q}}\overline{G}^{RBS}$.

Proposition III.14.8 *Let $\mathbf{P}_1, \dots, \mathbf{P}_m$ be a set of representatives of Γ-conjugacy classes of rational parabolic subgroups of* $\mathbf{G}$. *Then the G-orbits of the right action on $\overline{\Gamma\backslash G}^{RBS}$ are given by*

$$\overline{\Gamma\backslash G}^{RBS} = \Gamma\backslash G \cup \coprod_{i=1}^{m} \Gamma_{M_{\mathbf{P}_i}} \backslash M_{\mathbf{P}_i} K = \Gamma\backslash G \cup \coprod_{i=1}^{m} \Gamma_{M_{\mathbf{P}_i}} N_{P_i} A_{\mathbf{P}_i} \backslash G.$$

Proof. It follows from the definition that under the left multiplication, for any rational parabolic subgroup $\mathbf{P}$, the stabilizer of the boundary component $e(\mathbf{P})$ in Γ is equal to Γ_P, and the action factors through the map $\Gamma_P \to \Gamma_{M_{\mathbf{P}}}$. Then the first decomposition in the proposition is clear, and the second decomposition follows from the identification $M_{\mathbf{P}} K = N_P A_{\mathbf{P}} \backslash G$.

Proposition III.14.9 *The identity map on G extends to a continuous map ${}_{\mathbb{Q}}\overline{G}^{BS} \to {}_{\mathbb{Q}}\overline{G}^{RBS}$, and hence $\overline{\Gamma\backslash G}^{BS}$ dominates the compactification $\overline{\Gamma\backslash G}^{RBS}$.*

Proof. By [GJT, Lemma 3.28], it suffices to show that if an unbounded sequence y_j in G converges in ${}_{\mathbb{Q}}\overline{G}^{BS}$, then it also converges in ${}_{\mathbb{Q}}\overline{G}^{RBS}$. This is clear from the definitions of the topologies of ${}_{\mathbb{Q}}\overline{G}^{BS}$ and ${}_{\mathbb{Q}}\overline{G}^{RBS}$.

III.14.10 Summary and comments. The construction of the reductive Borel–Serre compactification $\overline{\Gamma\backslash G}^{RBS}$ is similar to $\overline{\Gamma\backslash G}^{BS}$ and $\overline{\Gamma\backslash X}^{RBS}$. But an important difference between $\overline{\Gamma\backslash G}^{RBS}$ and $\overline{\Gamma\backslash G}^{BS}$ is that the right G-action on $\Gamma\backslash G$ extends continuously to the former but not the latter. This can be naturally explained by the relation between $\overline{\Gamma\backslash G}^{RBS}$ and the subgroup compactification $\overline{\Gamma\backslash G}^{sb}$ in §III.18. As in the case of $\overline{\Gamma\backslash G}^{BS}$, the compactification $\overline{\Gamma\backslash G}^{RBS}$ also gives a compactification of $\Gamma\backslash G/H$, where H is a nonmaximal compact subgroup of G.

11

Properties of Compactifications of Locally Symmetric Spaces

In this chapter, we study relations between the many compactifications of locally symmetric spaces studied in the previous chapters and structures of the corners of the Borel–Serre compactification $\overline{\Gamma\backslash X}^{BS}$.

For a not necessarily Hermitian locally symmetric space $\Gamma\backslash X$, we have studied the following compactifications: the Satake compactifications, the Borel–Serre compactification, the reductive Borel–Serre compactification, the Tits compactification. The relations between these compactifications are easy to determine. Basically, the Borel–Serre compactification and the Tits compactification use complementary factors of the Langlands decomposition of parabolic subgroups, and hence are complementary to each other, while the Borel–Serre compactification dominates the reductive Borel–Serre compactification, which in turn dominates all the Satake compactifications.

On the other hand, when $\Gamma\backslash X$ is a Hermitian locally symmetric space, there are two more compactifications: the Baily–Borel compactification, and the toroidal compactifications. Since both the Borel–Serre compactification and the toroidal compactifications resolve the singularities of the Baily–Borel compactification in different senses, it is natural and important to compare these two compactifications. Though they are very different, they have nontrivial common quotients, for example, the Baily–Borel compactification. It turns out that the greatest common quotient is a new compactification, which is often different from the Baily–Borel compactification. All these relations will be studied in §III.15.

As mentioned earlier in §I.18 and Part II, the Oshima compactification was introduced to study structures of eigenfunctions of all invariant differential operators through their boundary values, and the Oshima compactification can be self-glued from the maximal Satake compactification. In §III.16, we self-glue the Borel–Serre compactification into a closed real analytic manifold, called the *Borel–Serre–Oshima compactification*. Both the real analytic structure of the Borel–Serre compactification and the fact that it can be embedded into a closed analytic manifold might be useful to a similar approach to study automorphic forms, for example, the meromorphic continuation of Eisenstein series. The Borel–Serre-Oshima compactification $\overline{\Gamma\backslash X}^{BSO}$

has been used in [Wes1] [Wes2] to study cohomology groups of the arithmetic subgroup Γ.

III.15 Relations between the compactifications

In this section, we examine relations between the compactifications of locally symmetric spaces studied earlier.

Let $\mathbf{G}$ be a semisimple linear algebraic group defined over $\mathbb{Q}$ of positive $\mathbb{Q}$-rank, and Γ an arithmetic subgroup. Then $\Gamma\backslash X$ is a noncompact locally symmetric space. Since X is a symmetric space of noncompact type, all the compactifications in Proposition III.15.2 are defined. It should be pointed out that the Borel–Serre compactification $\overline{\Gamma\backslash X}^{BS}$ and the reductive Borel–Serre compactification $\overline{\Gamma\backslash X}^{RBS}$ can be defined for a more general class of reductive groups $\mathbf{G}$. In fact, the Borel–Serre compactification $\overline{\Gamma\backslash X}^{BS}$ can be defined for an even larger class of spaces of S-type (see [BS2]).

This section is organized as follows. First, we assume that $\Gamma\backslash X$ is not necessarily Hermitian and summarize relations between all the compactifications discussed earlier in Proposition III.15.2. Relations between compactifications of $\Gamma\backslash G$ are discussed in Proposition III.15.3. Then we consider the case of $\Gamma\backslash X$ Hermitian. There are two more compactifications, and relations between all compactifications are stated in Proposition III.15.4. A relation between the Borel–Serre compactification $\overline{\Gamma\backslash X}^{BS}$ and the toroidal compactifications $\overline{\Gamma\backslash X}_{\Sigma}^{tor}$ was given by the Harris–Zucker conjecture. Some details about the proof of this conjecture are given in III.15.6 and III.15.7. A relation between the reductive Borel–Serre compactification and the toroidal compactifications is briefly described in III.15.8.

III.15.1 Recall that given two compactifications $\overline{\Gamma\backslash X}^1$, $\overline{\Gamma\backslash X}^2$, if the identity map on $\Gamma\backslash X$ extends to a continuous map $\overline{\Gamma\backslash X}^1 \to \overline{\Gamma\backslash X}^2$, then $\overline{\Gamma\backslash X}^1$ is said to dominate $\overline{\Gamma\backslash X}^2$, and this extended map is called the *dominating map*. If the dominating map is bijective, the compactifications are called *isomorphic*. Otherwise, $\overline{\Gamma\backslash X}^1$ is said to strictly dominate $\overline{\Gamma\backslash X}^2$.

A compactification $\overline{\Gamma\backslash X}^3$ dominated by both $\overline{\Gamma\backslash X}^1$ and $\overline{\Gamma\backslash X}^2$ is called a *common quotient* (CQ) of $\overline{\Gamma\backslash X}^1$ and $\overline{\Gamma\backslash X}^2$. There always exists a *greatest common quotient* (GCQ) of any two compactifications $\overline{\Gamma\backslash X}^1$ and $\overline{\Gamma\backslash X}^2$, denoted by $\overline{\Gamma\backslash X}^1 \wedge \overline{\Gamma\backslash X}^2$.

Proposition III.15.2 *The compactifications of* $\Gamma\backslash X$ *satisfy the following relations:*

1. *The Borel–Serre compactification* $\overline{\Gamma\backslash X}^{BS}$ *dominates the reductive Borel–Serre compactification* $\overline{\Gamma\backslash X}^{RBS}$*, and the inverse images of the dominating map over boundary points are nilmanifolds* $\Gamma_{N_P}\backslash N_P$.

2. *The reductive Borel–Serre compactification* $\overline{\Gamma\backslash X}^{RBS}$ *dominates the maximal Satake compactification* $\overline{\Gamma\backslash X}^{S}_{\max}$, *and the inverse image of the dominating map over the boundary points are quotients of Euclidean spaces. They are isomorphic if and only if the* $\mathbb{Q}$*-rank of* $\mathbf{G}$ *is equal to its* $\mathbb{R}$*-rank.*
3. *The maximal Satake compactification* $\overline{\Gamma\backslash X}^{S}_{\max}$ *dominates all other Satake compactifications* $\overline{\Gamma\backslash X}^{S}_{\tau}$, *if* $\overline{X}^{S}_{\tau}$ *is rational and* $\overline{\Gamma\backslash X}^{S}_{\tau}$ *is defined. And the inverse images of the dominating maps are Satake compactifications of lower-dimensional locally symmetric spaces. There are finitely many nonisomorphic Satake compactifications, and they are partially ordered with respect to the domination relation. In general, the minimal Satake compactifications are not unique.*
4. *Hence, both the Borel–Serre compactification* $\overline{\Gamma\backslash X}^{BS}$ *and the reductive Borel–Serre compactification* $\overline{\Gamma\backslash X}^{RBS}$ *dominate all the Satake compactifications* $\overline{\Gamma\backslash X}^{S}_{\max}$, $\overline{\Gamma\backslash X}^{S}_{\tau}$.
5. *When the* $\mathbb{Q}$*-rank of* $\mathbf{G}$ *is greater than or equal to* 2, *the GCQ of* $\overline{\Gamma\backslash X}^{BS}$ *and* $\overline{\Gamma\backslash X}^{T}$ *is given by*

$$\overline{\Gamma\backslash X}^{BS} \wedge \overline{\Gamma\backslash X}^{T} = \Gamma\backslash X \cup \{\infty\},$$

the one point compactification, and hence

$$\overline{\Gamma\backslash X}^{RBS} \wedge \overline{\Gamma\backslash X}^{T} = \Gamma\backslash X \cup \{\infty\}, \quad \overline{\Gamma\backslash X}^{S}_{\tau} \wedge \overline{\Gamma\backslash X}^{T} = \Gamma\backslash X \cup \{\infty\}.$$

On the other hand, when the $\mathbb{Q}$*-rank of* $\mathbf{G}$ *is equal to 1,*

$$\overline{\Gamma\backslash X}^{T} = \overline{\Gamma\backslash X}^{RBS} = \overline{\Gamma\backslash X}^{S}_{\tau},$$

and they are all equal to the end compactification, which is obtained by adding one point to each end of $\Gamma\backslash X$.

Proof. (1) follows from the original definition of $\overline{\Gamma\backslash X}^{RBS}$ or Proposition III.10.7. (2) follows from Proposition III.11.12 and the description of the dominating map ${}_{\mathbb{Q}}\overline{X}^{RBS} \to {}_{\mathbb{Q}}\overline{X}^{S}_{\max}$ in Proposition III.11.8.

All statements in (3) except the inverse images of the boundary points are contained in Proposition I.4.35 and the general procedure in §III.3 on how to pass from the geometrically rational Satake compactification $\overline{X}^{S}_{\tau}$ to $\overline{\Gamma\backslash X}^{S}_{\tau}$. To show that the inverse images of $\overline{\Gamma\backslash X}^{S}_{\tau_1} \to \overline{\Gamma\backslash X}^{S}_{\tau_2}$ are Satake compactifications, we note that the inverse images of the map $\overline{X}^{S}_{\tau_1} \to \overline{X}^{S}_{\tau_2}$ over the boundary are Satake compactifications of symmetric spaces of lower dimension. In fact, for any boundary component X_P in $\overline{X}^{S}_{\tau_2}$, the inverse image of X_P splits as a product $X_P \times \overline{X_{P'}}^{S}$, where $X_P \times X_{P'}$ is the largest boundary component of $\overline{X}^{S}_{\tau_1}$ that is mapped into X_P (see the discussions of the dominating map $\overline{X}^{S}_{\tau_1} \to \overline{X}^{S}_{\tau_2}$ in Proposition I.4.35). Since Γ induces discrete subgroups acting on X_P and $X_{P'}$, it defines a locally symmetric space $\Gamma_{X_{P'}}\backslash X_{P'}$, whose Satake compactification induced from $\overline{X_{P'}}^S$ is the inverse image over the points in $\Gamma_{X_P}\backslash X_P \subset \overline{\Gamma\backslash X}^S_{\tau_2}$.

(4) follows from Proposition III.11.12 and (3). (5) follows from Proposition III.12.7.

Proposition III.15.3 *Compactifications of $\Gamma\backslash G$ and $\Gamma\backslash X$ are related as follows.*

1. *The Borel–Serre compactification $\overline{\Gamma\backslash G}^{BS}$ dominates the reductive Borel–Serre compactification $\overline{\Gamma\backslash G}^{RBS}$, and the inverse images of the dominating maps over the boundary points are nilmanifolds $\Gamma_{N_P}\backslash N_P$, where $\mathbf{P}$ are rational parabolic subgroups.*
2. *The right K-action on $\Gamma\backslash G$ extends to a continuous action on $\overline{\Gamma\backslash G}^{BS}$, and the quotient $\overline{\Gamma\backslash G}^{BS}/K$ is isomorphic to $\overline{\Gamma\backslash X}^{BS}$.*
3. *The right K-action on $\Gamma\backslash G$ extends to a continuous action on $\overline{\Gamma\backslash G}^{RBS}$, and the quotient $\overline{\Gamma\backslash G}^{RBS}/K$ is isomorphic to $\overline{\Gamma\backslash X}^{RBS}$.*

Proof. (1) follows from Proposition III.14.9, (2) is given in Proposition III.13.8, and (3) is given in Corollary III.14.7.

Proposition III.15.4 *Suppose that $\Gamma\backslash X$ is a Hermitian locally symmetric space. Its compactifications satisfy the following relations:*

1. *The Baily–Borel compactification $\overline{\Gamma\backslash X}^{BB}$ is isomorphic, as a topological compactification, to a minimal Satake compactification $\overline{\Gamma\backslash X}^S_\tau$ in the partially ordered set of Satake compactifications, where the highest weight μ_τ is connected only to the last distinguished root in the sense of Proposition I.5.18.*
2. *Therefore, the Borel–Serre compactification $\overline{\Gamma\backslash X}^{BS}$ and the reductive Borel–Serre compactification $\overline{\Gamma\backslash X}^{RBS}$ dominate the Baily–Borel compactification $\overline{\Gamma\backslash X}^{BB}$.*
3. *Every toroidal compactification $\overline{\Gamma\backslash X}^{tor}_\Sigma$ dominates the Baily–Borel compactification, and the dominating map is a complex analytic map. When the toroidal compactification $\overline{\Gamma\backslash X}^{tor}_\Sigma$ is a projective variety, the dominating map $\overline{\Gamma\backslash X}^{tor}_\Sigma \to \overline{\Gamma\backslash X}^{BB}$ is an algebraic map (i.e., a morphism between algebraic varieties).*
4. *For any toroidal compactification $\overline{\Gamma\backslash X}^{tor}_\Sigma$, the GCQ $\overline{\Gamma\backslash X}^{tor}_\Sigma \wedge \overline{\Gamma\backslash X}^{BS}$ is a compactification called the intermediate compactification and denoted by $\overline{\Gamma\backslash X}^{int}$, which always dominates $\overline{\Gamma\backslash X}^{BB}$ but is not necessarily isomorphic to it.*
5. *For any toroidal compactification $\overline{\Gamma\backslash X}^{tor}_\Sigma$, the GCQ $\overline{\Gamma\backslash X}^{tor}_\Sigma \wedge \overline{\Gamma\backslash X}^{RBS}$ is isomorphic to the Baily–Borel compactification $\overline{\Gamma\backslash X}^{BB}$.*

Proof. (1) follows from Corollary I.5.29. (2) follows from (1) and Proposition III.15.2(4). (3) follows from the construction of $\overline{\Gamma\backslash X}^{tor}_\Sigma$ outlined in §III.7. (4) is proved in [Ji3] and an outline is given in III.15.6 below. (5) is also given in [Ji3].

In Proposition III.15.4.4, precise conditions are given in [Ji3] for when the greatest common quotient $\overline{\Gamma\backslash X}^{tor}_{\Sigma} \wedge \overline{\Gamma\backslash X}^{BS}$, or the intermediate compactification $\overline{\Gamma\backslash X}^{int}$, is equal to $\overline{\Gamma\backslash X}^{BB}$.

Remark III.15.5 In this book, we have not discussed functorial properties of compactifications of $\Gamma\backslash X$, i.e., if $i : \Gamma_1\backslash X_1 \to \Gamma_2\backslash X_2$ is an embedding and $\overline{\Gamma_2\backslash X_2}$ is a compactification of $\Gamma_2\backslash X_2$, then the closure of $i(\Gamma_1\backslash X_1)$ is a compactification. The question is what this induced compactification is. A similar question concerns the compactifications of products of $\Gamma\backslash X$.

III.15.6 It was conjectured by Harris and Zucker in [HZ, Conjecture 1.5.8] that the GCQ $\overline{\Gamma\backslash X}^{tor}_{\Sigma} \wedge \overline{\Gamma\backslash X}^{BS}$ is always equal to $\overline{\Gamma\backslash X}^{BB}$. The above result shows that this conjecture is not true, and the GCQ is given by the intermediate compactification $\overline{\Gamma\backslash X}^{int}$, which is often different from $\overline{\Gamma\backslash X}^{BB}$.

If $\Gamma\backslash X$ is a Hilbert modular surface, then the boundary $\partial\overline{\Gamma\backslash X}^{BS}$ is a union of rank-two torus bundles over a circle, one bundle for each end of $\Gamma\backslash X$, while $\partial\overline{\Gamma\backslash X}^{tor}_{\Sigma}$ is a union of cycles of rational curves $\mathbb{C}P^1$, one cycle for each end of $\Gamma\backslash X$, whose length depends on Σ. Then it is conceivable that $\overline{\Gamma\backslash X}^{BS}$ is completely incompatible with $\overline{\Gamma\backslash X}^{tor}_{\Sigma}$ at each end, and hence $\overline{\Gamma\backslash X}^{BS} \wedge \overline{\Gamma\backslash X}^{tor}_{\Sigma}$ is the compactification obtained by adding one point to every end of $\Gamma\backslash X$, which is exactly $\overline{\Gamma\backslash X}^{BB}$. Therefore, the conjecture is true in this case.

On the other hand, for a Picard modular surface $\Gamma\backslash X = \Gamma\backslash B^2$, where B^2 is the unit ball in $\mathbb{C}^2$, $\overline{\Gamma\backslash X}^{tor}_{\Sigma}$ is unique and $\partial\overline{\Gamma\backslash X}^{tor}_{\Sigma}$ is a union of elliptic curves, one for each end of $\Gamma\backslash X$, while $\partial\overline{\Gamma\backslash X}^{BS}$ is a union of circle bundles over the elliptic curves that appear in $\partial\overline{\Gamma\backslash X}^{tor}_{\Sigma}$. Since $\overline{\Gamma\backslash X}^{BB}$ is obtained by adding one point to each end of $\Gamma\backslash X$, these three compactifications fit into a tower

$$\overline{\Gamma\backslash X}^{BS} \underset{\neq}{\longrightarrow} \overline{\Gamma\backslash X}^{tor} \underset{\neq}{\longrightarrow} \overline{\Gamma\backslash X}^{BB}.$$

So the GCQ $\overline{\Gamma\backslash X}^{BS} \wedge \overline{\Gamma\backslash X}^{tor}_{\Sigma}$ is equal to $\overline{\Gamma\backslash X}^{tor}_{\Sigma}$, which strictly dominates $\overline{\Gamma\backslash X}^{BB}$, and the conjecture is false in this case.

In every case where the conjecture fails, this phenomenon in the Picard modular surface is present, i.e., $\overline{\Gamma\backslash X}^{BS}$ dominates $\overline{\Gamma\backslash X}^{tor}_{\Sigma}$ near some boundary components of $\overline{\Gamma\backslash X}^{tor}_{\Sigma}$, but $\overline{\Gamma\backslash X}^{tor}_{\Sigma}$ is strictly bigger than $\overline{\Gamma\backslash X}^{BB}$.

Besides this partial dominance of $\overline{\Gamma\backslash X}^{BS}$ over $\overline{\Gamma\backslash X}^{tor}_{\Sigma}$ possibly near some boundary components and the fact that they both are bigger than $\overline{\Gamma\backslash X}^{BB}$, the compactifications $\overline{\Gamma\backslash X}^{BS}$ and $\overline{\Gamma\backslash X}^{tor}_{\Sigma}$ are incompatible.

III.15.7 To prove this incompatibility, we proceed in two steps. The fibers (or inverse images) in $\overline{\Gamma\backslash X}^{BS}$ over boundary points in $\overline{\Gamma\backslash X}^{BB}$ are families of nilmanifolds over lower-dimensional locally symmetric spaces. Fix a fiber and assume that the base has positive dimension. In the first step, we use the incompatibility between the geodesic action in $\overline{\Gamma\backslash X}^{BS}$ and the torus action in $\overline{\Gamma\backslash X}_{\Sigma}^{tor}$ to show that every horizontal section of this bundle collapses to a point in any common quotient of $\overline{\Gamma\backslash X}^{BS}$ and $\overline{\Gamma\backslash X}_{\Sigma}^{tor}$. In this step, the relation in Proposition III.7.13 between the refined horospherical decomposition and the realization of the Siegel domain of the third kind plays an essential role in both the proof and understanding the proof.

In the second step, we show that the fundamental group of the base manifold acts ergodically on the fibers of the bundle (see Remark III.7.11), and hence the fibers have to collapse also in any common quotient because of the Hausdorff property. In this argument, the fact that the base locally symmetric space has positive dimension is crucial in order to get the nontrivial fundamental group that acts ergodically.

Remark III.15.8 By Proposition III.15.4.5, the identity map on $\Gamma\backslash X$ cannot be extended to a continuous map from $\overline{\Gamma\backslash X}^{RBS}$ to $\overline{\Gamma\backslash X}_{\Sigma}^{tor}$. On the other hand, it is shown in [GT1] that for any compact subset C of $\Gamma\backslash X$, there is a continuous map from $\overline{\Gamma\backslash X}^{RBS}$ to $\overline{\Gamma\backslash X}_{\Sigma}^{tor}$ which is equal to the identity on C.

III.16 Self-gluing of Borel–Serre compactification into Borel–Serre–Oshima compactification

In this section, we study the structure of the boundary faces of $\overline{\Gamma\backslash X}^{BS}$ and show that 2^r copies of $\overline{\Gamma\backslash X}^{BS}$ can be self-glued into a closed real analytic manifold by the method of §II.1, where r is the $\mathbb{Q}$-rank of $\mathbf{G}$, which is equal to the maximum of $\dim A_{\mathbf{P}}$ for all rational parabolic subgroups $\mathbf{P}$ of $\mathbf{G}$. By Remark II.1.3, this is the least number of copies of $\overline{\Gamma\backslash X}^{BS}$ needed to glue into a closed manifold.

Since this space is similar to the Oshima compactification of X, it is called the *Borel–Serre-Oshima compactification* of $\Gamma\backslash X$ and denoted by $\overline{\Gamma\backslash X}^{BSO}$. Since $\overline{\Gamma\backslash X}^{BSO}$ admits a $(\mathbb{Z}/2\mathbb{Z})^r$-action and the quotient is $\overline{\Gamma\backslash X}^{BS}$, functions on $\overline{\Gamma\backslash X}^{BS}$ can be lifted to $(\mathbb{Z}/2\mathbb{Z})^r$-invariant functions on the closed manifold $\overline{\Gamma\backslash X}^{BSO}$. This point of view has been used to study the trace formula for Hecke operators on the cohomology groups of Γ in [Wes2].

This section is organized as follows. The closure of each boundary face is a real analytic manifold with corners III.16.2, which is used to show that the rank of ${}_{\mathbb{Q}}\overline{X}^{BS}$ as a manifold with corners is equal to the $\mathbb{Q}$-rank of $\mathbf{G}$ in III.16.3. The boundary hypersurfaces are shown to be embedded in III.16.4. To self-glue ${}_{\mathbb{Q}}\overline{X}^{BS}$ into a smooth manifold without corners, we need a partition of the set of its boundary hypersurfaces, which is given in III.16.7. The self-gluing of ${}_{\mathbb{Q}}\overline{X}^{BS}$ is given in III.16.8, and of $\overline{\Gamma\backslash X}^{BS}$ in III.16.9. An application of $\overline{\Gamma\backslash X}^{BSO}$ is briefly mentioned in III.16.11.

III.16.1 We start by determining the corners and boundary faces of ${}_{\mathbb{Q}}\overline{X}^{BS}$. Then we show that the boundary faces satisfy the general conditions in §II.1 for self-gluing.

Lemma III.16.2 *For every rational parabolic subgroup* $\mathbf{Q}$*, the closure* $\overline{e(\mathbf{Q})}$ *of the boundary face* $e(\mathbf{Q}) = N_Q \times X_{\mathbf{Q}}$ *in* ${}_{\mathbb{Q}}\overline{X}^{BS}$ *is a (closed) boundary face of codimension* $\dim A_{\mathbf{Q}}$ *and is hence a real analytic submanifold with corners.*

Proof. From the definition of convergence of sequences of boundary points in ${}_{\mathbb{Q}}\overline{X}^{BS}$,

$$\overline{e(\mathbf{Q})} = e(\mathbf{Q}) \cup \coprod_{\mathbf{P} \subset \mathbf{Q}} e(P).$$

Therefore, $\overline{e(\mathbf{Q})}$ is covered by the corners $X(\mathbf{P})$ for all $\mathbf{P} \subseteq \mathbf{Q}$. In each corner $X(\mathbf{P})$,

$$\overline{e(\mathbf{Q})} \cap X(\mathbf{P}) = \coprod_{\mathbf{P} \subseteq \mathbf{R} \subseteq \mathbf{Q}} e(\mathbf{R}) \subset X \cup \coprod_{\mathbf{P} \subseteq \mathbf{R}} e(\mathbf{R}).$$

By Lemmas III.9.8 and III.9.9, in the decomposition

$$X(\mathbf{P}) = N_P \times \overline{A_{\mathbf{P},\mathbf{Q}}} \times \overline{A_{\mathbf{Q}}} \times X_{\mathbf{P}},$$

the intersection $\overline{e(\mathbf{Q})} \cap X(\mathbf{P})$ is given by

$$\overline{e(\mathbf{Q})} \cap X(\mathbf{P}) = N_P \times \overline{A_{\mathbf{P},\mathbf{Q}}} \times \{o_Q\} \times X_{\mathbf{P}},$$

which is clearly a real analytic submanifold with corners in $X(\mathbf{P})$ of codimension $\dim A_{\mathbb{Q}}$. This implies that $\overline{e(\mathbf{Q})}$ is a real analytic submanifold with corners in ${}_{\mathbb{Q}}\overline{X}^{BS}$ of codimension $\dim A_{\mathbf{Q}}$.

Corollary III.16.3 *The rank of* ${}_{\mathbb{Q}}\overline{X}^{BS}$ *as a manifold with corners is equal to the* $\mathbb{Q}$*-rank of the algebraic group* $\mathbf{G}$.

Proof. Recall from §II.1 that the rank of a manifold with corners is the maximal codimension of boundary faces, which is equal to $rk_{\mathbb{Q}}(\mathbf{G}) = \max\{\dim A_{\mathbf{P}} \mid \mathbf{P}$ runs over rational parabolic subgroups$\}$ by the above lemma.

Lemma III.16.4 *The boundary hypersurfaces of* ${}_{\mathbb{Q}}\overline{X}^{BS}$ *are the closures* $\overline{e(\mathbf{Q})}$*, where* $\mathbf{Q}$ *are rational parabolic subgroups of rank 1,* $\dim A_{\mathbf{Q}} = 1$*, i.e.,* $\mathbf{Q}$ *are proper maximal rational parabolic subgroups, and they are embedded in the sense defined in §II.1.*

Proof. The first statement clearly follows from Lemma III.16.2. To prove the second statement, we note that for every rational parabolic subgroup $\mathbf{P}$ of rank i, i.e., $\dim A_P = i$, there are exactly i maximal proper rational parabolic subgroups containing $\mathbf{P}$. In fact, this fact follows from the one-to-one correspondence between subsets of the set of simple roots $\Delta(P, A_{\mathbf{P}})$ and rational parabolic subgroups containing $\mathbf{P}$. This implies that every point in $e(\mathbf{P})$, which has rank i by Lemma III.16.2, is contained in exactly i different boundary hypersurfaces. This proves that all boundary hypersurfaces are embedded.

Lemma III.16.5 *Let* $\mathbf{Q}_1, \mathbf{Q}_2$ *be two rational parabolic subgroups and* $\mathbf{P} = \mathbf{Q}_1 \cap \mathbf{Q}_2$. *If* $\mathbf{P}$ *is not a rational parabolic subgroup, the boundary faces* $\overline{e(\mathbf{Q}_1)}$, $\overline{e(\mathbf{Q}_2)}$ *are disjoint. Otherwise,*

$$\overline{e(\mathbf{Q}_1)} \cap \overline{e(\mathbf{Q}_2)} = \overline{e(P)}.$$

Proof. This follows from the equation

$$\overline{e(\mathbf{Q}_i)} = \coprod_{\mathbf{P} \subseteq \mathbf{Q}_i} e(P).$$

Lemma III.16.6 *The boundary faces of* ${}_{\mathbb{Q}}\overline{X}^{BS}$ *are locally finite.*

Proof. For any rational parabolic subgroup $\mathbf{P}$ and its corner $X(\mathbf{P})$, there are only finitely many boundary faces $\overline{e(\mathbf{Q})}$ with

$$X(\mathbf{P}) \cap \overline{e(\mathbf{Q})} \neq \emptyset.$$

In fact, let $\mathbf{Q}_1, \ldots, \mathbf{Q}_n$ be the finitely many rational parabolic subgroups containing $\mathbf{P}$. Since

$$X(\mathbf{P}) = X \cup \coprod_{j=1}^{n} e(\mathbf{Q}_j)$$

and

$$\overline{e(\mathbf{Q})} = \coprod_{\mathbf{R} \subseteq \mathbf{Q}} e(\mathbf{R}),$$

it is clear that $\overline{e(\mathbf{Q})}$ has nonempty intersection with $X(\mathbf{P})$ if and only if $\mathbf{Q}$ contains one of $\mathbf{Q}_1, \ldots, \mathbf{Q}_n$ and hence is one of them. Since $X(\mathbf{P})$ is an open subset and ${}_{\mathbb{Q}}\overline{X}^{BS}$ is covered by these corners, every point has a neighborhood that meets only finitely many boundary faces.

Lemma III.16.7 *Let* r *be the* $\mathbb{Q}$*-rank of* $\mathbf{G}$. *Then there exists a partition of the set* $\mathcal{H}_{{}_{\mathbb{Q}}\overline{X}^{BS}}$ *of boundary hypersurfaces of* ${}_{\mathbb{Q}}\overline{X}^{BS}$ *into* r *parts,*

$$\mathcal{H}_{{}_{\mathbb{Q}}\overline{X}^{BS}} = \coprod_{j=1}^{r} \mathcal{H}_j,$$

such that for every j, *the hypersurfaces in* $\mathcal{H}_j$ *are disjoint.*

Proof. Fix a minimal rational parabolic subgroup $\mathbf{P}$ of $\mathbf{G}$. Then $\dim A_{\mathbf{P}} = r$. By the one-to-one correspondence between subsets of the simple roots $\Delta(P, A_{\mathbf{P}})$ and parabolic subgroups containing $\mathbf{P}$, there are exactly r maximal rational parabolic subgroups containing $\mathbf{P}$.

Denote the maximal rational parabolic subgroups containing $\mathbf{P}$ by $\mathbf{Q}_1, \dots, \mathbf{Q}_r$. For any other minimal rational parabolic subgroup $\mathbf{P}'$, it is known that there exists an element $g \in \mathbf{G}(\mathbb{Q})$ such that $\mathbf{P}' = g\mathbf{P}g^{-1}$. Under this conjugation, the maximal parabolic subgroups containing $\mathbf{P}$ are mapped to the maximal parabolic subgroups containing $\mathbf{P}'$. Denote them by $\mathbf{Q}'_1 = g\mathbf{Q}_1 g^{-1}, \dots, \mathbf{Q}'_r = g\mathbf{Q}_r g^{-1}$. We claim that this numbering of the maximal parabolic subgroups containing $\mathbf{P}'$ is independent of the choice of the element $g \in \mathbf{G}(\mathbb{Q})$. In fact, g is unique up to an element of $\mathbf{P}(\mathbb{Q})$. Since the conjugation by an element of $\mathbf{P}(\mathbb{Q})$ leaves all $\mathbf{Q}_1, \dots, \mathbf{Q}_r$ stable, the claim is proved.

Now define $\mathcal{H}_j$ to be the set of the boundary hypersurfaces $\overline{e(\mathbf{Q}'_j)}$ for all minimal rational parabolic subgroups $\mathbf{P}'$. Since every maximal rational parabolic subgroup contains a minimal rational parabolic subgroup, $\mathcal{H}_1, \dots, \mathcal{H}_r$ forms a partition of the set of boundary hypersurfaces of ${}_{\mathbb{Q}}\overline{X}^{BS}$.

By Lemma III.16.5, the hypersurfaces in each $\mathcal{H}_j$ are disjoint, since no two $\mathbf{Q}'_j$ contain a rational parabolic subgroup.

The manifold ${}_{\mathbb{Q}}\overline{X}^{BS}$ has infinitely many boundary faces, since there are infinitely many rational parabolic subgroups of $\mathbf{G}$. On the other hand, the following is true.

Theorem III.16.8 *Let r be the $\mathbb{Q}$-rank of $\mathbf{G}$ as above. Then 2^r copies of the Borel–Serre partial compactification ${}_{\mathbb{Q}}\overline{X}^{BS}$ can be glued into a closed analytic manifold by the methods in §II.1. This closed analytic manifold is denoted by ${}_{\mathbb{Q}}\overline{X}^{BSO}$ and admits a $(\mathbb{Z}/2\mathbb{Z})^r$-action whose quotient is equal to ${}_{\mathbb{Q}}\overline{X}^{BS}$.*

Proof. By Lemma III.16.7, the set of boundary hypersurfaces $\mathcal{H}_{{}_{\mathbb{Q}}\overline{X}^{BS}}$ admits a partition $\mathcal{H}_1, \dots, \mathcal{H}_r$ such that the hypersurfaces in each $\mathcal{H}_j$ are disjoint. Proposition II.1.2 or Proposition II.1.9 shows that 2^r copies of ${}_{\mathbb{Q}}\overline{X}^{BS}$ can be glued into a closed analytic manifold. Since ${}_{\mathbb{Q}}\overline{X}^{BS}$ is a real analytic manifold with corners by Proposition III.9.16, it follows from Proposition II.1.4 that ${}_{\mathbb{Q}}\overline{X}^{BSO}$ is an analytic manifold. Proposition II.1.5 gives the action of $(\mathbb{Z}/2\mathbb{Z})^r$.

Corollary III.16.9 *When Γ is a torsion-free arithmetic subgroup, 2^r copies of $\overline{\Gamma\backslash X}^{BS}$ can be glued into a closed analytic manifold, denoted by $\overline{\Gamma\backslash X}^{BSO}$, which admits a $(\mathbb{Z}/2\mathbb{Z})^r$-action whose quotient is equal to $\overline{\Gamma\backslash X}^{BS}$.*

Proof. By Proposition III.9.16, Γ acts on ${}_{\mathbb{Q}}\overline{X}^{BS}$ by real analytic diffeomorphism. By Proposition II.1.4, this Γ-action extends to ${}_{\mathbb{Q}}\overline{X}^{BSO}$. Then the quotient of ${}_{\mathbb{Q}}\overline{X}^{BSO}$

by Γ is a compact closed analytic manifold consisting of 2^r copies of $\overline{\Gamma\backslash X}^{BS}$. By Proposition II.1.5, the $(\mathbb{Z}/2\mathbb{Z})^r$-action on ${}_{\mathbb{Q}}\overline{X}^{BSO}$ commutes with Γ and hence descends to the quotient by Γ, which is clearly the union of 2^r copies of $\overline{\Gamma\backslash X}^{BS}$ with the real analytic structure given in Theorem III.9.18.

Remark III.16.10 The above corollary can also be proved directly without using ${}_{\mathbb{Q}}\overline{X}^{BSO}$. In fact, under the Γ-action, the partition $\mathcal{H}_1, \ldots, \mathcal{H}_r$ of $\mathcal{H}_{{}_{\mathbb{Q}}\overline{X}^{BS}}$ is preserved and hence induces a partition into r parts of the collection of the hypersurfaces of $\overline{\Gamma\backslash X}^{BS}$. Since $\overline{\Gamma\backslash X}^{BS}$ is a compact real analytic manifold with corners and hence has only finitely many boundary faces, Proposition II.1.2 or II.1.9 shows that 2^r copies of $\overline{\Gamma\backslash X}^{BS}$ can be glued into a compact closed analytic manifold.

Remarks III.16.11

(1) A different construction of $\overline{\Gamma\backslash X}^{BSO}$ as a C^∞-manifold has been independently given by Weselmann in [Wes1] in the adelic case and has been used by him in [Wes2] to compute the trace of the Hecke operators on cohomology groups of Γ. Briefly, since $(\mathbb{Z}/2\mathbb{Z})^r$ acts on $\overline{\Gamma\backslash X}^{BSO}$ with quotient equal to $\overline{\Gamma\backslash X}^{BS}$, differential forms on $\Gamma\backslash\overline{X}^{BS}$ can be identified with $(\mathbb{Z}/2\mathbb{Z})^r$-invariant differential forms on the closed manifold $\overline{\Gamma\backslash X}^{BSO}$.

(2) By similar methods, we can show that 2^r copies of $\overline{\Gamma\backslash G}^{BS}$ can be self-glued into a closed analytic manifold $\overline{\Gamma\backslash G}^{BSO}$. See [BJ3] for details.

III.16.12 Summary and comments. The compactification $\overline{\Gamma\backslash X}^{BSO}$ was motivated by the Oshima compactification $\overline{X}^{O}$. Since $\overline{X}^{O}$ has been used to study eigenfunctions of invariant different operators through the theory of differential equations with regular singularities, it is reasonable to expect that $\overline{\Gamma\backslash X}^{BSO}$ might be useful to study Eisenstein series, which are joint eigenfunctions on $\Gamma\backslash X$ of all invariant operators. For such a purpose, the property that $\overline{\Gamma\backslash X}^{BSO}$ is a closed real analytic manifold is important.

12

Subgroup Compactifications of $\Gamma\backslash G$

To construct compactifications of a symmetric space X, a natural method is to find a compact G-space and a G-equivariant embedding $i : X \to Z$ such that the closure of $i(X)$ in Z is a G-compactification of X. For example, the Satake compactifications $\overline{X}^S_\tau$ and the Furstenberg compactifications $\overline{X}^F_P$ are all defined this way. On the other hand, none of the compactifications of a locally symmetric space $\Gamma\backslash X$ has been constructed this way in the earlier chapters. One basic point of this chapter is that compactifications of $\Gamma\backslash X$ can also be studied via compactifications of the homogeneous space $\Gamma\backslash G$ that can be obtained by embeddings into compact spaces. On the other hand, as pointed out in §III.13.1, compactifications of $\Gamma\backslash G$ are also important in themselves.

In the first part of this chapter, we construct a compactification of $\Gamma\backslash G$ by embedding $\Gamma\backslash G$ into a compact G-space when Γ is equal to its own normalizer. In fact, the compact ambient G-space is the space $\mathcal{S}(G)$ of closed subgroups of G, and the compactification of $\Gamma\backslash G$ is called the *subgroup compactification* and denoted by $\overline{\Gamma\backslash G}^{sb}$. A slight modification also applies to $\Gamma\backslash X$ and gives a corresponding subgroup compactification $\overline{\Gamma\backslash X}^{sb}$. The subgroup compactifications $\overline{\Gamma\backslash G}^{sb}$ and $\overline{\Gamma\backslash X}^{sb}$ are dominated by the reductive Borel–Serre compactifications $\overline{\Gamma\backslash G}^{RBS}$ and $\overline{\Gamma\backslash X}^{RBS}$ respectively, and they are isomorphic to each other under certain conditions. Therefore, this reconstructs $\overline{\Gamma\backslash X}^{RBS}$ via an embedding into a compact space and taking the closure.

In the study of reduction theories of arithmetic groups, the identification of $SL(n, \mathbb{Z})\backslash SL(n, \mathbb{R})$ with the space $\mathcal{L}(\mathbb{R}^n)$ of *unimodular lattices* in $\mathbb{R}^n$ plays an important role. In the second part of this chapter, we use the space of lattices in $\mathbb{R}^n$ together with generalizations such as the space of closed subgroups in $\mathbb{R}^n$ and lattices in flags in $\mathbb{R}^n$ to study compactifications of $SL(n, \mathbb{Z})\backslash SL(n, \mathbb{R})$ and $SL(n, \mathbb{Z})\backslash SL(n, \mathbb{R})/SO(n)$. It turns out that we can get several Satake compactifications by scaling differently the lattices in $\mathbb{R}^n$ and making use of lattices in flags of subspaces in $\mathbb{R}^n$.

More specifically, in §III.17, we introduce maximal discrete subgroups of G, which form an important class of discrete subgroups that are equal to their own normalizers. Then we define a compactification of the corresponding homogeneous space $\Gamma\backslash G$ using the space $\mathcal{S}(G)$ of closed subgroups of G. In §III.18, we identify the boundary limit groups of $\overline{\Gamma\backslash G}^{sb}$ and relate them to $\overline{\Gamma\backslash G}^{RSB}$. In §III.19, we first use the space of closed subgroups $\mathcal{S}(\mathbb{R}^n)$ of $\mathbb{R}^n$ to compactify $\mathrm{SL}(n,\mathbb{Z})\backslash\mathrm{SL}(n,\mathbb{R})$ and hence $\mathrm{SL}(n,\mathbb{Z})\backslash\mathrm{SL}(n,\mathbb{R})/\mathrm{SO}(n)$. The resulting compactification is not isomorphic to any Satake compactification. In order to obtain Satake compactifications of $\mathrm{SL}(n,\mathbb{Z})\backslash\mathrm{SL}(n,\mathbb{R})/\mathrm{SO}(n)$, we need to scale the lattices and to introduce lattices in flags, which correspond to different scales of layers of lattices.

III.17 Maximal discrete subgroups and space of subgroups

In this section, we recall the definition and properties of maximal discrete subgroups of the Lie group G, and the space $\mathcal{S}(G)$ of closed subgroups of G, which will be used to construct the subgroup compactification $\overline{\Gamma\backslash G}^{sb}$ in the next section.

More specifically, in III.17.1, we introduce a more general notion of arithmetic subgroups for the purpose of considering maximal discrete subgroups. Maximal discrete subgroups are introduced in III.17.2. Existence of maximal arithmetic subgroups is proved in III.17.3. Several examples are given in III.17.4. Self-normalizing subgroups are defined in III.17.5. Maximal arithmetic subgroups are shown to be self-normalizing in III.17.7. This property is important for defining the subgroup compactification in the next section. A map i_Γ from $\Gamma\backslash G$ to the space $\mathcal{S}(G)$ of closed subgroups is defined in III.17.8. This map is shown to be proper in III.17.9. For a self-normalizing arithmetic group, i_Γ is a proper embedding (III.17.12). The closure of $\Gamma\backslash G$ under this embedding defines the subgroup compactification (III.17.13). A more precise version of Proposition III.17.9 is given in Proposition III.17.11, which determines the limit subgroups or the boundary points of the subgroup compactification. Proposition III.17.11 is proved in two steps: the first step is given by Proposition III.17.15, and the second step in III.17.16, which is illustrated through the example $\mathbf{G} = \mathrm{SL}(2)$.

III.17.1 In the previous sections, we always assume that the arithmetic subgroup is a subgroup of $\mathbf{G}(\mathbb{Q})$. In this chapter, we will call any discrete subgroup Γ of G *arithmetic* if it is commensurable with an arithmetic subgroup of $\mathbf{G}(\mathbb{Q})$. Clearly, such a Γ acts properly on X, and $\Gamma\backslash X$ has finite volume. A fundamental set for Γ can also be constructed by Siegel sets as in Proposition III.2.19. If $\mathbf{G}$ is of adjoint type, then there is no difference since any discrete subgroup commensurable with an arithmetic subgroup in $\mathbf{G}(\mathbb{Q})$ is automatically contained in $\mathbf{G}(\mathbb{Q})$ (see [Bo2]).

We will also assume in this chapter that all normal $\mathbb{Q}$-subgroups of $\mathbf{G}$ have strictly positive $\mathbb{Q}$-rank. Then G has no compact factor of strictly positive dimension.

Definition III.17.2 *A discrete subgroup Γ is called a maximal if it is not properly contained in any discrete subgroup. If Γ is also arithmetic, Γ is called a maximal arithmetic subgroup.*

Lemma III.17.3 *Under the above assumption on* $\mathbf{G}$, *any arithmetic subgroup* Γ *of* G *is contained in a maximal arithmetic subgroup.*

Proof. It is shown in [KM] (see also [Bo5]) that the volume of $\Gamma' \backslash X$ has a strictly positive uniform lower bound for all discrete subgroups Γ' of G. This implies that the length of any chain $\Gamma_0 \subset \Gamma_1 \subset \cdots$ of discrete subgroups starting with $\Gamma_0 = \Gamma$ has a uniform upper bound. Hence Γ is contained in some maximal discrete subgroup, which is clearly also arithmetic.

Note that in the above lemma, the maximal arithmetic subgroups containing Γ may not be unique. The point of this lemma is that it implies the existence of maximal arithmetic subgroups.

III.17.4 Examples of maximal arithmetic subgroups are given in [Bo3] and [Al1] [Al2]. In particular, $\mathrm{SL}(n, \mathbb{Z})$ and $Sp(n, \mathbb{Z})$ are maximal in $\mathrm{SL}(n, \mathbb{R})$ and $Sp(n, \mathbb{R})$ respectively. More generally, if $\mathbf{G}$ is split over $\mathbb{Q}$, then an arithmetic subgroup associated with an admissible Chevalley lattice is maximal ([Bo3, Theorem 7]). Such examples can also be defined in a split k-group if k is a number field with class number one (loc. cit.). For other examples, see [Al1] [Al2] [Bon1]–[Bon4].

Definition III.17.5 *A subgroup* Γ *of* G *is called self-normalizing if it is equal to its own normalizer* $\mathcal{N}(\Gamma) = \{g \in G \mid {}^g\Gamma = \Gamma\}$..

Lemma III.17.6 *Let* $\Gamma \subset G$ *be discrete, of finite covolume. Then the normalizer* $\mathcal{N}(\Gamma)$ *in* G *is discrete, and hence* Γ *is of finite index in* $\mathcal{N}(\Gamma)$.

Proof. Let M be the (ordinary) closure of $\mathcal{N}(\Gamma)$ in G. It is the real locus of an algebraic subgroup $\mathbf{M}$ of $\mathbf{G}$. Its identity component M^0 centralizes Γ. But Γ is Zariski dense in G [Bo2]; hence M^0 is reduced to the identity, and hence $M = \mathcal{N}(\Gamma)$ is discrete.

Corollary III.17.7 *If* Γ *is a maximal arithmetic subgroup, then it is self-normalizing.*

Proof. By the above lemma, $\mathcal{N}(\Gamma)$ is a discrete subgroup. Since it contains Γ, it is equal to Γ by the assumption that Γ is maximal.

III.17.8 Let $\mathcal{S}(G)$ be the space of closed subgroups of G. Clearly G acts on $\mathcal{S}(G)$ by conjugation. Recall from Proposition I.17.2 that the space $\mathcal{S}(G)$ may be endowed with a topology under which it is a compact Hausdorff G-space, i.e., the G-action is continuous.

Let $\Gamma \subset G$ be a discrete subgroup. Define

$$i_\Gamma : \Gamma \backslash G \to \mathcal{S}(G), \quad \Gamma g \to \Gamma^g = g^{-1} \Gamma g.$$

Since the right multiplication on $\Gamma \backslash G$ corresponds to the conjugation on $\mathcal{S}(G)$, this map is G-equivariant.

If $\Gamma = \mathcal{N}(\Gamma)$, it is clearly injective. If moreover Γ is arithmetic, then we shall show that i_Γ is a homeomorphism of $\Gamma\backslash G$ onto its image. The proof uses the reduction theory in Proposition III.2.19, and the main point is the following proposition.

Proposition III.17.9 *Assume that Γ is arithmetic. Let g_j be a divergent sequence in $\Gamma\backslash G$. Assume that the sequence Γ^{g_j} converges in $\mathcal{S}(G)$ to a closed subgroup Γ_∞. Then Γ_∞ is not discrete. In other words, the map i_Γ is proper.*

We can of course replace g_j by any element of Γg_j. Combining this observation with Proposition III.2.19 and passing to a subsequence if necessary, we can assume the following:

Assumption III.17.10 *For some $\mathbf{P}_i$, g_j can be written as $g_j = n_j a_j m_j$, where $n_j \in N_{P_i}$, $a_j \in A_{\mathbf{P}_i}$, $m_j \in M_{\mathbf{P}_i}K$ such that*

1. *$n_j \to n_\infty$ in N_{P_i},*
2. *$m_j \to m_\infty$ in $M_{\mathbf{P}_i}K$,*
3. *$a_j^\alpha \to +\infty$ for all $\alpha \in \Phi(P_i, A_{\mathbf{P}_i})$.*

This assumption implies that g_j converges to (n_∞, m_∞) in ${}_{\mathbb{Q}}\overline{X}^{BS}$. Under these assumptions, Proposition III.17.9 follows from the following more precise result.

Proposition III.17.11 *Under the assumption in Assumption III.17.10, the sequence of subgroups Γ^{g_j} converges in $\mathcal{S}(G)$ to the group $m_\infty^{-1} N_P \Gamma_P m_\infty$.*

Before proving this proposition, we draw the following conclusion.

Corollary III.17.12 *Assume $\Gamma = \mathcal{N}(\Gamma)$ and that Γ is arithmetic. Then $i_\Gamma : \Gamma\backslash G \to \mathcal{S}(G)$ is a homeomorphism of $\Gamma\backslash G$ onto the image $i_\Gamma(\Gamma\backslash G)$. In particular, when Γ is a maximal arithmetic subgroup, i_Γ is an embedding.*

Proof. We need to prove that Γg_j converges to Γg if and only if Γ^{g_j} converges to Γ^g.

Assume that $\Gamma g_j \to \Gamma g$. Since $G \to \Gamma\backslash G$ is a covering map, we can choose g_j such that $g_j \to g$, whence $\Gamma^{g_j} \to \Gamma^g$.

On the other hand, suppose that $\Gamma^{g_j} \to \Gamma^g$. It follows from Proposition III.17.11 that we can assume g_j to be bounded. Passing to a subsequence if necessary, we may assume that $g_j \to g' \in G$. Then $\Gamma^{g_j} \to \Gamma^{g'}$, and therefore $\Gamma^g = \Gamma^{g'}$; hence $g'g^{-1} \in \mathcal{N}(\Gamma) = \Gamma$ and $\Gamma g_j \to \Gamma g$.

Definition III.17.13 Assume that Γ is arithmetic and equal to its own normalizer $\mathcal{N}(\Gamma)$. We denote the closure $\Gamma\backslash G$ in $\mathcal{S}(G)$ under the embedding i_Γ by $\overline{\Gamma\backslash G}^{sb}$ and call it the subgroup compactification of $\Gamma\backslash G$.

Remark III.17.14 If Γ is not a maximal arithmetic subgroup, choose a maximal arithmetic subgroup containing $\Gamma_{\max}$. Then Γ is of finite index in $\Gamma_{\max}$, and the composition of the covering $\Gamma\backslash X \to \Gamma_{\max}\backslash X$ and of the embedding $\Gamma_{\max}\backslash X \hookrightarrow$

$\mathcal{S}(G)$ gives a map $\Gamma\backslash G \to \mathcal{S}(G)$, which is a finite covering map onto its image of degree $[\Gamma_{\max} : \Gamma]$. To use $\mathcal{S}(G)$ to define a compactification, consider the diagonal map

$$\Gamma\backslash G \to (\Gamma\backslash G \cup \{\infty\}) \times \mathcal{S}(G), \quad \Gamma g \mapsto (\Gamma g, g^{-1}\Gamma_{\max} g),$$

where $\Gamma\backslash G \cup \{\infty\}$ is the one-point compactification. Clearly this is an embedding of $\Gamma\backslash G$ into a compact G-space.

The first step in proving Proposition III.17.11 is the following proposition concerning the subgroup Γ_P of Γ. The second part deals with the complement $\Gamma - \Gamma_P$.

Proposition III.17.15 *For a rational parabolic subgroup P and a sequence g_j in G satisfying Assumption III.17.10, the sequence of subgroups $g_j^{-1}\Gamma_P g_j$ converges to $m_\infty^{-1} N_P \Gamma_{M_\mathbf{P}} m_\infty$ in $\mathcal{S}(G)$.*

Proof. The proof consists of two steps. The first step is to show that if the limit exists, it must be contained in $m_\infty^{-1} N_P \Gamma_{M_\mathbf{P}} m_\infty$. The second step shows that the limit exists.

Since Γ_P is contained in $N_P M_\mathbf{P}$ and $\Gamma_{M_\mathbf{P}}$ is the image of Γ_P under the projection $P = N_P A_\mathbf{P} M_\mathbf{P} \to M_\mathbf{P}$, $\Gamma_P \subseteq N_P \Gamma_{M_\mathbf{P}}$. This implies that $g_j^{-1}\Gamma_P g_j \subset g_j^{-1} N_P \Gamma_{M_\mathbf{P}} g_j$. Since $g_j = n_j a_j m_j$, we have

$$\begin{aligned} g_j^{-1} N_P \Gamma_{M_\mathbf{P}} g_j &= m_j^{-1} a_j^{-1} N_P \Gamma_{M_\mathbf{P}} n_j a_j m_j \\ &= m_j^{-1} N_P \Gamma_{M_\mathbf{P}} (a_j^{-1} n_j a_j) m_j. \end{aligned} \tag{III.17.1}$$

Since n_j is bounded, $a_j^{-1} n_j a_j \to id$. This implies that $g_j^{-1} N_P \Gamma_{M_\mathbf{P}} g_j$ converges to $m_\infty^{-1} N_P \Gamma_{M_P} m_\infty$ and hence $\lim_{j\to+\infty} g_j^{-1}\Gamma_P g_j \subset m_\infty^{-1} N_P \Gamma_{M_P} m_\infty$.

We next show that all elements of $m_\infty^{-1} N_P \Gamma_{M_\mathbf{P}} m_\infty$ are limits of sequences of points in $g_j^{-1}\Gamma_P g_j$. For any $\gamma \in \Gamma_{M_\mathbf{P}}$, we claim that $m_\infty^{-1} N_P \gamma m_\infty$ is contained in $\lim_{j\to+\infty} g_j^{-1}\Gamma_P g_j$. Since $\gamma \in \Gamma_{M_\mathbf{P}}$ is arbitrary, this claim implies that $\lim_{j\to+\infty} g_j^{-1}\Gamma_P g_j \supseteq m_\infty^{-1} N_P \Gamma_{M_\mathbf{P}} m_\infty$ and completes the proof of the proposition.

To prove the claim, we note that for every $\gamma \in \Gamma_{M_\mathbf{P}}$, there exists an element $n \in N_P$ such that $n\gamma \in \Gamma_P$, where n is not necessarily in Γ_{N_P}. This implies that $\Gamma_{N_P} n\gamma \subset \Gamma_P$. Using $g_j = n_j a_j m_j$ again, we have

$$\begin{aligned} g_j^{-1}\Gamma_{N_P} n\gamma g_j &= m_j^{-1} a_j^{-1} n_j^{-1} \Gamma_{N_P} n\gamma n_j a_j m_j \\ &= m_j^{-1} (a_j^{-1} n_j^{-1} a_j)(a_j^{-1}\Gamma_{N_P} a_j)(a_j^{-1} n a_j)\gamma(a_j^{-1} n_j a_j) m_j. \end{aligned} \tag{III.17.2}$$

Since n_j is bounded, $a_j^{-1} n_j a_j$ and $a_j^{-1} n_j^{-1} a_j \to id$. Similarly, $a_j^{-1} n a_j \to id$. Since Γ_{N_P} is a cocompact lattice in N_P, there exists a relatively compact open neighborhood G of 1 in N_P such that $N_P = C \cdot \Gamma_{N_P}$, whence also

$$(a_j^{-1} \cdot C \cdot a_j) \cdot (a_j^{-1} \cdot \Gamma_{N_P} \cdot a_j) = N_P.$$

But the $a_j^{-1} \cdot C \cdot a_j$ form a fundamental set of neighborhoods of 1, hence any $n \in N_P$ is a limit of a sequence $a_j^{-1} \cdot \gamma_j \cdot a_j$ $(\gamma_j \in \Gamma_{N_P})$.

III.17.16 *Proof of Proposition III.17.11* Write $\Gamma = \bigcup_{\gamma \in \Gamma/\Gamma_P} \gamma\Gamma_P$, where γ runs over a set of representatives of Γ/Γ_P. Then

$$g_j^{-1}\Gamma g_j = \bigcup_{\gamma \in \Gamma/\Gamma_P} g_j^{-1}\gamma\Gamma_P g_j = g_j^{-1}\Gamma_P g_j \cup \bigcup_{\gamma \in \Gamma/\Gamma_P, \gamma \notin \Gamma_P} g_j^{-1}\gamma\Gamma_P g_j.$$

In view of Proposition III.17.15, it suffices to prove that the sequence of subsets

$$\bigcup_{\gamma \in \Gamma/\Gamma_P, \gamma \notin \Gamma_P} g_j^{-1}\gamma\Gamma_P g_j \tag{III.17.3}$$

in G goes to infinity. By assumption, $n_j \to n_\infty$ and $m_j \to m_\infty$. For simplicity, we assume that $n_j = id$, $m_j = id$, and hence $g_j = a_j$.

Let

$$G = KP = KM_{\mathbf{P}}A_{\mathbf{P}}N_P \cong KM_{\mathbf{P}} \times A_{\mathbf{P}} \times N_P$$

be the decomposition of G induced from the Langlands decomposition of P. For any $g \in G$, write $g = (m_P(g), a_P(g), n_P(g)) \in KM_{\mathbf{P}} \times A_{\mathbf{P}} \times N_P$, and call $a_P(g)$ the $A_{\mathbf{P}}$-component of g. The idea is to show that the $A_{\mathbf{P}}$-component of the elements of the set in equation (III.17.3) uniformly goes to infinity. We will use the fundamental representations of $\mathbf{G}$ defined over $\mathbb{Q}$ [Bo4, §14] [BT, §12] to prove this.

Let $\mathbf{P}_0$ be a minimal rational parabolic subgroup of $\mathbf{G}$ contained in $\mathbf{P}$. For any simple $\mathbb{Q}$-root $\alpha \in \Delta(A_{\mathbf{P}_0}, P_0)$, there is a strongly rational representation $(\pi_\alpha, \mathbf{V}_\alpha)$ of $\mathbf{G}$ whose highest weight λ_α is orthogonal to $\Delta(A_{\mathbf{P}_0}, P_0) - \{\alpha\}$, and $\langle \lambda_\alpha, \alpha \rangle > 0$. Then the weight space of λ_α is invariant under the maximal parabolic subgroup $P_{0,\Delta-\{\alpha\}}$ [BT, §12.2]. Fix an inner product $\| \ \|$ on $\mathbf{V}_\alpha(\mathbb{R})$ that is invariant under K, and with respect to which $A_{\mathbf{P}_0}$ is represented by self-adjoint operators. Let e_0 be a unit vector in the weight space of λ_α. Let $P_{0,\Delta-\{\alpha\}} = M_{0,\Delta-\{\alpha\}}A_{0,\Delta-\{\alpha\}}N_{0,\Delta-\{\alpha\}}$ be the Langlands decomposition of $P_{0,\Delta-\{\alpha\}}$. Then for any $p \in M_{0,\Delta-\{\alpha\}}N_{0,\Delta-\{\alpha\}}$,

$$\pi_\alpha(p)e_0 = \pm e_0.$$

The Langlands decomposition $P_{0,\Delta-\{\alpha\}} = M_{\mathbf{P}_{0,\Delta-\{\alpha\}}}A_{\mathbf{P}_{0,\Delta-\{\alpha\}}}N_{P_{0,\Delta-\{\alpha\}}}$ induces the decomposition of G:

$$G = KM_{\mathbf{P}_{0,\Delta-\{\alpha\}}}A_{\mathbf{P}_{0,\Delta-\{\alpha\}}}N_{P_{0,\Delta-\{\alpha\}}} \cong KM_{\mathbf{P}_{0,\Delta-\{\alpha\}}} \times A_{\mathbf{P}_{0,\Delta-\{\alpha\}}} \times N_{P_{0,\Delta-\{\alpha\}}}.$$

For any $g \in G$, denote the $A_{P_{0,\Delta-\{\alpha\}}}$-component by $a_{\Delta-\{\alpha\}}(g)$. Then

$$\|\pi_\alpha(g)e_0\| = a_{\Delta-\{\alpha\}}(g)^{\lambda_\alpha},$$

where λ_α is restricted to the subgroup $A_{\mathbf{P}_{0,\Delta-\{\alpha\}}} \subseteq A_{\mathbf{P}_0}$. If the $\mathbb{Q}$-parabolic subgroup $\mathbf{P}$ is contained in $\mathbf{P}_{0,\Delta-\{\alpha\}}$, then $M_{\mathbf{P}}N_P \subseteq M_{\mathbf{P}_{0,\Delta-\{\alpha\}}}N_{P_{0,\Delta-\{\alpha\}}}$, and hence

$$\|\pi_\alpha(g)e_0\| = a_P(g)^{\lambda_\alpha}.$$

Now we follow the computations in [JM, pp. 505–506] [Bo6, pp. 550–551]. For any $p \in \Gamma_P$, let

$$g = a_j^{-1}\gamma p a_j.$$

Since $\mathbf{P}_0$ is a minimal $\mathbb{Q}$-parabolic subgroup contained in $\mathbf{P}$, we can write $\mathbf{P} = \mathbf{P}_{0,I}$, where $I \subseteq \Delta = \Delta(P_0, A_{P_0})$. For any $\alpha \in \Delta - I$, there are two cases: (1) $\gamma \in P_{0,\Delta-\{\alpha\}}$, (2) $\gamma \notin P_{0,\Delta-\{\alpha\}}$.

In case (1), $g \in P_{0,\Delta-\{\alpha\}}$, and hence by [BS2, Prop. 1.2], $g \in M_{\mathbf{P}_{0,\Delta-\{\alpha\}}} N_{P_{0,\Delta-\{\alpha\}}}$, and hence

$$a_P(g)^{\lambda_\alpha} = 1.$$

In case (2), using the *Bruhat decomposition* of $\mathbf{G}$ over $\mathbb{Q}$ [Bo4, §11.4] [JM, Lemma 10.11], write $\gamma^{-1} = uwtmv$, where $u \in N'_w$, $t \in A_{P_0}$, $m \in M_{P_0}$, $v \in N_{P_0}$, and w is from a set of fixed representatives of the $\mathbb{Q}$-Weyl group of $\mathbf{G}$. Consider the element $w^{-1}g$ and its $A_{\mathbf{P}}$-component $a_P(w^{-1}g)$. Then the computations in [JM, pp. 505-506] (also [Bo6, p. 551]) show that there exists a positive constant δ that depends only on Γ and the fundamental representation π_α such that

$$a_P(w^{-1}g)^{\lambda_\alpha} \geq a_j^\alpha \delta.$$

This implies that when $j \to +\infty$, $w^{-1}g$ and hence g go to infinity uniformly with respect to an arbitrary choice of $p \in \Gamma_P$.

Note that $P = P_{0,I} = \cap_\alpha P_{0,\Delta-\{\alpha\}}$, where $\alpha \in \Delta - I$. Hence for any $\gamma \notin \Gamma_P$, there exists at least one $\alpha \in \Delta - I$ such that $\gamma \notin P_{0,\Delta-\{\alpha\}}$. Since the set of w is a fixed finite set, this implies that when $j \to +\infty$, the subset of G defined in equation (III.17.3) goes to infinity. This completes the proof of the proposition.

Remarks III.17.17

(1) When $G = \mathrm{SL}_2(\mathbb{R})$ and Γ is of finite index in $\mathrm{SL}_2(\mathbb{Z})$, the second part of the proof showing the divergence to infinity can also be seen as follows. In this case X is equal to the upper half-plane $\{x + iy \mid y > 0\}$. For any $z \in X$, $\mathrm{Im}\ \gamma z$ is uniformly bounded from above, when γ runs over $\Gamma - \Gamma_\infty$. This is related to the fact that the Eisenstein series

$$E_\infty(z, s) = \sum_{\gamma \in \Gamma_\infty \backslash \Gamma} (\mathrm{Im}\ \gamma z)^s$$

converges uniformly for s with $\mathrm{Re}\ s > 1$.

(2) Besides the fact that the proof of the above proposition is related to the convergence of Eisenstein series, the limit subgroups $N_P \Gamma_{M_{\mathbf{P}}}$ are exactly the subgroups that leave invariant the constant term of the Eisenstein series along the parabolic subgroup P. These connections together with its natural definition make $\overline{\Gamma\backslash G}^{sb}$ an interesting compactification.

III.17.18 Summary and comments. The subgroup compactification $\overline{\Gamma\backslash G}^{sb}$ was motivated by the fact that lattices in $\mathbb{R}^n$ have been used in studying the space $SL(n,\mathbb{Z})\backslash SL(n,\mathbb{R})$, the reduction theory of $SL(n,\mathbb{Z})$, and the reduction theory of general arithmetic subgroups. In fact, $SL(n,\mathbb{Z})\backslash SL(n,\mathbb{R})$ can be identified with the moduli space of unimodular lattices in $\mathbb{R}^n$. For example, Mahler's compactness criterion for subsets of the space of lattices in $\mathbb{R}^n$ played an important role in compactness of $\Gamma\backslash G$ or equivalently $\Gamma\backslash G$. In this section, instead of abelian lattices in $\mathbb{R}^n$, we mapped $\Gamma\backslash G$ into the space of (nonabelian) lattices in G. In §III.19, we will use (abelian) lattices in $\mathbb{R}^n$ to study compactifications of $SL(n,\mathbb{Z})\backslash SL(n,\mathbb{R})$.

III.18 Subgroup compactification of $\Gamma\backslash G$ and $\Gamma\backslash X$

In the previous section, we have defined the subgroup compactification $\overline{\Gamma\backslash G}^{sb}$ for any maximal arithmetic subgroup Γ. In this section, we determine the limit subgroups on the boundary of $\overline{\Gamma\backslash G}^{sb}$ and relate the compactification $\overline{\Gamma\backslash G}^{sb}$ to the reductive Borel–Serre compactification $\overline{\Gamma\backslash G}^{RBS}$. By considering the space of K-orbits in $\mathcal{S}(G)$, we obtain a subgroup compactification of $\Gamma\backslash X$ and relate it to the reductive Borel–Serre compactification $\overline{\Gamma\backslash X}^{RBS}$ of $\Gamma\backslash G/K$. This gives a construction of $\Gamma\backslash G/K$ via the embedding method.

This section is organized as follows. We first show that the reductive Borel–Serre compactification $\overline{\Gamma\backslash G}^{RBS}$ dominates the subgroup compactification $\overline{\Gamma\backslash G}^{sb}$ in III.18.1. To determine G-orbits in $\overline{\Gamma\backslash G}^{sb}$, we introduce a Γ_M-equivalence relation on $\mathbb{Q}$-parabolic subgroups in III.18.2. The G-orbits are determined in III.18.4. Under some conditions on $\mathbf{G}$ and Γ, $\overline{\Gamma\backslash G}^{sb}$ is shown to be isomorphic to $\overline{\Gamma\backslash G}^{RBS}$ in III.18.6. Examples in which all these conditions are satisfied are given in III.18.7. The above discussions concern $\Gamma\backslash G$. To get a compactification of $\Gamma\backslash X$ using an embedding into a compact space, we need to divide $\mathcal{S}(G)$ by K (III.18.9). The induced compactification of $\Gamma\backslash X$ is given in III.18.10.

Proposition III.18.1 *Assume that Γ is a maximal arithmetic subgroup. Then the identity map on $\Gamma\backslash G$ extends to a continuous map from $\overline{\Gamma\backslash G}^{RBS}$ to $\overline{\Gamma\backslash G}^{sb}$ that is surjective and equivariant with respect to the right G-action.*

Proof. Let $\mathbf{P}$ be a rational parabolic subgroup. Let g_j be an unbounded sequence in $\Gamma\backslash G$ converging to $m_\infty \in \Gamma_{M_\mathbf{P}}\backslash X_\mathbf{P}$ in $\overline{\Gamma\backslash G}^{RBS}$. Since Γ_{N_P} is a cocompact lattice in N_P, we can choose a lift $\tilde{g}_j$ in G such that in the decomposition $\tilde{g}_j = (n_j, a_j, m_j) \in N_P \times A_P \times (M_\mathbf{P} K)$, the component n_j is bounded, and the component m_j converges to a lift $\tilde{m}_\infty$ of m_∞ in $M_\mathbf{P} K = \hat{e}(P)$.

By the definition of the convergence in $\overline{\Gamma\backslash G}^{RBS}$, we know that for all $\alpha \in \Phi(P, A_\mathbf{P})$, $(a_j)^\alpha \to +\infty$. Then by Proposition III.17.15, $\tilde{g}_j$ converges in $\overline{\Gamma\backslash G}^{sb}$ to $\tilde{m}_\infty^{-1} N_P \Gamma_{M_\mathbf{P}} \tilde{m}_\infty$, i.e., the subgroup $\tilde{g}_j^{-1}\Gamma\tilde{g}_j$ converges to $\tilde{m}_\infty^{-1} N_P \Gamma_{M_\mathbf{P}} \tilde{m}_\infty$. Since

$g_j^{-1}\Gamma g_j = \tilde{g}_j^{-1}\Gamma\tilde{g}_j$ and the limit $\tilde{m}_\infty^{-1}N_P\Gamma_{M_\mathbf{P}}\tilde{m}_\infty = m_\infty^{-1}N_P\Gamma_{M_\mathbf{P}}m_\infty$ does not depend on the choice of the lift $\tilde{m}_\infty$ in G, g_j converges in $\overline{\Gamma\backslash G}^{sb}$. This shows that every unbounded sequence in $\Gamma\backslash G$ that is convergent in $\Gamma\backslash\overline{G}^{RBS}$ also converges in $\overline{\Gamma\backslash G}^{sb}$. Since both $\Gamma\backslash\overline{G}^{RBS}$ and $\overline{\Gamma\backslash G}^{sb}$ are metrizable compactifications of $\Gamma\backslash G$, by [GJT, Lemma 3.28], the identity map on $\Gamma\backslash G$ extends to a continuous map from $\Gamma\backslash\overline{G}^{RBS}$ to $\overline{\Gamma\backslash G}^{sb}$, which is automatically surjective, and the extended map is G-equivariant with respect to the right G-action.

Definition III.18.2 Two rational parabolic subgroups $\mathbf{P}_1,\ \mathbf{P}_2$ are called Γ_M-equivalent if there exists $g \in G$ such that

$$g^{-1}N_{P_1}\Gamma_{M_{\mathbf{P}_1}}g = N_{P_2}\Gamma_{M_{\mathbf{P}_2}}.$$

Since N_{P_i} is the identity component of $N_{P_i}\cdot\Gamma_{M_{\mathbf{P}_i}}$, the normalizer of $N_{P_i}\cdot\Gamma_{M_{\mathbf{P}_i}}$ is contained in the normalizer of N_{P_i}, hence in P_i by [BT, Proposition 3.1]. This implies that if $\mathbf{P}_1,\ \mathbf{P}_2$ are Γ_M-equivalent, $\mathbf{P}_1,\ \mathbf{P}_2$ are conjugate under G and hence also under $\mathbf{G}(\mathbb{Q})$. On the other hand, if $\mathbf{P}_1,\ \mathbf{P}_2$ are Γ-conjugate, they are clearly Γ_M-equivalent. Let $\mathbf{P}_1,\ \dots,\ \mathbf{P}_k$ be a set of representatives of the Γ-conjugacy classes of rational parabolic subgroups. Then there exists a subset of $\mathbf{P}_1,\ \dots,\ \mathbf{P}_k$ that are representatives of the Γ_M-equivalent classes of rational parabolic subgroups. For simplicity, assume that they are given by $\mathbf{P}_1,\ \dots,\ \mathbf{P}_l$ for some $l \leq k$.

Lemma III.18.3 *For every i, the normalizer $\mathcal{N}(N_{P_i}\Gamma_{M_{\mathbf{P}_i}})$ of $N_{P_i}\Gamma_{M_{\mathbf{P}_i}}$ in $M_{\mathbf{P}_i}K$ is equal to the normalizer $\mathcal{N}(\Gamma_{M_{\mathbf{P}_i}})$ of $\Gamma_{M_{\mathbf{P}_i}}$ in $M_{\mathbf{P}_i}$.*

Proof. If an element normalizes $N_{P_i}\Gamma_{M_{\mathbf{P}_i}}$, then it normalizes its identity component N_{P_i}. It follows from [BT, Proposition 3.1] that the normalizer of N_{P_i} in G is equal to P_i. In fact, Proposition 3.1 in [BT] shows that there is a rational parabolic subgroup $\mathbf{P}$ that contains the normalizer of N_{P_i} and whose unipotent radical is equal to N_{P_i}. Clearly, such a parabolic subgroup has to be exactly equal to $\mathbf{P}_i$. Then it is clear that the normalizer of $N_{P_i}\Gamma_{M_{\mathbf{P}_i}}$ in $M_{\mathbf{P}_i}K$ is contained in $M_{\mathbf{P}_i}$ and hence is equal to the normalizer $\mathcal{N}(\Gamma_{P_i})$ of $\Gamma_{M_{\mathbf{P}_i}}$ in $M_{\mathbf{P}_i}$.

Theorem III.18.4 *Assume Γ to be a maximal arithmetic subgroup. Then*

$$\overline{\Gamma\backslash G}^{sb} = \Gamma\backslash G \cup \coprod_{i=1}^{l} \mathcal{N}(\Gamma_{M_{\mathbf{P}_i}}) \backslash M_{\mathbf{P}_i}K$$

is the decomposition of $\overline{\Gamma\backslash G}^{sb}$ into G-orbits.

Proof. It follows from Propositions III.17.15 and III.18.1 that under the map

$$\Gamma\backslash\overline{G}^{RBS} \to \overline{\Gamma\backslash G}^{sb},$$

the image, denoted by $b(P_i)$, of the boundary component $\Gamma_{M_{\mathbf{P}_i}}\backslash M_{\mathbf{P}_i}K$ of $\Gamma\backslash\overline{G}^{RBS}$ consists of subgroups of the form $m^{-1}N_{P_i}\Gamma_{M_{\mathbf{P}_i}}m$, where $m \in M_{\mathbf{P}_i}K$, $i = 1, \ldots, k$. By Lemma III.17.5, the normalizer of $N_{P_i}\Gamma_{M_{\mathbf{P}_i}}$ in $M_{\mathbf{P}_i}K$ is equal to the normalizer $\mathcal{N}(\Gamma_{M_{\mathbf{P}_i}})$ of $\Gamma_{M_{\mathbf{P}_i}}$ in $M_{\mathbf{P}_i}$, and hence we obtain that the image $b(P_i)$ can be identified with $\mathcal{N}(\Gamma_{M_{\mathbf{P}_i}})\backslash M_{\mathbf{P}_i}K$ through the map $m \to m^{-1}N_{P_i}\Gamma_{M_{\mathbf{P}_i}}m$.

Since each boundary component $\Gamma_{M_{\mathbf{P}_i}}\backslash M_{\mathbf{P}_i}K$ of $\Gamma\backslash\overline{G}^{RBS}$ is a G-orbit (Proposition III.14.8) and the map $\Gamma\backslash\overline{G}^{RBS} \to \overline{\Gamma\backslash G}^{sb}$ is G-equivariant (Proposition III.18.1), the image $b(P_i)$ is also a G-orbit. In fact, for $m^{-1}N_{P_i}\Gamma_{M_{\mathbf{P}_i}}m \in b(P_i)$ and $g \in G$,

$$g \circ m^{-1}N_{P_i}\Gamma_{M_{\mathbf{P}_i}}m = g^{-1}m^{-1}N_{P_i}\Gamma_{M_{\mathbf{P}_i}}mg.$$

From this it is clear that two image sets $b(P_i)$, $b(P_j)$ are equal if and only if P_i and P_j are Γ_M-equivalent. This gives the disjoint decomposition of $\overline{\Gamma\backslash G}^{sb}$ in the theorem, and shows that the decomposition is exactly the decomposition into the disjoint G-orbits on $\overline{\Gamma\backslash G}^{sb}$.

Proposition III.18.5 *For any arithmetic subgroup Γ, let Γ' be a maximal discrete group containing Γ. Then the projection map $\Gamma\backslash G \to \Gamma'\backslash G$ extends to a continuous map $\Gamma\backslash\overline{G}^{RBS} \to \overline{\Gamma'\backslash G}^{sb}$.*

Proof. We note that the quotient map $\Gamma\backslash G \to \Gamma'\backslash G$ extends to a continuous map $\Gamma\backslash\overline{G}^{RBS} \to \Gamma'\backslash\overline{G}^{RBS}$. Then the proposition follows from Proposition 5.12.

Proposition III.18.6 *Suppose that* **G** *is a semisimple algebraic group defined over $\mathbb{Q}$, $\Gamma \subset \mathbf{G}(\mathbb{Q})$ is a maximal arithmetic subgroup, and the Γ-conjugacy relation on the set of all rational parabolic subgroups induces the same relation as the Γ_M-equivalence relation in Definition III.18.2. If for every rational parabolic subgroup* **P**, *$M_{\mathbf{P}}$ is semisimple and has no compact factor of positive dimension, and its subgroup $\Gamma_{M_{\mathbf{P}}}$ is also maximal, then $\overline{\Gamma\backslash G}^{sb}$ is G-equivariantly isomorphic to the reductive Borel–Serre compactification $\Gamma\backslash\overline{G}^{RBS}$.*

Proof. Let $\mathbf{P}_1, \ldots, \mathbf{P}_k$ be a set of representatives of Γ-conjugacy classes of proper rational parabolic subgroups. By assumption, they are also representatives of the Γ_M-relation. Since $\mathcal{N}(\Gamma_{M_{\mathbf{P}_i}}) = \Gamma_{M_{\mathbf{P}_i}}$, by Theorem III.18.4, the boundary of $\overline{\Gamma\backslash G}^{sb}$ is equal to $\bigcup_{i=1}^{k}\Gamma_{M_{\mathbf{P}_i}}\backslash M_{\mathbf{P}_i}K$, which is also the boundary of $\Gamma\backslash\overline{G}^{RBS}$. This implies that the continuous map from $\Gamma\backslash\overline{G}^{RBS}$ to $\overline{\Gamma\backslash G}^{sb}$ in Proposition III.18.1 is bijective. Since both compactifications are Hausdorff, they are homeomorphic, and the homeomorphism is equivariant with respect to the right G-action.

Remark III.18.7 Examples in which all the conditions in the above theorem are satisfied include $G = \mathrm{SL}(n, \mathbb{R})$, $Sp(n, \mathbb{R})$, $\Gamma = \mathrm{SL}(n, \mathbb{Z})$, $Sp(n, \mathbb{Z})$.

Remark III.18.8 If Γ is maximal, but other conditions are not satisfied, then $M_{\mathbf{P}_i}$ is in general only reductive. Let $M'_{\mathbf{P}_i}$ be the derived group of $M_{\mathbf{P}_i}$, and $\mathcal{C}(M_{\mathbf{P}_i})$ the center of $M_{\mathbf{P}_i}$. Then $\mathcal{N}(\Gamma_{M_{\mathbf{P}_i}})$ contains $\mathcal{C}(M_{\mathbf{P}_i})$, and $\mathcal{C}(M_{\mathbf{P}_i})\backslash\mathcal{N}(\Gamma_{M_{\mathbf{P}_i}})$ is a discrete subgroup of $M'_{\mathbf{P}_i}$, and $\mathcal{N}(\Gamma_{M_{\mathbf{P}_i}})\backslash M_{\mathbf{P}_i}K$ is equal to $(\mathcal{C}(M_{\mathbf{P}_i}))\backslash\mathcal{N}(\Gamma_{M_{\mathbf{P}_i}})\backslash M'_{\mathbf{P}_i}K$. This shows that the boundary faces of $\overline{\Gamma\backslash G}^{sb}$ are analogous to the boundary faces in the maximal Satake compactification $\overline{\Gamma\backslash X}^{S}_{\max}$ in §III.11.

III.18.9 Since the group G acts continuously on $\mathcal{S}(G)$ by conjugation, the maximal compact subgroup K acts continuously on $\mathcal{S}(G)$, and the quotient $\mathcal{S}(G)/K$ by K is a compact Hausdorff.

Assume that Γ is a maximal arithmetic subgroup. The embedding $i_\Gamma : \Gamma\backslash G \to \mathcal{S}(G)$ induces an embedding

$$i_\Gamma : \Gamma\backslash X = \Gamma\backslash G/K \to \mathcal{S}(G)/K, \quad \Gamma g K \mapsto K g^{-1}\Gamma g^{-1} K.$$

The closure of $i_\Gamma(\Gamma\backslash X)$ in $(G)/K$ is called the *subgroup compactification* of $\Gamma\backslash X$ and denoted by $\overline{\Gamma\backslash X}^{sb}$.

Proposition III.18.10 *Assume that Γ is a maximal arithmetic subgroup. Then the reductive Borel–Serre compactification $\overline{\Gamma\backslash X}^{RBS}$ dominates the subgroup compactification $\overline{\Gamma\backslash X}^{sb}$. If* **G** *is semisimple, then the maximal Satake compactification $\overline{\Gamma\backslash X}^{S}_{\max}$ also dominates $\overline{\Gamma\backslash X}^{sb}$. Furthermore, if all the rational parabolic subgroups of* **G** *of the same type, i.e., conjugate under* $\mathbf{G}(\mathbb{Q})$, *are Γ-conjugate, then $\overline{\Gamma\backslash X}^{S}_{\max}$ is isomorphic to $\overline{\Gamma\backslash X}^{sb}$.*

Proof. The first statement follows from the corresponding result for $\Gamma\backslash G$ in Proposition III.18.1. The second statement follows from Theorem III.18.4. By assumption, all rational parabolic subgroups of G of the same type are Γ-conjugate. This implies that two rational parabolic subgroups are Γ_M-equivalent if and only if they are Γ-conjugate. Then the last statement follows from the equality

$$\mathcal{N}(\Gamma_P)M_{\mathbf{P}}/K_{\mathbf{P}} = X_P$$

for any rational parabolic subgroup **P**.

III.18.11 Summary and comments. The relations between the subgroup compactification $\overline{\Gamma\backslash G}^{sb}$ and other comactifications such as the reductive Borel–Serre compactifications and the Satake compactifications and the form of the limit subgroups in the boundary of $\overline{\Gamma\backslash G}^{sb}$ indicate that this compactification is natural. As mentioned earlier, the left-G multiplication on $\Gamma\backslash G$ does not extend continuously to $\overline{\Gamma\backslash G}^{BS}$, but does extend continuously to $\overline{\Gamma\backslash G}^{RBS}$. A natural explanation is given by the relation between $\overline{\Gamma\backslash G}^{RBS}$ and $\overline{\Gamma\backslash G}^{sb}$, since the left-$G$ multiplication on $\Gamma\backslash G$ clearly extends continuously to $\overline{\Gamma\backslash G}^{sb}$.

III.19 Spaces of flags in $\mathbb{R}^n$, flag lattices, and compactifications of $\mathbf{SL}(n, \mathbb{Z})\backslash\mathbf{SL}(n, \mathbb{R})$

In the reduction theory of arithmetic groups, the space $\mathcal{L}(\mathbb{R}^n)$ of lattices in $\mathbb{R}^n$ was used crucially in Mahler's compactness criterion (see [Bo4]). Since $\mathrm{SL}(n, \mathbb{Z})\backslash\mathrm{SL}(n, \mathbb{R})$ can be identified with the space $\mathcal{L}(\mathbb{R}^n)$ of lattices in $\mathbb{R}^n$ of covolume 1, it is natural to study compactification of this homogeneous space by embedding it into the compact space $\mathcal{S}(\mathbb{R}^n)$, which consists of closed subgroups of $\mathbb{R}^n$. By dividing out by K on the right, we obtain a compactification of $\Gamma\backslash X = \mathrm{SL}(n, \mathbb{Z})\backslash\mathrm{SL}(n, \mathbb{R})/\mathrm{SO}(n)$. It turns out that this compactification is not a Satake compactification or other familiar one.

To obtain Satake compactifications of $\Gamma\backslash X$ using the identification with the space $\mathcal{L}(\mathbb{R}^n)$, we need to scale the lattices and use groups that do not belong to the space $\mathcal{S}(\mathbb{R}^n)$. To obtain the maximal Satake compactification, we need to scale successively at different rates and consider spaces of flags in $\mathbb{R}^n$ and lattices in these flags. These give explicit examples of the decomposition of the Satake compactifications of $\mathrm{SL}(n, \mathbb{Z})\backslash\mathrm{SL}(n, \mathbb{R})$ into $G = \mathrm{SL}(n, \mathbb{R})$-orbits and the fibration of each orbit over the *flag variety*.

As mentioned in the preface, this chapter, in particular this section, was motivated by [Mac], where attempts were outlined to describe the reductive Borel–Serre and the Borel–Serre compactifications of $\mathrm{SL}(n, \mathbb{Z})\backslash\mathrm{SL}(n, \mathbb{R})/\mathrm{SO}(n)$ and more generally $\Gamma\backslash\mathrm{SL}(n, \mathbb{R})/\mathrm{SO}(n)$, $\Gamma \subset \mathrm{SL}(n, \mathbb{Z})$, in terms of lattices in $\mathbb{R}^n$.

This section is organized as follows. The spaces $\mathcal{S}(\mathbb{R}^n)$, $\mathcal{L}(\mathbb{R}^n)$ of closed subgroups and lattices in $\mathbb{R}^n$ are introduced in III.19.1. Then $\mathrm{SL}(n, \mathbb{Z})\backslash\mathrm{SL}(n, \mathbb{R})$ is identified with $\mathcal{L}(\mathbb{R}^n)$ in III.19.2. The embedding of $\mathrm{SL}(n, \mathbb{Z})\backslash\mathrm{SL}(n, \mathbb{R})$ in $\mathcal{S}(\mathbb{R}^n)$ gives the lattice compactification (III.19.3). When $n \geq 3$, this compactification is not isomorphic to any of the earlier compactifications (III.19.5). (See III.19.6 for some informal explanations.) To realize some Satake compactifications using lattices in $\mathbb{R}^n$, we need to scale the lattices suitably. Different scalings lead to the sublattice compactification in III.19.9 and the suplattice compactification in III.19.11. They are isomorphic to nonmaximal Satake compactifications. To obtain the maximal Satake compactification, we need to introduce flags of vector subspaces and flag lattices (III.19.13). Basics of the space $\mathcal{FL}(\mathbb{R}^n)$ of flag lattices are studied in III.19.15–III.19.18. The flag-lattice compactification is defined in III.19.19, which is shown to be isomorphic to the reductive Borel–Serre compactification of $\mathrm{SL}(n, \mathbb{Z})\backslash\mathrm{SL}(n, \mathbb{R})$ in III.19.20. The maximal Satake compactification of $\mathrm{SL}(n, \mathbb{Z})\backslash\mathrm{SL}(n, \mathbb{R})/\mathrm{SO}(n)$ is identified with a flag-lattice compactification in III.19.22.

III.19.1 Let $\mathcal{S}(\mathbb{R}^n)$ be the space of closed subgroups of $\mathbb{R}^n$. As shown in Proposition I.17.2 in §I.17, it is a compact Hausdorff space. The group $\mathrm{SL}(n, \mathbb{R})$ acts on $\mathcal{S}(\mathbb{R}^n)$ by right multiplication: for any $g \in \mathrm{SL}(n, \mathbb{R})$ and $H \in \mathcal{S}(\mathbb{R}^n)$,

$$H \cdot g = Hg,$$

where g acts as a linear transform and Hg is clearly an (additive) abelian subgroup.

Let $\mathbb{Z}^n$ be the standard lattice in $\mathbb{R}^n$. Then for any $g \in \mathrm{SL}(n,\mathbb{R})$, $\mathbb{Z}^n g$ is a unimodular lattice in $\mathbb{R}^n$, and the stabilizer of $\mathbb{Z}^n$ is $\mathrm{SL}(n,\mathbb{Z})$. Let $\mathcal{L}(\mathbb{R}^n)$ be the space of unimodular lattices in $\mathbb{R}^n$. Then we have the following result.

Proposition III.19.2 *The homogeneous space* $\mathrm{SL}(n,\mathbb{Z})\backslash\mathrm{SL}(n,\mathbb{R})$ *can be canonically identified with* $\mathcal{L}(\mathbb{R}^n)$ *under the map*

$$\mathrm{SL}(n,\mathbb{Z})g \mapsto \mathbb{Z}^n g.$$

Since $\mathcal{L}(\mathbb{R}^n)$ is naturally contained in $\mathcal{S}(\mathbb{R}^n)$, we obtain the following result.

Proposition III.19.3 *The map*

$$i : \mathrm{SL}(n,\mathbb{Z})\backslash\mathrm{SL}(n,\mathbb{R}) \to \mathcal{S}(\mathbb{R}^n), \quad \mathrm{SL}(n,\mathbb{Z})g \mapsto \mathbb{Z}^n g,$$

is an $\mathrm{SL}(n,\mathbb{R})$*-equivariant embedding with respect to the right* $\mathrm{SL}(n,\mathbb{R})$*-action, and the closure of* $i(\mathrm{SL}(n,\mathbb{Z})\backslash\mathrm{SL}(n,\mathbb{R}))$ *is an* $\mathrm{SL}(n,\mathbb{R})$*-Hausdorff compactification, denoted by* $\overline{\mathrm{SL}(n,\mathbb{Z})\backslash\mathrm{SL}(n,\mathbb{R})}^{la}$ *and called the lattice compactification.*

Proof. We first show that the map i is an embedding. For a sequence $\mathrm{SL}(n,\mathbb{Z})g_j$ in $\mathrm{SL}(n,\mathbb{Z})\backslash\mathrm{SL}(n,\mathbb{R})$ converging to $\mathrm{SL}(n,\mathbb{Z})g_\infty$ for some $g_\infty \in \mathrm{SL}(n,\mathbb{R})$, we can choose suitable representatives g_j, g_∞ such that $g_j \to g_\infty$. This clearly implies that $\mathbb{Z}^n g_j \to \mathbb{Z}^n g_\infty$ in $\mathcal{S}(\mathbb{R}^n)$.

On the other hand, we need to show that if $\mathbb{Z}^n g_j$ converges to $\mathbb{Z}^n g_\infty$, then $\mathrm{SL}(n,\mathbb{Z})g_j$ converges to $\mathrm{SL}(n,\mathbb{Z})g_\infty$ in $\mathrm{SL}(n,\mathbb{Z})\backslash\mathrm{SL}(n,\mathbb{R})$. In fact, we claim that g_j is bounded. Assume the claim first. By passing to a subsequence, we can assume that $g_j \to g'_\infty$ for some $g'_\infty \in \mathrm{SL}(n,\mathbb{R})$. By the continuity of the map i in the previous paragraph, $\mathbb{Z}^n g_j \to \mathbb{Z}^n g'_\infty$. Hence $\mathbb{Z}^n g_\infty = \mathbb{Z}^n g'_\infty$, which implies that $\mathrm{SL}(n,\mathbb{Z})g_\infty = \mathrm{SL}(n,\mathbb{Z})g'_\infty$, and hence $\mathrm{SL}(n,\mathbb{Z})g_j$ converges to $\mathrm{SL}(n,\mathbb{Z})g_\infty$ in $\mathrm{SL}(n,\mathbb{Z})\backslash\mathrm{SL}(n,\mathbb{R})$.

To prove the claim, let $\mathbf{P}$ be the minimal rational parabolic subgroup of $\mathrm{SL}(n)$ consisting of upper triangular matrices. Since there is only one $\mathrm{SL}(n,\mathbb{Z})$-conjugacy class of minimal rational parabolic subgroups of $\mathrm{SL}(n)$, a Siegel set associated with $\mathbf{P}$ projects surjectively to $\Gamma\backslash G$. If g_j is not bounded, then a suitable representative of g_j has the horospherical coordinates with respect to $\mathbf{P}$,

$$g_j = (n_j, a_j, m_j) \in N_P \times A_{\mathbf{P}} \times M_{\mathbf{P}}K, \tag{III.19.4}$$

satisfying the following conditions:

1. n_j, m_j are bounded,
2. $a_j = \mathrm{diag}(d_{j,1}, \dots, d_{j,n})$ is unbounded.

It is clear that $\mathbb{Z}^n a_j$ does not converge to a discrete subgroup in $\mathbb{R}^n$, and it follows that $\mathbb{Z}^n g_j$ does not converge to any discrete subgroup either, in particular, not to $\mathbb{Z}^n g_\infty$. This contradiction proves the claim.

Remark III.19.4 The action of $\mathrm{SL}(n,\mathbb{R})$ on $\mathcal{S}(\mathbb{R})$ is a right action instead of the usual (left) group action in order to be consistent with the left action used to define the quotient $\Gamma\backslash G = \mathrm{SL}(n,\mathbb{Z})\backslash\mathrm{SL}(n,\mathbb{R})$ used in this book. If we use the quotient $\mathrm{SL}(n,\mathbb{R})/\mathrm{SL}(n,\mathbb{Z})$, then the action of $\mathrm{SL}(n,\mathbb{R})$ on $\mathcal{S}(\mathbb{R}^n)$ is the left action and $\mathrm{SL}(n,\mathbb{R})/\mathrm{SL}(n,\mathbb{Z}) \cong \mathcal{L}(\mathbb{R}^n)$.

Since the embedding $i : \mathrm{SL}(n,\mathbb{Z})\backslash\mathrm{SL}(n,\mathbb{R}) \to \mathcal{S}(\mathbb{R}^n)$ is equivariant with respect to the right K-action, the lattice compactification $\overline{\mathrm{SL}(n,\mathbb{Z})\backslash\mathrm{SL}(n,\mathbb{R})}^{la}$ admits the right action of K and the quotient by K defines a compactification of $\mathrm{SL}(n,\mathbb{Z})\backslash\mathrm{SL}(n,\mathbb{R})/\mathrm{SO}(n)$, also called the *lattice compactification* of the locally symmetric space $\mathrm{SL}(n,\mathbb{Z})\backslash\mathrm{SL}(n,\mathbb{R})/\mathrm{SO}(n)$, and denoted by $\overline{\mathrm{SL}(n,\mathbb{Z})\backslash\mathrm{SL}(n,\mathbb{R})/\mathrm{SO}(n)}^{la}$.

Proposition III.19.5 *If $n \geq 3$, the compactification of $\overline{\mathrm{SL}(n,\mathbb{Z})\backslash\mathrm{SL}(n,\mathbb{R})/\mathrm{SO}(n)}^{la}$ is not isomorphic to any Satake compactification, the Borel–Serre compactification, or the reductive Borel–Serre compactification of $\mathrm{SL}(n,\mathbb{Z})\backslash\mathrm{SL}(n,\mathbb{R})/\mathrm{SO}(n)$. If $n = 2$, it is isomorphic to the Satake compactifications and the reductive Borel–Serre compactification of $\mathrm{SL}(n,\mathbb{Z})\backslash\mathrm{SL}(n,\mathbb{R})/\mathrm{SO}(n)$.*

Proof. Let $\mathbf{P}$ be the minimal rational parabolic subgroup consisting of upper triangular matrices as in the proof of the previous proposition. Then the positive Weyl chamber $A_{\mathbf{P}}^+$ is given by

$$A_{\mathbf{P}}^+ = \{a = \mathrm{diag}(d_1, \dots, d_n) \mid d_1 > d_2 > \cdots > d_n, d_1 \dots d_n = 1\}.$$

Consider any sequence a_j satisfying $a_j^\alpha \to +\infty$, for all $\alpha \in \Phi(P, A_{\mathbf{P}})$. Clearly the conditions are equivalent to the following:

1. $a_j = \mathrm{diag}(d_{j,1}, \dots, d_{j,n})$ is not bounded.
2. for any pair $i \leq n-1$, $d_{j,i}/d_{j,i+1} \to +\infty$.

In any of the compactifications mentioned in the proposition, the image of $a_j K$ in $\mathrm{SL}(n,\mathbb{Z})\backslash\mathrm{SL}(n,\mathbb{R})/\mathrm{SO}(n)$ will converge to the same boundary point for all sequences a_j satisfying the above conditions. But this is not the case with the lattice compactification of $\mathrm{SL}(n,\mathbb{Z})\backslash\mathrm{SL}(n,\mathbb{R})/\mathrm{SO}(n)$. In fact, we can find two such sequences a_j such that $\mathbb{Z}^n a_j$ converges to different limits in $\mathcal{S}(\mathbb{R}^n)$. For simplicity, assume $n = 3$. The first choice is $a_j = \mathrm{diag}(j, 1, j^{-1})$, and the lattice $\mathbb{Z}^3 a_j$ converges to the subgroup $\{(0, x_2, x_3) \mid x_2 \in \mathbb{Z}, x_3 \in \mathbb{R}\}$. The second choice is $a_j = \mathrm{diag}(j, j^{-\frac{1}{3}}, j^{-\frac{2}{3}})$, and the lattice $\mathbb{Z}^3 a_j$ converges to a different subgroup $\{(0, x_2, x_3) \mid x_2 \in \mathbb{R}, x_3 \in \mathbb{R}\}$. This proves the first statement in the proposition.

When $n = 2$, $\dim A_{\mathbf{P}} = 1$, and a sequence $a_j = \mathrm{diag}(a_{j,1}, a_{j,2})$ goes to infinity if and only if $a_{j,1} \to +\infty$, $a_{j,2} = a_{j,1}^{-1} \to 0$, and the sequence of lattices $\mathbb{Z}^2 a_j$ converges to $\{(0, x_2) \mid x_2 \in \mathbb{R}\}$. In this case, all the Satake compactifications and the reductive Borel–Serre compactification are isomorphic to the one point compactification.

To prove the second statement in the proposition, we note that $X_{\mathbf{P}}$ consists of only one point and any unbounded sequence y_j in $\Gamma\backslash X$ converging in $\overline{\Gamma\backslash X}^{RBS}$ has a lift $\tilde{y}_j = (n_j, a_j) \in N_P \times A_{\mathbf{P}}$ such that (1) n_j is bounded, (2) $a_j = \mathrm{diag}(a_{j,1}, a_{j,1}^{-1})$, $a_{j,1} \to +\infty$. Since $a_j^{-1} n_j a_j \to e$, it follows that

$$\mathbb{Z}^2 n_j a_j = \mathbb{Z}^2 a_j (a_j^{-1} n_j a_j) \to \{(0, x_2) \mid x_2 \in \mathbb{R}\}$$

in $\mathcal{S}(\mathbb{R}^2)$. This implies that $\overline{\Gamma\backslash X}^{RBS}$ dominates $\overline{\Gamma\backslash X}^{sb}$ and hence they are isomorphic.

III.19.6 One explanation for the difference between the lattice compactification $\mathrm{SL}(n, \mathbb{Z})\backslash\mathrm{SL}(n, \mathbb{R})/\mathrm{SO}(n)$ and the Satake compactifications in the above proposition is that in the Satake compactifications, the limit of the sequence $a_j K$ depends on the behaviors of the roots, while in the lattice compactification, the limit depends on the values of the weights, which are given by the diagonal elements $d_{j,1}, \dots, d_{j,n}$. In order to recover the Satake compactifications from the spaces of lattices and closed subgroups, we need to scale the lattices suitably so that the behaviors of the weights do not affect the limits. It will turn out that two minimal Satake compactifications of $\mathrm{SL}(n, \mathbb{Z})\backslash\mathrm{SL}(n, \mathbb{R})/\mathrm{SO}(n)$ can be obtained this way. To obtain the maximal Satake compactification, we need to apply successive scaling and consider lattices in flags.

III.19.7 To discuss scaling of lattices, we recall some facts about the reduced basis of lattices in $\mathbb{R}^n$. The basic reference is [Si3, Chap. III, §5].

For any lattice Λ in $\mathbb{R}^n$, we can choose a reduced basis of Λ as follows. Recall that a set of vectors $c^{(1)}, \dots, c^{(k)}$ of Λ is called a *primitive set* if whenever any $\mathbb{R}$-linear combination $\lambda_1 c^{(1)} + \cdots + \lambda_k c^{(k)}$ belongs to Λ, then all the coefficients λ_i are in $\mathbb{Z}$.

Consider all primitive sets of one vector b in Λ, and choose $b^{(1)}$ such that the norm $\|b^{(1)}\|$ is minimal, which is denoted by $\lambda_1(\Lambda)$ and called the first minimum. Suppose $b^{(1)}, \dots, b^{(k)}$ have been chosen. Consider the set of $b \in \Lambda$ such that $b^{(1)}, \dots, b^{(k)}, b$ is primitive, and choose $b^{(k+1)}$ such that the norm $\|b^{(k+1)}\|$ is minimal among this set. Denote this norm by $\lambda_{k+1}(\Lambda)$. By induction, we get a reduced basis $b^{(1)}, \dots, b^{(n)}$ of Λ.

It should be pointed out that the reduced basis is not necessarily unique, but the first minimum $\lambda_1(\Lambda)$ depends only on Λ.

III.19.8 We define a different embedding by scaling the lattice so that the first minimum is equal to 1:

$$i_m : \mathrm{SL}(n, \mathbb{Z})\backslash\mathrm{SL}(n, \mathbb{R}) \to \mathcal{S}(\mathbb{R}^n), \quad \mathrm{SL}(n, \mathbb{Z})g \mapsto \lambda_1(\mathbb{Z}^n g)^{-1}\mathbb{Z}^n g.$$

It is clearly $\mathrm{SL}(n, \mathbb{R})$-equivariant, and the closure of $i_m(\mathrm{SL}(n, \mathbb{Z})\backslash\mathrm{SL}(n, \mathbb{R}))$ in $\mathcal{S}(\mathbb{R}^n)$ is a compactification of $\mathrm{SL}(n, \mathbb{Z})\backslash\mathrm{SL}(n, \mathbb{R})$, called the *sublattice compactification*, denoted by $\overline{\mathrm{SL}(n, \mathbb{Z})\backslash\mathrm{SL}(n, \mathbb{R})}^{subla}$. The quotient by $K = \mathrm{SO}(n)$ on the right gives a compactification of $\mathrm{SL}(n, \mathbb{Z})\backslash\mathrm{SL}(n, \mathbb{R})/K$, also called the *sublattice compactification* and denoted by

$$\overline{\mathrm{SL}(n,\mathbb{Z})\backslash\mathrm{SL}(n,\mathbb{R})/\mathrm{SO}(n)}^{subla}.$$

Let P be the minimal parabolic subgroup of $\mathrm{SL}(n,\mathbb{R})$ consisting of upper triangular matrices, and

$$\mathfrak{a}_P^+ = \{H = (t_1,\dots,t_n) \mid t_1+\cdots+t_n = 0, t_1 > t_2 > \cdots > t_n\}$$

the positive Weyl chamber above. Then the simple roots are $\alpha_1,\dots,\alpha_{n-1}$ given by

$$\alpha_1(H) = t_1 - t_2, \dots, \alpha_{n-1}(H) = t_{n-1} - t_n.$$

Let τ be an irreducible, faithful representation such that its highest weight μ_τ is connected only to α_{n-1}. Since $\mathrm{SL}(n)$ splits over $\mathbb{Q}$, the associated Satake compactification $\overline{X}_\tau^S$ is geometrically rational. Let $\overline{\mathrm{SL}(n,\mathbb{Z})\backslash\mathrm{SL}(n,\mathbb{R})/\mathrm{SO}(n)}_\tau^S$ be the induced Satake compactification.

Proposition III.19.9 *With the above notation, the sublattice compactification of* $\overline{\mathrm{SL}(n,\mathbb{Z})\backslash\mathrm{SL}(n,\mathbb{R})/\mathrm{SO}(n)}^{supla}$ *is isomorphic to the minimal Satake compactification* $\overline{\mathrm{SL}(n,\mathbb{Z})\backslash\mathrm{SL}(n,\mathbb{R})/\mathrm{SO}(n)}_\tau^S$.

Proof. Let $\mathbf{P}$ be the minimal rational parabolic subgroup of $\mathrm{SL}(n,\mathbb{C})$ consisting of upper triangular matrices. Then any standard rational parabolic subgroup containing $\mathbf{P}$ is a subgroup of block upper triangular matrices of block sizes $n_i - n_{i-1}$, where $n_0 = 0 < n_1 < \cdots < n_k = n$ is an increasing sequence between 0 and n. Denote the sequence by Π and the parabolic subgroup by $\mathbf{P}_\Pi$. Then the Langlands decomposition of P_Π is given by

$$A_{\mathbf{P}_\Pi} = \{a = \mathrm{diag}(d_1,\dots,d_n) \mid d_{n_{i-1}+1} = \cdots = d_{n_i}, \quad i = 1,\dots,k\}, \tag{III.19.5}$$

$$M_{\mathbf{P}_\Pi} = \{D = \mathrm{diag}(D_1,\dots,D_k) \mid D_i \in \pm\mathrm{SL}(n_i - n_{i-1},\mathbb{R}), \det D_1 \cdots \det D_k = 1\},$$

i.e., $M_{\mathbf{P}_\Pi}$ consists of block diagonal matrices of sizes $n_i - n_{i-1}$. Hence there is a projection

$$\pi_- : M_{\mathbf{P}_\Pi} \to \mathrm{SL}(n_k - n_{k-1},\mathbb{R}), \quad \mathrm{diag}(D_1,\dots,D_k) \mapsto \pm D_k, \tag{III.19.6}$$

where the sign $\pm$ is chosen so that $\pm D_k \in \mathrm{SL}(n_k - n_{k-1},\mathbb{R})$.

The minimal Satake compactification satisfies

$$\overline{\mathrm{SL}(n,\mathbb{Z})\backslash\mathrm{SL}(n,\mathbb{R})/\mathrm{SO}(n)}_\tau^S = \mathrm{SL}(n,\mathbb{Z})\backslash\mathrm{SL}(n,\mathbb{R})/\mathrm{SO}(n) \cup \coprod_{i=1}^{n-1} \mathrm{SL}(i,\mathbb{Z})\backslash\mathrm{SL}(i,\mathbb{R})/\mathrm{SO}(i). \tag{III.19.7}$$

Since every rational parabolic subgroup is $\mathrm{SL}(n,\mathbb{Z})$-conjugate to a standard parabolic subgroup above, the reduction theory for $\mathrm{SL}(n,\mathbb{Z})$ implies that for any sequence y_j in $\mathrm{SL}(n,\mathbb{Z})\backslash\mathrm{SL}(n,\mathbb{R})/\mathrm{SO}(n)$ going to infinity, by passing to a subsequence, we can assume that there exist a standard parabolic subgroup $\mathbf{P}_\Pi$ and a suitable lift $\tilde{y}_j$ such that in the horospherical

$$\tilde{y}_j = (n_j, a_j, m_j K_{P_\Pi}) \in N_{P_\Pi} \times A_{\mathbf{P}_\Pi} \times X_{\mathbf{P}_\Pi},$$

the coordinates satisfy

1. n_j is bounded.
2. $m_j \to m_\infty$ in $M_{\mathbf{P}_\Pi}$.
3. For every $i = 2, \dots, k$, $d_{j,n_{i-1}}/d_{j,n_i} \to +\infty$, where $a_j = (d_{j,1}, \dots, d_{j,n})$.

Clearly, $\lambda_1(\mathbb{Z}^n a_j)^{-1}\mathbb{Z}^n a_j$ converges to the discrete subgroup $\mathbb{Z}e_{n_{k-1}+1} + \dots + \mathbb{Z}e_n$, denoted by Λ'_∞, where $e_1, \dots, e_n$ is the standard basis of $\mathbb{R}^n$. Since $n_j a_j m_j = a_j n_j^{a_j} m_j$ and $n_j^{a_j} \to e$, it follows that

$$\lambda_1(\mathbb{Z}^n \tilde{y}_j)^{-1}\mathbb{Z}^n \tilde{y}_j \to \Lambda'_\infty m_\infty.$$

Let π_- be the projection in equation (III.19.6). Then

$$\Lambda'_\infty m_\infty = \Lambda'_\infty \pi_-(m_\infty).$$

This implies that the sequence y_j converges to the K-orbit $\Lambda'_\infty \pi_-(m_\infty)K$ in the compactification $\overline{\mathrm{SL}(n,\mathbb{Z})\backslash\mathrm{SL}(n,\mathbb{R})/K}^{supla}$, and the limit depends only on $\pi_-(m_\infty)K$ and the integer n_{k-1} in Π.

But the sequence y_j also converges in $\overline{\mathrm{SL}(n,\mathbb{Z})\backslash\mathrm{SL}(n,\mathbb{R})/K}^S_\tau$ and the limit is equal to the image of $\pi_-(m_\infty)\mathrm{SO}(n_k - n_{k-1})$ in $\mathrm{SL}(n_k - n_{k-1},\mathbb{Z})\backslash\mathrm{SL}(n_k - n_{k-1},\mathbb{R})/\mathrm{SO}(n_k - n_{k-1})$ in equation (III.19.7). Since every sequence in $\mathrm{SL}(n,\mathbb{Z})\backslash\mathrm{SL}(n,\mathbb{R})/K$ that converges to a point $\mathrm{SL}(n_k - n_{k-1},\mathbb{Z})m'\mathrm{SO}(n_k - n_{k-1})$ in $\overline{\mathrm{SL}(n,\mathbb{Z})\backslash\mathrm{SL}(n,\mathbb{R})/K}^S_\tau$ is a combination of finite sequences of the form above satisfying

$$\mathrm{SL}(n_k - n_{k-1},\mathbb{Z})\pi_-(m_\infty)\mathrm{SO}(n_k - n_{k-1}) = \mathrm{SL}(n_k - n_{k-1},\mathbb{Z})m'\mathrm{SO}(n_k - n_{k-1}),$$

it follows that the two compactifications in the proposition are isomorphic.

III.19.10 The Satake compactification in the above proposition is not the Satake compactification associated with the standard representation, since the highest weight of the standard representation is connected only to α_1 instead of α_{n-1}.

To realize the minimal Satake compactification of $\mathrm{SL}(n,\mathbb{Z})\backslash\mathrm{SL}(n,\mathbb{R})/K$ associated with the standard representation, we need a different scaling on the lattices.

For any lattice Λ, consider all bases $v_1, \dots, v_n$ such that

$$\|v_1\| \le \dots \le \|v_n\|. \tag{III.19.8}$$

Define

$$\nu_i(\Lambda) = \min_{v_1,\dots,v_n} \|v_i\|, \quad i = 1, \dots, n,$$

where $v_1, \dots, v_n$ ranges over all the bases satisfying the condition in equation (III.19.8). Clearly, for any $i = 1, \dots, n$,

$$\lambda_i(\Lambda) \geq \nu_i(\Lambda) \geq \lambda_1(\Lambda) > 0,$$

and $\nu_i(\Lambda)$ is realized by some basis. It is known that $\nu_i(\Lambda)$ and $\lambda_i(\Lambda)$ are uniformly bounded in terms of each other [Si3, Lemma 2, p. 98].

By definition, $\nu_n(\Lambda)$ depends only on Λ. By scaling using $\nu_n(\Lambda)$, we get a different embedding

$$i^m : \mathrm{SL}(n, \mathbb{Z})\backslash\mathrm{SL}(n, \mathbb{R}) \to \mathcal{S}(\mathbb{R}^n), \quad \mathrm{SL}(n, \mathbb{Z})g \to (\nu_n(\mathbb{Z}g))^{-1}\mathbb{Z}^n g.$$

The closure of $i^m(\mathrm{SL}(n, \mathbb{Z})\backslash\mathrm{SL}(n, \mathbb{R}))$ in $\mathcal{S}(\mathbb{R}^n)$ is called the *suplattice compactification*, denoted by $\overline{\mathrm{SL}(n, \mathbb{Z})\backslash\mathrm{SL}(n, \mathbb{R})}^{supla}$, and its quotient by $K = \mathrm{SO}(n)$ on the right gives a compactification of $\mathrm{SL}(n, \mathbb{Z})\backslash\mathrm{SL}(n, \mathbb{R})/\mathrm{SO}(n)$, also called the *suplattice compactification* and denoted by

$$\overline{\mathrm{SL}(n, \mathbb{Z})\backslash\mathrm{SL}(n, \mathbb{R})/\mathrm{SO}(n)}^{supla}.$$

Proposition III.19.11 *The compactification $\overline{\mathrm{SL}(n, \mathbb{Z})\backslash\mathrm{SL}(n, \mathbb{R})/\mathrm{SO}(n)}^{sup-la}$ is isomorphic to the standard Satake compactification of the locally symmetric space* $\mathrm{SL}(n, \mathbb{Z})\backslash\mathrm{SL}(n, \mathbb{R})/\mathrm{SO}(n)$, *which is the minimal Satake compactification associated with the standard representation of* $\mathrm{SL}(n, \mathbb{R})$, *i.e., the identity representation* $\mathrm{SL}(n, \mathbb{R}) \to \mathrm{SL}(n, \mathbb{R})$.

Proof. We use the notation of the proof of Proposition III.19.9. Consider a sequence y_j in $\mathrm{SL}(n, \mathbb{Z})\backslash\mathrm{SL}(n, \mathbb{R})/\mathrm{SO}(n)$ and its lift $\tilde{y}_j$ such that in the horopsherical decomposition with respect to a standard parabolic subgroup $\mathbf{P}_\Pi$,

$$\tilde{y}_j = (n_j, a_j, m_j K_{P_\Pi}) \in N_{P_\Pi} \times A_{\mathbf{P}_\Pi} \times X_{\mathbf{P}_\Pi},$$

the coordinates satisfy the following conditions:

1. n_j is bounded.
2. $m_j \to m_\infty$ in $M_{\mathbf{P}_\Pi}$.
3. For every $i = 2, \dots, k$, $d_{j,n_{i-1}}/d_{j,n_i} \to +\infty$, where $a_j = (d_{j,1}, \dots, d_{j,n})$.

Clearly, $\nu_n(\mathbb{Z}^n a_j)^{-1}\mathbb{Z}^n a_j$ converges to the closed subgroup $\mathbb{Z}e_1 + \cdots + \mathbb{Z}e_{n_1} + \mathbb{R}e_{n_1+1} + \cdots + \mathbb{R}e_n$, denoted by Λ''_∞. As in the proof of Proposition III.19.9, $\nu_n(\mathbb{Z}^n \tilde{y}_j)^{-1}\mathbb{Z}^n \tilde{y}_j$ converges to $\Lambda''_\infty m_\infty$.

Let π^+ be the projection of $M_{\mathbf{P}_\Pi}$ to the first factor $\mathrm{SL}(n_1, \mathbb{R})$ as in equation (III.19.6). Then

$$\Lambda''_\infty m_\infty = \Lambda''_\infty \pi^+(m_\infty).$$

This implies that the sequence y_j converges to $\Lambda''_\infty \pi^+(m_\infty)K$ in the compactification $\overline{\mathrm{SL}(n,\mathbb{Z})\backslash\mathrm{SL}(n,\mathbb{R})/K}^{sup-la}$, and the limit only depends on $\pi^+(m_\infty)K$ and the integer n_1 in Π.

As in the proof of Proposition III.19.9 again, this implies that the standard Satake compactification of $\mathrm{SL}(n,\mathbb{Z})\backslash\mathrm{SL}(n,\mathbb{R})/\mathrm{SO}(n)$ is isomorphic to the suplattice compactification.

III.19.12 After realizing these two minimal Satake compactifications, a natural question is how to use some scaling to realize the maximal Satake compactification of $\mathrm{SL}(n,\mathbb{Z})\backslash\mathrm{SL}(n,\mathbb{R})/\mathrm{SO}(n)$. The above results suggest that we need to do scalings at different levels. This will lead us out of the space $\mathcal{S}(\mathbb{R}^n)$. In fact, we need to introduce flags of linear subspaces in $\mathbb{R}^n$ and lattices in them.

Definition III.19.13

1. *A flag in $\mathbb{R}^n$ is a strictly increasing sequence of linear subspaces* $\mathbf{F}: V_0 = \{0\} \subset V_1 \subset \cdots \subset V_k = \mathbb{R}^n$.
2. *A lattice Λ in a flag $\mathbf{F}$ is a collection $\Lambda^1, \dots, \Lambda^k$ such that Λ^i is a lattice in the quotient V_i/V_{i-1}. Such a lattice is also called a flag lattice. The lattice Λ is called a unimodular lattice if each Λ^i is a unimodular lattice.*

The flag with $k = 1$ is trivial and consists of only $\{0\} \subset \mathbb{R}^n$, and a lattice in the flag is a lattice in $\mathbb{R}^n$. On the other hand, when $k = n$, the flag is full, and each quotient V_i/V_{i-1} is of dimension 1, and there is only one unimodular lattice in this full flag.

For any flag $\mathbf{F}$, its stabilizer in $\mathrm{SL}(n,\mathbb{R})$ is defined to be

$$\{g \in \mathrm{SL}(n,\mathbb{R}) \mid gV_i = V_i, i = 1, \dots, k\}.$$

Proposition III.19.14 *The stabilizer of any nontrivial flag $\mathbf{F}$ is a (proper) parabolic subgroup of* $\mathrm{SL}(n,\mathbb{R})$, *denoted by $P_{\mathbf{F}}$.*

Proof. Let $e_1 = (1, 0, \dots, 0), \dots, e_n = (0, \dots, 0, 1)$ be the standard basis of $\mathbb{R}^n$. For any sequence $n_1 < n_2 < \cdots < n_k = n$, let V_i be the subspace spanned by $e_1, \dots, e_{n_i}$. Then $\{0\} \subset V_1 \subset \cdots \subset V_k = \mathbb{R}^n$ is called a standard flag. Clearly, the stabilizer of this standard flag is the subgroup of upper block triangular matrices with block sizes given by $n_1, n_2 - n_1, \dots, n_k - n_{k-1}$, which is a standard parabolic subgroup of $\mathrm{SL}(n,\mathbb{R})$. Since any two bases can be mapped to each other up to suitable scalar multiples by an element in $\mathrm{SL}(n,\mathbb{R})$, any flag is the image under some element of $\mathrm{SL}(n,\mathbb{R})$ of some standard flag and the above result implies that its stabilizer is conjugate to a standard parabolic subgroup.

Lemma III.19.15 *For any flag $\mathbf{F}$, let $P_{\mathbf{F}} = N_{P_{\mathbf{F}}} A_{P_{\mathbf{F}}} M_{P_{\mathbf{F}}}$ be the Langlands decomposition with respect to the maximal compact subgroup $K = \mathrm{SO}(n)$. Then $M_{\mathbf{P}_{\mathbf{F}}}$ acts transitively on the set of lattices in the flag $\mathbf{F}$ of any fixed covolume, where by the fixed covolume, we mean that the covolume of each lattice Λ^i is fixed.*

Proof. It suffices to consider the case that $\mathbf{F}$ is a standard flag as in the proof of the previous proposition. Then the elements of $M_{P_{\mathbf{F}}}$ consists of block diagonal matrices, each of which is a square matrix of size $n_i - n_{i-1}$ and determinant ± 1, and the proposition is clear in this case.

Define

$$\mathcal{FL}(\mathbb{R}^n) = \{(\mathbf{F}, \Lambda) \mid \mathbf{F} \text{ is a flag }, \Lambda \text{ is a unimodular lattice in } \mathbf{F}\}. \qquad \text{(III.19.9)}$$

Since the lattice Λ uniquely determines the flag $\mathbf{F}$, we will often identify the pair $(\mathbf{F}, \Lambda)$ with Λ for convenience. The right action of $\mathrm{SL}(n, \mathbb{R})$ on $\mathbb{R}^n$ extends to a right action on flags and lattices, and hence to the space $\mathcal{FL}(\mathbb{R}^n)$.

Clearly $\mathrm{SL}(n, \mathbb{Z})\backslash\mathrm{SL}(n, \mathbb{R}) \cong \mathcal{L}(\mathbb{R}^n)$ is included in $\mathcal{FL}(\mathbb{R}^n)$ as the subset when the flag $\mathbf{F}$ is trivial. We are going to define a topology on $\mathcal{FL}(\mathbb{R}^n)$ to make it into a compact Hausdorff space and hence a compactification of $\mathrm{SL}(n, \mathbb{Z})\backslash\mathrm{SL}(n, \mathbb{R})$, which will be shown to be isomorphic to the reductive Borel–Serre compactification $\overline{\mathrm{SL}(n, \mathbb{Z})\backslash\mathrm{SL}(n, \mathbb{R})}^{RBS}$, whose quotient by $\mathrm{SO}(n)$ on the right gives a compactification of $\mathrm{SL}(n, \mathbb{Z})\backslash\mathrm{SL}(n, \mathbb{R})/\mathrm{SO}(n)$ that is isomorphic to the maximal Satake compactification.

III.19.16 The topology of $\mathcal{FL}(\mathbb{R}^n)$ is defined in terms of convergent sequences as follows.

Given two flags $\mathbf{F}_1 : V_0 = \{0\} \subset V_1 \subset \cdots \subset V_k = \mathbb{R}^n$ and $\mathbf{F}_2 : W_0 = \{0\} \subset W_1 \subset W_2 \subset \cdots \subset W_l = \mathbb{R}^n$, we say that $\mathbf{F}_2$ is a refinement of $\mathbf{F}_1$ if every linear subspace V_i of $\mathbf{F}_1$ appears in $\mathbf{F}_2$, i.e., there exist n_i, $i = 1, \ldots, k$, such that

$$V_i = W_{n_i}, \quad i = 1, \ldots, k.$$

Then

$$\{0\} = V_{i-1}/W_{n_{i-1}} \subset W_{n_{i-1}+1}/W_{n_{i-1}} \subset \cdots \subset V_i/W_{n_{i-1}} = V_i/V_{i-1} \qquad \text{(III.19.10)}$$

is a flag in V_i/V_{i-1} induced from $\mathbf{F}_2$. Since for every $j = n_{i-1}, n_{i-1} + 1, \ldots, n_i$,

$$\frac{W_{j+1}/W_{n_{i-1}}}{W_j/W_{n_{i-1}}} = \frac{W_{j+1}}{W_j},$$

a lattice Λ in $\mathbf{F}_2$ defines a lattice in the induced flag in V_i/V_{i-1}.

Clearly, every flag is a refinement of the trivial flag, and the complete flags have no refinements.

The idea of the topology is that for any sequence of lattices in a flag $\mathbf{F}$, if it is unbounded, it can converge only to a lattice in a refined flag.

III.19.17 First, we define how an unbounded sequence of lattices Λ_j in $\mathbb{R}^n$ converges to a lattice in a nontrivial flag $\mathbf{F}$.

Given a flag $\mathbf{F} : \{0\} \subset V_1 \subset \cdots \subset V_k = \mathbb{R}^n$, let $n_i = \dim V_i$, $i = 1, \ldots, k$. Let $\Lambda_\infty : \Lambda_\infty^1, \ldots, \Lambda_\infty^k$ be a unimodular lattice in the flag $\mathbf{F}$. Then the sequence Λ_j of lattices in $\mathbb{R}^n$ converges to $(\mathbf{F}, \Lambda_\infty)$ if and only if the following conditions are satisfied:

1. $\nu_{i+1}(\Lambda_j)/\nu_i(\Lambda_j) \to +\infty$ if $i = n_1, \dots, n_k$, and converges to a finite number otherwise.
2. For each $i = n_1, \dots, n_k$, $\nu_i(\Lambda_j)^{-1}\Lambda$ converges to a closed subgroup of $\mathbb{R}^n$ whose linear span is equal to V_i, and whose image in V_i/V_{i-1} is a lattice that is a multiple of the unimodular lattice Λ^i_∞.

Given a flag $\mathbf{F}_1 : \{0\} \subset V_1 \subset \cdots \subset V_k = \mathbb{R}^n$ and a refinement $\mathbf{F}_2$ and a lattice Λ_∞ in it, and a sequence of lattices Λ_j in $\mathbf{F}_1$, then $(\mathbf{F}_1, \Lambda_j)$ converges to $(\mathbf{F}_2, \Lambda_\infty)$ if and only if for any $i = 1, \dots, k$, the sequence of lattices Λ^i_j in V_i/V_{i-1} converges in the sense described in the previous paragraph to the lattice in the flag in V_i/V_{i-1} in equation (III.19.10), which is induced from the lattice Λ_∞ in $\mathbf{F}_2$.

Proposition III.19.18 *The space $\mathcal{FL}(\mathbb{R}^n)$ with the above topology is a compact Hausdorff space.*

Proof. It suffices to show that every sequence in $\mathcal{FL}(\mathbb{R}^n)$ has a convergent subsequence. As in the definition of the topology of $\mathcal{FL}(\mathbb{R}^n)$ above, it suffices to show that every sequence of lattices Λ_j in $\mathbb{R}^n$ has a convergent subsequence $\Lambda_{j'}$ in $\mathbf{FL}(\mathbb{R}^n)$, i.e., there exist a flag $\mathbf{F}$ and a lattice Λ_∞ in it such that $\Lambda_{j'}$ converges to $(\mathbf{F}, \Lambda_\infty)$.

Suppose the minimal norm $\nu_1(\Lambda_j)$ is uniformly bounded away from zero. Then by the Mahler compactness criterion [Bo4, Corollary 1.9], Λ_j is a relatively compact family in the space $\mathcal{L}(\mathbb{R}^n)$ of unimodular lattices, and hence a subsequence converges to a lattice in $\mathbb{R}^n$. If not, by passing to a subsequence, we can assume that $\nu_1(\Lambda_j) \to 0$. By passing to a further subsequence, we can assume that for any i, $\nu_i(\Lambda_j)/\nu_1(\Lambda_j)$ goes either to $+\infty$ or a finite positive number. Let i_1 be the smallest integer such that the limit is equal to $+\infty$. Clearly, $1 < i_1 \le n$, since the lattice Λ_i is unimodular and hence $\nu_n(\Lambda_j)/\nu_1(\Lambda_j) \to +\infty$. Let $b^{(1)}_j, \dots, b^{(n)}_j$ be a reduced basis of Λ_j. Then by passing to a further subsequence, we obtain that

1. for any $i \le i_1 - 1$, $\nu_1(\Lambda_j)^{-1}b^{(i)}$ converges to a nonzero vector;
2. for any $i \ge i_1$, $\|\nu_1(\Lambda_j)^{-1}b^{(i)}\| \to +\infty$.

This implies that $\nu_1(\Lambda_j)^{-1}\Lambda$ converges to a discrete subgroup whose linear span contains it as a lattice. Denote this linear subspace by V_1, which has dimension $i_1 - 1$.

By passing to a further subsequence, we assume that for any $i > i_1$, $\nu_i(\Lambda)/\nu_{i_1}(\Lambda)$ goes either to $+\infty$ or a finite positive number. Let i_2 be the least integer such that this limit is $+\infty$. Next, consider the subgroups $\nu_{i_2}(\Lambda_j)^{-1}\Lambda_j$. Then as in the previous paragraph, after passing to a further subsequence, it converges to a closed subgroup S that contains V_1. Denote the linear span of S by V_2. Then the image of S in the quotient V_2/V_1 is a lattice, which can be scaled to be unimodular. By induction, we get a flag $\mathbf{F} : \{0\} \subset V_1 \subset V_2 \subset \cdots \subset V_k = \mathbb{R}^n$ and a lattice Λ_∞ in $\mathbf{F}$ such that a subsequence $\Lambda_{j'}$ converges to $(\mathbf{F}, \Lambda_\infty)$ in the topology of $\mathcal{FL}(\mathbb{R}^n)$ defined above. This proves the compactness.

The uniqueness of the limit of a convergent sequence Λ_j is clear. In fact, the limit has to be of the form constructed in the previous paragraph, and the construction leads to a unique flag and the lattice in it.

III.19.19 Under the embedding

$$\mathrm{SL}(n,\mathbb{Z})\backslash\mathrm{SL}(n,\mathbb{R}) \cong \mathcal{L}(\mathbb{R}^n) \to \mathcal{FL}(\mathbb{R}^n),$$

the proof of the above proposition shows that $\mathrm{SL}(n,\mathbb{Z})\backslash\mathrm{SL}(n,\mathbb{R})$ is an open dense subset, and hence $\mathcal{FL}(\mathbb{R}^n)$ is a compactification of $\mathrm{SL}(n,\mathbb{Z})\backslash\mathrm{SL}(n,\mathbb{R})$, called the *flag-lattice compactification*, denoted by $\overline{\mathrm{SL}(n,\mathbb{Z})\backslash\mathrm{SL}(n,\mathbb{R})}^{fl}$. The right $\mathrm{SL}(n,\mathbb{R})$-action clearly extends to the compactification.

Proposition III.19.20 *The flag-lattice compactification* $\overline{\mathrm{SL}(n,\mathbb{Z})\backslash\mathrm{SL}(n,\mathbb{R})}^{fl}$ *is isomorphic to* $\overline{\mathrm{SL}(n,\mathbb{Z})\backslash\mathrm{SL}(n,\mathbb{R})}^{RBS}$*, the reductive Borel–Serre compactification.*

Proof. First we define a bijective map from $\overline{\mathrm{SL}(n,\mathbb{Z})\backslash\mathrm{SL}(n,\mathbb{R})}^{RBS}$ to $\overline{\mathrm{SL}(n,\mathbb{Z})\backslash\mathrm{SL}(n,\mathbb{R})}^{fl}$ extending the identity map on $\mathrm{SL}(n,\mathbb{Z})\backslash\mathrm{SL}(n,\mathbb{R})$.

For each increasing sequence $\Pi : n_0 = 0 < n_1 < n_2 < \cdots < n_k = n$, we associate with the sequence the standard rational parabolic subgroup $\mathbf{P}_\Pi$ of $\mathrm{SL}(n,\mathbb{C})$ consisting of block upper triangular matrices of sizes $n_i - n_{i-1}$. Then

$$\Gamma_{M_{\mathbf{P}_\Pi}}\backslash M_{\mathbf{P}_\Pi} = \prod_{i=1}^{k} \mathrm{SL}(n_i - n_{i-1},\mathbb{Z})\backslash\mathrm{SL}(n_i - n_{i-1},\mathbb{R}).$$

Let V_i be the linear span of $e_1,\dots,e_i$ as above, and $\mathbf{F}_\Pi : \{0\} \subset V_1 \subset \cdots \subset V_k = \mathbb{R}^n$ the flag associated with Π. The standard lattice $\mathbb{Z}^n$ in $\mathbb{R}^n$ induces a standard lattice $\mathbb{Z}^n_\Pi$ in the flag $\mathbf{F}_\Pi$. Since $M_{\mathbf{P}_\Pi}$ acts transitively on the set of unimodular lattices in $\mathbf{F}_\Pi$ and the stabilizer of $\mathbb{Z}^n_\Pi$ in $M_{\mathbf{P}_\Pi}$ is equal to $\Gamma_{M_{\mathbf{P}_\Pi}}$, it follows that the set of unimodular lattices in $\mathbf{F}_\Pi$ can be identified with $\Gamma_{M_{\mathbf{P}_\Pi}}\backslash M_{\mathbf{P}_\Pi}$ under the map

$$\Gamma_{M_{\mathbf{P}_\Pi}} m \mapsto \mathbb{Z}^n_\Pi m.$$

Note that any flag in $\mathbf{F}$ in $\mathbb{R}^n$ is equivalent to a standard one $\mathbf{F}_\Pi$ under the right action of $\mathrm{SL}(n,\mathbb{R})$. This implies that each of the $\mathrm{SL}(n,\mathbb{R})$-orbits in $\mathcal{FL}(\mathbb{R}^n)$ can be identified with $N_{\mathbf{P}_\Pi} A_{\mathbf{P}_\Pi} \Gamma_{M_{\mathbf{P}_\Pi}}\backslash\mathrm{SL}(n,\mathbb{R})$, which is a boundary component of $\overline{\mathrm{SL}(n,\mathbb{Z})\backslash\mathrm{SL}(n,\mathbb{R})}^{RBS}$. In fact, since every rational parabolic subgroup of $\mathrm{SL}(n)$ is $\mathrm{SL}(n,\mathbb{Z})$-conjugate to a standard rational parabolic subgroup $\mathbf{P}_\Pi$, we obtain that

$$\begin{aligned}\overline{\mathrm{SL}(n,\mathbb{Z})\backslash\mathrm{SL}(n,\mathbb{R})}^{RBS} &= \mathrm{SL}(n,\mathbb{Z})\backslash\mathrm{SL}(n,\mathbb{R}) \cup \coprod_\Pi \Gamma_{M_{\mathbf{P}_\Pi}}\backslash M_{\mathbf{P}_\Pi}\mathrm{SO}(n)\\ &= \mathrm{SL}(n,\mathbb{Z})\backslash\mathrm{SL}(n,\mathbb{R}) \cup \coprod_\Pi N_{P_\Pi} A_{\mathbf{P}_\Pi} \Gamma_{M_{\mathbf{P}_\Pi}}\backslash\mathrm{SL}(n,\mathbb{R}),\end{aligned}$$

where Π runs over all the increasing sequences between 1 and n. Therefore we have obtained the desired identification:

$$\overline{\mathrm{SL}(n,\mathbb{Z})\backslash\mathrm{SL}(n,\mathbb{R})}^{RBS} \cong \mathcal{FL}(\mathbb{R}^n), \quad \Gamma_{M_{\mathbf{P}_\Pi}} g \mapsto \mathbb{Z}^n_\Pi g.$$

By [GJT, Lemma 3.28], it suffices to show that if an unbounded sequence y_j in $\mathrm{SL}(n,\mathbb{Z})\backslash\mathrm{SL}(n,\mathbb{R})$ converges in the reductive Borel–Serre compactification $\overline{\mathrm{SL}(n,\mathbb{Z})\backslash\mathrm{SL}(n,\mathbb{R})}^{RBS}$, then it also converges to the same limit point in $\overline{\mathrm{SL}(n,\mathbb{Z})\backslash\mathrm{SL}(n,\mathbb{R})}^{fl}$.

By definition, there exists a standard rational parabolic subgroup $\mathbf{P}_\Pi$ such that a suitable lift $\tilde{y}_j$ of y_j has a horospherical decomposition

$$\tilde{y}_j = (n_j, a_j, m_j) \in N_{P_\Pi} \times A_{\mathbf{P}_\Pi} \times M_{\mathbf{P}_\Pi} K,$$

where the coordinates $n_j \in N_{P_\Pi}$, $a_j = \mathrm{diag}(d_{j,1}, \dots, d_{j,n}) \in A_{\mathbf{P}_\Pi}$, $m_j \in M_{\mathbf{P}_\Pi} K$ satisfy the following conditions:

1. n_j is bounded.
2. $m_j \to m_\infty$ some element $m_\infty \in M_{\mathbf{P}_\Pi} K$.
3. For any $i = 1, \dots, k$, $s, t \in \{n_{i-1}+1, \dots, n_i\}$, $d_{j,s} = d_{j,t}$, and $d_{j,n_{i-1}}/d_{j,n_i} \to +\infty$.

It is clear that the sequence of lattices $\mathbb{Z}^n a_j$ converges to the standard flag lattice $\mathbb{Z}^n_\Pi$ in the flag $\mathbf{F}_\Pi$. Write

$$n_j a_j m_j = a_j n_j^{a_j} m_j.$$

Since $n_j^{a_j}$ converges to the identify element, and $m_j \to m_\infty$, it follows that $\mathbb{Z}^n \tilde{y}_j$ converges to the flag lattice $\mathbb{Z}^n_\Pi m_\infty$, the boundary point in $\mathcal{FL}(\mathbb{R}^n)$ corresponding to m_∞ in $\overline{\mathrm{SL}(n,\mathbb{Z})\backslash\mathrm{SL}(n,\mathbb{R})}^{RBS}$.

III.19.21 We can use the flag-lattice compactification $\overline{\mathrm{SL}(n,\mathbb{Z})\backslash\mathrm{SL}(n,\mathbb{R})}^{fl}$ to get another realization of $\overline{\mathrm{SL}(n,\mathbb{Z})\backslash\mathrm{SL}(n,\mathbb{R})/\mathrm{SO}(n)}^{S}_{\max}$, the maximal Satake compactification. In fact, $\mathrm{SO}(n)$ acts on the right of $\overline{\mathrm{SL}(n,\mathbb{Z})\backslash\mathrm{SL}(n,\mathbb{R})}^{fl}$, and the quotient contains $\mathrm{SL}(n,\mathbb{Z})\backslash\mathrm{SL}(n,\mathbb{R})/\mathrm{SO}(n)$ as an open dense subset. We call the resulting compactification $\overline{\mathrm{SL}(n,\mathbb{Z})\backslash\mathrm{SL}(n,\mathbb{R})}^{fl}/\mathrm{SO}(n)$ also the *flag-lattice compactification* of $\mathrm{SL}(n,\mathbb{Z})\backslash\mathrm{SL}(n,\mathbb{R})/\mathrm{SO}(n)$ and denote it by

$$\overline{\mathrm{SL}(n,\mathbb{Z})\backslash\mathrm{SL}(n,\mathbb{R})/\mathrm{SO}(n)}^{fl}.$$

Proposition III.19.22 *The compactification $\overline{\mathrm{SL}(n,\mathbb{Z})\backslash\mathrm{SL}(n,\mathbb{R})/\mathrm{SO}(n)}^{fl}$ is isomorphic to $\overline{\mathrm{SL}(n,\mathbb{Z})\backslash\mathrm{SL}(n,\mathbb{R})/\mathrm{SO}(n)}^{S}_{\max}$, the maximal Satake compactification.*

Proof. Since $\mathrm{SL}(n)$ splits over $\mathbb{Q}$, by Proposition III.11.12, the maximal Satake compactification $\overline{\mathrm{SL}(n,\mathbb{Z})\backslash\mathrm{SL}(n,\mathbb{R})/\mathrm{SO}(n)}^{S}_{\max}$ is isomorphic to the reductive Borel–Serre compactification $\overline{\mathrm{SL}(n,\mathbb{Z})\backslash\mathrm{SL}(n,\mathbb{R})/\mathrm{SO}(n)}^{RBS}$. By Proposition III.14.7, the quotient $\overline{\mathrm{SL}(n,\mathbb{Z})\backslash\mathrm{SL}(n,\mathbb{R})}^{RBS}$ by $\mathrm{SO}(n)$ on the right gives the compactification

$\overline{\mathrm{SL}(n, \mathbb{Z})\backslash\mathrm{SL}(n, \mathbb{R})/\mathrm{SO}(n)}^{RBS}$. Hence, the proposition follows from the previous one.

III.19.23 Summary and comments. For the special homogeneous space $\mathrm{SL}(n, \mathbb{Z})\backslash \mathrm{SL}(n, \mathbb{R})$ of the form $\Gamma\backslash G$, its compactifications via the space of lattices $\mathcal{L}(\mathbb{R}^n)$ and the space of flag-lattices $\mathcal{FL}(\mathbb{R}^n)$ continue the method of studying locally symmetric spaces via lattices in $\mathbb{R}^n$, i.e., the geometry of numbers. The realizations of the Satake compactifications of $\mathrm{SL}(n, \mathbb{Z})\backslash\mathrm{SL}(n, \mathbb{R})$ and $\mathrm{SL}(n, \mathbb{Z})\backslash\mathrm{SL}(n, \mathbb{R})/\mathrm{SO}(n)$ in this section give natural explanations of the boundary points. The orbits in the boundary of the Satake compactifications of $\mathrm{SL}(n, \mathbb{Z})\backslash\mathrm{SL}(n, \mathbb{R})$ can also be identified via natural flag spaces. Using additional structures on lattices in $\mathbb{R}^n$ and flag lattices, one can also compactify $\Gamma\backslash\mathrm{SL}(n, \mathbb{R})$, where Γ is a principal congruence subgroup, $Sp(n, \mathbb{Z})\backslash Sp(n, \mathbb{R})$, and other related spaces.

13

Metric Properties of Compactifications of Locally Symmetric Spaces $\Gamma\backslash X$

From the point of view of Riemannian geometry, locally symmetric spaces form a very special and important class of Riemannian manifolds. They enjoy some rigid and extremal properties such as those in the Mostow strong rigidity [Mos], the Margulis superrigidity [Mag] [Zi], and the minimality of the entropy [BCG]. In these studies, the large-scale geometry and compactifications of X play a crucial role.

In the earlier chapters, the metric aspects of the locally symmetric space $\Gamma\backslash X$ did not play an important role in compactifications of $\Gamma\backslash X$. In this chapter, we study metric properties of the compactifications discussed earlier. Besides the above motivations from the geometry, another important reason for such studies comes from the question on extension of the period map for degenerating families of algebraic varieties. Furthermore, geometry at infinity and compactifications of $\Gamma\backslash X$ are important in understanding the spectral analysis on $\Gamma\backslash X$, in particular, the spectral theory of automorphic forms.

In §III.20, we study the class of eventually distance-minimizing (EDM) geodesics in $\Gamma\backslash X$. They are clearly contained in the class of all geodesics of $\Gamma\backslash X$ going to infinity, and the inclusion is strict unless $rk_{\mathbb{Q}}(\mathbf{G}) = rk_{\mathbb{R}}(\mathbf{G}) = 1$. Then we use these EDM geodesics to define the geodesic compactification $\Gamma\backslash X \cup \Gamma\backslash X(\infty)$. After classifying the EDM geodesics, we show that the geodesic compactification is isomorphic to the Tits compactification $\overline{\Gamma\backslash X}^T$ defined in §III.12. Then we define the Gromov compactification $\overline{\Gamma\backslash X}^T$ in §III.20.16 and show that it is also isomorphic to the Tits compactification $\overline{\Gamma\backslash X}^T$ and hence to $\Gamma\backslash X \cup \Gamma\backslash X(\infty)$ as well.

In §III.21, we study the rough geometry of $\Gamma\backslash X$ and show that it is up to finite distance a metric cone over the quotient $\Gamma\backslash\Delta_{\mathbb{Q}}(\mathbf{G})$ of the Tits building $\Delta_{\mathbb{Q}}(\mathbf{G})$ by Γ, which is a finite complex and is the boundary of $\overline{\Gamma\backslash X}^T$. Then we prove the Siegel conjecture on comparison of the restriction of two metrics on Siegel sets.

In §III.22, we introduce the notion of hyperbolic compactifications and study the question of when a holomorphic map from the punctured disk $D^{\times}$ to a Hermitian locally symmetric space $\Gamma\backslash X$ can be extended over the puncture to a map from the disk D to compactifications of $\Gamma\backslash X$.

In §III.23, we show how the boundaries of compactifications of $\Gamma\backslash X$ can occur naturally as parameter spaces for the generalized eigenfunctions of the continuous spectrum of $\Gamma\backslash X$. When the $\mathbb{Q}$-rank of $\mathbf{G}$ or $\Gamma\backslash X$ is equal to 1, we introduce the notion of *scattering geodesics* and *sojourn times*, and study the *Poisson relation*, which connects the sojourn times of scattering geodesics and the frequencies of oscillation of the *scattering matrices*, which form an important part of the continuous spectrum.

III.20 Eventually distance-minimizing geodesics and geodesic compactification of $\Gamma\backslash X$

For a symmetric space X of noncompact type, we recall the geodesic compactification in §I.2. In fact, the construction also works for any simply connected nonpositively curved Riemannian manifold M, and this geodesic compactification plays an important role in the study of geometry and analysis on M. A natural question is whether such a geodesic compactification can be constructed for nonsimply connected Riemannian manifolds.

In this section, we recall a general method in [JM, §9] on constructing such a geodesic compactification, and then apply it to the locally symmetric space $\Gamma\backslash X$ to define the geodesic compactification $\Gamma\backslash X \cup \Gamma\backslash X(\infty)$. We also define the Gromov compactification $\overline{\Gamma\backslash X}^G$. Then we show that the geodesic compactification is isomorphic to the Tits compactification $\overline{\Gamma\backslash X}^T$ defined in §III.12 and the Gromov compactification $\overline{\Gamma\backslash X}^G$.

Specifically, for any Riemannian manifold M, EDM geodesics and rays, and the set $M(\infty)$ of equivalence classes of EDM geodesics are defined in III.20.1. If M is noncompact, then $M(\infty)$ is nonempty (III.20.2). In order to use $M(\infty)$ to define a compactification of M (III.20.5), we introduce two assumptions in III.20.3 and III.20.4. To apply this method to compactify $\Gamma\backslash X$, we classify EDM geodesics of $\Gamma\backslash X$ in III.20.8. The geodesic compactification $\Gamma\backslash X \cup \Gamma\backslash X(\infty)$ is defined in III.20.12, and is shown to be isomorphic to the Tits compactification $\overline{\Gamma\backslash X}^T$ in Proposition III.20.15 and isomorphic to the Gromov compactification $\overline{\Gamma\backslash X}^G$ in III.20.17.

As mentioned earlier, all geodesics in this section are oriented and of unit speed.

Definition III.20.1 *Let M be a complete Riemannian manifold, and d the Riemannian distance function. A geodesic $\gamma : \mathbb{R} \to M$ is called eventually distance-minimizing (EDM) if there exists some t_0 such that for all $t_1, t_2 \geq t_0$,*

$$d(\gamma(t_1), \gamma(t_2)) = |t_1 - t_2|.$$

A half-geodesic $\gamma : [0, +\infty) \to M$ is called a distance-minimizing ray (or simply a ray) if for all $t_1, t_2 \geq 0$,

$$d(\gamma(t_1), \gamma(t_2)) = |t_1 - t_2|.$$

Since M is complete, every ray extends to an EDM geodesic, and every EDM geodesic gives a ray after a suitable shift in parameter.

Two EDM geodesics (or rays) γ_1, γ_2 in M are called *equivalent* if

$$\lim_{t\to+\infty} \sup d(\gamma_1(t), \gamma_2(t)) < +\infty.$$

Let $M(\infty)$ be the set of equivalence classes of geodesics in M.

If M is simply connected and negatively curved, then every geodesic in M is distance minimizing, and $M(\infty)$ is equal to the earlier definition in §I.2, the sphere at infinity.

Lemma III.20.2 *A (complete) Riemannian manifold M is noncompact if and only if $M(\infty)$ is nonempty.*

Proof. If $M(\infty)$ is nonempty, let γ be an EDM geodesic. Clearly, the sequence of points $\gamma(n)$ does not have a convergent subsequence, and hence M is noncompact. On the other hand, if M is noncompact, let x_j be a sequence in M going to infinity. Since M is complete, there exists a geodesic γ_j with $\gamma_j(0) = x_0$ and $x_j = \gamma_j(t_j)$ for $t_j = d(x_0, x_j)$. After passing to a subsequence, there exists a geodesic γ_∞ such that $\gamma_j(t)$ converges to $\gamma_\infty(t)$ uniformly for t in compact subsets. It can be checked easily that for all $t_1, t_2 \geq 0$, $d(\gamma_\infty(t_1), \gamma_\infty(t_2)) = |t_2 - t_1|$, i.e., γ_∞ is EDM.

From now on, we assume that M is a noncompact complete Riemannian manifold. To define a compactification of M by adding $M(\infty)$ at infinity, we need to make two assumptions:

Assumption III.20.3 *There exists a compact subset ω_0 in M, called a base compact subset, such that every point x in M is connected to ω_0 by a ray, i.e., there exists a ray $\gamma : [0, \infty) \to M$ such that $\gamma(0) \in \omega_0$, and $x = \gamma(t)$ for some t.*

Assumption III.20.4 *Suppose Assumption III.20.3 holds. Let ω be any compact subset containing ω_0 and let y_j, y'_j be any two sequences of points in M going to infinity with $d(y_j, y'_j)$ bounded. Let γ_j, γ'_j be any two sequences of rays connecting y_j, y'_j to ω respectively and $\gamma_j(t) \to \gamma_\infty(t)$, $\gamma'_j(t) \to \gamma'_\infty(t)$ uniformly for t in compact subsets, where $\gamma_\infty, \gamma'_\infty$ are some rays. Then γ_∞ and γ'_∞ are equivalent.*

Both assumptions hold when M is simply connected and nonpositively curved. In fact, the base compact subset ω_0 can be chosen to consist of any point $x \in M$, and the second assumption follows from the comparison with the Euclidean space. On the other hand, it is easy to construct examples of nonsimply connected manifolds or manifolds that are not nonpositively curved such that both assumptions are not satisfied [JM, 9.15].

Proposition III.20.5 *Let M be a complete Riemannian manifold satisfying Assumptions III.20.3 and III.20.4. Then there is a canonical topology on $M \cup M(\infty)$ with respect to which it is a compact Hausdorff compactification of M, called the geodesic compactification of M.*

Proof. We will describe the topology only in terms of convergent sequences of unbounded sequences in M. For details and the proof for the compactness and Hausdorff property of the topology, see [JM, §9]. Let ω be any compact subset containing the base compact subset ω_0 in Assumption III.20.3. For any unbounded sequence $y_j \in M$, let γ_j be a (DM) ray connecting y_j to ω, $\gamma_j(t_j) = x_j$. For an equivalence class of EDM geodesics $[\gamma] \in M(\infty)$, the unbounded sequence $y_j \in M$ converges to $[\gamma]$ if and only if for any sequence of rays γ_j chosen above, there is a subsequence $\gamma_{j'}$ such that $\gamma_{j'}(t)$ converges to a geodesic $\gamma_\infty(t)$ uniformly for t in compact subsets for some $\gamma_\infty \in [\gamma]$. Convergence of sequences of points of $M(\infty)$ can also be described explicitly.

In the above proposition, Assumption III.20.3 is clearly needed to get the rays γ_j, and Assumption III.20.4 is needed to show that the limit of y_j is independent of the choice of these rays γ_j. When M is simply connected and nonpositively curved, $M \cup M(\infty)$ is the geodesic compactification defined earlier in §I.2.

III.20.6 To apply the above general construction to $M = \Gamma\backslash X$, we need to determine all EDM geodesics in $\Gamma\backslash X$.

For any rational parabolic subgroup $\mathbf{P}$, $a \in A_{\mathbf{P}}$, $H \in \mathfrak{a}_{\mathbf{P}}^{+}(\infty)$, $n \in N_P$, $z \in X_{\mathbf{P}}$, define

$$\tilde{\gamma}(t) = (n, a \exp tH, z) \in X. \tag{III.20.1}$$

It follows from the formula for the invariant metric of X in the next lemma that $\tilde{\gamma}$ is a geodesic in X (see [Bo6, Prop. 1.6, Cor. 1.7] [Bo7]).

Lemma III.20.7 *For any rational parabolic subgroup* $\mathbf{P}$, *let* $X = N_P \times A_{\mathbf{P}} \times X_{\mathbf{P}}$ *be the rational horospherical decomposition associated with* P *in §III.1.*

1. *Let* dx^2, da^2, *and* dz^2 *be the invariant metrics on* X, $A_{\mathbf{P}}$, *and* $X_{\mathbf{P}}$ *respectively induced from the Killing form. Then at the point* $(n, z, a) \in X$,

$$dx^2 = dz^2 + da^2 + \sum_{\alpha \in \Phi(P, A_{\mathbf{P}})} a^{-2\alpha} h_\alpha(z),$$

where $h_\alpha(z)$ *is a metric on the root space* $\mathfrak{g}_\alpha$ *that depends smoothly on* $z \in X_{\mathbf{P}}$.

2. *For any two points* $(n_1, z_1, a_1), (n_2, z_2, a_2) \in X$,

$$d_X((n_1, z_1, a_1), (n_2, z_2, a_2)) \geq d_A(a_1, a_2),$$

where $d_X(\cdot, \cdot)$ *and* $d_A(\cdot, \cdot)$ *are the distance functions on* X *and* $A_{\mathbf{P}}$ *respectively.*

The projection of these geodesics gives all EDM geodesics in $\Gamma\backslash X$.

Proposition III.20.8 *With the above notation, the projection in* $\Gamma\backslash X$ *of* $\tilde{\gamma}$ *in equation (III.20.1) is an EDM geodesic, and every EDM geodesic in* $\Gamma\backslash X$ *is of this form for some rational parabolic subgroup* $\mathbf{P}$.

To determine whether a geodesic in X is projected to an EDM geodesic in $\Gamma\backslash X$, we use the following fact. Let D be the Dirichlet domain for Γ acting on X with center $x_1 \in X$:

$$D = \{x \in X \mid d(x, x_1) < d(x, \gamma x_1) \text{ for all } \gamma \in \Gamma, \gamma \neq e\}.$$

Lemma III.20.9

1. *The Dirichlet domain D is a fundamental domain for the Γ-action and is star shaped, i.e., for any $x \in D$, the geodesic from x_1 to x belongs to D.*
2. *Let $x_1' \in \Gamma\backslash X$ be the projection of $x_1 \in X$. A geodesic ray $\gamma(t)$, $t \geq 0$, in $\Gamma\backslash X$ with $\gamma(0) = x_1'$ is DM if and only if its lift $\tilde{\gamma}(t)$ to X with $\tilde{\gamma}(0) = x_1$ belongs to the Dirichlet domain D with center x_1.*

Proof. The basic point is the formula for the distance function $d_{\Gamma\backslash X}$ in terms of the distance function d_X:

$$d_{\Gamma\backslash X}(\pi(p), \pi(q)) = \inf_{\gamma\in\Gamma} d_X(p, \gamma q),$$

where $\pi : X \to \Gamma\backslash X$ is the canonical projection.

To understand the shape of the Dirichlet fundamental domain in terms of the horospherical decomposition, we need the following result (see [JM, Proposition 10.8] for proof).

Proposition III.20.10 *For any rational parabolic subgroup $\mathbf{P}$ and compact subsets $U \subset N_P$, $V \subset X_{\mathbf{P}}$, there exists a positive number $t_0 = t_0(\Gamma, U, V)$ such that for any $a_0 \in A_{\mathbf{P},t_0}$, $a_1 \in A_{\mathbf{P}}^+ \cdot a_0$, $n, n' \in U$, $z, z' \in V$, and $\gamma \in \Gamma - \Gamma_P$, the following inequality holds:*

$$d((n, a_1, z), (n', a_0, z')) < d((n, a_1, z), \gamma \cdot (n', a_0, z')).$$

Proposition III.20.11 *Assume that Γ is a neat arithmetic subgroup. For any $(n_0, z_0) \in N_P \times X_{\mathbf{P}}$ and any sequence $y_j = (n_j, a_j, z_j) \in N_P \times A_{\mathbf{P}} \times X_{\mathbf{P}}$ with $n_j \to n_0$, $z_j \to z_0$, and $a_j^\alpha \to +\infty$ for all $\alpha \in \Phi(P, A_{\mathbf{P}})$, there exist compact neighborhoods U of n_0 and V of z_0, and $j_0 \geq 1$ such that when $j \geq j_0$, the Dirichlet domain D_j for Γ with center (n_j, a_j, z_j) contains $U \times \{a_k\} \times V$ when $k \gg j$.*

Using this proposition, Lemma III.20.9, and the classification of geodesics in §I.2, we can show Proposition III.20.8. See [JM, Theorem 10.18] for details.

Using the classification of EDM geodesics, we can prove the following results. See [JM, §11] for details of their proofs.

Proposition III.20.12 *Assume that Γ is a neat arithmetic subgroup. Then $\Gamma\backslash X$ satisfies the assumptions in III.20.3 and III.20.4. Hence the geodesic compactification $\Gamma\backslash X \cup \Gamma\backslash X(\infty)$ is defined.*

Remark III.20.13 One can also show directly that the projection of $\tilde{\gamma}(t) = (n, a \exp tH, z)$ is an EDM geodesic by using the precise reduction theory in Proposition III.2.21 (see [JM, Proposition 10.5]). Then the validity of Assumption III.20.3 for $\Gamma\backslash X$ also follows from the precise reduction theory.

Proposition III.20.14 *Let $\gamma(t)$ be an EDM geodesic. Then as $t \to +\infty$, $\gamma(t)$ converges to a point in the Tits compactification $\overline{\Gamma\backslash X}^T$. Two EDM geodesics in $\Gamma\backslash X$ converge to the same point in $\overline{\Gamma\backslash X}^T$ if and only if they are equivalent. Hence, there is a bijection between $\Gamma\backslash X(\infty)$ and $\Gamma\backslash\Delta_{\mathbb{Q}}(\mathbf{G})$.*

The first statement is clear from the classification of EDM geodesics in Proposition III.20.8 and the definition of $\overline{\Gamma\backslash X}^T$. It is also clear that every boundary point of $\overline{\Gamma\backslash X}^T$ is the limit of an EDM geodesic. For the proof of the second statement, see [JM, Proposition 11.3].

As a corollary of the above proposition, we obtain the following identification.

Proposition III.20.15 *The identity map on $\Gamma\backslash X$ extends to a continuous map $\overline{\Gamma\backslash X}^T \to \Gamma\backslash X \cup \Gamma\backslash X(\infty)$. Hence, by the previous proposition, $\overline{\Gamma\backslash X}^T$ is isomorphic to $\Gamma\backslash X \cup \Gamma\backslash X(\infty)$.*

III.20.16 In §I.17, we introduced the Gromov compactification of X. In fact, as pointed out there, the construction outlined there works for any complete Riemannian manifold M (see [BGS]). Briefly, let $C^0(M)$ be the space of continuous functions, and $\tilde{C}(M)$ the quotient by constant functions. For every point $x \in M$, $d(\cdot, x)$ is the distance measured from x. Let $[d(\cdot, x)]$ be its image in $\tilde{C}(M)$. Then the map

$$i : M \to \tilde{C}(M), \quad x \mapsto [d(\cdot, x)]$$

is an embedding, and the closure $\overline{i(M)}$ is a Hausdorff compactification of M, called the *Gromov compactification* and denoted by $\overline{\Gamma\backslash X}^G$.

Proposition III.20.17 *The Gromov compactification $\overline{\Gamma\backslash X}^G$ is isomorphic to the Tits compactification $\overline{\Gamma\backslash X}^T$, and hence also isomorphic to the geodesic compactification $\Gamma\backslash X \cup \Gamma\backslash X(\infty)$.*

See [JM, §12] for details of the proof.

III.20.18 Summary and comments. The structure of geodesics in a Riemannian manifold M is an important part of the geometry of the Riemannian manifold M. For a simply connected nonpositively curved Riemannian manifold M, the geodesic compactification $M \cup M(\infty)$ is well known and has played an important role in understanding the geometry and analysis of M. For a nonsimply connected manifold, we could not consider all geodesics nor all geodesics that go to infinity. It turns out that the right condition on geodesics is the EDM condition. The identification of the set of equivalence classes of such EDM geodesics with the quotient $\Gamma\backslash\Delta_{\mathbb{Q}}(\mathbf{G})$ is similar to the relation between the sphere at infinity $X(\infty)$ and the spherical Tits building $\Delta(G)$ in §I.2.

III.21 Rough geometry of $\Gamma\backslash X$ and Siegel conjecture on metrics on Siegel sets

In this section, we first study locally symmetric spaces as metric spaces in the category in which we identify two metric spaces when the Hausdorff distance between them is finite. Then we prove a conjecture of Siegel on comparison of two metrics on Siegel domains.

Specifically, we first recall some general facts about Hausdorff distance (III.21.1) and the tangent cones at infinity of noncompact Riemannian manifolds (III.21.3). Then we use the precise reduction theory for arithmetic groups to determine the tangent cone at infinity of $\Gamma\backslash X$ (III.21.7), (III.21.14). A key step is to understand a skeleton of $\Gamma\backslash X$ (III.21.10). The Siegel conjecture on comparison of metrics on Siegel sets is stated in III.21.16, and proved in III.21.17.

Definition III.21.1 *If X, Y are two subsets of a metric space (Z, d), then the Hausdorff distance $d_Z^H(X, Y)$ between X, Y in Z is defined as follows:*

$$d_Z^H(X, Y) = \inf\{\varepsilon \mid d(x, Y), d(X, y) \leq \varepsilon \text{ for all } x \in X, y \in Y\}.$$

If X, Y are any two metric spaces, then the Hausdorff distance $d_H(X, Y)$ between them is defined by

$$d_H(X, Y) = \inf_Z d_Z^H(f(X), f(Y)),$$

where Z is a metric space, and $f : X \to Z$, $f : Y \to Z$ are isometric embeddings.

Definition III.21.2 Let (M_n, d_n, x_n), $n \geq 1$, be a sequence of pointed metric spaces, where d_n is the distance function of M_n and x_n is a basepoint in M_n. Then (M_n, d_n, x_n) is defined to converge to a pointed metric space $(M_\infty, d_\infty, x_\infty)$ in the sense of Gromov–Hausdorff if for all $R > 0$, the Hausdorff distance between the metric ball $B(x_n, R)$ in M_n and the metric ball $B(x_\infty, R)$ in M_∞ goes to zero as $n \to \infty$.

Definition III.21.3 Let (M, d) be a metric space. For any $t > 0$, $\frac{1}{t}d$ defines another metric on M. Let $x_0 \in M$ be a basepoint. If the Gromov–Hausdorff limit $\lim_{t\to\infty}(M, \frac{1}{t}d, x_0)$ exists, then it is a metric cone and called the tangent cone at infinity of M, denoted by $T_\infty M$. This limit is clearly independent of the choice of the basepoint x_0.

Definition III.21.4 A metric space (M, d) is called a length space if the distance between any two points in M is equal to the minimum of the lengths of all curves joining them.

If (M, g) is a complete Riemannian manifold and d_g is the induced distance function, then (M, d_g) is a length space. If $T_\infty M$ exists, it is also a length space (see [Gro2, 3.8]).

Lemma III.21.5 *Let M be a topological space with a distance function d defined locally, i.e., when x, y belong to a small neighborhood, $d(x, y)$ is defined. Then there is a canonical length structure associated with d as in [Gro2, 1.4].*

Proof. It is shown in [Gro2, pp. 1–2] that a distance function canonically defines a length structure. Since the dilation is defined locally, the same argument works for a locally defined distance function.

III.21.6 We will use these general results to determine the tangent cone at infinity $T_\infty \Gamma\backslash X$.

Let $\mathbf{P}_1, \dots, \mathbf{P}_k$ be representatives of Γ-conjugacy classes of rational parabolic subgroups of $\mathbf{G}$. Recall the precise reduction theory in Proposition III.2.21. For each $\mathbf{P}_i$, let $A_{\mathbf{P}_i,T}$ be the shifted cone as in equation (III.1.16). Fix a $T \gg 0$ and identify $\coprod_{i=0}^n A_{\mathbf{P}_i,T}$ with the subset $\coprod_{i=0}^n A_{\mathbf{P}_i,T} x_0$ in $\Gamma\backslash X$. Then the Riemannian distance function on $\Gamma\backslash X$ induces a distance function on the subspace $\coprod_{i=1}^k A_{\mathbf{P}_i,T}$, denoted by d_{ind}. Note that we cannot exclude right away the possibility that some points x, y in one $A_{\mathbf{P}_i,T}$ may be connected by a distance-minimizing curve not entirely contained in $A_{\mathbf{P}_i,T}$.

Lemma III.21.7 *If the tangent cone at infinity $T_\infty(\coprod_{i=1}^k A_{\mathbf{P}_i,T}, d_{ind})$ of the subspace $(\coprod_{i=1}^k A_{\mathbf{P}_i,T}, d_{ind})$ exists, then $T_\infty \Gamma\backslash X$ also exists and is equal to $T_\infty(\coprod_{i=1}^k A_{\mathbf{P}_i,T}, d_{ind})$.*

Proof. From the precise reduction reduction theory in Proposition III.2.21, it is clear that the Hausdorff distance between $(\coprod_{i=1}^k A_{\mathbf{P}_i,T}, d_{ind})$ and $\Gamma\backslash X$ is finite. Then the lemma follows easily.

III.21.8 We define another length structure on $\coprod_{i=1}^k A_{\mathbf{P}_i,T}$ in order to study this induced distance function d_{ind}.

Identify $A_{\mathbf{P}_i,T}$ with a cone in the Lie algebra $\mathfrak{a}_{\mathbf{P}_i}$ through the exponential map and endow it with the metric defined by the Killing form. Denote this metric by d_S, called the simplicial metric. Then $(A_{\mathbf{P}_i,T}, d_S)$ is a cone over $\mathfrak{a}^+_{\mathbf{P}_i}(\infty) = A^+_{\mathbf{P}_i}(\infty)$, where $A^+_{\mathbf{P}_i}(\infty)$ is the open simplex in the Tits building $\Delta_{\mathbb{Q}}(\mathbf{G})$ associated with $\mathbf{P}_i$. In fact, with a suitable simplicial metric on $A^+_{\mathbf{P}_i}(\infty)$, $(A_{\mathbf{P}_i,T}, d_S)$ is a metric cone over $A^+_{\mathbf{P}_i}(\infty)$.[1]

We now glue these metric cones $(A_{\mathbf{P}_i,T}, d_S)$ together to get a local distance function on $\coprod_{i=1}^k A_{\mathbf{P}_i,T}$. Since $A_{\mathbf{P}_i,T}$ is a translate of the positive chamber $A^+_{\mathbf{P}_i}$, $(A_{\mathbf{P}_i,T}, d_S)$ is isometric to $(A^+_{\mathbf{P}_i}, d_S)$. Identify $(A_{\mathbf{P}_i,T}, d_S)$ with $(A^+_{\mathbf{P}_i}, d_S)$. Let $(\overline{A^+_{\mathbf{P}_i}}, d_S)$ be the closure of $(A^+_{\mathbf{P}_i}, d_S)$ in $(A_{\mathbf{P}_i}, d_S)$. Any face F of the polyhedral

[1] The metric on $A^+_{\mathbf{P}_i}(\infty)$ induced from the distance function of $A_{\mathbf{P}_i}$ defined by the Killing form is not a simplicial metric, since by definition, $A^+_{\mathbf{P}_i}(\infty)$ is a part of the unit sphere in $A_{\mathbf{P}_i}$.

cone $\overline{A^+_{\mathbf{P}_i}}$ is the chamber $\overline{A^+_{\mathbf{P}_F}}$ of a rational parabolic subgroup $\mathbf{P}_F$ containing $\mathbf{P}_i$. The group $\mathbf{P}_F$ is Γ-conjugate to a unique representative $\mathbf{P}_j$ above. Identify $\overline{A^+_{\mathbf{P}_j}}$ with the face F. Gluing all the spaces $\overline{A^+_{\mathbf{P}_i}}$ together using this face relation gives a topological space $\cup_{i=1}^k \overline{A^+_{\mathbf{P}_i}} / \sim$. Suppose that $\overline{A^+_{\mathbf{P}_j}}$ is glued onto a face of $\overline{A^+_{\mathbf{P}_i}}$. Since both metrics on $\overline{A^+_{\mathbf{P}_i}}$ and $\overline{A^+_{\mathbf{P}_j}}$ are induced from the Killing form, they coincide on $\overline{A^+_{\mathbf{P}_j}}$. Therefore, all the metric spaces $(\overline{A^+_{\mathbf{P}_i}}, d_S)$ are compatible and can be glued together to give a locally defined distance function on $\cup_{i=1}^k \overline{A^+_{\mathbf{P}_i}} / \sim$. By Lemma III.21.5, there is an induced length function on $\cup_{i=1}^k \overline{A^+_{\mathbf{P}_i}} / \sim$, denoted by l_S.

As a topological space, $\cup_{i=1}^k \overline{A^+_{\mathbf{P}_i}} / \sim$ is a cone over the finite complex $\Gamma\backslash\Delta_{\mathbb{Q}}(\mathbf{G})$; and as a set, $\cup_{i=1}^k \overline{A^+_{\mathbf{P}_i}} / \sim$ has a disjoint decomposition $\coprod_{i=1}^k A^+_{\mathbf{P}_i}$ that can be identified with $\coprod_{i=1}^k A_{\mathbf{P}_i,T}$. Therefore, the length space $(\cup_{i=1}^k \overline{A^+_{\mathbf{P}_i}} / \sim, l_S)$ defines a length structure on $\coprod_{i=1}^k A_{\mathbf{P}_i,T}$, denoted by $(\coprod_{i=1}^k A_{\mathbf{P}_i,T}, l_S)$.

An important property of this length space $(\coprod_{i=1}^k A_{\mathbf{P}_i,T}, l_S)$ is the following lemma. The basic idea of the proof is that since all the minimal rational parabolic subgroups $\mathbf{P}$ (and hence their positive chambers $A^+_{\mathbf{P}}$) are conjugate, there is no short-cut in $(\coprod_{i=1}^k A_{\mathbf{P}_i,T}, l_S)$ connecting two points in one chamber $\overline{A^+_{\mathbf{P}}}$ by going through other chambers.

Lemma III.21.9 *For any $i = 0, \dots, n$, $A_{\mathbf{P}_i,T}$ is a convex subspace of the length space $(\coprod_{i=1}^k A_{\mathbf{P}_i,T}, l_S)$; in other words, for any two points in $A_{\mathbf{P}_i,T}$, any curve that connects x, y and realizes the distance between them is contained in $A_{\mathbf{P}_i,T}$ and is hence a straight line segment contained entirely in $A_{\mathbf{P}_i,T}$. In particular, on each $A_{\mathbf{P}_i,T}$, $l_S = d_S$, where d_S is the simplicial distance on $A_{\mathbf{P}_i,T}$ defined by the Killing form.*

Proposition III.21.10 *There is a continuous map $\varphi : \Gamma\backslash X \to (\coprod_{i=1}^k A_{\mathbf{P}_i,T}, l_S)$ that restricts to the identity map on the subset $\coprod_0^n A_{\mathbf{P}_i,T}x_0$ when it is identified with $\coprod_{i=1}^k A_{\mathbf{P}_i,T}$.*

The basic idea is to retract $\Gamma\backslash X$ to the *skeleton* $\coprod_{i=1}^k A_{\mathbf{P}_i,T}x_0$, which is further identified with $\coprod_{i=1}^k A_{\mathbf{P}_i,T}$. For proofs of this and other statements, see [JM, §5].

Lemma III.21.11 *For any $i = 1, \dots, n$, the induced metric d_{ind} on $A_{\mathbf{P}_i,T} = A_{\mathbf{P}_i,T}x_0$ is equal to the simplicial metric d_S on $A_{\mathbf{P}_i,T}$. Hence, for any two points $x, y \in \coprod_{i=1}^k A_{\mathbf{P}_i,T}$,*

$$d_{ind}(x, y) \geq l_S(x, y).$$

Using the precise reduction in Proposition III.2.21, we obtain the bound in the other direction.

Lemma III.21.12 *There exists a finite constant c such that for any* $x, y \in \coprod_{i=1}^{k} A_{\mathbf{P}_i,T}$,

$$d_{ind}(x, y) \leq l_S(x, y) + c.$$

Since the tangent cone at infinity of $(\coprod_{i=1}^{k} A_{\mathbf{P}_i,T}, l_S)$ exists and is equal to itself, the above two lemmas imply the following results.

Lemma III.21.13 *The tangent cone at infinity* $T_\infty(\coprod_{i=1}^{k} A_{\mathbf{P}_i,T}, d_{ind})$ *exists and is equal to* $(\coprod_{i=1}^{k} A_{\mathbf{P}_i,T}, l_S)$.

Proposition III.21.14 *The tangent cone at infinity* $T_\infty(\Gamma\backslash X)$ *exists and is equal to* $(\coprod_{i=1}^{k} A_{\mathbf{P}_i,T}, l_S)$*, and hence equal to a metric cone over the complex* $\Gamma\backslash\Delta_{\mathbb{Q}}(\mathbf{G})$.

Proof. Since the Hausdorff distance between $\Gamma\backslash X$ and $(\coprod_{i=1}^{k} A_{\mathbf{P}_i,T}, d_{ind})$ is finite, the result follows from the above lemma.

III.21.15 In the remainder of this section, we apply the above results to prove the Siegel conjecture mentioned earlier. In [Si2, §10], Siegel stated the following conjecture.

Conjecture III.21.16 *For any rational parabolic subgroup* $\mathbf{P}$ *and a Siegel set* $U \times A_{\mathbf{P},t} \times V$*, where* $U \subset N_P$, $V \subset X_{\mathbf{P}}$ *are bounded subsets, there exists a positive constant C such that for every pair of points* $p, q \in U \times A_{\mathbf{P},t} \times V$*, and every* $\gamma \in \Gamma$,

$$d_X(p, \gamma q) \geq d_X(p, q) - C,$$

where d_X *is the distance function of X; or equivalently,*

$$d_{\Gamma\backslash X}(\pi(p), \pi(q)) \leq d_X(p, q) \leq d_{\Gamma\backslash X}(p, q) + C,$$

where $\pi : X \to \Gamma\backslash X$ *is the canonical projection and* $d_{\Gamma\backslash X}$ *is the distance function of* $\Gamma\backslash X$.

Proposition III.21.17 *The Siegel conjecture is true.*

Proof. For any $p, q \in X$, $d_{\Gamma\backslash X}(\pi(p), \pi(q)) = \min\{d_X(p, \gamma q) \mid \gamma \in \Gamma\}$. Hence $d_{\Gamma\backslash X}(\pi(p), \pi(q)) \leq d_X(p, q)$.

To prove the other inequality, we can assume that $\mathbf{P}$ is equal to some $\mathbf{P}_i$, since it is conjugate to some $\mathbf{P}_i$, and any Siegel set associated with $\mathbf{P}$ is also mapped to a Siegel set. Let $U_i \times A_{\mathbf{P}_i,T} \times V_i$ be the Siegel set in the disjoint decomposition in the precise reduction in Proposition III.2.21. Assume that $p, q \in U_i \times A_{\mathbf{P}_i,T} \times V_i$. Write the horospherical coordinates, $p = (n_p, \exp H_p, z_p)$, $q = (n_q, \exp H_q, z_q)$. Since U_i, V_i are bounded,

$$d_X(p, q) = d_{A_{\mathbf{P}_i}}(\exp H_p, \exp H_q) + O(1),$$

where $O(1)$ denotes some uniformly bounded quantity. Under the map ϕ in Proposition III.21.10,

$$\varphi(p) = \exp H_p, \quad \varphi(q) = \exp H_q.$$

By Lemmas III.21.11 and III.21.12,

$$d_{\Gamma\backslash X}(\pi(p), \pi(q)) = d_{l_S}(\exp H_p, \exp H_q) + O(1) = d_{A_{\mathbf{P}_i}}(\exp H_p, \exp H_q) + O(1),$$

and hence

$$d_{\Gamma\backslash X}(\pi(p), \pi(q)) \geq d_X(p, q) - C$$

for some constant independent of p, q.

In general, for any $p, q \in U \times A_{\mathbf{P}_i,t} \times V$, there exist $p', q' \in U_i \times A_{\mathbf{P}_i,T} \times V_i$ such that

$$d_X(p, p') = O(1), \quad d_X(q, q') = O(1),$$

which implies that

$$d_{\Gamma\backslash X}(\pi(p), \pi(p')) = O(1), \quad d_{\Gamma\backslash X}(\pi(q), \pi(q')) = O(1).$$

Then the equality $d_{\Gamma\backslash X}(\pi(p), \pi(q)) \geq d_X(p, q) - C$ follows from $d_{\Gamma\backslash X}(\pi(p'), \pi(q')) \geq d_X(p', q') - C$.

Remark III.21.18 An independent proof of the Siegel conjecture is given in [Leu1]. The special case $\Gamma = SL(n, \mathbb{Z})$ was proved earlier in [Din]. See [Ab] and [AbM] for generalizations of the Siegel conjecture and related results.

III.21.19 Summary and comments. Up to a finite Hausdorff distance, the locally symmetric space $\Gamma\backslash X$ is a cone over the quotient $\Gamma\backslash\Delta_{\mathbb{Q}}(\mathbf{G})$, a finite simplicial complex. Though this does not catch many fine properties of $\Gamma\backslash X$, for example, local topology near infinity, this does give a global picture, or rather the shape, of $\Gamma\backslash X$. The Siegel conjecture fits well into this framework.

III.22 Hyperbolic compactifications and extension of holomorphic maps from the punctured disk to Hermitian locally symmetric spaces

In this section, we study the Borel–Serre compactification $\overline{\Gamma\backslash X}^{BS}$, the reductive Borel–Serre compactification $\overline{\Gamma\backslash X}^{RBS}$, and the Satake compactifications $\overline{\Gamma\backslash X}^{S}$ from the point of view of metric spaces.

This section is organized as follows. First we recall the big Picard theorem and relate it to extension of holomorphic maps from the punctured disk and metric properties of the natural compactification $\mathbb{C}P^1$ of $\mathbb{C}P^1 \setminus \{0, 1, \infty\}$ (III.22.1). A general extension problem is formulated in III.22.2, and an answer is given in III.22.3. Motivated by this, the notion of hyperbolic compactifications is introduced in III.22.5, which is shown to be invariant under quasi-isometries. Existence and other properties of hyperbolic compactifications are given in III.22.9 and III.22.10. In III.22.11, the reductive Borel–Serre compactification $\overline{\Gamma\backslash X}^{RBS}$ is shown to be hyperbolic. As a corollary, the Baily–Borel compactification is also hyperbolic (III.22.12), which in turn implies the Borel extension theorem of holomorphic maps from the punctured disk (III.22.13). On the other hand, the Borel–Serre compactification is not hyperbolic (III.22.14). Similarly, the toroidal compactifications are not hyperbolic either (III.22.15). To show that the Baily–Borel compactification is extremal in certain senses among complex analytic compactifications of $\Gamma\backslash X$ (III.22.18 and III.22.19), we state a generalization on extension of holomorphic maps from the punctured disk (III.22.16). To show the difference between the geodesic compactification on the one hand and the Satake compactifications and the reductive Borel–Serre compactification on the other, we introduce the notion of asymptotic compactifications in III.22.21. Then the geodesic compactification is shown to be an asymptotic compactification (III.22.22), but the Satake compactifications are not (III.22.23).

III.22.1 In complex analysis of one variable, there is a well-known result called the *big Picard theorem*, which says that near an essentially singular point of a meromorphic function, its values can miss at most three points of $\mathbb{C}P^1$. This result can be expressed in terms of metric properties of $\mathbb{C}P^1\setminus\{0, 1, \infty\}$, which is of the form $\Gamma\backslash\mathbf{H}$. Let $D = \{z \in \mathbb{C} \mid |z| < 1\}$ be the unit disk, and $D^\times = D \setminus \{0\}$ the punctured disk. Then the big Picard theorem is equivalent to the statement that every holomorphic map

$$f : D^\times \to \mathbb{C}P^1 \setminus \{0, 1, \infty\}$$

can be extended across the puncture to a holomorphic map

$$f : D \to \mathbb{C}P^1.$$

This is related to the fact that the *Kobayashi pseudometric* on $\mathbb{C}P^1 \setminus \{0, 1, \infty\}$ is a metric. Briefly, the Kobayashi metric is the maximal pseudometric such that (1) it coincides with the Poincaré metric for the unit disc, (2) and it is distance-decreasing under holomorphic maps. (See [Kob2] for details and the definition of the Kobayashi pseudometric.)

This result can be generalized to the following situation. Let Y be a complex space and M a complex subspace of Y whose closure $\overline{M}$ is compact.

Problem III.22.2 *Does every holomorphic map $f : D^\times \to M$ extend to a holomorphic map $f : D \to Y$?*

In the above example, $Y = \mathbb{C}P^1$ and $M = \mathbb{C}P^1 \setminus \{0, 1, \infty\}$, and the answer is positive. Though the answer is negative in general, it holds under some conditions.

Proposition III.22.3 *The answer to the above extension problem is positive if the following two conditions hold:*

1. *M is a hyperbolic manifold, i.e., the Kobayashi pseudometric d_M of M is a metric.*
2. *For any two sequences $p_j, q_j \in M$ with $p_j \to p_\infty$, $q_j \to q_\infty$ in Y, if $d_M(p_j, q_j) \to 0$, then $p_\infty = q_\infty$.*

For the proof of this proposition, see [Kob2]. The conditions say that the compactification $\overline{M}$ of M in Y is small in a certain sense. If these conditions are satisfied, the embedding $M \hookrightarrow Y$ is called a *hyperbolic embedding with respect to the metric* d_M.

III.22.4 As mentioned earlier, many important spaces $\Gamma\backslash X$ arise as moduli spaces of certain varieties or structures in algebraic geometry and are often noncompact. A natural, important problem is to understand how these objects degenerate, or their moduli points go to the infinity of $\Gamma\backslash X$. This is equivalent to understanding whether holomorphic maps

$$f : D^\times \to \Gamma\backslash X$$

extend and the senses in which they extend.

Motivated by the above results, we introduce the following definition.

Definition III.22.5 *Let (Y, d) be a metric space. A compactification $\overline{Y}$ is called a hyperbolic compactification with respect to the metric d if the following condition is satisfied: Let p_j, q_j be any two sequences in Y that converge to two boundary points $p, q \in \overline{Y} - Y$ respectively. If $d(p_j, q_j) \to 0$, then $p = q$.*

Intuitively, a hyperbolic compactification is small in the sense that the boundary points will not separate sequences of points that are close in the distance function d. Clearly, the smallest compactification, the one-point compactification $Y \cup \{\infty\}$, is always hyperbolic, since all sequences of points that go to infinity have the same limit. This intuition also tells us that if a compactification is smaller, it has a better chance of being hyperbolic.

Lemma III.22.6 *Let d_1, d_2 be two metrics on Y, and let $\overline{Y}$ be a compactification of Y. If $\overline{Y}$ is hyperbolic with respect to d_1 and there exists a constant $C > 0$ such that*

$$d_1(x, y) \leq C d_2(x, y), \quad \textit{for all } x, y \in Y,$$

then $\overline{Y}$ is also hyperbolic with respect to d_2. Therefore, if d_1, d_2 are quasi-isometric, i.e., there exists $C > 1$ such that

$$C^{-1} d_2(x, y) \leq d_1(x, y) \leq C d_2(x, y), \quad \textit{for all } x, y \in Y,$$

then $\overline{Y}$ is hyperbolic with respect to d_2 if and only if it is hyperbolic with respect to d_1.

Proof. This follows from the definition of hyperbolic compactifications and the fact that if $d_2(p_j, q_j) \to 0$, then $d_1(p_j, q_j) \to 0$.

III.22.7 Suppose that X is a Hermitian symmetric space. It is known that X is hyperbolic in the sense that the Kobayashi pseudometric is a metric. Let $d_{X,k}$ be the distance function associated with the Kobayashi metric. Let d_X be the distance of the invariant metric. It is also known that the metric $d_{X,k}$ is invariant under G, in particular under Γ. Denote the induced metric on $\Gamma\backslash X$ by $d_{\Gamma\backslash X,k}$. Similarly, denote the metric on $\Gamma\backslash X$ induced from d_X by $d_{\Gamma\backslash X}$.

The discussions in [Bo6, 3.1–3.3] imply the following equivalence.

Lemma III.22.8 *Assume that X is a Hermitian symmetric space. Then the two distance functions $d_{k,X}$ and d_X are quasi-isometric, and hence $d_{k,\Gamma\backslash X}$ and $d_{\Gamma\backslash X}$ are also quasi-isometric.*

If the reference to a distance d is clear, it will be omitted, and a hyperbolic compactification $\overline{Y}$ with respect to d will simply be referred to as a hyperbolic compactification of Y.

Lemma III.22.9 *If a compactification $\overline{Y}^2$ of Y is dominated by a hyperbolic compactification $\overline{Y}^1$, then $\overline{Y}^2$ is also hyperbolic.*

Proof. Let p_j, q_j be any two sequences in Y converging to p, q in $\overline{Y}^2$ respectively and $d(p_n, q_n) \to 0$. Since $\overline{Y}^1$ is compact, we can choose subsequences $p_{j'}, q_{j'}$ such that $p_{j'} \to p'$, and $q_{j'} \to q'$ in $\overline{Y}^1$. Since $\overline{Y}^1$ dominates $\overline{Y}^2$, p' is mapped to p under the dominating map, and q' is mapped to q. Since $\overline{Y}^1$ is hyperbolic by assumption, $p' = q'$. This implies that $p = q$, and hence $\overline{Y}^2$ is hyperbolic.

The following two general facts are also known. See [Ji4, §2] for proofs.

Proposition III.22.10

1. *Every locally compact metric space (Y, d) has a unique largest hyperbolic compactification.*
2. *For any compactification $\overline{Y}$ of a locally compact metric space (Y, d), there is a unique largest hyperbolic compactification among all the compactifications dominated by $\overline{Y}$. This compactification is called the hyperbolic reduction of the compactification of $\overline{Y}$ and denoted by $Red_H(\overline{Y})$.*

In the following, when we discuss the hyperbolicity of compactifications of $\Gamma\backslash X$, we always use the invariant metric.

Proposition III.22.11 *The reductive Borel–Serre compactification $\overline{\Gamma\backslash X}^{RBS}$ is hyperbolic.*

Proof. Let p_j, q_j be two sequences in $\Gamma\backslash X$ converging to boundary points p, q respectively. Assume that $d(p_j, q_j) \to 0$. We need to prove that $p = q$.

We prove this by contradiction. Suppose $p \neq q$. Let $\mathbf{P}$ be a rational parabolic subgroup such that a point $z \in X_\mathbf{P}$ is projected to p. For any $\varepsilon > 0$, let $B_{X_\mathbf{P}}(z, \varepsilon)$ be the metric ball in $X_\mathbf{P}$ or radius ε and center z, and similarly $B_{A_\mathbf{P}}(a, \varepsilon)$ a ball in $A_\mathbf{P}$.

Since $p_j \to p$, there exists a list $\tilde{p}_j$ of p_j such that in the horospherical decomposition $\tilde{p}_j = (n_j, a_j, z_j)$, the coordinates satisfy the conditions (1) n_j is bounded, (2) $z_j \to z$ in $X_\mathbf{P}$, (3) $a_j^\alpha \to +\infty$ for all $\alpha \in \Phi(P, A_\mathbf{P})$.

On the other hand, $q_j \to q \neq p$. Hence there exist $\varepsilon_0 > 0$ and j_0 such that when $j \geq j_0$, for any $j \geq j_0$ and any lift $\tilde{q}_j$ of q_j, the horospherical coordinates of $\tilde{q}_j = (n'_j, a'_j, z_j)$ satisfy the condition

$$(a'_j, z'_j) \notin B_{A_\mathbf{P}}(a_j, 2\varepsilon_0) \times B_{X_\mathbf{P}}(z, 2\varepsilon_0).$$

Otherwise, we could extract a subsequence of $\tilde{q}_j$ converging to z in ${}_\mathbb{Q}\overline{X}^{RBS}$. By the formula of the metric of X in terms of the horospherical decomposition in Lemma III.20.7, we have

$$d_X(\tilde{p}, \tilde{q}) \geq \max\{d_{X_\mathbf{P}}(z_p, z_q), d_{A_\mathbf{P}}(a_p, a_q)\},$$

where $\tilde{p} = (n_p, a_p, z_p), \tilde{q} = (n_q, a_q, z_q) \in N_P \times A_\mathbf{P} \times X_\mathbf{P}$. When $j_0 \gg 1$, we can assume that $z_j \in B_{X_\mathbf{P}}(z, \varepsilon_0)$ when $j \geq j_0$. This implies that

$$d_X(\tilde{p}_j, \tilde{q}_j) \geq \varepsilon_0,$$

and hence

$$d_{\Gamma\backslash X}(p_j, q_j) \geq \varepsilon_0.$$

This contradicts the assumption $d(p_j, q_j) \to 0$.

Corollary III.22.12 *The Baily–Borel compactification $\overline{\Gamma\backslash X}^{BB}$ is hyperbolic.*

Proof. This follows from the fact that $\overline{\Gamma\backslash X}^{RBS}$ dominates $\overline{\Gamma\backslash X}^{BB}$ in Proposition III.15.4, Lemma III.22.9, and Proposition III.22.11.

Together with Proposition III.22.3, Corollary III.22.12 implies the following result.

Corollary III.22.13 *Let $D = \{z \in \mathbb{C} \mid |z| < 1\}$ be the unit disk, and $D - \{0\}$ the punctured disk. Then every holomorphic map $f : D - \{0\} \to \Gamma\backslash X$ extends to a holomorphic map $\tilde{f} : D \to \overline{\Gamma\backslash X}^{BB}$.*

This extension property implies the great Picard theorem and also has important applications to the theory of variation of Hodge structures (see [Gri] [GS]). Corollary III.22.13 was first proved in [Bo6], and related results were also obtained in [KO]. Another proof of Corollary III.22.13 was given in [GS, p. 99].

Corollary III.22.14 *The Borel–Serre compactification $\overline{\Gamma\backslash X}^{BS}$ is not hyperbolic, and its hyperbolic reduction $Red_H(\overline{\Gamma\backslash X}^{BS})$ is the reductive Borel–Serre compactification $\overline{\Gamma\backslash X}^{RBS}$.*

Proof. By definition, $\overline{\Gamma\backslash X}^{BS}$ dominates $\overline{\Gamma\backslash X}^{RBS}$. Since $\overline{\Gamma\backslash X}^{RBS}$ is hyperbolic, it suffices to prove that every two points in the inverse image of the dominating map over any boundary point in $\overline{\Gamma\backslash X}^{RBS}$ are limits of two sequences of points p_j, q_j with $d(p_j, q_j) \to 0$. This is clear from the expression for the invariant metric in Proposition III.20.7 and the definition of $\overline{\Gamma\backslash X}^{BS}$.

This result gives another reason for the name of the reductive Borel–Serre compactification.

As recalled in §III.7, the toroidal compactifications all dominate the Baily–Borel compactification.

Proposition III.22.15 *The hyperbolic reduction of any toroidal compactification $\overline{\Gamma\backslash X}_{\Sigma}^{tor}$ is the Baily–Borel compactification $\overline{\Gamma\backslash X}^{BB}$.*

The idea of the proof is similar to that of Corollary III.22.14. See [Ji4, Theorem 5.3].

To state an application of this proposition, we need the following generalization in [KK2] of the big Picard theorem in Proposition III.22.3.

Proposition III.22.16 *Let W be a compact complex analytic space, and $Y \subset W$ a hyperbolically embedded complex subspace, i.e., the closure of M in W is a hyperbolic compactification. Let A be a closed complex subspace of a complex manifold Z. If the singularities of A are normal crossings, then every holomorphic map $f : Z - A \to Y$ extends to a holomorphic map $\tilde{f} : W \to Z$.*

In the proposition, if we take $Z = D$, the unit disc, A the origin, this is a reformulation of the big Picard theorem in Proposition III.22.3.

Proposition III.22.17 *Assume that $\Gamma\backslash X$ is Hermitian. Let Y be a complex space hyperbolically embedded in a compact complex space W. Then every holomorphic map $f : \Gamma\backslash X \to Y$ extends to a holomorphic map $\overline{\Gamma\backslash X}^{BB} \to W$.*

Proof. Let $\overline{\Gamma\backslash X}_{\Sigma}^{tor}$ be a toroidal compactification that is smooth and whose boundary $\partial\overline{\Gamma\backslash X}_{\Sigma}^{tor} - \Gamma\backslash X$ is a union of divisors with normal crossings. By the previous proposition, the holomorphic map $f : \Gamma\backslash X \to Y$ extends to a holomorphic map $\tilde{f} : \overline{\Gamma\backslash X}^{BB} \to W$. Since the embedding of Y in W is hyperbolic, the map $\tilde{f}$ factors through $Res_H(\overline{\Gamma\backslash X}_{\Sigma}^{tor})$. By Proposition III.22.15, $Res_H(\overline{\Gamma\backslash X}_{\Sigma}^{tor}) = \overline{\Gamma\backslash X}^{BB}$, and hence $\tilde{f}$ induces the desired extension.

Corollary III.22.18 *Assume that $\Gamma\backslash X$ is Hermitian. If $\overline{\Gamma\backslash X}$ is a hyperbolic compactification that is also a complex analytic space, then it is dominated by the Baily–Borel compactification, i.e., the Baily–Borel compactification is the largest complex analytic hyperbolic compactification.*

This corollary shows that among all the hyperbolic complex analytic compactifications of $\Gamma\backslash X$, $\overline{\Gamma\backslash X}^{BB}$ is the maximal one. On the other hand, as mentioned earlier in §III.4, the Baily–Borel compactification $\overline{\Gamma\backslash X}^{BB}$ is isomorphic to a minimal Satake compactification. The following result shows that it is also minimal in another sense.

Corollary III.22.19 *Assume that $\Gamma\backslash X$ is Hermitian. If $\overline{\Gamma\backslash X}'$ is a complex analytic compactification such that the boundary $\overline{\Gamma\backslash X}' - \Gamma\backslash X$ is a union of divisors with normal crossings, then $\overline{\Gamma\backslash X}'$ dominates $\overline{\Gamma\backslash X}^{BB}$.*

Proof. By Corollary III.22.12, the inclusion $\Gamma\backslash X \to \overline{\Gamma\backslash X}^{BB}$ is a hyperbolic embedding. Then the proposition follows from Proposition III.22.16.

III.22.20 Finally, we mention another general metric compactification, which gives another explanation of the geodesic compactification $\Gamma\backslash X \cup \Gamma\backslash X(\infty) = \overline{\Gamma\backslash X}^T$.

Definition III.22.21 *A compactification $\overline{Y}$ of a metric space (Y, d) is an asymptotic compactification if it satisfies the following condition: for any two sequences p_j, q_j in Y converging to p, q in $\overline{Y}$ respectively, if $\limsup_{j\to+\infty} d(p_j, q_j) < +\infty$, then $p = q$.*

Clearly, if $\overline{Y}$ is a hyperbolic compactification, it is an asymptotic compactification. But the converse is not true.

The asymptotic compactifications have a functorial property in the category of locally compact metric spaces with morphisms given by ε-Hausdorff approximations. Similarly, we can define the asymptotic reduction of any compactification $\overline{Y}$.

Proposition III.22.22 *The Tits compactification $\overline{\Gamma\backslash X}^T$ is an asymptotic compactification.*

Proof. We claim that the geodesic compactification $\Gamma\backslash X \cup \Gamma\backslash X(\infty)$ is an asymptotic compactification. By Proposition III.20.15, $\overline{\Gamma\backslash X}^T$ is isomorphic to $\Gamma\backslash X \cup \Gamma\backslash X(\infty)$ and is hence an asymptotic compactification. In fact, it follows from the precise reduction theory in Proposition III.2.21 that there exists a compact base subspace ω_0 such that for any two sequences $p_j, q_j \in \Gamma\backslash X$ going to infinity, there exist rays $\gamma_i(t), \delta_j(t)$ connecting ω_0 with p_j, q_j respectively; furthermore, if $\gamma_j(t) \to \gamma(t)$ and $\delta_j(t) \to \delta(t)$, and $d(p_j, q_j)$ is bounded, then $\gamma \sim \delta$, i.e., $\limsup_{t\to+\infty} d(\gamma(t), \delta(t)) < +\infty$, which implies that p_j, q_j converge to the same point in $\Gamma\backslash X \cup \Gamma\backslash X(\infty)$.

Proposition III.22.23 *The Borel–Serre compactification $\overline{\Gamma\backslash X}^{BS}$ is never an asymptotic compactification, and its asymptotic reduction is equal to the end compactification, which is equal to the one-point compactification if the $\mathbb{Q}$-rank of* **G** *is greater than or equal to* 2. *Similarly, the reductive Borel–Serre compactification $\overline{\Gamma\backslash X}^{RBS}$ and the Satake compactifications $\overline{\Gamma\backslash X}_{\tau}^{S}$ are not asymptotic compactifications unless both the $\mathbb{Q}$- and $\mathbb{R}$- ranks of* **G** *are equal to* 1.

See [Ji4, §6] for details of the above definition and the proofs of the propositions.

III.22.24 Summary and comments. The notions of hyperbolic embeddings and hyperbolic compactifications are closely related to the notion of Kobayashi hyperbolic spaces (see [Kob1] [Kob2] for details). The approach of studying compactifications of $\Gamma\backslash X$ from the point of view of metric spaces simplifies several results, for example, the Borel extension on holomorphic maps from the punctured disk. It also gives a different perspective on sizes of compactifications.

III.23 Continuous spectrum, boundaries of compactifications, and scattering geodesics of $\Gamma\backslash X$

In this section, we discuss relations between the continuous spectrum and the geometry at infinity of $\Gamma\backslash X$. The basic reference is [JM, §13].

This section is organized as follows. The decomposition principle on invariance of the continuous spectrum under compact perturbations is stated in III.23.1. The generalized eigenfunctions and parametrization of the generalized eigenspace of $\mathbb{R}^n$ are discussed in III.23.2. For noncompact $\Gamma\backslash X$, the generalized eigenfunctions are given by Eisenstein series (III.23.3, III.23.4). Then two distinguished subspaces, the discrete subspace $L^2_{dis}(\Gamma\backslash X)$, and the continuous subspace $L^2_{con}(\Gamma\backslash X)$, are introduced, and the spectral decomposition of $L^2_{con}(\Gamma\backslash X)$ is given in III.23.6 and III.23.7. This decomposition is interpreted as a parametrization of the generalized eigenspace of $\Gamma\backslash X$ in terms of the boundaries of the reductive Borel–Serre compactification $\overline{\Gamma\backslash X}^{RBS}$ and the geodesic compactification $\Gamma\backslash X \cup \Gamma\backslash X(\infty)$ (III.23.8). To refine relations between the continuous spectrum and the geometry at infinity, we recall the Poisson relation for functions on a circle in III.23.10, which relate the spectrum of the circle to the lengths of (closed) geodesics. To obtain a generalization of the Poisson relation to noncompact manifolds, we introduce scattering geodesics in III.23.11. The sojourn time of scattering geodesics is introduced in III.23.13. For $\Gamma\backslash X$ of $\mathbb{Q}$-rank 1, the structure of scattering geodesics is studied in III.23.15; scattering matrices of $\Gamma\backslash X$ are introduced in III.3.16, and the generalized Poisson relation between the sojourn times of scattering geodesics and scattering matrices is stated in III.23.17.

Let Δ be the Laplace operator of $\Gamma\backslash X$. It is well known that when $\Gamma\backslash X$ is compact, the spectrum of Δ (also called the spectrum of $\Gamma\backslash X$) is discrete. On the other hand, when $\Gamma\backslash X$ is noncompact, its spectrum is no longer discrete. The continuous spectrum as a set depends only on the geometry at infinity. In fact, we have the following *decomposition principle*:

Proposition III.23.1 *For any complete noncompact Riemannian manifold M, its singular spectrum, in particular the continuous spectrum, does not change under any compact perturbations and hence depends only on the geometry at infinity.*

In general, more refined spectral data such as the spectral measure and generalized eigenfunctions will change under any compact perturbations.

A natural question is how the continuous spectrum is related to the geometry at infinity.

We will discuss two such problems: (1) parametrization of the continuous spectrum in terms of boundaries of compactifications, (2) the Poisson relation between the scattering matrices of the continuous spectrum and the sojourn times of scattering geodesics.

III.23.2 The basic example is $M = \mathbb{R}^n$. In this case, it follows from the Fourier transform that the spectrum is purely continuous and is equal to $[0, +\infty)$, and for each $\lambda > 0$, the *generalized eigenfunctions* are given by $e^{i\sqrt{\lambda}\langle x, \chi\rangle}$, where $\chi \in S^{n-1}$, whose superpositions

$$\int_{S^{n-1}} e^{i\sqrt{\lambda}\langle x,\chi\rangle} f(\chi) d\chi$$

give functions in the *generalized eigenspace* of $\mathbb{R}^n$ with eigenvalue λ, where $f \in L^1(S^{n-1})$. Clearly, these generalized eigenfunctions $e^{i\sqrt{\lambda}\langle x,\chi\rangle}$ are linearly independent. Since the unit sphere S^{n-1} can be identified with the sphere at infinity $\mathbb{R}^n(\infty)$ in the geodesic compactification, the boundary of the compactification $\mathbb{R}^n \cup \mathbb{R}^n(\infty)$ is the parameter space of a basis of generalized eigenfunctions.

III.23.3 When $\Gamma\backslash X$ is noncompact, it has a *continuous spectrum*, and the *generalized eigenfunctions* of the continuous spectrum are given by Eisenstein series. We first recall several basic facts about Eisenstein series and the spectral decomposition of $L^2(\Gamma\backslash X)$ following [La], [Ar2], [MW], and [OW2].

For any rational parabolic subgroup $\mathbf{P}$ of $\mathbf{G}$, let $\mathfrak{a}_\mathbf{P}^*$ be the dual of $\mathfrak{a}_\mathbf{P}$. For any L^2-eigenfunction φ of the Laplace operator Δ on the boundary locally symmetric space $\Gamma_{M_\mathbf{P}}\backslash X_\mathbf{P}$, we define an *Eisenstein series* $E(\mathbf{P}|\varphi, \Lambda)$, $\Lambda \in \mathfrak{a}_\mathbf{P}^* \otimes \mathbb{C}$, as follows:

$$E(\mathbf{P}|\varphi, \Lambda : x) = \sum_{\gamma\in\Gamma_P\backslash\Gamma} e^{(\rho_\mathbf{P}+\Lambda)(H_\mathbf{P}(\gamma x))}\varphi(z_\mathbf{P}(\gamma x)),$$

where $\rho_\mathbf{P}$ is the half sum of the positive roots in $\Phi(P, A_\mathbf{P})$ with multiplicity equal to the dimension of the root spaces, and

$$x = (n_\mathbf{P}(x), z_\mathbf{P}(x), \exp H_\mathbf{P}(x)) \in N_\mathbf{P} \times X_\mathbf{P} \times A_\mathbf{P} = X,$$

the horospherical decomposition with respect to $\mathbf{P}$.

When $\mathrm{Re}(\Lambda) \in \rho_\mathbf{P} + \mathfrak{a}_\mathbf{P}^{*+}$, the above series converges uniformly for x in compact subsets of X (see [La, Lemma 4.1]). It is a theorem of Langlands [La, Chap. 7] [MW, Chap. 4] [OW2, Chap. 6] that $E(\mathbf{P}|\varphi, \Lambda)$ can be meromorphically continued as a function of Λ to the whole complex space $\mathfrak{a}_{\mathbf{P}}^* \otimes \mathbb{C}$, and $E(\mathbf{P}|\varphi, \Lambda)$ is regular when $\mathrm{Re}(\Lambda) = 0$.

The Eisenstein series $E(\mathbf{P}|\varphi, \Lambda)$ are clearly Γ-invariant and hence define functions on $\Gamma\backslash X$.

Lemma III.23.4 *If φ has eigenvalue ν, i.e., $\Delta\varphi = \nu\varphi$, then for any $\Lambda \in \sqrt{-1}\mathfrak{a}_{\mathbf{P}}^*$,*

$$\Delta E(\mathbf{P}|\varphi, \Lambda) = (\nu + |\rho_{\mathbf{P}}|^2 + |\Lambda|^2) E(\mathbf{P}|\varphi, \Lambda).$$

For any $f \in L^2(\sqrt{-1}\mathfrak{a}_{\mathbf{P}}^*)$, define a function $\hat{f}$ on $\Gamma\backslash X$ by

$$\hat{f}(x) = \int_{\sqrt{-1}\mathfrak{a}_{\mathbf{P}}^*} f(\Lambda) E(\mathbf{P}|\varphi, \Lambda : x) d\Lambda.$$

It is known that $\hat{f} \in L^2(\Gamma\backslash X)$ (see [OW1, pp. 328–329] for example). For every such pair of $\mathbf{P}$ and φ, denote the span in $L^2(\Gamma\backslash X)$ of all such functions $\hat{f}$ above by $L^2_{\mathbf{P},\varphi}(\Gamma\backslash X)$.

These subspaces induce a decomposition of $L^2(\Gamma\backslash X)$. Denote the subspace of $L^2(\Gamma\backslash X)$ spanned by L^2-eigenfunctions of the Laplace operator Δ by $L^2_{dis}(\Gamma\backslash X)$, called the *discrete subspace*, and the orthogonal complement of $L^2_{dis}(\Gamma\backslash X)$ in $L^2(\Gamma\backslash X)$ by $L^2_{con}(\Gamma\backslash X)$, called the continuous subspace. Then

$$L^2(\Gamma\backslash X) = L^2_{dis}(\Gamma\backslash X) \oplus L^2_{con}(\Gamma\backslash X).$$

The subspace $L^2_{con}(\Gamma\backslash X)$ can be decomposed into the subspaces $L^2_{\mathbf{P},\varphi}(\Gamma\backslash X)$ [La, Chap. 7] [OW1, Theorem 7.5].

Proposition III.23.5 *With the notation as above,*

$$L^2_{con}(\Gamma\backslash X) = \sum_{\mathbf{P}} \sum_{\varphi} L^2_{\mathbf{P},\varphi}(\Gamma\backslash X),$$

where $\mathbf{P}$ sums over all proper rational parabolic subgroups of $\mathbf{G}$, and φ is over eigenfunctions of $\Gamma_{M_{\mathbf{P}}}\backslash X_{\mathbf{P}}$ that form a basis of $L^2_{dis}(\Gamma_{M_{\mathbf{P}}}\backslash X_{\mathbf{P}})$.

Proposition III.23.6 *For any $\lambda > 0$, the generalized eigenspace of $\Gamma\backslash X$ with eigenvalue λ is spanned by $E(\mathbf{P}|\varphi, \Lambda)$, where $\mathbf{P}$ is a parabolic subgroup, φ an L^2 eigenfunction of $\Gamma_{M_{\mathbf{P}}}\backslash X_{\mathbf{P}}$ with eigenvalue ν satisfying $\nu \leq \lambda - |\rho_{\mathbf{P}}|^2$, and $\Lambda \in \sqrt{-1}\mathfrak{a}_{\mathbf{P}}^*$ satisfying $|\Lambda|^2 = \lambda - |\rho_{\mathbf{P}}|^2 - \nu$.*

The sum of the subspaces in Proposition III.23.5 is not direct, and the Eisenstein series $E(\mathbf{P}|\varphi, \Lambda)$ for different $\mathbf{P}$ are not linearly independent. For example, if two rational parabolic subgroups $\mathbf{P}_1, \mathbf{P}_2$ are conjugate under Γ, then the subspaces $\sum_\varphi L^2_{\mathbf{P}_1,\varphi}(\Gamma\backslash X)$, $\sum_\varphi L^2_{\mathbf{P}_2,\varphi}(\Gamma\backslash X)$ are equal to each other. Besides this relation, there is another relation coming from the functional equations of Eisenstein series. It turns out that these are all the relations between them after all (see [Ar2, Main Theorem, p. 256] [OW1, Proposition 7.4] [JM, Proposition 13.14]).

Proposition III.23.7 *For any rational parabolic subgroup* $\mathbf{P}$ *and an* L^2*-eigenfunction* φ *on* $\Gamma_{M_\mathbf{P}}\backslash X_\mathbf{P}$*, identify* $L^2(\sqrt{-1}\mathfrak{a}_\mathbf{P}^{*+})$ *with a subspace of* $L^2(\Gamma\backslash X)$ *through the map*

$$f \to \hat{f} = \int_{\sqrt{-1}\mathfrak{a}_\mathbf{P}^{*+}} f(\Lambda)E(\mathbf{P}|\varphi, \Lambda)d\Lambda.$$

Then the continuous subspace $L^2_{con}(\Gamma\backslash X)$ *is equal to the following direct sum:*

$$\sum_{i=1}^{k}\sum_{\varphi_{i,j}} \oplus L^2(\sqrt{-1}\mathfrak{a}_{\mathbf{P}_i}^{*+}),$$

where $\mathbf{P}_1, \dots, \mathbf{P}_k$ *is a set of representatives of* Γ*-conjugacy classes of rational parabolic subgroups of* $\mathbf{G}$*, and* $\varphi_{i,j}$ *is over an orthonormal basis of eigenfunctions of* $L^2_{dis}(\Gamma_{M_{\mathbf{P}_i}}\backslash X_{\mathbf{P}_i})$*.*

See [JM, Proposition 13.14] for proof of the above proposition.

III.23.8 The above proposition says that the generalized eigenspaces of $\Gamma\backslash X$ are naturally parametrized by the boundaries of the geodesic compactification $\Gamma\backslash X \cup \Gamma\backslash X(\infty)$ and the reductive Borel–Serre compactification $\overline{\Gamma\backslash X}^{RBS}$. In fact, consider the generalized eigenspace of eigenvalue λ. For each rational parabolic subgroup $\mathbf{P}_i$ above and a boundary eigenfunction $\varphi_{i,j}$, let $\nu_{i,j}$ be the eigenvalue of $\varphi_{i,j}$. Then the generalized eigenfunctions are

$$E(\mathbf{P}_i|\varphi_{i,j}, \Lambda),$$

where the parameter Λ belongs to the spherical section

$$\{\Lambda \in \mathfrak{a}_{\mathbf{P}_i}^* \mid |\Lambda|^2 = \lambda - \nu_{i,j} - |\rho_{\mathbf{P}_i}|^2\},$$

which can be identified with $\mathfrak{a}_{\mathbf{P}_i}^+(\infty)$ if $\lambda > \nu_{i,j} + |\rho_\mathbf{P}|^2$. By Proposition III.20.15,

$$\overline{\Gamma\backslash X}(\infty) = \coprod_{i=1}^{m} \mathfrak{a}_{\mathbf{P}_i}^+(\infty).$$

Since the eigenfunctions $\varphi_{i,j}$ span $L^2_{dis}(\Gamma_{M_{\mathbf{P}_i}}\backslash X_{\mathbf{P}_i})$, the boundary locally symmetric spaces $\Gamma_{M_{\mathbf{P}_i}}\backslash X_{\mathbf{P}_i}$ are also needed. By Proposition III.10.10,

$$\overline{\Gamma\backslash X}^{RBS} = \Gamma\backslash X \cup \coprod_{i=1}^{m} \Gamma_{M_{\mathbf{P}_i}}\backslash X_{\mathbf{P}_i}.$$

It is in this sense that the boundaries of $\Gamma\backslash X \cup \Gamma\backslash X(\infty)$ and $\overline{\Gamma\backslash X}^{RBS}$ are the parameter spaces for the generalized eigenspaces of $\Gamma\backslash X$.

Remark III.23.9 To emphasize the relations between geodesics and the continuous spectrum, we point out that the boundary of the reductive Borel–Serre compactification $\overline{\Gamma\backslash X}^{RBS}$ can also be identified with certain classes of EDM geodesics (see [JM, §14]).

III.23.10 In the rest of this section, we study relations between the spectral measure of the continuous spectrum and the geodesics going to infinity.

First we recall relations between the *length spectrum* of closed geodesics and the eigenvalues of a compact Riemannian manifold.

The classical *Poisson relation*

$$\sum_{n \in \mathbb{Z}} \hat{f}(2\pi n) = \sum_{n \in \mathbb{Z}} f(n), \tag{III.23.2}$$

where $f \in C_0^\infty(\mathbb{R})$, and $\hat{f}$ its Fourier transform, $\hat{f}(t) = \int_{\mathbb{R}} f(x) e^{-itx} dx$, can be regarded as an identity between the length spectrum of closed geodesics and the eigenvalues of $\mathbb{R}/\mathbb{Z}$. It asserts that the singularities of the Fourier transform of the spectral measure are given by the delta functions at the lengths of the closed geodesics in $\mathbb{R}/\mathbb{Z}$ (see [Ji8] for more details).

This relation has been generalized to compact Riemannian manifolds in [DG]. It states that the singularities of the Fourier transform of the spectral measure are supported at the lengths of closed geodesics. If the metric is generic, the singularities can be described explicitly.

For a noncompact locally symmetric space $\Gamma\backslash X$, the spectral measure of the continuous spectrum is essentially determined by the scattering matrices, and the relevant geodesics are not closed geodesics, but a certain class of geodesics going to infinity defined below.

Definition III.23.11 A geodesic $\gamma(t)$ in $\Gamma\backslash X$ is called a scattering geodesic if it is EDM in both directions, i.e., there exist t_+, t_- such that for any $t_1, t_2 \geq t_+$ or $t_1, t_2 \leq t_-$,

$$d(\gamma(t_1), \gamma(t_2)) = |t_1 - t_2|.$$

Then it follows from the definition of $\Gamma\backslash X \cup \Gamma\backslash X(\infty)$ that both limits $\lim_{t \to +\infty} \gamma(t)$ and $\lim_{t \to -\infty} \gamma(t)$ exist in $\Gamma\backslash X \cup \Gamma\backslash X(\infty)$.

III.23.12 When the $\mathbb{Q}$-rank r of $\mathbf{G}$ is greater than 1, $\Gamma\backslash X$ has continuous spectra of all dimensions up to r. It is reasonable that the scattering geodesics are related to the one-dimensional spectrum. For this reason, in the following, we assume that the $\mathbb{Q}$-rank of $\mathbf{G}$ is equal to 1. Then topologically, the ends of $\Gamma\backslash X$ are topological cylinders and correspond to Γ-conjugacy classes of rational parabolic subgroups of $\mathbf{G}$.

For a closed geodesic, a natural invariant is its length. On the other hand, the length of a scattering geodesic is infinite and we need a finite renormalization. It turns out that there is a canonical renormalization, called the sojourn time.

Let $\mathbf{P}_1, \dots, \mathbf{P}_k$ be a set of representatives of Γ-conjugacy classes of rational parabolic subgroups of $\mathbf{G}$. Since the $\mathbb{Q}$-rank of $\mathbf{G}$ is equal to 1, by the precise reduction theory in Proposition III.2.21, there exist bounded sets $U_i \subset N_{P_i}$, $V_i \subset X_{\mathbf{P}_i}$ such that for $T \gg 1$, $U_i \times A_{\mathbf{P}_i,T} \times V_i$ is mapped injectively into $\Gamma\backslash X$ and its image $\pi(U_i \times A_{\mathbf{P}_i,T})$ covers one end, where $\pi : X \to \Gamma\backslash X$ is the projection map. Hence there exists a smooth submanifold with boundary ω_T in $\Gamma\backslash X$ such that $\Gamma\backslash X$ admits the disjoint decomposition

$$\Gamma\backslash X = \omega_T \cup \coprod_{i=1}^{k} \pi(U_i \times A_{\mathbf{P}_i,T} \times V_i).$$

When $\Gamma\backslash X$ is a Riemann surface, this is the decomposition into cusp neighborhoods and a compact core. Since the $\mathbb{Q}$-rank of $\mathbf{G}$ is equal to 1, the picture is similar. Clearly, ω_T depends on the (height) parameter T.

Definition III.23.13

1. *For any pair of rational parabolic subgroups $\mathbf{P}_i, \mathbf{P}_j$, a scattering geodesic γ is called a scattering geodesic from the end of $\mathbf{P}_i$ to the end of $\mathbf{P}_j$ if for $t \ll 0$, $\gamma(t) \in \pi(U_i \times A_{\mathbf{P}_i,T} \times V_i)$, and for $t \gg 0$, $\gamma(t) \in \pi(U_j \times A_{\mathbf{P}_j,T} \times V_j)$.*
2. *For any scattering geodesic γ between the ends of $\mathbf{P}_i$ and $\mathbf{P}_j$, choose $T \gg 1$ such that the intersection $\gamma \cap \pi(U_i \times A_{\mathbf{P}_i,T} \times V_i)$ is of the form*

$$\gamma((-\infty, t_-) = \pi(\{(n_i, a_i \exp tH_i, z_i) \in N_{P_i} \times A_{\mathbf{P}_i} \times X_{\mathbf{P}_i} \mid t < t_-\},$$

and $\gamma \cap \pi(U_j \times A_{\mathbf{P}_j,T} \times V_j)$ is of a similar form $\gamma([t_+, +\infty))$. Then the sojourn time of γ is defined to be $t_+ - t_- - 2\log T$.

Intuitively, the height parameter T of the Siegel sets is chosen high enough so that the scattering geodesic shoots out directly to infinity once it enters the cusp regions. By the classification of EDM geodesics in Proposition III.20.8, this is possible. Clearly, when T is bigger, $t_+ - t_-$ is bigger, and the subtraction $-2\log T$ adjusts the changes, and the sojourn time is well-defined (see [JZ1, §2]). In fact, the sojourn time can be interpreted in terms of the Bruhat decomposition in [JZ2]. Basically, it is the time that γ spends around the compact core of $\Gamma\backslash X$.

Remark III.23.14 It should be pointed that in defining the sojourn time, the Riemannian metric on X is defined by a suitable multiple of the Killing form such that if $\alpha \in \Phi(P_i, A_{\mathbf{P}_i})$ is the short root, then its norm is equal to 1. Since all $\mathbf{P}_i$ are conjugate under $\mathbf{G}(\mathbb{Q})$, this is possible.

Proposition III.23.15 *For any pair of rational parabolic subgroups $\mathbf{P}_i, \mathbf{P}_j$, the set of sojourn times between the ends of $\mathbf{P}_i, \mathbf{P}_j$ is discrete. For each sojourn time, the set of scattering geodesics with the given sojourn time form a smooth family parametrized by a common finite cover of the two boundary locally symmetric spaces $\Gamma_{M_{\mathbf{P}_i}}\backslash X_{\mathbf{P}_i}$ and $\Gamma_{M_{\mathbf{P}_j}}\backslash X_{\mathbf{P}_j}$.*

The discreteness of the sojourn time is proved in [JZ2], and the parametrization of the scattering geodesics is proved in [JZ1, Theorem 1, p. 442]

For an Eisenstein series $E(P_i, \phi, \Lambda)$, its constant term along P_j is defined to be

$$E_{P_j}(P_i, \phi, \Lambda)(x) = \int_{\Gamma_{N_{P_j}} \backslash N_{P_j}} E(P_i, \phi, \Lambda)(nx)dn,$$

with the total measure of $\Gamma_{N_{P_j}} \backslash N_{P_j}$ normalized to be 1. For simplicity, identify $\mathfrak{a}_{\mathbf{P}_j}$ with $\mathbb{R}$ so that the positive chamber corresponds to the positive half-axis. Then in the horospherical coordinate decomposition $x = (n, e^H, z) \in N_{P_j} \times A_{\mathbf{P}_j} \times X_{\mathbf{P}_j}$,

$$E_{P_j}(P_i, \phi, \Lambda)(e^H z) = \delta_{ij} e^{(\rho+\Lambda)(H)} \phi(z) + e^{(\rho-\Lambda)(H)} c(\Lambda)(\phi)(z),$$

where $\rho = \rho_{\mathbf{P}_j}$, and the operator $c(\lambda)$ is called the scattering operator. If ϕ ranges over a suitable orthonormal basis of $L^2_{dis}(\Gamma_{M_{\mathbf{P}_i}} \backslash X_{P_i})$, it becomes a matrix, called the *scattering matrix.* It plays a crucial role in the meromorphic continuation of the Eisenstein series and the spectral decomposition of $L^2(\Gamma \backslash X)$ (see [La] [Ar2] [OW1] [MoW] for details). The density of the continuous spectrum can also be expressed in terms of the scattering matrix.

Proposition III.23.16 *The singular supports of the Fourier transform of the scattering matrix are supported on the sojourn times of the scattering geodesics.*

The structures of the singularities are also be described, and the parameter space of the scattering geodesics in Proposition III.23.15 naturally enters into the formula. This is a generalization of the Poisson relation to noncompact Riemannian manifolds. See [JZ1, Theorem 2] for details.

III.23.17 Summary and comments. The spectral theory of locally symmetric spaces has been studied intensively and extensively from many different points of view. See [Sel1] [Sel2] [La] [Bo10] [MoW]. The results in this section indicate some geometric aspects of the spectral theory: parametrization of generalized eigenspaces and boundaries of compactifications of $\Gamma \backslash X$, and the Poisson relation between the scattering geodesics and the scattering matrices of $\Gamma \backslash X$.

References

[Abe] H. Abels, *Reductive groups as metric spaces*, in *Groups: topological, combinatorial and arithmetic aspects*, pp. 1–20, in *London Math. Soc. Lecture Note*, Ser., 311, Cambridge Univ. Press, 2004.

[AbM] H. Abels, G. Margulis, *Coarsely geodesic metrics on reductive groups*, in *Modern dynamical systems and applications*, pp. 163–183, Cambridge Univ. Press, 2004.

[Ale] V. Alexeev, *Complete moduli in the presence of semiabelian group action*, Ann. of Math. 155 (2002) 611–708.

[AlB1] V. Alexeev, M. Brion, *Stable reductive varieties. I. Affine varieties*, Invent. Math. 157 (2004) 227–274.

[AlB2] V. Alexeev, M. Brion, *Stable reductive varieties. II. Projective case*, Adv. Math. 184 (2004) 380–408.

[Al1] N. D. Allan, *The problem of the maximality of arithmetic groups*, in *Algebraic groups and discontinuous subgroups*, Proc. Symp. Pure Math IX, (A.Borel and G.D. Mostow eds.) AMS 1966.

[Al2] N. D. Allan, *Maximality of some arithmetic groups*, in *Revista Colombiana de Matematicas*, Monografias Matematicas, vol. 7 Sociedad Colombiana de Matematicas, Bogota 1970 iv+57 pp.

[Anc] A. Ancona, *Negatively curved manifolds, elliptic operators, and the Martin boundary*, Ann. of Math. 125 (1987) 495–536.

[And] M. Anderson, *The Dirichlet problem at infinity for manifolds of negative curvature*, J. Differential Geom. 18 (1983) 701–721.

[AS] M.T. Anderson, R. Schoen, *Positive harmonic functions on complete manifolds of negative curvature*, Ann. of Math. 121 (1985) 429–461.

[AJ] J.P. Anker, L. Ji, *Heat kernel and Green function estimates on noncompact symmetric spaces*, Geom. Funct. Analy. 9 (1999) 1035–1091.

[Ar1] J. Arthur, *The trace formula and Hecke operators*, in *Number theory, trace formulas and discrete groups (Oslo, 1987)*, Academic Press, Boston, MA, 1989, pp. 11–27.

[Ar2] J. Arthur, *Eisenstein series and the trace formula*, in Proc. of Symp. in Pure Math., vol. 33 (1979) part 1, pp. 253–274.

[Ar3] J. Arthur, *A Trace formula for reductive groups I*, Duke Math. J. 45 (1978) 911–952.

[As1] A. Ash, *Small-dimensional classifying spaces for arithmetic subgroups of general linear groups*, Duke Math. J. 51 (1984) 459–468.

[As2] A. Ash, *Cohomology of congruence subgroups* $\mathrm{SL}(n, Z)$, Math. Ann. 249 (1980) 55–73.

[As3] A. Ash, *On eutactic forms*, Canad. J. Math. 29 (1977) 1040–1054.

[As4] A. Ash, *Deformation retracts with lowest possible dimension of arithmetic quotients of self-adjoint homogeneous cones*, Math. Ann. 225 (1977) 69–76.

[AsB] A. Ash, B.Borel, *Generalized modular symbols, Cohomology of arithmetic groups and automorphic forms*, pp. 57–75, Lecture Notes in Math., vol. 1447, Springer, 1990.

[AM] A. Ash, M. McConnell, *Cohomology at infinity and the well-rounded retract for general linear groups*, Duke Math. J. 90 (1997) 549–576.

[AMRT] A. Ash, D. Mumford, M. Rapoport, Y. Tai, *Smooth compactifications of locally symmetric varieties*, Math. Sci. Press, Brookline, 1975.

[AR] A. Ash, L. Rudolph, *The modular symbol and continued fractions in higher dimensions*, Invent. Math. 55 (1979) 241–250.

[At] M. Atiyah, *Convexity and commuting Hamiltonians*, Bull. Lond. Math. Soc. 14 (1982) 1–15.

[ABS] H. Azad, M. Barry, G. Seitz, *On the structure of parabolic subgroups*, Comm. Algebra 18 (1990) 551–562.

[Ba1] W. Baily, *Satake's compactification of V_n*, Amer. J. Math. 80 (1958) 348–364.

[Ba2] W. Baily, *On the Hilbert-Siegel modular space*, Amer. J. Math. 846–874.

[Ba3] W. Baily, *Introductory lectures on automorphic forms*, Princeton University Press, 1973.

[Ba4] W. Baily, *Special arithmetic groups and Eisenstein series*, in *Complex analysis and algebraic geometry*, pp. 307–317, Iwanami Shoten, Tokyo, 1977.

[Ba5] W. Baily, *On compactifications of orbit spaces of arithmetic discontinuous groups acting on bounded symmetric domains*, in *Algebraic Groups and Discontinuous Subgroups*, Proc. Sympos. Pure Math., pp. 281–295, Amer. Math. Soc., 1966.

[Ba6] W. Baily, *Fourier-Jacobi series*, in *Algebraic Groups and Discontinuous Subgroups*, Proc. Sympos. Pure Math., pp. 296–300, Amer. Math. Soc., 1966.

[Ba7] W. Baily, *On the orbit spaces of arithmetic groups*, in *Arithmetical Algebraic Geometry*, pp. 4–10, Harper & Row, 1965.

[BB1] W. Baily, A. Borel, *Compactification of arithmetic quotients of bounded symmetric domains*, Ann. of Math. 84 (1966) 442–528.

[BB2] W. Baily, A. Borel, *On the compactification of arithmetically defined quotients of bounded symmetric domains*, Bull. Amer. Math. Soc. 70 (1964) 588–593.

[Bal1] W. Ballmann, *Lectures on spaces of nonpositive curvature*, Birkhäuser Verlag, 1995.

[Bal2] W. Ballmann, *Manifolds of nonpositive sectional curvature and manifolds without conjugate points*, in *Proceedings of the International Congress of Mathematicians*, (Berkeley, Calif., 1986) pp. 484–490, Amer. Math. Soc., 1987.

[Bal3] W. Ballmann, *Nonpositively curved manifolds of higher rank*, Ann. of Math. 122 (1985) 597–609.

[BBE] W. Ballmann, M. Brin, P. Eberlein, *Structure of manifolds of nonpositive curvature. I*, Ann. of Math. 122 (1985) 171–203.

[BBS] W. Ballmann, M. Brin, R. Spatzier, *Structure of manifolds of nonpositive curvature. II*, Ann. of Math. 122 (1985) 205–235.

[BGS] W. Ballmann, M. Gromov, V. Schroeder, *Manifolds of nonpositive curvature*, Prog. in Math., vol. 61, Birkhäuser, Boston, 1985.

[BFS1] E. van den Ban, M. Flensted-Jensen, H. Schlichtkrull, *Harmonic analysis on semisimple symmetric spaces: a survey of some general results*, in *Representation theory and automorphic forms*, pp. 191–217, Proc. Sympos. Pure Math., vol. 61, Amer. Math. Soc., 1997.

[BFS2] E. van den Ban, M. Flensted-Jensen, H. Schlichtkrull, *Basic harmonic analysis on pseudo-Riemannian symmetric spaces*, in *Noncompact Lie groups and some of their applications*, pp. 69–101, Kluwer Acad. Publ., 1994.

[BN] P. Bayer, J. Neukirch, *On automorphic forms and Hodge theory*, Math. Ann. 257 (1981) 137–155.

[Ben1] Y. Benoist, *Convexes divisibles. II*, Duke Math. J. 120 (2003) 97–120.

[Ben2] Y. Benoist, *Propriétés asymptotiques des groupes linéaires*, Geom. Funct. Anal. 7 (1997) 1–47.

[Bes] M. Bestvina, *Local homology properties of boundaries of groups*, Michigan Math. J. 43 (1996) 123–139.

[BG] M. Bestvina, G. Mess, *The boundary of negatively curved groups*, J. Amer. Math. Soc. 4 (1991) 469–481.

[BCG] G. Besson, G. Courtois, S. Gallot, *Entropies et rigidités des espaces localement symétriques de courbure strictement négative*, Geom. Funct. Anal. 5 (1995) 731–799.

[Bet] F. Betten, *Causal compactification of compactly causal spaces*, Trans. Amer. Math. Soc. 355 (2003) 4699–4721.

[BO] F. Betten, G. Ólafsson, *Causal compactification and Hardy spaces for spaces of Hermitian type*, Pacific J. Math. 200 (2001) 273–312.

[BDP] E. Bifet, C. De Concini, C. Procesi, *Cohomology of regular embeddings*, Adv. Math. 82 (1990) 1–34.

[BFG] D. Blasius, J. Franke, F. Grunewald, *Cohomology of S-arithmetic subgroups in the number field case*, Invent. Math. 116 (1994) 75–93.

[Bon1] A. Bondarenko, *Classification of maximal arithmetic subgroups of indefinite groups of type* (C_l), Dokl. Akad. Nauk Belarusi 38 (1994) 29–32.

[Bon2] A. Bondarenko, *On the maximality of arithmetic subgroups of indefinite orthogonal groups of type* (D_l), Math. USSR-Sb. 69 (1991) 417–430.

[Bon3] A. Bondarenko, *Classification of maximal arithmetic subgroups of indefinite orthogonal groups of type* (B_l), Mat. Sb. (N.S.) 127 (1985) 72–91.

[Bon4] A. Bondarenko, *On the classification of maximal arithmetic subgroups in split groups*, Mat. Sb. (N.S.) 102 (1977) 155–172.

[Bo1] A. Borel, *Density properties for certain subgroups of semisimple groups without compact factors*, Ann. of Math. 72 (1960) 179–188.

[Bo2] A. Borel, *Density and maximality of arithmetic subgroups*, J. Reine Angew. Math. 224 (1966) 78–89.

[Bo3] A. Borel, *Reduction theory for arithmetic groups*, Proc. of Symp. in Pure Math., Vol. 9 (1969) pp. 20–25.

[Bo4] A. Borel, *Introduction aux groupes arithmétiques*, Hermann, Paris, 1969.

[Bo5] A. Borel, *Sous-groupes discrets de groupes semi-simples*, Séminaire Bourbaki, Exp. 358 (1968/69) Lect. Notes Math. 179 (1971) 199–216.

[Bo6] A. Borel, *Some metric properties of arithmetic quotients of symmetric spaces and an extension Theorem*, J. of Diff. Geom. 6 (1972) 543–560.

[Bo7] A. Borel, *Stable real cohomology of arithmetic groups*, Ann. Sci. Ec. Norm. Super., 7 (1974) 235–272.

[Bo8] A. Borel, *Notes of a seminar on compactifications of symmetric spaces at IAS, Princeton*, 1987.

[Bo9] A. Borel, *Linear algebraic groups*, 2nd enlarged edition, GTM 126, Springer 1991.

[Bo10] A. Borel, *Automorphic forms on* $SL_2(R)$, Cambridge Tracts in Mathematics 130, Cambridge University Press, 1997.

[Bo11] A. Borel, *Semisimple groups and Riemannian symmetric spaces*, Texts & Readings in Math. vol. 16, Hindustan Book Agency, New Delhi, 1998.

[Bo12] A. Borel, *Essays in the history of Lie groups and algebraic groups*, History of Mathematics 21, A.M.S. 2001.

[Bo13] A. Borel, *Ensembles fondamentaux pour les groupes arithmétiques*, Colloque sur la théorie des groupes algébriques, Bruxelles, 1962, pp. 23–40.

[Bo14] A. Borel, *Linear algebraic groups*, in *Algebraic Groups and Discontinuous Subgroups*, Proc. Sympos. Pure Math., vol. 9 (1965) pp. 3–19.

[BHC] A. Borel, Harish-Chandra, *Arithmetic subgroups of algebraic groups*, Ann. of Math. 75 (1962) 485–535.

[BJ1] A. Borel, L. Ji, *Compactifications of symmetric and locally symmetric spaces*, Math. Research Letters 9 (2002) 725–739.

[BJ2] A. Borel, L. Ji, *Compactifications of symmetric spaces*, to appear in J. Diff. Geom.

[BJ3] A. Borel, L. Ji, *Compactifications of locally symmetric spaces*, to appear in J. Diff. Geom.

[BJ4] A. Borel, L. Ji, *Lectures on compactifications of symmetric and locally symmetric spaces*, in *Lie Theory: unitary representations and compactifications of symmetric spaces*, ed. by J.P. Anker and B. Orsted, Birkhäuser, 2005, pp. 69–137.

[BS1] A. Borel, J-P. Serre, *Théorèmes de finitude en cohomologie galoisienne*, Comm. Math. Helv. 39, 1964, 111–164; A. Borel C.P. II 362–415.

[BS2] A. Borel, J.P. Serre, *Corners and arithmetic groups*, Comment. Math. Helv. 48 (1973) 436–491. A. Borel C.P. III, 244–299.

[BS3] A. Borel, J.P. Serre, *Cohomologie d'immeubles et groups S-arithmétiques*, Topology 15 (1976) 211–232.

[BT] A. Borel, J. Tits, *Groupes réductifs*, Publ. Sci. IHES, 27 (1965) 55–150.

[BW] A. Borel, N. Wallach, *Continuous cohomology, discrete subgroups, and representations of reductive groups*, Second edition, *Mathematical Surveys and Monographs*, vol.67, Amer. Math. Soc., 2000, xviii+260 pp.

[Bu1] N. Bourbaki, *General topology*, Springer-Verlag, 1989.

[Bu2] N. Bourbaki, *Groupes et algèbres de Lie*, Ch. 4–6, Hermann, Paris, 1868.

[Bu3] N. Bourbaki, *Éléments de Mathématique, Livre VI, Intégration*, Hermann, Paris, 1963.

[Bre] B. Breckner, *A note on the triviality of the Bohr-compactification of Lie groups*, Studia Univ. Babeş-Bolyai Math. 44 (1999) 11–24.

[Bri1] M. Brion, *Group completions via Hilbert schemes*, J. Algebraic Geom. 12 (2003) 605–626.

[Bri2] M. Brion, *The behaviour at infinity of the Bruhat decomposition*, Comment. Math. Helv. 73 (1998) 137–174.

[Bri3] M. Brion, *Spherical varieties*, in *Proc. of the International Congress of Mathematicians*, Vol. 1, 2, pp. 753–760, Birkhäuser, 1995.

[BLV] M. Brion, D. Luna, T. Vust, *Espaces homogènes sphériques*, Invent. Math. 84 (1986) 617–632.

[Br1] K. Brown, *Buildings*, Springer-Verlag, 1989.

[Br2] K. Brown, *Cohomology of groups*, Graduate Texts in Mathematics, 87. Springer-Verlag, New York, 1994.

[BMP] R. Bruggeman, R. Miatello, I. Pacharoni, *Density results for automorphic forms on Hilbert modular groups*, Geom. Funct. Anal. 13 (2003) 681–719.

[BMW] R. Bruggeman, R. Miatello, N. Wallach, *Resolvent and lattice points on symmetric spaces of strictly negative curvature*, Math. Ann. 315 (1999) 617–639.

[BuM1] U. Bunke, M. Olbrich, *The spectrum of Kleinian manifolds*, J. Funct. Anal. 172 (2000) 76–164.

[BuM2] U. Bunke, M. Olbrich, *Group cohomology and the singularities of the Selberg zeta function associated to a Kleinian group*, Ann. of Math. 149 (1999) 627–689.

[BuM3] U. Bunke, M. Olbrich, *Gamma-cohomology and the Selberg zeta function*, J. Reine Angew. Math. 467 (1995) 199–219.

[BuM4] U. Bunke, M. Olbrich, *Selberg zeta and theta functions. A differential operator approach*, Akademie-Verlag, Berlin, 1995. 168 pp.

[BuM5] U. Bunke, M. Olbrich, *The wave kernel for the Laplacian on the classical locally symmetric spaces of rank one, theta functions, trace formulas and the Selberg zeta function*, Ann. Global Anal. Geom. 12 (1994) 357–405.

[BuMo] M. Burger, N. Monod, *Bounded cohomology of lattices in higher rank Lie groups*, J. Eur. Math. Soc. 1 (1999) 199–235.

[BuW] J. Burgos, J. Wildeshaus, *Hodge modules on Shimura varieties and their higher direct images in the Baily-Borel compactification*, Ann. Sci. E'cole Norm. Sup. 37 (2004) 363–413.

[BuS1] K. Burns, R. Spatzier, *On the topological Tits buildings and their classifications*, Publ. Sci. IHES 65 (1987) 5–34.

[BuS2] K. Burns, R. Spatzier, *Manifolds of nonpositive curvature and their buildings*, Inst. Hautes Itudes Sci. Publ. Math. 65 (1987) 35–59.

[By] A. Bytsenko, *Heat-kernel asymptotics of locally symmetric spaces of rank one and Chern-Simons invariants*, in *Quantum gravity and spectral geometry*, Nuclear Phys. B Proc. Suppl. 104 (2002) 127–134.

[Cas1] W. Casselman, *Canonical extensions of Harish-Chandra modules to representations of G*, Canad. J. Math. 41 (1989) 385–438.

[Cas2] W. Casselman, *Geometric rationality of Satake compactifications*, in *Algebraic groups and Lie groups*, pp. 81–103, Austral. Math. Soc. Lect. Ser., 9, Cambridge Univ. Press, Cambridge, 1997.

[Cass1] J. Cassels, *An introduction to the geometry of numbers*, Corrected reprint of the 1971 edition, in *Classics in Mathematics*, Springer-Verlag, 1997. viii+344 pp.

[Cass2] J. Cassels, *Rational quadratic forms*, in *London Mathematical Society Monographs*, 13. Academic Press, Inc., 1978. xvi+413 pp.

[Cat1] E. Cattani, *On the partial compactification of the arithmetic quotient of a period matrix domain*, Bull. Amer. Math. Soc. 80 (1974) 330–333.

[Cat2] E. Cattani, *Mixed Hodge structures, compactifications and monodromy weight filtration*, in *Topics in transcendental algebraic geometry*, pp. 75–100, Ann. of Math. Stud., vol. 106, Princeton Univ. Press, 1984.

[Cat3] E. Cattani, *Homogeneous spaces and Hodge theory*, *Conference on differential geometry on homogeneous spaces*, Rend. Sem. Mat. Univ. Politec. Torino 1983, Special Issue, 1–16 (1984).

[CK] E. Cattani, A. Kaplan, *Extension of period mappings for Hodge structures of weight two*, Duke Math. J. 44 (1977) 1–43.

[CKS1] E. Cattani, A. Kaplan, W. Schmid, *Some remarks on L^2 and intersection cohomologies*, in *Hodge theory*, 32–41, Lecture Notes in Math., vol. 1246, Springer, 1987.

[CKS2] E. Cattani, A. Kaplan, W. Schmid, *Variations of polarized Hodge structure: asymptotics and monodromy*, in *Hodge theory*, pp. 16–31, Lecture Notes in Math., vol. 1246, Springer, 1987.

[CKS3] E. Cattani, A. Kaplan, W. Schmid, *L^2 and intersection cohomologies for a polarizable variation of Hodge structure*, Invent. Math. 87 (1987) 217–252.

[Ch1] C.L. Chai, *Arithmetic minimal compactification of the Hilbert-Blumenthal moduli spaces*, Ann. of Math. 131 (1990) 541–554.

[Ch2] C.L. Chai, *Arithmetic compactification of the Siegel moduli space*, in *Theta functions—Bowdoin 1987*, Part 2, pp. 19–44, Proc. Sympos. Pure Math., vol. 49, Part 2, Amer. Math. Soc., 1989.

[Ch3] C.L. Chai, *Siegel moduli schemes and their compactifications over C*, in *Arithmetic geometry*, pp. 231–251, Springer, 1986.

[Ch4] C.L. Chai, *Compactification of Siegel moduli schemes*, London Mathematical Society Lecture Note Series, vol. 107, Cambridge University Press, 1985. xvi+326 pp.

[CL] R. Charney, R. Lee, *Cohomology of the Satake compactification*, Topology 22 (1983) no. 4 389–423.

[CY] S.Y. Cheng, S.T. Yau, *Differential equations on Riemannian manifolds and their geometric applications*, Comm. Pure Appl. Math. 28 (1975) 333–354.

[Chr1] U. Christian, *Über die Uniformisierbarkeit nicht-elliptischer Fixpunkte Siegelscher Modulgruppen*, J. Reine Angew. Math. 219 (1965) 97–112.

[Chr2] U. Christian, *Zur Theorie der Modulfunktionen n-ten Grades. II*, Math. Ann. 134 (1958) 298–307.

[Chr3] U. Christian, *Zur Theorie der Modulfunktionen n-ten Grades*, Math. Ann. 133 (1957) 281–297.

[CoG1] J. Conze, Y. Guivarc'h, *Densité d'orbites d'actions de groupes linéaires et propriétés d'equidistribution de marches allatoires*, in *Rigidity in dynamics and geometry*, pp. 39–76, Springer, 2002.

[CoG2] J. Conze, Y. Guivarc'h, *Limit sets of groups of linear transformations*, in *Ergodic theory and harmonic analysis (Mumbai, 1999)*, Sankhya Ser. A 62 (2000) 367–385.

[Con] J.H. Conway, *Understanding groups like* $\Gamma_0(N)$, in *Groups, difference sets, and the Monster*, pp. 327–343, Ohio State Univ. Math. Res. Inst. Publ., vol. 4, de Gruyter, 1996.

[CoS] J.H. Conway, N. Sloane, *Sphere packings, lattices and groups*, Third edition, Springer-Verlag, 1999. lxxiv+703 pp.

[DGMP] C. De Concini, M. Goresky, R. MacPherson, C. Procesi, *On the geometry of quadrics and their degenerations*, Comment. Math. Helv. 63 (1988) 337–413.

[DP1] C. De Concini, C. Procesi, *Complete symmetric varieties*, in *Invariant Theory*, Springer Lecture Notes, vol. 996, pp. 1–44, 1983.

[DP2] C. De Concini, C. Procesi, *Complete symmetric varieties. II. Intersection theory*, in *Algebraic groups and related topics*, pp. 481–513, Adv. Stud. Pure Math., vol. 6, North-Holland, 1985.

[DP3] C. De Concini, C. Procesi, *Cohomology of compactifications of algebraic groups*, Duke Math. J. 53 (1986) 585–594.

[DS] C. De Concini, T.A. Springer, *Compactification of symmetric varieties*, Transform. Groups 4 (1999) 273–300.

[Deg] D. de George, *Length spectrum for compact locally symmetric spaces of strictly negative curvature*, Ann. Sci. Icole Norm. Sup. (4) 10 (1977) 133–152.

[DW] D. de George, N. Wallach, *Limit formulas for multiplicities in* $L^2(\Gamma\backslash G)$, Ann. Math. 107 (1978) 133–150.

[Dei1] A. Deitmar, *Geometric zeta functions of locally symmetric spaces*, Amer. J. Math. 122 (2000) 887–926.

[Dei2] A. Deitmar, *Equivariant torsion of locally symmetric spaces*, Pacific J. Math. 182 (1998) 205–227.

[DH1] A. Deitmar, W. Hoffmann, *On limit multiplicities for spaces of automorphic forms*, Canad. J. Math. 51 (1999) 952–976.

[DH2] A. Deitmar, W. Hoffmann, *Spectral estimates for towers of noncompact quotients*, Canad. J. Math. 51 (1999) 266–293.

[De] M. Demazure, *Limites de groupes orthogonaux ou symplectiques*, preprint, 1980.

[Dim] M. Dimitrov, *Compactifications arithmétiques des variétés de Hilbert et formes modulaires de Hilbert pour $\Gamma_1(\mathfrak{c}, \mathfrak{n})$*, Geometric aspects of Dwork theory. Vol. I, II, pp. 527–554, Walter de Gruyter, 2004.

[Din] J. Ding, *A proof of a conjecture of C. L. Siegel*, J. Number Theory 46 (1994) 1–11.

[Do1] Do Ngoc Diep, *On the Langlands type discrete groups. III. The continuous cohomology*, Acta Math. Vietnam. 22 (1997) 379–394.

[Do2] Do Ngoc Diep, *On the Langlands type discrete groups. II. The theory of Eisenstein series*, Acta Math. Vietnam. 16 (1991) 77–90.

[Do3] Do Ngoc Diep, *On the Langlands type discrete groups. I. The Borel-Serre compactification*, Acta Math. Vietnam. 12 (1987) 41–54.

[Dr] V. Drinfeld, *Elliptic modules*, Mat. Sb. (N.S.) 94 (1974) 594–627.

[DG] J. Duistermaat, V. Guillemin, *The spectrum of positive elliptic operators and periodic bicharacteristics*, Invent. Math. 29 (1975) 39–79.

[Dy1] E.B. Dynkin, *Markov process and problems in analysis*, Proc. of ICM at Stockholm, 1962, pp. 36–58 (Amer. Math. Soc. Transl Ser. 2, vol. 31, pp. 1–24.)

[Dy2] E.B. Dynkin, *Nonnegative eigenfunctions of the Laplace-Beltrami operator and the Brownian motion in some symmetric spaces*, Izv. Akad. Nauk. SSSR 30 (1966) 455–478.

[Eb] P. Eberlein, *Geometry of nonpositively curved manifolds*, Chicago Lectures in Mathematics, University of Chicago Press, 1996.

[EO] P. Eberlein, B. O'Neill, *Visibility manifolds*, Pacific J. Math. 46 (1973) 45–109.

[EGM] J. Elstrodt, F. Grunewald, J. Mennicke, *Groups acting on hyperbolic spaces*, Springer-Verlag, 1998.

[EL] S. Evens, J. Lu, *On the variety of Lagrangian subalgebras. I*, Ann. Sci. Ecole Norm. Sup. 34 (2001) 631–668.

[Fa] G. Faltings, *Arithmetische Kompaktifizierung des Modulraums der abelschen Varietäten*, in *Workshop Bonn 1984*, 321–383, Lecture Notes in Math., vol. 1111, Springer, 1985.

[FC] G. Faltings, C.L. Chai, *Degeneration of abelian varieties*, Springer-Verlag, Berlin, 1990.

[Fl] M. Flensted-Jensen, *Analysis on non-Riemannian symmetric spaces, CBMS Regional Conference Series in Mathematics*, vol. 61, 1986. x+77 pp.

[Fr] E. Freitag, *Hilbert modular forms*, Springer-Verlag, 1990.

[FrH] E. Freitag, C. Hermann, *Some modular varieties of low dimension*, Adv. Math. 152 (2000) 203–287.

[FK] D. Fuks, V. Rokhlin, *Beginner's course in topology*, Springer-Verlag, 1984.

[FH] W. Fulton, J. Harris, *Representation theory. A first course*, Grad. Texts in Math., vol. 129, Springer-Verlag, New York, 1991.

[Ful] W. Fulton, *Introduction to toric varieties*, Annals of Mathematics Studies, vol. 131, Princeton University Press, 1993.

[FM] W. Fulton, R. MacPherson, *A compactification of configuration spaces*, Ann. of Math. (2) 139 (1994) 183–225.

[Fu1] H. Furstenberg, *A Poisson formula for semi-simple Lie groups*, Ann. of Math. 72 (1963) 335–386.

[Fu2] H. Furstenberg, *Boundaries of Riemannian symmetric spaces*, in *Symmetric spaces* ed. by W. Boothby and G. Weiss, Dekker, 1972, pp. 359–377.

[Fu3] H. Furstenberg, *Boundary theory and stochastic processes on homogeneous spaces*, Proc. of Symp. in Pure Math. of AMS, vol. XXVI, 1973, pp. 193–229.

[Fu4] H. Furstenberg, *Boundaries of Lie groups and discrete subgroups*, in *Actes du Congrès International des Mathématiciens (Nice, 1970)*, Tome 2, pp. 301–306, Gauthier-Villars, 1971.

[Fu5] H. Furstenberg, *Random walks and discrete subgroups of Lie groups*, in *Advances in Probability and Related Topics*, Vol. 1 pp. 1–63, Dekker, 1971.

[Fu6] H. Furstenberg, *Poisson boundaries and envelopes of discrete groups*, Bull. Amer. Math. Soc. 73 (1967) 350–356.

[Fu7] H. Furstenberg, *Translation-invariant cones of functions on semi-simple Lie groups*, Bull. Amer. Math. Soc. 71 (1965) 271–326.

[GaR] H. Garland, M.S. Raghunathan, *Fundamental domains for lattices in* (R)*-Rank 1 semisimple Lie groups*, Ann. of Math. 92 (1970) 279–326.

[Ga] P. Garrett, *Holomorphic Hilbert modular forms*, Wadsworth & Brooks/Cole, 1990.

[Gek1] E. Gekeler, *Satake compactification of Drinfeld modular schemes*, *Proceedings of the conference on p-adic analysis*, pp. 71–81, 1986.

[Gek2] E. Gekeler, *Compactification du schéma des modules de Drinfeld*, in *Séminaire de théorie des nombres*, 1985–1986, Exp. No. 16, 7 pp., Univ. Bordeaux I, Talence, 1986.

[Ger1] P. Gerardin, *Harmonic functions on buildings of reductive split groups*, in *Operator algebras and group representations*, Vol. I (Neptun, 1980) Monogr. Stud. Math., 17, Pitman, 1984, pp.208–221.

[Ger2] P. Gerardin, *Géométrie des compactifications des espaces hermitiens localement symétriques*, in *Several complex variables (Hangzhou, 1981)*, pp. 57–66, Birkhäuser Boston, 1984.

[GW] S. Giulini, W. Woess, *The Martin compactification of the Cartesian product of two hyperbolic spaces*, J. Reine Angew. Math. 444 (1993) 17–28.

[Gol] B. Goldbarb, *Novikov conjectures for arithmetic groups with large actions at infinity*, K-Theory 11 (1997) 319–372.

[GHM] M. Goresky, G. Harder, R. MacPherson, *Weighted cohomology*, Invent. Math. 116 (1994) 139–213.

[GHMN] M. Goresky, G. Harder, R. MacPherson, A. Nair, *Local intersection cohomology of Baily-Borel compactifications*, Compositio Math. 134 (2002) 243–268.

[Go] M. Goresky, *Compactifications and cohomology of modular varieties*, Lecture notes for Clay Mathematics Institute summer school on *harmonic analysis, trace formula, and Shimura varieties*, 2003.

[GM1] M. Goresky, R. MacPherson, *Lefschetz numbers of Hecke correspondences*, in *The zeta functions of Picard modular surfaces*, ed. by R. Langlands and D. Ramakrishnan, Univ. Montreal, Montreal, QC, 1992, pp. 465–478.

[GM2] M. Goresky, R. MacPherson, *The topological trace formula*, Crelle's J. 560 (2003) 77–150.

[GT1] M. Goresky, Y. Tai, *Toroidal and reductive Borel-Serre compactifications of locally symmetric spaces*, Amer. J. Math. 121 (1999) 1095–1151.

[GT2] M. Goresky, Y. Tai, *The moduli space of real abelian varieties with level structure*, Compositio Math. 139 (2003) 1–27.

[GT3] M. Goresky, Y. Tai, *Anti-holomorphic multiplication and a real algebraic modular variety*, J. Differential Geom. 65 (2003) 513–560.

[Gra1] D. Grayson, *Reduction theory using semistability. II* Comment. Math. Helv. 61 (1986) 661–676.

[Gra2] D. Grayson, *Reduction theory using semistability*, Comment. Math. Helv. 59 (1984) 600–634.

[Gri] P. Griffiths, *Periods of integrals on algebraic manifolds: Summary of main results and discussion of open problems*, Bull. Amer. Math. Soc. 76 (1970) 228–296.

[GS] P. Griffiths, W. Schmid, *Recent developments in Hodge theory: a discussion of techniques and results*, in *Discrete subgroups of Lie groups and applications to moduli*, Oxford Univ. Press, 1975, pp. 31–127.

[Gro1] M. Gromov, *Asymptotic Invariants of Infinite Groups*, in *Geometric Group Theory*, Vol. 2 (Sussex, 1991) Cambridge Univ. Press, 1993, pp. 1–295.

[Gro2] M. Gromov, *Structures Métriques pour les Variétés Riemanniennes*, CEDIC, Paris, 1981.

[Grs1] J. Grosche, *Über die Erzeugbarkeit paramodularer Gruppen durch Elemente mit Fixpunkten*, J. Reine Angew. Math. 293/294 (1977) 86–98.

[Grs2] J. Grosche, *Über die Fundamentalgruppen von Quotientenräumen Siegelscher und Hilbert-Siegelscher Modulgruppen*, Nachr. Akad. Wiss. Göttingen Math.-Phys. Kl. II 9 (1976) 119–142.

[Grs3] J. Grosche, *Uber die Fundamentalgruppen von Quotientenräumen Siegelscher Modulgruppen*, J. Reine Angew. Math. 281 (1976) 53–79.

[Gru] P. Gruber, *Geometry of numbers*, in *Handbook of convex geometry*, Vol. A, B, pp. 739–763, North-Holland, 1993.

[GruL] P. Gruber, C. Lekkerkerker, *Geometry of numbers*, Second edition, North-Holland Publishing Co., 1987. xvi+732 pp.

[Gu] V. Guillemin, *Sojourn times and asymptotic properties of the scattering matrix*, RIMS Kyoto Univ. 12 (1977) 69–88.

[GJT] Y. Guivarc'h, L. Ji, J.C. Taylor, *Compactifications of symmetric spaces*, in *Progress in Math.*, vol. 156, Birkhäuser, Boston, 1998.

[GT] Y. Guivarc'h, J.C. Taylor, *The Martin compactification of the polydisc at the bottom of the positive spectrum*, Colloq. Math. 60/61 (1990) 537–546.

[Ha] W. Hammond, *Chern numbers of 2-dimensional Satake compactifications*, J. London Math. Soc. (2) 14 (1976) 65–70.

[Har1] G. Harder, *Eisenstein cohomology of arithmetic groups and its applications to number theory*, in *Proceedings of the International Congress of Mathematicians, Vol. I, II*, pp. 779–790, Math. Soc. Japan, Tokyo, 1991.

[Har2] G. Harder, *Some results on the Eisenstein cohomology of arithmetic subgroups of* GL_n, in *Cohomology of arithmetic groups and automorphic forms*, pp. 85–153, Lecture Notes in Math., vol. 1447, Springer, Berlin, 1990.

[Har3] G. Harder, *Eisenstein cohomology of arithmetic groups. The case* GL_2, Invent. Math. 89 (1987) 37–118.

[Har4] G. Harder, *On the cohomology of* $SL(2, O)$, in *Lie groups and their representations* (Proc. Summer School on Group Representations of the Bolyai Janos Math. Soc., Budapest, 1971) pp. 139–150, 1975.

[Har5] G. Harder, *On the cohomology of discrete arithmetically defined groups*, in *Discrete subgroups of Lie groups and applications to moduli*, pp. 129–160, Oxford Univ. Press, 1975.

[Har6] G. Harder, *A Gauss-Bonnet formula for discrete arithmetically defined groups*, Ann. Sci. Icole Norm. Sup. 4 (1971) 409–455.

[Hae1] J.Harer, *Stability of the homology of the mapping class groups of orientable surfaces*, Ann. of Math. 121 (1985) 215-249.

[Hae2] J.Harer, *The cohomology of the moduli space of curves*, in *Theory of moduli* , pp. 138-221, Lecture Notes in Math., 1337, Springer, 1988.

[HC] Harish-Chandra, *Automorphic forms on semisimple Lie groups*, Lect. Notes in Math. vol. 62, Springer-Verlag, 1968.

[Had] J. Hadamard, *Non-Euclidean geometry in the theory of automorphic functions*, ed. by J.Gray and A.Shenitzer, Amer. Math. Soc., 1999.

[Has1] M. Harris, *Automorphic forms and the cohomology of vector bundles on Shimura varieties*, in *Automorphic forms, Shimura varieties, and L-functions*, Vol. II , pp. 41–91, Perspect. Math., vol. 11, Academic Press, 1990.

[Has2] M. Harris, *Functorial properties of toroidal compactifications of locally symmetric varieties*, Proc. London Math. Soc. (3) 59 (1989) 1–22.

[HaT] M. Harris, R. Taylor, *The geometry and cohomology of some simple Shimura varieties*, Annals of Mathematics Studies, vol. 151, Princeton University Press, 2001. viii+276 pp.

[HZ1] M. Harris, S. Zucker, *Boundary cohomology of Shimura varieties. III. Coherent cohomology on higher-rank boundary strata and applications to Hodge theory*, Mim. Soc. Math. Fr. (N.S.) No. 85 (2001) vi+116 pp.

[HZ2] M. Harris, S. Zucker, *Boundary cohomology of Shimura varieties II. Hodge theory at the boundary*, Invent. Math. 116 (1994) 243–307.

[HZ3] M. Harris, S. Zucker, *Boundary cohomology of Shimura varieties. I. Coherent cohomology on toroidal compactifications*, Ann. Sci. Ecole Norm. Sup. (4) 27 (1994) 249–344.

[Hav1] W.Harvey, *Geometric structure of surface mapping class groups*, in *Homological group theory* (Proc. Sympos., Durham, 1977), pp. 255-269, London Math. Soc. Lecture Note Ser., 36, Cambridge Univ. Press, 1979.

[Hav2] W.Harvey, *Boundary structure of the modular group*, in *Riemann surfaces and related topic*, pp. 245–251, Ann. of Math. Stud., 97, Princeton Univ. Press, Princeton, 1981.

[Hav3] W.Harvey, *Modular groups and representation spaces*, in *Geometry of group representations*, pp. 205–214, Contemp. Math., vol. 74, Amer. Math. Soc., 1988.

[HT] B. Hassett, Y. Tschinkel, *Geometry of equivariant compactifications of* G_a^n, Internat. Math. Res. Notices 22 (1999) 1211–1230.

[Hat1] K. Hatada, *Hecke correspondences and Betti cohomology groups for smooth compactifications of Hilbert modular varieties of dimension* ≤ 3, Proc. Japan Acad. Ser. A Math. Sci. 72 (1996) 189–193.

[Hat2] K. Hatada, *On the local zeta functions of compactified Hilbert modular schemes and action of the Hecke rings*, Sci. Rep. Fac. Ed. Gifu Univ. Natur. Sci. 18 (1994) 1–34.

[Hat3] K. Hatada, *On the action of Hecke rings on homology groups of smooth compactifications of Siegel modular varieties and Siegel cusp forms*, Tokyo J. Math. 13 (1990) 191–205.

[Hat4] K. Hatada, *Correspondences for Hecke rings and (co-)homology groups on smooth compactifications of Siegel modular varieties*, Tokyo J. Math. 13 (1990) 37–62.

[Hat5] K. Hatada, *Correspondences for Hecke rings and l-adic cohomology groups on smooth compactifications of Siegel modular varieties*, Proc. Japan Acad. Ser. A Math. Sci. 65 (1989) 62–65.

[Htt1] T. Hattori, *Geometry of quotient spaces of* $\mathrm{SO}(3)\backslash\mathrm{SL}(3,\mathbf{R})$ *by congruence subgroups*, Math. Ann. 293 (1992) 443–467.

[Htt2] T. Hattori, *Collapsing of quotient spaces of* $\mathrm{SO}(n)\backslash\mathrm{SL}(n,\mathbf{R})$ *at infinity*, J. Math. Soc. Japan, 47 (1995) 193–225

[He1] H. He, *An analytic compactification of symplectic group*, J. Differential Geom. 51 (1999) 375–399.

[He2] H. He, *Compactification of classical groups*, Comm. Anal. Geom. 10 (2002) 709–740.

[HeS] G. Heckman, H. Schlichtkrull, *Harmonic analysis and special functions on symmetric spaces*, Perspectives in Mathematics, vol. 16, Academic Press, 1994, xii+225 pp.

[Hel1] S. Helgason, *Geometric analysis on symmetric spaces*, in *Mathematical Surveys and Monographs*, vol. 39. Amer. Math. Soc., 1994. xiv+611 pp.

[Hel2] S. Helgason, *Groups and geometric analysis. Integral geometry, invariant differential operators, and spherical functions*, in *Mathematical Surveys and Monographs*, vol. 83, Amer. Math. Soc., 2000, xxii+667 pp.

[Hel3] S. Helgason, *Differential geometry, Lie groups, and symmetric spaces*, Pure and Applied Math. vol. 80, Academic Press, 1978

[Hel4] S. Helgason, *Invariant differential equations on homogeneous manifolds*, Bull. Amer. Math. Soc. 83 (1977) 751–774.

[Hel5] S. Helgason, *A duality for symmetric spaces with applications to group representations*, Advances in Math. 5 (1970) 1–154.

[Her1] R. Hermann, *Compactification of homogeneous spaces. I*, J. Math. Mech. 14 (1965) 655–678.

[Her2] R. Hermann, *Compactification of homogeneous spaces. II*, J. Math. Mech. 15 (1966) 667–681.

[Her3] R. Hermann, *Geometric aspects of potential theory in the bounded symmetric domains*, Math. Ann. 148 (1962) 349–366.

[Her4] R. Hermann, *Geometric aspects of potential theory in the bounded symmetric domains. II*, Math. Ann. 151 (1963) 143–149.

[Her5] R. Hermann, *Geometric aspects of potential theory in symmetric spaces. III*, Math. Ann. 153 (1964) 384–394.

[Her6] R. Hermann, *Compactifications of homogeneous spaces and contractions of Lie groups*, Proc. Nat. Acad. Sci. U.S.A. 51 (1964) 456–461.

[Hi1] F. Hirzebruch, *The Hilbert modular group, resolution of the singularities at the cusps and related problems*, Séminaire Bourbaki, 231ème année (1970/1971) Exp. No. 396, pp. 275–288. Lecture Notes in Math., Vol. 244, Springer, 1971.

[Hi2] F. Hirzebruch, *Hilbert modular surfaces*, Enseignement Math. 19 (1973) 183–281.

[Hi3] F. Hirzebruch, *Automorphe Formen und der Satz von Riemann-Roch*, in *Symposium internacional de topologma algebraica International symposium on algebraic topology*, pp. 129–144, Universidad Nacional Autsnoma de Mixico and UNESCO, 1958.

[Hir1] H. Hironaka, *Resolution of singularities of an algebraic variety over a field of characteristic zero. I, II*, Ann. of Math. 79 (1964) 109–203; ibid., 79 (1964) 205–326.

[Hir2] H. Hironaka, *On resolution of singularities (characteristic zero)*, in *Proc. Internat. Congr. Mathematicians (Stockholm, 1962)*, pp. 507–521, 1963.

[Ho1] R. Holzapfel, *Geometry and arithmetic around Euler partial differential equations*, Reidel Publishing Company, 1986.

[Ho2] R. Holzapfel, *The ball and some Hilbert problems*, Birkhäuser, 1995.

[HW1] W. Hoffman, S. Weintraub, *Cohomology of the boundary of Siegel modular varieties of degree two, with applications*, Fund. Math. 178 (2003) 1–47.

[HW2] W. Hoffman, S. Weintraub, *The Siegel modular variety of degree two and level three. Trans*, Amer. Math. Soc. 353 (2001) 3267–3305.

[HW3] W. Hoffman, S. Weintraub, *Cohomology of the Siegel modular group of degree two and level four*, Mem. Amer. Math. Soc. 133 (1998) 59–75.

[Ho] R. Howe, *Lecture notes on compactifications of symmetric spaces*, Yale University.

[HOW] J. Huang, T. Oshima, N. Wallach, *Dimensions of spaces of generalized spherical functions*, Amer. J. Math. 118 (1996) 637–652.

[Hu] B. Hunt, *The geometry of some special arithmetic quotients*, Lect. Notes in Math., vol. 1637, Springer-Verlag, 1996.

[HKW] K. Hulek, C. Kahn, S. Weintraub, *Moduli spaces of abelian surfaces: compactification, degeneration and theta functions*, Walter de Gruyter, 1993.

[HS1] K. Hulek, G. Sankaran, *The nef cone of toroidal compactifications of* $\mathbf{A}_4$, Proc. London Math. Soc. 88 (2004) 659–704.

[HS2] K. Hulek, G. Sankaran, *The geometry of Siegel modular varieties*, in *Higher dimensional birational geometry*, pp. 89–156, Adv. Stud. Pure Math., vol. 35, Math. Soc. Japan, 2002.

[HS3] K. Hulek, G. Sankaran, *Degenerations of* $(1, 3)$ *abelian surfaces and Kummer surfaces*, in *Algebraic geometry: Hirzebruch 70*, pp. 177–192, Contemp. Math., vol. 241, Amer. Math. Soc., 1999.

[Ig1] J. Igusa, *On the desingularization of Satake compactifications*, in *Algebraic Groups and Discontinuous Subgroups*, Proc. Sympos. Pure Math., Amer. Math. Soc., pp. 301–305, 1966.

[Ig2] J. Igusa, *A desingularization problem in the theory of Siegel modular functions*, Math. Ann. 168 (1967) 228–260.

[Ig3] J. Igusa, *On the theory of compactifications*, AMS Summer Institute on Algebraic Geometry (Woods Hole) 1964, Lecture notes.

[IP1] A. Iozzi, J. Poritz, *The moduli space of boundary compactifications of* $\mathrm{SL}(2, R)$, Geom. Dedicata 76 (1999) 65–79.

[IP2] A. Iozzi, J. Poritz, *Boundary compactifications of* $\mathrm{SL}(2, R)$ *and* $\mathrm{SL}(2, C)$, Forum Math. 11 (1999) 385–397.

[Is] S. Ishikawa, *Symmetric subvarieties in compactifications and the Radon transform on Riemannian symmetric spaces of noncompact type*, J. Funct. Anal. 204 (2003) 50–100.

[Iv1] N.Ivanov, *Algebraic properties of mapping class groups of surfaces*, in *Geometric and algebraic topology*, pp.15-35, Banach Center Publ., 18, PWN, Warsaw, 1986.

[Iv2] N.Ivanov, *Attaching corners to Teichmller space*, Leningrad Math. J. 1 (1990) 1177-1205.

[Iv3] N.Ivanov, *Complexes of curves and Teichmller spaces*, Math. Notes 49 (1991) 479-484.

[Iv4] N.Ivanov, *Mapping class groups*. in *Handbook of geometric topology*, pp. 523–633, North-Holland, 2002.

[Ja] I. James, *Remarkable mathematicians: from Euler to von Neumann*, Cambridge University Press, 2002.

[Ji1] L. Ji, *Satake and Martin compactifications of symmetric spaces are topological balls*, Math. Res. Letters 4 (1997) 79–89.

[Ji2] L. Ji, *The trace class conjecture for arithmetic groups*, J. Diff. Geom. 48 (1998) 165–203.

[Ji3] L. Ji, *The greatest common quotient of Borel-Serre and the toroidal compactifications*, Geometric and Functional Analysis, 8 (1998) 978–1015.

[Ji4] L. Ji, *Metric compactifications of locally symmetric spaces*, International J. of Math. 9 (1998) 465–491.

[Ji5] L. Ji, *Spectral theory and geometry of locally symmetric spaces*, in *Proc. of International Congress of Chinese mathematicians*, AMS/IP Stud. Adv. Math., 20, Amer. Math. Soc., Providence, RI, 2001, pp. 261–270.

[Ji6] L. Ji, *Lectures on locally symmetric spaces and arithmetic groups*, in *Lie groups and automorphic forms*, International Press and Amer. Math. Soc., 2005, pp. 87–146.

[Ji7] L. Ji, *Introduction to symmetric spaces and their compactifications*, in *Lie Theory: unitary representations and compactifications of symmetric spaces*, ed. by J.P.Anker and B.Orsted, Birkhäuser, 2005, pp. 1–67.

[Ji8] L. Ji, *Scattering and geodesics of locally symmetric spaces*, in *Geometry and nonlinear partial differential equations*, AMS/IP Studies in Advanced Mathematics, vol. 29, 2002, ed. by S. Chen and S.T. Yau, pp. 39–52.

[Ji9] L. Ji, *Buildings and their applications in geometry and topology*, preprint 2005.

[JL] L. Ji, J. Lu, *New realizations of the maximal Satake compactifications of Riemannian symmetric spaces of noncompact type*, Lett. in Math. Physics 69 (2004) 139–145.

[JM] L. Ji, R. MacPherson, *Geometry of compactifications of locally symmetric spaces*, Ann. Inst. Fourier (Grenoble) 52 (2002) no. 2, 457–559.

[JZ1] L. Ji, M. Zworski, *Scattering matrices and scattering geodesics*, Ann. Scient. de Éc. Norm. Sup. 34 (2001) 441–469.

[JZ2] L. Ji, M. Zworski, *Correction and supplements to "Scattering matrices and scattering geodesics,"* Ann. Scient. Éc. Norm. Sup. 35 (2002) 897–901.

[Jun] H. Junghenn, *Distal compactifications of semigroups*, Trans. Amer. Math. Soc. 274 (1982) 379–397.

[Jur] J. Jurkiewicz, *Torus embeddings, polyhedra, k^*-actions and homology*, Dissertationes Math. (Rozprawy Mat.) vol. 236, 1985.

[Kan1] S. Kaneyuki, *Compactification of parahermitian symmetric spaces and its applications. II. Stratifications and automorphism groups*, J. Lie Theory 13 (2003) 535–563.

[Kan2] S. Kaneyuki, *Compactification of parahermitian symmetric spaces and its applications. I. Tube-type realizations*, in *Lie theory and its applications in physics*, pp. 63–74, World Sci. Publishing, 2000.

[Kan3] S. Kaneyuki, *On the causal structures of the Šilov boundaries of symmetric bounded domains*, in *Prospects in complex geometry*, pp. 127–159, Lecture Notes in Math., vol. 1468, Springer, 1991.

[Kan4] S. Kaneyuki, *On orbit structure of compactifications of para-Hermitian symmetric spaces*, Japan. J. Math. (N.S.) 13 (1987) 333–370.

[KaS] S. Kaneyuki, M.Sudo, *On Šilov boundaries of Siegel domains*, J. Fac. Sci. Univ. Tokyo Sect. I 15 (1968) 131–146.

[Ka] F.I. Karpelevic, *The geometry of geodesics and the eigenfunctions of the Beltrami-Laplace operator on symmetric spaces*, Trans. Moscow Math. Soc. 14 (1965) 51–199.

[KaK] M. Kashiwara, A. Kowata, K. Minemura, K. Okamoto, T. Oshima, M. Tanaka, *Eigenfunctions of invariant differential operators on a symmetric space*, Ann. of Math. 107 (1978) 1–39.

[KaU] K. Kato, S. Usui, *Borel-Serre spaces and spaces of SL(2)-orbits*, in *Algebraic geometry 2000*, Azumino (Hotaka) 321–382, Adv. Stud. Pure Math., 36, Math. Soc. Japan, Tokyo, 2002.

[Kat] S. Katok, *Fuchsian groups*, University of Chicago Press, 1992.

[Kaus] I. Kausz, *A modular compactification of the general linear group*, Doc. Math. 5 (2000) 553–594.

[KM] D. Kazhdan, G.A. Margulis, *A proof of Selberg's hypothesis* (Russian) Math. Sbornik N.S. 75 (1968) 163–168.

[Ke1] J. Keller, *Wave Propagation*, Proc. of ICM 1994 in Zürich, Birkhäuser, Switzerland, pp. 106–119.

[Ke2] Kelly, *General topology*, Graduate Texts in Math. vol. 27, Springer, 1955.

[KKMS] G. Kempf, F. Knudsen, D. Mumford, B. Saint-Donat, *Toroidal embeddings. I*, Lecture Notes in Mathematics, Vol. 339, Springer-Verlag, 1973. viii+209 pp.

[Ki] P. Kiernan, *On the compactifications of arithmetic quotients of symmetric spaces*, Bull. Amer. Math. Soc. 80 (1974) 109–110.

[KK1] P. Kiernan, S. Kobayashi, *Comments on Satake compactification and Picard theorem*, J. Math. Soc. Japan 28 (1976) 577–580.

[KK2] P. Kiernan, S. Kobayashi, *Satake compactification and extension of holomorphic mappings*, Invent. Math. 16 (1972) 237–248.

[Kn1] F. Knop, *Homogeneous varieties for semisimple groups of rank one*, Compositio Math. 98 (1995) 77–89.

[Kn2] F. Knop, *The Luna-Vust theory of spherical embeddings*, in *Proc. of the Hyderabad Conference on Algebraic Groups*, pp. 225–249, Manoj Prakashan, Madras, 1991.

[KL] F. Knop, H. Lange, *Some remarks on compactifications of commutative algebraic groups*, Comment. Math. Helv. 60 (1985) 497–507.

[Kob1] S. Kobayashi, *Hyperbolic complex spaces*, Grundlehren der Mathematischen Wissenschaften, vol. 318, 1998. xiv+471 pp.

[Kob2] S. Kobayashi, *Hyperbolic manifolds and holomorphic mappings*, Marcel Dekker Inc., 1970.

[KO] S. Kobayashi, T. Ochiai, *Satake compactification and the great Picard theorem*, J. Math. Soc. Japan 23 (1971) 340–350.

[Kom1] B. Komrakov, *The lattice of geometric compactifications of symmetric Riemannian spaces*, (Russian) Uspehi Mat. Nauk 28 (1973) 217–218.

[Kom2] B. Komrakov, *Compactifications of simply connected Riemannian spaces of nonpositive curvature*, (Russian) Vesci Akad. Navuk BSSR Ser. Fiz.-Mat. Navuk 1973, no. 3, 45–52, 136.

[Ko1] A. Koranyi, *Poisson integrals and boundary components of symmetric spaces*, Inv. Math. 34 (1976) 19–35.

[Ko2] A. Koranyi, *A survey of harmonic functions on symmetric spaces*, Proc. of Symp. in Pure Math. vol. XXXV, 1979, pp. 323–344.

[Ko3] A. Koranyi, *Harmonic functions on symmetric spaces*, in *Symmetric spaces* ed. by W. Boothy and G. Weiss, 1973, Marcel Dekker, 1972, pp. 380–412.

[Ko4] A. Koranyi, *Admissible limit sets of discrete groups on symmetric spaces of rank one*, in *Topics in Geometry: in memory of Joseph D'Atri*, ed. by S. Gindikin, Birkhäuser, 1996, pp. 231–240.

[Ko5] A. Koranyi, *Admissible convergence in Cartan-Hadamard manifolds*, J. of Geometric Analysis 11 (2001) 233–239.

[Ko6] A. Koranyi, *Compactifications of symmetric spaces and harmonic functions*, in *Analyse harmonique sur les groupes de Lie*, II, pp. 341–366, Lecture Notes in Math., vol. 739, Springer, 1979.

[Ko7] A. Koranyi, *Remarks on the Satake-Furstenberg compactifications*, to appear in *Pure and Appl. Math. Quarterly*.

[Ko8] A. Koranyi, *Generalizations of Fatou's theorem to symmetric spaces*, Rice Univ. Studies 56, no. 2, (1971) 127–136.

[Ko9] A. Koranyi, *Boundary behavior of Poisson integrals on symmetric spaces*, Trans. Amer. Math. Soc. 140 (1969) 393–409.

[Ko10] A. Koranyi, *Harmonic functions on Hermitian hyperbolic space*, Trans. Amer. Math. Soc. 135 (1969) 507–516.

[KW] A. Koranyi, J. Wolf, *Realization of Hermitian symmetric spaces as generalized half planes*, Ann. of Math. 81 (1965) 265–288.

[Kr] S. Krantz, *Function theory of several complex variables*, John Wiley & Sons, 1982.

[KuM] S. Kudla, J. Millson, *Geodesic cyclics and the Weil representation. I. Quotients of hyperbolic space and Siegel modular forms*, Compositio Math. 45 (1982) 207–271.

[Kug] M. Kuga, *Fiber varieties over a symmetric space whose fibers are abelian varieties*, in *Algebraic Groups and Discontinuous Subgroups*, pp. 338–346, Amer. Math. Soc., 1966.

[Ku] K. Kuratowski *Topology*, Academic Press, 1966.

[Kus1] G.F. Kusner, *Irreducible projective compactifications of noncompact symmetric spaces*, Trudy Sem. Vektor. Tenzor. Anal. No. 21, (1983) 153–190.

[Kus2] G.F. Kusner, *Karpelevic's compactification is homeomorphic to a sphere* (in Russian) Trudy Sem. Vector. Tenzor. Anal. 19 (1979) 95–111. (See MR 82b:53069 for a summary).

[Kus3] G.F. Kusner, *The compactification of noncompact symmetric Riemannian spaces*, (Russian) Trudy Sem. Vektor. Tenzor. Anal. 16 (1972) 99–152.

[Kus4] G.F. Kusner, *A certain compactification of noncompact symmetric Riemannian spaces*, Dokl. Akad. Nauk SSSR 190 (1970) 1282–1285.

[Kz] A. Kuznetsov, *Laumon's resolution of Drinfeld's compactification is small*, Math. Res. Lett. 4 (1997) 349–364.

[Laf1] L. Lafforgue, *Pavages des simplexes, schémas de graphes recollés et compactification des* $\mathrm{PGL}_r^{n+1}/\mathrm{PGL}_r$, Invent. Math. 136 (1999) 233–271.

[Laf2] L. Lafforgue, *Une compactification des champs classifiant les chtoucas de Drinfeld*, J. Amer. Math. Soc. 11 (1998) 1001–1036.

[Laf3] L. Lafforgue, *Compactification de l'isoginie de Lang et diginirescence des structures de niveau simple des chtoucas de Drinfeld*, C. R. Acad. Sci. Paris Sir. I Math. 325 (1997) no. 12, 1309–1312.

[La] R. Langlands, *On the functional equations satisfied by Eisenstein series*, in *Lecture Notes in Mathematics*, Vol. 544. Springer-Verlag, 1976, v+337 pp.

[LR] R. Langlands, D. Ramakrishnan, *The zeta functions of Picard modular surfaces*, les publications CRM, Montreal, 1992.

[Lan1] E. Landvogt, *Some functorial properties of the Bruhat-Tits building*, J. Reine Angew. Math. 518 (2000) 213–241.

[Lan2] E. Landvogt, *A compactification of the Bruhat-Tits building*, Lecture Notes in Mathematics, vol. 1619, Springer-Verlag, Berlin, 1996.

[Lar] M. Larsen, *Arithmetic compactification of some Shimura surfaces*, in *The zeta functions of Picard modular surfaces*, pp. 31–45, Montreal, 1992.

[LMP] A. Lau, P. Milnes, J. Pym, *Compactifications of locally compact groups and quotients*, Math. Proc. Cambridge Philos. Soc. 116 (1994) 451–463.

[LL] J. Lawson, Y. Lim, *Compactification structure and conformal compressions of symmetric cones*, J. Lie Theory 10 (2000) 375–381.

[Le] M. Lee, *Mixed automorphic forms, torus bundles, and Jacobi forms*, Lecture Notes in Mathematics, vol. 1845, Springer-Verlag, 2004. x+239 pp.

[LeS1] R. Lee, J. Schwermer, *Geometry and arithmetic cycles attached to* $\mathrm{SL}_3(Z)$. *I*, Topology 25 (1986) 159–174.

[LeS2] R. Lee, J. Schwermer, *Cohomology of arithmetic subgroups of* SL_3 *at infinity*, J. Reine Angew. Math. 330 (1982) 100–131.

[LeW1] R. Lee, S. Weintraub, *On certain Siegel modular varieties of genus two and levels above two*, in *Algebraic topology and transformation groups*, pp. 29–52, Lecture Notes in Math., 1361, Springer, 1988.

[LeW2] R. Lee, S. Weintraub, *Topology of the Siegel spaces of degree two and their compactifications*, in *Proceedings of the 1986 topology conference*, Topology Proc. 11 (1986) 115–175.

[LeW3] R. Lee, S. Weintraub, *Cohomology of* $\mathrm{Sp}_4(Z)$ *and related groups and spaces*, Topology 24 (1985) 391–410.

[LeW4] R. Lee, S. Weintraub, *Cohomology of a Siegel modular variety of degree* 2, in *Group actions on manifolds*, pp. 433–488, Contemp. Math., vol. 36, Amer. Math. Soc., 1985.

[Leh1] J. Lehner, *Discontinuous groups and automorphic functions*, Amer. Math. Soc., Math. Surveys, No. VIII, 1964.

[Leh2] J. Lehner, Review of [Si4], MR0074534 (17,602d) Amer. Math. Soc.

[LM] F. Lescure, L. Meersseman, *Compactifications equivariantes non kählériennes dun groupe algebrique multiplicatif*, Ann. Inst. Fourier (Grenoble) 52 (2002) 255–273.

[Leu1] E. Leuzinger, *Tits geometry, arithmetic groups, and the proof of a conjecture of Siegel*, J. Lie Theory 14 (2004) 317–338.

[Leu2] E. Leuzinger, *On polyhedral retracts and compactifications of locally symmetric spaces*, Differential Geom. Appl. 20 (2004) 293–318.

[Leu3] E. Leuzinger, *Kazhdan's property (T)* L^2*-spectrum and isoperimetric inequalities for locally symmetric spaces*, Comment. Math. Helv. 78 (2003) 116–133.

[Leu4] E. Leuzinger, *Geodesic rays in locally symmetric spaces*, Diff. Geom. Appl. 6 (1996) 55–65.

[Leu5] E. Leuzinger, *An exhaustion of locally symmetric spaces by compact submanifolds with corners*, Invent. Math. 121 (1995) 389–410.

[Leu6] E. Leuzinger, *On the trigonometry of symmetric spaces*, Comment. Math. Helv. 67 (1992) 252–286.

[Leu7] E. Leuzinger, *On the Gauss–Bonnet formula for locally symmetric spaces of noncompact type*, Enseign. Math. 42 (1996) 201–214.

[Li] J. Li, *Nonvanishing theorems for the cohomology of certain arithmetic quotients*, J. Reine Angew. Math. 428 (1992) 177–217.

[LiM] J. Li, J. Millson, *On the first Betti number of a hyperbolic manifold with an arithmetic fundamental group*, Duke Math. J. 71 (1993) 365–401.

[LiS1] J-S. Li, J. Schwermer, *On the Eisenstein cohomology of arithmetic groups*, Duke Math. J. 123 (2004) 141–169.

[LiS2] J-S. Li, J. Schwermer, *Automorphic representations and cohomology of arithmetic groups*, in *Challenges for the 21st century* (Singapore, 2000) World Sci. Publishing, 2001, pp. 102–137.

[LiS3] J-S. Li, J. Schwermer, *Constructions of automorphic forms and related cohomology classes for arithmetic subgroups of* G_2, Compositio Math. 87 (1993) 45–78.

[LP] P. Littelmann, C. Procesi, *Equivariant cohomology of wonderful compactifications*, in *Operator algebras, unitary representations, enveloping algebras, and invariant theory*, pp. 219–262, Progr. Math., 92, Birkhäuser Boston, 1990.

[LoM] N. Lohoue, S. Mehdi, *The Novikov-Shubin invariants for locally symmetric spaces*, J. Math. Pures Appl. (9) 79 (2000) 111–140.

[Lo1] E. Looijenga, *Compactificatioms defined by arrangements I: The ball quotient case*, Duke Math. J. 118 (2003) 151–187.

[Lo2] E. Looijenga, *Compactificatioms defined by arrangements II: locally symmetric varieties of type IV*, Duke Math. J. 119 (2003) 527–588.

[Lo3] E. Looijenga, *L^2-cohomology of locally symmetric varieties*, Compositio Math. 67 (1988) 3–20.

[Lo4] E. Looijenga, *Cellular decompositions of compactified moduli spaces of pointed curves*, in *The moduli space of curves*, pp. 369–400, Progr. Math., 129, Birkhäuser Boston, 1995.

[Lo5] E. Looijenga, *Smooth Deligne-Mumford compactifications by means of Prym level structures*, J. Algebraic Geom. 3 (1994) 283–293.

[Lo6] E. Looijenga, *L^2-cohomology of locally symmetric varieties*, Compositio Math. 67 (1988) 3–20.

[Lo7] E. Looijenga, *New compactifications of locally symmetric varieties*, in *Proceedings of the 1984 Vancouver conference in algebraic geometry*, pp. 341–364, CMS Conf. Proc., 6, Amer. Math. Soc., 1986.

[LoR] E. Looijenga, M. Rapoport, *Weights in the local cohomology of a Baily-Borel compactification*, in *Complex geometry and Lie theory*, pp. 223–260, Proc. Sympos. Pure Math., 53, Amer. Math. Soc., 1991.

[Lot] J. Lott, *Heat kernels on covering spaces and topological invariants*, J. Differential Geom. 35 (1992) 471–510.

[LMR] A. Lubotzky, S. Mozes, M. Raghunathan, *The word and Riemannian metrics on lattices of semisimple groups*, Inst. Hautes Etudes Sci. Publ. Math. No. 91 (2000) 5–53.

[Lu1] D. Luna, *Sur les plongements de Demazure*, J. Algebra 258 (2002) 205–215.

[Lu2] D. Luna, *Variétés sphériques de type A*, Publ. Math. Inst. Hautes Etudes Sci. No. 94, (2001) 161–226.

[Lu3] D. Luna, *Toute variété magnifique est sphérique*, Transform. Groups 1 (1996) 249–258.

[LV] D. Luna, Th. Vust, *Plongements d'espaces homogènes*, Comment. Math. Helv. 58 (1983) 186–245.

[Mac] R. MacPherson, *Seminar notes on locally symmetric spaces and their compactifications*, taken by M. Goresky, 1985, 1989, 1990.

[MM1] R. MacPherson, M. McConnell, *Explicit reduction theory for Siegel modular threefolds*, Invent. Math. 111 (1993) 575–625.

[MM2] R. MacPherson, M. McConnell, *Classical projective geometry and modular varieties*, in *Algebraic analysis, geometry, and number theory*, (Baltimore, MD, 1988) 237–290, Johns Hopkins Univ. Press, Baltimore, MD, 1989.

[MP] R. MacPherson, C. Procesi, *Making conical compactifications wonderful*, Selecta Math. (N.S.) 4 (1998) 125–139.

[MR] C. Maclachlan, A. Reid, *The arithmetic of hyperbolic 3-manifolds*, GTM vol. 219, Springer-Verlag, 2003.

[Ma] W. Magnus, *Noneuclidean tesselations and their groups*, Academic Press, 1974.

[Mag] G. Margulis, *Discrete subgroups of semisimple Lie groups*, Springer-Verlag, 1991.

[Mar] R.S. Martin, *Minimal Positive Harmonic Functions*, Trans. Amer. Math. Soc. 49 (1941) 137–172

[Mas] B. Massmann, *Equivariant Kähler compactifications of homogeneous manifolds*, J. Geom. Anal. 2 (1992) 555–574.

[MaM1] Y. Matsushima, S. Murakami, *On certain cohomology groups attached to Hermitian symmetric spaces*, Osaka J. Math. 2 (1965) 1–35.

[MaM2] Y. Matsushima, S. Murakami, *On vector bundle valued harmonic forms and automorphic forms on symmetric Riemannian manifolds*, Ann. of Math. 78 (1963) 365–416.

[Mc1] M. McConnell, *Cell decompositions of Satake compactifications*, in *Geometric combinatorics (Kotor, 1998)*, Publ. Inst. Math. (Beograd) (N.S.) 66(80) (1999) 46–90.

[Mc2] M. McConnell, *Classical projective geometry and arithmetic groups*. Math. Ann. 290 (1991) 441–462.

[Me1] R. Melrose, *Geometric Scattering Theory*, Cambridge Univ Press, New York, 1995

[Me2] R. Melrose, *A private letter on self-gluing manifolds with corners*, 1998.

[MW1] R. Miatello, N. Wallach, *The resolvent of the Laplacian on locally symmetric spaces*, J. Differential Geom. 36 (1992) 663–698.

[MW2] R. Miatello, N. Wallach, *Kuznetsov formulas for products of groups of R-rank one*, in *Festschrift in honor of I. I. Piatetski-Shapiro on the occasion of his sixtieth birthday, Part II*, pp. 305–320, Israel Math. Conf. Proc., 3, 1990.

[MW3] R. Miatello, N. Wallach, *Kuznetsov formulas for real rank one groups*, J. Funct. Anal. 93 (1990) 171–206.

[Mi1] J. Milne, *Canonical models of (mixed) Shimura varieties and automorphic vector bundles*, in *Automorphic forms, Shimura varieties, and L-functions*, Vol. I (Ann Arbor, MI, 1988) ed. by L. Clozel and J. Milne, Academic Press, 1990, pp. 283–414.

[Mi2] J. Milne, *Shimura varieties and motives*, in *Motives* (Seattle, WA, 1991) Proc. Sympos. Pure Math., 55, Part 2, Amer. Math. Soc., Providence, RI, 1994, pp. 447–523.

[Mil1] J. Millson, *On the first Betti number of a constant negatively curved manifold*, Ann. of Math. 104 (1976) 235–247.

[Mil2] J. Millson, *Intersection numbers of cycles on locally symmetric spaces and Fourier coefficients of holomorphic modular forms in several complex variables*, in *Theta functions*, Part 2, pp. 129–142, Proc. Sympos. Pure Math., 49, Part 2, Amer. Math. Soc., 1989.

[MiR] J. Millson, M. Raghunathan, *Geometric construction of cohomology for arithmetic groups. I*, Proc. Indian Acad. Sci. Math. Sci. 90 (1981) 103–123.

[MoW] C. Moeglin, J. Waldspurger, *Spectral decomposition and Eisenstein series*, in *Cambridge Tracts in Mathematics*, vol. 113, Cambridge University Press, 1995. xxviii+338 pp.

[Mok1] N. Mok, *Metric rigidity theorems on Hermitian locally symmetric manifolds*, World Scientific, 1989.

[Mok2] N. Mok, *Compactification of complete Kähler surfaces of finite volume satisfying certain curvature conditions*, Ann. of Math. 129 (1989) 383–425.

[MZ] N. Mok, J. Zhang, *Compactifying complete Kähler-Einstein manifolds of finite topological type and bounded curvature*, Ann. of Math. 129 (1989) 427–470.

[Mo1] C.C. Moore, *Compactifications of symmetric spaces I*, Amer. J. Math. 86 (1964) 201–218.

[Mo2] C.C. Moore, *Compactifications of symmetric spaces II: The Cartan domains*, Amer. J. Math. 86 (1964) 358–378.

[Mor] L. Moret-Bailly, *Compactifications, hauteurs et finitude*, in *Seminar on arithmetic bundles: the Mordell conjecture (Paris, 1983/84)*, Astérisque No. 127 (1985) 113–129.

[MS] H. Moscovici, R. Stanton, *R-torsion and zeta functions for locally symmetric manifolds*, Invent. Math. 105 (1991) 185–216.

[Mos] G. Mostow, *Strong rigidity of locally symmetric spaces*, Princeton University Press, 1973.

[MT] G. Mostow, T. Tamagawa, *On the compactness of arithmetically defined homogeneous spaces*, Ann. of Math. 76 (1962) 446–463.

[Mul1] W. Muller, *Weyl's law for the cuspidal spectrum of* SL_n, C. R. Math. Acad. Sci. Paris 338 (2004) 347–352.

[Mul2] W. Muller, *The trace class conjecture in the theory of automorphic forms. II*, Geom. Funct. Anal. 8 (1998) 315–355.

[Mul3] W. Muller, *The trace class conjecture in the theory of automorphic forms*, Ann. of Math. 130 (1989) 473–529.

[Mul4] W. Muller, *Spectral theory and geometry*, *First European Congress of Mathematics, Vol. I (Paris, 1992)*, pp. 153–185, Progr. Math., 119, Birkhäuser, 1994.

[Mum1] D. Mumford, *A new approach to compactifying locally symmetric varieties*, in *Discrete subgroups of Lie groups and applications to moduli* (Internat. Colloq., Bombay, 1973) pp. 211–224, Oxford Univ. Press, Bombay, 1975

[Mum2] D. Mumford, *Tata lectures on theta I*, in *Progress in Mathematics*, vol. 28, Birkhäuser, 1983.

[Mum3] D. Mumford, *Hirzebruch's proportionality theorem in the noncompact case*, Invent. Math. 42 (1977) 239–272.

[Nad] A. Nadel, *On complex manifolds which can be compactified by adding finitely many points*, Invent. Math. 101 (1990) 173–189.

[NT] A. Nadel, H. Tsuji, *Compactification of complete Kähler manifolds of negative Ricci curvature*, J. Differential Geom. 28 (1988) 503–512.

[Nam1] Y. Namikawa, *Toroidal compactification of Siegel spaces*, Lecture Notes in Mathematics, 812. Springer, Berlin, 1980

[Nam2] Y. Namikawa, *A new compactification of the Siegel space and degeneration of Abelian varieties. I*, Math. Ann. 221 (1976) 97–141.

[Nam3] Y. Namikawa, *A new compactification of the Siegel space and degeneration of Abelian varieties. II*, Math. Ann. 221 (1976) 201–241.

[Nam4] Y. Namikawa, *Toroidal degeneration of Abelian varieties*, *Complex analysis and algebraic geometry*, pp. 227–237, Iwanami Shoten, 1977.

[Ne1] Y. Neretin, *Pencils of geodesics in symmetric spaces, Karpelevich boundary, and associahedron-like polyhedra*, in *Lie groups and symmetric spaces*, Amer. Math. Soc. Transl. Ser. 2, 210, 2003, pp. 225–255.

[Ne2] Y. Neretin, *Geometry of* $\mathrm{GL}_n(\mathbb{C})$ *at infinity: hinges, complete collineations, projective compactifications, and universal boundary*, in *The orbit method in geometry and physics (Marseille, 2000)*, Progr. Math., vol. 213, Birkhäuser, 2003, pp. 297–327.

[Ne3] Y. Neretin, *Hinges and the Study-Semple-Satake-Furstenberg-De Concini-Procesi-Oshima boundary*, in *Kirillov's seminar on representation theory*, Amer. Math. Soc. Transl. Ser. 2, vol. 181, 1998, pp. 165–230.

[Ni] E. Nisnevich, *Arithmetic and cohomology invariants of semisimple group schemes and compactifications of locally symmetric spaces*, Functional Anal. Appl. 14 (1980) 65–66.

[NR] M. Nori, M. Raghunathan, *On conjugation of locally symmetric arithmetic varieties*, in *Proceedings of the Indo-French Conference on Geometry*, pp. 111–122, Hindustan Book Agency, Delhi, 1993.

[Od] T. Oda, *Convex bodies and algebraic geometry: An introduction to the theory of toric varieties*, Springer-Verlag, Berlin, 1988.

[OO] G. Olafsson, B. Orsted, *Causal compactification and Hardy spaces*, Trans. Amer. Math. Soc. 351 (1999) 3771–3792.

[Ol1] M. Olbrich, *L^2-invariants of locally symmetric spaces*, Doc. Math. 7 (2002) 219–237.

[Ol2] M. Olbrich, *Hodge theory on hyperbolic manifolds of infinite volume*, in *Lie theory and its applications in physics, III*, pp. 75–83, World Sci. Publishing, 2000.

[Os1] T. Oshima, *A realization of Riemannian symmetric spaces*, J. Math. Soc. Japan, 30 (1978) 117–132.

[Os2] T. Oshima, *A realization of semisimple symmetric spaces and construction of boundary value maps*, Adv. Studies in Pure Math. vol 14, *Representations of Lie groups*, 1986, pp. 603–650.

[Os3] T. Oshima, *Fourier analysis on semisimple symmetric spaces*, in *Non-commutative harmonic analysis*, in Lect. Notes in Math. 880, pp. 357–369, Springer-Verlag, 1981.

[Os4] T. Oshima, *Harmonic analysis on semisimple symmetric spaces*, Sugaku Expositions 12 (2002) 151–170.

[OsS1] T. Oshima, J. Sekiguchi, *Eigenspaces of invariant differential operators on an affine symmetric spaces*, Invent. Math. 57 (1980) 1–81.

[OsS2] T. Oshima, J. Sekiguchi, *The restricted root system of a semisimple symmetric pair*, Adv. Studies Pure Math. 4, Group representations and systems of differential equations, 1984, 433–497.

[OW1] M.S. Osborne, G. Warner, *The theory of Eisenstein systems*, Academic Press, 1981, New York.

[OW2] M.S. Osborne, G. Warner, *The Selberg trace formula II: partition, reduction, truncation*, Pacific Jour. Math. 106 (1983) 307–496.

[Pe1] P. Perry, *A Poisson summation formula and lower bounds for resonances in hyperbolic manifolds*, Int. Math. Res. Not. 34(2003) 1837–1851.

[Pe2] P. Perry, *Spectral theory, dynamics, and Selberg's zeta function for Kleinian groups, Dynamical, spectral, and arithmetic zeta functions*, 145–165, Contemp. Math., vol. 290, Amer. Math. Soc., 2001.

[Pe3] P. Perry, *Asymptotics of the length spectrum for hyperbolic manifolds of infinite volume*, Geom. Funct. Anal. 11 (2001) 132–141.

[Pe4] P. Perry, *The Selberg zeta function and a local trace formula for Kleinian groups*, J. Reine Angew. Math. 410 (1990) 116–152.

[PS] I. Piatetski-Shapiro, *Geometry of classical domains and theory of automorphic functions*, Gordon and Breach, New York, 1969.

[Pi1] R. Pink, *On compactification of Drinfeld moduli schemes*, in *Moduli spaces, Galois representations and L-functions*, Surikaisekikenkyusho Kokyuroku No. 884 (1994) 178–183.

[Pi2] R. Pink, *On l-adic sheaves on Shimura varieties and their higher direct images in the Baily-Borel compactification*, Math. Ann. 292 (1992) 197–240.

[Pi3] R. Pink, *Arithmetical compactification of mixed Shimura varieties*, Bonner Mathematische Schriften, vol. 209, 1990. xviii+340 pp.

[PR] V. Platonov, A. Rapinchuk, *Algebraic groups and Number theory*, Academic Press, 1994.

[Po] H. Poincaré, *Papers on Fuchsian groups*, Springer-Verlag, 1985.

[Pro] C. Procesi, *The toric variety associated to Weyl chambers*, in *Mots*, pp. 153–161, Lang. Raison. Calc., Hermès, Paris, 1990.

[Ra1] M. Raghunathan, *Discrete subgroups of Lie groups*, Springer-Verlag, 1972.

[Ra2] M. Raghunathan, *A note on quotients of real algebraic groups by arithmetic subgroups*, Invent. math. 4 (1968) 318–335.

[Ra3] M. Raghunathan, *The first Betti number of compact locally symmetric spaces*, in *Current trends in mathematics and physics*, pp. 116–137, Narosa, New Delhi, 1995.

[Ri1] R. Richardson, *On orbits of algebraic groups and Lie groups*, Bull. Australian Math. Soc. 25 (1982) 1–28.

[Ri2] R. Richardson, *Orbits, invariants and representations associated to involutions of reductive groups*, Inv. Math. 66, 1982, 287–312.

[Rol1] J. Rohlfs, *Projective limits of locally symmetric spaces and cohomology*, J. Reine Angew. Math. 479 (1996) 149–182.

[Rol2] J. Rohlfs, *Lefschetz numbers for arithmetic groups*, in *Cohomology of arithmetic groups and automorphic forms*, pp. 303–313, Lecture Notes in Math., vol. 1447, Springer, 1990.

[Rol3] J. Rohlfs, *Maximal arithmetically defined groups and related group schemes*, in *Proceedings of the 5th School of Algebra*, pp. 59–73, Soc. Brasil. Mat., Rio de Janeiro, 1978.

[RSc] J. Rohlfs, J. Schwermer, *An arithmetic formula for a topological invariant of Siegel modular varieties*, Topology 37 (1998) 149–159.

[RSg] J. Rohlfs, T. Springer, *Applications of buildings*, in *Handbook of incidence geometry*, pp. 1085–1114, North-Holland, 1995.

[RSp1] J. Rohlfs, B. Speh, *Pseudo-Eisenstein forms and cohomology of arithmetic groups. II*, in *Algebraic groups and arithmetic*, pp. 63–89, Tata Inst. Fund. Res., 2004.

[RSp2] J. Rohlfs, B. Speh, *Pseudo-Eisenstein forms and cohomology of arithmetic groups. I*, Manuscripta Math. 106 (2001) 505–518.

[RSp3] J. Rohlfs, B. Speh, *Boundary contributions to Lefschetz numbers for arithmetic groups. I*, in *Cohomology of arithmetic groups and automorphic forms*, pp. 315–332, Lecture Notes in Math., vol. 1447, Springer, 1990.

[RSp4] J. Rohlfs, B. Speh, *On limit multiplicities of representations with cohomology in the cuspidal spectrum*, Duke Math. J. 55 (1987) 199–211.

[Roo] G. Roos, *Volume of bounded symmetric domains and compactification of Jordan triple systems*, in *Lie groups and Lie algebras*, pp. 249–259, Math. Appl., 433, Kluwer Acad. Publ., 1998.

[Ros] W. Rossman, *The structure of semisimple symmetric spaces*, Can. J. Math. 31, 1979, 157–180.

[Sal1] M. Salvai, *On the volume and energy of sections of a circle bundle over a compact Lie group, Differential geometry (Valencia, 2001)*, pp. 262–267, World Sci. Publishing, 2002.

[Sal2] M. Salvai, *On the Laplace and complex length spectra of locally symmetric spaces of negative curvature*, Math. Nachr. 239/240 (2002) 198–203.

[San1] G. Sankaran, *Abelian surfaces with odd bilevel structure*, in *Number theory and algebraic geometry*, pp. 279–300, London Math. Soc. Lecture Note Ser., vol. 303, Cambridge Univ. Press, 2003.

[San2] G. Sankaran, *Moduli of polarised abelian surfaces*, Math. Nachr. 188 (1997) 321–340.

[San3] G. Sankaran, *Fundamental group of locally symmetric varieties*, Manuscripta Math. 90 (1996) 39–48.

[Sap1] L. Saper, *Tilings and finite energy retractions of locally symmetric spaces*, Comment. Math. Helv. 72 (1997) 167–202.

[Sap2] L. Saper, *Geometric rationality of equal-rank Satake compactifications*, Math. Res. Lett. 11 (2004) 653–671.

[Sap3] L. Saper, *On the cohomology of locally symmetric spaces and of their compactifications*, in *Current developments in mathematics, 2002*, pp. 219–289, Int. Press, 2003.

[SaS] L. Saper, M. Stern, *L_2-cohomology of arithmetic varieties*, Ann. of Math. 132 (1990) 1–69.

[SaZ] L. Saper, S. Zucker, *An introduction to L^2-cohomology*, in *Several complex variables and complex geometry*, Part 2, pp. 519–534, Proc. Sympos. Pure Math., 52, Part 2, Amer. Math. Soc., 1991.

[Sar] P. Sarnak, *The arithmetic and geometry of some hyperbolic three-manifolds*, Acta Math. 151 (1983) 253–295.

[Sat1] I. Satake, *On representations and compactifications of symmetric spaces*, Ann. of Math. 71 (1960) 77–110.

[Sat2] I. Satake, *On compactifications of the quotient spaces for arithmetically defined discontinuous groups*, Ann. of Math. 72 (1960) 555–580.

[Sat3] I. Satake, *On the compactification of Siegel spaces*, J. India Math. Soc. 20 (1956) 259–281.

[Sat4] I. Satake, *A note on holomorphic imbeddings and compactification of symmetric domains*, Amer. J. Math., 90 (1968) 231–247.

[Sat5] I. Satake, *On the arithmetic of tube domains (blowing-up of the point at infinity)*, Bull. Amer. Math. Soc. 79 (1973) 1076–1094.

[Sat6] I. Satake, *On Siegel's modular functions*, in *Proceedings of the international symposium on algebraic number theory, (Tokyo & Nikko, 1955)*, pp. 107–129, Science Council of Japan, Tokyo, 1956.

[Sat7] I. Satake, *Review of [AMRT] in Math. Review*, MR0457437 (56 # 15642).

[Sat8] I. Satake, *Algebraic structures of symmetric domains*, Princeton University Press, 1980.

[Sat9] I. Satake, *Compactifications, past and present*, Sugaku 51 (1999) 129–141.

[Sat10] I. Satake, *Equivariant holomorphic embeddings of symmetric domains and applications*, in *Research on automorphic forms and zeta functions*, Surikaisekikenkyusho Kokyuroku No. 1002 (1997) 19–31.

[Sat11] I. Satake, *On the rational structures of symmetric domains. I*, in *International Symposium in Memory of Hua Loo Keng, Vol. II*, pp. 231–259, Springer, 1991.

[Sat12] I. Satake, *On the rational structures of symmetric domains. II. Determination of rational points of classical domains*, Tohoku Math. J. 43 (1991) 401–424.

[Sat13] I. Satake, *Linear imbeddings of self-dual homogeneous cones*, Nagoya Math. J. 46 (1972) 121–145.

[Sat14] I. Satake, *On some properties of holomorphic imbeddings of symmetric domains*, Amer. J. Math. 91 (1969) 289–305.

[Sat15] I. Satake, *On modular imbeddings of a symmetric domain of type (IV)*, in *Global Analysis*, pp. 341–354, Univ. Tokyo Press, 1969.

[Sat16] I. Satake, *Holomorphic imbeddings of symmetric domains into a Siegel space*, Amer. J. Math. 87 (1965) 425–461.

[Sch] H. Schlichtkrull, *Hyperfunctions and harmonic analysis on symmetric spaces*, Birkhäuser, 1984.

[ScT] P. Schneider, J. Teitelbaum, *p-adic boundary values*, in *Cohomologies p-adiques et applications arithmfiques*, Astérisque No. 278 (2002) 51–125.

[Schm1] W. Schmid, *Boundary value problems for group invariant differential equations*, in *The mathematical heritage of Elie Cartan (Lyon, 1984)*, Astérisque 1985, pp. 311–321.

[Schm2] W. Schmid, *Variation of Hodge structure: the singularities of the period mapping*, Invent. Math. 22 (1973) 211–319.

[Shw1] J. Schwermer, *On Euler products and residual Eisenstein cohomology classes for Siegel modular varieties*, Forum Math. 7 (1995) 1–28.

[Shw2] J. Schwermer, *Eisenstein series and cohomology of arithmetic groups: the generic case*, Invent. Math. 116 (1994) 481–511.

[Shw3] J. Schwermer, *On arithmetic quotients of the Siegel upper half space of degree two*, Compositio Math. 58 (1986) 233–258.

[Shw4] J. Schwermer, *Holomorphy of Eisenstein series at special points and cohomology of arithmetic subgroups of* $\mathrm{SL}_n(Q)$, J. Reine Angew. Math. 364 (1986) 193–220.

[Shw5] J. Schwermer, *Cohomology of arithmetic groups, automorphic forms and L-functions*, in *Cohomology of arithmetic groups and automorphic forms* (Luminy-Marseille, 1989) Lecture Notes in Math., 1447, Springer, Berlin, 1990, pp. 1–29.

[Sen] S. Senthamarai Kannan, *Remarks on the wonderful compactification of semisimple algebraic groups*, Proc. Indian Acad. Sci. Math. Sci. 109 (1999) 241–256.

[Sel1] A. Selberg, *Collected papers. Vol. I*, Springer-Verlag, 1989. vi+711 pp.

[Sel2] A. Selberg, *Harmonic analysis and discontinuous groups in weakly symmetric Riemannian spaces with applications to Dirichlet series*, J. Indian Math. Soc. (N.S.) 20 (1956) 47–87.

[Sel3] A. Selberg, *On discontinuous groups in higher-dimensional symmetric spaces*, in *Contributions to function theory (internat. Colloq. Function Theory, Bombay, 1960)*, pp. 147–164, Tata Institute of Fundamental Research, Bombay.

[Ser1] J-P. Serre, *Cohomologie galoisienne*, LNM 5, 5th edition, Springer 1994.

[Ser2] J-P. Serre, *Arithmetic groups*, in *Homological group theory*, pp. 105–136, London Math. Soc. Lecture Note Ser., 36, Cambridge Univ. Press, 1979.

[Ser3] J-P. Serre, *Groupes discrets-compactifications*, *Colloque sur les Fonctions Sphériques et la Théorie des Groupes*, Exp. No. 6 4 pp., Inst. Ilie Cartan, Univ. Nancy, 1971.

[Ser4] J-P. Serre, *Cohomologie des groupes discrets*, in *Prospects in mathematics*, pp. 77–169. Ann. of Math. Studies, No. 70, Princeton Univ. Press, 1971.

[Ser5] J-P. Serre, *Trees*, Springer-Verlag, 1980.

[Sha] J. Shah, *Projective degenerations of Enriques' surfaces*, Math. Ann. 256 (1981) 475–495.

[Shi] G. Shimura, *Introduction to the arithmetic theory of automorphic functions*, Princeton University Press, 1971.

[Si1] C.L. Siegel, *Symplectic Geometry*, Academic Press, 1964

[Si2] C.L. Siegel, *Zur reduktionstheorie quadratischer Formen*, Publ. of Math. Soc. Japan, vol. 5, 1959; Des. Werke III, pp. 274–327.

[Si3] C.L. Siegel, *Lectures on the geometry of numbers*, Springer, 1989.

[Si4] C.L. Siegel, *Zur Theorie der Modulfunktionen n-ten Grades*, Comm. Pure Appl. Math. 8 (1955) 677–681.

[Si5] C.L. Siegel, *Einführung in die Theorie der Modulfunktionen n-ten Grades*, Math. Ann. 116 (1939) 617–657.

[Sil] J. Silverman, *The arithmetic of elliptic curves*, GTM 106, Springer-Verlag, 1986.

[Sin] D. Sinha, *Manifold-theoretic compactifications of configuration spaces*, Selecta Math. (N.S.) 10 (2004) 391-428.

[SiY] Y. Siu, S.T. Yau, *Compactification of negatively curved complete Kähler manifolds of finite volume*, in *Seminar on Differential Geometry*, pp. 363–380, Ann. of Math. Stud., vol. 102, Princeton Univ. Press, 1982.

[So1] C. Soule, *Perfect forms and the Vandiver conjecture*, J. Reine Angew. Math. 517 (1999) 209–221.

[So2] C. Soule, *The cohomology of* $\mathrm{SL}_3(Z)$, Topology 17 (1978) 1–22.

[Sp1] T.A. Springer, *Algebraic groups with involutions*, Canadian Math. Soc. Conference Proceedings 6, 1986, 461–471.

[Sp2] T.A. Springer, *Combinatorics of B-orbits in a wonderful compactification*, in *Algebraic groups and arithmetic*, pp. 99–117, Tata Inst. Fund. Res., Mumbai, 2004.

[Sp3] T.A. Springer, *Intersection cohomology of* $B \times B$*-orbit closures in group compactifications*, J. Algebra 258 (2002) 71–111.

[Sp4] T.A. Springer, *Linear algebraic groups*, Second edition, in *Progress in Mathematics*, vol. 9, Birkhäuser Boston, 1998, xiv+334 pp.

[St] M. Stern, L^2*-index theorems on locally symmetric spaces*, Invent. Math. 96 (1989) 231–282.

[Ste1] H. Sterk, *Compactifications of the period space of Enriques surfaces. II*, Math. Z. 220 (1995) 427–444.

[Ste2] H. Sterk, *Compactifications of the period space of Enriques surfaces. I*, Math. Z. 207 (1991) 1–36.

[Str1] E. Strickland, *Computing the equivariant cohomology of group compactifications*, Math. Ann. 291 (1991) 275–280.

[Str2] E. Strickland, *An algorithm related to compactifications of adjoint groups*, in *Effective methods in algebraic geometry*, pp. 483–489, Progr. Math., vol. 94, Birkhäuser Boston, 1991.

[Str3] E. Strickland, *A vanishing theorem for group compactifications*, Math. Ann. 277 (1987) 165–171.

[Sul] D. Sullivan, *The Dirichlet problem at infinity for a negatively curved manifold*, J. Differential Geom. 18 (1983) 723–732.

[Sun] T. Sunada, *Closed geodesics in a locally symmetric space*, Tahoku Math. J. 30 (1978) 59–68.

[Ta1] J.C. Taylor, *The Martin compactification associated with a second order strictly elliptic partial differential operator on a manifold*, in *Topics in probability and Lie groups: boundary theory*, ERM Proceedings and Lect. Notes, vol. 28, AMS, pp. 153–202.

[Ta2] J.C. Taylor, *Compactifications determined by a polyhedral cone decomposition of* $\mathbb{R}^n$, in *Harmonic analysis and discrete potential theory*, ed. M.A. Picardello, Plenum Press, New York, 1992.

[Ta3] J.C. Taylor, *The Martin compactification of a symmetric space of noncompact type at the bottom of the positive spectrum: an introduction*, in *Potential theory*, pp. 127–139, de Gruyter, 1992.

[Tc1] A. Tchoudjem, *Cohomologie des fibrés en droites sur les compactifications des groupes re'ductifs*, Ann. Sci. E'cole Norm. Sup. 37 (2004) 415–448.

[Tc2] A. Tchoudjem, *Cohomologie des fibrés en droites sur la compactification magnifique d'un groupe semi-simple adjoint*, C. R. Math. Acad. Sci. Paris 334 (2002) 441–444.

[Te] J. Teitelbaum, *The geometry of p-adic symmetric spaces*, Notices Amer. Math. Soc. 42 (1995) 1120–1126.

[Ter1] C. Terng, *Recent progress in submanifold geometry*, in *Differential geometry*, pp. 439–484, Proc. Sympos. Pure Math., 54, Part 1, Amer. Math. Soc. 1993.

[Ter2] C. Terng, *Isoparametric submanifolds and their Coxeter groups*, J. Differential Geom. 21 (1985) 79–107.

[Th] M. Thaddeus, *Complete collineations revisited*, Math. Ann. 315 (1999) 469–495.

[Tho1] G. Thorbergsson, *A survey on isoparametric hypersurfaces and their generalizations*, in *Handbook of differential geometry*, Vol. I, pp. 963–995, North-Holland, 2000.

[Tho2] G. Thorbergsson, *Isoparametric foliations and their buildings*, Ann. of Math. 133 (1991) 429–446.

[Thu] W.Thurston, *Travaux de Thurston sur les surfaces*, Sminaire Orsay. Astérisque, vol. 66-67. Société Mathématique de France, Paris, 1979. 284 pp.

[Tim] D. Timashev, *Equivariant compactifications of reductive groups*, Sb. Math. 194 (2003) 589–616.

[Ti1] J. Tits, *Buildings of spherical type and BN-pairs*, Lect. Notes in Math. 386, Springer-Verlag, 1974

[Ti2] J. Tits, *On Buildings and Their Applications*, Proc. ICM, Vancouver, 1974, pp. 209–220

[Ts] H. Tsuji, *Complete negatively pinched Kähler surfaces of finite volume*, Tohoku Math. J. 40 (1988) 591–597.

[Ul] A. Ulyanov, *Polydiagonal compactification of configuration spaces*, J. Algebraic Geom. 11 (2002) 129–159.

[Uz1] T. Uzawa, *Compactifications of symmetric varieties and applications to representation theory*, in *Representation theory and noncommutative harmonic analysis (Japanese)*, Surikaisekikenkyusho Kokyuroku No. 1082 (1999) 137–142.

[Uz2] T. Uzawa, *Compactifications of symmetric varieties*, in *Representation theory of groups and homogeneous spaces (Japanese)*, Surikaisekikenkyusho Kokyuroku No. 1008 (1997) 81–100.

[Uz3] T. Uzawa, *On equivariant completions of algebraic symmetric spaces*, in *Algebraic and topological theories*, pp. 569–577, Kinokuniya, Tokyo, 1986.

[Vd] G. van der Geer, *Hilbert modular surfaces*, Springer-Verlag, 1988.

[VT1] M. van der Put, J. Top, *Algebraic compactification and modular interpretation*, in *Drinfeld modules, modular schemes and applications*, pp. 141–166, World Sci. Publishing, 1997.

[VT2] M. van der Put, J.Top, *Analytic compactification and modular forms*, in *Drinfeld modules, modular schemes and applications*, pp. 113–140, World Sci. Publishing, 1997.

[Va] V.S. Varadarajan, *Harmonic analysis on real reductive groups*, Springer Lect. Notes in Math. 576, 1977.

[Ve1] T. Venkataramana, *Lefschetz properties of subvarieties of Shimura varieties*, in *Current trends in number theory*, pp. 265–270, Hindustan Book Agency, New Delhi, 2002.

[Ve2] T. Venkataramana, *On cycles on compact locally symmetric varieties*, Monatsh. Math. 135 (2002) 221–244.

[Ve3] T. Venkataramana, *Restriction maps between cohomologies of locally symmetric varieties*, in *Cohomology of arithmetic groups, L-functions and automorphic forms*, pp. 237–251, Tata Inst. Fund. Res. Stud. Math., vol.15, 2001.

[Ve4] T. Venkataramana, *Cohomology of compact locally symmetric spaces*, Compositio Math. 125 (2001) 221–253.

[Vu1] T. Vust, *Plongements d'espaces symétriques algébriques: une classification*, Ann. Scuola Norm. Sup. Pisa Cl. Sci. 17 (1990) 165–195.

[Vu2] T. Vust, *Opération de groupes réductifs dans un type de cônes presque homogènes*, Bull. Soc. Math. France 102, 1974, 317–333.

[Wal1] N. Wallach, *Real reductive groups.I* , Academic Press, 1988, xx+412 pp.

[Wal2] N. Wallach, *Real reductive groups. II*, Academic Press, 1992, xiv+454 pp.

[Wal3] N. Wallach, *Automorphic forms*, Notes by Roberto Miatello, in *New developments in Lie theory and their applications*, Progr. Math., vol. 105, 1–25, Birkhäuser Boston, 1992.

[Wal4] N. Wallach, *The powers of the resolvent on a locally symmetric space*, in *Algebra, groups and geometry*, Bull. Soc. Math. Belg. Sir. A 42 (1990) 777–795.

[Wal5] N. Wallach, *Square integrable automorphic forms and cohomology of arithmetic quotients of* $\mathrm{SU}(p, q)$, Math. Ann. 266 (1984) 261–278.

[Wal6] N. Wallach, *The spectrum of compact quotients of semisimple Lie groups*, in *Proc. of the International Congress of Mathematicians (Helsinki, 1978)*, pp. 715–719, 1980.

[Wal7] N. Wallach, *An asymptotic formula of Gelfand and Gangolli for the spectrum of* $G\backslash G$, J. Differential Geometry 11 (1976) 91–101.

[Wal8] N. Wallach, *On the Selberg trace formula in the case of compact quotient*, Bull. Amer. Math. Soc. 82 (1976) 171–195.

[Wal9] N. Wallach, *Harmonic analysis on homogeneous spaces*, Marcel Dekker, 1973. xv+361 pp.

[Wan1] W. Wang, *A note on the logarithmic canonical bundle of a smooth noncompact locally symmetric Hermitian variety*, Comment. Math. Univ. St. Paul. 48 (1999) 1–5.

[Wan2] W. Wang, *On the smooth compactification of Siegel spaces*, J. Differential Geom. 38 (1993) 351–386.

[Wan3] W. Wang, *On the moduli space of principally polarized abelian varieties*, in *Mapping class groups and moduli spaces of Riemann surfaces*, pp. 361–365, Contemp. Math., vol. 150, Amer. Math. Soc., 1993.

[War] G. Warner, *Harmonic analysis on semisimple Lie groups I*, Springer-Verlag, New York, 1972.

[Wer1] A. Werner, *Compactification of the Bruhat-Tits building of PGL by seminorms*, Math. Z. 248 (2004) 511–526.

[Wer2] A. Werner, *Arakelov intersection indices of linear cycles and the geometry of buildings and symmetric spaces*, Duke Math. J. 111 (2002) 319–355.

[Wer3] A. Werner, *Compactification of the Bruhat-Tits building of PGL by lattices of smaller rank*, Doc. Math. 6 (2001) 315–341.

[Wes1] U. Weselmann, *Twisted topological trace formula*, notes of a talk taken by M.Goresky, Oberwolfach, Germany, 1997.

[Wes2] U. Weselmann, in preparation.

[Wh] H. Whitney, *Elementary structure of real algebraic varieties*, Ann. Math. 66. 1957, 545–566.

[Wi] J. Wildeshaus, *Mixed sheaves on Shimura varieties and their higher direct images in toroidal compactifications*, J. Algebraic Geom. 9 (2000) 323–353.

[Wolf] M.Wolf, *The Teichmüller theory of harmonic maps*, J. Differential Geom. 29 (1989) 449-479.

[Wo1] J. Wolf, *Flag manifolds and representation theory*, in *Geometry and representation theory of real and p-adic groups*, pp. 273–323, Progr. Math., vol. 158, Birkhäuser Boston, 1998.

[Wo2] J. Wolf, *Spaces of constant curvature*, Third edition, Publish or Perish, 1974, xv+408 pp.

[Wo3] J. Wolf, *Fine structure of Hermitian symmetric spaces*, in *Symmetric spaces: short courses presented at Washington University*, ed. by W. Boothby and G. Weiss, Marcel Dekker, 1972, pp. 271–357.

[Wo4] J. Wolf, *The action of a real semisimple group on a complex flag manifold. I. Orbit structure and holomorphic arc components*, Bull. Amer. Math. Soc. 75 (1969) 1121–1237.

[WK] J. Wolf, A. Koranyi, *Generalized Cayley transformations of bounded symmetric domains*, Amer. J. Math. 87 (1965) 899–939.

[Wolp] S.Wolpert, *Geometry of the Weil-Petersson completion of Teichmüller space*

[Xu] Q. Xu, *Representations and compactifications of locally compact groups*, Math. Japon. 52 (2000) 323–329.

[Ya] I. Yaglom, *Felix Klein and Sophus Lie: evolution of the idea of symmetry in the nineteenth century*, Birkhäuser, Boston, 1988.

[Ye1] S. Ye, *Compactification of Kähler manifolds with negative Ricci curvature*, Invent. Math. 106 (1991) 13–25.

[Ye2] S. Ye, *Compactification of complete Kähler surfaces with negative Ricci curvature*, Invent. Math. 99 (1990) 145–163.

[Yo1] M. Yoshida, *Fuchsian differential equations: with special emphasis on the Gauss-Schwarz theory*, Friedr. Vieweg & Sohn, 1987.

[Yo2] M. Yoshida, *Hypergeometric functions, my love: modular interpretations of configuration spaces*, Friedr. Vieweg & Sohn, 1997.

[Zi] R. Zimmer, *Ergodic theory and semisimple groups*, Birkhäuser, Boston, 1984.

[Zu1] S. Zucker, *L_2 cohomology of warped products and arithmetic groups*, Invent. Math. 70 (1982) 169–218.

[Zu2] S. Zucker, *Satake compactifications*, Comment. Math. Helv. 58 (1983) 312–343.

[Zu3] S. Zucker, *On the reductive Borel-Serre compactification: L^p-cohomology of arithmetic groups*, Amer. J. Math. 123 (2001) 951–984.

[Zu4] S. Zucker, *L^2-cohomology of Shimura varieties*, in *Automorphic forms, Shimura varieties, and L-functions*, Vol. II (Ann Arbor, MI, 1988) Perspect. Math., 11, ed. by L. Clozel and J. Milne Academic Press, 1990, pp. 377–391.

[Zu5] S. Zucker, *On the boundary cohomology of locally symmetric varieties*, Vietnam J. Math. 25 (1997) 279–318.

[Zu6] S. Zucker, *L^p-cohomology: Banach spaces and homological methods on Riemannian manifolds*, in *Differential geometry: geometry in mathematical physics and related topics*, pp. 637–655, Proc. Sympos. Pure Math., vol. 54, Part 2, Amer. Math. Soc., 1993.

[Zu7] S. Zucker, *An introduction to L^2-cohomology*, in *Several complex variables and complex geometry*, Part 2 pp. 519–534, Proc. Sympos. Pure Math., 52, Part 2, Amer. Math. Soc., 1991.

[Zu8] S. Zucker, *L^p-cohomology and Satake compactifications*, *Prospects in complex geometry*, pp. 317–339, Lecture Notes in Math., vol. 1468, Springer, 1991.

[Zu9] S. Zucker, *L^2-cohomology and intersection homology of locally symmetric varieties. III*, in *Actes du Colloque de Thiorie de Hodge*, Astérisque No. 179–180 (1989) 245–278.

[Zu10] S. Zucker, *L_2-cohomology and intersection homology of locally symmetric varieties. II*, Compositio Math. 59 (1986) 339–398.

[Zu11] S. Zucker, *L_2-cohomology and intersection homology of locally symmetric varieties*, in *Singularities*, Part 2 pp. 675–680, Proc. Sympos. Pure Math., vol. 40, Amer. Math. Soc., 1983.

[Zu12] S. Zucker, *Hodge theory and arithmetic groups*, in *Analysis and topology on singular spaces, II, III*, pp. 365–381, Astérisque, 101–102, Soc. Math. France, 1983.

Notation Index

Index